International Conference on Advances in Mechanical Engineering-2022 (ICAME-2022)

IOP Conference Series: Materials Science and Engineering
Volume 1259

Nagpur, India
26 – 28 May 2022

ISBN: 978-1-7138-6422-6
ISSN: 1757-8981

Printed from e-media with permission by:

Curran Associates, Inc.
57 Morehouse Lane
Red Hook, NY 12571

Some format issues inherent in the e-media version may also appear in this print version.

Institute of Physics
Dirac House, Temple Back
Bristol BS1 6BE UK

Phone: 44 1 17 929 7481
Fax: 44 1 17 920 0979

techtracking@iop.org

Additional copies of this publication are available from:

Curran Associates, Inc.
57 Morehouse Lane
Red Hook, NY 12571 USA
Phone: 845-758-0400
Fax: 845-758-2633
Email: curran@proceedings.com
Web: www.proceedings.com

TABLE OF CONTENTS

IOP Conf. Series: Materials Science and Engineering 1259 (2022) 011001 doi:10.1088/1757-899X/1259/1/011001

PREFACE

Advances in Mechanical Engineering in both fundamentals and applied research are going to facilitate the new digital era of automated machines & manufacturing processes to expedite local socio-economic development.

Advances in Mechanical Engineering in both fundamentals and applied research are going to facilitate the new digital era of automated machines & manufacturing processes to expedite local socio-economic development. These advances are going to be integrated in the near future for development of new manufacturing processes and new digital automated world. They will be important part of the automotive technology, advance machining processes, renewable energy sources, IOT based mechanical controls, aviation technology & electric vehicles etc. However, advance design and analysis methods, bio medical implant's by additive manufacturing, advance materials, MEMS, advances in propulsions and power concepts, advance aerodynamics and CFD, FEA optimization, vibration and acoustics, virtual instrumentation with IOT, robotics are key challenges, and they need to be solved through advances in Mechanical engineering through focused and integrated efforts with the use of smart technologies. These advance technologies performs an important role in the 'Digital Technology Transformation' to integrate environmental friendly Mechanical systems (e.g. Industry Automation, Advance Energy conversion, Robotics, New Modeling and simulation methods etc.) for the sustainable development. The necessity for 'Advance Mechanical systems' is broadly understood and shared, as the world starts to embrace a transition towards new methods and technology for sustainable development in Industry 4.0. As the Editors of the Proceeding of ICAME-2022 in IOP Conference series Material Science & Engineering Journal (ISSN: 1757-899X), we wish to express sincere gratitude to the authors for diligent research dissemination as well as to the reviewers for providing constructive and critical comments in improving the quality of the selected articles. The objective of this volume is to present advancements in advance technologies for environment friendly automation systems and their management, policies for making an energy transition towards sustainable societies. The ICAME-2022 is a major multidisciplinary sustainable engineering conference organized with the objective of bringing together academia, industry practitioners and engineering graduates to analyze scientific studies focusing on advance mechanical system design and analysis with IOT controls and smart material concepts to integrate the relationship among them. The ICAME-2022 is envisioned to analyze the increasing penetration of environment friendly technologies with energy efficiency and management as central topics with effective participation from the academicians, industry practitioners, governmental policy makers, as well as of community service providers. The ICAME-2022 has been jointly organized by G H Raisoni College of Engineering (GHRCE) Nagpur & G H Raisoni Institute of Engineering Technology (GHRIET)Nagpur, India on 27-28 May 2022. As editor(s) of this ICAME-2022, we want to tribute deep sense of appreciations to the international Publishing agency IOP Science for acceptance of our presented papers in IOP Conference Series, Material Science & Engineering Journal. We wish to express our sincere appreciation to the authors, plenary speakers, presenters, participants, supporters, reviewers, technical committee members as well as the entire team of G. H. Raisoni College of Engineering(GHRCE), Nagpur & G H Raisoni Institute of Engineering & Technology(GHRIET), Nagpur(INDIA). We are very grateful to Shri Sunil Raisoni, Chairperson of the Raisoni Group of Institutions; Dr. Sachin Untawale, Director, GHRCE, Nagpur; Dr. Vivek Kapur, Director, GHRIET, Nagpur, Dr. P.V. Walke, Dr. S.B. Jaju, Dr. R.S. Shelke, Dr. Vinod Bhaiswar, Dr. Abhay Khalatkar, Dr. Dheeraj Deshmukh, Dr. A. S.Raut, Mr. Amol Pitale, Dr.A.K. Mahalle, Dr. Manish Gupta, Dr. V. W. Khond, Mr. T.C. Parshiwankar, Dr. A. S. Shahare and the ICAME-2022 organizing committee members for contributing effectively in making the ICAME-2022 and publication of the volume in IOP Proceedings successful. We strongly believe that the ICAME-2022 has provided an effective platform to integrate academia, industry practitioners and engineering graduates to share their research and innovations for the sustainable development of the society.

IOP Conf. Series: Materials Science and Engineering 1259 (2022) 011002 doi:10.1088/1757-899X/1259/1/011002

Peer Review Statement

All papers published in this volume have been reviewed through processes administered by the Editors. Reviews were conducted by expert referees to the professional and scientific standards expected of a proceedings journal published by IOP Publishing Publishing.

- **Type of peer review:** Double Anonymous
- **Conference submission management system:** Easychair
- **Number of submissions received:** 121
- **Number of submissions sent for review:** 121
- **Number of submissions accepted:** 84
- **Acceptance Rate (Submissions Accepted / Submissions Received × 100):** 69.4
- **Average number of reviews per paper:** 2
- **Total number of reviewers involved:** 50
- **Contact person for queries:**
  Name: Dr. Rupesh Shelke
  Email: rupesh.shelke@raisoni.net
  Affiliation: Professor and Head, Mechanical Engineering Department, G H Raisoni College of Engineering, Nagpur, Maharashtra, India

IOP Conf. Series: Materials Science and Engineering 1259 (2022) 012001 doi:10.1088/1757-899X/1259/1/012001

Observational Analysis of Stationary C.I. Engine Output and Emission Evaluation for Variance of Induction Pressure Aspects

Dipak C Talele[1,2,*], Dheeraj S Deshmukh[3], Manish S Deshmukh[4], Sheetal D Deshmukh[5] and Vilas S Patil[6]

[1]PhD Research Scholar, U.I C.T. K.B.C. North Maharashtra University, Jalgaon, Maharashtra State, India
[2]Assistant Professor, Department of Mechanical Engineering, S.S.B.T.'s, C.O.E.T., Bambhori, Jalgaon, Maharashtra State, India
[3]Associate Professor, Department of Mechanical Engineering, G. H. Raisoni College of Engineering, Nagpur, Maharashtra State, India.
[4]Associate Professor, Department of Mechanical Engineering, AISSMS College of Engineering, Pune - 01, MS, India
[5]Assistant Professor, Food Technology Department, L.I.T., RTMNU, Nagpur, Maharashtra State, India.
[6]Retired Professor, U.I C.T. K.B.C. North Maharashtra University, Jalgaon, Maharashtra State, India.

*Corresponding Author Email: taleledeepak190687@gmail.com

Orchid Ids:- [1]https://orcid.org/ 0000-0003-1907-2904, [3]https://orcid.org/ 0000-0001-6074-3111, [4]https://orcid.org/ 0000-0001-8981-2643, [5]https://orcid.org/ 0000-0003-4708-1437,

Abstract. In the current environment, improved engine performance, reduced fuel consumption and reduced emissions are important goals for automotive manufacturers. To understand the factors involved in the compartment assemblies in Compression Ignition (C I) engines it is crucial to concentrate on such goals. Pressure rating, fuel atomization, fuel pressure, fuel quality, air pressure rating, air temperature, pressure and other design components such as piston, Air induction system, and assembly chamber and more to work. It is evident from the literature review and research work carried out to date that the airing system is best method of promoting C I operation of the engine. The design of the Air suction system could be achieved in this way that it takes into account the objectives of entire system. Effective air suction system improvement needs least energy utilization and most extreme use of air intake pressure for decrease in poisonous productions at outlet of C I engines. Furthermore, powerful use of fuel vitality. In present work, air suction arrangement, exclusively less air inlet pressure during suction stroke of four stroke cycle engine concern is topic of explicit enthusiasm for plan. Advancement of air induction system with air induction pressure regulator, flow distribution, action is studied and useful information is developed through proper intake manifold arrangement for better fuel efficiency.

1. Introduction

Compression Ignition engine is key part in a vehicle. An engine combust fuel in cylinder, then produces power so named as internal combustion engine. C I engine is important amongst other engines because of higher thermal efficiencies. Different vehicles that is light as well as heavy duty diesel engines could be utilized. Mostly I.C. Engines, the air intake network consists of a number of interlinking tubing. The section of the engine which always provides the air to the cavity being identified as air inlet/intake system. Primary objective of to stream air conduit will be to feed air to the air intake manifold of its engine head. Based on the air circulation inside chamber, the fuel injection is mixed to air in the Diesel engine. Fuel combustion is controlled by air flow this is a crucial parameter since air is traditional provided into the cylinder via air intake manifold through higher velocity. C I engine, airflow causes fuel- air mix as well as combustion rates. Due to the growing air density, a kinetic energy of that same air creates turbulence as well as the rapid fuel air mixture. Owing to higher cooling of cylinder walls caused by increased turbulence heat lost to atmosphere is reduced very important achievement. Air inducted in suction captivates heat from the cylinder parapets this causes reduction in ignition delay-period and thereby surges thermal efficiency of the C I engines [1 & 2].

2. Review of literature

Important operating parameters for internal combustion engine include engine speed, load, backpressure, induction pressure, and basic dependent variable fuel consumption rate. One of the important engine operating parameters is air induction pressure during the suction stroke of the internal combustion engine.[1] The advancement of an air inlet system by correct inlet air pressure controller for engine performance enhancement with an adequate air supply system for stationary diesel engine applications is the main task of project work. Understanding of engine operating parameters and additional system parameters included in the proper combustion thereby power output of C I engine is crucial, to obtain energy efficiency.[2] Certain engine operating conditions include compression ratio, liquid fuel atomization, Diesel fuel injection pressure, diesel fuel efficiency, diesel ignition rate, fuel & air ratio, supply air pressure, cylinder temperature depending on piston configuration, air induction method, combustion chamber configuration, related parameters disturbing the engine performance are vital to be considered. Air induction system development is one of the important factors that decide the air supply condition for proper combustion process achievement, to enhance C.I. engine performance.[3] Production of the air intake manifold should be carried out in such a way as to take into account the overall goals of the system. Effective air induction design process needs proper fuel efficiency and optimum air pressure usage to minimize CI emission levels engine as well as for the efficient use of fuel oil. [4]

2.1. International Status

With the increasingly stringent requirements of the day-to-day environmental expansion and the international maritime company over the ship's NOx and suite emissions. The efficiency of diesel engines has now become a major source of research both domestically and overseas. Efficiency is currently mainly optimized by engine, in-engine as well as first-engine filtering methods. After filtration, the engine basically consists with Lean Trap (LNT) as well as Selective Catalytic Reduction; Key part of pre-engine as well as engine filtration is to reap benefits by ignition process enhancement, also to monitor the output of cylinder pollutants including such as oxygen-rich ignition, exhaust gas recycle (EGR) and diesel fuel efficiency.[5]

At moment, engine treatment method is indeed nascent of diesel engines, making it much more difficult to bring it into common use. Engine optimization technique, based on the condition of air or oxygen, enhances combustion performance and decreasing carbide emissions, however cylinder rising temperatures favour NO growth; exhaust gas recycle (EGR) or air-intake humidity or even other steps favour NOx gas emissions; That's also, nevertheless, just at cost of the economies of diesel engines. And the most sophisticated engine filtration technologies shakes the "NO-Soot" pollution barrier curve. Predicated also on kinetic mechanism of its combustion process, and inside cylinder NO production is primarily estimated using three parameters: a cylinder inside temperature, the concentrations of reaction as well as the duration of a high-temperature reaction phase. NO is generated as much as nitrogen is available. The existence of N_2 in an inlet combustion environment which doesn't have N_2 could be probably eliminated; so, as the production of NOx is almost equal to zero.[6]

The segregation of N_2 from its air needed for nitrogen-free ignition and also the amount of non-N_2 gases to mere oxygen may avoid over-dramatic ignition of fuel oil with high percentage oxygen. Combustion & its optimization tools are essential components of nitrogen-free ignition. Work focuses on the viability and effective adaptation steps of O_2 and CO_2 in the climate, considering the running characteristics of diesel engines. Several related studies on combustion technologies in such environments have been conducted nationwide and abroad, but in particular how the application of various types of combustible coal stoves, such as pulverized coal and NOx, is prioritized and minimized by emissions of many pollutants such as dust. Depending upon the learning, next suitable, observational analysis on the combustion efficiency of vessels engine throughout the O_2 and CO_2 conditions was performed with a test object including the "4135ACa" form of C.I. The research subject in "ZS195" is the form of diesel cylinder engine, as well as the actual procedure of that same single-cylinder engine relates to the O_2 & CO_2 conditions engine.[6]

Mathematical Simulation of 3 types of inlet system is performed and the flow field of the inlet manifold is evaluated. Responses indicate intake air end positioned in the centre of the regulator will significantly improve input consistency in order to enhance intake system efficiency and ignition quality. As per the theory of CFD, 3-D numerical simulations of the intake air system of its engine was carried out, and that is an accurate and possible analytical method. The analysis reveals pressure drop and stream consistency of an air intake are 2 key assessment metrics, but Strategy Three satisfies such two design requirements. [7,8].

2.2. *Current status in India*

The economically growing countries around the world, such as India, is currently experiencing three main problems: rising energy consumption, climate pollution and waste dumping. It contributes to a rising energy usage due to rapid population growth, industrial development as well as modern lifestyles. Besides, C.I. engines are being used as a key source of income and employment for people for mobility, manufacturing, agriculture including small-scale energy production. And in present situation, several C.I. engines use fossil fuels that are increasingly diminishing, prompting researchers to build and propagate renewable resources that provide alternatives that fossil fuels. In either case, a bio-gas twin-fuel mode is detrimental to the diesel engine issue.

Few researchers also tried many different dual-fuel C.I. engines incorporating biogas. For reference, Duck and Vattanavichian have investigated to identify the efficiency of bio-gas insertion with in emission & performance criterions of Diesel engine. Increased carbon monoxide (CO) and incomplete combustion Hydro-carbon productions at particular energy consumption were detected in two fuel mode. The diesel options contributed for 43 to 49 per cent of the overall load output. Macarevisenai et al. Enhanced CO and HC pollutants with fall-in Brake thermal efficiency in bio-gas dual-fuel engines are being recorded. It's Bora et al. anticipating variations in the impact of the compression ratio (CR) on the efficiency of dual-fuel bio-gas engines. Thermal efficiency rose at 18, 17.5, 17 as well as 16 compression ratios (CR), or 20.0, 18.3, 17.1 and 16.4 per cent respectively. It's Tipayawang et al. Bio-gas is being used in a small-scale C.I. engine for lengthy-term test Performance. They found short piston wear that it was not sufficient to decline engine output during extended service. [9]

Amberita utilized bio-gas for C.I. engines to fuel 70 per cent of methane as well as could observe started saving to 87.5 per cent of diesel. Barrick & Murugan included biogas also in C.I. engine and then observed a 6.2 per cent decline in thermal performance, a decrease in nitrogen oxide (NOx) as well as a rise in HC and CO emissions, and also in engine exhaust. Barrick and Murugan pilot experimented with enhancements in performance and release of exhaust gases characteristics

IOP Conf. Series: Materials Science and Engineering 1259 (2022) 012001 doi:10.1088/1757-899X/1259/1/012001

compared to the average fuel injection time and 23 degrees prior to outer dead center was observed. Verma et al. The influence of the proportion of methane mostly on efficiency of dual fuels of 3 distinct bio-gas compositions has been studied. Forth herein were shown by bio-gas diesel type activity with 93 percent methane. [9]

Performance improvement approaches are implemented for better output with the dual-fuel engines. Thermal barrier coating (TBC) has been used by investigators to improve thermal insulation with in combustion chamber, leading to combustion temperature rise and thus improvement in performance as well as emission characteristics. Idin et al. Thermal barrier coating (TBC) has been used to improve the performance of diesel engines using 88 percent zirconium oxide, 4 percent magnesium oxide, and 8 percent aluminium oxide pistons and valves derived from cottonseed oil biodiesel. Coated engines (CE) and Brake thermal efficiency (BTE) have shown a lower saving in HC and CO contaminants, while Knox releases have also augmented compared to others in engines. [10]

Jena et al. Zirconia (YZZ) coating has been investigated to stabilize the effect of yttria on engine performance and exhaust pollutants .Since a diesel additives, ferric chloride (0.2 g L^{-1}) has been used, that increased Brake thermal efficiency (BTE) about 2.7 percent and lowered emissions like HC , CO, as well as smoke opacity that exclude emissions of nitric oxide (NO).Sivakumar & Kumar were experimenting with YZZ coating diesel engines and saw an improvement in brake power of 3-5 percent with drop in fuel efficiency, HC and CO emissions. From Mahapatra et al. Used in steam-injected YSZ coated diesel engines. In contrast to CE without vapour injection, Brake thermal efficiency (BTE) increased with reduced BTE. [11]

3. Experimental Setup Used

Table 1. Testing Setup Specifications for experimentation.

Sr. No.	Parameter	Specification
01	Make	Kirlosker, one cylinder, 4-stroke C I engine
02	Out Put Power	5 [H.P]
03	Speed	1500 [R.P.M.]
04	Stroke length	110 [mm]
05	Bore diameter	80[mm]
06	Loading type	Electrical resistance type load, with copper element and load Procedure
07	Moment arm	0.2 [meter]
08	Air Box Orifice dia.	25[mm]
09	Cd for orifice	0.64

Figure 1. Single Cylinder Kirlosker Make Stationary C.I. Engine Experimental Setup

4. Result and Discussion

The following observations are made during the review of the literature and experimental work carried out are listed below-

4.1. Thermal explanation of the intake air pressure effect mostly on output of the 4-S C.I. engine.

Investigate the effects of pressure of air induction on C.I. engine Output enhancement through effective utilization of energy in fresh air for emission control and possible energy recovery activity is so far as the basic aim of this work. Engine performance improvement with the aid of the air induction mechanism to attain overall system objectives, the system needs an interpretation of integrated system output elements. Throughout this context of an engine system, the aims are to improve fuel efficiency and lowering engine emission levels. As per these targets, the significant element for enhancing engine overall efficiency is the intake air system design for minimal back pressure requirements, see Figure 2. As the indicator diagram shows during the suction process from point e to a, the work is done per cycle and the air induction pressure drops during the suction process are cumulative rise in negative region of the drawing as shown by the blank region below the atmospheric pressure during suction stroke. The regular reduction in the level of air induction pressure at the input, it is observed that the normal surge in the value of the fuel consumption is significant at the outcome. It specifies that a minor reduction in the value of air induction pressure severely surges fuel consumption.

More emphasis has been put on developing aspects of C.I. engines for its output analysis and development during previous studies. Deployable aspects of maintenance, primarily relevant to the C.I. engine induction system, which would be the effect of its decrease of air induction pressure mostly on engine, is explicitly oriented and evaluated to examine the actual result on engine output in that same context.

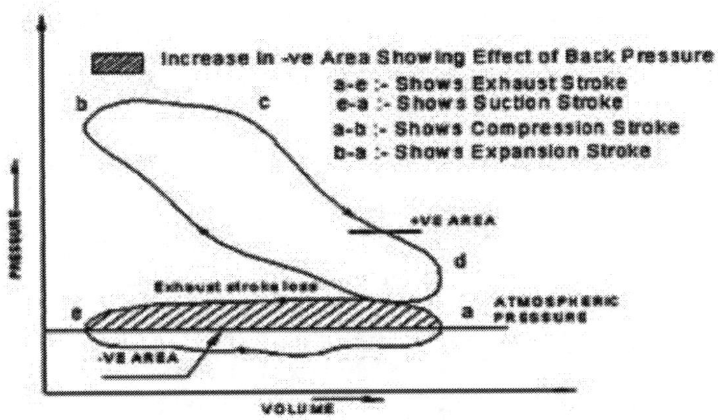

Figure 2. Explanation of the effect of air induction pressure using indicator diagram

In this investigation, it is felt necessary to study cause and effect analysis for air induction pressure phenomena. Consequently, opportunity to improve the quality requirements from each intake system part and its arrangement are very essential so as to achieve a positive induction pressure frequently during different operating situations of engine. High induction pressure for efficient utilization of fuel energy under wide range of working conditions for engine, not much high induction pressure is needed for efficient utilization of fuel energy under wide range of working conditions for engine.

4.2. Design aspects of intake manifold

The air intake manifold's primary objective is to uniformly disperse air within the manifold runner, which again is necessary for even an efficient design of its intake manifold. The configuration of the intake side has a significant impact on the engine's volumetric performance. An haphazard circulation of air results in much less volumetric efficiency, loss of power, and higher fuel consumption. This

IOP Conf. Series: Materials Science and Engineering 1259 (2022) 012001 doi:10.1088/1757-899X/1259/1/012001

same filling with cylinders besides air could be positively or negatively affected depending mostly on magnitude and frequency of pressure waves within the intake side. Such pressure waves' magnitude and frequency depend mostly on configuration of its inlet manifold, engine speed, as well as valve timing. Unstable design of the method of induction and even the effect of it's manifold for charging is highly dependent mostly on engine speed. Owing to air input within the intake side, which would be a feature of pulses that differ. Thus, in engine manifolds, such pulses must be fine-tuned to deliver the right power.

4.3. Air motion aspects inside engine cylinder

Air movement from inside of the engine cylinder involves the movement of air, fuel, including exhaust gas inside the cylinders mostly during compression stroke, fuel combustion process and expansion stroke of cycle. It is necessary to surge fuel air mixture, to raise combustion rapidity and efficacy, and also to raise fuel-air vaporization within the chamber. Air motion contains swirling motion, tumble movement, squish movement, and turbulence. Planned rotating movement of the cylinder axis inside the cylinder is regarded called "swirl". Planned angular velocity is induced by the geometrical arrangement of its inlet port & valves and its opening method. The design of the intake manifold on the lateral part to make the inlet stream into another cylinder creates swirl. Prepared rotational movement perpendicular to either axis of cylinder inside the cylinder is referred to as "Tumble". That gas combination filling the volume at just the outside periphery of a cylinder is pushed radially inner even as outside volume is compact to approximately nil because as piston reaches the TDC.

The radial inward movement of the gas combination is termed "Squish." Flows in the C.I. engines are turbulent. An engine is due to the high-speed flow inside, into, then out of the cylinder. Thermal rate of transfer including such as evaporation, heat transfer, blending & rate of combustion within the engine are increased by the scale factor. Elevated turbulence mostly throughout suction stroke is indeed due to the turbulence including its inlet jet alone results in the conversion of that same non-turbulent flow of inlet stream with turbulence. Even then, turbulence is minimized as the flow rate slows down near the Bottom Dead Core. Again, near the Top Dead Core, turbulence rises as swirls, squishes and tumbles rise during compression.

Turbulence is becoming more uniform in composition all of the way thru the cylinder due to swirl. Elevated turbulence near the Top Dead Centre when combustion takes place is very favourable for combustion. It splits up and stretches the front flames a lot faster than those of a laminar flame. Air-fuel is expended in a really small time, but auto-ignition as well as knocking are prevented.

4.4. Impact of inclination of air intake manifold on the C.I. engine Output

That Inlet Manifold does a critical role for engine efficiency. Air moves inside of cylinder either from the suction port or from the suction manifold Air moves into its cylinder either from the intake port or from the intake manifold. In the present research, inlet manifolds are made of soft steel with different angle inclinations. Manifolds are produced at angle 300, 600, 900 & standard. It has been seen in most situations that, while exiting the intake port, its jet of air impinges on either the cylinder wall then diverts back, enabling small and wide vertices to form inside the cylinder.

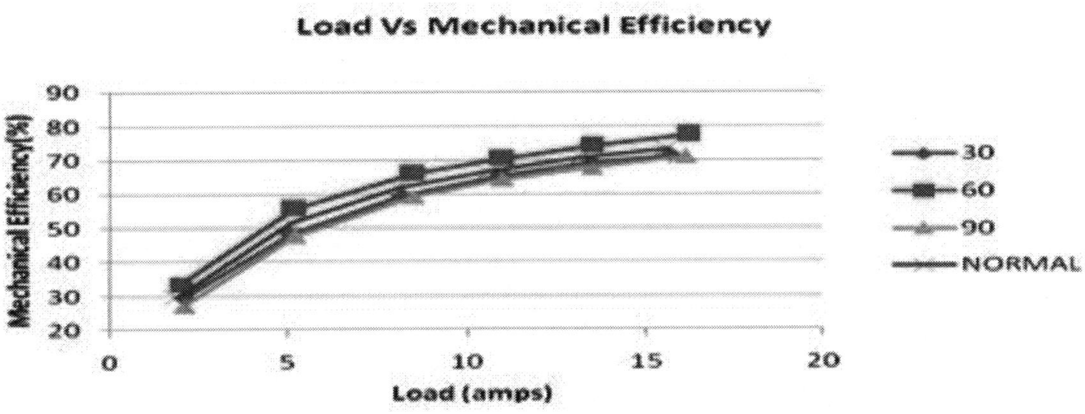

Figure 3. Load Versus Mechanical efficiency at various manifold design at 160 bar injection pressure

Refer Figure 3, it is visible hat mechanical efficiency is more at 600 than different manifold designs.

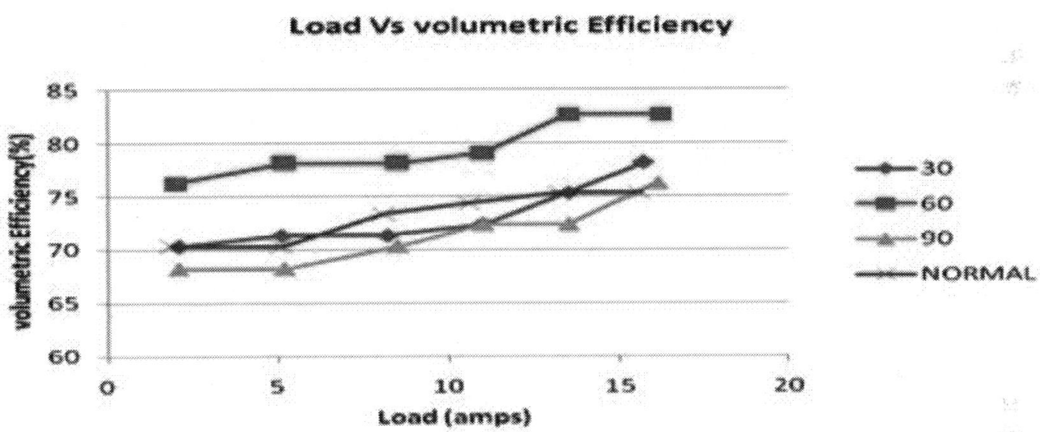

Figure 4. Load Versus Volumetric efficiency at different manifold designs at 160 bar injection pressure

Refer Figure 4, it is seen that volumetric efficiency is good at 60^0 than various manifold angles.

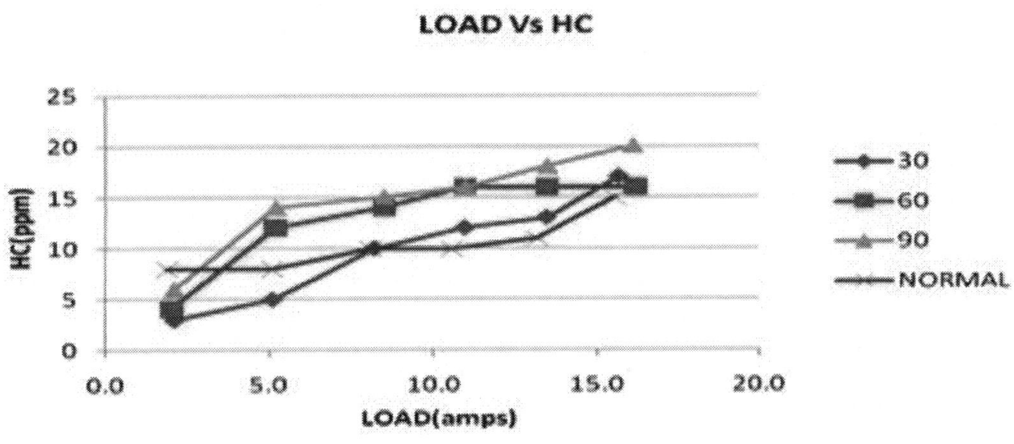

Figure 5. Load Versus HC (ppm) release at various manifold angles at 160 bar injection pressure

Refer Figure 5, it is visible that an HC (ppm) emission is fewer at normal manifold inclination than manifold angles.

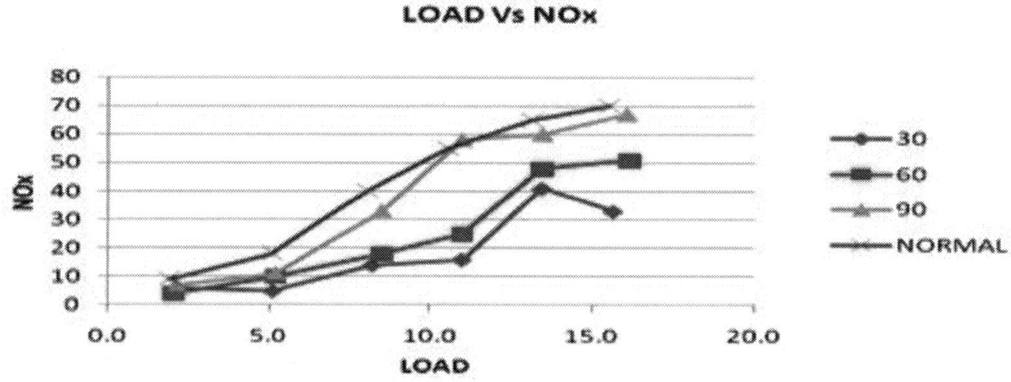

Figure 6. Load Versus NOx emissions at various manifold designs at 160 bar injection pressure

Refer Figure 6, this is visible oxides of Nitrogen emission is fewer at 30^0 compare to manifold designs. Inlet manifold design angles considerably impacts flow arrangement in the cylinder. Review of air suction system advancements affects performance and fuel efficiency. Improved performance is found instead of the normal one with various the manifold designs. Cumulative fuel injection pressure effects ample combustion of fuel that consequences in a decrease in effluence level.

4.5. Effects of inside threaded manifold of engine output

Turbulence is accomplished in the intake side with various varieties of steady pitch threaded from inside. The influence of different internal threads. Acme, buttress and knuckle threads by constant pitch are considered.

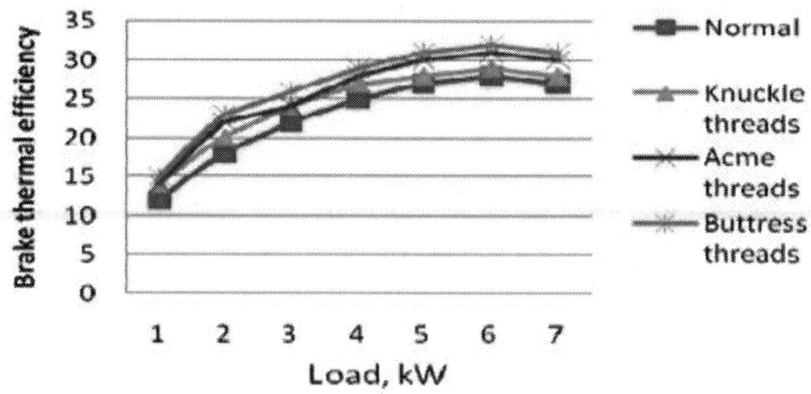

Figure 7. Load Versus Brake thermal efficiency at various threads from inside

Brake thermal efficiency versus load for C I engine by inlet manifolds having 3-different threads from inside. ACME, buttress and knuckle threads are equated to the engine with typical inlet manifold and is shown in Figure 7. As per the figure, it is clearly seen that the brake thermal efficiencies are increasing with an addition of load for formations that are under calculation. It is also perceived that the engine by inlet manifolds with internal buttress threads has the highest thermal efficiency. This is because the inlet manifold with buttress threads found a better swirl coefficient and swirl ratio equated to the inlet manifold by ACME and knuckle threads.

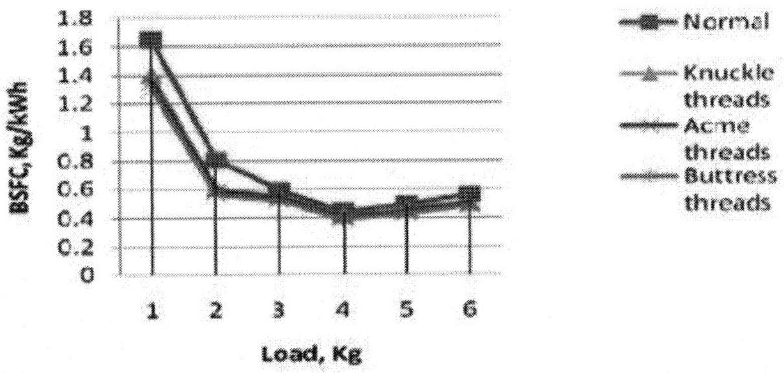

Figure 8. Load Versus Brake Specific Fuel Consumption at various internal threads

Brake Specific Fuel Consumption (BSFC) tests engine efficiency. In fact, Engine performance and BSFC are inversely proportion interrelated. Engine performance is better at Lower BSFC. Technicians use BSFC instead of thermal efficiency since there is no conventional concept of thermal efficiency more than Just a across the globe as Figure 8 indicates. BSFC Vs load for 3 main types of internal threads upon this intake manifold of the diesel engine viz. ACME, buttress and knuckle threads are compared to the standard inlet manifold motor. The Brake Specific Fuel Consumption (BSFC) decreases with an increase in load for all the modifications under consideration. From figure above, it has been shown that the internal threaded inlet manifold engine provides less precise Brake Specific Fuel Consumption (BSFC) than some other internal threads.
The decline in Brake Specific Fuel Consumption (BSFC) may be due to an increased swirl created in the inlet manifold inside the buttress compared with the standard inlet manifold engine.

5. Conclusion

Design aspects for inlet manifold arrangement are noteworthy in C I engine. Enhanced homogenization of the air-fuel blend improves the burning of fuel due to a swirl in the C.I. engine chamber. Appropriate design of an air induction system is utilized to generate a churn that is valuable to surge the diesel engine combustion efficiency. Swirl air is dependent on engine speed. As the engine speed is increased, swirls also raise. As the swirl rises, the rate of evaporation as well as the air-fuel mixture increase, which contributes to thorough combustion. Better output than the standard could be achieved by adjusting the inclination of the manifold. Throughout the engine, the internal buttress threads upon this inlet manifold get a maximum upsurge in thermal efficiency also decline in BSFC than the usual inlet manifold is reported. Obviously, it depends on the output characteristics of the engine, the buttress threads on the intake manifold are known as the most favorable configuration. So as to achieve the better performance of engine, the least fuel utilization, the maximum thermal efficiency as well as the smallest amount of particulate emissions, the airflow field produced within the engine cylinder becomes useful, that is responsible by the nature of the air induction system. Consequently, so as to obtain a positive induction pressure during the various working conditions of the engine, suitable structural features of each induction system part and their configuration are very important. Instead of altering the engine and fuel, exhaust emissions can be minimized efficiently without adversely affecting the engine's efficiency through using fuel energy efficiently in a range of engine operating environments using full induction pressure.

Acknowledgment
Authors are thankful Institutes for providing technical support, library facility also thankful to contemporaries for valuable inputs.

References

[1] Thamaraikanan R, Anish M and Kanimozhi B. "Design and Analysis of an Intake Manifold in

an IC Engine", Applied Mechanics and Materials. Vol.766, pp.1021-1027, 2015.

[2] Wang C Y, He Z X and Wang Q. "Internal Flow Field Analysis of Engine Intake Manifold Based on RE and CFD", Advanced Materials Research, vol. 1079, pp. 930-934, 2015.

[3] Rahiman A, AD M S and Ramis M K. "CFD Analysis of Flow Field Development in a Direct Injection Diesel Engine with Different Manifolds", American Journal of Fluid Dynamics, Vol.4, No.3, pp.102-113,2014.

[4] Saravanan D, Gokhale A and Karthikeyan N. "Design and Development of a Flow Based Dual Intake Manifold System", SAE International Journal of Engines, Vol.8, No1., pp. 68-74,2014.

[5] Qinming Tan, Yihuai Hu, Xusheng Zhang, and Hongsheng Zhang "A Study on the Combustion Performance of Diesel Engines with O2 and CO2 Suction" in Hindawi Publishing Corporation Journal of Chemistry Volume 2016, Article ID 1258314, 7 pages http://dx.doi.org/10.1155/2016/1258314

[6] Bambang Wahono, and Harutoshi Ogai "Combustion Model and Control Parameter Optimization Methods for Single Cylinder Diesel Engine" in Hindawi Publishing Corporation Journal of Optimization Volume 2014, Article ID 135163, 9 pages http://dx.doi.org/10.1155/2014/135163.

[7] David C and Pascal C. "Fluid dynamic modeling of junctions internal combustion engine inlet and exhaust systems", Journal of Thermal Science, Vol. 19, No.5, pp.410 —418,2010.

[8] Jianmin Xu,"Flow analysis of engine intake manifold based on computational fluid dynamics", IOP Conf. Series: Journal of Physics: Conf. Series 916 (2017) 012043 doi :10.1088/1742-6596/916/1/012043

[9] Jay V. Shah, Prof. P. D. Patel, D. J. Jotava "Effect of Intake Manifold Inclination on Performance and Emission Parameters of 4-Stroke Single Cylinder C.I. Engine: A Technical Review", International Journal of Engineering Research & Technology (IJERT), Vol. 3 Issue 1, January – 2014, ISSN: 2278-0181, pp. 2310-2313

[10] G.Sivakumar and S.Senthil Kumar "Investigation on effect of Yttria Stabilized Zirconia coated piston crown on performance and emission characteristics of a diesel engine" Alexandria Engineering Journal Volume 53, Issue 4, December 2014, Pages 787-794

[11] Jena SP, Acharya SK, Das HC, Patnaik PP and Bajpai S "Investigation of the effect of FeCl3 on combustion and emission of diesel engine with thermal barrier coating", Sustainable Environment Research (2017), doi: 10.1016/j.serj.2017.10.002.

[12] Ceviz M and Akim M. "Design of a new SI engine intake manifold with variable length plenum", Energy Conversion and Management, Vol. 51, No11., pp. 2239 — 2244,2010.

[13] M.L.S Deva Kumar, S.Drakshayani and K.Vijaya Kumar Reddy "Effect of Fuel Injection Pressure on Performance of Single Cylinder Diesel Engine at Different Intake Manifold Inclinations" International Journal of Engineering and Innovative Technology (IJEIT) Volume 2, Issue 4, October 2012, ISSN: 2277-3754, pp. 20-28

[14] Dr. Sambhe Rajeshkumar U. and Dr. Shrirao Pankaj N."Performance Analysis of Single Cylinder (DI) Diesel Engine by Air Swirl Induction With Internally Threaded Inlet Manifolds", International Journal of Mechanical Engineering And Technology (IJMET), ISSN 0976 – 6340 (Print) ISSN 0976 – 6359 (Online) Volume 5, Issue 7, July (2014), pp. 144-150

[15] Dr. Hiregoudar Yerrennagoudaru and Shiva Prasad Desai "Effect of Inlet Air Swirl on Four Stroke Single Cylinder Diesel Engine Performance", International Journal of Recent Development in Engineering and Technology, ISSN 2347-6435 (Online) Volume 2, Issue 6, June 2014) pp. 95-103

[16] Deshmukh D.S., Deshmukh M.S., and Modak J.P. "Experimental Investigation of Effects of Operating Variables on a C.I. Engine Performance", International Journal of Industrial Engineering and Technology (IJIET). ISSN 0974-3146 Volume 2, Number 1 (2010), pp. 197-206.

[17] Patil A.A., Deshmukh D.S., Navale L.G. and Patil V.S. "Experimental Investigation of a C.I. Engine Operating Parameters for Energy Efficient Exhaust System Development" International Journal of Innovations in Mechanical & Automobile Engineering (IJIMAE), ISSN 2249-2968 (Print), (2011), Issue – I, Vol. -II, pp. 60-64.

[18] Deshmukh D.S., Modak J.P. and Nayak K.M. "Experimental Analysis of Backpressure Phenomenon Consideration for C.I. Engine Performance Improvement" International Powertrains, Fuels & Lubricants Meeting, Rio De Janeiro, Brazil, (2010) SAE Paper No. 2010-01-1575.

[19] Deshmukh M.S., Deshmukh D.S. and Modak J.P. "Experimental Investigation of Diesel Particulate Filter System on Single cylinder Stationary C.I. Engine" at ICEEE-2009, International Conference organized by Mulshi Institute of Technology and Research, Sambhave, Mulshi, Pune, (2009).

[20] Deshmukh D.S., Deshmukh M.S., and Modak J.P. "Design Considerations for Diesel Particulate Filter for Enhancement of Engine Performance and Emission Control" International Journal of Engineering Research and Technology. ISSN 0974-3154 Volume 3, Number 3 (2010), pp. 761—774.

[21] Deshmukh D.S., Deshmukh M.S., and Modak J.P. "Experimental Investigation of the Effect of Backpressure on an I.C. Engine Operating Performance" international journal of emerging technologies and applications in engineering, technology and sciences (IJ-ETA-ETS) (ISSN: 0974-3588 I APRIL 2010), (2010), Page no. 23-28.

[22] Deshmukh Dheeraj S., Deshmukh M.S., Patil P D., and Patel S.U. "Experimental Investigation on an Internal Combustion Engine for Operating Performance Enhancement" International Journal of Innovations in Mechanical & Automobile Engineering (IJIMAE), ISSN 2249-2968 (Print), Nov.-.2012, Issue – III, pp. 136-156.

[23] Talele Dipak C., Dr. Deshmukh Dheeraj S. and Rajput Swapnil A, "Possible Causes of Induction Pressure Reduction in a Four Stroke Compression Ignition Engine: A Review" in International Engineering Research Journal (IERJ) Special Issue 3 Page 183-186, 2016, ISSN 2395-1621.

[24] Arvindkumar K; Adhithiyan N; Darsak V S and Dinesh C "Optimisation of Intake Manifold Design Using Fibre Reinforced Plastic" International Journal of Scientific & Engineering Research, Volume 5, Issue 4, April-2014 922, ISSN 2229-5518

[25] Rajkumar, S, Hussain, R, Ponnusamy and Ragupathi. (2018). "Design and Analysis of Diesel Engine Intake Manifold Using CFD Simulation".

[26] Chinchanikar Sourabh Shrikrishna "Design, Analysis and Optimization of an Intake Manifold of two-cylinder Stationary Diesel Engine" International Research Journal of Engineering and Technology (IRJET) e-ISSN: 2395-0056, p-ISSN: 2395-0072, Volume: 05 Issue: 02 | Feb-2018 www.irjet.net

[27] Solanki P.M., Deshmukh D.S., Diware V.R., Deshmukh M.S. "Experimental investigation of thermoelectric generator system", Materials Today: Proceedings, Volume 47, Part 11, 2021, Pages 3012-3017, ISSN 2214-7853, https://doi.org/10.1016/j.matpr.2021.05.478.

[28] Deshmukh Sheetal D, Gabhane Shubham and Deshmukh Dheeraj S "Osmotic dehydration of carrot strips and modelling" , Published under licence by IOP Publishing Ltd, Journal of Physics: Conference Series, Volume 1913, International Conference on Research Frontiers in Sciences (ICRFS 2021) 5th-6th February 2021, Nagpur, India, doi:10.1088/1742-6596/1913/1/012093, Journal of Physics: Conference Series, 1913 (2021) 012093, Page No. 1-10, June 2021

[29] Kulkarni Mahesh.V., Deshmukh D.S. and Shekhawat S.P, "An innovative design approach of hot water storage tank for solar water heating system using artificial neural network", Materials Today: Proceedings, Volume 46, Part 11, 2021, Pages 5400-5405, ISSN 2214-7853, https://doi.org/10.1016/j.matpr.2020.09.058.

[30] Patil, P.P., Dr.Deshmukh D.S., Dr.Vaidya A.M and Paul, I.D., "Experimental Investigation of

New Innovative Flat Plate Collector for Solar Water Heater" (May 17, 2019). Proceedings of the Second International Conference on Emerging Trends in Science & Technologies For Engineering Systems (ICETSE-2019), SSRN: https://ssrn.com/abstract=3509071 or http://dx.doi.org/10.2139/ssrn.3509071

[31] Patil Pundlik N., Dehmukh Dheeraj S. and Patil Vilas S., "Analysis of the effect of LPG on the Performance and Frictional Power Loss for SI Engine", International Journal of Innovative Technology and Exploring Engineering (IJITEE), ISSN: 2278-3075, Volume-8, Issue-9S3, July 2019, Page No. 1406 -1409, Retrieval Number: I33000789S319/2019©BEIESP DOI: 10.35940/ijitee.I3300.0789S319, Published By: Blue Eyes Intelligence Engineering & Sciences Publication,

[32] Pimpale Shailesh, Deshmukh Manish, Shelke Rajesh and Deshmukh Dheeraj, published a research paper titled "Experimental Investigation on Hip Implant Materials Development through Analytical and Finite Element Analysis: 3D Modelled Computed Tomography", Biointerface Research in Applied Chemistry, Platinum Open Access Journal (ISSN: 2069-5837, Volume 12, Issue 3, 2022, 4103 – 4125, online available 15 August 2021, https://biointerfaceresearch.com/, https://doi.org/10.33263/BRIAC123.41034125

[33] Shailesh S. Pimpale, Manish S. Deshmukh, Rajesh T. Shelke and Dheeraj S. Deshmukh, published a research paper titled "Biomaterial Properties of Femur Implant on Acetabulum Erosion: A Review", Journal of Biomimetics, Biomaterials and Biomedical Engineering, ISSN: 2296- 9845, Volume 51, Page No. 39-62, June 2021, Online available 14 June 2021, @ Trans Tech Publications Ltd., Switzerland, ESCI Indexed, https://doi.org/10.4028/www.scientific.net/jbbbe.51.39

[34] Dr. Bhaskar B. S., Dr. Deshmukh M. S., Dr. Choudhary S.K. and Dr. Deshmukh D. S. "Heat Transfer Through Porous Materials (Aluminum Foam) Empirical Optimization of a Heat Exchanger", International Journal of Mechanical Engineering, ISSN: 0974-5823, Vol. 7 No. 1, Page No. 47 – 55, January, 2022, Copyrights @Kalahari

IOP Conf. Series: Materials Science and Engineering 1259 (2022) 012002 doi:10.1088/1757-899X/1259/1/012002

Effect of helmholtz resonator shape on suppression of thermo-acoustic instability

N N Deshmukh, A Ansari, A Phalak, J D'sa 'M Chaudhari' A Mathew
Department of Mechanical Engineering
Fr. C. Rodrigues Institute of Technology, Vashi
Navi Mumbai, Maharashtra, India - 400703

Email: afzal.ansari@fcrit.ac.in, nilajd@aero.iitb.ac.in

Abstract. In combustion chambers thermo-acoustic instabilities arise due to the coupling between the unsteady heat release and pressure fluctuations. The instabilities created cause serious structural vibration, flame blow out, increase in heat transfer and damage, thereby reducing the life span and performance of jet engines, gas turbines and industrial burners. Therefore, there is need to develop a simple technique for effective control of these instabilities. The present work involves experimental study of thermo-acoustic instabilities occurring in a Rijke tube and their suppression by using Helmholtz resonators of different shape (a passive method of suppression). The combustor model is replaced by Rijke tube which is an experimental setup extensively employed for research purpose. Rijke tube is provided with a co-axial pre-mixed gas burner as the source of heat which could be placed at any desired position. When a heat source with proper air to fuel ratio is placed at about one fourth of the tube length, 3rd mode of instability is obtained and a loud noise of frequency 681 Hz is produced. A comparative study of thermo-acoustic instability suppression using cylindrical, spherical and conical Helmholtz resonators has been performed. Spherical resonator has been found to be the most effective one imparting a maximum damping effect of 20 dB and an average damping effect of 10.5 dB.

1. Introduction

Thermo-acoustic instability is a critical phenomenon taking place in the gas turbine engines in aerospace and industrial applications in which lean premixed air-fuel mixture is used. In a combustor, the phenomenon of resonance takes place when the acoustic pressure waves superimpose over the heat waves. The occurrence of resonance magnifies the amplitude of the vibration of the fluid by several times. If this phenomenon is not checked, the oscillations induced in the fluid is transmitted to the adjoining structural components resulting in vibration of the solid structures and finally leading to ruinous structural failures causing loss of lives and property. Hence, it is really important to search for various methods and implement them to abate the thermo-acoustic instability in order to ascertain the structural integrity. The growth of thermo-acoustic instability is characterized by the Rayleigh Index which explains the occurrence of thermo-acoustic instabilities depending upon the phase angle between pressure fluctuations related to acoustic waves and unsteady heat release [1]. Li and Zhao [2] have discussed the growth of instability which depends on parameters like inlet velocity, temperature of the heat source and the location at which the heat source is placed. Several methods have been proposed and implemented to suppress the thermo-acoustic instabilities. They can be broadly categorized into active methods and passive methods. Methods involving an external energy source to suppress the instability are known as active methods, whereas, the methods where no external energy source is needed to abate the instability are passive methods. The active techniques are much more complicated than passive methods. Extensive use of sensors and actuators is involved in active methods. Experiments have been carried out by Deshmukh and Sharma [3] using active methods in which micro-jets of air at different positions along the Rijke tube have been used to suppress the instability. Further, investigations have been carried out by Deshmukh and Sharma [4] to study the effects of pointed injection and diffused injection over a premixed flame on the sound pressure level. Deshmukh et al. [5] have studied the effect of volume and length of the neck of a Helmholtz resonator on the suppression of thermo-acoustic instabilities and have concluded that for a longer neck length the standing wave developed inside the Rijke tube has to travel a longer distance resulting in less reduction in sound pressure level, hence shorter neck length was found to be more effective than longer neck length. They have also concluded that larger volume with same neck length become ineffective and doesn't contribute for reduction in sound pressure level. Wang et al. [6] have studied the effectiveness of distributed Helmholtz resonator and have concluded that instead of using a Helmholtz resonator with single neck the use of multiple small holes with an equivalent open area can increase the damping bandwidth of the resonator. Tang [7] has theoretically and experimentally investigated the effect of taper in the geometry of the resonator neck and found that providing taper in the neck significantly improves the sound absorption capacity of the resonator. Bourquard and Noiray [8] have studied the stability of the coupled system analytically and experimentally. A model for their acoustic impedance is derived and experimentally validated. In a second part, a thermos-acoustic instability is mimicked by an electro-acoustic feedback loop in a rectangular cavity, to which the dampers are added. The length of the dampers can be adjusted, so that the system can be studied for tuned and detuned conditions. In their study they have shown for tuned dampers, the best stabilization is achieved at the exceptional points. Selamet et al. [9] have studied the effect of fibrous material on the acoustic performance of a concentric circular Helmholtz resonator. They have studied the effect of density and the thickness of the fibrous material in the cavity on the resonance frequency and the transmission loss. They have concluded that both the resonance frequency and the peak value of the transmission loss decrease with increasing absorbent thickness. Hosseini et al. [10] have studied the effects of superimposition between thermal and acoustic waves in a burner and heat exchanger setup and have also tested and proposed a method to break the resonance and suppress the instability.

Since active methods employ complex control techniques and sophisticated sensors and actuators, to overcome the drawback related to complexity of the mechanism, passive methods are preferred over the active ones. The passive methods involve use of Helmholtz resonators, perforated liners and baffles. Among the several methods, use of Helmholtz resonator is widely accepted in most of the applications. Bourquard and Noiray [11] studied the abation of instabilities emerging out of thermo-acoustic effects by the use of dampers on the combustion chamber walls which could decouple the acoustic waves from the heat waves. The combustion chamber is modelled as a cavity featuring a linearly unstable thermo-acoustic mode. The analytical model has been experimentally validated on a simple rectangular cavity, where a feedback loop involving a loudspeaker and microphones have been used to model and suppress the instability arising from the resonance between acoustic pressure waves and heat waves. The analytical predictions of the stabilization capabilities of a Helmholtz resonator are in coherence with the experimental results. Gupta and Gupta [12] have proposed that different results were obtained when different shapes of resonators were used. Cylindrical, triangular and spherical resonators were mounted on an exhaust pipe to study the effects of the different shape of resonators. The maximum efficient resonator was found to be the cylindrical shaped Helmholtz resonator. Selamet and Lee [13] have performed a series of experiments to analyse the performance of a circular resonator with extended neck. It was concluded that the resonance decreases with the increase in neck length. Komkin et al. [14] have generated sound field with the help of a loud speaker and placed a Helmholtz resonator at the end wall of a circular duct to study its absorption characteristics. Jo et al. [15] discussed the effect of acoustic cavities with large orifice area on damping. Dowd and Meadows [16] have studied the influence of the introduction of a metal plate on the acoustic properties of the Helmholtz resonator. Positioning of the porous inert media near anti-nodes provided the better damping compared to other positions. The results showed that the use of PIM (porous inert media) extended the operational limits of the combustor compared to the conventional combustor without PIM coating. Yu et al. [17] have studied the phenomena of maximization of energy reduction in a Helmholtz resonator and related it to the internal resistance of a Helmholtz resonator and its influence on the acoustic suppression. Zhang et al. [18] devised a Helmholtz resonator whose neck area could be tuned in real-time and studied its damping effect. Paschereit et al. [19] have achieved unstable thermo-acoustic modes by modifying the acoustic boundary conditions and controlled them by the use of a swirl stabilizer. Noiray et al. [20] have studied the passive method of suppressing instabilities in a multi-point injection configuration. A perforated plate on which multiple conical flames are anchored has been used to study the suppression characteristics. Zhao and Li [21] have compiled and reviewed data related to acoustic dampers which have been used to attenuate the instability arising in combustion related to aerospace industry.

From the review of the already existing literature in the field of thermo-acoustic instability suppression, it can be concluded that not much research has been carried out to compare the effects of different shape of Helmholtz resonators on thermo-acoustic instability abation. In this paper efforts have been taken to compare the effects of change in the shape of Helmholtz resonator on reduction of instabilities arising from the superimposition of acoustic and thermal waves. Three resonators of same volume but different shapes viz. spherical, cylindrical and conical have been designed and fabricated and their performances have been compared. Raw signals in terms of voltage vs time are acquired and further subjected to FFT in order to acquire the voltage variation with respect to frequency using the NI-DAQ and LabVIEW software. The data are further post processed using the calibration factor of the pressure transducer and presented in decibels (dB) as sound pressure level (SPL). This paper gives an overall comprehensive study of the effects of shape of Helmholtz resonators of different shapes. As an outcome of the study, for the same volume, spherical resonator has been found to be the most efficient one which has helped to suppress the instability by a maximum magnitude of around 20 dB and an average magnitude of 10.5 dB.

2. Experimental set-up and Instrumentation

The experimental setup consists of a horizontal Rijke tube having an internal diameter of 80 mm, length of 800 mm and thickness of 5 mm. The L/d ratio has been maintained as 10 in order to ensure 1D combustor effect. Nine equidistant holes of M6 are made on the wall of Rijke tube for the purpose of mounting the Helmholtz resonator. A co-axial cylindrical gas burner of 30 mm diameter is used as heat source. A lean mixture of air and LPG is prepared in a pre-mixer after metering appropriate amount using rotameters. Figure 1 shows the schematic layout of the experimental setup.

On the wall of the Rijke tube, 9 pressure taps are provided at an interval of $x/L = 0.1$. These pressure taps are used for measuring the wall pressures. The wall pressure measurement is performed using RS 395-257, ultra-low differential pressure transducer. It has a range of 0 – 60 mm of water and accuracy of ± 2%. A 4-channel C series dynamic signal acquisition module, NI 9234 from National Instruments, having an operating temperature range between −400 to 700°C) has been used to acquire the temperature and acoustic. NI USB 6211, a multifunctional DAQ device, has been used for the measurement of pressure. The data was acquired at the rate of 50000 Hz and 8124 samples were acquired using NI LabVIEW.

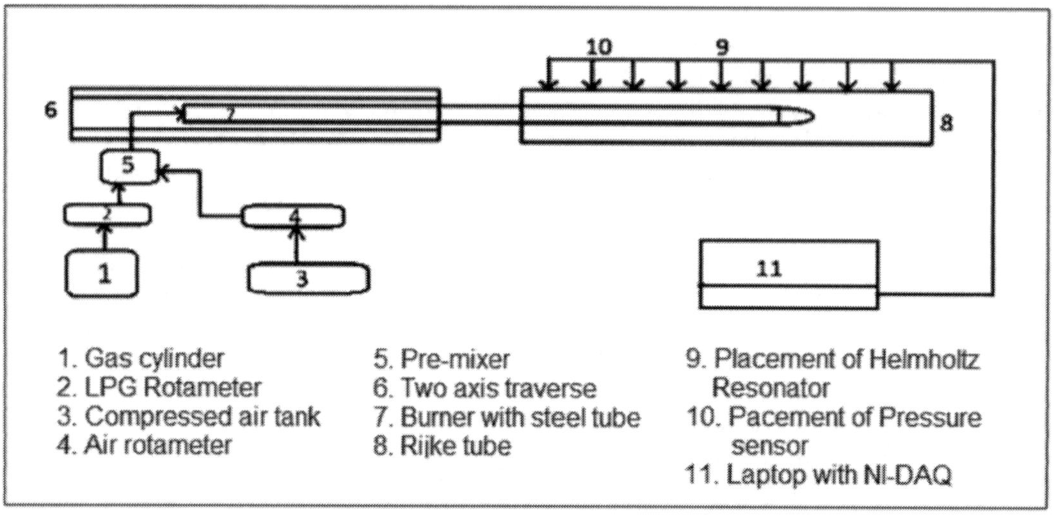

Figure 1. Schematic diagram of experimental setup.

2.1. Design, Modelling and Fabrication of Helmholtz Resonator

The present study involved three different types of Helmholtz resonators based on their shape viz. spherical, conical and cylindrical. In preliminary study, it was observed that 3rd mode of thermo-acoustic instability exist for wide range of air/fuel ratios and burner positions. For suppressing third mode of instability, the Helmholtz resonators of different shapes were designed keeping constant volume ($V = 2.11\times10^{-5}$ m^3). The model of the spherical, conical and cylindrical Helmholtz resonators and their sectional views are shown in Figure 2(a), figure 2(b), figure 3(a), figure 3(b), figure 4(a) and figure 4(b) respectively. The radii of spherical, conical and cylindrical resonators are 17.2 mm, 20.1 mm and 24.1 mm respectively. The 3D models of resonators were developed in Autodesk Inventor 2018 and exported as an STL file into Fraktory 3D printing software. The Helmholtz resonators were fabricated using 3D printing machine as a one piece except cylindrical resonator. Cylindrical resonator was fabricated in two parts and press fitted to avoid the leakages. To sustain Helmholtz resonators at high temperature, PLA (Polylactic Acid) was used as a 3D printing material. The melting point of PLA is 130°C – 180°C, whereas maximum surface temperature of Rijke tube reaches up to 80°C. Figure 3, 4 and 5 show the final view of the resonator after the fabrication. Each resonator is fitted with a M6 grub screw to the neck with outer diameter 10 mm and length of 10 mm. This assembly was fitted with external thread on the holes drilled and threaded on the wall of the Rijke tube.

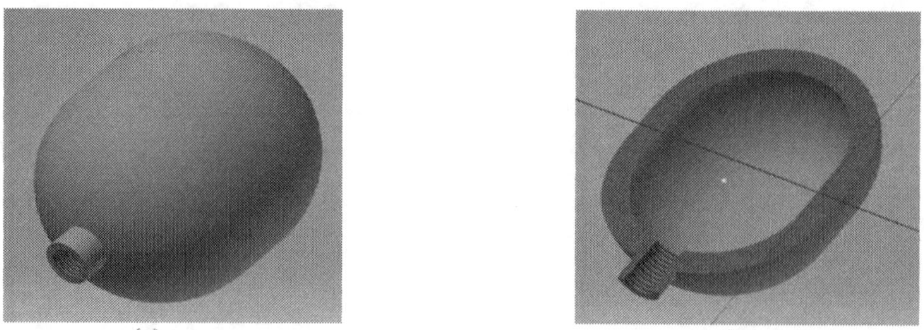

(a) (b)

Figure 2. Spherical Helmholtz resonator with (a) 3D view and (b) Sectional view.

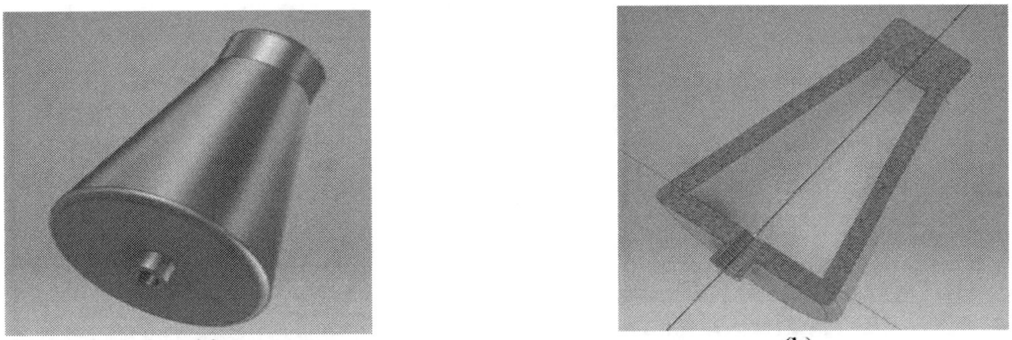

(a) (b)

Figure 3. Conical Helmholtz resonator with (a) 3D view and (b) Sectional view.

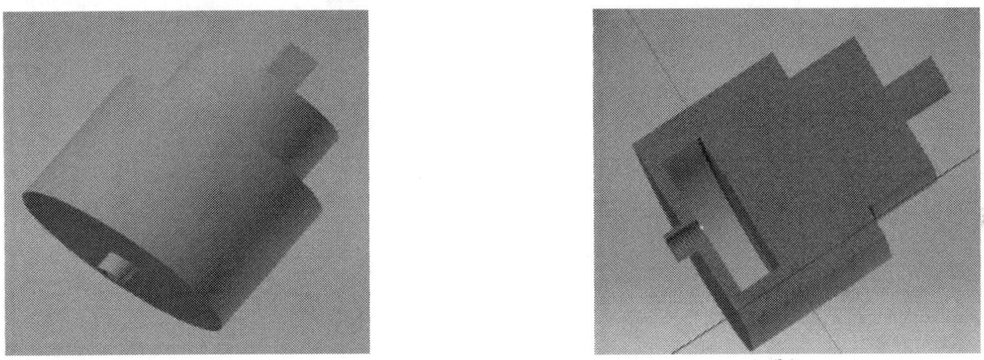

(a) (b)

Figure 4. Cylindrical Helmholtz resonator with (a) 3D view and (b) Sectional view.

2.2. Experimental Methodology

To evaluate the performance of Helmholtz resonator of different shapes, experiments were carried out for 9.1 LPM of air supply and 0.25 LPM of fuel supply. The stoichiometric air to fuel ratio for LPG is 28 and experiments were carried out at an air to fuel ratio of 36.4 which is leaner than the stoichiometric ratio. The experiments were carried out without Helmholtz resonator and with Helmholtz resonator to compare the results. Three types of resonators with spherical, conical and cylindrical shapes which were fabricated using a Fused Deposition Modelling Rapid Prototyping process were used for carrying out the experiments. The burner was positioned at L/4 distance from both ends of the Rijke tube and proper air to fuel ratio was maintained with the help of the rotameters in order to achieve the third mode of instability. Using LabVIEW, measurements of wall pressure were carried out. Figure 5 indicates block diagram for wall pressure measurements using NI LabVIEW.

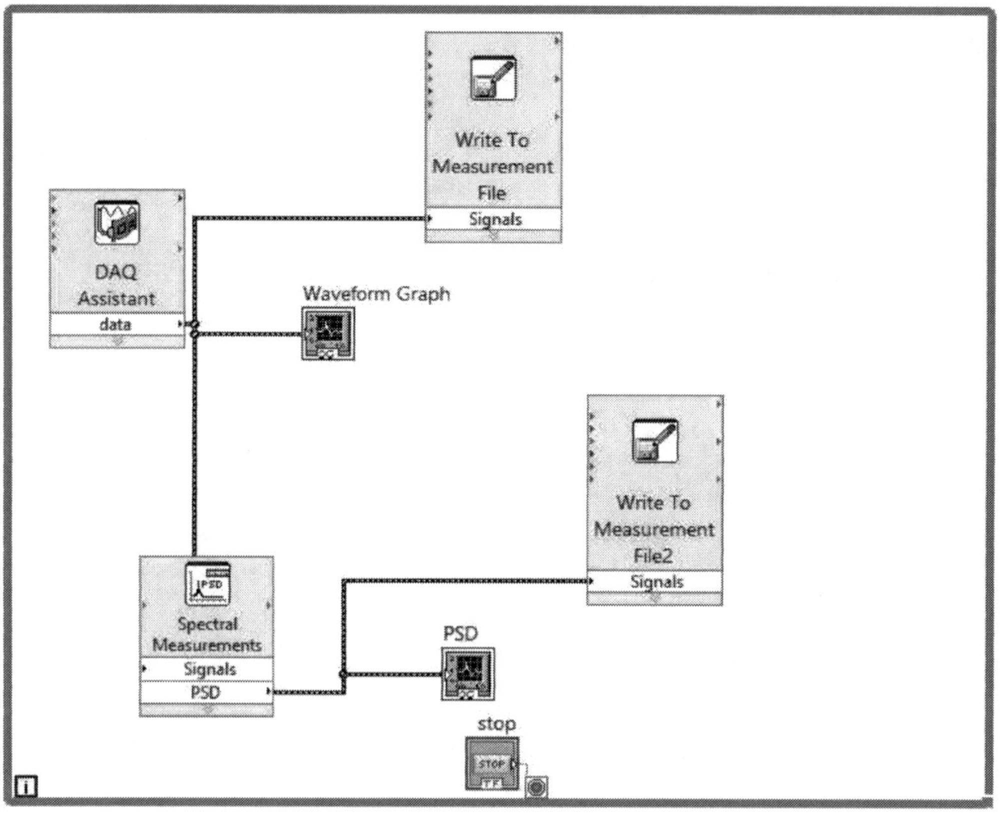

Figure 5. Block diagram of LabVIEW VI.

Pressure readings were taken by attaching an ultra-low differential pressure transducer to pressure tap of 1 mm diameter on the wall of Rijke tube through PVC tube. Before starting the experiments, the transducer was calibrated using a manometer and calibration factor was determined. For the calibration purpose, one end of pressure transducer was connected to pressure tap and the other end was vented to the atmosphere. The excitation power was provided externally by a supply of 8 Volt DC. The output of pressure transducer is obtained in volts, which was converted to Pascal by dividing with the calibration factor. The pressure reading obtained is summation of hydrodynamic pressure and acoustic pressure. The final sound pressure level in dB was calculated using the relation given in equation (1).

$$\mathrm{SPL} = 10\log_{10}\frac{\mathrm{P}^2}{\mathrm{P}_{\mathrm{ref}}^2} \tag{1}$$

Where, $\mathrm{P}_{\mathrm{ref}} = 20\ \mu\mathrm{Pa}$

3. Results and Discussion

The 3rd mode of thermo-acoustic instabilities were observed at a frequency of 681 Hz. The location of Helmholtz Resonator along the length of the Rijke tube affects the output sound pressure intensity. It was observed that there is a significant suppression of sound pressure level for 3rd mode at 5th hole as it was close to anti-node position. The excitation voltage used was 8V and the corresponding calibration factor of the sensor was found out to be 0.00256 V/Pa. The output of the pressure transducer provides raw signal in volts, which is converted to Pascal by dividing with calibration factor. Without Helmholtz resonator and with Helmholtz resonator are the two sets of readings that are shown for 3rd mode (681 Hz) in time versus pressure for different resonators.

3.1. Comparison based on time domain signal

Figure 6, 7 and 8 show wall pressure variation at 3 locations for spherical, cylindrical and conical Helmholtz resonators. The maximum wall pressure difference (without HR and with HR) was 65 Pa for spherical Helmholtz resonator, whereas for cylindrical and conical Helmholtz resonators it was found to be around 45 Pa and 40 Pa respectively. The lowest pressure fluctuation was around 360 Pa for spherical Helmholtz resonator, whereas around 430 Pa for cylindrical and conical Helmholtz resonators. This shows the effectiveness of spherical Helmholtz resonator compared to other two resonators.

3.2. Comparison based on FFT analysis

The FFT (Fast Fourier Transformation) of the obtained raw voltage signal was performed in LabVIEW software. The output obtained was in the form of voltage with respect to frequency and then it was converted to decibel (dB).

Figure 9, 10 and 11 show frequency plot obtained by FFT analysis for all three Helmholtz resonators. Comparative plots clearly show the suppression of thermo-acoustic instabilities up to 20 dB for spherical Helmholtz resonator. The developed standing wave inside the Rijke tube enters in to the Helmholtz resonator and shape of Helmholtz resonator helps to suppress the maximum amplitude at the resonance. Spherical Helmholtz resonator allows the standing wave to move inside the resonator freely compared to cylindrical and conical resonator. All plots show that the peak sound pressure level decreases without rise in other harmonics. With addition of Helmholtz resonator of different shape, no noticeable changes in frequency plot are observed. Therefore with addition of Helmholtz resonator of small volume, thermo-acoustic instability can be eliminated at the design stage of combustor.

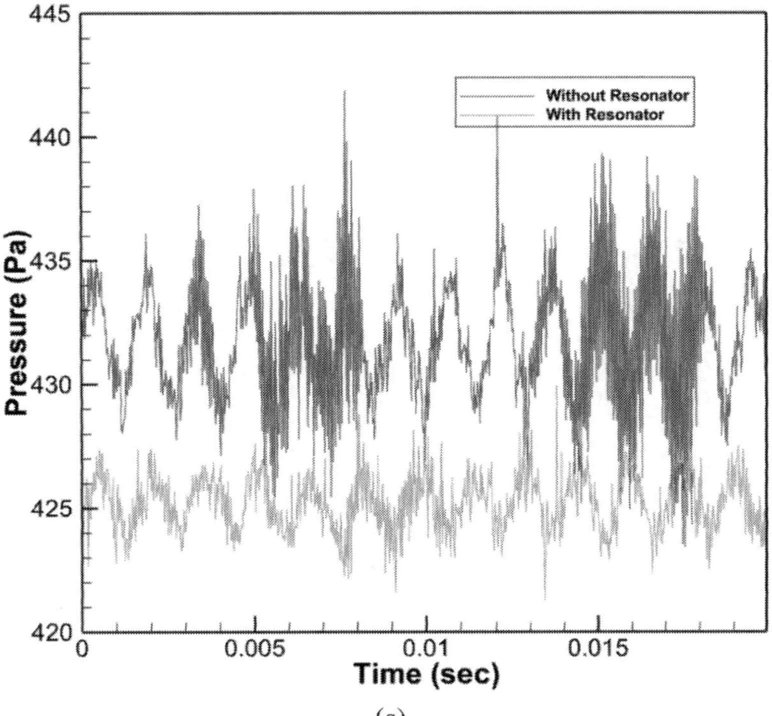

(a)

IOP Conf. Series: Materials Science and Engineering 1259 (2022) 012002 doi:10.1088/1757-899X/1259/1/012002

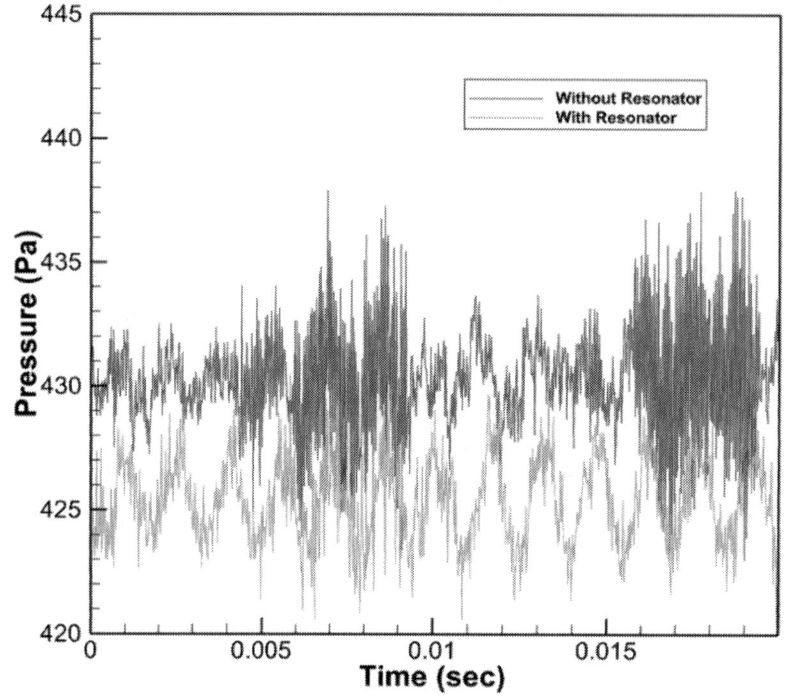

(b)

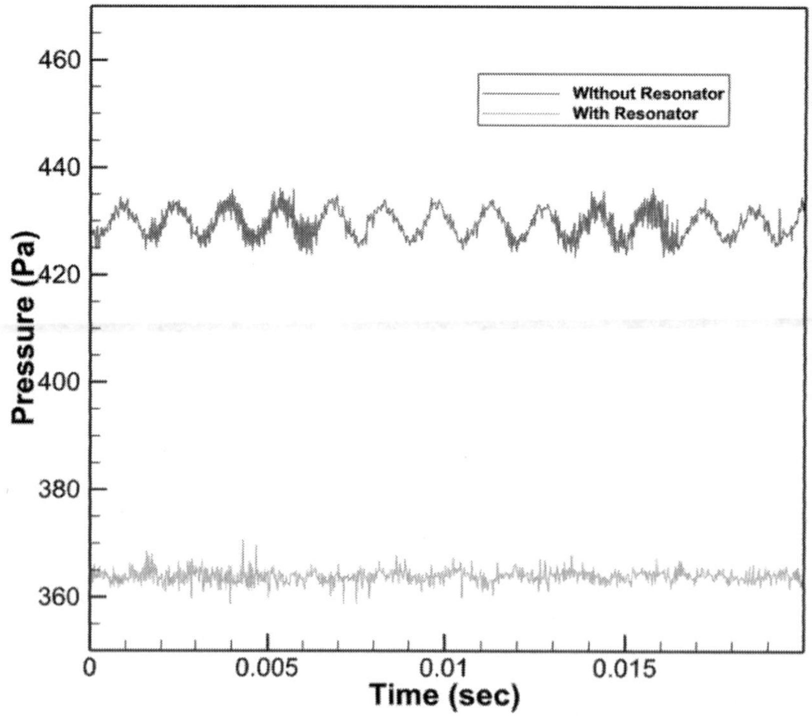

(c)

Figure 6. Time domain signal of pressure transducer using spherical resonator at (a) x/L = 0.2, (b) x/L = 0.6 and (c) x/L = 0.9.

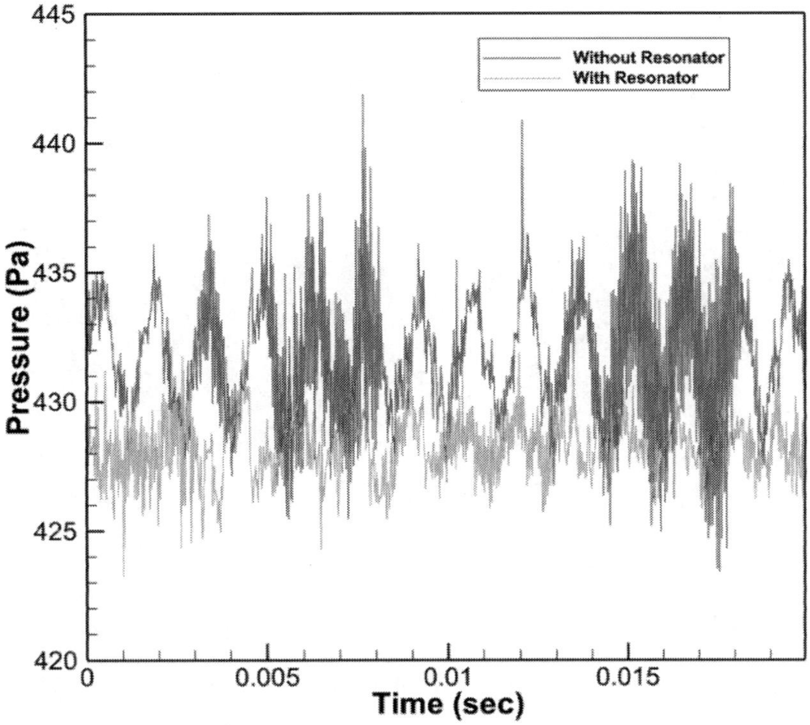
Without Resonator
With Resonator
445
440
435
430
425
420
Pressure (Pa)
0
0.005
0.01
0.015
Time (sec)

(a)

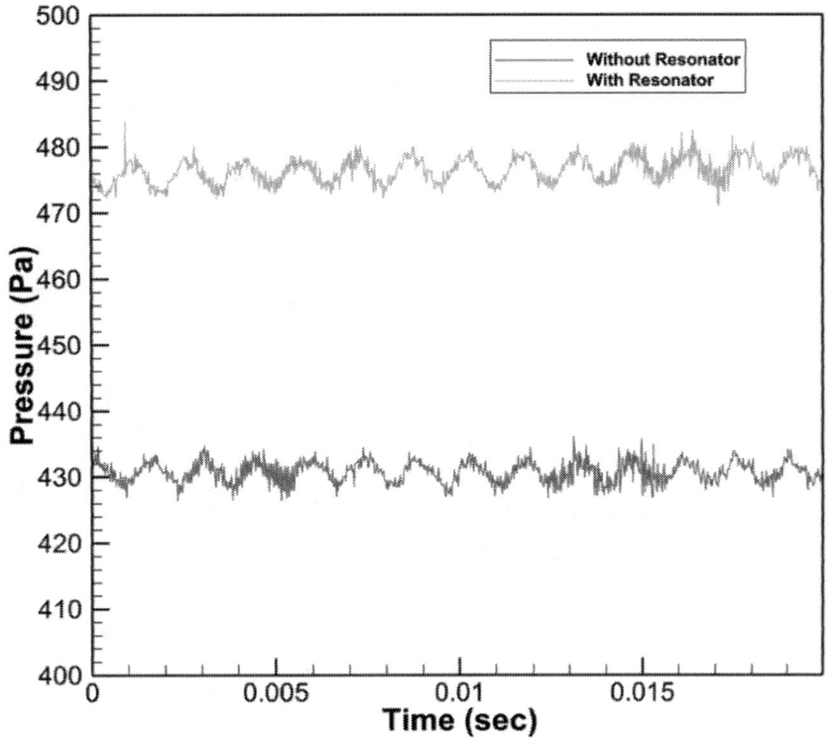
Without Resonator
With Resonator
500
490
480
470
460
450
440
430
420
410
400
Pressure (Pa)
0
0.005
0.01
0.015
Time (sec)

(b)

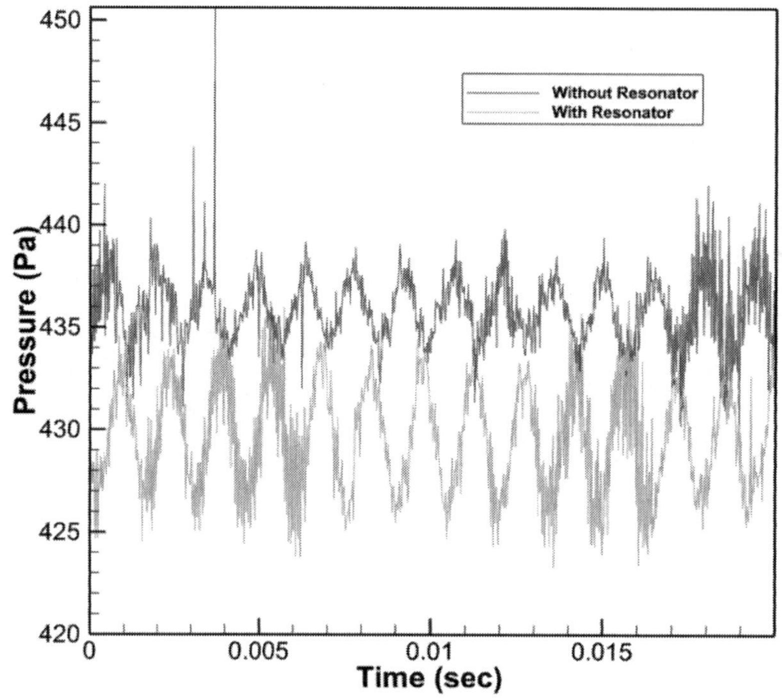

(c)

Figure 7. Time domain signal of pressure transducer using cylindrical resonator at (a) x/L = 0.2, (b) x/L = 0.6 and (c) x/L = 0.9.

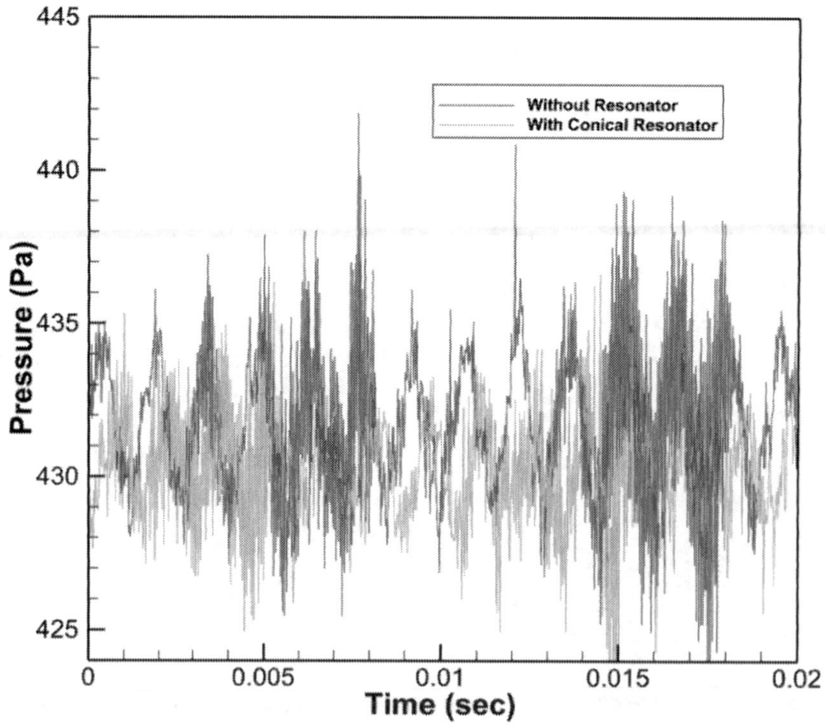

(a)

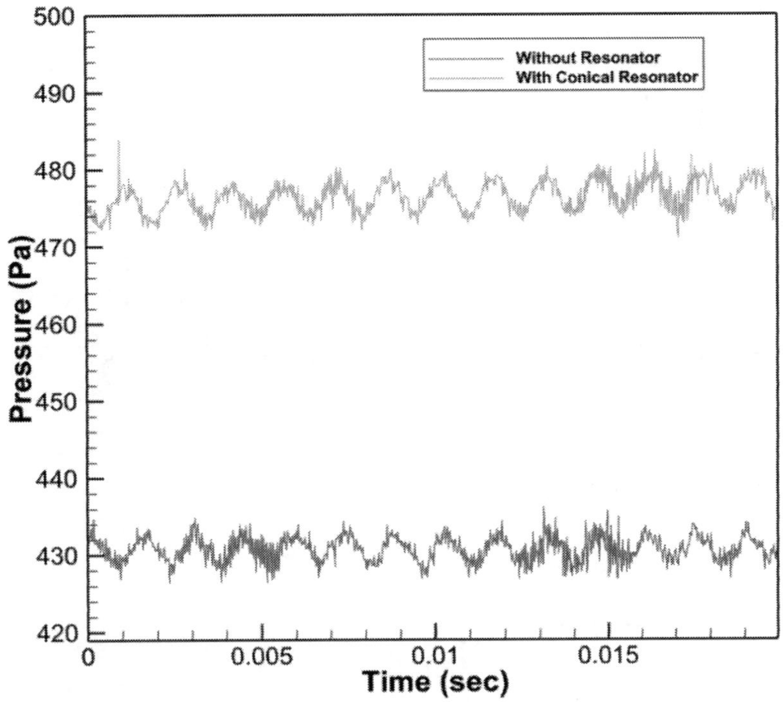

(b)

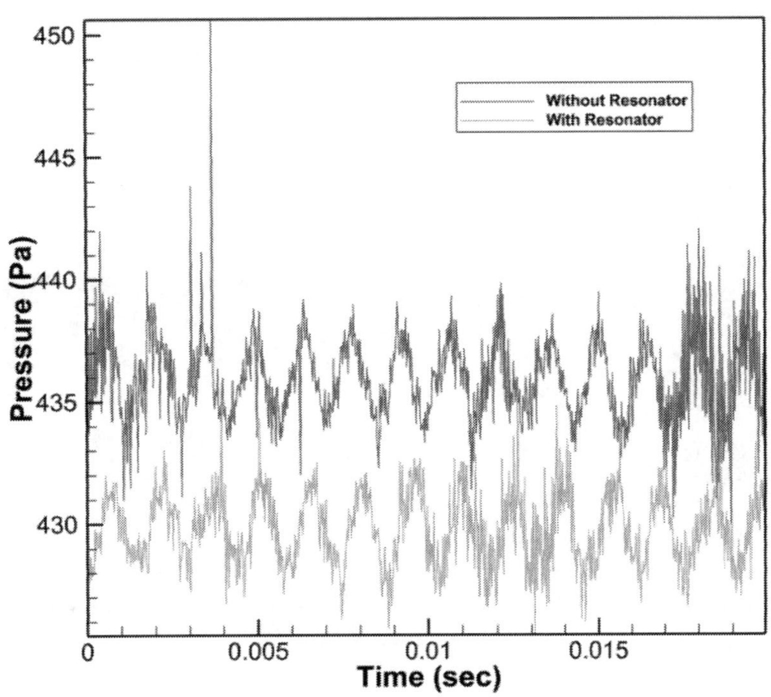

(c)

Figure 8. Time domain signal of pressure transducer using conical resonator at (a) x/L = 0.2, (b) x/L = 0.6 and (c) x/L = 0.9.

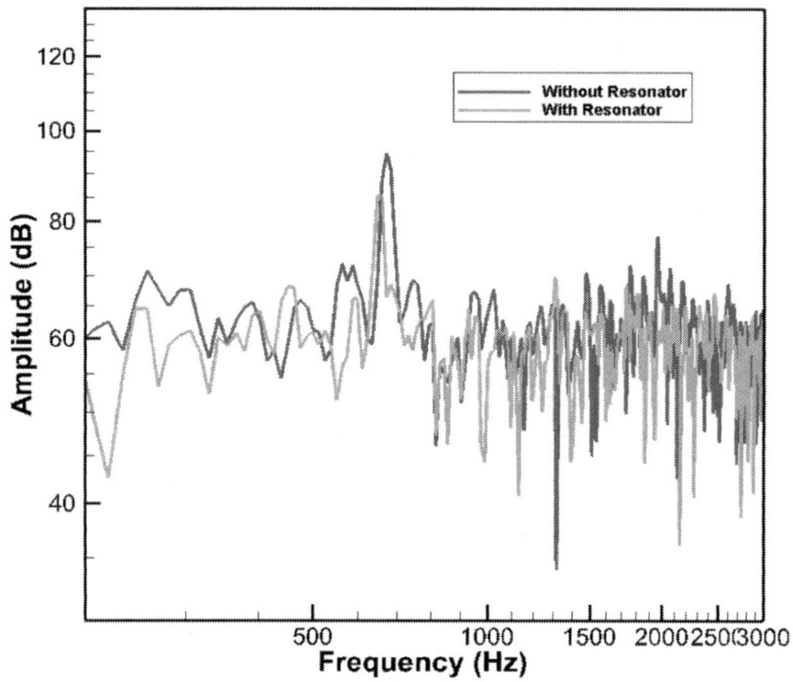
Without Resonator
With Resonator
120
100
80
60
40
Amplitude (dB)
500
1000
1500
2000
2500
3000
Frequency (Hz)

(a)

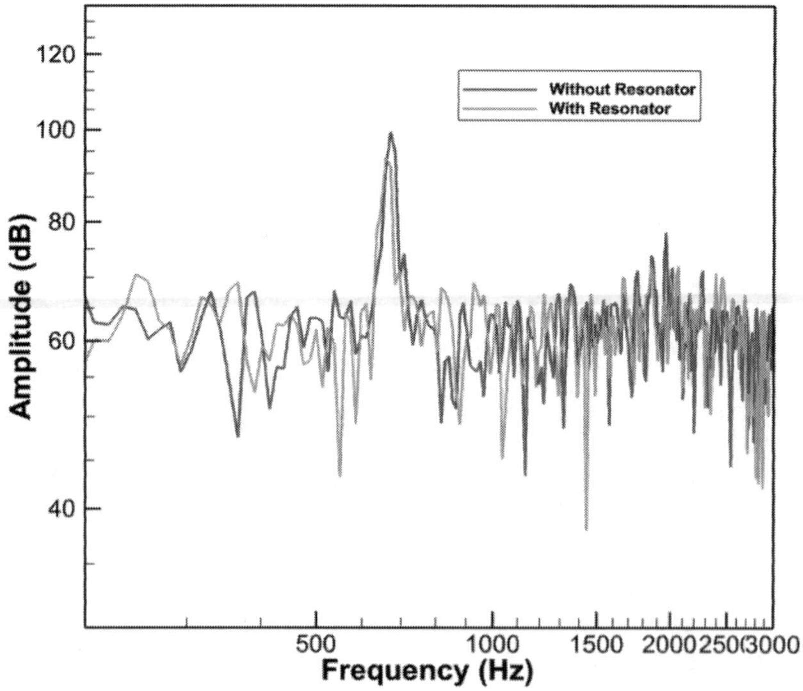
Without Resonator
With Resonator
120
100
80
60
40
Amplitude (dB)
500
1000
1500
2000
2500
3000
Frequency (Hz)

(b)

IOP Conf. Series: Materials Science and Engineering 1259 (2022) 012002 doi:10.1088/1757-899X/1259/1/012002

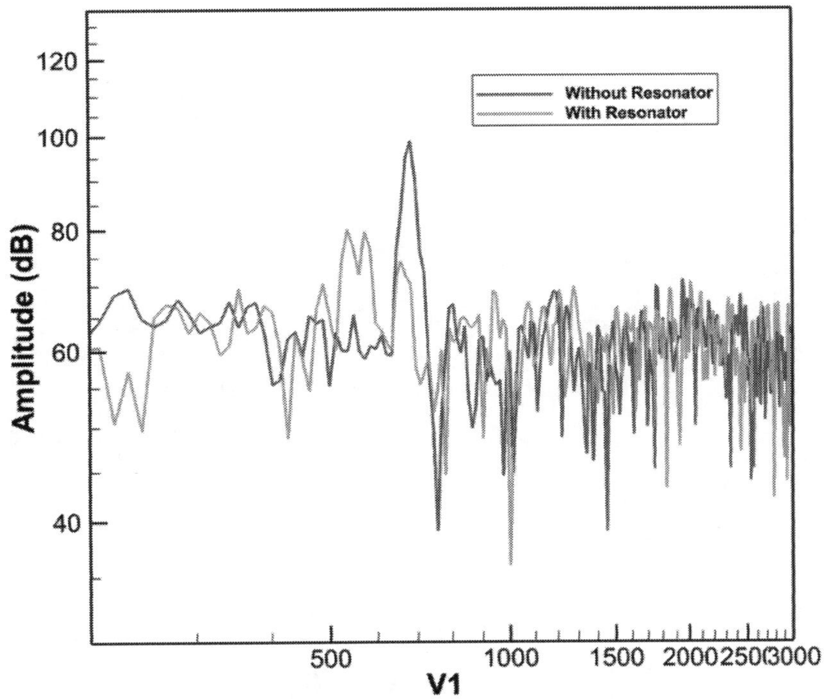

(c)

Figure 9. Frequency plot for spherical resonator at (a) x/L = 0.2, (b) x/L = 0.6 and (c) x/L = 0.9.

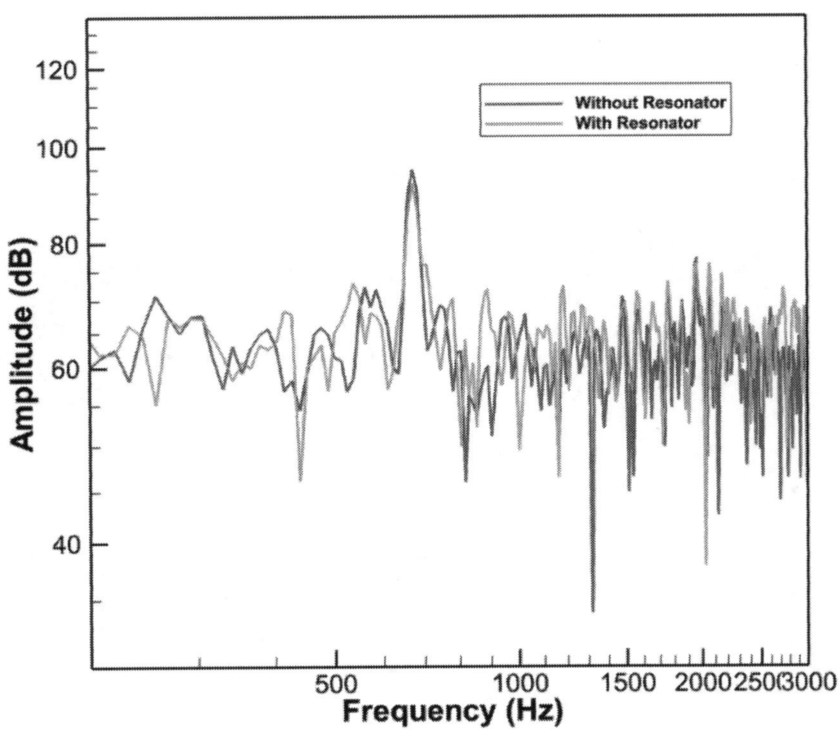

(a)

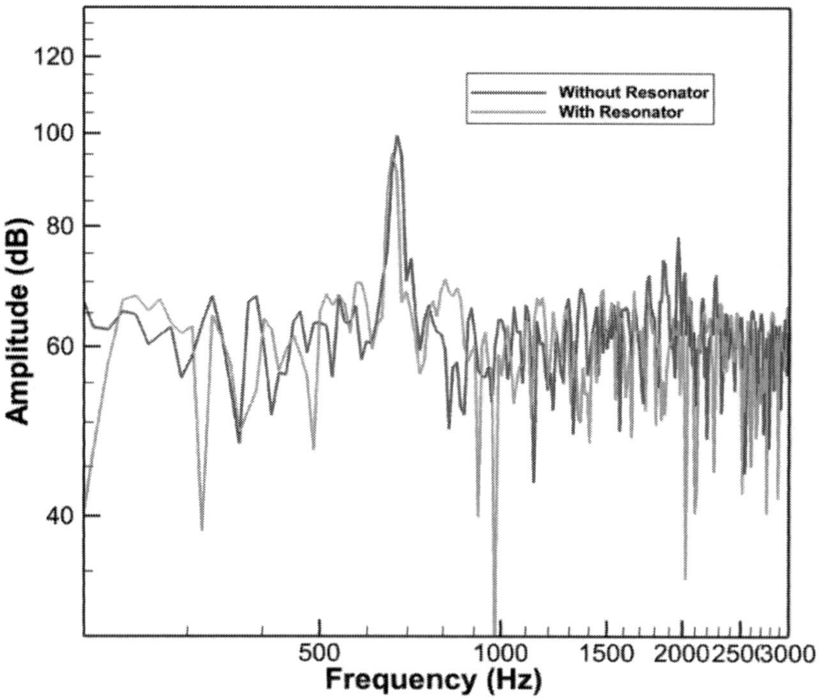

(b)

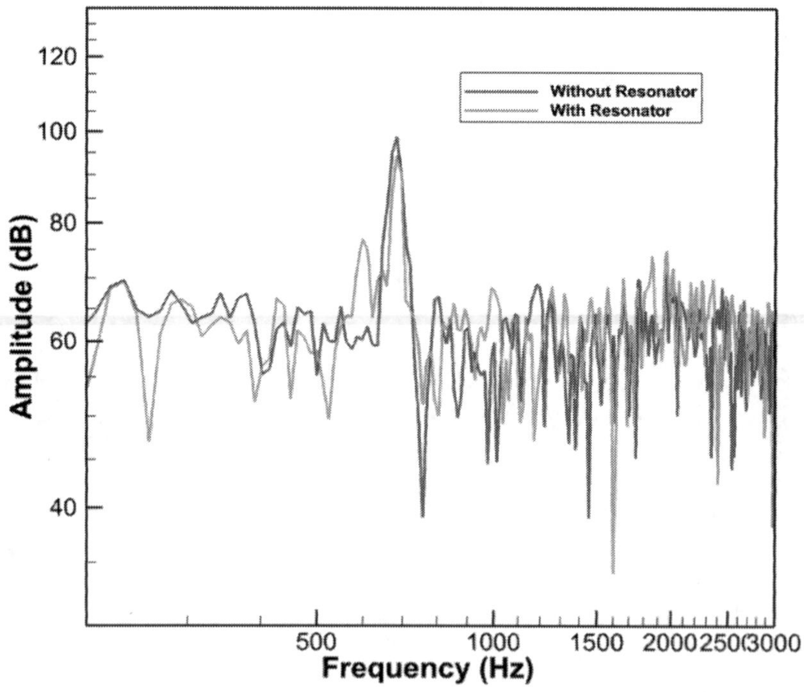

(c)

Figure 10. Frequency plot for cylindrical resonator at (a) x/L = 0.2, (b) x/L = 0.6 and (c) x/L = 0.9.

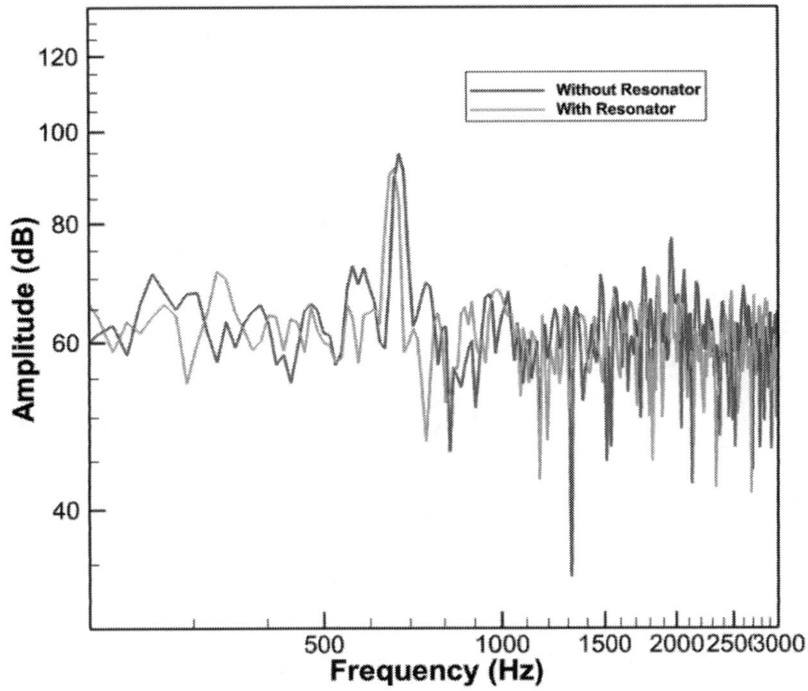
Without Resonator
With Resonator
120
100
80
60
40
Amplitude (dB)
500
1000
1500
2000
2500
3000
Frequency (Hz)

(a)

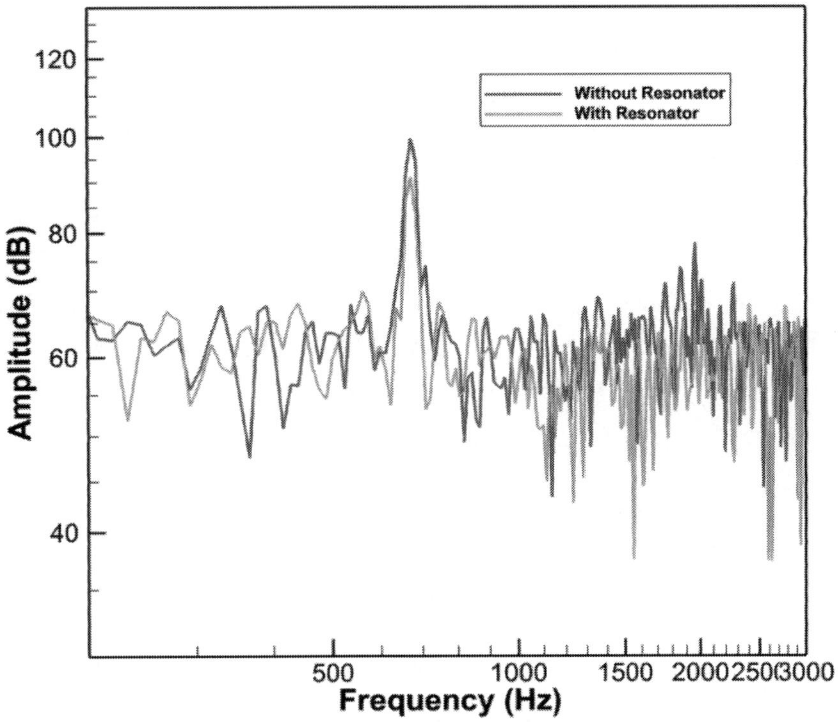
Without Resonator
With Resonator
120
100
80
60
40
Amplitude (dB)
500
1000
1500
2000
2500
3000
Frequency (Hz)

(b)

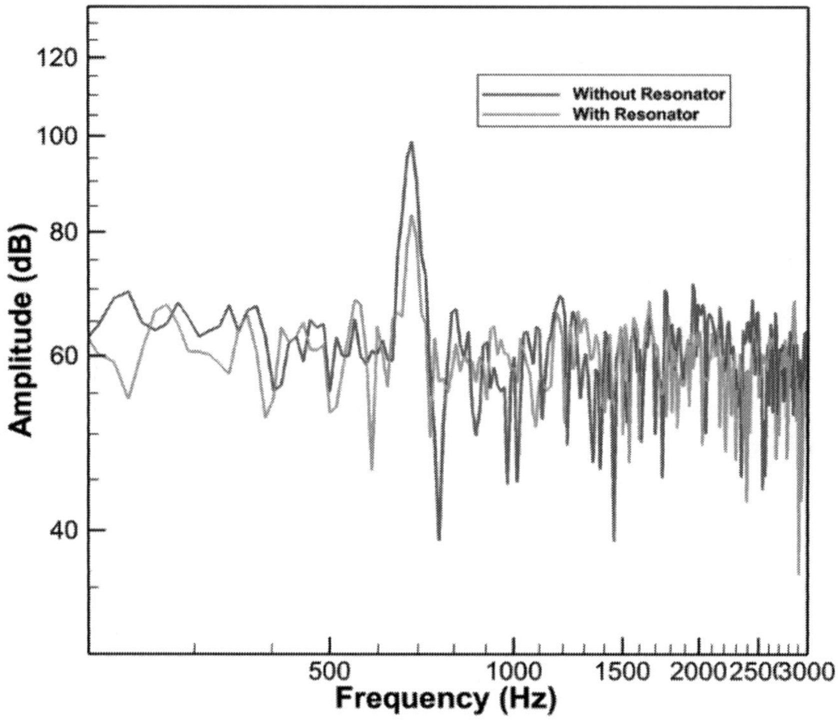

(c)

Figure 11. Frequency plot for conical resonator at (a) x/L = 0.2, (b) x/L = 0.6 and (c) x/L = 0.9.

Figure 11 shows amplitude (dB) was reduced in the range of 3 dB to 18 dB at different holes. The maximum sound pressure reduction was observed as 18 dB at 9th hole. Due to presence of back pressure in the pre-mixer, frequency shift was observed at some of the positions. The comparison of reduction in the decibel levels for all the three resonators is shown in figure 12.

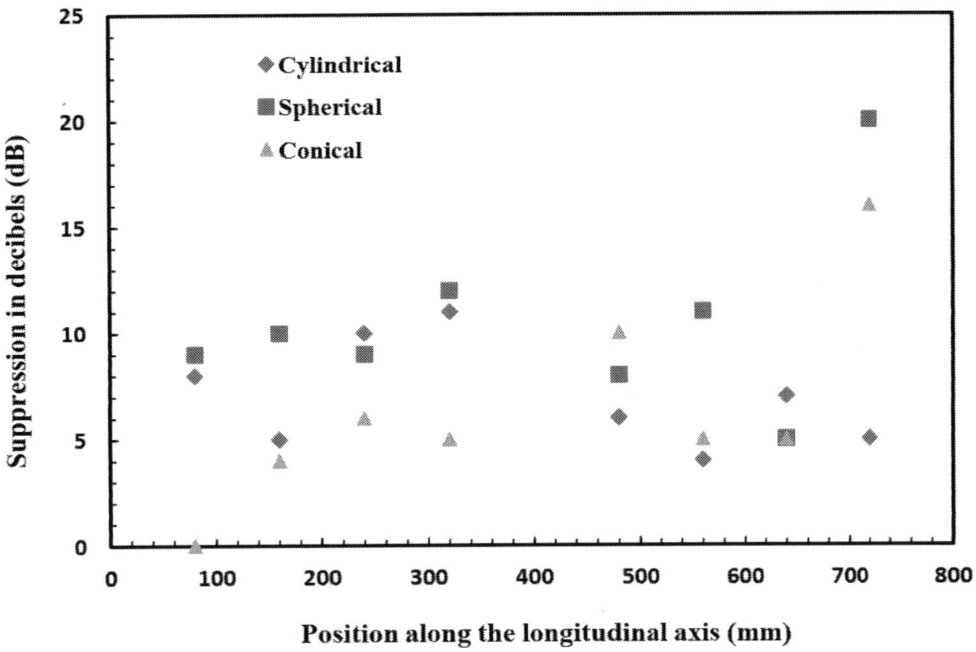

Figure 12. Performance comparison of various shaped resonators.

The maximum reduction of around 20 dB in sound pressure level is observed for the spherical resonator. Since the maximum values of decrease in sound pressure level are not achieved at the points of anti-node, the average values of reduction in sound pressure level are estimated experimentally. The average suppression in SPL is found out to be around 10.5 dB for spherical resonator, 7 dB for cylindrical resonator and 6.375 dB for conical resonator.

Figure 13 shows frequency plot for resonators of different shapes to understand effectiveness of resonators over a wide range of frequency. From plot, it is clear that spherical resonator is very effective for suppressing thermo-acoustic instability. There is no frequency shift or rise in amplitude at any other frequency due to mounting of resonator on the combustor.

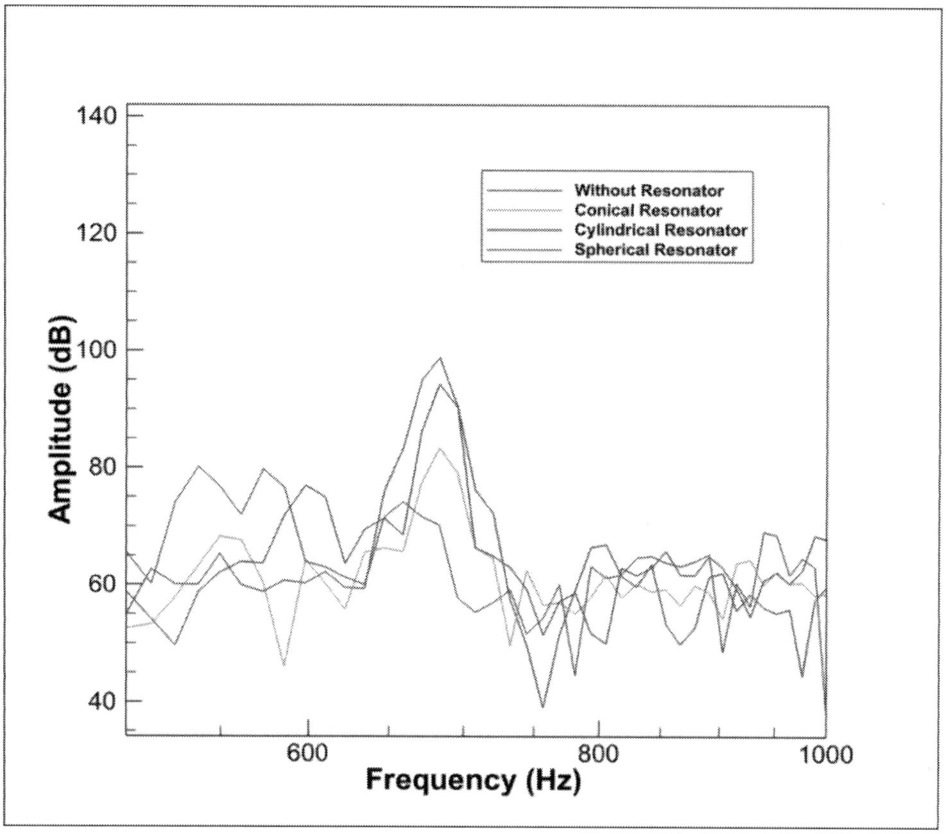

Figure 13. Frequency Plot for resonators of different shapes.

3.3. Comparison of Transmission Loss

The performance of the resonators have been compared on the basis of transmission loss of the different shaped resonators. The transmission loss has been calculated using the relation given in equation (2).

$$TL = 10\log_{10}\left[1+\left(\frac{A_c}{2A_p}\frac{\tan kl_c + (A_v/A_c)\tan kl}{1-(\frac{A_v}{A_c})\tan kl.\tan kl_c}\right)^2\right] \quad (2)$$

Where,

A_c = Neck Area
A_p = Pipe Area
A_v = Cavity Area
k = wave Number = $\frac{2\pi}{\lambda}$ (where λ is the wavelength of the acoustic waves)
l_c = Neck length
l = Cavity length

Based on the above relation the transmission losses for spherical, cylindrical and conical resonators have been determined to be 4.935 dB, 1.327 dB and 0.00541 dB respectively. The results have been compared in table 1. The comparison shows that transmission losses are highest in spherical resonator followed by cylindrical resonator and conical resonator. For calculating the transmission loss for the resonators following parameters have been used:

Table 1. Comparison of Transmission Losses.

Resonator Type	Neck Diameter d_c	Neck Length l_c	Pipe Diameter (m) d_p	Cavity Diameter (m) d	Cavity length (m) l	Transmission Loss (dB) TL
Spherical	0.006	0.01	0.079	0.0343	0.024	4.935
Cylindrical	0.006	0.01	0.079	0.04822	0.011	1.327
Conical	0.006	0.01	0.079	0.040186	0.05	0.005

For the calculation of transmission loss for spherical resonator, the effective cavity length has been taken as 24 mm instead of 34.3 mm as the cross section area of the sphere decreases at sections away from the centre. From the above analysis the spherical resonator has been found to be most effective followed by cylindrical resonator and conical resonator.

4. Conclusion

The thermo-acoustic instability occurs at 3rd mode with frequency around 681 Hz. The various geometries of various Helmholtz resonators are designed using 3rd mode of frequency. The passive method of subduing the thermo-acoustic instability in a Rijke tube, by means of Helmholtz resonator has been successfully developed and implemented. The maximum damping achieved by spherical resonator is 20 dB, cylindrical resonator is 11 dB and conical resonator is 18 dB. But since the maximum values of damping in the sound pressure level are not exactly at the antinode points, the average damping capacity has been taken for comparing the performance of the resonators. The comparison of average damping capacity shows the reduction in sound pressure level to be around 10.5 dB for spherical resonator, 7 dB for cylindrical resonator and 6.375 dB for conical resonator. The same trend of damping is shown by the transmission loss analysis for the three resonators which shows maximum transmission loss for spherical resonator followed by cylindrical resonator and conical resonator. From the results obtained it has been concluded that thermo-acoustic instability of 3rd mode of frequency can be best attenuated by the use of spherical Helmholtz resonator. While designing Helmholtz resonator, the shape is very important as the volume of all resonators were kept same to compare the result. It has also been observed that for effective control of thermo-acoustic instability, factors like the location of heat source, shape of Helmholtz resonator and placement of Helmholtz resonator play a key role. With proper design, elimination of thermo-acoustic instability is possible at design stage. This will help in improving the performance and life of combustor.

References

[1] Rayleigh 1878 The explanation of certain thermoacoustical phenomena *Nature* **18** 319–21

[2] Li S and Zhao D 2013 Heat flux and acoustic power in a convection-driven T-shaped thermoacoustic system *Energy Convers. Manag.* **75** 336–47

[3] Deshmukh N N and Sharma S D 2017 Experiments on heat content inside a rijke tube with suppression of thermo-acoustics instability *Int. J. Spray Combust. Dyn.* **9** 85–101

[4] Deshmukh N N and Sharma S D 2016 Effect of pointed and diffused air injection on premixed flame confined in a rijke tube *Cogent Eng.* **3** 1–12

[5] Deshmukh N N, Ansari A, Degwekar S, Unnikrishnan R and Paul B T 2021 Effects of volume and neck length of helmholtz resonator on thermo-acoustic instability *Int. Conf. Nascent Technol. Eng.*

[6] Wang J, Rubini P, Qin Q and Houston B 2019 A model to predict acoustic resonant frequencies of distributed helmholtz resonators on gas turbine engines *Appl. Sci.* **9**

[7] Tang S K 2005 On Helmholtz resonators with tapered necks *J. Sound Vib.* **279** 1085–96

[8] Bourquard C and Noiray N 2019 Stabilization of acoustic modes using helmholtz and quarter-wave resonators tuned at exceptional points *J. Sound Vib.* **445** 288–307

[9] Selamet A, Xu M B, Lee I J and Huff N T 2005 Helmholtz resonator lined with absorbing material *J. Acoust. Soc. Am.* **117** 725–33

[10] Hosseini N, Kornilov V N, Teerling O J, Arteaga I L and Goey P 2018 Evaluating thermos-acoustic properties of heating appliances considering the burner and heat exchanger as acoustically active elements *Combust. Flame* **191** 486–95

[11] Poinsot T, Bourienne F, Candel S, Esposito E and Lang W 1989 Suppression of combustion instabilities by active control *J. Propuls. Power* **5** 14–20

[12] Gupta A K and Gupta N 2016 Development of shape of helmholtz resonator cavity for attenuation of low frequency noise of pure reactive muffler *Int. J. Sci. Dev. Res.* **1** 37–42

[13] Selamet A and Lee I 2003 Helmholtz resonator with extended neck *Acoust. Soc. Am.* **113**

[14] Komkin A I, Mironov M A and Bykov A I 2017 Sound absorption by a helmholtz resonator *Acoust. Phys.* **63** 385–92

[15] Jo S, Choi Y and Kim H J 2019 Evaluation of the damping capacity according to the geometric and the number of resonator with thermal environment using a rijke tube *Aerosp. Sci. Technol.* **88** 1–8

[16] Dowd C and Meadows J 2018 Thermo-acoustic instability model with porous media: linear stability analysis and the impact of porous media *J. Eng. Gas Turbines Power* **141** 1–13

[17] Yu G, Li D and Li C 2007 Internal resistance optimization of a helmholtz resonator in noise control of small enclosures *14th Int. Congr. Sound Vib.*

[18] Zhang Z, Zhao D, Han N, Wang S and Li J 2015 Control of combustion instability with a tunable helmholtz resonator *Aerosp. Sci. Technol.* **41** 55–62

[19] Paschereit C O, Gutmark E and Weisenstein W 1998 Control of thermoacoustic instabilities and emissions in an industrial-type gas-turbine combustor *Symp. Combust.* **27** 1817–24

[20] Noiray N, Durox D, Schuller T and Candel S 2007 Passive control of combustion instabilities involving premixed flames anchored on perforated plates *Proc. Combust. Inst.* **31** I 1283–90

[21] Zhao D and Li X Y 2014 A review of acoustic dampers applied to combustion chambers in aerospace industry *Prog. Aerosp. Sci.* **37** 2213–15

EXPERIMENTAL INVESTIGATION ON HEAT TRANSFER PERFORMANCE OF ROTATORY DESICCANT DEHUMIDIFIER

Zafar Alam[1*] and Taliv Hussain[2]

[1,2] Mechanical Engineering Department, Aligarh Muslim University, Aligarh, India. 202002

[*] Author corresponding email: zafaralamalig003@gmail.com

Abstract. Dry air plays an essential role in achieving human comfort and in other applications like product drying, food storage, etc. one of the best methods to produce dry air is by employing a desiccant dehumidifier. A rotatory desiccant wheel with silica gel is used in the present work. Current work involves the optimization of the desiccant wheel performance in terms of airflow rates along with ambient temperatures. Optimization of desiccant wheel is suggested by maximum value of various performance parameters involved in this study. Required conditions have been set up by the pre-treatment section installed before the desiccant wheel. At the regeneration section two electric heater (250 watt each) is fitted through which the desiccant material is regenerated. This work is conducted for humid and hot conditions and found that the optimum functioning of desiccant wheel dehumidifier is found at 31°C ambient temperature and at the inlet 3.5 m/s velocity of air.

Keywords: humidity, silica gel, dry air, pre-treatment, dehumidifier.

Nomenclature

h	specific enthalpy (kJ kg^{-1})
c_p	air specific heat (kJ kg^{-1} K^{-1})
MRC	moisture removal Capacity (kg h^{-1})
$\dot{V}$	flow rate (volumetric) (m^3 h^{-1})
SER	energy ratio sensible (dimensionless)
DW	desiccant wheel
RR	Renewal Rate (Kg h^{-1})
DCOP	coefficient of performance of dehumidification (dimensionless)
T	temperature (°C)
Greek symbols:	
ρ	air density (kg m^{-3})
Δh_{vs}	water vaporization latent heat (kJ kg^{-1})
ω	air moisture ratio (kg kg^{-1})
Subscripts:	
P	process
deh	dehumidification
reg	regeneration

IOP Conf. Series: Materials Science and Engineering 1259 (2022) 012003 doi:10.1088/1757-899X/1259/1/012003

1. Introduction

Because the amount of moisture in the atmosphere varies depending on the temperature, higher air temperatures may absorb more water, whereas lower air temperatures cannot absorb the same or more water. This also applies to compressed air systems, and if human bodies are an example of how excess water and moisture can impair the performance, consider how a compressed air system would be affected. To keep up the comfort conditions by controlling temperature and stickiness, customary forced air system is most generally utilized where the dehumidification of air is accomplished by bringing the temperature underneath the dew point in the cooling loop to consolidate water fume. One of the cheapest way to achieve it by employing desiccant dehumidifier.

Jeong et al. (2010) [1] proposed results indicated that an ideal speed of rotational was present which amplified the moisture removal and 94% enhanced COP when compared along with ordinary vapour pressure fridge. Also, formed and broke down the idea of using the vapours heat (50°C) from power device or cooling framework as the warmth source in a 4 segment dehumidification desiccant framework which prompted extensive energy savings. Rafique et al. (2018) [2] given the impacts of encompassing conditions and different boundaries on the exhibition of the framework. The framework execution bends that aids to decide coefficient of execution (COP) and the air source conditions of the framework are strained for varied scope of renewal flow rate of air (1.5–4.5 kg $m^{-2}s^{-1}$), surrounding recovery temperature (55–85 °C), air proportions of moistness (0.010–0.026 kg kg^{-1}) surrounding temperature of air (25–40 °C), and air flow rate (1.5–8.0 kg $m^{-2}s^{-1}$). Fong et al. (2020) [3] proposed activity of the DW at a advanced speed for energy recuperation just deprived of heat recovery offered 9.5% progress in the sunlight based division and decrease of 11.5% in the all-out essential energy utilization, when contrasted with the previous one when the occasional impact was additionally considered. Muthu et al. (2021) [4] used novel composite wheel designs improved the act of these amalgam wheels by up to 40 %. Similarly, planned set of different proposal of revolving desiccant wheel hybrid the transport phenomena taking place and method based CFD code numerically rsults were simulated. Hussain et al. (2021) improved the precooling of spherical food product by suggesting optimum value of airflow rate and corresponding cooling time. They suggested that cooling is improved up to critical speed i.e., 2 m/s for the spherical food product orange [5]. Hussain et al. (2020) optimized the COP of the VCRS system by using steel cellulose and wire mesh condensers. They suggested that COP in case of cellulose condensers is more than the steel wire mesh condensers due to evaporative cooling [6]. Alam and Alam & Hussain (2022) optimized the performance of desiccant-based dehumidifier based on process air flow rates and process air ambient temperature on the basis of regeneration rate and other performance parameters [7-8].

The current project aims to enhance and optimize the functioning of a solid desiccant dehumidifier concerning airflow velocities and ambient temperatures. Because this study only looked at hot and humid situations, the temperature range that was considered was 30°C to 35°C. At the inlet section and the renewal portion of the DW, air flow rates were changed (i.e., 2.8, 3.5, 4.5, and 5.2 m s^{-1}).

2. Experimental setup

Experiments were done to calculate the functioning of a desiccant-based hybrid structure, as depicted in Figure 1. Various experiments have been carried out at various airflow rates and Process air inlet temperatures. The adsorption procedure was started immediately after regeneration. The blower processes the process air prior to leaving it through with the silica-based dehumidifier system. The humidity of the process (inlet) air falls and the temperature of the air at process increases after going out of the DW, and the process (inlet) air travels through evaporator coil of hybrid VCRS system.

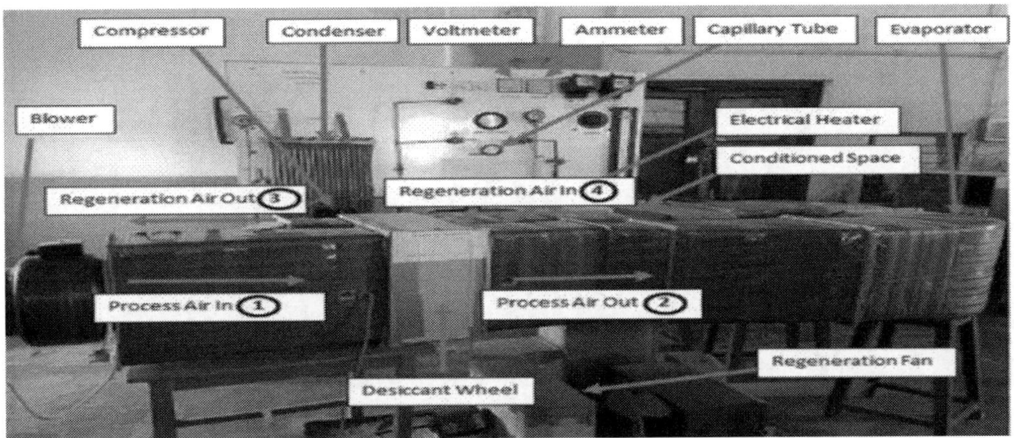

Figure 1. Actual Experimental setup used in present work

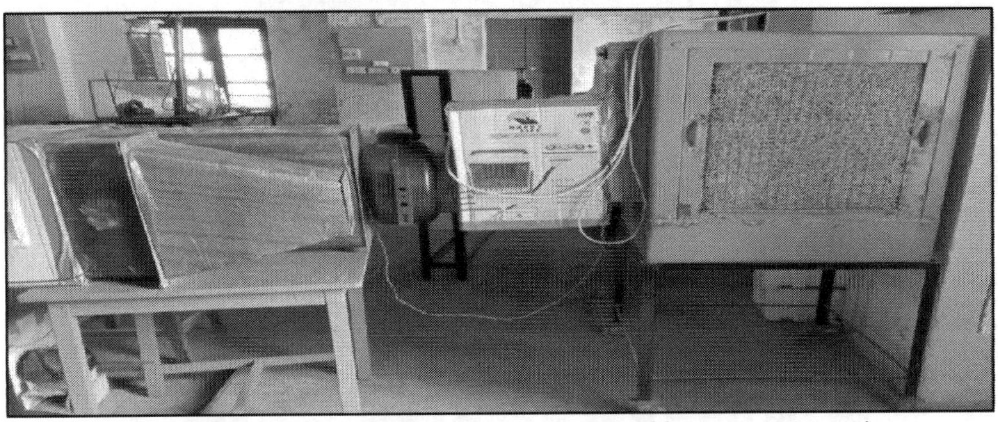

Figure 2. Shows Actual Experimental setup with pre-treatment section

Figure 3. shows electric heater in the pre-treatment section

Figure 2 shows the process air pre-treatment section coupled with the actual experimental setup and Figure 3 shows the electric heater rods (500 watts each) in the pre-treatment section. Through Direct evaporative dehumidifier we can set the dampness in the incoming air also the required temperature is maintained by electric heater rods. The main purpose of adding pre-treatment section with the experimental setup is to maintain the humidity in the incoming air. Figure 4 indicates the setup line diagram.

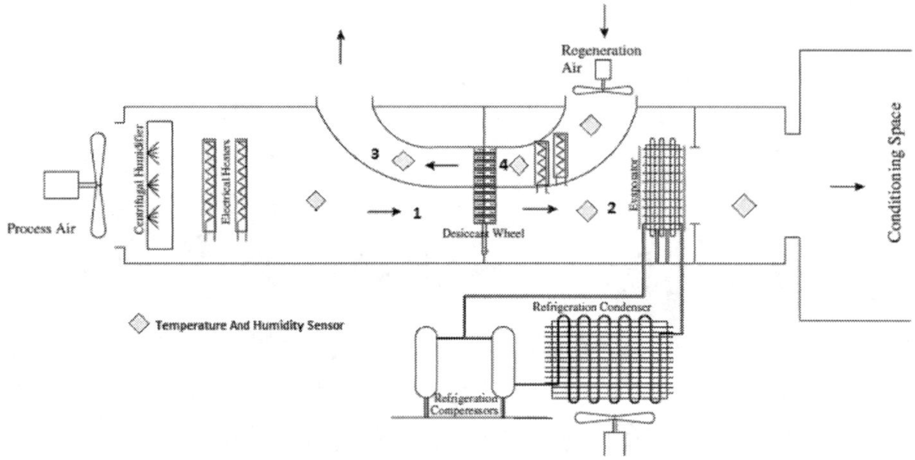

Figure 4. Line Illustration of the Investigational unit.

2.1. *Evaluating Instruments and Device*

Data Logger is accustomed to measure the relative humidity and temperature (dry bulb) of the air. An anemometer is accustomed to measure velocity of air.

Table.1 signifies the measuring tools.

Apparatus	Power	Sampling Rate	Range	Precision
Anemometer	DC, 3.0 V	--	0 - 30 m/s	±5%
Data Logger	220V Standard supply Single Phase	5 seconds to 255 minutes	0 to 100 %RH -10 to +85 °C	±2.5% ±0.5°C

2.2. *Desiccant specifications*

Table 2. Shows the desiccant specifications

Aspect	Size
Volume fraction of desiccant, (Φ)	0.48
Silica gel specific heat, c_d (J kg^{-1}K^{-1})	921
silica gel thermal conductivity of, k_d (W m^{-1}K^{-1})	0.175
Altitude of flow route, 2a (m)	0.0018
Pitch of flow route, 2b (m)	0.0032
matrix materials Specific heat, c_m (J kg^{-1}K^{-1})	1030
Wheel dia, D (m)	0.37
matrix solid density, ρ_m (kg m^{-3})	625
Porousness, ε	0.4
silica gel density, ρ_d(kg m^{-3})	1129
channel wall Width, δ (m)	0.00034

Air stream route to the total area of one passage (Ratio), (A_r)	0.844
Opening radius, r (m)	11×10^{-10}
Wheel Distance, L_w (m)	0.1

2.3. Blower Specification

Table 3. Shows the Typical Specification of the blower used in the experiment

Voltage (V)	Frequency (Hz)	Power (W)	Speed (RPM)	Air flow (m3 h^{-1})	Noise (dB)	Operating Temperature (°C)
115/120	60	220	2600	1050	68	-30 to +60

2.4. Duct Specifications

Table 4. Shows the Typical Specification of the Fabricated Duct in the experiment.

Material	Thickness of sheet	Insulation Material	Insulation thickness
Galvanized Iron sheet	Gauge 18/ 1.214 mm	PU Foam	2.5 cm

3. Methodology

A "humidity pump" has been compared to the activity of a desiccant dehumidifier. A desiccant dehumidifier transports latent heat moisture starting with one airstream then onto the next, similar to how a heat pump distributes good warmth from one airstream to the next. A desiccant dehumidifier can also be considered a "dampness concentrator" because it eliminates dampness as of the cycle stream of air and transports it to a considerably lesser recurrence stream of air, concentrating the process dampness into the reactivation air. The temperature of the cycle air rises as it is dehumidified. On the other hand, the reactivation air is humidified, so its temperature drops as it dehumidifies.

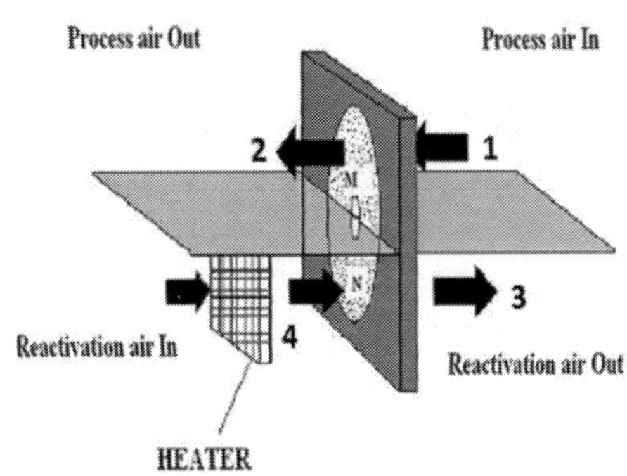

Figure 5. Sketch of the desiccant unit studied in this work

3.1. Functioning parameters

3.1.1 The Moisture Removal Capacity (MRC): It denotes the DW mass flow rate of dampness removed. [9].

$$MRC = \rho_1 \times \dot{V}_P \times (\omega_1 - \omega_2) \qquad (1)$$

IOP Conf. Series: Materials Science and Engineering 1259 (2022) 012003 doi:10.1088/1757-899X/1259/1/012003

3.1.2 The Coefficient of Performance of Dehumidification (DCOP): characterizes the proportion of power (thermal) used for dehumidification of air compared to thermal power used during renewal process [10].

$$\text{DCOP} = \frac{\rho_1 \times \dot{V}_P \times \Delta h_{vs} \times (\omega_1 - \omega_2)}{\rho_1 \times \dot{V}_{reg} \times C_p \times (T_4 - T_1)} = \frac{\dot{V}_P \times \Delta h_{vs} \times (\omega_1 - \omega_2)}{\dot{V}_{reg} \times (h_4 - h_1)} \quad (2)$$

"The latent heat of vaporization of water $\boldsymbol{\Delta h_{vs}}$ has been assessed by means of the following empirical cubic function" [11].

$$\Delta h_{vs} = -0.614342 \times 10^{-4} \times T_1^3 + 0.0158927 \times 10^{-2} \times T_1^2 - 0.236418 \times 10 \times T_1 + 0.250079 \times 10^4 \quad (3)$$

3.1.3 Energy Ratio (Sensible) (SER): On the process side, it epitomizes the fraction among the power (thermal) used to heat the air through the wheel and the thermal energy used for the renewal procedure [12].

$$\text{SER} = \frac{\rho_1 \times \dot{V}_P \times C_p \times (T_2 - T_1)}{\rho_1 \times \dot{V}_{reg} \times C_p \times (T_4 - T_1)} \quad (4)$$

3.1.4 Regeneration rate (RR): indicates rate of flow (mass) of moisture removed from process section by the renewal section by the wheel [13].

$$\text{RR} = \rho_1 \times \dot{V}_R \times (\omega_3 - \omega_4) \quad (5)$$

4. Results & discussions

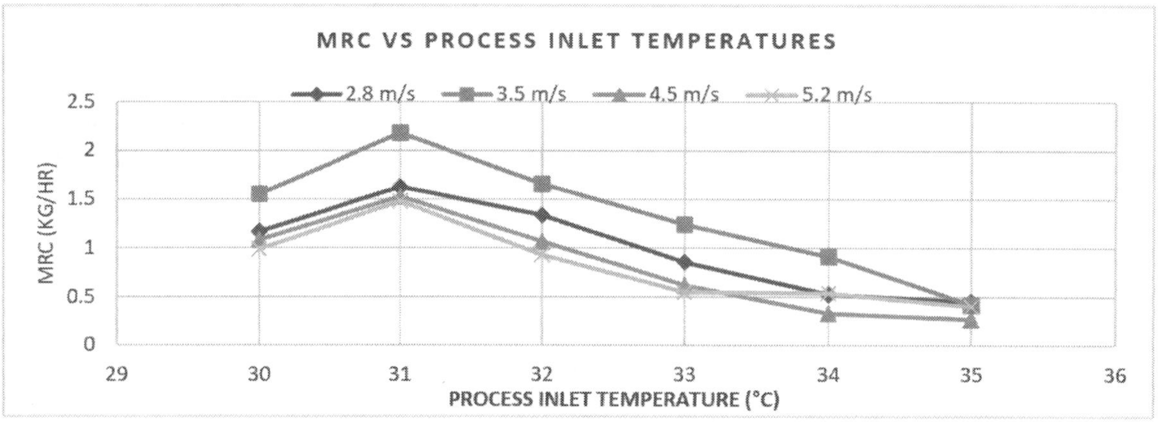

Figure 6. shows the variation of MRC for same Process and Regeneration Inlet Velocities

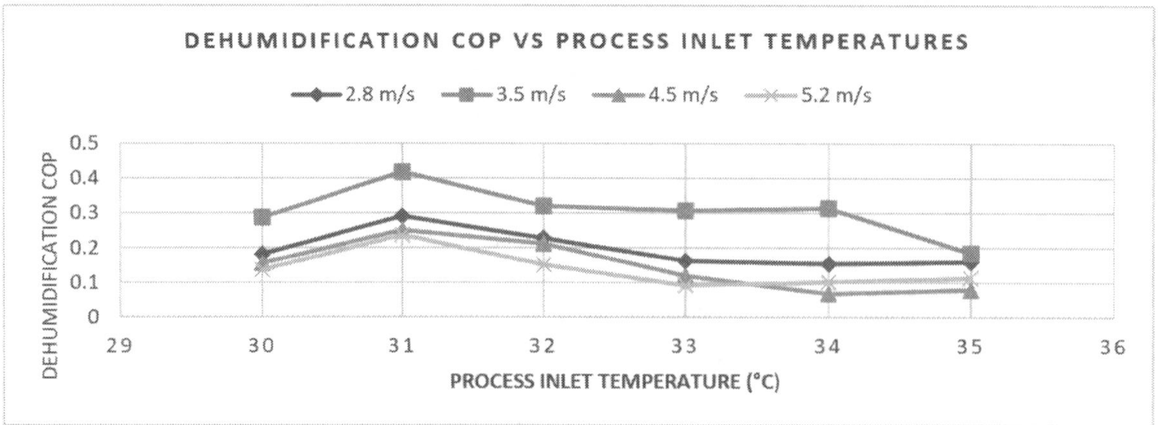

Figure 7. shows the difference of DCOP for same Process and Renewal Inlet Velocities

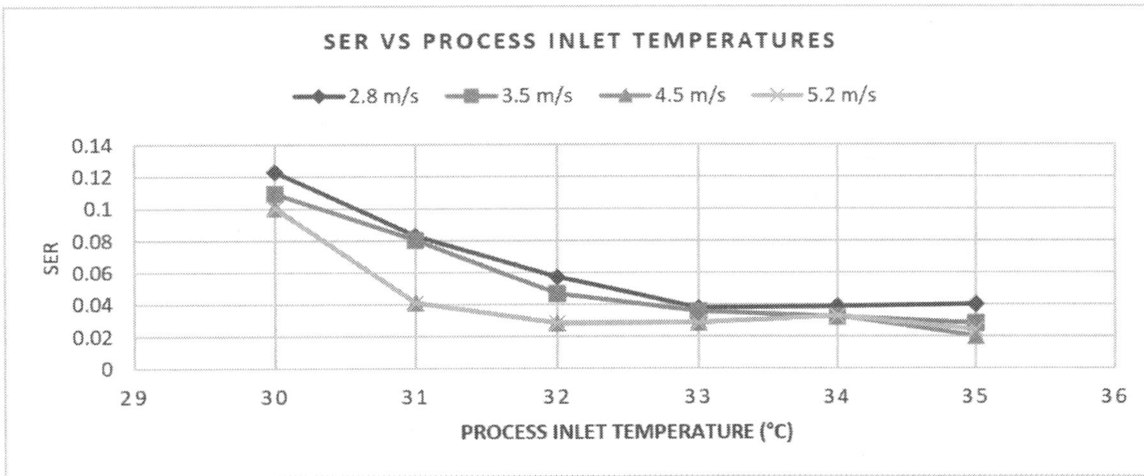

Figure 8. shows the variation of SER for same Process and Regeneration Inlet Velocities.

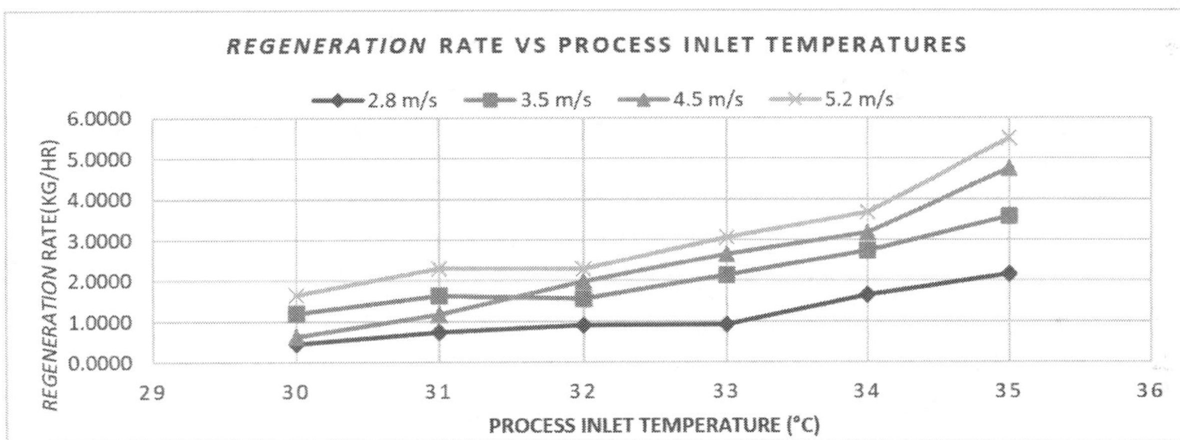

Figure 9. Displays the dissimilarity of Renewal rate for same Process and Regeneration Inlet velocities.

Moisture removal Capacity increases by rise in Process inlet Temperature of Air as indicated in figure 6. As this one reduces contact time among the process air and surface of DW, it further reduces with relatively higher flow rates. Moisture removal capacity also decreases after certain Process air inlet Temperature. It is encouraged via less temperatures as the adsorption procedure is exothermic. Hence, the process air temperature rise, results in MRC reduction. In this case Maximum value of MRC is obtained at 3.5 ms^{-1} process and renewal velocity (air inlet) and at 31°C Process inlet Temperature.

It is clearly observed from the Figure 7 that DCOP first increases then decreases. DCOP increases with the rise in humidity proportion at air inlet of the process side. In this Case Optimum values of DCOP is obtained at 3.5 ms^{-1} process and recovery inlet air velocity and at 31°C air inlet Temperature at the process section of the DW.

Figure 8 indicates variation of sensible energy ratio with different renewal and process flow rates of air via air inlet temperatures at the process side. With humidity ratio at process air inlet, Outlet temperature at the process sector increases significantly. While the DW eliminates larger amount of vapor, greater will be the adsorption heat released that raises the temperature variation at process inlet and outlet section of the desiccant wheel. As wheel removes lesser amount of water vapor, lesser will be heat of adsorption released which further decreases the SER with increase in Process inlet temperature.

Regeneration rate increases with increase in flow rate of air and air inlet temperature at the process side of the DW as indicated in Figure 9.

5. Error analysis

Kline and McClintock [14] explained the root sum square approach, that was used to accomplish the error research in this work, and the relative error values obtained for the factors investigated are displayed in Table 5.

Table 5. displays the uncertainty of the Calculated Performance Factors.

Factors	Error (%)
Moisture Removal Capacity (MRC)	7.89 %
Dehumidification COP (DCOP)	2.02 %
Sensible Energy Ratio (SER)	3.13 %
Regeneration Rate (RR)	4.76 %

6. Conclusions

1. Temperature of regeneration is directly proportional to corresponding moisture removal capacity and regeneration rate. Rise in temperature of regeneration gives relatively higher MRC and Regeneration Rate. For relatively higher flow rates, the MRC decreases due to less time of contact between the surface of desiccant and incoming air.
2. DCOP is directly proportional to the process inlet air humidity ratio. Therefore, it increases with increases in process air humidity ratio and decrease with decrease in process inlet humidity ratio.
3. With Increase in MRC of the DW, SER increases and vice versa. As MRC increases the extent of humidity removed from the process air increases, that leads to rise in temperature variation among process outlet and inlet of the desiccant wheel because of growth in heat of adsorption released. With decreases in MRC the SER decreases, as the wheel removes lesser amount of dampness from the process air leads to lesser heat of adsorption released.
4. Regeneration rate is directly proportional to the regeneration air inlet velocities and regeneration temperatures but inversely proportional to the rise in ambient moisture, due to rise in vapour pressure of regeneration air, it decreases with increase in ambient moisture.
5. For varying regeneration and process air inlet velocities optimum values of these performance parameters is found on 3.5 m/s process and renewal rate of incoming air & at 31°C process air inlet temperature.
6. For hot and humid conditions these optimized values of air flow rates and ambient temperature can be used to generate the dry air for various applications like hybrid VCRS system, food storage, pharmaceutical storage etc.

Acknowledgements

Current experimental work was funded by TEQIP-III, MHRD Government of India

References

[1] Jeong, J., Yamaguchi, S., Saito, K. and Kawai, S., 2010, "Performance analysis of four – partition desiccant wheel and hybrid dehumidification air conditioning system", International Journal of Refrigeration, Vol. 33, pp. 496-509.
[2] Rafique, M. M., Rehman, S., Alhems, L. M., & Ali Shakir, M. (2018). A Liquid Desiccant Enhanced Two Stage Evaporative Cooling System—Development and Performance Evaluation of a Test Rig. Energies, 11(1), 72.
[3] Fong, K. F., & Lee, C. K. (2020). Solar desiccant cooling system for hot and humid region–A new perspective and investigation. Solar Energy, 195, 677-684.
[4] Muthu, S., Sekarapandian, N., & Ashok, K. (2021, April). Performance Enhancement of Rotary Desiccant Wheel using Novel Homogenous Composite Desiccant Designs. In IOP Conference Series: Materials Science and Engineering (Vol. 1123, No. 1, p. 012055). IOP Publishing.
[5] Hussain, T., Kamal, M. A., Alam, Z., Hafiz, A., & Ahmad, A. (2021). Experimental and numerical investigation of spherical food product during forced convection cooling. Measurement: Food, 3, 100006.

[6] Hussain, T., Singh, A. K., Mittal, A., Verma, A., & Alam, Z. (2020). Performance evaluation of vapor compression refrigeration system by varying air flow rates in air-cooled and evaporatively cooled condensers. International Journal of Energy for a Clean Environment, 21(1).

[7] Alam, Z., & Hussain, T. (2022). Experimental Exploration on Performance Advancement of Solid Desiccant Dehumidifier. IOP Conference Series: Materials Science and Engineering, 1224(1), 012011. doi:10.1088/1757-899x/1224/1/012011.

[8] Alam, Z., & Hussain, T. (2022). Experimental Analysis and Performance Evaluation of Solid Desiccant Dehumidifier Using Silica Gel. International Journal of Analytical, Experimental and Finite Element Analysis, 9(1).

[9] Slayzak, S. J., & Ryan, J. P. (2000). Desiccant Dehumidification Wheel Test Guide. National Renewable Energy Laboratory. NERL/TP-550-26131.

[10] Ge, T. S., Ziegler, F., & Wang, R. Z. (2010). A mathematical model for predicting the performance of a compound desiccant wheel (A model of compound desiccant wheel). Applied Thermal Engineering, 30(8-9), 1005-1015.

[11] Angrisani, G., Minichiello, F., Roselli, C., & Sasso, M. (2011). Experimental investigation to optimise a desiccant HVAC system coupled to a small size cogenerator. Applied Thermal Engineering, 31(4), 506-512.

[12] Henning, H. M., Erpenbeck, T., Hindenburg, C., & Santamaria, I. S. (2001). The potential of solar energy use in desiccant cooling cycles. International journal of Refrigeration, 24(3), 220-229.

[13] Angrisani, G., Capozzoli, A., Minichiello, F., Roselli, C., & Sasso, M. (2011). Desiccant wheel regenerated by thermal energy from a microcogenerator: experimental assessment of the performances. Applied Energy, 88(4), 1354-1365.

[14] Kline, S. J. (1953). Describing uncertainty in single sample experiments. Mech. Engineering, 75, 3-8.

A Technical Review of Available Desiccant Design Techniques

Ajay Sharma[1], Shubham Kumar Mishra[1], Ashutosh Kumar Verma[1], Laxmikant Yadav[1]

[1]Department of Mechanical Engineering, NIT Hamirpur, Hamirpur-177005, India.

skm22121997@gmail.com ; ajaysharma54.me@gmail.com

Authors details:

1. Ajay Sharma – **ajaysharma54.me@gmail.com**, ORCID: 0000-0003-3333-0383
2. Shubham Kumar Mishra - skm22121997@gmail.com, ORCID: 0000-0002-3535-0352
3. Ashutosh Kumar Verma – Ashutosh.verma742@gmail.com. ORCID: 0000-0002-8281-8839
4. Dr. Laxmikant Yadav – laxmikant.yadav2149@gmail.com, ORCID: 0000-0002-6772-4797

Abstract. Desiccant wheel system is a topic of research because of increased requirement in HVAC industry and in dehumidification industry. In HVAC industry desiccant cooling system operating on regenerating heat supply from renewable source as solar are making their impact and desiccant wheel plays important role as main part in dehumidification industry. It's been approximately two decades since the desiccant wheel has made around in market for large level. Many designs and technologies are already available for desiccant wheels and some are available within the research and experiments only. This article provides a better understanding of available research and designs for desiccant wheel systems. Review for difference of liquid and solid desiccants operated cooling systems, various regenerating technologies used for desiccant different desiccant material and different materials for subtract of desiccant wheel, different construction design in terms of number of sectors and direction of air flowing in the desiccant sections prepared and discussed here. Apart from this various simulation models of desiccant systems studied based on mathematical models are studied here. Based on these reviews an effective conclusion is given for desiccant operated systems. At the end of the review future possibilities for advancement in desiccant industries are mentioned.

Keywords: *Desiccant, Sustainable technology, Cooling, Technical review.*

1. Introduction

The human working habitat has changed as a result of today's rapid technological advancements, resulting in advance in the utilisation of vapour compression-based chilling systems. The tropical region has the highest number of rapidly developing cities. The transaction to the "glass box" workspace has resulted in greater temperature discomfort, which has a damaging effect on occupant health and performance. To attain pleasant interior conditions, this leads to a large emphasis on conventional mechanical cooling, which is neither ecologically acceptable nor energy efficient. [1]. For building sectors total amount of electricity consumption because of space cooling stands for nearly 16 percentage. Since 1990, space cooling power consumption has more than tripled, posing big problem for electricity supplying system, particularly during high demand spans and intense hot weathers [2]. Desiccant cooling systems are adaptable to a variety of climates. While there are high-efficiency air conditioners on the market, most people opt for models that are less efficient because of the low cost of these systems. Resident air conditioning accounts for 70% of all available air conditioning [3]. Since 2000, requirement for air conditioning and cooling has increased at a rate of 4-percentages each year on average, that is double fast as requirement of residential and industrial hot water and lighting. In addition to highly efficient equipment, work has been done to realise the requirement of earth environment friendly refrigeration systems, which mandates the use of low global warming potential refrigerants, organic refrigerants, without refrigerant systems should be used. Solar operated desiccant system is an alternate solution to the global energy deficit problem at the refrigeration level for desiccant air conditioning regeneration. It functions as an excellent humidity control system, which is an essential aspect in human comfort. The devices incorporated with the solar operated dehumidifier are a dehumidifier to remove water content of air, heat exchange to transfer heat without changing the phase of liquid water here and to cool down the air evaporative coolers are used. A desiccant wheel is harmless to environment and used for de-moisturize the air, allowing evaporative cooling or other cooling alternatives to help decrease the air temperature [4]. There are many literatures are available for desiccant wheel system defining it's use for dehumidification and cooling system. Outcome from experiments done by Fong and Lee [5] shows that high rotational speed for desiccant wheel required for warm and humid environments, this experiment done for various operating conditions for the desiccant wheel. For desiccant wheel system the material used for adsorbing water content with different properties has great impact to optimize the output of dehumidifier wheel system. By demonstrating that the ideal desiccant material has a type V adsorption isotherm, Zheng, Ge, and Wang [6] categorized all novel materials employed in the field of desiccant refrigeration. There is heat as well as mass transfer takes place between the input air for process section and the adsorbent desiccant material. This heat and conventional moisture transfer is both happened for both sides, from air to adsorbent and from adsorbent to air. One other novel use of desiccant coated heat exchanger is that by passing cooling water or any natural refrigerant from the desiccant material the moisture can be removed along with the cooling effect from the input process air at same time [7].

In this paper a proper review of literature study of work done by many researchers is given on many elements of desiccant technology in order to increase the availability of data on its uses. Desiccant cooling systems can be used to enhance classic cooling heat ventilation and air conditioning as evaporative cooling by increasing their environmental usefulness, as demonstrated in this research. Comparing with the generally used mechanical cooling systems, it has the ability to optimise indoor air level, reduce expenses and power consumption, and safeguard the environment. The desiccant-based cooling system uses basic energies such as sustainable sun energy generated power and waste heat from industry when energy supply shortage and climate degradation are global concerns. It also demonstrates the usage of combined desiccant and other cooling cycles, that are good options for both air-conditioning needs but also good for energy saving and climate change concerns. The current research also looks at the most prevalent lower temperature regenerated adsorbent (near-ambient). This also demonstrates the various materials used to remove the desiccant wheel. This publication will assist new scholars in getting a short overview of previous developments in this field.

1.1. Desiccant wheel system

Low rotational speed wheel with heat and mass transfer between air and material to adsorb water moisture is the definition of a desiccant wheel. On thin sheet of matrix material adsorbent material is coated or impregnated and then a sheet is corrugated and sandwiched between to sheets to create flow channel. There are possibilities to generate and use different shapes of flow channel as triangular, sinusoidal, rectangular etc. Because of low pressure drop and optimum flow sinusoidal is generally used and, in our study, it is considered. According to the air stream that passes through the rotary dehumidifier wheel, there are process and regeneration sector. Process air (adsorption sector) is the section where dehumidification takes place and humidification of air happened in regeneration sector of the rotary dehumidifier.

Based on various type of frame of adsorbent material solid desiccant system differentiate as follows:

1. Stationary bed: A stationary or fixed bed is packing the desiccant material. It dehumidifies the input air as it flows through it. To regenerate the desiccant, the process air is frequently changed with heated regeneration air.
2. Rotating dehumidifier wheel: Desiccant is coated or impregnated in the wheel's substrate. One part of the rotary adsorbent wheel dehumidifies the process air as it revolves, while the other is heated and regenerated. A continuous dehumidification process is possible as a result of this.
3. Cross air flow cooled: All of the process flow channel is coated with the adsorbent material and the secondary flow channel that's where the cooling fluid continues to flow. To regenerate the desiccant, the cooling fluid is frequently exchanged with heated fluid.

In desiccant wheel technologies thermally activated desiccants are used. As shown in Figure 1 Thermally activated desiccant is made up of a desiccant substance that goes through a cyclic process of direct dehumidification of moist air followed by thermal regeneration.

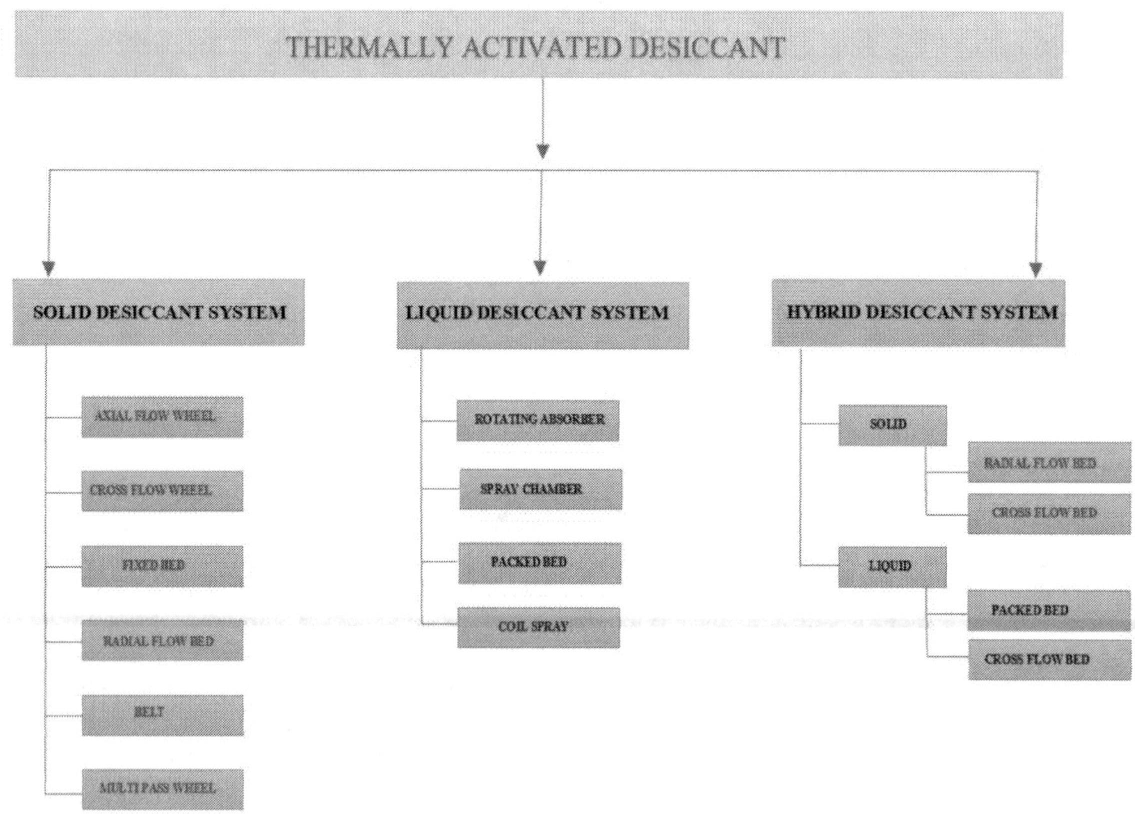

Figure. 1. Thermally activated rotary dehumidification system [8]

There is solid, liquid and Hybrid (combination of both state) type of desiccant systems working in desiccant-based cooling & dehumidification system. There are various types in each form of system which are as shown in the classification, Axial flow system in solid desiccant system and spray chamber in liquid desiccant system are widely used. Main research of purpose of desiccant systems are for cooling purpose, but there is difference of working for both solid desiccant-based cooling system and liquid desiccant operated chilling system. This difference is in terms of substance medium, cooling techniques, substrate material and number of sectors discussed here.

1.1.1. Solid Desiccant Wheel Cooling System

1. Solid desiccant-based system with conventional form of energy used for regeneration
   In this system there is solid desiccant wheel used for de-moisturizing of input air, along with blower, heat exchanger, gear motor, process fan use. As shown in Figure 2 input wet air from one side of wheel entered and processed within the desiccant wheel. Dry air is generated as the adsorbent remove moisture from the input air in rotary desiccant

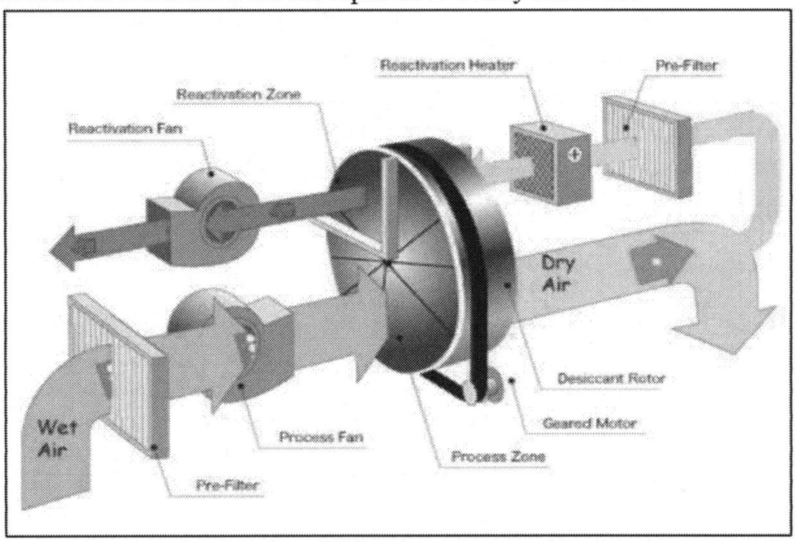

Figure 2. Desiccant Wheel Dehumidification system with electric source of regeneration [9]

wheel. Regeneration of air is done for removing water from adsorbent and this is achieved by electrical energy source. By passing the input air of regeneration section over the electric heater the ait gets heated and then it is passed from the flow channels of the desiccant wheel.

2. Solid desiccant-based system with sun energy used for regeneration

Figure 3. Desiccant Wheel Dehumidification system with Solar source of regeneration [10]

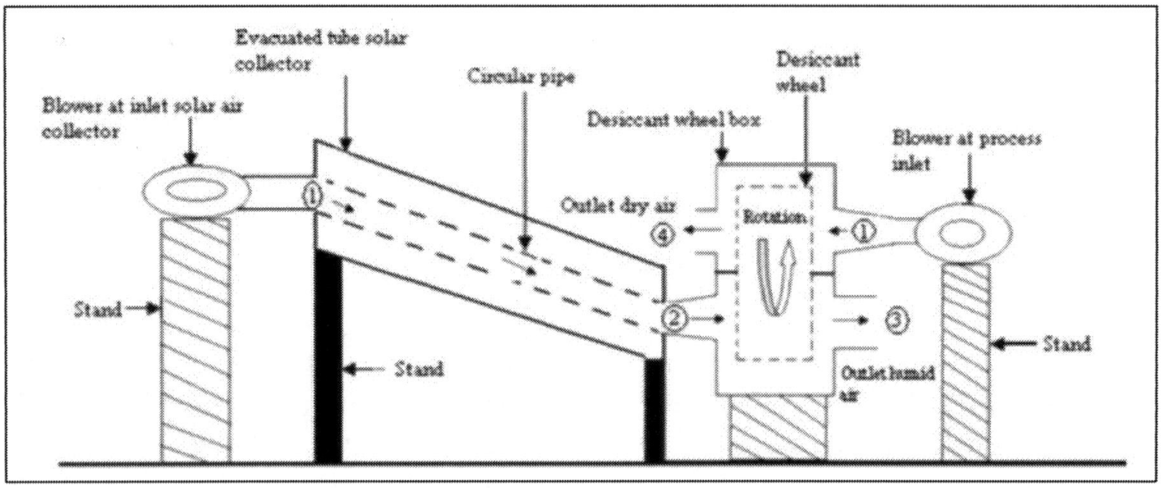

A solar energy operated evacuated tube heat collector with a surface area of 4.44 m^2 is used for generating the regeneration air. At temperatures between 45 and 75 degrees Celsius, the desiccant wheel is regenerated. The evacuated tube has a transparent tube and an aluminium-nickel covered absorber tube, as shown in Figure 3. [10]. The system's remaining functions are the same as a general solid desiccant wheel system, with dry and cool air produced at the process section's output.

1.1.2. Liquid Desiccant Cooling System

Because of the increasing expense of conventional energy fuels and other climate change issues, liquid desiccant-based cooling systems are among the solutions to reduce the cost of energy requirement of air handling systems without affecting air quality used for residential and industrial purpose and convenience. Which is not only environmentally safe (no use of ozone-depleting refrigerants), but also makes very good utilisation of alternative and new energy sources. The operation of a liquid desiccant cooling system differs from that of a solid desiccant cooling system. Liquid desiccant in form of mist is sprayed in the compartment made for dehumidification, air is passed through this compartment and this will abstract moisture from input air. Further this high moisture contained liquid desiccant is transferred for regeneration in other compartment and there in this compartment water contains from liquid desiccant is removed by supplying heat. Apart from regeneration and process section the remaining process for cooling is approximately same as output process air is used for effective cooling by decreasing its temperature using evaporative cooling.

1. Liquid desiccant-based cooling system by utilizing solar powered energy or other renewable energy

Figure 4 is showing examined liquid desiccant dehumidifiers that is regnerated using solar powered enenrgy. Performance of this system founds better for greenhouse gas project and study estimated and compared the cost with the conventional system.

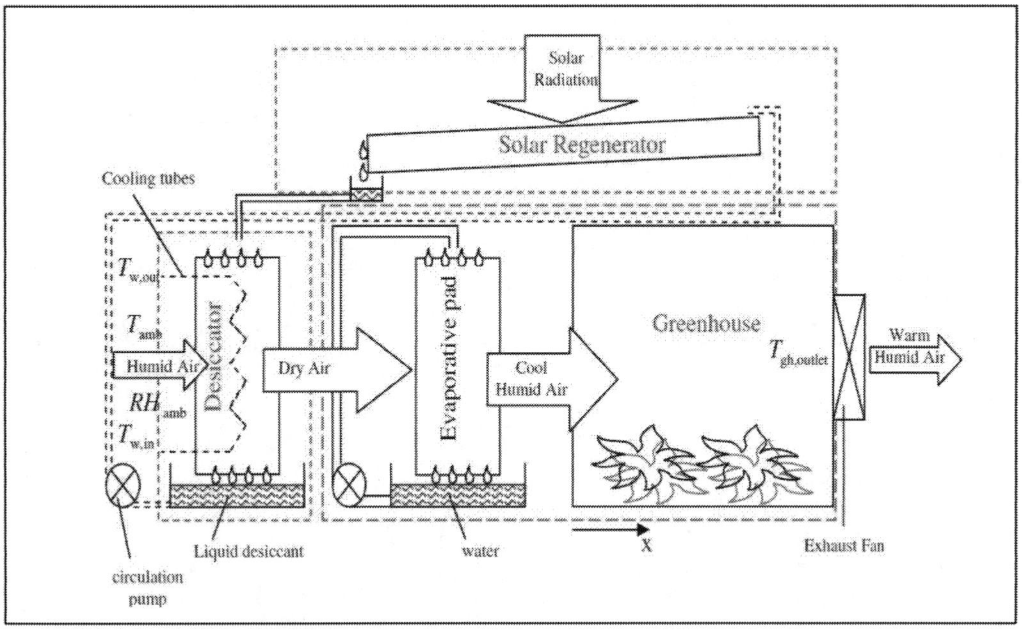

Figure 4. Liquid desiccant-based chiller with Solar powered source of regeneration [11]

A system with solar powered regeneration and general system with electricity energy used for regeneration are tested and compared for $MgCl_2$ adsorbent. When compared to traditional evaporative cooling, the liquid desiccant system desiccant system reduces average daily high temperatures by 5.5-7.5°C in the hotter months, which is enough to keep vegetable growing all year.[11]

2. Hybrid solid and liquid desiccant cooling system regenerated using sun powered energy
The standard vapour compression refrigeration technique is combined with a liquid desiccant dehumidifier in a hybrid solid and liquid chiller. It is a better option to use for high temperature and high humid type of conditions. One of the benefits of this system is that diverse types of energy, especially low sun powered thermal energy, may be incorporated, lowering the system's electricity consumption. In this paper authors have simulated the system for Kuala Lumpur city of Shri Lanka in TRNSYS. [12]

Figure 5 shows the implicit requirement is treated by a liquid-based dehumidification subsystem, while the perceptive requirement is handled through a cooling subsystem. The suggested system consists substitute systems as, the LiCl-H_2O solution liquid desiccant system, which adsorb water from the input air.

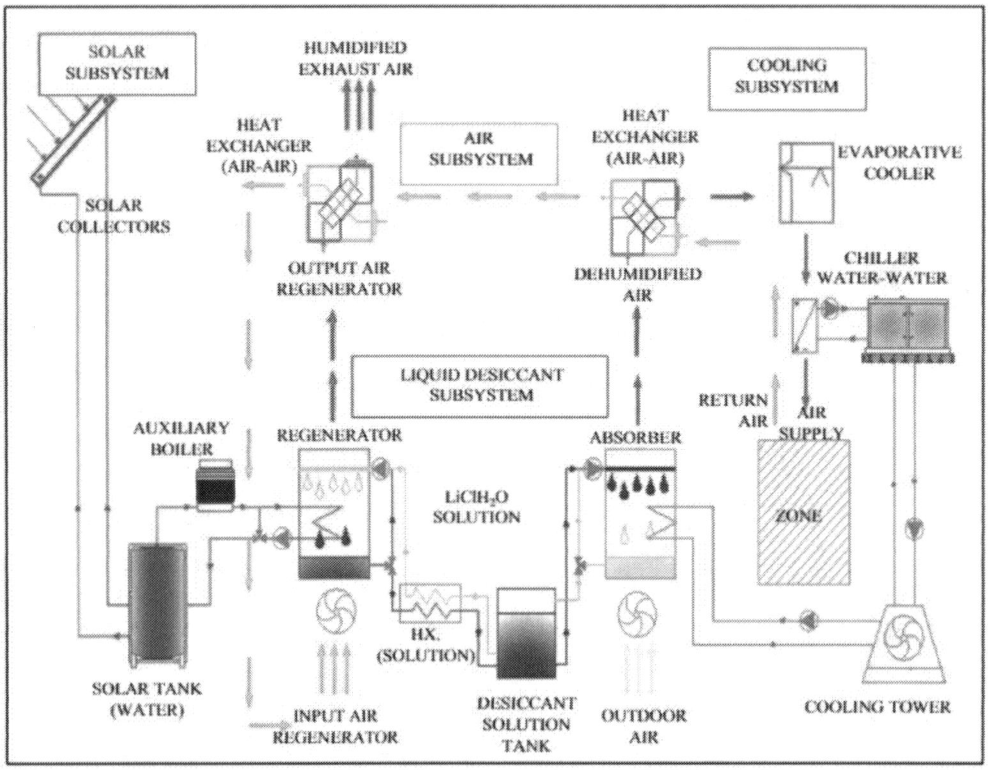

Figure 5. Simulation of hybrid solid and liquid desiccant cooling system regenerated using sun powered energy [13]

The solar powered flat plate collector's regeneration system to get the need of temperature in the regeneration air and the liquid vapour compression chiller to handles the implicit load. At last, an absorber, additional components such as regulators and pumps, heat exchangers to work between air mediums, a regenerator, a solution exchanger [13]. The restoration temperature for adsorbent material along with the LiCl mass percentage in the dehydrator tank, and the direct sunlight area are the most critical characteristics in the scenario studied.

3. Liquid desiccant-based cooling by conventional source of energy

Figure 6 implies about research of examine liquid desiccant-based dehumidification worked for calcium chloride, Lithium chloride and triethylene glycol. There is low effectiveness for this liquid desiccant chiller but this can be eliminated by using better heat exchanger to transfer heat between liquids [14]. Apart from this the system working is as similar to the other liquid desiccant-based system.

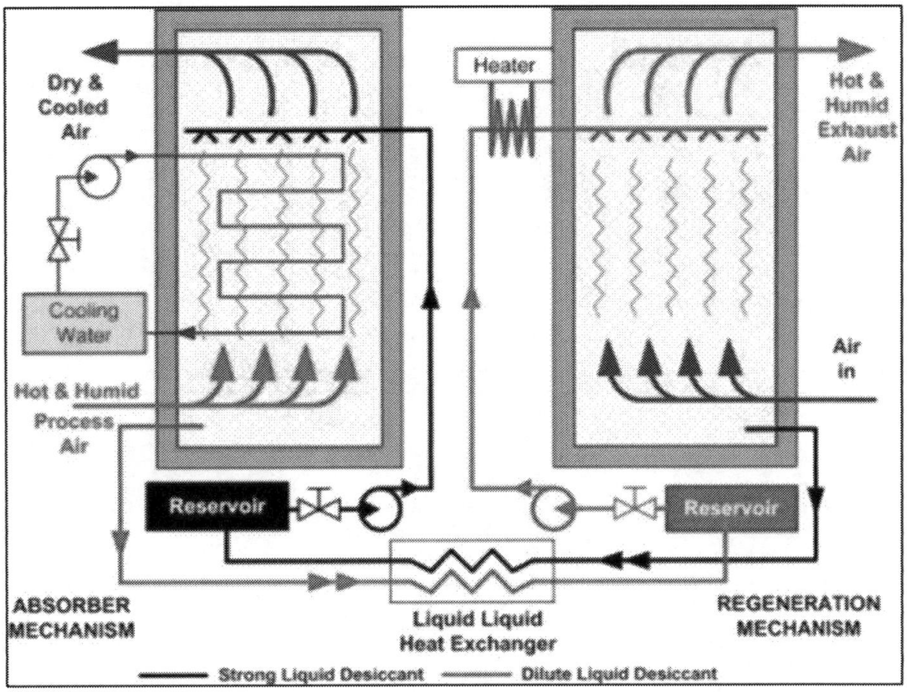

Figure 6. A liquid desiccant dehumidification system with liquid-liquid Heat Exchanger [14]

2. Cooling Techniques Used for Processed Air

For cooling in desiccant system there is mainly two techniques that are used, Passive cooling and Evaporative cooling. Passive cooling involves improving indoor thermal comfort in a building with little to no dependence on 'man-made' energy. Passive cooling techniques help achieve both, reduced demand for cooling as well as lower energy consumption for cooling. Comfort cooling technique involves 'cooling the human body' to attain comfort. Human body metabolizes energy and releases it in form of heat and sweat. Higher air speeds hells increasing the efficiency of sweat evaporation, which leads to thermal comfort. Comfort ventilation is ideal for climate zones that have high levels of relative humidity through the year ranging between 50-90%, like tropical rainforests.

Comfort ventilation is applicable when still air temperatures seem to be too warm due to low speed and its inability to carry out excess moisture which causes discomfort. It essentially helps remove excess moisture through proper ventilation. Evaporative cooling, in its wider definition, refers to any process in which the sensible heat in an airflow is transferred for the latent heat of droplets of water or wetted surfaces. Sensible heat is increases in air temperature without a change in state of working fluid, whereas latent heat refers to phase shifts between different states of working fluid. This means that when water meets hot air, it changes its phase from liquid to gaseous/vaporous form; thereby lowering the air temperature. Direct evaporative cooling systems involve passing air over water, which could be in the form of a water body or even drip channels. The air, cooled and humidified by the evaporation of water, is then introduced into the building.

3. Different Of Desiccant System Based on Mathematical Model

There is various material which shows dehumidification characteristics to adsorb water molecule from the input air. Similarly, there is different substrate or matrix materials available for descant wheel to use in appropriate condition. Classification for desiccant material for solid, liquid and advanced form is shown in Figure 7.

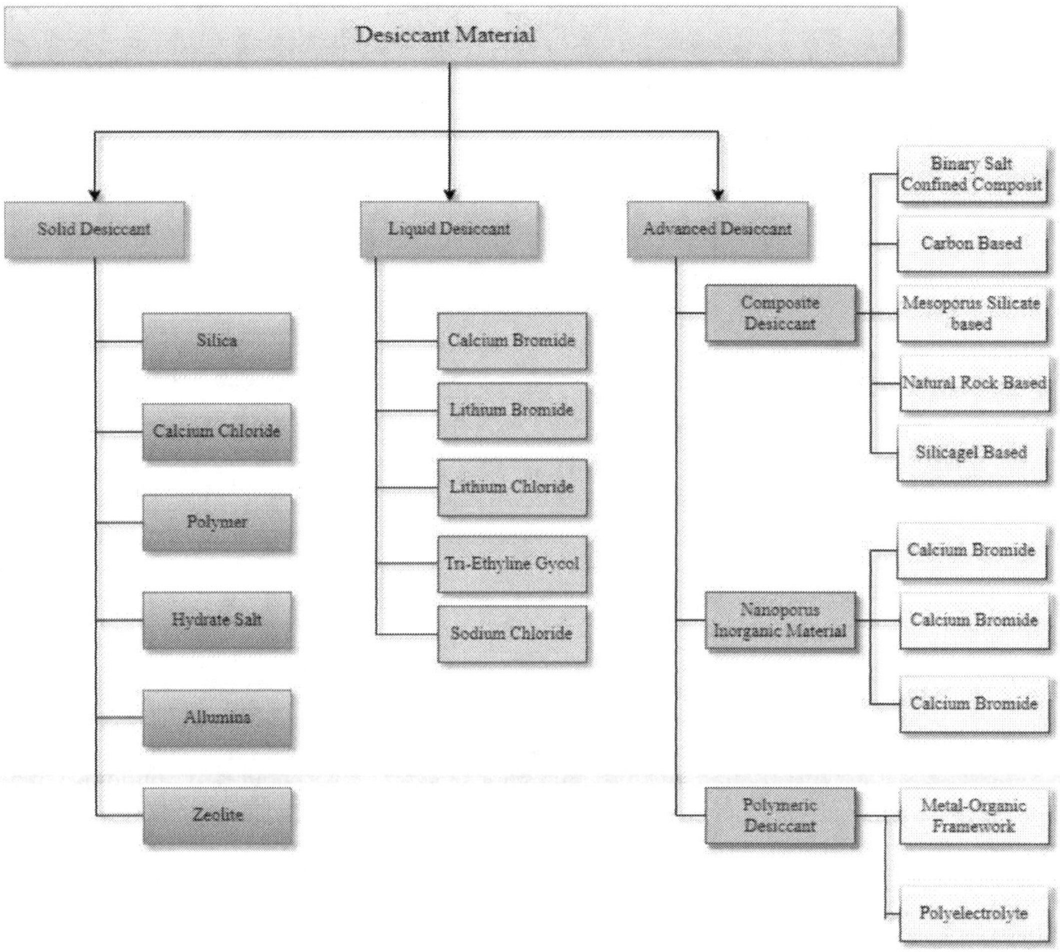

Figure 7. Different desiccant material used in desiccant adsorbent and desiccant cooling system

A desiccant is a solid or liquid substance that is used to absorb or adsorb water vapour the air by absorption or adsorption [16]. According to Abdulrahman Th. Mohammad et al [17], Desiccant materials increase dehumidifier performance in desiccant-based chilling systems by boosting moisture removing capacity during adsorption operations, which may be improved by mixing diverse desiccant materials into composite materials. Absorbents for liquid desiccant-based chilling systems include CaCl, LiBr, sodium chloride, and triethylene glycol [18]. S. Bouzenada et al [19] used Calcium chloride (CaCl2) in their research and discovered that while CaCl2 is the lowest and most easily accessible liquid adsorbent material, its vapour pressure at a specific temperature is considerably greater than the other materials.

Cortes et al. [21] and Rafique et al. [22] investigated the body made and behaviour patterns of various adsorbent materials such as alumina, composites, carbon black, clay, silica gel, molecular sieve polymers and metal silicates which are generally used as desiccant in desiccant wheel dehumidification systems. Calcium chloride, Silica gel, Clay, and Molecular sieve are all seen in Figure 8. Yadav et al. [23] evaluated several types of solid desiccant materials for generating regenerative heat using renewable sun powered energy, and found that silica gel is the most effective desiccant solution for food process drying. Zhang et al. [24] conducted experiments on several materials-based dehumidifier wheels and came to the conclusion that 3A silica gel can be regenerated at room temperature. However, when compared to other improved desiccants, this advanced material did not exhibit a significant improvement in dehumidification capability.

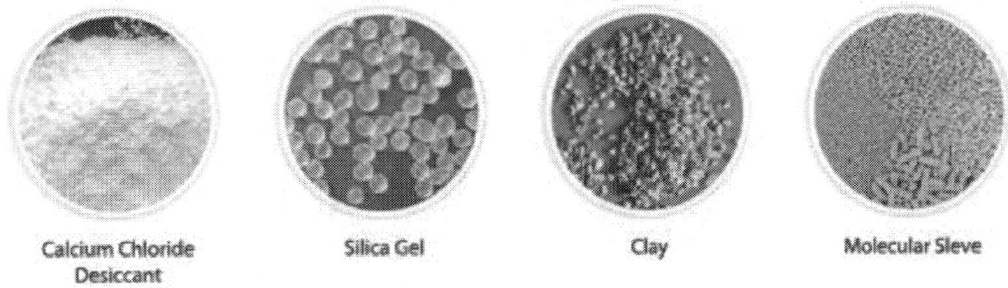

Figure 8. Images of Solid Desiccant materials

As desiccant material plays important role to decide the de-moisturising capacity and increase in temperature for air, Matrix material of desiccant also contribute to this. Substrate or matrix material is the material on which desiccant is impregnated or coated. So, in the air passages as air passes over the desiccant material a small amount of moisture is adsorbed by the matrix material and due to temperature increment in the desiccant material, heat also transferred to matrix material. There are different materials used as matrix material of desiccant wheel i.e., Copper, Aluminum, Brass, Magnesium alloy, Stainless steel, Titanium alloy, Teflon, Fiberglass. Different materials show different properties and accordingly they used. As copper is having high thermal conductivity so it dissipates heat at fast speed than others, that is useful for industrial conditions for higher temperature involvement. Paper fiber glass is mainly used matrix material because of its flexibility and less sensitivity for temperature. Also, fiber glass is easily compatible with different desiccants and desiccant material can be coated on fiber glass.

4. Different Geometry Shapes of Flow Channels in Desiccant Wheel

The design and shape appearance element channels have a significant influence on the transport process and performance of the desiccant wheel. The majority of desiccant wheel manufacturers now use sinusoidal channels. This study used a mathematical model of heat transfer rate for desiccant and water to assess the influence of flow form and geometrical dimentions of the matrix on the operation of a dehumidifier wheel.

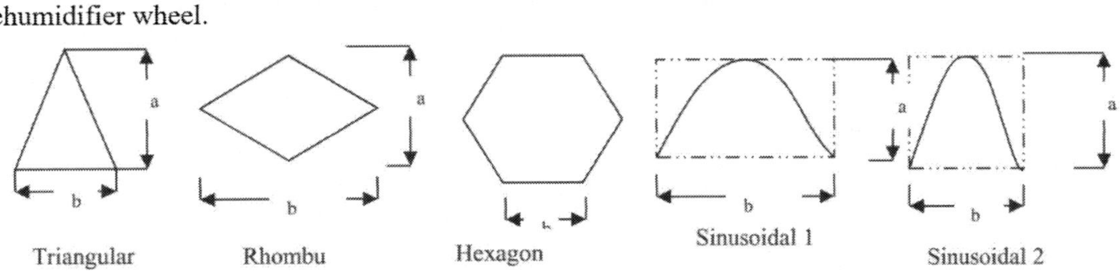

Figure 9. Different air flow channel Shapes used for desiccant wheel [25]

The different shapes of the channel which are taken into consideration in this study of counter flow desiccant wheel system are shown in Figure 9. Different characteristics will exist for the flow through these types of channels in terms of hydraulic diameter and, as a result, the Nusselt number will be different and also the friction factor [25]. In this investigation, the area of all flow channel is kept constant at 3.89 mm2, it is the area of a standard dehumidifier wheel sinusoidal flow channel.

According to the data given in Figure 10, rectangular, sinusoidal and triangular cross-sections dehumidify outshine circular, hexagonal, and square cross-sections. The best three shapes out of all possess a lower main diameter thus as a consequence, greater velocities and heat transfer coefficients.

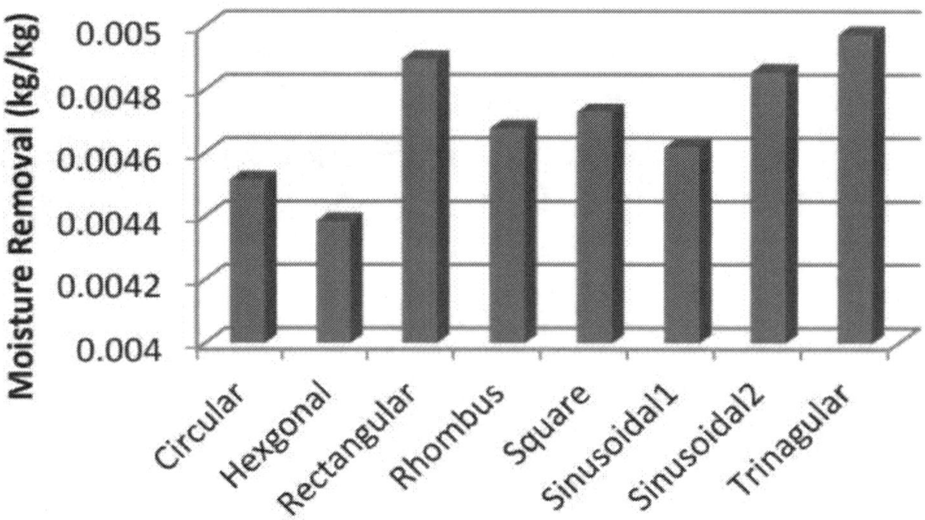

Figure 10. Different channel shapes geometry used in desiccant wheel [25]

Moisture removal in triangular type of shape is maximum but this comes with high pressure drop because of the flow through the channel. The sinusoidal channel possesses optimum moisture removal with lower value of pressure drop in the flow passage. Similarly, D.b Jani et al. [26] shows comparison of various kind of flow passages in ANSYS. And the results shows that the sinusoidal section has the optimum output as compared to all other shapes.

5. Difference of Desiccant Wheel on The Basis of Number of Sectors

There are two basic sectors in desiccant wheel process and regeneration sector. There is one more section purge section provided, which increase effectiveness of regeneration sector. Purge section tends to remove dust particles and dehumidify the wheel so that it can be used efficiently in the process section. Avadhesh and Bajpai [27] have compared dehumidification wheel with two and four sectors. They have developed a heat and mass transfer mathematical model which examines two and four sectors with different angles, same is shown in the form of Table 1 and Table 2. The higher moisture content elimination (MR) efficiency for both two and four sectors is the same, although two sector may be attained at double the RPH of four sector. Additionally, use of hot purge sector and cool purge sector will increase relative moisture removal (MR) efficiency.

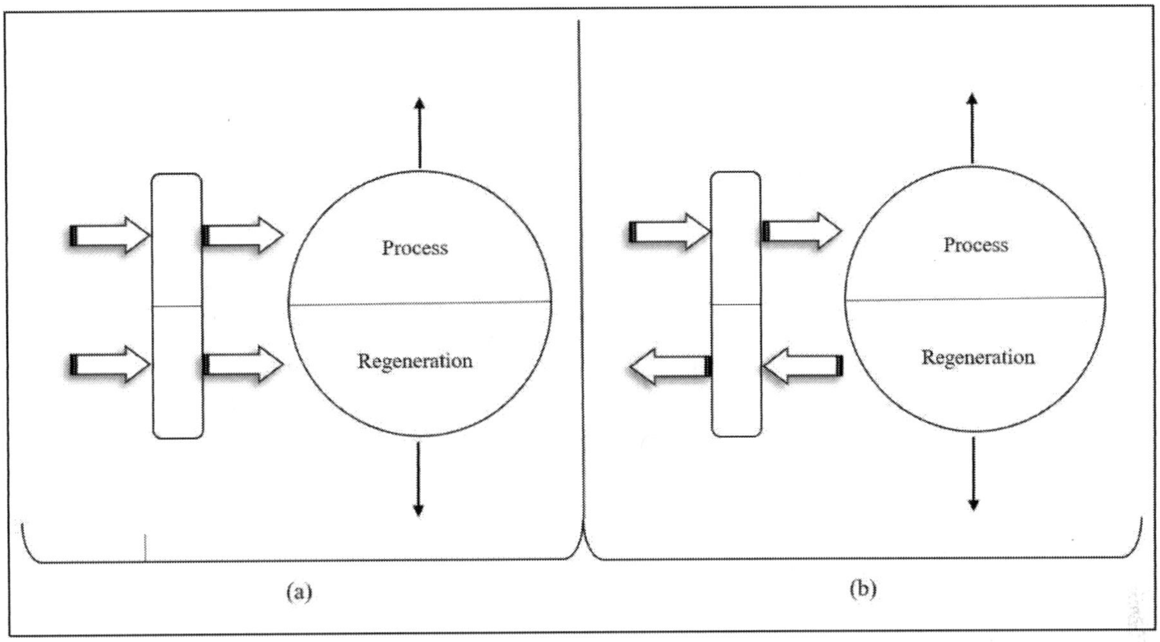

Figure 11. (a) Counter flow and (b) Parallel flow in Desiccant wheel

Based on different output examined by researchers, Counter flow in desiccant wheel performs better than of parallel flow. This is because heat and mass transfer acts better for counter flow, also optimum utilization by using heat exchangers is possible in counter flow. Apart from this purge section is also useful for increasing effectiveness of desiccant wheel system.

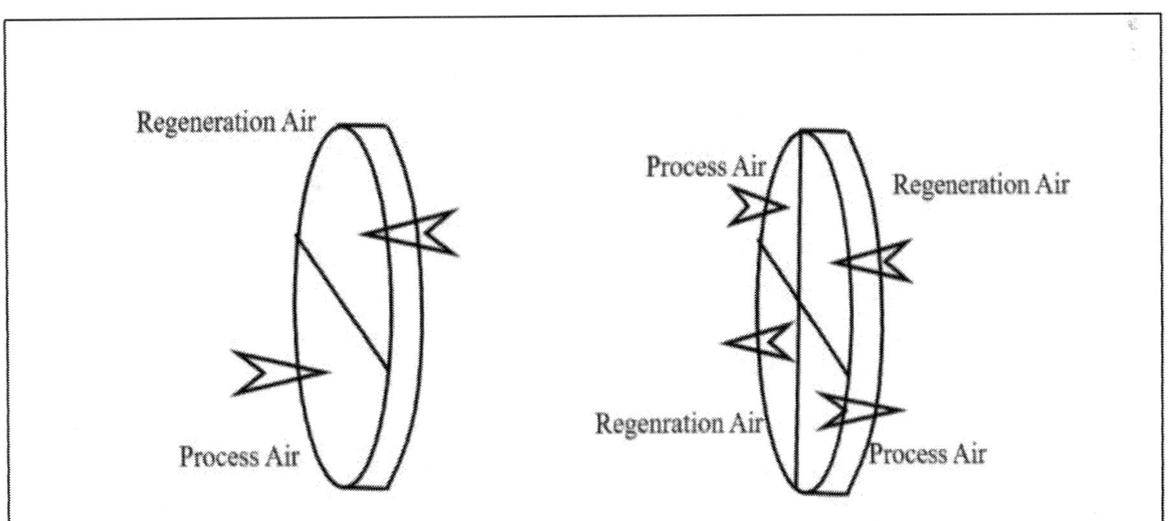

(a) (b)

Figure 12. Diagram of a) Two sector b) Four sector rotating dehumidifier wheel

IOP Conf. Series: Materials Science and Engineering 1259 (2022) 012004 doi:10.1088/1757-899X/1259/1/012004

Table 1. Comparison of water content removal efficiency and temperature change for various two and four different sector adsorbent wheel

		Process Sector angle	Regeneration Sector angle	Max. removal Efficiency	Max. temp. difference	RPH
Case-1	Two Sector	180°	180°	25.28%	7.79°	25
	Four Sector	90° - two PS	90° - two RS			50
Case-2	Two Sector	200°	160°	21.36%	6.59°	25
	Four Sector	100° - Two PS	80° - Two RS			50
Case-3	Two Sector	220°	140°	17.00%	5.29°	30
	Four Sector	110° - Two PS	70° - Two RS			60

Use of purge section in the dehumidifier wheel shows increment in performance. By author effect of different between two types of purge portion hot and cold purge portion. Adsorbent, regeneration, cooled & heated purge portion are the different sections of the dehumidifier wheel. The heat is exchanged between these two purge regions that used for reducing losses. Adsorption of process air occurs in the process or adsorption sector of the wheel, while humidification of regeneration air occurs in the regeneration sector.

The four different sector dehumidifier wheels have a larger parallel moisture removal (MR) effectiveness than the two different sector adsorbent wheels at low revolution per hour and low regenerative orientation. The maximum moisture removal efficiency is the same for both the two-sector and four-sector dehumidifier operated mechanisms; however, the two-sector dehumidifier wheel accomplishes it at double the rph of the four-sector desiccant wheel.

Table 2. Comparison of moisture eliminating efficiency and temperature change for various two and four different sector adsorbent wheel with purge sector included [27]

Two Sector arrangement with hot and cold purge		Sector Angle	Max. Moisture removal	Max. temp. difference	RPH
	Process	160°	0.217	6.68° C	15
Case-1	Purge 1 & 2	20°	0.410	12.92° C	35
	Regeneration	160°	0.577	19.11° C	50
	Process	140°	0.216	6.64° C	15
Case-2	Purge 1 &2	40°	0.4047	12.71° C	35
	Regeneration	140°	0.5659	18.74° C	50

When compared to low regeneration temperatures, the purge sector performs better especially at high regeneration temperatures. The desiccant technologies with a larger purge and a lesser regenerating portion found to be optimum at low revolution per hour than the desiccant technologies with a smaller purge and a bigger regeneration portion.

6. Conclusion

Desiccant-based air chilling and handling technologies are a feasible replacement to conventional chilling systems, particularly when latent loads are high. This method enables for the cooling cycle to be powered by low-temperature industrial waste heat or solar electricity. This has gained the market of desiccant based chilling technologies in past 20 years. From review of desiccant technologies following points are clear

- Counter flow is more beneficial than of parallel flow.
- Sinusoidal and triangular shape for flow channel is most efficient to use, but sinusoidal performs better in case for low pressure difference.
- Purge sector increase effectiveness of dehumidifier wheel system.
- Silica gel is most used desiccant material in solid dehumidifier wheel technologies.
- Glycols are generally used liquid desiccant for liquid dehumidifier technologies.
- Renewable resources are best alternative to use for regeneration un both liquid and solid dehumidifier technologies.
- More simulation and research required to generate accurate models to guess the behaviour of desiccant technologies working under various environmental conditions.

7. Future Aspect of Desiccant Wheel System

Desiccant based technologies are environment friendly and due to this it is replacing the conventional cooling system. The economic issue of cost, environmental issues of harmful effects and social effect of marketing of desiccant needs to be examine and these topics should be covered and explained with a diversify researches. Evaporative chilling technologies included for cooling of air in desiccant technologies, should be employed to design low cost, more dependable, and cooling systems with low space. Desiccant technologies need to targeted for industrial purpose because of good return of energy and money in terms of hot and humid climatic zones. The advancement of the components of adsorbent-based wheel system is in the,

1. Cooling sources
2. Desiccant materials
3. Desiccant adsorbers
4. Regeneration sources & storage

But apart from above points other important desiccant systems should expand applications in the following:

- Improve and increase competition between desiccant producer industries.
- To improve the COP there is requirement of increase the material properties.
- There is requirement of change the system of air handling units.
- Learning the negative effects of high-water content on the indoor air level and on the indoor appliances

References

[1] Huang, S. M., Zuo, X., Li, J. J. E., Li, S. F. Y., Bay, B. H., & Ong, C. N. (2012). Metabolomics studies show dose-dependent toxicity induced by SiO2 nanoparticles in MRC-5 human fetal lung fibroblasts. *Advanced healthcare materials*, *1*(6), 779-784.

[2] IEA (2021), Cooling, IEA, Paris https://www.iea.org/reports/cooling

[3] Birol, F. (2018). The future of cooling: opportunities for energy-efficient air conditioning. *International Energy Agency*.

[4] Bhabhor, K. K., & Jani, D. B. (2021). Performance analysis of desiccant dehumidifier with different channel geometry using CFD. *Journal of Building Engineering*, *44*, 103021.

[5] Fong, K. F., & Lee, C. K. (2018). New perspectives in solid desiccant cooling for hot and humid regions. *Energy and Buildings*, *158*, 1152-1160.

[6] Zheng, X., Ge, T. S., & Wang, R. Z. (2014). Recent progress on desiccant materials for solid desiccant cooling systems. *Energy*, *74*, 280-294.

[7] Saeed, A., & Al-Alili, A. (2017). A review on desiccant coated heat exchangers. *Science and Technology for the Built Environment*, *23*(1), 136-150.

[8] Cortés, F. B., Chejne, F., Carrasco-Marín, F., Pérez-Cadenas, A. F., & Moreno-Castilla, C. (2012). Water sorption on silica-and zeolite-supported hygroscopic salts for cooling system applications. *Energy Conversion and Management*, *53*(1), 219-223.

[9] Patnaik, S., Lenz, T. G., & Löf, G. O. G. (1990). Performance studies for an experimental solar open-cycle liquid desiccant air dehumidification system. *Solar Energy*, *44*(3), 123-135.

[10] Bhabhor, K. K., & Jani, D. B. (2019). Progressive development in solid desiccant cooling: A review. *International Journal of Ambient Energy*, 1-24.

[11] Lychnos, G., & Davies, P. A. (2012). Modelling and experimental verification of a solar-powered liquid desiccant cooling system for greenhouse food production in hot climates. *Energy*, *40*(1), 116-130.

[12] Coca-Ortegón, A., Prieto, J., & Coronas, A. (2016). Modelling and dynamic simulation of a hybrid liquid desiccant system regenerated with solar energy. *Applied Thermal Engineering*, *97*, 109-117.

[13] Belguith, S., Meddeb, Z., & Ben Slama, R. (2021). Performance analysis of desiccant cooling system using polyacrylic acid sodium salt desiccant wheel. *Science and Technology for the Built Environment*, *27*(10), 1368-1380.

[14] Jain, S., & Bansal, P. K. (2007). Performance analysis of liquid desiccant dehumidification systems. *International journal of refrigeration*, *30*(5), 861-872.

[15] Sphaier, L. A., & Worek, W. M. (2006). Comparisons between 2-D and 1-D formulations of heat and mass transfer in rotary regenerators. *Numerical Heat Transfer, Part B: Fundamentals*, *49*(3), 223-237.

[16] Gao, Z., Mei, V. C., & Tomlinson, J. J. (2005). Theoretical analysis of dehumidification process in a desiccant wheel. *Heat and mass transfer*, *41*(11), 1033-1042.

[17] Mohammad, A. T., Mat, S. B., Sulaiman, M. Y., Sopian, K., & Al-abidi, A. A. (2013). Survey of hybrid liquid desiccant air conditioning systems. *Renewable and Sustainable Energy Reviews*, *20*, 186-200.

[18] Narayanan, R., Saman, W., White, S., & Goldsworthy, M. (2010). *Modeling and parametric analysis of silica-gel desiccant wheels*. CQUniversity.

[19] Bouzenada, S., Frainkin, L., & Léonard, A. (2017). Experimental investigation on vapor pressure of desiccant for air conditioning application. *Procedia Computer Science*, *109*, 817-824.

[20] Yadav, A., & Bajpai, V. K. (2013). Analysis of various designs of a desiccant wheel for improving the performance using a mathematical model. *Journal of Renewable and Sustainable Energy*, *5*(2), 023110.

[21] Mohammad, A. T., Mat, S. B., Sopian, K., & Al-abidi, A. A. (2016). Survey of the control strategy of liquid desiccant systems. *Renewable and Sustainable Energy Reviews*, *58*, 250-258.

[22] Rafique, M. M., Gandhidasan, P., Rehman, S., & Al-Hadhrami, L. M. (2015). A review on desiccant based evaporative cooling systems. *Renewable and Sustainable Energy Reviews*, *45*, 145-159.

[23] Yadav, A., & Bajpai, V. K. (2012). Experimental comparison of various solid desiccants for regeneration by evacuated solar air collector and air dehumidification. *Drying Technology*, *30*(5), 516-525.

[24] Zhang, L. Z., Fu, H. X., Yang, Q. R., & Xu, J. C. (2014). Performance comparisons of honeycomb-type adsorbent beds (wheels) for air dehumidification with various desiccant wall materials. *Energy*, *65*, 430-440.

[25] Narayanan, R. (2017). Investigation of geometry effects of channels of a silica-gel desiccant wheel. *Energy Procedia*, *110*, 20-25.

[26] Bhabhor, K. K., & Jani, D. B. (2021). Performance analysis of desiccant dehumidifier with different channel geometry using CFD. *Journal of Building Engineering*, *44*, 103021.

[27] Yadav, A., & Bajpai, V. K. (2013). The performance of solar powered desiccant dehumidifier in India: An experimental investigation. *International Journal of Sustainable Engineering*, *6*(3), 239-257.

Variation in the enhancement of heat transfer with different fluids using twisted tape inserts

Shrikant Arunrao Thote[1], Prof. (Dr.) Netra Pal Singh[2]

[1]Research Scholar, Department of Mechanical Engineering, Oriental University, Indore, (M.P.) - 453555, India.
[2]Professor, Department of Mechanical Engineering, Oriental University, Indore, (M.P.) - 453555, India.

shrikantathote@gmail.com, netrapaljadon@gmail.com

Abstract. This research intends to look into the transformation of heat, factors affecting the friction, and characteristics of the efficacy ratio for the ratio of twist as 3.0, 4.4, and 6.0, independently of a concentric pipe heat exchanger with and without trapezoidal section twisted tape insert. For different Reynold's Number (Re) ranges of 200-2000, 800-2000, and 2000-12000, experimentation was accomplished to find the transformation of heat energy and aspect of factors of the friction of coaxial tubes. The investigated data from the tube without an insert were compared to the conventional correspondence to assure that the investigation output was valid. The findings of the trapezoidal section twisted tape tube has been in contrast to the simple tube's results. The consequences reveal that the trapezoidal section twisted tape appreciably improves heat transmission along with friction coefficient. The Trapezoidal section twisted tape enhancement transmission of heat was once additionally judged to be practical because the realized output ratio is maximum than unity. The twisted tape has a greater Nusselt Number (Nu), frictional influence, and overall efficacy attainment than a plain tube, by the research. According to the tapes that have been tested, the tapes with the lowest twist ratio 3.0 are greater fine than the tapes with the other ratio of twist, as a consequence of the fact that the tape with the lowest ratio of twist is the most environmentally friendly at supplying heat augmentation.
Keywords: Heat Exchanger, Insert Twisted Tape, Thermal Performance, Friction Factor, Nanofluid, Heat Transfer Enhancement, Wings and cuts, Laminar flow, Flow regimes.

1. Introduction

Heat exchangers are utilized in a variety of sectors, including coolants, computer ICs, automotive, solar heat collecting systems, and so on. The heating and cooling procedures consume an excessive quantity of energy, which contributes to worsening climate change [1]. Globally, governments have launched programs to keep our planet sustainable and have also urged business groups to engage in the Campaign for sustainable Energy [2]. Optimizing Heat Transfer (HT) will help in the reduction of energy waste in a broader area of applications. Because of the rising consciousness for power savings, many researchers have been involved in HT enhancement, because an enlargement in HT rate leads to expanded machine performance [3], [4], this is a requirement for all thermal application systems. HT improvement methodology is categorized into several classes: compound, active, and passive [5]. [6] Experimentally studied the effect of Twisted Tape (TT) in laminar flow. The partitioning of tube float ensuing in greater float velocities, discount in hydraulic diameter growing the warmth switch quantum, fluid with helically twisting action growing drift route and secondary fluid action enhancing convective HT, and helically twisting fluid action growing glide direction and secondary fluid action enhancing convective HT are the most important motives for TT inserts bettering of HT. [7] The heat transmission and stress drop features of laminar glide in a circular tube with typically spaced TT inserts linked with rods were studied. They discovered that TT inserts attached to a rod in a tube had a 40% lower stress drop than a tube with non-stop TT and that the former has the preferable thermohydraulic achievement. They also looked at how the diameter of the connecting rod and the breadth of the tape inserts influenced HT and strain loss. The discovery was that narrow tape thing widths resulted in poor thermo-hydraulic performance, while a narrower connecting rod had the best results. [8] Extended their study to the HT behaviors in a tube with regularly spaced TT inserts in laminar and turbulent flows. They examined that the HT-coefficient and friction aspects were significantly reduced when compared to tubes with non-stop TT. A similar type of effort was carried out by [9]–[11], and others, in which the degree of HT enhancement and friction factor (FF) are investigated for various types of TT inserts and their diverse layouts, as well as various types of fluid ducts. Even though extraordinary attempts have been made to improve heat transmission through the use of TTs, the underlying science for improving HT has received little attention in the literature. The most recent research provides a physical explanation for HT enhancement as a function of twist-ratio and Re [12].

By reducing the DR from 0.4 to 0.13 the enhancement of heat transfer occurs by 25.2% and also for the heat exchangers with vortex generators with N = 14 and DR = 0.26, the highest $\eta = 2.25$ (thermal efficiency) obtained at Re = 5000. The output of rotated inclined elliptical (RIE) inserts on heat exchanger pipes for heat transfer and the thermal performance factor of double-pipe were investigated [13]. RIE turbulators improve heat transfer by up to 30.7 percent when compared to non-rotated elliptic (NRIE) turbulators, and for the RIE turbulators with $\beta = 90^{\circ}$, dp/b=0.250, and $\alpha = 25^{\circ}$ the highest thermal performance factor of 2.23 was achieved. When NRIE vortex generators with dp=1.5 mm are used, the heat transfer coefficient is improved by 59.95 percent when compared to typical non-rotated louvered strips. It was discovered that using DPIE turbulators increased the Nusselt number by 217.4 percent when compared to the tube without vortex generators. The maximum thermal efficiency parameter of 1.849 was obtained for the DPIE vortex generators with d/b = 0.25 and $\alpha = 25$. A 14.0 percent increase in friction factor was also observed for the DPIE insert with d/b (0.25), and the heat transfer increased by 39.4 percent for the DPIE insert with d/b (0.25) [14]. The separation of the thermal boundary layer is the primary cause of heat transfer enhancement [15].

By extending the area with consideration of sink/source and chemical reaction, the steady boundary layer of a two-dimensional magnetohydrodynamic of a viscous magneto micro polar was investigated. They discovered a rise in velocity concerning the Soret, Richardson, and Grashof numbers. In comparison to the micropolar fluid, the Newtonian fluid had a higher velocity [16]. Mohammad Alkasassbeh et al. [17] investigated and derived a method and claimed that the new derived method outperformed the old method when working on similar types of problems. They also obtained better numerical method properties. R Djebali et al. [18] investigated that, due to cold fluid transverse entry and flow intrusion under the ceiling, the heated cavity not be analyzed as a part of an infinitely hot and cold plate for the phenomena of the boundary layer that occurred along with the isothermal vertical wall.

The influence of secondary flow intensity on heat transfer intensity and friction factor was investigated, as well as twenty-two distinct architectures of circle tube-finned heat exchangers with and without VGs. Circle tubes and vortex generators create secondary flow intensity, which is quantified using the dimensionless measure Se. In a circle tube-finned heat exchanger, Se may accurately represent the secondary flow intensity created by tubes and VGs. Nu and Se have a one-to-one connection, which means Nu and Se have a power exponential function relationship. This means that in a circle tube-finned heat exchanger, the volumetrically averaged Se is the single determining factor of Nu [19]. The computational results revealed that the hydrothermal features of the SCHE are greatly influenced by the flow patterns in the presence of VGs. When the Oil/MWCNT rates are increased in the presence of VGs, negative turbulent velocities are generated in large amounts, promoting good agitation of nanofluid particles and therefore increased great transfer rates. This heat exchanger can be used in a variety of applications, including residential heating and cooling, agricultural food drying, and water desalination [20].

In shell and tube heat exchangers, the connection between increased fouling resistance and heat transfer losses is exponential. The exchange areas are exaggerated in design issues that estimate the

IOP Conf. Series: Materials Science and Engineering 1259 (2022) 012005 doi:10.1088/1757-899X/1259/1/012005

total heat transfer coefficient without factoring in the fouling variables peculiar to the heat transport fluid, therefore overestimating the costs of heat recovery. In shell and tube heat exchangers, the LMTD defines the severity of fouling resistance. Because of the exponential link between exchanger heat loss and fouling buildup, higher LMTD values worsen heat losses more than lower ones [21].

1.1. Basic Terminologies

1.1.1 Thermo Hydraulic Performance

The thermohydraulic overall performance of an insert is referred to be appropriate for a given Re with a wide variety if the HT-coefficient improves appreciably with a little expansion in FF. A thermal-hydraulic overall achievement estimate is normally used to examine the overall performance of more than a few inserts in a particular fluid that goes with the flow situation.

1.1.2 Nu

It is the ratio of gradients of temperature by conduction to convection at the wall surface.

1.1.3 Prandtl Number (Pr)

It is evaluated by the ratio of momentum-diffusivity to heat-diffusivity in molecules.

1.1.4 Pitch

It is the interval measured parallel to the axis between two points in the same plane. Tape that is twisted.

1.1.5 Ratio of a twist (p)

It is the ratio of the length of the pitch to tube inner diameter is known as the twist ratio [12].

1.2. TT

Due to its consistent efficiency and ease of installation, the TT is common in a passive HT-improvement technology. The use of TTs to improve the thermal-hydraulic output of tubes is becoming more widespread [11].

1.2.1 Plain TT

Inserting a swirling flow into the inner test tube of a concentric dual-pipe heat-exchanger with different ratios of twist as p = 3.0, 4.4, and 6.0 may be used to assess the HT-coefficient and flow-friction properties of the heat exchanger. The increase in the rate of heat transmission with the TT inserts is spotted to be significantly governed by tape-induced for swirl or vortex motion [22].

1.2.2 Modified TT

HT can be improved by experimenting with different twist geometries and twist ratios. Inserting half-length TT into the heat exchangers in an inner test tube can also induce swirling flow. Tape-induced -swirl or vortex motion is shown to have a dynamic influence on the heat transmission rate of TT- inserts. The HT coefficient rises by 40% when half-length twisted-tape-inserts are used instead of basic heat-exchangers [23].

2. Material and Experimental Setup

2.1 Physical Model

Figure 2 depicts the experimental structure, which is denoted through distinct colors: blue for cold fluid traits and crimson for hot fluid characteristics. Two concentric tubes, one internal and the other outer, make up the arrangement. Hot fluid runs via a copper tube with an interior diameter of 2.85 cm and a size of 200 cm, whilst fluid flows on the contrary route through the annulus.

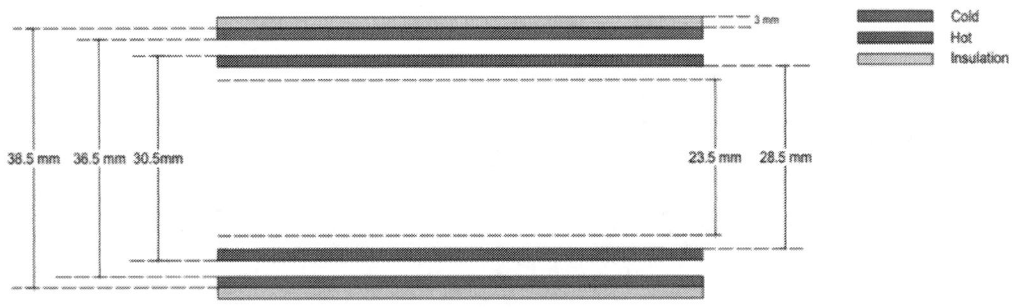

Figure 1. Heat Exchanger Dimensions

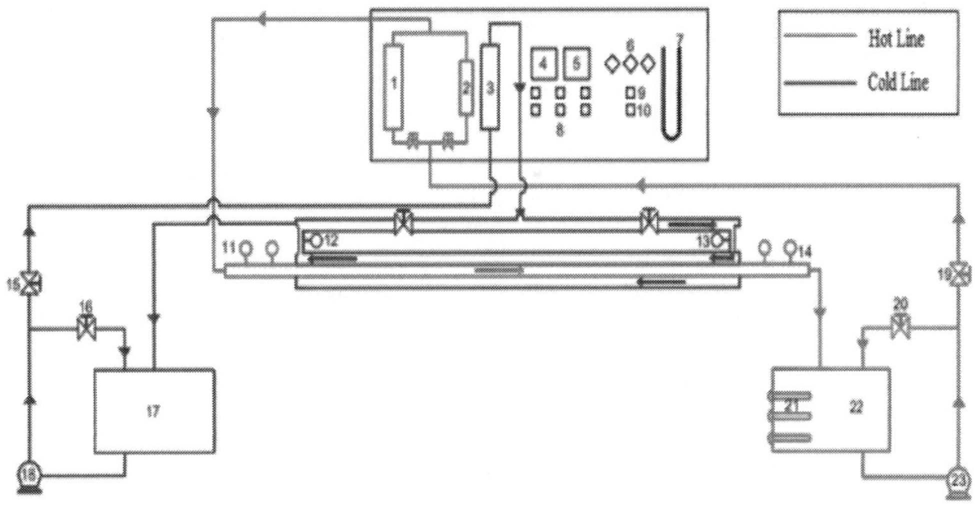

1. Hot Fluid Meter
2. Hot Fluid Meter
3. Cold Fluid Meter
4. Temperature Indicator
5. Temperature Controller
6. Heat Indicator
7. U- Tube Meter
8. Switches
9. Ammeter
10. Voltmeter
11. RTD Hot Fluid In
12. RTD Cold Fluid Out
13. RTD Cold Fluid In
14. RTD Hot Fluid Out
15. Cold Fluid Control Valve
16. Cold Fluid Bypass Valve
17. Cold Fluid Sump
18. Cold Fluid Pump
19. Hot Fluid Control Valve
20. Hot Fluid Bypass Valve
21. Heater
22. Hot Fluid Tank
23. Hot Fluid Pump

Figure 2. Structural outline of Heat exchanger

The temperature sensors (PT 100) are used to measure the temperature of the hot and cold fluid at the input and outflow with an accuracy of 10%. As indicated in Figure 2, 04 temperature sensors (two on each end) monitor the temperature of the hot fluid at the input and outlet, whereas 02 temperature sensors (one on either side) measure the temperature of cold fluid. Three 1 KW heaters are used to heat the fluid in a tank. The temperature on the panel is controlled by a temperature controller, as indicated by the temperature- indicator mounted on the panel. The temperature of the hot fluid at the inlet is maintained at 55°C, while the cold fluid temperature at the inlet is vindicated at 28°C. The rate of flow of cold fluid is kept constant at ten LPM to keep the Re constant between 2000 and 12000, while the rate of flow of hot fluid is regulated between two and seven LPM with a 0.5 LPM increment to keep the Re constant between 2000 and 12000. A U tube manometer watched the reduction in pressure when it achieve the equilibrium conditions in the case of the plane tube, while RTDs assessed the input and output temperatures of hot and cold fluid.

2.2 TT dimensions

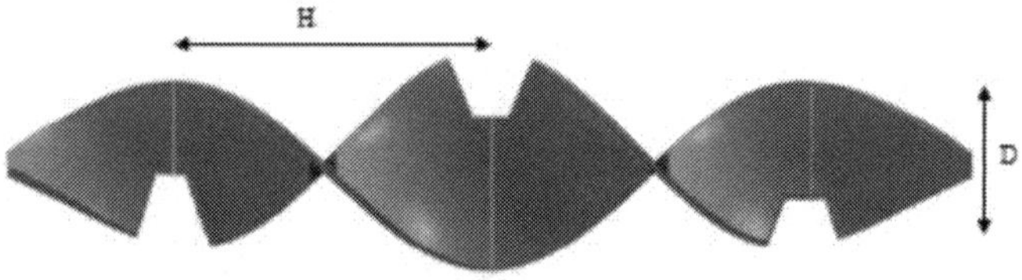

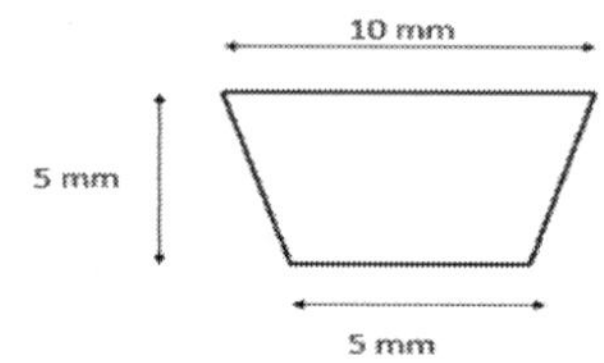

Figure 3 (a) and (b). Inserted TT and Dimensions of the Trapezoidal-Section

Figures 3 (a) and (b) show TT inserts with trapezoidal–cuts at the top and backside on alternate sides. The twisted tape is constructed of 0.15 cm thick and 2.35 cm wide strips of aluminum with the ratio of twist as 6.0, 4.4, and 3.0 (p = H/D, where H= 10.8 cm, 7.92 cm, and 5.4 cm and D=1.8 cm, respectively). The twist ratio is the proportion of twist diameter to twist size. The full-length TT is used with the scaling for trapezoidal-cut as 0.5 cm depth, 0.5 cm base, and 1 cm breadth at the top. The cuts on the insert are alternated on the bottom and top to enhance the mixing of fluid alongside the look at the segment's walls. The pressure drop is recorded again after inserting the circular tube into the U tube manometer, and the hot and cold fluid temperatures at the intake and output were measured by RTDs and also correlated to the values obtained by the plane tube.

3. Research Methodology

The tube hydraulic diameter affects both the FF and the average quantity of Nu. Qc may be stated as: cold water absorbs the heat in a uniform heat flow condition expressed as below:

$$Qc = m \times C_p \times (T_o - T_i) \tag{1}$$

Under the condition of the state of equilibrium, the HT rate via the applied part can be described as follows:

$$Qc = h_{avg} \times A\,(Tw - Ty) \tag{2}$$

Tw= provided as the average wall temperature;

$$\textstyle\sum_{i=1}^{n} Tiw/n \tag{3}$$

Where, n = the thermocouples are used to determine the wall temperature.

$$T_y = \frac{To+Ti}{2}$$

Solving (1) and (2) gives;

$$h_{avg} \times A \times (Tw - T_y) = m \times C_p \times (To - Ti) \tag{4}$$

As a result, the average coefficient of HT may be described as follows:

$$h = \frac{m \times Cp \times (To-Ti)}{A \times (Tw-Ty)} \tag{5}$$

The average coefficient of HT may be useful to evaluate the mean Nu, as illustrated below:

$$Nu = \frac{havg \ x \ dh}{k} \tag{6}$$

To calculate the Re below mentioned expression is utilized;

$$Re = \frac{\rho U dh}{\mu} \tag{7}$$

The hydraulic diameter (dh = d and it is the inside diameter of the pipe) is defined as the typical dimension for internal-flow circumstances:

$$d_h = \frac{4A}{P} \tag{8}$$

The hydraulic diameter of heat exchangers (Figure 2) can be stated as follows:

$$dh = do - di \tag{9}$$

The following co-relations of Nu were made to attain the HT and for the tube with and without insert for the three different fluids [22], [23].

$$Nu = 0.023\ x\ Re^{0.8}\ x\ Pr^{0.3} \quad (10)$$

For Tube with Trapezoidal-cut twisted tape insert

$$Nu = 0.1197\ x\ Re^{0.82}\ x\ Pr^{0.33}\ x\ p^{-0.77} \quad (11)$$

$$Nu = 0.005\ x\ Re^{1.14} x\ Pr^{0.4} x\ p^{-0.5} \quad (12)$$

The value of the Prandtl number is 3

4. Performance Evaluation

The thermo-hydraulic behavior of an insert is primarily governed by the flow conditions, aside from the insert designs (turbulent or laminar). The primary findings of all studies concerning various inserts and flow are listed below [24]:

4.1 Insert tape twisted when the flow is laminar

The thermal resistance in the laminar flow is not confined to any specific place; when the flow is laminar TT mixes and inserts efficiently in the bulk flow and performs exceptionally than the other inserts. Fluid characteristics, such as the Pr on the other hand, affect TT performance. If the Pr is big, such as Pr = 30, TT and other inserts, such as a wire-coil, will not perform in terms of thermo-hydraulic performance. Twisted short-length tape out-performs full-length-TT in terms of continuous pumping power. TT can be used to increase heat transmission by creating the heat-exchangers with laminar flow [5], [25].

4.2 Insert tape twisted when the flow is turbulent

Inside the turbulent flow, the TT effectively upgrades to the precise range of Re, but not beyond that. TT fails in turbulent flow as compared to wire coils because it causes a high-pressure drop and stops the flow [26]. As a result, the "thermo-hydraulic performance" of TT in turbulent flow is lower than that of a wire coil. As a consequence, it can be said that the wire coil is a good alternative for the design of a compact heat exchanger in a turbulent flow. When compared to a TT that is the entire length, the turbulent flow with a short piece of TT generates favorable thermo-hydraulic behavior [27].

4.3 Laminar flow in the wire coil

The HT rate of the coil in the wire is greatly boosted in a laminar flow. The performance, on the other hand, is determined by the Pr. When the Pr is large, the performance coil in the wire is favorable because the layer of thermal boundary is thin when correlating to the layer of hydrodynamic boundary, and the coil readily penetrates the thick layer of the boundary. As a result, both pressure and heat transmission are decreased greatly [28].

4.4 Turbulent flow in the wire coil

The HT has been significantly enhanced when the flow is turbulent due to the wire coil. The turbulent flow is superior to the laminar flow in terms of performance. The wire-coil beats TT when the flow is turbulent in terms of thermo-hydraulic performance [29], [30].

5. Results and Discussion

Table 1. Nusselt Number (water used as working fluid)

Re	Nusselt Number (Nu)			
	Plain Tube	**p = 3.0**	**p = 4.4**	**p = 6.0**
2000	14.338942	37.5842	27.877999	22.125472
3000	19.833078	52.4083	38.55821	30.536749
4000	24.965548	66.3444	48.603661	38.448117
5000	29.844831	79.6655	58.202046	46.007393

6000	34.531394	92.5122	67.45857	53.297436
7000	39.063536	104.9773	76.440146	60.370941
8000	43.467544	117.3248	85.19287	67.264211
9000	47.762508	129.0012	93.750306	74.003681
10000	51.962868	140.642	102.13792	80.609403
11000	56.079904	152.0747	110.37563	87.097073
12000	60.12265	163.3216	118.47944	93.479287

Table 2. Nusselt Number (air used as working fluid)

Re	Nusselt Number (Nu)			
	Plain Tube	**p = 3.0**	**p = 4.4**	**p = 6.0**
800	4.36	8.96394	7.8643901	6.8804657
900	4.36	10.25201	8.907338	7.7677947
1000	4.36	11.33541	9.9665501	8.6689611
1100	4.36	12.89635	11.040611	9.5827609
1200	4.36	14.18019	12.128349	10.508196
1300	4.36	15.61745	13.228773	11.444425
1400	4.36	16.99215	14.341037	12.390727
1500	4.36	18.16344	15.464408	13.34648
1600	4.36	19.61928	16.598244	14.311136
1700	4.36	21.19899	17.74198	15.284214
1800	4.36	22.49878	18.895111	16.265285
1900	4.36	23.89126	20.057185	17.253966
2000	4.36	25.09683	21.227796	18.249909

Table 3. Nusselt Number (oil used as working fluid)

Re	Nusselt Number (Nu)			
	Plain Tube	**p = 3.0**	**p = 4.4**	**p = 6.0**
200	4.36	18.2777	15.2732	13.2941
300	4.36	28.5772	23.77782	20.577
400	4.36	39.3778	32.6962	28.2143
500	4.36	50.567	41.93546	36.1263
600	4.36	62.076	51.43875	44.2644
700	4.36	73.8578	61.16737	52.5955
800	4.36	85.8785	71.09317	61.0955
900	4.36	98.1118	81.19462	69.7458
1000	4.36	110.5373	91.45474	78.5321
1100	4.36	123.1383	101.8597	87.4424
1200	4.36	135.9009	112.3982	96.467
1300	4.36	148.8134	123.0604	105.598
1400	4.36	161.8659	133.8382	114.827
1500	4.36	175.0486	144.7244	124.15
1600	4.36	188.357	155.7128	133.559

1700	4.36	201.7815	166.7978	143.052
1800	4.36	215.317	177.9745	152.623
1900	4.36	228.9583	189.2385	162.269
2000	4.36	242.7005	200.5858	171.986

5.1 Variation of Nusselt Number (Nu) concerning Reynold's Number (Re)

5.1.1 Fluid used as water

Figure 4 depicts the variation of Nu with Re for plain tube with and without insert for the ratio of twist 6.0, 4.4, and 3.0 respectively. The Nu for trapezoidal section twisted tape is bigger than the plain tube, because of twisted tape along with the length of tube the flow is spiral and it mix-up with the flow, hence enhancing the heat transfer rate.

5.1.2 Fluid used as air

The Nu seems to grow substantially as the Re rises, indicating an increase in convective HT. Across the tested range, the Nu rises as the Re rises (laminar regime). Furthermore, as seen in figure 5, larger H/D ratios are associated with lower Nu.

5.1.3 Fluid used as oil

Figure 6 indicates that the fluctuation occurs for the Nu concerning Re with and without insert for the used ratio of twist. The flow is spiraling towards the direction of flow and it disturbs the complete flow, hence increase in heat transfer rate due to obtained Nu is higher than the plain tube.

For the plane tube, the Nu is steady in the laminar flow regime and may be computed as Nu =48/11 =4.36. Tape-inserts improve the Nu by a factor of 35-40, which is higher for lower twist ratios.

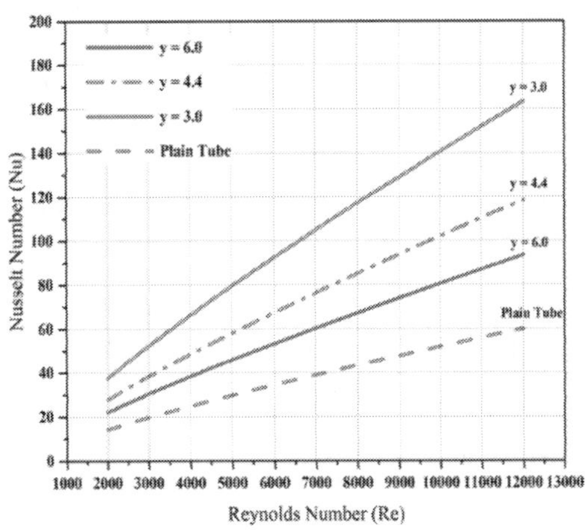

Figure 4. Comparison of Nusselt Number concerning Reynold's Number (water used as working fluid)

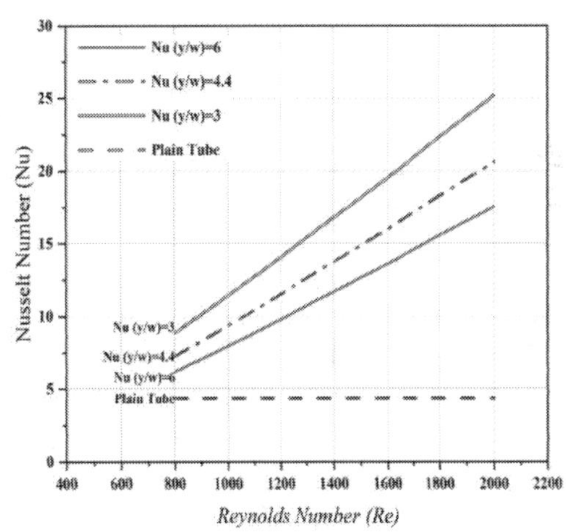

Figure 5: Comparison of Nusselt Number concerning Reynold's Number (air used as working fluid)

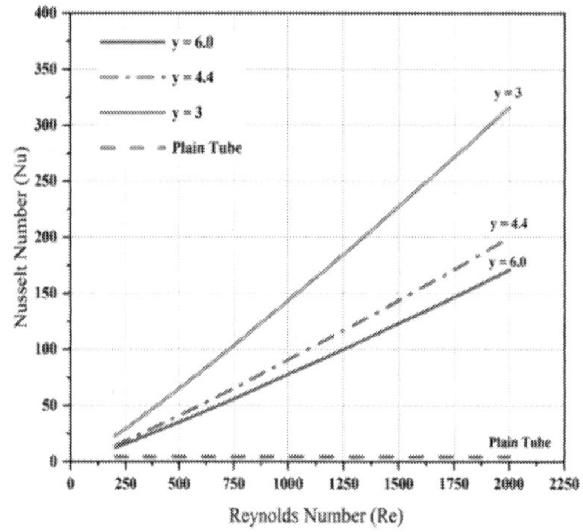

Figure 6: Comparison of Nusselt Number concerning Reynold's Number (oil used as working fluid)

5.2 Variation of Friction Factor (f) concerning Reynold's Number (Re)

Figure 7, 8 and 9 depicts the friction factor (f) worsen Reynolds Number (Re) for an insert and without insert. The friction factor is lowest for the tube without an insert and highest for the Trapezoidal-Cut-Twisted-Tape insert for lowest ratio twist (y = 3). Though the friction factor reduces with increases of Re, increasing the flow of swirls with a lower twist ratio produces the maximum friction factor parameter by employing a twisted tape insert with a twist ratio of 3.

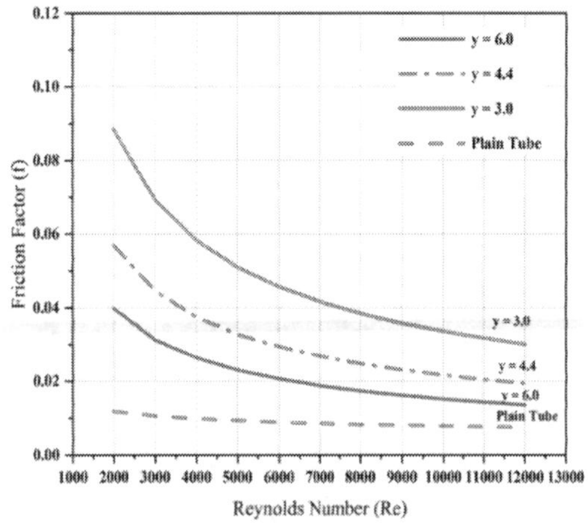

Figure 7: Comparison of Friction Factor concerning Reynold's Number (water used as working fluid)

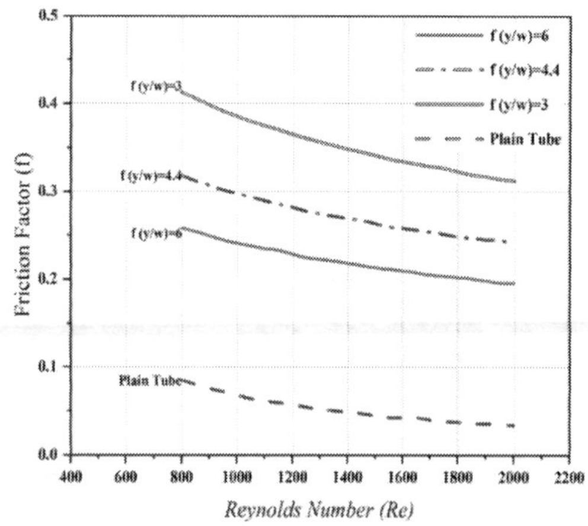

Figure 8: Comparison of Friction Factor concerning Reynold's Number (air used as working fluid)

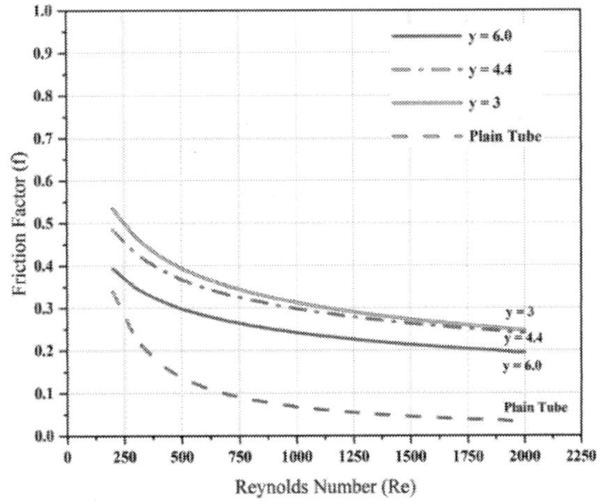

Figure 9: Comparison of Friction Factor concerning Reynold's Number (oil used as working fluid)

5.3 Variation of Efficiency (η) concerning Reynold's Number (Re)

5.3.1 Fluid used as water

Figure 10 indicate the efficiency for the Trapezoidal-Cut-Twisted-Tape insert for the ratio of twist 6.0, 4.4, and 3.0 respectively. It indicates the efficiency is larger than unity, implementing of modification for total saving of energy is prudent.

5.3.2 Fluid used as air

Figure 11 shows the change in the thermal performance with Reynolds Number for the inserts with the ratio of twist 6.0, 4.4, and 3.0 respectively and it indicates the efficiency for all the cases are higher than unity, it means the use of enhancement is valuable towards the total energy saving.

5.3.3 Fluid used as oil

The efficiency (thermal) variation with Reynolds Number is shown in Figure 12 and it can be seen that Trapezoidal-Cut-Twisted-Tape with lower twist ratio (y = 3) gives better thermal performance than the other twist ratio used.

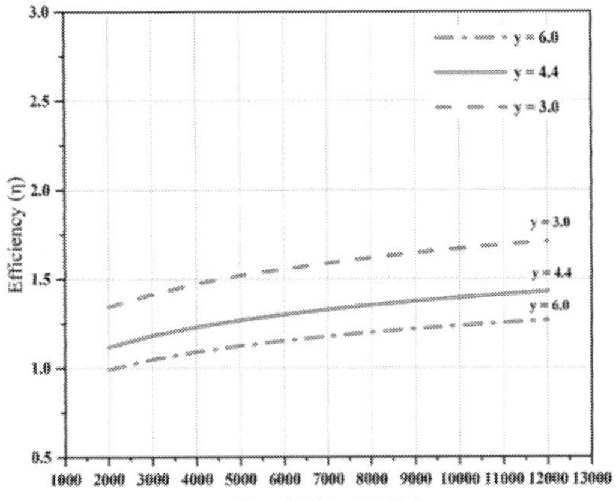

Figure 10: Comparison of Efficiency concerning Reynold's Number (water used as working fluid)

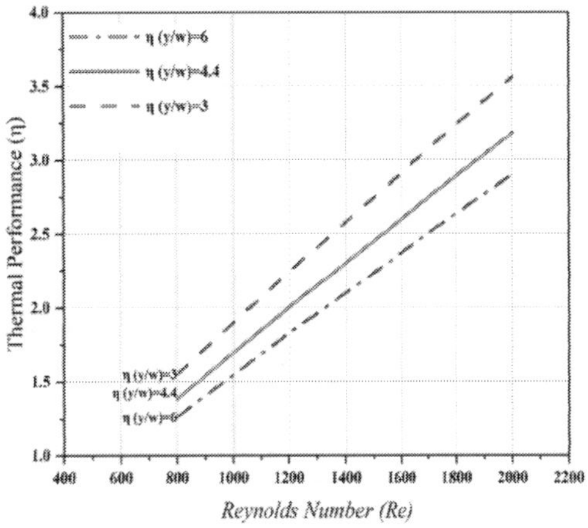

Figure 11: Comparison of Efficiency concerning Reynold's Number (air used as working fluid)

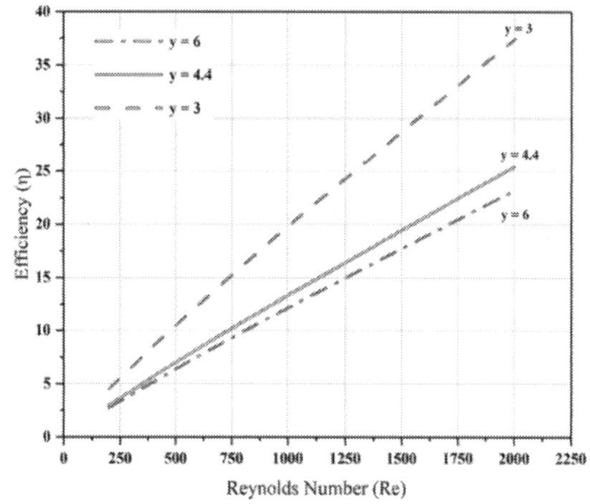

Figure 12: Comparison of Efficiency concerning Reynold's Number (oil used as working fluid)

6. Conclusion

The heat transmission and FF capabilities of a copper-based concentric tube by inserting with a trapezoidal-cut- twisted-tape were experimentally examined for the ratio of twist of 6.0, 4.4, and 3.0 for different ranges of Re as 2000-12000, 800-2000, and 200-2000. The HT-coefficient and FF characteristics improve in value when the twist-ratio decreases. Trapezoidal-cut-twisted-tapes boost HT rates by 26%, 40%, and 51%, for the ratio of twist as 6.0, 4.4, and 3.0 as compared to a plain tube. Above all, the output ratio produced by trapezoidal-cut-twisted-tape is more than one, suggesting that this change has the potential to save energy. HT and flow characteristics of TT-inserts with varying twist-ratios and Re are studied. When the twist-ratio drop, the Re rises, and heat transmission increases.

When the twist ratio lowers owing to a rise in radial convection, HT increases. The smaller ratio of twist enhances HT due to the enhanced transverse motion of fluid in core flow commencing by the higher degree of the helical channel rotating flow.

When the Re rises owing to axial convection, the heat transmission is accelerated. By disturbing the barrier layer generated by more convective HT from the wall to the fluid, an increase in Re improves heat transfer.

References

[1] Bergles A E 2001 The Implications and Challenges of Enhanced Heat Transfer for the Chemical Process Industries. *Chem. Eng. Res. Des.*, 79, 437-444.
[2] Zhang J, Zhu X, Mondejar M E and Haglind F 2019 A review of heat transfer enhancement techniques in plate heat exchangers. *Renew. Sustain. Energy Rev.*,101, 305-328.
[3] Rohsenow W M, Hartnett J P and Cho Y I 1998 Handbook of heat transfer. *Int. J. Heat Mass Transf.*
[4] YILMAZ İ H and GÖKSU T T 2019 Düz tüp içerisinde burulmuş şerit kullanarak ısı geçişinin iyileştirilmesi. *Bitlis Eren Üniversitesi Fen Bilim. Derg.*, 8, 251-260.
[5] Ahmad S, Abdullah S and Sopian K 2020 A review on the thermal performance of nanofluid inside circular tube with twisted tape inserts. *Adv. Mech. Eng.*, 12, 1-26.
[6] Manglik R M and Bergles A E 1993 Heat Transfer and Pressure Drop Correlations for Twisted-Tape Inserts in Isothermal Tubes: Part I—Laminar Flows. *J. Heat Transfer*, 115, 881-889.
[7] Saha S K, Gaitonde U N and Date A W 1989 Heat transfer and pressure drop characteristics of laminar flow in a circular tube fitted with regularly spaced twisted-tape elements. *Exp. Therm. Fluid Sci.*, 2, 310-322.

[8] Eiamsa-ard S, Thianpong C and Promvonge P 2006 Experimental investigation of heat transfer and flow friction in a circular tube fitted with regularly spaced twisted tape elements. *Int. Commun. Heat Mass Transf.*, 33, 1225-1233.

[9] Jaisankar S, Radhakrishnan T K and Sheeba K N 2009 Experimental studies on heat transfer and friction factor characteristics of thermosyphon solar water heater system fitted with spacer at the trailing edge of twisted tapes. *Appl. Therm. Eng.*, 29, 1224-1231.

[10] Ayub Z H and Al-Fahed S F 1993 The effect of gap width between horizontal tube and twisted tape on the pressure drop in turbulent water flow. *Int. J. Heat Fluid Flow*, 14, 64-67.

[11] Rahimi M, Shabanian S R and Alsairafi A A 2009 Experimental and CFD studies on heat transfer and friction factor characteristics of a tube equipped with modified twisted tape inserts. *Chem. Eng. Process. Process Intensif.*, 48, 762-770.

[12] Savekar A, Jangid D, Gurjar M, Patil V and Sewatkar C M 2015 Analysis of Heat Transfer in Pipe with Twisted Tape Inserts. *Proceedings of the 2 nd International Conference on Fluid Flow, Heat and Mass Transfer Ottawa, Ontario, Canada*, 143-151.

[13] Nakhchi M E, Hatami M and Rahmati M 2021 Experimental evaluation of performance intensification of double-pipe heat exchangers with rotary elliptical inserts. *Chem. Eng. Process. - Process Intensif.*, 169, 108615.

[14] Nakhchi M E, Hatami M and Rahmati M 2021 Experimental investigation of performance improvement of double-pipe heat exchangers with novel perforated elliptic turbulators. *Int. J. Therm. Sci.*, vol. 168, 107057.

[15] Nakhchi M E, Hatami M and Rahmati M 2021 Effects of CuO nano powder on performance improvement and entropy production of double-pipe heat exchanger with innovative perforated turbulators. *Adv. Powder Technol.*, 32, 3063-3074.

[16] Warke A S, Ramesh K, Mebarek-Oudina F and Abidi A 2022 Numerical investigation of the stagnation point flow of radiative magnetomicropolar liquid past a heated porous stretching sheet. *J. Therm. Anal. Calorim.*, 147, 6901-6912.

[17] Alkasassbeh M, Omar Z, Mebarek-Oudina F, Raza J and Chamkha A 2019 Heat transfer study of convective fin with temperature-dependent internal heat generation by hybrid block method. *Heat Transf. Res.*, 48, 1225-1244.

[18] Djebali R, Mebarek-Oudina F and Rajashekhar C 2021 Similarity solution analysis of dynamic and thermal boundary layers: further formulation along a vertical flat plate. *Phys. Scr.*, 96, 085206.

[19] Guan Y, Hu W, Zhang Y, Song K, and Wang L 2020 Numerical Study of the Intensity Correlation between Secondary Flow and Heat Transfer of Circle Tube-Finned Heat Exchanger with Vortex Generators. *Comput. Model. Eng. Sci.*, 123, 237-256.

[20] Maouedj R, Menni Y, Inc M, Chu Y M, Ameur H and Lorenzini G 2021 Simulating the Turbulent Hydrothermal Behavior of Oil/MWCNT Nanofluid in a Solar Channel Heat Exchanger Equipped with Vortex Generators. *Comput. Model. Eng. Sci.*, 126, 855-889.

[21] Musonye F S, Ndiritu H and Kinyua R 2021 Performance Assessment of Heat Exchangers for Process Heat Integration. *Energy Eng.*, 118, 211-224.

[22] Dhamane N B, Nalawade D B, and Dange M M 2014 Performance Analysis of Wavy Twisted Tape Insert for Heat Transfer in a Circular Tube. *Int. J. Technol. Res. Eng.*, 1, 25-28.

[23] Chowdhuri M, Hossain R and Sarkar M 2011 An experimental investigation of turbulent flow heat transfer through tube with rod-pin insert. *Int. J. Eng. Sci. Technol.*, 3, 76-81.

[24] Johar G and Hasda V 2010 Experimental Studies On Heat Transfer Augmenatation Using Modified Reduced Width Twisted Tapes (RWTT) As Inserts For Tube Side Flow Of Liquids. National Institute of Technology, Rourkela.

[25] Bartwal A, Gautam A, Kumar M, Mangrulkar C K and Chamoli S 2018 Thermal performance intensification of a circular heat exchanger tube integrated with compound circular ring–metal wire net inserts. *Chem. Eng. Process. - Process Intensif.*, 124, 50-70.

[26] Abolarin S M, Everts M and Meyer J P 2019 Heat transfer and pressure drop characteristics of alternating clockwise and counter clockwise twisted tape inserts in the transitional flow regime. *Int. J. Heat Mass Transf.*, 133, 203-217.

[27] Dewan A, Mahanta P, Raju K S and Kumar P S 2004 Review of passive heat transfer augmentation techniques. *Proc. Inst. Mech. Eng. Part A J. Power Energy*, 218, 509-527.

[28] Jedsadaratanachai W, Jayranaiwachira N and Promvonge P 2015 3D numerical study on flow structure and heat transfer in a circular tube with V-baffles. *Chinese J. Chem. Eng.*, 23, 342-349.

[29] Zheng N, Liu P, Shan F, Liu J, Liu Z and Liu W 2016 Numerical studies on thermo-hydraulic characteristics of laminar flow in a heat exchanger tube fitted with vortex rods. *Int. J. Therm.*

IOP Conf. Series: Materials Science and Engineering 1259 (2022) 012005 doi:10.1088/1757-899X/1259/1/012005

Sci., 100, 448-456.
[30] Dhumal A H, Kerkal G M and Pawale K T 2017 Heat Transfer Enhancement for Tube in Tube Heat Exchanger Using Twisted Tape Inserts. *Int. J. Adv. Eng. Res. Sci.*, 4, 89-92.

Acknowledgments

The authors are grateful to SGP Instruments & Control, Ambernath (E), District: - Thane (M.S.) -India for allowing the experimentation work.

Subscripts

avg	Average
i	inside
o	outside

Nomenclature

L	Tube length, m
d	Diameter of the tube, m
Re	Reynolds Number
Nu	Nusselt Number
Pr	Prandtl Number
y	Twist ratio
A	Area, m^2
Cp	Specific heat, J/kg-K
m	Flow rate of fluid, kg/sec
T	Fluid temperature, °C
h	Coefficient of heat transfer, W/m^2-K
k	Thermal conductivity, W/m-K
ρ	Fluid density, kg/m^3
μ	Dynamic viscosity, kg/m-sec
U	Fluid velocity, m/sec
dh	Hydraulic diameter, m
Ty	Average temperature of fluid, °C
Tw	Average wall temperature, °C

Heat Integration and Heat Exchanger Network Design with Aspen Energy Analyze

K R PETHE[1]; P M DABHADE[1]; V J KOLEKAR[1]; M D SARDARE*

School of chemical engineering, MIT Academy of Engineering, Alandi, Pune, India

Email Id: mamtasardare@gmail.com; mdsardare@mitaoe.ac.in;
https://orcid.org/**0000-0003-0205-841X**

Abstract. The considerable amount of energy used in industrial processes is typically used for heating and cooling. The synthesis of heat ex-changer networks (HEN) is a critical topic in this industry in order to reduce energy and capital costs in chemical plants. Heating and cooling systems in industry must be designed efficiently, which can be done using technologies like pinch analysis to create an optimal heat recovery network. Heat pinch analysis is a systematic method for designing thermally efficient systems. A designer can determine the minimum heating and cooling requirements as well as the maximum heat recovery (MHR) potential by locating a thermodynamic bottleneck, or the pinch point for heat recovery. This Paper uses Aspen Energy Analyzer for HEN Design.
Heat integration network design is a cost-effective strategy that helps industries to cut costs by reducing the number of external utilities needed and increase profitability by lowering energy consumption. The HENs' composite curves, grand composite curves, grid representation, and target reports were created using the Aspen Energy Analyzer. The software suggested ten designs, which were then optimized based on total annualized cost. By comparing it to other optimized designs, the best design is determined.
Key Words: Pinch technology, heat integration, HEN, pinch point, Aspen Energy Analyzer, Hint

1.Introduction

The products which are getting produced in industry or the processes which are required for crude oil to produce fuel, requires an amount of energy and the cost of this production depends on this amount of consumed energy in the plant. The more the amount of energy consumed in the process plant, higher the cost for the production will be required and vice versa. So from this we got to know that the energy consumption is important factor in any process plant. This energy consumption factor should always be considered in any process to reduce the cost and to increase the profit of the industry. So minimization of the energy consumption is highly important in any process plant. If we see then it is difficult to determine the minimum amount of energy required by a process plant by using old method which do not include pinch technology. Pinch Technology is a methodology which is used to reduce energy consumption by establishing thermodynamically feasible energy targets (or lowest energy consumption) and realizing them through the optimization of heat recovery systems, energy supply sources, and process operating conditions. Process integration, heat-integration, energy-integration, and pinch analysis are other terms for the same thing. Since pinch technology was discovered in the 1970s and was first used by industries in the 1980s, it is safe to assume that process plants built and designed prior to 1980 were not subjected to pinch analysis. As a result, we can conclude that the amount of energy required in such facilities was likely greater than the amount required.

The majority of petroleum refineries and petrochemical firms used a lot of energy. In most chemical processing industries, this has been a major source of worry. Because the plants were developed in an era when energy was cheap, employing traditional design energy (TDE) approaches such as mass and energy balances, laws of thumb, sound engineering judgment, and the designers' creative talent, which are not cost effective. Pinch analysis was not used in the construction of some refineries and petrochemical industries prior to the 1980s. As a result, a significant amount of fuel is consumed, CO_2 is released into the atmosphere, and operational costs are high. This raises severe concerns about the need for a better process design, which is referred to as Process Integration.

1.1 Overview of the kaduna refinery and petrochemical company's crude distillation unit :

Crude oil is extracted from the earth, which contains a wide range of chemicals such as gases, water, and dirt (Minerals). If crude oil is to be processed without fouling or corrosion in process equipment, it must first be pretreated. Crude desalting and Field separation are two methods for pretreatment of petroleum. The Crude Distillation Unit is part of the Kaduna Refining and Petrochemical Company's [3] (KRPC) Fuels department, where local crude is distilled into naphtha, gasoline, kerosene, diesel, and bottom residue. Raw crude oil is pumped to the CDU after settling and dewatering at the tank farm. It passes through a heat exchanger train, the desalter (for removal of salt and sediments), the pre-flash column (for removal of lighter ends) and the crude furnace where it is heated up, then to the fractionating column where the crude is separated into its components. The vapours are removed from the top, condensed and sent to saturated gas concentration unit (SGCU) for further separation and production of LPG or cooking gas while the liquids are withdrawn from the sides, based on the boiling point range.

2. Materials and Methods:

2.1 *Materials:*

The materials used include Pinch analysis software (Aspen Energy Analyzer), Stream data of the crude distillation unit of Kaduna refinery and petrochemical company and Computer set.

2.2 Method:

The Aspen Energy Analyzer version 8.6 process tool was employed to perform a detailed and accurate pinch analysis of the heat exchanger networks on the Crude distillation unit in Kaduna refinery. To do this, the thermal data obtained from [3] the data extraction were fed as input to the software to construct the composite curve, balanced composite curve, grand composite curve and grid representation of all networks. The following pinch rules were employed in order to achieve the minimum energy targets for the crude preheating process.

(1) Heat must not be transferred across the pinch

(2) There must be no external cooling above the pinch and no external heating below the pinch (heaters must be placed above and coolers below the pinch).

Violating any of these rules will lead to cross-pinch heat transfer resulting in an increase in the energy requirement beyond the target. Any heat transfer across the pinch is excess heat which is wasted, and expressed as a pinch penalty.

2.2.1 Steps in Pinch Analysis

The pinch analysis of the industrial ammonia plant was achieved through the following steps:

a) Data Extraction

This involves extraction of the data from the process flow diagram to form thermal problem table. The data to be extracted is stated in the materials required for the work.

b) The Generation of a Thermal Data Table

The data table was formed from the extracted data by Aspen energy analyzer.

c) The application of Aspen Energy Analyzer Software Version 8.6

The extracted data were fed as input into Aspen energy analyzer [2] to complete the thermal data table.

IOP Conf. Series: Materials Science and Engineering 1259 (2022) 012006 doi:10.1088/1757-899X/1259/1/012006

d) Analysis of the Composite Curves and Grid Diagrams to meet the Objective of the Work is to formulate and complete the thermal data table, equation (1) was used:

$Q = M * Cp * \Delta T$ ... (1)

Where: M - mass flow rate of the stream (kg/hr)

Cp - specific heat capacity of the process streams (kJ/kg 0C)

ΔT - temperature difference between inlet temperatures and the outlet temperature of each stream (^{0}C)

$CP = M * Cp$ ... (2)

Where CP is the heat capacity flow-rate and measured as enthalpy change per unit temperature difference (kJ/C-h).

3.Results and Discussions

The results of the crude distillation unit of the Kaduna refinery and petrochemical firm in Kaduna, Nigeria, are presented in this section. Table 1 shows the stream specification and data collection for cold and hot steams from the crude distillation unit of the Kaduna refinery and petrochemical firm in Kaduna, Nigeria.

Table 1: Stream Data of CDU unit of Kaduna Refinery

Stream	Mass Flow (Kg/s)	Specific Heat (KJ/kg. °C)	Supply Temperature (°C)	Target Temperature (°C)	HTC (W/m2.^{0}C)	M.Cp (KJ/°C.s)
H1	48.67	2.81	319	244	45	136.7627
H2	20.13	2.2	73	40	680	44.286
H3	32.41	2.83	347	45	643	91.7203
H4	144.57	2.74	264	180	648	396.1218
H5	22.47	2.78	297	110	652	62.4666
H6	22.56	2.8	248	50	596	63.168
H7	26.41	2.2	73	40	589	58.102
H8	22.39	2.77	232	120	646	62.0203
H9	146.24	2.69	167	70	700	393.3856
H10	128.54	2.68	147	73	420	344.4872
C1	178	2.1	30	232	740	373.8
C2	170	2.79	232	343	752	474.3
C3	158.24	2.7	226	232	784	427.248
C4	170	2.82	343	398	810	479.4

3.1 Outcome of Aspen Energy Analyzer

Figure 1 shows the minimum cooling and heating duty requirement, pinch temperature and area targets for given data at ΔTmin of 20 °C.

Energy Targets	
Heating [kJ/h]	2.021e+008
Cooling [kJ/h]	2.464e+008

Area Targets	
Counter Current [m2]	3.153e+004
1-2 Shell & Tube [m2]	3.442e+004

Pinch Temperatures	
Hot	Cold
264.0 C	244.0 C

Number of Units Targets	
Total Minimum	15
Minimum for MER	18
Shells	28

Cost Index Targets	
Capital [Cost]	6.864e+006
Operating [Cost/s]	0.2530
Total Annual [Cost/s]	0.3104

Figure 1: Results Extracted from AEA for ΔTmin=20°C

Based on the requirement utilities were added to meet heating and cooling duty; shown in Figure

Name	Load [kJ/h]	Cost Index [Cost/s]	Losses [kJ/h]	Outlet T [C]
Cooling Water	2.464e+008	1.454e-002	0.0000	25.00
Fired Heat (1000)	2.021e+008	0.2385	0.0000	400.0

Figure 2: Utility requirements

3.2 Composite curve

Figure 3 shows the composite diagram for given data the upper curve represents the hot streams (Red) composite-curve while the lower curve represents the cold streams (Blue) composite-curve. The part of the hot stream in composite-curve that is extended beyond the start of the cold stream, cannot be cooled by process to process heat transfer. Therefore the minimum cooling utility of 246369976.2 kJ/h is required. The part of the cold stream composite-curve that is extended beyond the start of the hot stream cannot be heated by process-to-process heat transfer. Therefore the minimum hot utility of 202055718.960 kJ/h is required. The point where the two curves are closest is the pinch point and the corresponding temperature is the pinch temperature (254 ^{0}C). Also from Figure 3 the hot pinch temperature was found to be 264 ^{0}C and the cold pinch temperature was found to be 244 ^{0}C.

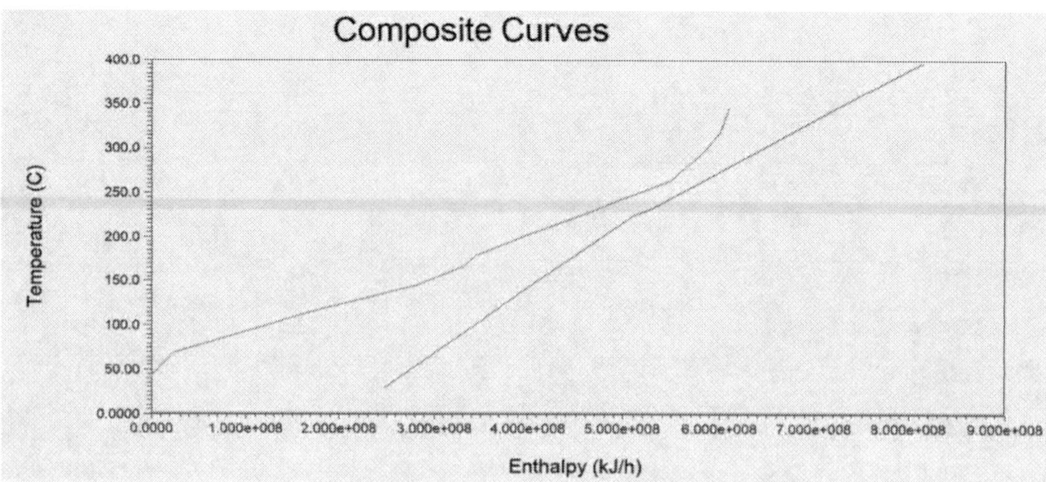

Figure 3: Composite curves at initial at ΔTmin=20°C

The Grand-Composite Curve (GCC) in Figure 4 depicts the heat accessible at various temperature intervals as well as the process's net heat flow (which is zero at the pinch). ΔTmin is 20 degrees Celsius. GCC tells us how much net heating and cooling is needed, as well as the temperatures at which it is necessary.

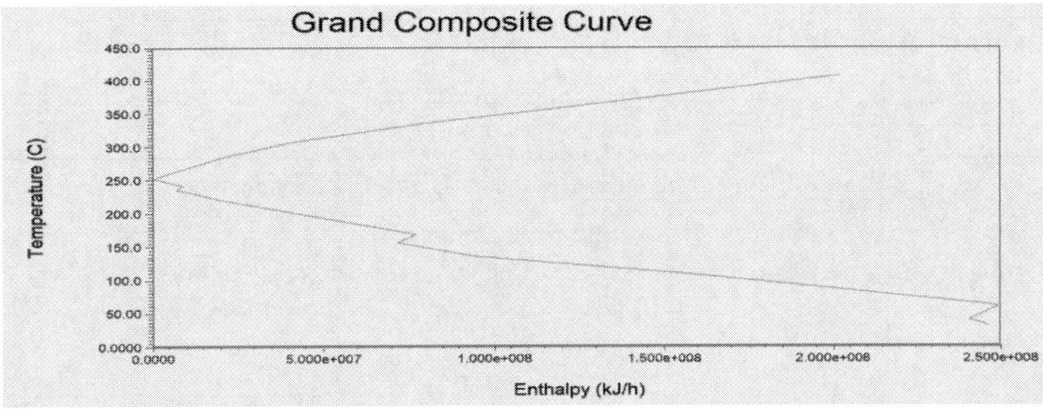

Figure 4: Grand-Composite curve at initial at ΔTmin=20°C

3.3 Heat Exchanger Network Analysis

In Aspen Energy Analyzer Heat, exchanger can be done manually and automatically by recommend design. For given data, software had generated 10 designs. These designs were optimized and the best design was proposed. The table 2 shows 10 recommended designs and table 3 shows optimized designs.

Table 2: Initial designs on the basis of total annualized cost

Design	Total Cost index	Area(m^2)	Units	Shell	Capital Cost index	Heating(KJ/h) $*10^8$	Cooling(KJ/h) $*10^8$	Op. cost index
A_Design1	0.3337	34540.16	26	111	8932733	1.94	2.38	0.2424
A_Design2	0.3399	36955.18	30	116	9545279	1.94	2.3	0.2424
A_Design3	0.3322	33459.54	27	103	8588364	1.95	2.4	0.2445
A_Design8	0.325	31098.14	26	86	7832795	1.96	2.4	0.245
A_Design5	0.3371	35962.27	33	115	9274535	1.94	2.38	0.2424
A_Design6	0.3251	30916.63	29	92	7897664	1.9	2.4	0.2445
A_Design7	0.3253	31334.81	27	102	8119088	1.93	2.38	0.2423

IOP Conf. Series: Materials Science and Engineering 1259 (2022) 012006 doi:10.1088/1757-899X/1259/1/012006

A_Design8	0.325	31098.14	26	86	7832795	1.96	2.4	0.245
A_Design9	0.3277	32509.85	25	92	8186415	1.95	2.39	0.2441
A_Design10	0.3311	33881.99	28	103	8681802	1.94	2.38	0.2424

Table 3: Optimized design on the basis of total annualized cost

Optimized Designs								
Design	Total Cost index	Area(m^2)	Units	Shell	Capital Cost index	Heating(KJ/h) $*10^8$	Cooling(KJ/h) $*10^8$	Op. cost index
A_ Design 1- O	0.3097	21647.18	26	70	5655604	2.01	2.46	0.2519
A_ Design 2- O	0.306	21444.02	30	65	5573383	1.99	2.43	0.2491
A_ Design 3-O	0.3098	23983.4	27	72	6149838	1.97	2.42	0.247
A_Design4 -O	0.31	24618.56	28	69	6251974	1.97	2.41	0.2462
A_Design5 -O	0.311	21136.23	33	68	5540424	2.03	2.48	0.2544
A_Design6 -O	0.3096	22677.77	29	70	5874039	1.99	2.44	0.2496
A_Design7 -O	0.31	22711.9	27	73	5881098	2	2.44	0.2499
A_ Design 8-O	**0.2534**	**8493.02**	**26**	**29**	**6637165**	**1.48**	**1.92**	**0.186**
A_Design9 -O	0.3129	22455.5	25	62	5661801	2.04	2.48	0.2551
A_Design10-O	0.3084	21671.24	28	66	5605156	2	2.44	0.2511

Out of 10 optimized designs design 8 is most efficient on because less area and utility requirements. Also from alpha plot (Figure 5) which is a plot of operating cost vs capital cost for all designs it can be seen that design 8 is most economical. However, it has overshooted the no. of targeted units which also increased the capital cost

IOP Conf. Series: Materials Science and Engineering 1259 (2022) 012006 doi:10.1088/1757-899X/1259/1/012006

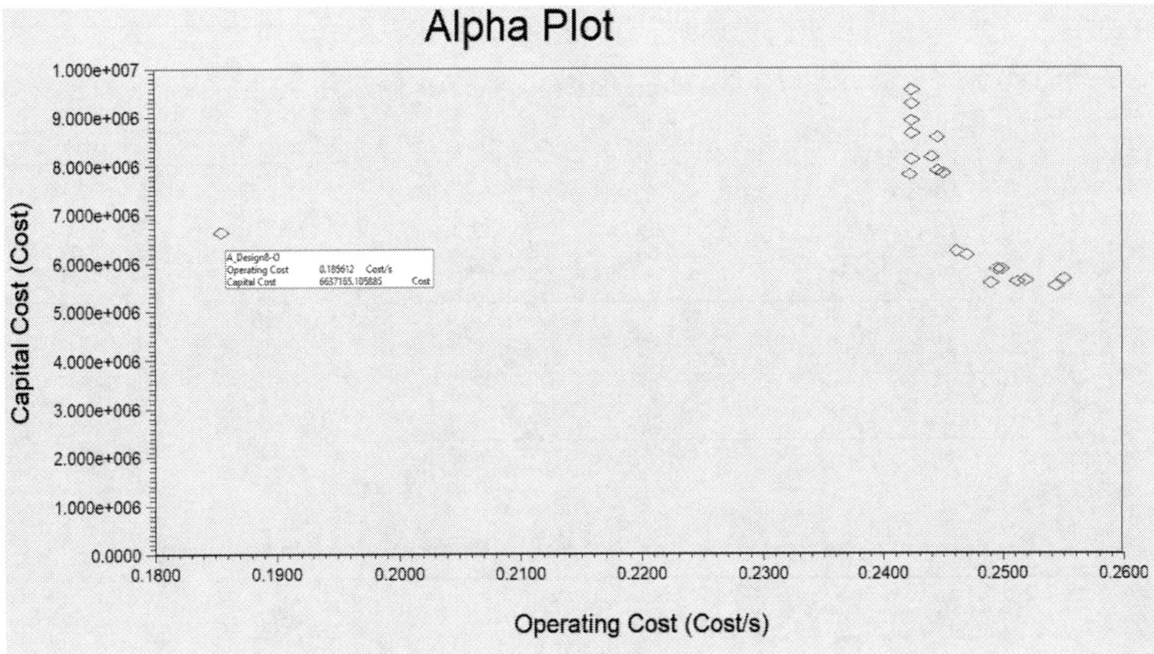

Figure 5: Alpha Plot

Table 4 and 5 shows the performance of design 8 which is relative to the targets. It can be seen Number of units has overshooted by 44.44% whereas capital cost has overshooted by 38.0641%.

Table 4: Network cost Indexes relative to target

Network Cost Indexes (design 8)		
	Cost Index	% of Target
Heating(Cost/s)	0.174282195	73.0780
Cooling(Cost/s)	1.13E-02	77.9205
Operating(Cost/s)	0.185611593	73.3563
Capital(Cost)	6637165.106	138.0641
Total Cost(Cost/s)	0.253402186	86.4145

Table 5: Network Performance relative to target

Network Performance (design 8)		
	HEN	% of Target
Heating(KJ/h)	147658192.3	73.0780
Cooling(KJ/h)	191973081.4	77.9205
Number of Units	26	144.4444
Number of Shells	29	93.5484
Total Area (m2)	8493.022322	41.6728

Table 6 shows the comparison of a various parameter of previous work [3] on the same data of Kaduna refinery. It can be seen that the pinch temperature is different for same ΔTmin. Although the heating duty is almost same but there is a vast difference in cooling duty this can also be justified by the composite diagrams of present study (Fig.3) and Barambu Nafiu Umar et al. (Fig.6). There is a difference in the lower part of the Composite diagram which explains the difference in cooling duties of two studies. Moreover, same results to present study were obtained when an open source software HINT was used to generate composite curves and target report (Fig.7a and 7b).

Heat recovery of $4.1784*10^8$ KJ/h occurs in the suggested design. The hot utility reduction is approximately 26.9% of the overall target heat load, whereas the cold utility reduction is approximately 22.8% of the initial target load. The minimum number of units for heat transfers, such as heaters, coolers, and heat exchangers, increases, resulting in a 38.1 % increase in capital cost.

Table 6: Comparison with past work

Parameter	Units	Barambu Nafiu Umar et al. (2017)	Present study
ΔTmin	^0C	20	20
Pinch point	^0C	222	254
Heating utility	MW	57.65	56.1266
Cooling utility	MW	2.638	68.4361

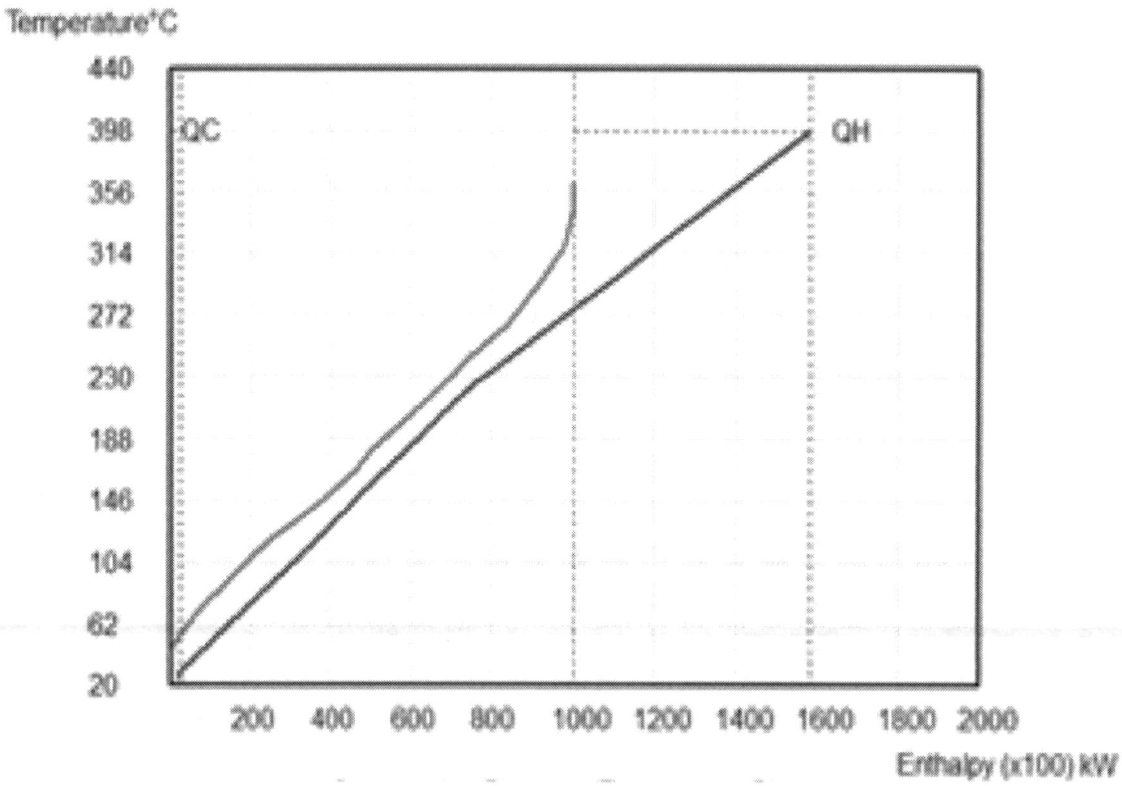

Figure 6: Composite curves of Barambu Nafiu Umar et al. (2017)

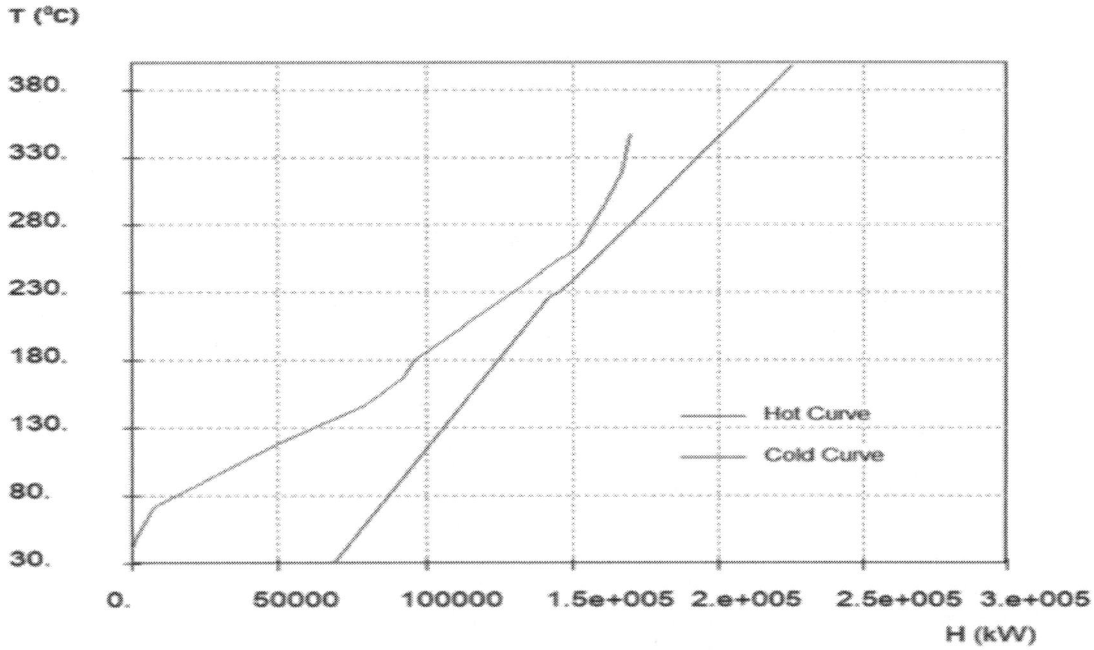

Figure 7a: Composite curves obtained from HINT

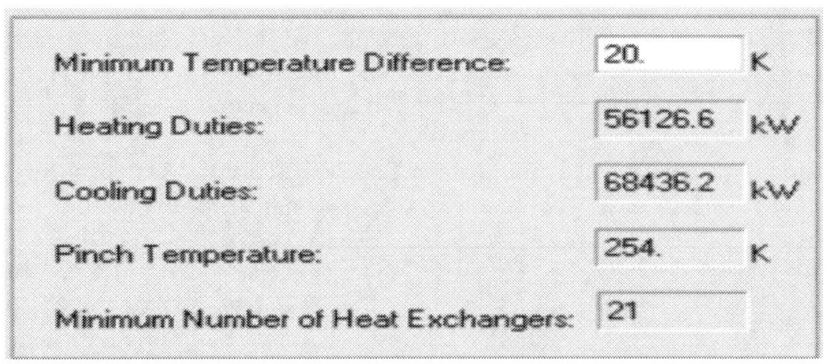

Figure 7b: Target report obtained from HINT

4. Conclusion:

The pinch temperatures are found to be 264°C (hot pinch) and 244°C (cold pinch). The cold utility requirement is $2.464*10^8$ [kJ/h] (Cooling Water) and hot utility requirement is $2.021*10^8$ [kJ/h] (Fired Heat).

The Crude Distillation Unit of the Kaduna Refinery [3] was subjected to a pinch analysis utilizing Aspen Energy Analyzer. Heat recovery of $4.1784*10^8$ KJ/h occurs in the suggested design. The hot utility reduction is approximately 26.9% of the overall target heat load, whereas the cold utility reduction is approximately 22.8% of the initial target load. The minimum number of units for heat transfers, such as heaters, coolers, and heat exchangers, increases, resulting in a 38.1 % increase in capital cost above the initial goal capital cost, since capital cost investment is a one-time investment, it will be helpful in the long run.

The difference in the pinch point and cooling utility between present study and Barambu Nafiu Umar et al. can be because of difference in software's. HINT and Aspen Energy Analyser are giving similar results whereas HENSAD is showing variation in some parameters.

5. References:

[1] Robin Smith, Chemical Process Design, McGraw-Hill, New York (1995)

[2] Iregbu, Precious O. Yiga, Francis, Tijani, Olamilekan E. ,Pinch Analysis of Heat Exchanger Networks of an Industrial Ammonia Plant for Fertilizer Production; International Research Journal of Advanced Engineering and Science,(2020)

[3] Barambu Nafiu Umar , El-Nafaty Usman Aliyu, Saeed Isa Ahmed; Energy Integration Of Crude Distillation Plant Using Pinch Analysis (A Case Study Of Kaduna Refinery And Petrochemical Company, Nigeria); International Journal of Innovative Research and Advanced Studies (IJIRAS), (2017)

[4] Mohammed Nasiru Bello, Sadiq Ali Zangeri, Pinch Analysis of Vacuum Distillation Unit of Kaduna Refinery and Petrochemical Company, IOSR Journal of Engineering, (2021)

[5] Padole Manjusha, Sapkal Pradyumna2, Dawande S.D. , Nitin Kanse, Heat Integration of Crude Organic Distillation Unit, Research Journal of Engineering Sciences (2013)

[6] Ian C Kemp, Pinch Analysis and process integration, a user guide on process integration for effective use of energy, IChemE, Elsevier Limited, (2007)

[7] Bodo Linnhoff and John R. Flower, Synthesis of Heat Exchanger Networks, AIChe Journal, (1978)

[8] Bodo Linnhoff and E. Hindmarsh, the Pinch Design Method for Heat Exchanger Networks, Chemical Engineering Science, (1982)

[9] André L.H. Costa, Eduardo M. Queiroz; An extension of the problem table algorithm for multiple utilities targeting, Energy Conversion and Management, (2009)

[10] W.D Seider, J.D Seader, D.R Lewin, Product and Process Design Principles, (2003)

[11] Shoaib, A. M., Emam, E. A., Osman, W. S.; Industrial Application of Plant Crude Oil through Comparison of Heat Exchanger Network Synthesis Techniques; Journal of Petroleum and Mining Engineering ;(2021)

[12] Luqman Buchori 1 and Farah Marda Yesica Putri; Pinch Analysis for Styrene Production With Lummus/UOP Smart SM Technology; International Conference on Chemical and Material Engineering (ICCME 2021)

[13] Keivan Nemati-Amirkolaii, Hedi Romdhana and Marie-Laure Lameloise; Pinch Methods for Efficient Use of Water in Food Industry: A Survey Review; Sustainability 2019

[14] Nikhil M Gadave, Pramod P Kothmire; Thermo hydraulic evaluation performance of shell and tube heat exchanger with different tube geometries, International Journal of Engineering and Technology (2019)

[15] Renu K.Shastri[ab]Chinmaya P.Mohanty[c]; **Machinability investigation on Nimonic C263 alloy in electric discharge machine, Material today : proceeding, 2020**

[16] Deore H., Sardare M., Nemade P. Experimental Modeling and Evacuation of Cr(VI) from Wastewater by Using Nanostructured Ceria.Advances in Civil Engineering and Infrastructural Development. Lecture Notes in Civil Engineering, vol 87. Springer, Singapore. (2021)

Study of the Secant Pile Wall technique, its application and Sustainable Construction Using the Non-Biodegradable Waste as partial replacement for Concrete materials

Pallavi Chakole[1], **Shubham Date**[2]

[1].Department. of Civil Engineering , Yeshwantrao college of Engineering and technology
Nagpur, India
chakole121pallu@gmail.com
[2]. Civil Engineering Dept , Yeshwantrao Chavan College of Engineering
Nagpur, India
Shubhamdate9696@gmaill.com

Abstract. The planning and execution of the project of a Secant Pile wall are presented in this study. Secant pile walls are a type of embedded retaining wall that allows for excavation of great depths in water-logged terrain and soft sub stratum. To construct the walls, pilings are placed in a certain order. Soft piles are set first, then hard piles are laid on both sides, overlapping and cutting through the soft piles. This design can also be utilised as a cut-off wall for fine sands and very soft wet soils.

In light of the significant growth, the idea of using non-biodegradable trash in concrete is also incorporated. Many non-biodegradable wastes exist; this study includes research on the potential reuse of ceramic waste, plastic granules, and glass trash as partial replacements in concrete materials. These waste materials are collected in various percentages and replenished using the cited publications. The properties of concrete containing the appropriate waste material are investigated.

Keywords— Secant pile wall, Auguring, borehole, Augmentation, vibro-hammer

1. Introduction

India, a country having a wide diversity within, is emerging itself as one of the fastest developing country in the world. With the demand and pace of development in India, infrastructural development can be considered as the backbone of it. Our country have many major cities viz Mumbai, Delhi, Chennai, Kolkata, etc. where there is a demand of high rise infrastructure may fall in some challenging situation. There can be the problems orienting to the challenging condition such as soft soil of low bearing capacity, high water table level, congestion of space for construction, threat to adjacent structure. All this problems can be overcome by constructing secant wall. Secant pile is type of retaining wall which is used for deep excavation in soft soil and water charged area. The objective of a secant wall is to keep the ground or composition beneath it secure and safe despite the retaining structure's lateral and downward pressure.

A. Secant Pile wall

The secant pile technique entails the construction of overlapping concrete piles. Lower strength concrete is typically used to create alternate unreinforced or "female" piles. Standard reinforced concrete "masculine" piles make up the remaining piles. Construction of intersecting reinforced concrete piles results in secant pile walls & thus forming a wall like barrier to adjacent media. Steel rebar is used to reinforce secant piles. Or steel beams and are formed by drilling through mud or auguring. In an urban setting, the secant pile wall method is a low-cost option for constructing retaining walls forming basements, cut-and-cover tunnels, tanks, and other deep excavations.

2. Methodology

1. Building of a wall of instructions: An instruction wall is built to mark the location of the secant pile wall.
2. Casing setup: The casing is driven into the earth with a vibro-hammer, leaving around 1 meter of casing exposed.
3. Auguring of the main borehole: To construct a basic borehole, a drilling tool known as the auger, cuts and removes earth from the casing. The outside casing helps to stabilize the earth around the borehole. If the casing is insufficiently long, to arrive at the appropriate depth in the earth, the slurry of betonies is employed to hold the soil in place.
4. Concentering of the main borehole: The primary bored pile is built by pouring concrete into the borehole.
5. Augmentation of the second drilled hole: After the casing has been extracted from the two primary drilled piles with the vibro-hammer, to create a secondary borehole, the auger cuts and removes dirt between the two primary drilled piles. The secondary drilled hole runs parallel to the primary drilled piles.
6. Placement of steel casing: The cage made of steel is lifted and lowered into the Crane-driven secondary borehole.
7. Pouring concrete into the borehole number two: The borehole number two is constructed by pouring concrete into the borehole.
8. Repetition of the procedure: Steps 2 through 7 are continued until the secant pile wall is constructed in its entirety.

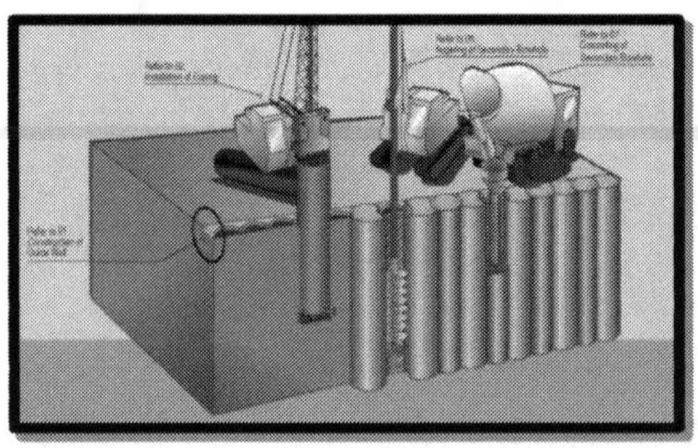

Figure 1. Procedure of Installation

3. Experimental program

There are various non-biodegradable wastes of which the decomposition and disposal is not an economical as well as environmental friendly task, and therefore it should be utilized effectively wherever possible. Based on the referred papers two non-biodegradable waste were selected. First was the ceramic waste which was partially replaced by the 10% of coarse aggregate of size between 10mm to 4.75mm and secondly 10% of uniformly distributed glass wastes of size varying from 4.75mm to 150 micron size concrete's fine aggregate is partially replaced.

Materials:

1. Cement: The cement employed in this investigation was 43-grade Portland Pozzolana Cement (OPC), which met IS 8112-89 specifications.

2. Fine aggregate: The sand was locally available and confirmed to be zone II with a specific gravity of 2.62. Sand was tested according to Indian Standard Specification IS: 383-1970.

3. Coarse aggregate: The coarse aggregate utilised had a specific gravity of 2.93 and was 20mm in size. The tests were carried out in accordance with Indian Standard Specification IS: 383-1970.

4. Ceramic waste: Waste ceramic waste available in your area is brought into desired size by crusher between 10mm to 4.75mm for replacement of Coarse aggregate.

5. Glass: Locally accessible waste glass was gathered. Glass trash is a very tough substance. Glass must be ground to the correct size before being added to the concrete. i.e. to desired size i.e. between 4.75mm to 150 micron for fine aggregate has been replaced.

Mix Design:

The proportions of the concrete mix without glass powder were calculated according to Indian Standard Specifications IS: 10262-1982For the M30 concrete grade, a mix design was created. A water-to-cement ratio of 0.55 was used to make the mixture. According to IS 10262-2009, the material mix ration is 1:2.05:3.22. Then there was the utilization of natural fine aggregate.

4. Analysis and Result

Property of concrete containing respective material and results obtained are shown in the below.

Table 1. Mix Design (Control Mix)

Water (liters)	Cement	Fine aggregate	Coarse aggregate
197	358.18 kg/m3	737.33 kg/m3	1153.27 kg/m3
0.55	1	2.05	3.22

Table 2. Partial Replacement Details

Material	Partially Replaced material	Replacement percentage	Quantity (kg/m3)
Ceramic Waste	Coarse aggregate	10%	115.327
Glass Waste	Fine aggregate	10%	73.733

Table 3. Properties of fresh concrete

S.R	Property	Standard (mm)	Ceramic waste (mm)	Glass (mm)
1	Slump value	150	140	155
2	Flow value	43	45	45

Table 4. Cement replaced by Ceramic waste

Sr no.	Material	7 days		14 days		28 days	
1	Control Mix	16.37	16.44	26.63	26.33	28.57	29.89
		15.82		26.09		29.18	
		17.13		26.27		31.92	
2	Ceramic waste concrete (10%)	14.36	13.33	20 51	21.24	29.28	29.56
		13.12		21.30		30.31	
		12.51		21.81		29.09	

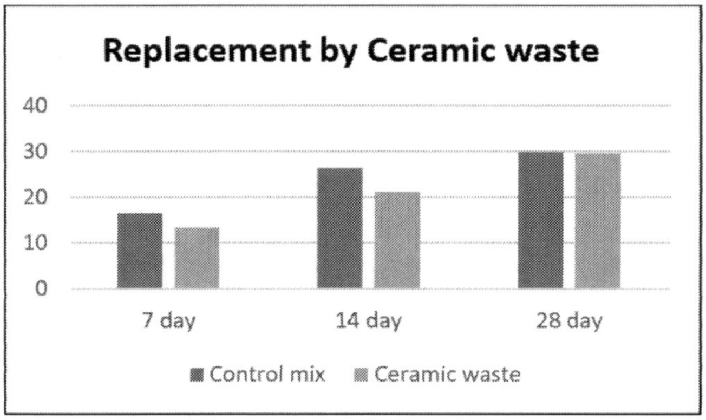

Table 5. Cement replaced by Glass waste

Sr no.	Material	7 days		14 days		28 days	
1	Control Mix	16.87	16.44	26.63	26.33	28.57	29.89
		15.82		26.09		29.18	
		12.13		26.27		31.92	
2	Glass powder waste (10%)	20.3	21.66	25.61	24.54	30.36	31.33
		21.5		24.32		31.89	
		23.18		23.69		31.74	

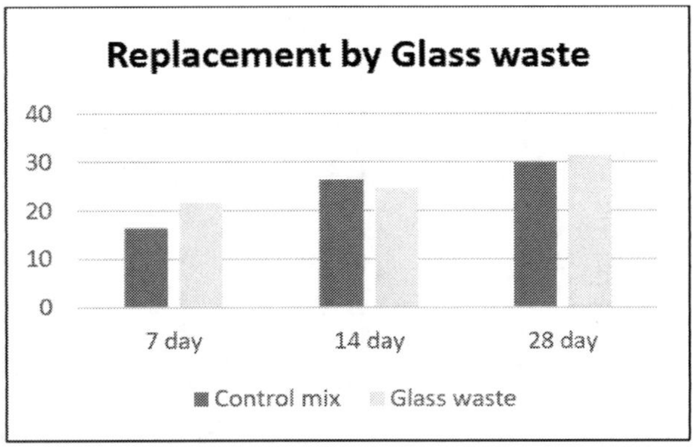

Table 6. Cement replaced by Plastic granules

Sr no.	Material	7 days		14 days		28 days	
1	Control Mix	16.87		26.63		28.57	
		15.82	16.44	26.09	26.33	29.18	29.89
		17.13		26.27		31.92	
2	Plastic granules (10%)	12.56		23.6		28.06	
		13.57	13.78	22.08	22.22	29.82	28.87
		15.11		20.98		28.73	

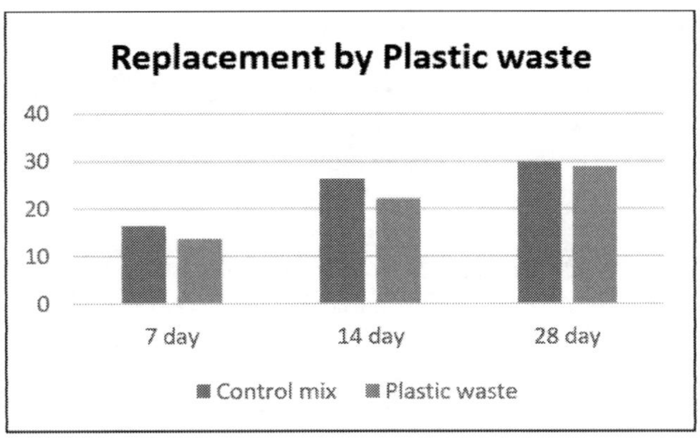

Table 7. Compressive strength test

No. of days	Control mix	Ceramic waste	Glass powder glass	Plastic granules waste
7 days	16.44	13.33	21.66	13.78
14 day	26.33	21.24	24.54	22.22
28 day	29.89	29.56	31.33	28.87

IOP Conf. Series: Materials Science and Engineering 1259 (2022) 012007 doi:10.1088/1757-899X/1259/1/012007

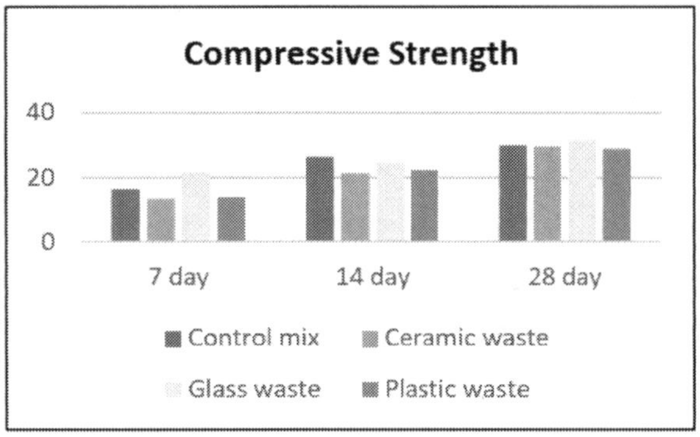

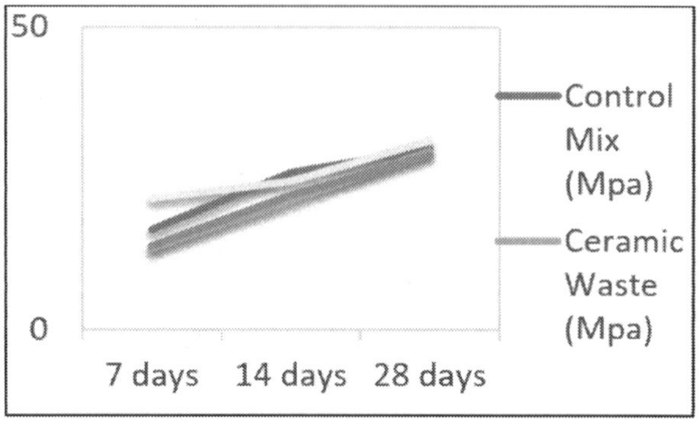

Figure 2. Preparation, casting and testing of blocks

5. Results and Conclusion

It has been noted that when a 10% replacement of coarse aggregate is used ceramic waste, the compressive strength at 7 day and 14 day is discovered to be reduced by about 23.33% and 22.4% respectively on average. There is a small decrease of 1.12% in compressive strength at 28 d at same replacement level.

While in another case, when fine aggregate is changed, it is discovered that by 10% glass waste, the compressive strength increases by roughly 31.75 percent on average after 7 days. However, at 14 and 28 days, there is a clear rise in compressive strength is only 5.95% and 3.71% at the same replacement level.

6. Conclusion

Secant pile wall can be used in geotechnical critical condition and for excavation of depths previously considered to be infeasible. It also increases construction alignment flexibility

It is observed that when coarse aggregate is replaced by 10% ceramic waste, the compressive strength at 7 d and 14 d is found to decrease by about 23.33% and 22.4% respectively on average. There is a small decrease of 1.12% in compressive strength at 28 d at same replacement level.

While in another case, when fine aggregate is replaced with 10% glass waste, the compressive strength at 7 days is shown to improve by around 31.75 percent on average. However, at the same replacement level, the improvement in compressive strength at 14 and 28 days is only 5.95 percent and 3.71 percent, respectively.

And in case of plastic waste aggregate, there is a slight decrease of 16.18 % and 15.60% in the 7th day ad 14th day compressive strength values, respectively. The 28th day compressive strength is also decreased by 3.41 %.

The studies and experimentation on the replacement of fine aggregate by non-biodegradable waste i.e. glass and ceramic waste shown that the strength values were not largely disturbed by the replacements made in the design. Hence it will be economical and environmental friendly to use such replacements in material for the construction of secant pile wall.

Reusing waste material wherever possible can save money and reduce harmful effects on environment. Efforts should be made to create and utilise materials derived from waste or by-products. By using by-products, you can save money and space when it comes to disposing of them. In our project we have used waste material in concrete by replacing coarse or fine aggregate in suitable proportion.

Concrete using glass waste has the highest compressive strength. Thus we can conclude that glass waste is acceptable among all the three materials, satisfying the eco-friendly and economic factor.

7. Reference

[1]Hadi Suroor, Mahi Galagoda, Chris McGhee (2008) " Design and construction of circular secant pile walls in soft clays".

[2]Andrew Cushing, Stephen Young (2000) " Foundation construction challenges at 100 11th avenue in manhattan – Secant pile wall".

[3]Dave Bradshow, P.Eng, (2011)Chief Estimator deep foundation contractors, Inc. "Deep Foundation Contractor Inc. Installs secant pile wall near Toronto waterfront".

[4]Hakan Elci (2014)"Utilization of crushed floor and wall tiles as aggregate in concrete production".

[5]Anna Halicka, Pawel Ogrodnik, Bartosz Zegardlo (2013)"Using ceramic sanitary ware waste as concrete aggregate".

[6]Shilpa Raju, Dr. P.R Kumar (2014)"Effect of using glass powder in concrete".

[7]S.P Gautam,Vikas Srivastava, V.C Agrawal(2014) "Use of glass waste as fine aggregate in concrete".

[8]Patil Hiteshkumar, Bavishkar Shrikant R, Dr Ahirrao Sarjerao P, Dr Anmol Saner(2017) "Polymer Concrete by using waste Plastic Granules".

[9] P.Suganthy, Dinesh Chandrasekar, Sathish Kumar.P.K(2013) "Utilization of pulverized plastic in cement concrete as fine aggregate."

[10] Brajesh Mishra, Ravi Shanker Mishra (2015) "A study on use of Plastic Waste Aggregate as partial replacement of natural coarse aggregate in cement concrete mix"

[11] Dr.K.Manjula Vani (2012) "Non biodegradables as building materials"

[12] Rob Jameson, Malcolm Drilling Company, Inc.,3503 Breakwater Court, Hayward, California, USA, Eric S.Lindquist (2012) "Secant Pile Shoring – Developments in Design and Construction"

[13] Nick Wharmby, Brian Perry Civil, Hamilton, New Zealand (2010) "Development of secant Pile Retaining Wall Construction in Urban New Zealand"

[14] Geotechnical Engineer Jorgen Johannessen, Norconsult AS NGM (2012) "A 25m Deep Secant Pile Retaining Shaft in Sensitive Clay".

[15] Dima M.Kannan, Sherif H.Aboukar, Amr S. EL-Dieb, Mahmoud M.Reda Taha(2012) "High performance concrete incorporating ceramic waste powder as large partial development of Portland cement".

Thermal Performance Improvement of Modified Hemispherical Cavity Receiver with Air Curtain

Abhishek Simer, Akanksha Maurya, Anoop Kumar

Department of Mechanical Engineering, National Institute of Technology-Hamirpur, HP, India

E-mail: 20mme001@nith.ac.in, akanksha@nith.ac.in, anoop@nith.ac.in

Abstract. Heat losses from solar cavity receivers degrade the efficiency of solar parabolic dish collector system. Specifically, increased convection losses emulate a significant contribution in decreasing the performance of the modified solar cavity receiver and hence, reducing these convection losses has become a matter of great concern. The Air curtain plays a crucial role in minimizing the convection losses. This work proposes an air curtain-based modified cavity receiver wherein a plane nozzle is placed on the face of receiver to produce air curtain effect. Air curtain will prevent hot air from escaping the modified cavity receiver and increase the stagnation zone inside it to reduce natural convection on the surfaces. Numerical analyses are performed considering different air velocities and receiver orientations to evaluate the effect of air curtains on the convection losses. The impact of plane air nozzle parameters like outlet temperature, inclination, and outlet velocity of the nozzle on convection losses are shown and these parameters' optimal value is evaluated. Our results state that the air curtain attached to the modified solar cavity receiver significantly improves performance by reducing the convection losses.

Keywords: Cavity receiver, air curtain, convection losses, Computational fluid dynamics

1. Introduction

The energy demand is growing exponentially day by day; looking at other side, non-renewable sources of energy are exhausting. Renewable sources of energy act as a good substitute for fulfilling energy demand; moreover, CSP are observed to be one of the most fitting system to harness solar energy. The concentrating solar power system consists of two major components; cavity receivers and collectors (or concentrators). The absorbed solar energy is focused on the cavity receiver due to which temperature of HTF increases, which can be used for further applications. For cavity receiver thermal efficiency plays a vital role in making solar power generating system more efficient. Geometry, orientation, ambient conditions like wind velocity and its direction, atmospheric temperature and characteristics of heat transfer fluid are deciding factors for its thermal performance. The receiver's performance is directly related to the thermal heat loss occurring in the solar cavity receiver. So for maximizing receiver's efficiency, these losses must be reduced. Thermal loss through the solar cavity receiver constitutes radiative loss, conductive loss and convective loss. Natural convection loss are responsible for a significant portion of total heat losses. So, it has become a matter of great concern to reduce these convection heat losses. The geometric parameters of the receiver and its inclination play a prime role in reducing natural convection losses. However, employing air curtain will act as a barrier or invisible wall between environment and aperture of cavity receiver to prevent hot air from escaping. Due to this, the stagnation zone inside the cavity will increase; hence there will be less heat loss by natural convection.

So many researches have been done in past to study heat loss through convection in solar cavity receivers. Reddy et al.(2016) did a numerical analysis to find the impact of speed of wind and its direction on convection losses in a modified solar cavity receiver. The thermal loss from the cavity receiver due to forced convection is constant and is greatest when receiver is inclined at 90°, regardless of its wind direction [1]. A numerical investigation on cylindrical solar cavity receiver was done by Lan Xiao et al.(2012), according to study, at high wind speeds, the inclination and buoyancy impact becomes less important [2]. Prakash et al.(2014) investigate the natural convection heat loss from a cavity receiver numerically. The study concluded that convectional heat loss decreased by 80% with the increment in angle of inclination [3]. Numerical and an experimental study on cylindrical solar cavity receiver using water as HTF was performed by Prakash et al.(2009) [4]. It stated that the convection losses increase with increased receiver's mean temperature. They also suggested that convective losses caused by the wind are greater than no wind conditions. Pentagon conical, helical conical and triangular conical cavity receiver are investigated numerically by Iss et al. (2022) for increasing the performance of cavity receiver. Results stated that at a temperature of 300K, most efficient conical cavity receiver was helical conical cavity receiver with maximum thermal performance of 66.68% [5]. Gebhart et al.(1975) experimentally study natural convection losses over a heated hemisphere and influence of heat transfer fluid flow for both inverted and upright orientations of the cavity [6]. Another experiment was performed by Lewandowski et al.(1996) to analyse the impact of heat transfer from an isothermal hemisphere cavity receiver in inverted position, considering water and dehydrated glycerine as working fluid [7]. Koenig et al.(1981) stated a model to estimate convective losses from solar cavity receivers under working temperature condition between 550°C and 900°C [8]. An experiment was performed on the modified solar cavity and semi-cavity receiver designs by Kaushika et al.(1993). It is suggested that other cavity designs were less efficient than modified cavity receivers and semi-cavity [9]. An investigation was done to analyze and compare the thermal efficiency of the solar cavity receivers for different designs like spherical, conical, hetero- conical, cylindrical and elliptical by Harris et al.(1985). The results reveal that considerable changes for power profiles of cavity receivers are caused by change in rim angle of concentrator and geometry of cavity receiver [10]. Taumoefolau et al.(2004) performed an investigation to study natural convection losses for different ranges of inclination changing from -90° to +90° with varying temperature 450°C to 650°C on an electrically heated cavity receiver [11]. The aspect ratio, Rayleigh number and inclination were found to be governing factors for flowing heat transfer fluid considering cubical, hemispherical, and spherical geometries in an investigation performed by Prakash et al.(2012) [12]. It was observed that as opening ratio is increased by some value, the natural convection loss increases too. Nusselt no. correlation was also developed involving the influence of cavity geometry and Rayleigh number. A secondary reflector and an extra quartz glass window were employed to increase the performance of the solar cavity receiver for model proposed by Chang et al.(2015) and results revealed that the cavity receiver can effectively limit heat loss [13]. Reddy et al.(2009) presented a numerical investigation on a CPC with cone and trumpet reflectors of solar dish modified cavity receiver. Convective loss at different operating temperatures and inclinations were also investigated and results revealed that natural convection losses are decreased by 19.81%, 29.23%and 19.16% [14]. A hollow receiver with an air curtain was investigated for natural convection loss by Jiabin Fang et al.(2019) for hexagonal cavity receivers numerically. A reduction of 28.6% was observed for natural convection losses under the optimum conditions [15]. An experimental investigation on design of cavity receiver was performed by Venkatachalam et al.(2019) to increase the performance of the solar cavity receiver [16]. Aspect ratio of cavity receiver was changed for deciding the design parameter. Result stated that as we increase aspect ratio, receiver's surface temperature got reduced and the performance of cavity receiver decreased. Optimal value of aspect ratio was discovered to be 0.8 for maximum thermal performance. A comparison of conical and spiral solar cavity receivers for a dish collector system is performed by Pavlovic et al.(2018). It was observed that there was an increment in thermal efficiency by 5.63% at 100°C and 40.45% at 200°C for conical cavity [17]. Two different types of cavity receiver designs were studied numerically in an analysis executed by Tharun et al.(2022).Two different deigns were conical and elliptical cavity receiver. The analysis was performed on different inclination, temperature and aspect ratio of the cavity receivers. Results showed that elliptical cavity receiver has more heat transfer coefficient as compare to other receiver [18]. An experimental and computational investigation was presented by Reynolds et al.(2004) to examine the heat loss and flow patterns of trapezoidal cavity [19]. Kribus et al.(1999) designed the solar thermal receiver to minimize the heat loss, considering aperture is divided in two stages [20]. Reddy et al.(2008) performed a comparative study for receivers of solar collector dish system. Numerical investigations were performed for cavity receiver designs such as hemispherical, semi-cavity receivers, and modified solar cavity receivers considering varying area ratios and inclinations. It was found that area ratio of 8 has minimum convection heat loss among different configurations [21]. An analytical model was presented by Clausing et al.(1981) for a cubical type receiver which helps in predicting the convective loss from solar cavity receiver. Result proved that transfer of energy through aperture is mostly caused by wind and buoyancy [22]. The impact of cavity shapes on thermal loss was investigated by Jilte et al.(2014) and Daabo et al.(2016), and it is suggested that the conical cavity has least convection heat losses among all the shapes [23], [24].A hot mirror coating was installed on the aperture of cavity receiver in an investigation performed by Ferrer et al. (2021). This investigation was carried out numerically. Result showed that as the opening size of the cavity was decreased, an increment in efficiency was observed when simulation is performed at a temperature approximately greater than 900K [25]. The impact of aspect ratio for cylindrical solar cavity receiver on thermal loss was studied by Shavazi et al.(2015) and the surface areas of the convection and stagnation zones were used to create correlations for accurate prediction of experimental findings [26]. Numerical and experimental investigations were performed to find the impact of emissivity, heat flux, aperture ratio, and orientation on radiative loss and natural convection heat losses by Wu et al.(2014), Shen et al.(2015), and Shen et al.(2016) [27], [28], [29]. According to the research, it was revealed that the aperture size and orientation have a significant influence on natural convection loss. There has been very few research to account for the effect of an air curtain on the convection loss of a solar cavity receiver. Based on past literature, no study has been stated to examine the impact of air curtains on modified hemispherical solar cavity receiver. Aperture region is covered to make a barrier to decrease thermal loss from the solar cavity receiver using air curtain. A plane air nozzle will be there on the one face of cavity receiver, which will create air barrier to separate the high-temperature zone and ambient zone. In the present work, convection losses of modified cavity receiver with air curtain is numerically studied with different nozzle inclinations θ_{noz}, air outlet temperature $T_{out,noz}$ and outlet air velocity $u_{out,noz}$. Fig [1] shows the schematic for the modified cavity receiver with an installed air-curtain.

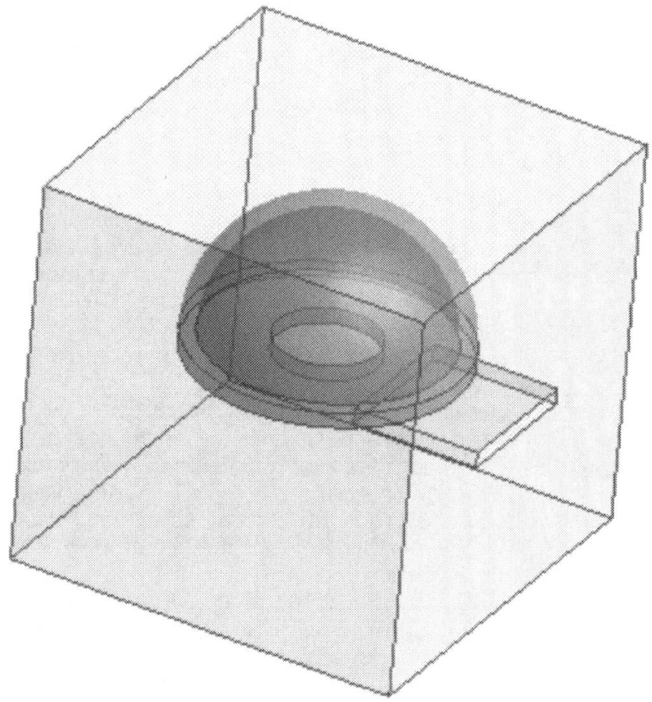

Figure 1. Modified hemispherical cavity receiver with air curtain.

2. Modelling

The illustration of a modified solar cavity receiver with an air curtain is as shown by Fig[1]. The receiver used in the current numerical study contains copper tubings, and the attached air curtain was installed with a plane air nozzle. The outer diameter of the hemispherical cavity receiver was considered to be equal to 34 cm. And the inner diameter of the cavity receiver was set to be equal to 30 cm. The diameter of aperture opening region was considered to be equal to 13 cm. The modified hemispherical geometry used for current numerical procedure was taken from numerical analysis performed by Kumar et al.(2008), numerical analysis was used to generate the modified hemispherical geometry used in the current numerical technique [21] . Air nozzle of width 20mm was used in the present model of modified cavity receiver's and outer hemispherical portion was insulated to a thickness of 20 mm. 3D, steady state and laminar flow conditions are used in ANSYS FLUENT setup module. For the numerical analysis, a pressure-based solver was used because this type of solver is employed in the process of finding solutions to problems involving laminar flow conditions. Energy, continuity and momentum equation are three basic equations used to perform numerical simulations.

Continuity Equation

$$\nabla \cdot v = 0 \tag{1}$$

Momentum Equation

$$v \cdot \nabla v = X - \frac{\nabla p}{\rho} + \nu \nabla^2 v \tag{2}$$

IOP Conf. Series: Materials Science and Engineering 1259 (2022) 012008 doi:10.1088/1757-899X/1259/1/012008

Energy Equation

$$\nabla \cdot (k_s \nabla T) = \rho \frac{Dh}{Dt} - \frac{Dp}{Dt} \tag{3}$$

2.1. Boundary Conditions

The outermost hemispherical surface of the modified solar cavity receiver was made adiabatic. The inside surface of hemisphere and the front side of the cavity receiver were made isothermal. The boundary condition for inlet pressure of 1 atm was set for the enclosing domain. The velocity outlet condition along with temperature was set for the plane air nozzle as shown in Table[1].

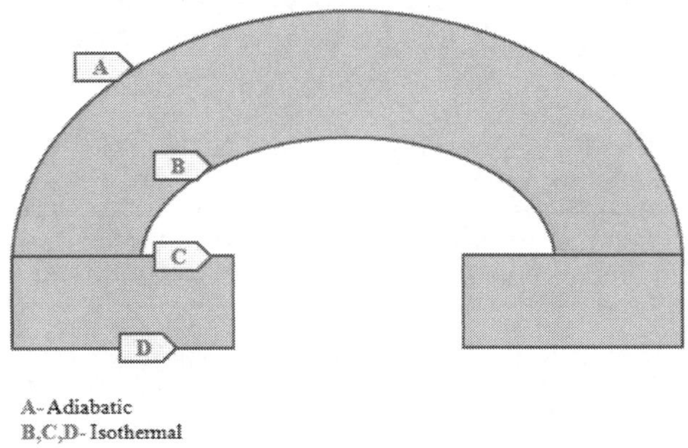

Figure 2. Boundary Conditions applied to modified cavity receiver.

Air curtain uses a plane air nozzle on the one face of the modified cavity receiver. This nozzle was inclined to some angle to produce an efficient air curtain to reduce the convection losses. The outlet temperature, velocity, and inclination of air nozzle are changed to get the different effects of air curtain. Parameters of the air curtain are represented in Table 1.

Table 1. Parameters of air nozzle to create different air curtain

Parameter	Quantity
L_{noz}	20 mm
θ_{noz}	0°, 7.5°, 15°, 30°
$T_{out,noz}$	400 K
$u_{out,noz}$	1 m/s, 2 m/s, 4 m/s, 6 m/s

2.1.1. Adiabatic Boundary Condition For the adiabatic boundary condition, the considered heat flux 0 was implemented on the outer hemispherical surface of the modified hemispherical cavity receiver.

2.1.2. Isothermal Boundary Condition For inner hemispherical surface and front side of modified solar cavity receiver,temperature was set to be constant and is equal to 673K.

2.1.3. Velocity Inlet At the entry of the air curtain, inlet velocity and temperature conditions were implemented.

2.1.4. Pressure Inlet Pressure inlet conditions was set for the enclosing cuboid and the pressure was taken as atmospheric pressure.

The current numerical procedure used a pressure-based solver, and laminar steady-state conditions were applied to the model. On putting these boundary condition to the model, the equations were solved using simulation methods. The aspect ratio of 8 was taken for the modified solar cavity receiver and the temperature of the fluid in the solar cavity receiver was supposed to be the same as that of surface temperature. Second order upwind scheme was employed for the spatial discretization and SIMPLEC system for pressure velocity coupling was utilized. After this, convergence criteria of 10^{-3} were taken for residual of the momentum, continuity, and energy equation.

3. Grid Independence

The computational model undergoes a grid independence test. Fig.[3] shows the mesh around the modified cavity receiver. A tetrahedral mesh was used for meshing purposes, a very fine mesh was considered close to the wall of the modified solar cavity receiver as the region is of great concern, and a less fine mesh was considered for the remaining region. Table 2 shows the convection losses with the no. of considered elements. On increasing number of elements more than 463743, convection losses were found to be stable. So the finalized grid contains 463743 elements.

Table 2. Grid Independence

S.NO.	No. of elements	Convection Loss (W)
1	316507	137.1
2	330265	159.27
3	365802	194.93
4	373650	198.71
5	400137	207.56
6	463743	207.89

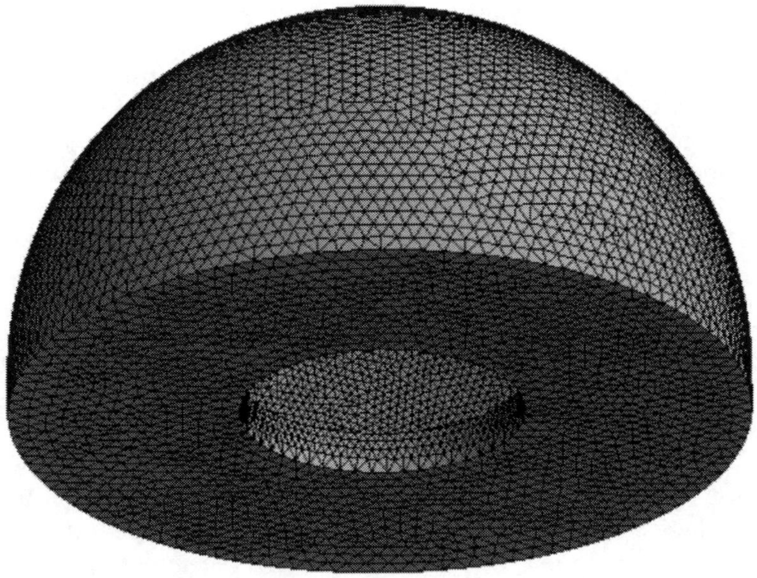

Figure 3. Meshing of modified cavity receiver.

4. Validation

The present numerical model is validated for a hemispherical shaped enclosure heated from below, while the outside surface of cavity receiver was kept at a lower temperature. Nusselt no. correlation was developed experimentally by Yasuaki et al.(1994) [30]. A steady-state, laminar condition was assumed for the hemispherical enclosure, and Nusselt no. for different Rayleigh no. was solved numerically for the experiment. According to correlation from the experimental procedure the Nusselt no. is proposed as $Nu = 0.185Ra^{0.25}$ for $10^6 \leq R \geq 6 \times 10^{10}$ and $6 \leq Pr \geq 13000$ by Yasuaki et al.(1994) [30]. Nusselt no. obtained by current numerical procedure and by experimental correlation are shown in Table 3. It has been observed that there was an insignificant difference between Nusselt no. obtained from experimental correlation and numerical procedure.

Table 3. Nusselt number validation

Sl. no.	Rayleigh no.	Nusselt no. from experimental corelation	Nusselt no. using current numerical procedure	Deviation
1	3.692×10^8	25.64	25.26	0.38
2	4.005×10^8	26.16	26.46	-0.3
3	4.308×10^8	26.65	27.84	-1.19
4	4.9237×10^8	27.55	30.58	-3.03

5. Results and Discussions

A computational model is suggested for natural convection losses for modified solar cavity receiver using different parameters of air curtain. The parameters include inclination θ_{noz}, outlet velocity $u_{out,noz}$ and temperature $T_{out,noz}$ of plane air nozzle are analyzed numerically to find the optimal nozzle inclination with outlet air temperature and outlet velocity. The inclination angle, θ_{noz} varies from 0° to 30° producing air curtain of velocity varying from 1m/s

IOP Conf. Series: Materials Science and Engineering 1259 (2022) 012008 doi:10.1088/1757-899X/1259/1/012008

to 6m/s along with an outlet temperature of 400K. Since, for the case of natural convection, velocity is approximately around 1m/s, the outlet velocity of air curtain varies from 1m/s to 6m/s. The modified solar cavity receiver's natural convection loss is found to be 207.89 W. Using air curtain; convection losses are investigated as stated:

At 0° inclination θ_{noz} of the plane air nozzle, convection losses are analysed at different velocity at 400K outlet temperature $T_{out,noz}$. It was found that convection losses are found minimum at 1m/s for the 400K outlet air temperature of the plane air nozzle Fig[4].

On providing an inclination θ_{noz} of 7.5° to the air curtain,natural convection losses are found at 400K of outlet air temperature of air curtain. The natural convection losses are minimum at 1m/s for 400K outlet temperature Fig[5].

For the case of 15° inclination θ_{noz} to the air curtain,at 400K outlet air temperature, minimum natural convection losses are at 1m/s outlet air velocity $u_{out,noz}$ Fig [6].

Further increasing the inclination, θ_{noz} of air curtain to 30° Fig [7], it was concluded that natural convection losses at 400K outlet air temperature are minimum at 1m/s outlet air velocity $u_{out,noz}$.

Looking at graphs discussed above the heat loss occurring due to convection are increasing firstly and then decrease thereafter. The inside volume of the hemisphere is divided into two zones viz. stagnant zone and convection zone. Approximately the upper half portion of the cavity receiver is acquired by the stagnant zone and the remaining portion constitute convection zone. As velocity of air coming out of cavity increases, an increment in convection losses can be seen due to increase in convection zone inside the modified cavity receiver. This increase in vicinity of convection zone promotes the convection losses occurring through the cavity receiver which in turn also reduces the stagnant zone inside the cavity. But as it reaches at certain outlet air velocity, the stagnant zone inside the cavity again started developing and covering more and more region. This phenomenon start reducing the convection heat losses further. This increment and decrement in region occupied by zones is followed at almost all the inclination of air curtain. From the analysis it was concluded that heat losses through natural convection rises firstly and then decreases with increase in air velocity of outlet of air nozzle for almost all the scenarios. In calculating the natural convection heat loss the difference in temperature between the wall and air plays an important part. In an air curtain, it is the outlet air velocity and inclination which decide the effectiveness of an air curtain. Natural convection losses rises at low outlet air velocity as the value of outlet air velocity is not able to create an effective air barrier and avoid the hot air to come out of solar cavity receiver. Furthermore it can be observed from the graph that when the outlet air velocity condition is sufficient to create an air barrier, the natural convection losses start decreasing.

It can be noted from Fig.[9] that natural convection losses decrease with increase in inclination of the air nozzle. As there is increment in angle of inclination of nozzle, the low velocity zone inside the modified solar cavity receiver increases and this increment will thereafter stretch the stagnation zone in the upper space of modified cavity receiver.

The result showed that all the air curtain with varying inclination reduce the convection losses. Due to increase in nozzle's inclination of the air curtain, the natural convection loss decreases.

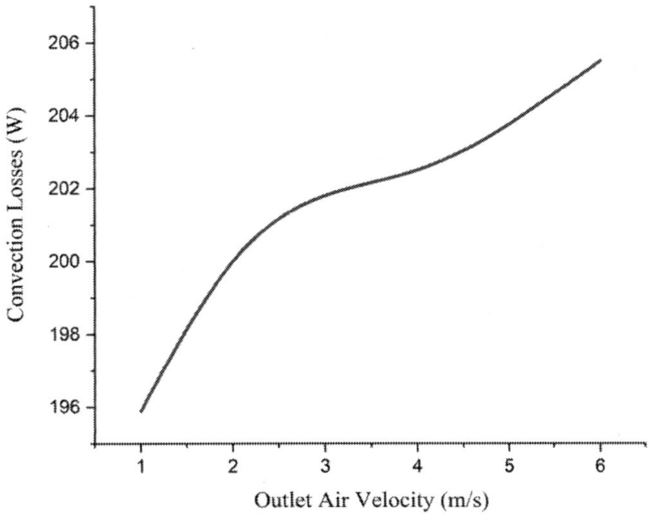

Figure 4. Convection losses for 0° inclination of nozzle of air curtain.

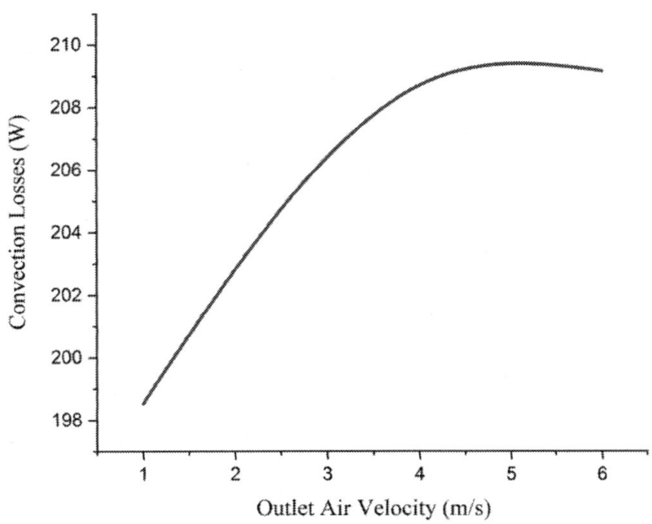

Figure 5. Convection losses for 7.5° inclination of nozzle of air curtain.

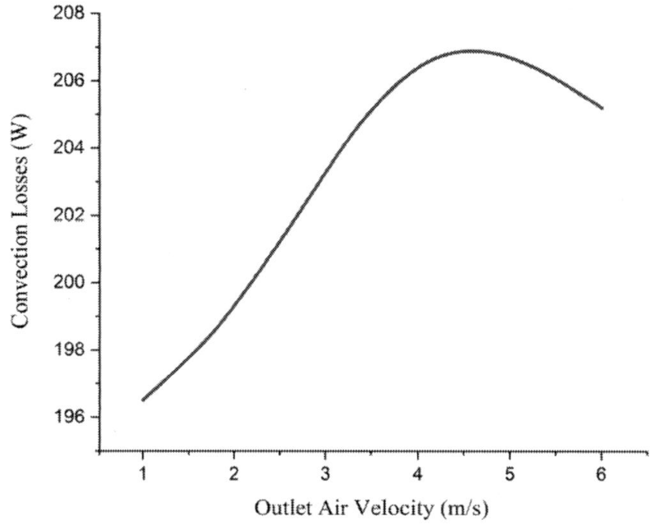

Figure 6. Convection losses for 15° inclination of nozzle of air curtain.

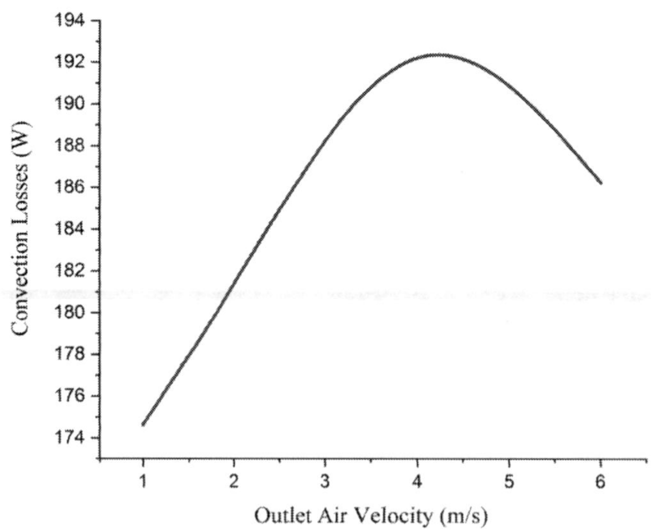

Figure 7. Convection losses for 30° inclination of nozzle of air curtain.

The temperature contours of the modified hemispherical solar cavity receiver along with air curtain is as shown in the Fig[8] which shows 3d view of modified cavity receiver along with air curtain at 15° inclination.

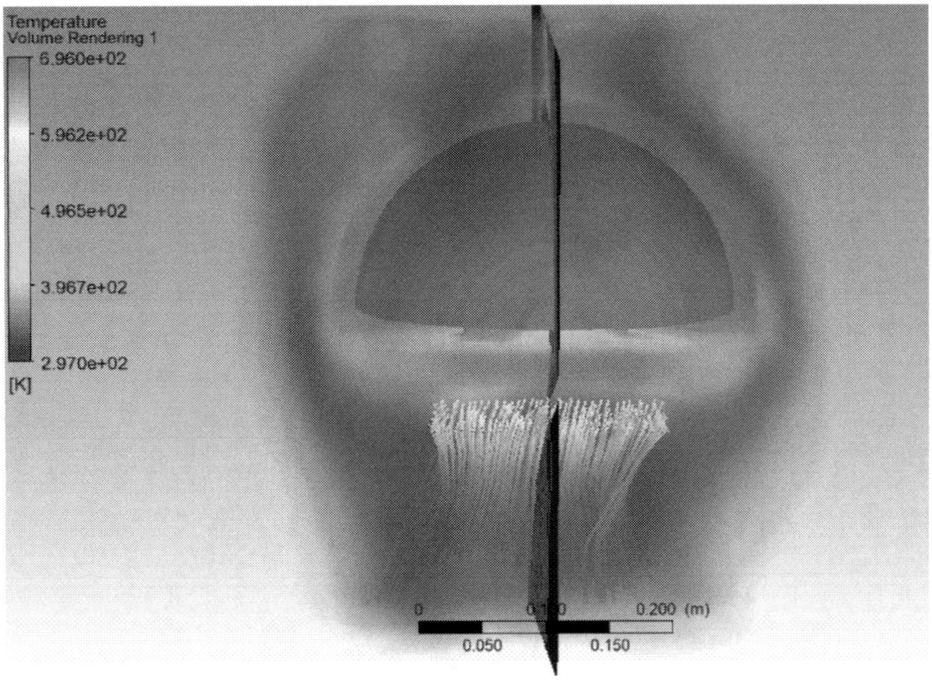

Figure 8. 3D design of modified cavity receiver.

6. Conclusion

Numerical investigation of modified cavity receiver with various air curtain parameter was done in this computational model. Plane air nozzle was equipped over the aperture of modified solar cavity receiver. Plane air nozzle's parameter consisting inclination θ_{noz}, outlet air temperature $T_{out;noz}$ and outlet velocity $u_{out;noz}$ are varied to get the optimal results .The chart[9] shows the impact of these parameter on natural convection loss for 400K outlet air temperature.The result concluded that different air curtains, to varying degree have an impact of decreasing the natural convection loss. Results in this paper state that natural convection loss decreases with inclination of air curtain.

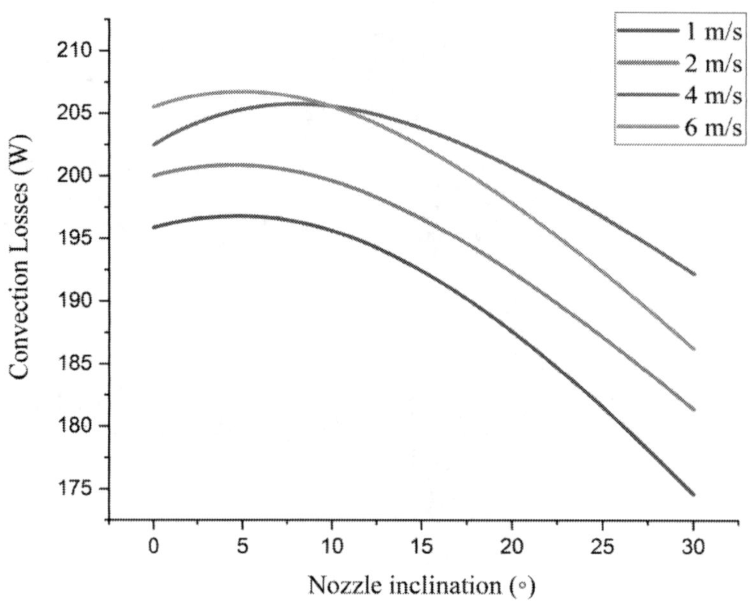

Figure 9. Relationships between air curtain nozzle's inclination angle and natural convective heat loss at different outlet air velocity.

For 400K outlet temperature $T_{out,noz}$ natural convection losses are minimum at 30° of inclination of air nozzle with 1m/s of outlet air velocity $u_{out,noz}$.The difference in temperature between the air and the wall perform a crucial role in the variation of the natural convective heat loss. An increase in natural convection heat loss of solar cavity receiver was observed on increasing the outlet air velocity $u_{out,noz}$of the air curtain. When the natural convection heat loss of modified hemispherical solar cavity receiver is compared with the natural convection heat loss of modified solar cavity receiver with air curtain,a reduction of 16.01% was achieved under the optimal condition of inclination θ_{noz} 30°,outlet air velocity $u_{out,noz}$ 1m/s and outlet air temperature $T_{out,noz}$ of 400K in the present study.

Nomenclature

Variables	Descriptions
h	Enthalpy of fluid (kJ/kg)
L_{noz}	Width of nozzle(m)
Pr	Prandtl Number
P	Pressure (Pa)
k_s	Thermal conductivity of insulation (W/mK)
v	Heat flux (W/m^2)
R	Reynolds Number
R_a	Rayleigh number
T	Temperature ($^{\circ}C$ or K)
$T_{out,noz}$	Outlet Air Temperature
$u_{out,noz}$	Outlet Air Velocity
ρ	Fluid density(kg/m^3)
X	Body force per unit volume
ν	Kinematic viscosity
N_u	Nusselt number
θ_{noz}	Nozzle inclination
Subscripts	
noz	nozzle
out	outlet
Abbrevation	
CSP	Concentrated Solar Power System
HTF	Heat Transfer Fluid
CPC	Compound Parabolic Concentrator
SIMPLEC	Semi-Implicit Method for Pressure Linked Equations-Consistent

References

[1] K. Reddy, G. Veershetty, and T. S. Vikram, "Effect of wind speed and direction on convective heat losses from solar parabolic dish modified cavity receiver," *Solar Energy*, vol. 131, pp. 183–198, 2016.

[2] L. Xiao, S.-Y. Wu, and Y.-R. Li, "Numerical study on combined free-forced convection heat loss of solar cavity receiver under wind environments," *International Journal of Thermal Sciences*, vol. 60, pp. 182–194, 2012.

[3] M. Prakash, "Numerical study of natural convection heat loss from cylindrical solar cavity receivers," *International Scholarly Research Notices*, vol. 2014, 2014.

[4] M. Prakash, S. Kedare, and J. Nayak, "Investigations on heat losses from a solar cavity receiver," *Solar Energy*, vol. 83, no. 2, pp. 157–170, 2009.

[5] O. O. Issa, V. Thirunavukkarasu, and P. Sudhakar, "Novel designs of cavity receiver for a solar parabolic dish concentrator," in *Journal of Physics: Conference Series*, vol. 2054, no. 1. IOP Publishing, 2021, p. 012036.

[6] Y. Jaluria and B. Gebhart, "On the buoyancy-induced flow arising from a heated hemisphere," *International Journal of Heat and Mass Transfer*, vol. 18, no. 3, pp. 415–431, 1975.

[7] W. M. Lewandowski, P. Kubski, J. M. Khubeiz, H. Bieszk, T. Wilczewski, and S. SZYMAŃSKI, "Theoretical and experimental study of natural convection heat transfer from isothermal hemisphere," *International journal of heat and mass transfer*, vol. 40, no. 1, pp. 101–109, 1996.

[8] A. Koenig and M. Marvin, "Convection heat loss sensitivity in open cavity solar receivers," *Final report, DOE contract No. EG77-C-04-3985, Department of Energy, Oak Ridge, Tennessee*, 1981.

[9] N. Kaushika, "Viability aspects of paraboloidal dish solar collector systems," *Renewable Energy*, vol. 3, no. 6-7, pp. 787–793, 1993.

[10] J. A. Harris and T. G. Lenz, "Thermal performance of solar concentrator/cavity receiver systems," *Solar energy*, vol. 34, no. 2, pp. 135–142, 1985.

[11] T. Taumoefolau, S. Paitoonsurikarn, G. Hughes, and K. Lovegrove, "Experimental investigation of natural convection heat loss from a model solar concentrator cavity receiver," *J. Sol. Energy Eng.*, vol. 126, no. 2, pp. 801–807, 2004.

[12] M. Prakash, S. Kedare, and J. Nayak, "Numerical study of natural convection loss from open cavities," *International Journal of Thermal Sciences*, vol. 51, pp. 23–30, 2012.

[13] H. Chang, C. Duan, K. Wen, Y. Liu, C. Xiang, Z. Wan, S. He, C. Jing, and S. Shu, "Modeling study on the thermal performance of a modified cavity receiver with glass window and secondary reflector," *Energy Conversion and Management*, vol. 106, pp. 1362–1369, 2015.

[14] K. Reddy and N. S. Kumar, "Convection and surface radiation heat losses from modified cavity receiver of solar parabolic dish collector with two-stage concentration," *Heat and Mass Transfer*, vol. 45, no. 3, pp. 363–373, 2009.

[15] J. Fang, N. Tu, J. F. Torres, J. Wei, and J. D. Pye, "Numerical investigation of the natural convective heat loss of a solar central cavity receiver with air curtain," *Applied Thermal Engineering*, vol. 152, pp. 147–159, 2019.

[16] T. Venkatachalam and M. Cheralathan, "Effect of aspect ratio on thermal performance of cavity receiver for solar parabolic dish concentrator: An experimental study," *Renewable energy*, vol. 139, pp. 573–581, 2019.

[17] S. Pavlovic, R. Loni, E. Bellos, D. Vasiljević, G. Najafi, and A. Kasaeian, "Comparative study of spiral and conical cavity receivers for a solar dish collector," *Energy Conversion and Management*, vol. 178, pp. 111–122, 2018.

[18] M. Tharun, S. Navin, K. Arjun Singh, A. Sampathkumar, and S. K. Natarajan, "Numerical comparison of elliptical and conical cavity receiver of solar parabolic dish collector system," in *Advances in Mechanical and Materials Technology*. Springer, 2022, pp. 607–618.

[19] D. Reynolds, M. Jance, M. Behnia, and G. Morrison, "An experimental and computational study of the heat loss characteristics of a trapezoidal cavity absorber," *Solar Energy*, vol. 76, no. 1-3, pp. 229–234, 2004.

[20] A. Kribus, P. Doron, R. Rubin, J. Karni, R. Reuven, S. Duchan, and E. Taragan, "A multistage solar receiver:: The route to high temperature," *Solar Energy*, vol. 67, no. 1-3, pp. 3–11, 1999.

[21] N. S. Kumar and K. Reddy, "Comparison of receivers for solar dish collector system," *Energy Conversion and Management*, vol. 49, no. 4, pp. 812–819, 2008.

[22] A. Clausing, "An analysis of convective losses from cavity solar central receivers," *Solar Energy*, vol. 27, no. 4, pp. 295–300, 1981.

[23] R. Jilte, S. Kedare, and J. Nayak, "Investigation on convective heat losses from solar cavities under wind conditions," *Energy Procedia*, vol. 57, pp. 437–446, 2014.

[24] A. M. Daabo, S. Mahmoud, and R. K. Al-Dadah, "The optical efficiency of three different geometries of a small scale cavity receiver for concentrated solar applications," *Applied energy*, vol. 179, pp. 1081–1096, 2016.

[25] K. Mohamad and P. Ferrer, "Thermal performance and design parameters investigation of a novel cavity receiver unit for parabolic trough concentrator," *Renewable Energy*, vol. 168, pp. 692–704, 2021.

[26] E. Abbasi-Shavazi, G. Hughes, and J. Pye, "Investigation of heat loss from a solar cavity receiver," *Energy Procedia*, vol. 69, pp. 269–278, 2015.

[27] S.-Y. Wu, F.-H. Guo, and L. Xiao, "Numerical investigation on combined natural convection and radiation heat losses in one side open cylindrical cavity with constant heat flux," *International Journal of Heat and Mass Transfer*, vol. 71, pp. 573–584, 2014.

[28] Z.-G. Shen, S.-Y. Wu, L. Xiao, D.-L. Li, and K. Wang, "Experimental and numerical investigations of combined free convection and radiation heat transfer in an upward-facing cylindrical cavity," *International Journal of Thermal Sciences*, vol. 89, pp. 314–326, 2015.

[29] Z.-G. Shen, S.-Y. Wu, and L. Xiao, "Numerical study of wind effects on combined convective heat loss from an upward-facing cylindrical cavity," *Solar Energy*, vol. 132, pp. 294–309, 2016.

[30] Y. Shiina, K. Fujimura, T. Kunugi, and N. Akino, "Natural convection in a hemispherical enclosure heated from below," *International Journal of Heat and Mass Transfer*, vol. 37, no. 11, pp. 1605–1617, 1994.

Effect of undercut slots on the overall performance of microchannel heatsink

M Nawaz Khan[1], M N Karimi[1] ,M A Khan[2], M Y Siddiqui[2], T Rehmaan[2]

[1] Department of Mechanical Engineering, Jamia Millia Islamia, New Delhi, India
[2] Department of Mechanical Engineering, Integral University, Lucknow, India

Email:khan.nawaz0786@gmail.com, ORCID:0000-0002-2251-2782

Abstract. In the current study, a numerical analysis has been performed to examine the performance characteristics of the slotted microchannel heat sink. Twelve equidistant slots are entrenched at the base wall of the channel. Four different shapes of slots are considered in this study. They are rectangular (case 1), trapezoidal (case 2), V (case 3) and curved shaped (case 4). Numerical simulations have been performed for these cases with Reynold's number ranging between 150 to 350 with a constant heat flux of 100 W/cm^2 and the results are compared with the plain channel. It is observed that cases 1 and 2 have higher overall thermal performance than the plain channel in which case 2 outperforms case 1 whereas the performance of cases 3 and 4 are lower than the plain microchannel. It is suggested from the analysis that the plain microchannel may be replaced by a trapezoidal shape slotted microchannel in further studies.
Keywords: Microchannel; Slots; Thermal Performance; Numerical Analysis

1. Introduction

The fast growth of the electronics industry over the last two decades has brought down the component size and necessitated a high heat transfer rate. It is quite difficult to achieve the cooling of these small-size components. The heat removal problem was previously solved by using naturally circulating air cooling. Tiny electronics equipment nowadays generates more heat, which cannot be dissipated by air cooling. The researchers have provided a solution to this problem by using an expanded surface concept in a microchannel heat sink. Micromachining technology is used to create these microchannels, which are primarily made of copper or silicon as the substrate material. Tuckerman and Pease[1] were the first to propose the breakthrough concept of heat removal through rectangular microchannels, which were able to reject heat flux up to 700 W/cm2 at a pressure drop of 1.94 bar. The greater surface area to volume ratio of microchannel heat sink results in a high heat transfer coefficient, and low coolant inventory requirement which are the major advantages of MCHS over conventional heat sinks.

2. Literature Review

A literature review has been performed to help comprehend the advancements in the field of microchannel heat sinks. The researchers modified the geometry of the microchannel heat sink cross-section in order to

augment the performance of the heat sink to satisfy the rising demand for heat transfer[2][3][4]. Jin et al.[5] enhanced the heat transmission in various types of microchannels using different configurations of pinfins. The impact of the hydraulic diameter for different shapes of microchannel cross-section on heatsink is analysed by Dalei et al [6]. Saravanan et al.[7] in their work reported the augmentation of heat transfer by incorporating square shaped pin fins. The efficiency of microchannel for various topologies and fins was investigated by Jadhav et al. [8] [9] who discovered that staggered arrangement performs best over the inline configuration of pinfins. Yadav et al.[10] investigated three distinct fin setups in microchannels and found that fins in the upstream region outperform the fins in the downside region and also revealed that a microchannel with a full finned layout is more successful than a partial finned arrangement. Qidwai et al. [11] propose the use of radial fin heat sink for single phase fluid. They found that for a higher Nusselt number a trade-off must be decided between channel depth and flow rate. Prajapati [12] in their work revealed the effect of the length of the fins and reported an enhancement of heat transfer upto 0.8 mm height of fin and a further increase in height retards the heat transmission whereas pressure drop always increases with height. Nawaz et al. [13] in their study reveals that descending height fins at inlet side and ascending arrangement at outlet side has the highest heat transfer capability also they impement the circular perforated pinfin in the rectangular microchannel heat sink to improve the performance of microchannel[14] .
According to the above-mentioned literature, investigators have put a lot of effort into the upliftment of heat transfer performance of heat sinks by implementing extended surfaces of various shapes and sizes along the length of the microchannel. However, this augmented heat transfer is offset by a larger pressure drop. The heat augmentation, combined with the higher pressure drop, has an impact on overall performance. From review, no or very little work has been done to investigate the impacts of undercut on the efficiency of microchannels. Therefore this work explores the influence of undercut slots etched on the base wall of the channel. Hence this work discloses the impact of slots on the capability of rectangular microchannel. To the best knowledge of the authors, no work has been performed on the given structure, which proves the novelty of this work.

3. Computational modelling

Figure 1 depicts the isometric view of the heat sink. Solid domain of microchannel is etched on the deoxidized copper base and water is employed as a coolant. Dimensions of a working domain are $L \times W \times H = 10 \times 0.2 \times 0.3$ mm whereas, the channel width (W_c) and height (H_c) are 100 μm and 200 μm. Twelve slots *(N)* are equidistantly spaced at the base of a microchannel. The structures of the slots considered in this study are rectangular, trapezoidal, curved and V-shaped as shown in figure 2. The maximum width of the slot is taken as 50% of the channel width whereas the depth of the slot is 10% of the channel height.

The assumptions which are considered for computational analysis are mentioned below:

1. Flow is assumed to be incompressible, laminar, 3d and steady-state.
2. At a solid and fluid interface, a no-slip boundary condition is considered.
3. Heat transfer due to radiation is neglected.
4. At the inlet, flow is in developing state.
5. The upper surface of the channel and the fluid domain is assumed to be adiabatic.
6. Thermophysical property of solid substrate remains constant.

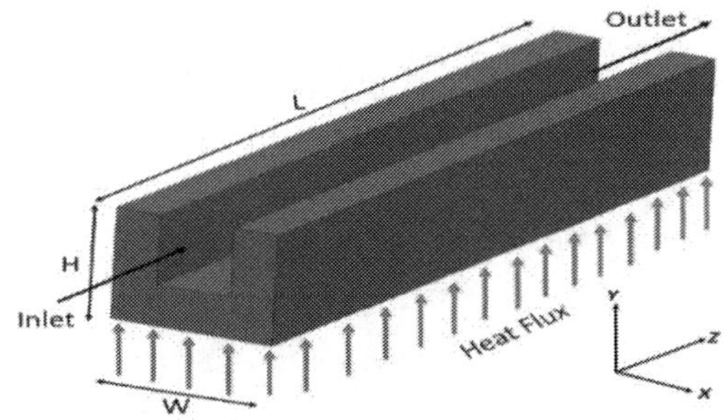

Fig. 1: Isometric view of a microchannel heat sink.

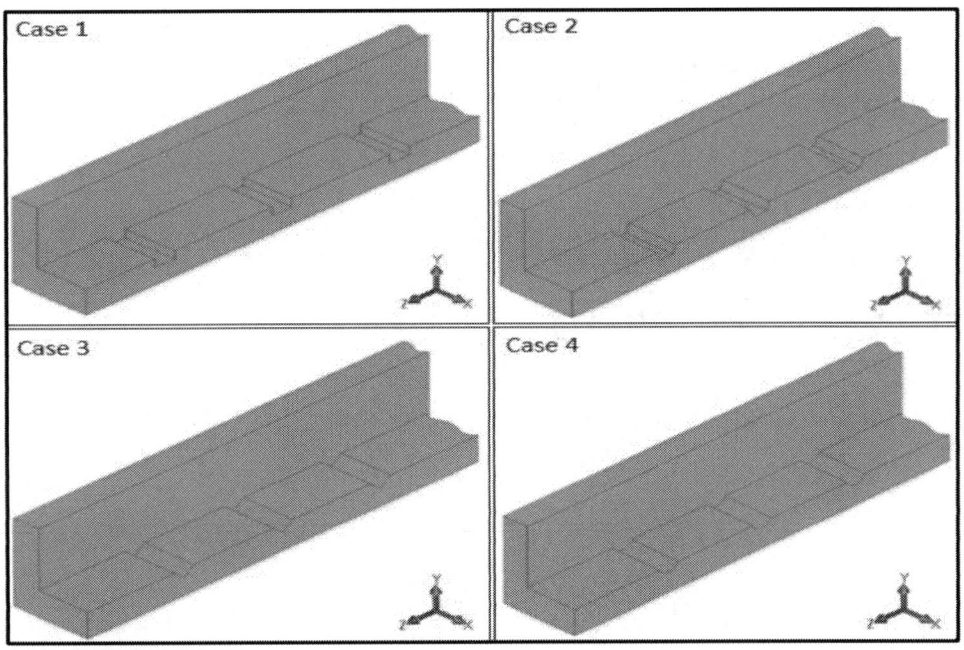

Fig 2: Regional perspective of different configurations of heatsink.

The appropriate general governing equations for solid substrate and working fluid were reduced to the following form using the aforementioned assumptions.

For solid domain

Continuity and Momentum equation

$$\nabla.(\vec{V}) = 0 \quad (1)$$

Energy equation

$$k_s \nabla^2 T = 0 \quad (2)$$

For liquid domain

Continuity equation

$$\nabla.(\rho_{wf}\vec{V}) = 0 \tag{3}$$

Momentum equation

$$\nabla.(\rho_{wf}\vec{V}.\nabla\vec{V}) = -\nabla P + \nabla.\mu_{wf}\left[(\nabla\vec{V} + \vec{V}^t) - \left(\frac{2}{3}\right)\nabla\vec{V}\right] + \rho_{wf}.\vec{g} \tag{4}$$

3.1. Boundary Conditions

The aforesaid set of governing equations is solved using Ansys Fluent by adopting the appropriate boundary conditions at various positions throughout the domain.

At x = 0 (inlet),

- Constant fluid temperature, $T = T_{wf,in} = 288\ K$
- Constant fluid velocity, $u = u_{wf,in},\ v = w = 0$

At x = L (outlet),

- Atmospheric Pressure, $P_{wf,out=}\ P_{atm}$

At y = 0 (heated base wall),

- a constant flux of 10^6 W/m^2 is provided.

At y = H (top surface),

- adiabatic boundary conditions are applied , $k_s\left(\frac{\partial T}{\partial y}\right)_{s,\Gamma} = k_{wf}\left(\frac{\partial T}{\partial y}\right)_{wf,\Gamma} = 0$

At z = 0, B (side walls),

- symmetric conditions, $\left(\frac{\partial}{\partial z}\right) = 0$

Coupled boundaries with no-slip conditions are defined at the interface

- $-k_s\left(\frac{\partial T}{\partial n}\right)_s = -k_{wf}\left(\frac{\partial T}{\partial n}\right)_{wf}$
- $u = v = w = 0$

3.2. Data reduction

Reynold's number is presented in this article by the following relation

$$Re = \frac{\rho_{wf,in}u_{in}D_h}{\mu_{wf,in}} \tag{5}$$

Where, $u_{in},\ \rho_{wf,in}$ is the velocity and density of water at the inlet

$$D_h \text{ (hydraulic diameter)} = \frac{2(W_{ch} \times H_{ch})}{W_{ch} + H_{ch}} \quad (6)$$

The expression of the average Nusselt number (Nu) is given as

$$Nu = \frac{q_\Gamma D_h}{k_{wf}(T_{\Gamma,m} - T_{wf,m})} \quad (7)$$

where, k_{wf} is the thermal conductivity of water; $T_{\Gamma,m}$ is the mean temperature at the interface and calculated by area-weighted approach whereas $T_{wf,m}$ is the mean temperature of coolant liquid and evaluated by the volume average approach; A_{HB} and A_Γ represents the area of heated base wall and interface respectively.

$$q_\Gamma = \frac{q\, A_{HB}}{A_\Gamma} \quad (8)$$

In the above equation, A_{HB} denotes the area of heated base wall and A_Γ is the area of solid liquid interface. The governing equations are solved with the help of using a commercial package of computational fluid dynamics (FLUENT). A simple numerical scheme for pressure velocity coupling with second-order spatial discretization for pressure and second-order upwind for momentum and energy has been considered to solve the set of differential equations. The value of residuals for the energy equation is set to 10^{-6} whereas for velocity and continuity equation, it is limited to 10^{-5} for the sake of convergence in the solution.

3.3. Grid size selection and validation

A grid test is required to determine the optimal grid size for avoiding mistakes caused by the coarseness of elements while not distorting the physics of the computational model and allowing us to complete the calculation in the shortest possible time. The model is discretized by adopting the unstructured mesh of hexahedral element by providing finer meshing at the perforation due to complex nature of flow at this particular region Nusselt number and pressure drop (ΔP) are the parameters determined for different numbers of elements and their values are provided in table 1. The table shows that the difference in values above 3258342 is small, hence the grid size of 3258342 elements has been used for further calculations to reduce computational time.

Table 1: Grid test

Number of elements	Nu	Difference%	ΔP (Pascal)	Difference%
960203	12.28	2.61	57152	3.79
1856651	12.15	1.56	58953	0.62
3258342	11.94	0.17	59316	0.01
4296604	11.92		59320	

To validate our current numerical scheme, the experimental analysis of Qu and Mudawwar[15] has been replicated by the approach of a present model. An exact similar single rectangular plane microchannel with the same initial and boundary conditions acting on it is created. To acquire the results, numerical simulations were performed using the current numerical scheme. Figure 3 represents the correlation of pressure drop and temperature distribution with the work of Qu and Mudawwar. The current numerical scheme for both examined parameters is in excellent accordance with the actual data, as shown in the figure, and may be used for further exploration of microchannel heat sinks.

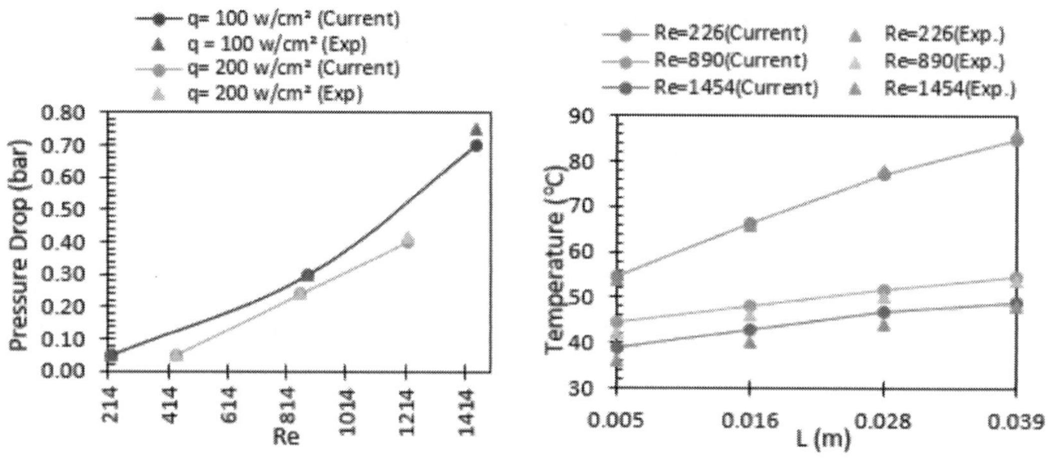

Fig 3: Comparison between results of the present scheme and Qu et al. for pressure drop and temperature.

4. Results

4.1. Analysis of Nusselt number:

To analyse the outcomes of etching slots at bottom wall a graph between Nusselt number and Reynold's number has been plotted and depicted in figure 4. From figure 4 (a), one may perceive that the value of the average Nu is increased with Reynold's number for all the cases. Different researchers corroborated the same observation in their investigation. The reason for this increase is that as Reynold's number rises, the discharge of the water rises, assisting to extract a considerable amount of heat. The maximum Nusselt number is associated with Case 2 followed by Case 1.

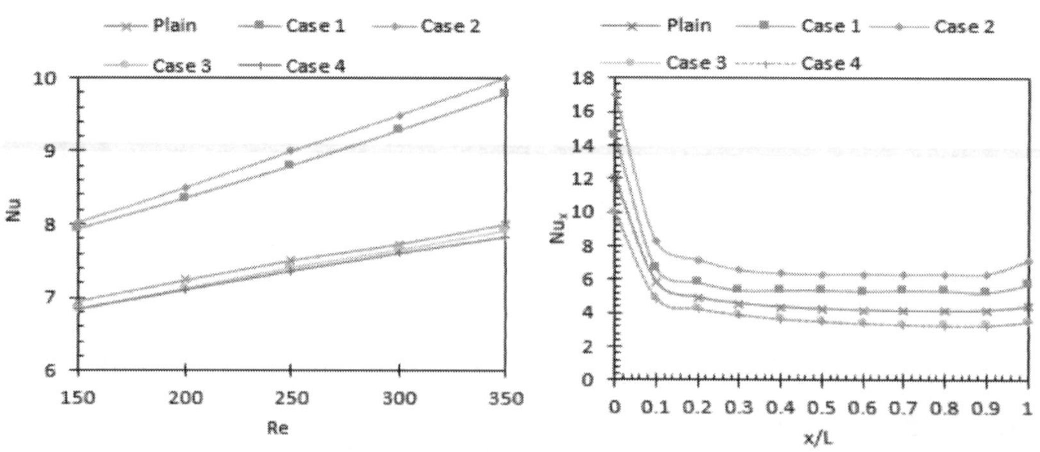

Fig 4: Variation of (a) Nusselt number vs Reynolds number, (b) Local Nusselt number at the planes perpendicular to the direction of flow (Re=250).

This increase in heat transfer capability in the above cases is due to the machining of the slot, which increases the interface area at the bottom wall. The same can also be acknowledged by the temperature contour of the channel depicted at the centre vertical plane as shown in figure 5. From the contour, it has been noticed that the temperature at the interface of case 2 is minimum which justifies the highest performance of case 2. Even interface area of case 1 is slightly elevated than case 2 although it has a lower Nusselt number. The rationale behind the underperformance of case 1 is associated with the complete entrapment of liquid in the slot. The amount of heat transfer is minimum for case 4. The slope of cases 3 and 4 are identical at Re up to the value of 200 and then case3 outperforms case4 with the value of Reynold's number

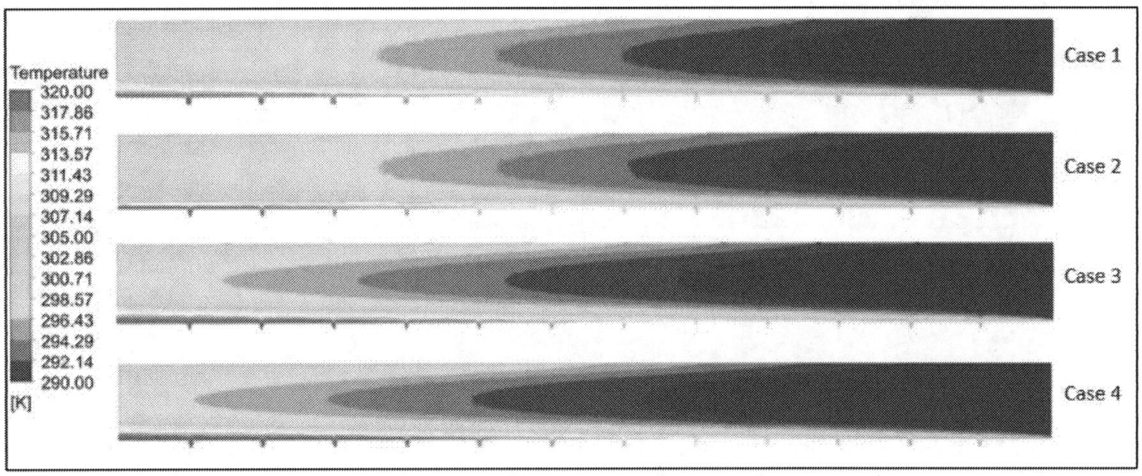

Fig 5: Temperature contours at a plane passing through the central vertical line of microchannel at Re 250.

The variation of local Nu is illustrated in figure 4 (b). The value of the local Nu is largest at an inlet, which is owing to the maximal local heatflux present at the inlet. Slope of the local Nu decreases with the length of the channel in all cases, which corresponds to a decrease in local heat transfer. The interesting results obtained from the figure are about cases 1 and 2. The slope of cases 1 and 2 are considerably constant after x/L= 0.4 to 0.9. This nature of the graph is the result of having slots of rectangular and trapezoidal shape at the base wall. The local Nusselt number increases at the location of these slots and hence abstains the tendency of diminishing slope across the length of channel.

4.2. Analysis of Pressure drop:

The effect of slots on the pressure drop across the microchannel is investigated and compared in this section to that of a plain microchannel. The relative pressure drop for all microchannel designs is shown in Figure 6 (a). With an increase in Reynold's number, a stronger resistance has been produced by the large quantity of water resulting in an elevation in pressure drop for all configurations. As expected the minimum pressure drop is recorded for the plain microchannel as it has the least heat transfer area which offers minimum flow resistance. This highest pressure drop is attributed to case 1. The complete entrapment of water in the rectangular slot with the maximum interface area mutually contributed to the highest pressure drop of case 1. After case 1, the maximum value of pressure drop is recorded for case 2 followed by cases 3 and 4. This curtailing trend of pressure drop is the outcome of the compound effect of reduction in resisting areas and the shape of the slots in each case respectively. The lines of pressure drop of different cases in the figure are diverging with the increase in Reynold's number and it is due to that the slots with larger areas contribute higher pressure at a larger Reynold's number.

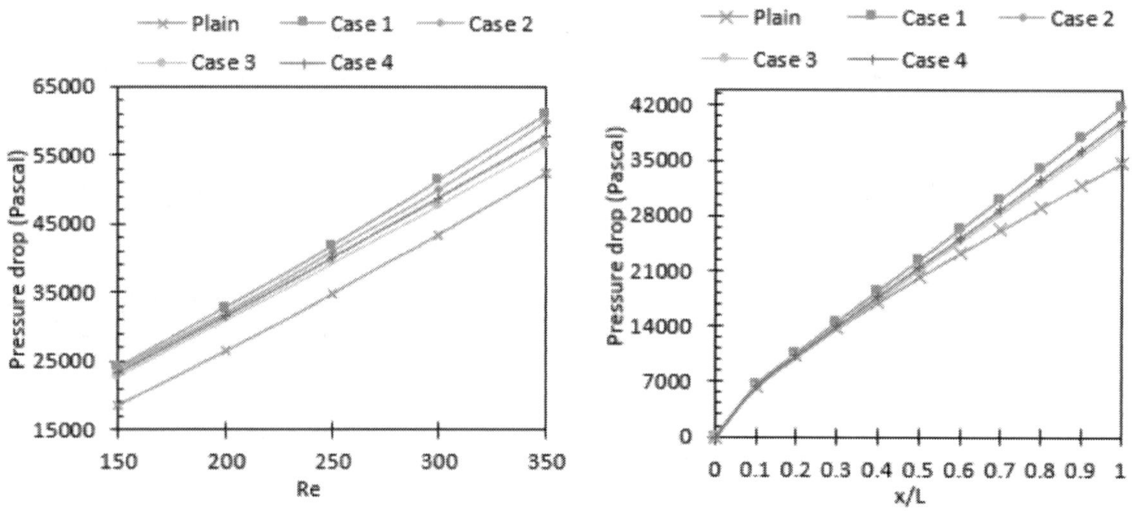

Fig 6:. (a) Variation of Total Pressure Drop vs Reynolds number, (b) Variation of Local Pressure Drop at sections along the direction of flow (Re=250).

Figure 6 (b) shows the fluctuation of the local pressure drop at Re =250 for distinct cases of a microchannel heat sink. The intensity of the local pressure drop is measured at ten equally spaced planes perpendicular to the direction of flow. It has been observed from the figure that initially the slope of the local pressure drop is indistinguishable for all arrangements. This similarity in local pressure drop is because of the absence of slots up to this region and also the flow is in the developing stage but from x/L = 0.1 the variations are observed and the maximum slope is recorded for case 1. The slope of the plain channel reduces as the flow approaches the exit of the channel whereas the slope of all other cases remains constant. This reduction in the slope of the plain channel at downstream is consequence of water heating at upstream of the channel resulting in the increase in density of the fluid and hence lowering the resistance and pressure drop whereas in cases 1 to 4, the presence of slots neutralises the effect of density increasing.

4.3. Overall performance (OTP):

It is essential to determine the OTP because the heat transfer potential of a microchannel heat sink is increased at the penalty of pressure drop when slots are entrenched at the base wall. The OTP is characterized to evaluate the corresponding relations between the enhancement of heat transfer and the increment in pressure drop.

$$\eta = \frac{Nu/Nu_o}{(\Delta P/\Delta P_o)^{1/3}}$$

Figure 7 shows relationship between OTP and the Reynold number. From the figure, it has been observed that the OTP of all the cases increases with the Reynolds number. The overall thermal performance of cases 1 and 2 for all the values of Reynold's number are greater than 1 which justifies the use of a microchannel with the rectangular and trapezoidal slot over the plain case. However, the maximum OTP is recorded for case 2 which suggests the adoption of trapezoidal slots in rectangular microchannel heat sinks. The values of overall thermal performance of cases 3 and 4 are lower than 1 hence the slots of triangular and curved shapes are not recommended for the microchannel heat sinks.

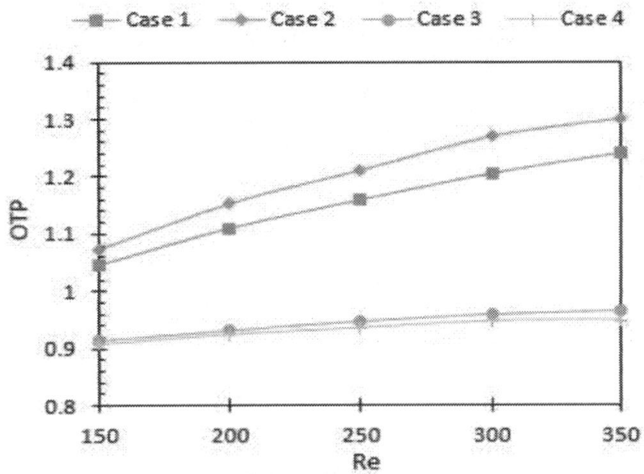

Fig 7: Variation of OTP with Re.

5. Conclusion

To improve the heat augmentation of microchannels various techniques have been used. Many researchers focused on the implementation of extended surfaces for heat transfer enhancement. This study depicts the performance of a microchannel heat sink with twelve equidistant slots of different shapes

Following are the important concluding points drawn from the analysis:

1. Only the slots of rectangular and trapezoidal shape on the base wall increase the heat transfer capability on the account of pressure drop relative to the plain microchannel. The maximum Nu is recorded for case 2 followed by case1 whereas the minimum is for case 4.
2. The maximum pressure drop is recorded for case 1 due to the complete entrapment of fluid in the slot whereas the minimum is associated with the case of the plain microchannel. In the cases of slotted microchannel case 4 has the least pressure drop succeeded by case 3.
3. Overall thermal performance portrays the mutual effect of heat transfer and pressure drop on the performance of the microchannel heat sink. Overall thermal performance is greater than 1 for cases 1 and 2 of the slotted microchannel. The value of overall thermal performance is maximum for case 2 followed by case 1 whereas cases 3 and 4 have performance values less than 1. Hence case 1 may be replaced plain microchannel in future studies of heat transfer enhancement

References

[1] Tuckerman and Pease, "High-Performance Heat Sinking for VLSI," *IEEE ELECTRON DEVICE Lett.*, vol. 2, no. 5, pp. 126–129, 1981, doi: 10.1177/0164027595174002.

[2] Y. Sui, P. S. Lee, and C. J. Teo, "An experimental study of flow friction and heat transfer in wavy microchannels with rectangular cross section," *Int. J. Therm. Sci.*, vol. 50, no. 12, pp. 2473–2482, 2011, doi: 10.1016/j.ijthermalsci.2011.06.017.

[3] L. Lin, Y. Y. Chen, X. X. Zhang, and X. D. Wang, "Optimization of geometry and flow rate distribution for double-layer microchannel heat sink," *Int. J. Therm. Sci.*, vol. 78, pp. 158–168, 2014, doi: 10.1016/j.ijthermalsci.2013.12.009.

[4] L. Lin, M. X. Deng, X. X. Zhang, and X. D. Wang, "Numerical analysis and parametric study of multilayered microchannel heat sinks," *Adv. Mech. Eng.*, vol. 7, no. 7, pp. 1–10, 2015, doi: 10.1177/1687814015594124.

[5] J. Zhao, S. Huang, and L. Gong, "Numerical Studies on Geometric Features of Microchannel Heat Sink with Pin Fin Structure," no. September, pp. 7–10, 2014.

[6] D. Jing, S. Song, and L. He, "Numerical studies on the hydraulic and thermal performances of microchannels with different cross-sectional shapes," *Int. J. Heat Mass Transf.*, vol. 128, pp. 1344–1350, 2019, doi: 10.1016/j.ijheatmasstransfer.2018.08.006.

[7] V. Saravanan and C. K. Umesh, "Numerical comparison for thermo-hydraulic performance of pin fin heat sink with micro channel pin fin heat sink," *Sadhana - Acad. Proc. Eng. Sci.*, vol. 43, no. 7, pp. 1–15, 2018, doi: 10.1007/s12046-018-0875-1.

[8] S. V. Jadhav and P. M. Pawar, "Performance analysis of microchannel with different pin fin layouts," *Int. J. Numer. Model. Electron. Networks, Devices Fields*, vol. 33, no. 2, pp. 1–19, 2019, doi: 10.1002/jnm.2697.

[9] S. V. Jadhav, P. M. Pawar, and B. P. Ronge, "Effect of pin-fin geometry on microchannel performance," *Chem. Prod. Process Model.*, vol. 14, no. 1, 2019, doi: 10.1515/cppm-2018-0016.

[10] V. Yadav, K. Baghel, R. Kumar, and S. T. Kadam, "Numerical investigation of heat transfer in extended surface microchannels," *Int. J. Heat Mass Transf.*, vol. 93, pp. 612–622, 2016, doi: 10.1016/j.ijheatmasstransfer.2015.10.023.

[11] M. O. Qidwai, M. M. Hasan, N. Z. Khan, and M. Ariz, "Performance evaluation of simple radial fin design for single phase liquid flow in microchannel heat sink," *Proc. Inst. Mech. Eng. Part E J. Process Mech. Eng.*, p. 095440892094426, Jul. 2020, doi: 10.1177/0954408920944269.

[12] Y. K. Prajapati, "Influence of fin height on heat transfer and fluid flow characteristics of rectangular microchannel heat sink," *Int. J. Heat Mass Transf.*, vol. 137, pp. 1041–1052, 2019, doi: 10.1016/j.ijheatmasstransfer.2019.04.012.

[13] M. N. Khan and M. N. Karimi, "Analysis of heat transfer enhancement in microchannel by varying the height of pin fins at upstream and downstream region," *J. Process Mech. Eng.*, vol. 234, no. 4, pp. 1–10, 2021, doi: 10.1177/0954408921992975.

[14] M. N. Khan, M. N. Karimi, and M. O. Qidwai, "Effect of circular perforated pin fin on heat transfer and fluid flow characteristics of rectangular microchannel heat sink," *Numer. Heat Transf. Part A Appl.*, pp. 1–15, Aug. 2022, doi: 10.1080/10407782.2022.2101809.

[15] W. Qu and I. Mudawar, "Experimental and numerical study of pressure drop and heat transfer in a single-phase micro-channel heat sink," *Int. J. Heat Mass Transf.*, vol. 45, no. 12, pp. 2549–2565, 2002, doi: 10.1016/S0017-9310(01)00337-4.

IOP Conf. Series: Materials Science and Engineering 1259 (2022) 012010 doi:10.1088/1757-899X/1259/1/012010

Utilization of Waste Heat from Condenser of Domestic Refrigerator for Water Heating

Wasim Babulal Patel[1] Pranav G. Charkha[2] Amol V. Joshi [3] Jitendra N. Wadadkar[4]

[1,2,3,4] *Department of Mechanical Engineering, G. H. R. I. B. M. Jalgaon, Maharashtra India*

1 *wpwasimpatel@gmail.com*
2 *pranav.charkha@raisoni.net*
3 *amol.joshi@raisoni.net*
4 *jitendra.wadadkar@raisoni.net*

Abstract. : Conservation of energy is very important factor in economic and financial development of country. Refrigeration system utilizes enormous amount and lost from refrigerator is also a major concern which ultimately rejected to atmosphere causing rise in temperature. On the other hand electricity is consumed to heat water. So it is necessary to utilize this rejected heat. In this paper the extracted heat from refrigeration is used to heat water by utilizing water cooled condenser without affecting COP of the system. The setup is installed on domestic refrigerator containing condensing coil, temperature sensors for reading and insulated tank attached to domestic refrigerator as it is widely used in almost every house.

Keywords: VCR System, Domestic refrigerator, Waste heat recovery, Modified condenser, Water heating

1. Introduction

Vapor-compression refrigeration (VCR) is widely used refrigeration cycle existing for use. The general application of VCR system is conditioning of hotels, private houses, offices, public buildings, cinema halls, automotive, and hospitals etc. [1, 4]. A domestic refrigerator operates on VCR cycle where heat is extracted from sealed chamber called evaporator and that heat is transfer or reject in surrounding through air cooled condenser [4]. The temperature in the evaporator is kept always below atmosphere is referred as refrigeration, so continuous heat extraction is there. This waste heat from condenser can be used to heat water simultaneously without effecting other parameter of domestic refrigerator [6, 7].

1.1 Vapor Compression Refrigeration System

VCR consist of Compressor Condenser Expansion device and evaporator as shown in the given figure [2]. Process 1-2 is compression of low pressure refrigerant to a high pressure and temperature vapor. Next process 2-3 is condensation of high temperature and high pressure refrigerant to low temperature and high pressure liquid. Moving ahead process 3-4 is expansion followed by evaporation process 4. Evaporator is the section where heat is extracted from the system and heat is rejected to atmosphere during Condensation.

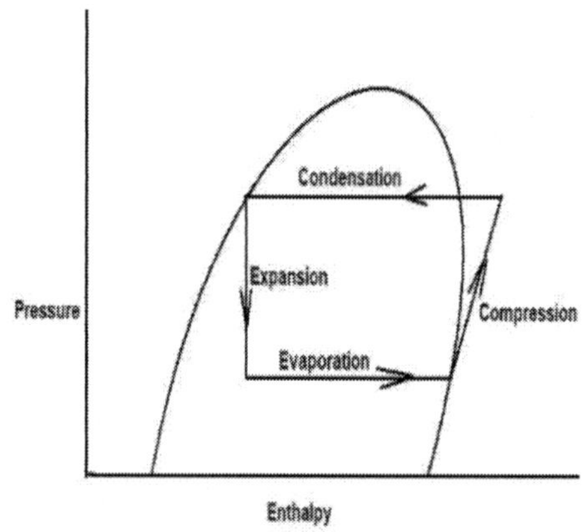

Figure 1. Pressure-Enthalpy curve

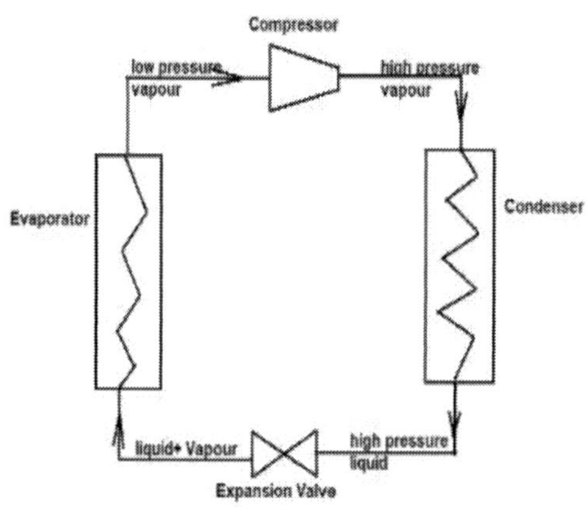

Figure 2. Vapour compression refrigeration cycle

2. Problem Definition and Objectives

A literature reviews of various papers and research has done and it can be conclude that helical shape and zig zag shape condensers and so many refrigerants are used for effect producing and heat rejection process increment in condenser and less focused was done on circular type coil in condenser research so we can define problem as heat can be recovered at higher rate from circular type coil condenser in cabin using R-134a refrigerant. The objectives of this paper and research is as follows

1. To use the rejected heat from condenser to heat water.
2. To promote green and clean energy.
3. To minimize use of electricity used for water heating

3. Literature Survey

[1]R. Vali, V. Reddy explained in their research paper based on performance investigation of refrigeration system. In this paper he used helical type of coil geometry of condenser heat exchanger with refrigerant R-134A and R-410A for finding out the best geometry suited for heat transfer to water. The conclusion was that the helical coil has better performance than old coil used in refrigeration by using R-410a than R-134a in terms of high refrigerating effect with better C.O.P. [2] D. Eriksson and R. Hashemi have researched on water generation methods using Vapor compression cycle by building a prototype. They came to a conclusion that lower relative hu midity of system fluid cause decrease in performance of VCR system. [3] Vinay Vishwanath, Rohan Jikar explained their research paper on Water Dispenser Using Air Conditioning system. The conclusion from the research is that there is no need of using heater for water heating. The function of water heater can be replaced by waste heat from condenser. [4] Sana Shaikh, Priyanka Rewate have presented research paper based on Fabrication of both Hot and Cold water dispenser. The use can be based on type of refrigerant used. There is no change in efficiency in comparison with system that using air cooled condenser. [5]Dhawal A. Khachane, Mr. D. Y. Dhande explained about waste Heat Recovery from refrigeration system. They proposed model [10] in which heat which is wasted from condenser coil (can be act as heat exchanger) in refrigeration cycle can be used to heat the water. Also, both water temperature equalized after 175 minute for heating 15 kg water. [6] Amol Vikas Joshi, Tejasbhai Pravinbhai Mali et al gave brief idea about waste heat recovery from domestic refrigeration system. They used compound air and water-cooling in refrigeration system. This system gives considerable improvement in COP and decrease in power consumption leads to conclusion that it is best way to use waste heat coming from condenser. [7] Dhananjay Parmar,Neelesh Soni presented research paper for utilizing waste heat in vapor compression refrigeration system. Heat is utilized by installing water chamber in between compressor and condenser. The product (hot water) is used for various purposes like bathing, washing in addition also used in fabricating industries, processing industries, diluting, for sterilization necessities. [8] S. White and D. Cleland had discussed on domestic refrigeration for refrigeration purpose and water heating. The objective of this paper was to reused heat for refrigeration and hot water, mainly required for food industry with modification of main parameters to make it reliable. Refrigerant which was used for this research is CO2.

4. Design Description

4.1 Component of System

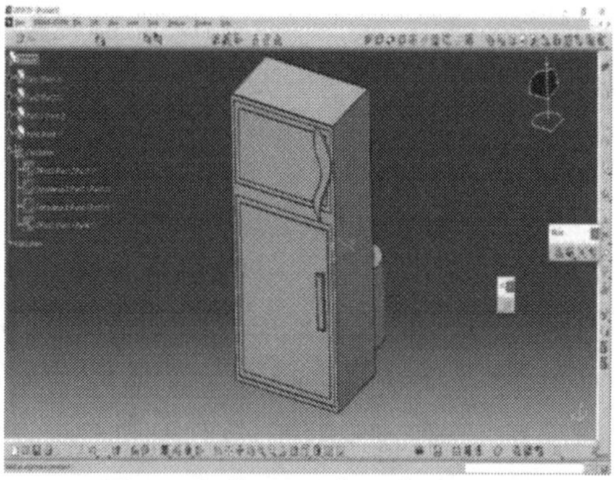

Figure 3. Construction of system

4.1.1 Compressor: The compressor is main component used to increase the pressure of refrigerants from the evaporator up to condenser pressure at which condenser is designed for handling.

4.1.2 Condenser: Condenser is used for heat released device in refrigerator. Heat is taken out from refrigerant and phase change of refrigerant is happen and the heat transmission fluid can be air or water. The process of condenser is same as that of evaporator except working Temperature (Both are heat exchanger)

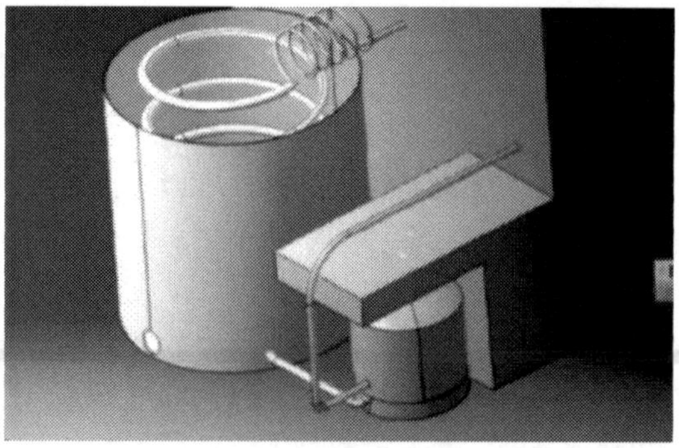

Figure 4. Design of Compressor, Receiver Tank and Condenser Coil

4.1.3 Expansion Valve: It lower the pressure of refrigerant to the evaporating pressure by expanding the refrigerant, the device also referred as throttling device or an expansion device use to undertake this procedure.

d. Evaporator: Evaporator has a function of converting low pressure liquid Refrigerant to low pressure vapor refrigerant by absorbing latent heat form chamber.

5. Experimental Set – Up

The setup compiles of domestic refrigerator with modification is used with following specification is to be used:

- Capacity in Liter: 190L
- Power Rating: Single phase 220-240 volt, 50 Hz, A.C. supply. 220 W
- Refrigerant type: 134 A (CF3CH2F)

IOP Conf. Series: Materials Science and Engineering 1259 (2022) 012010 doi:10.1088/1757-899X/1259/1/012010

- Condenser: air cool condenser with some amount modified water cool condenser

Figure 5. Experimental setup

5.1 Calculations

Attached tank measurements taken while constructing the set up

Radious = 12 cm = 0.12 m

Height = 26 cm = 0.26 m

Volume = $\Pi r^2 h$ = 3.14 * 0.12^2 * 0.26

= 11.756 * 10^{-3} m^3

Mass of circulating water = Mw = ρw * Vw = 11.756 kg

Cw = 4.187 kj/kg k

Temperature of water at initial level = T1 = 25°

Temperature of water at exit of cabin = T2 = 45°

Time = T = 55 min

Now we know by balancing of heat

Heat lost by refrigerant in condenser coil = Heat absorbed by water to be heated in cabin

Q = Mw * Cw * d T

Q = 11.75 * 4.187 * (45 – 25) = 983.9 kj

Since we have considered 1 ton of unit = 210 kj/min

Capacity of cabin = 983.9/ (210* 55) = 0.0851

6. Results and discussion

The coefficient of performance (COP) can be defined as the ratio of refrigerating effect to work of compressor. Readings are taken for time interval of 10 minute. COP for each readings are calculated as follows:-

IOP Conf. Series: Materials Science and Engineering 1259 (2022) 012010 doi:10.1088/1757-899X/1259/1/012010

- COP actual = Heat extracted in refrigerator / Work done by compressor.
- COP waste heat recovery = Heat extracted in refrigerator / (Work done by compressor - Waste heat recovery achieved)

It is observed that from experiment that

COP of domestic refrigerator without modification = 1.1

COP of domestic refrigerator after modification = 1.1851 for heat recovery through hot insulated tank. It is seen from calculations that COP enhance by 7.73 %. Heat recovered inside insulated tank used to heat water, whose temperature increased by 20 degree centigrade.

Table 1. Observation table

Time in minute	Temperature of hot insulated tank in ^{o}C
0	25
15	29
25	33
35	37
45	40
50	43
55	45

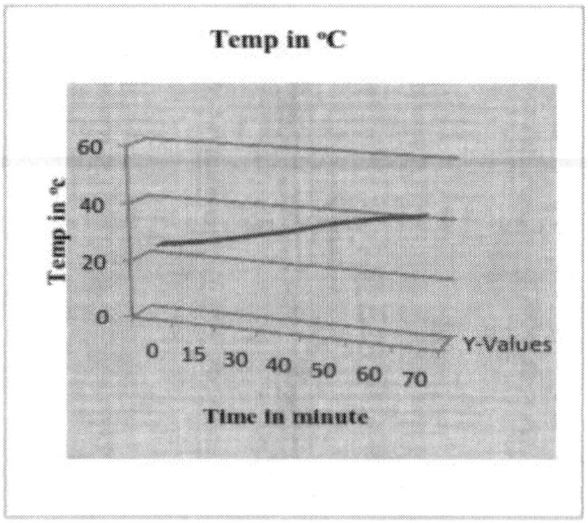

Figure 6 Temp vs Time plot

7. Conclusions

1. Heated water can be used for domestic purposes such as Bathing, hand washing, washing clothes, Dish washing etc.
2. There is no change in efficiency and it is same as air cooled condenser.
3. As waste heat is used for heating the water so no extra power required for system instead the electricity is saved used for heating of water by guizer and heater.

8. References

[1] Vali, R. H., & Reddy, V. M. (2015). Experimental investigation and performance evaluation of vapor compression refrigeration system with helical type condenser by using R-134A and R-410A refrigerants. International Journal of Engineering Research and Science & Technology, 2319-5991.

[2] Eriksson, D., & Hashemi, R. (2009). Evaluation of suitable methods for water generation. Skolan för industriell teknik och management, Kungliga Tekniska högskolan.

[3] Vishwanath, V., & Jikar, R. (2015). Water Dispenser System Using Air Conditioner. International Journal of Engineering Science Invention, 4, 21-32.

[4] Shaikh Sana, & Rewate Priyanka (2016). Fabrication of Hot and Cold Water Dispenser. International Journal of Current Trends in Engineering & Research (IJCTER), Volume 2(7), 1-6.

[5] Khachane Dhawal A., & Dhande D. Y. (2016).Development and Experimental Analysis of Waste Heat Recovery of Heat Pump. International Journal of Modern Engineering Research| Volume 6, 42-49.

[6] Mali, T. P., Saini, M., et al. (2017). Waste heat recovery in domestic refrigeration system in the application of water heating. Journal for Research| Volume, 3(01), 82-85.

[7] Parmar, D., & Soni, N. (2017). Review on study of waste heat utilization techniques in vapour compression refrigeration system. International Research Journal of Engineering and Technology, 4(7), 2033-39.

[8] White, S. D., Cleland, D. J., et al (1997). A heat pump for simultaneous refrigeration and water heating. In IPENZ Annual Conference 1997, Proceedings of: Engineering our nation's future; Volume 2; Papers presented in the technical programme of the IPENZ Annual Conference held in Wellington, February 7-10, 1997 (p. 117). Institution of Professional Engineers New Zealand.36-42.

[9] Kongre, D. U., Chiddarwar, et al (2013). Testing and performance analysis on air conditioner cum water dispenser. International Journal of Engineering Trends and Technology (IJETT), 4, 772-5.

[10] Amol Vikas Joshi, Devanand Tayade et al (2022) An Experimental Evaluation of FEA-Based Helical Compression Spring to esign the Shock Absorber. Smart Technologies for Energy, Environment and Sustainable development, Vol 2, Springer Proceedings in Energy, 299-308.

Numerical analysis of heat transfer from perforated fins of an induction motor housing

M B Bhambere[1, 2] **and S S Chaudhari**[1]

[1]Mechanical Engineering Department, Yeshwantrao Chavan College of Engineering, Nagpur, India.
[2]Mechanical Engineering Department, Shri Sant Gajanan Maharaj College of Engineering, Shegaon, India.

mechymadhao@gmail.com (ORCID ID: 0000-0001-7495-6097)

Abstract. The major goal of this research article is to demonstrate how perforation on the fin surface of an induction motor's housing improves cooling performance. Heat transfer study was also conducted between traditional fins and the suggested perforated fins housing. An Induction Motor (IM) overheats as a result of continuous operation, raising its temperature, lowering its performance and shortening its lifespan. The innovative designs and development of fins are necessary for effective cooling and enhancing the thermal performance of IM. In the present work, the thermal analysis of induction motor is carried out by means of providing the perforations on existing fins and varying perforation size from 4 mm to 10 mm diameters and number of perforation from 20 to 11, on fins and observed the steady state heat transfer and temperature variation. In this regard, the half part of IM is modelled because of symmetry and steady state thermal analysis is carried out for un-finned, solid fins and perforated fins using ANSYS software. Results show that, more reduction in temperature on the surface of IM is observed in case of perforated fins having 11 perforations of 10 mm diameter, as compared to the other cases.

Keywords: perforated fins, induction motor fins, heat transfer, fin tip temperature.

1. Introduction

The Induction Motors are frequently employed in both commercial and industrial application. They are usually basic in design, tough in build, maintenance-free, and inexpensive. Because induction motors are used in industry for lengthy periods of time at full load, they might overheat, which is undesirable. The temperature of the different internal and exterior sections of the motor rises as a result of overheating, and if this condition persists, the Induction motor fails. Fins are connected externally on the exterior surface of the body attached to the housing of an induction motor to prevent overheating and for appropriate heat dissipation from an electric motor. Many scholars have previously completed much of the work on thermal analysis of induction motors, which is briefly summarized here. In [1] studied the cooling of electric motors and mentioned that the mechanical strength is lost and significant differential thermal expansion is caused by the high temperatures obtained during this brief period of quick and intense heating. In [2] authors studied experimentally the thermal aspect and distribution of airflow over the housing of an induction motor and proposed a modified casing which shows a significant (10 degree celcius) decrease in surface temperature as compared to the commercial motor. In [3] done the Computational fluid Dynamics (CFD) and thermal models for thermal management of various components of electric motor and investigated the effect of fin geometry, fin pitch and air flow velocity on cooling performance of an electric motor. In [4] this paper discussed the issues related to convection cooling from the surface of the electric machine, turbulent cooling along the end winding and heat transfer across the air gap. This [5] research described how response surface approach was used to optimize fin design by creating a Matlab program and optimization tool. Even with just 10% of the original fin thickness transferring heat to the surrounding air, the cooling performance was the same. This work focuses on the investigation of machine ventilation and more conductive heat transfer analysis utilizing the commercial software package Motor-CAD, which is specialized to motor cooling optimization [6]. In [7] investigated three alternative methods for thermal analysis of electrical machines: Computational Fluid Dynamics (CFD), Finite Element Analysis (FEA), and Lumped Parameter Thermal Network (LPTN). The effects of guide vanes on the heat transfer coefficient across the un-finned frame of an externally cooled induction motor were studied in [8] this paper, which introduced a new method to guide vanes for an un-finned electric motor. In [9],The thermal performance of a BLDC motor was analyzed using CFD, and the effect of altering the fin geometry (using three heat sink designs) on cooling performance was investigated. The results reveal that introducing a proposed finned hosing can drop temperature by up to 15%.

In [10], direct winding heat exchanger (DWHX) for cooling of electric motor has been developed; The suggested thermal model was created and tested experimentally. In this [11] study and the correctness of the proposed thermal model has errors of roughly 4% to 6%. This [12] study detailed the analysis of heat removal utilizing a thermal resistance network technique to determine the best design(water jacket liquid cooling) for an induction motor cooling system. Here in [13] studied air flow over a finned surface, heat transfer between the finned surface and the cooling air. Finally, by altering the present fin design and making minor alterations to the frame's outside, the electric motor's cooling capacity might be increased. In [14], a numerical study was conducted by simulating a three-dimensional, viscous, steady turbulent flow model and examining the heat sink geometry, and it was discovered that by decreasing rib spacing and increasing rib height for the same flow rate, the pressure drop and heat transfer increases significantly. Researchers optimized the electric motor frame by studying convection disturbances at the outer section of the electric motor. The results reveal that in the improved design of the frame, the convection area increases and the temperature in the optimized system was significantly lower than in the current system [15]. The temperature distribution of the cross section of the housing was analyzed using a thermo-vision camera to compare the results of a computational and experimental research of heat transmission in the finned housing of an electric motor. The thermo-vision camera proved to be an efficient experimental instrument for validating numerical models of heat transfer in an electric motor's finned casing [16].

In [17] authors investigated the thermal management system for electric motors employing L-shaped heat pipes in an experimental setting. The motor surface temperature was lowered by 33.8 degrees Celsius, according to the results. The finite element approximation approach was used to investigate conductive and convective heat transport through a finned casing of an electric motor in [18] this work. There is also need for more investigation into the types and size of fins mounted to the motor casing. For the optimization of cooling fins design of a casted housing, a 2D technique with conjugate heat transfer was explored and utilized; this approach will minimize the time it takes to assess a huge number of alternative designs [19]. The influence of core loss distribution, airflow speed distribution, and thermal radiation was evaluated in this [20] study, which compared multiple thermal modeling methodologies for Totally Enclosed Fan Cooled (TEFC) electric motors. The thermal performance of a TEFC electric motor was numerically simulated and studied in [21]. The Taguchi technique was utilized to decrease the computing work required to calculate fin size, fin pitch, and fan cover position. The results reveal that the fin pitch ratio has a considerable impact on the average temperature increase of the winding, with the new recommended design resulting in a 4.8 degree Celsius reduction in the average temperature rise of the winding. This [22] research looked at the performance of electric motors as well as heat and temperature losses. A comparative analysis of several types of fins and materials has been conducted in this [23] work. Fins are designed in Solid-

IOP Conf. Series: Materials Science and Engineering 1259 (2022) 012011 doi:10.1088/1757-899X/1259/1/012011

works, and heat transfer analysis is performed using ANSYS 16. Finally, the optimal design and material model is recommended. The thermal study of an induction motor was carried out in this [24] article using the finite element software packages motor CAD and ANSYS, with the total number and height of the fins being varied. The results demonstrate that utilizing 44 fins with a 15 mm fin length gave the best results and the lowest temperature at the fin tip in a specific situation.

The goal of this research is to provide the findings of a numerical simulation of thermal analysis of an induction motor housing with fins using ANSYS software. Unfinned surface, solid fins, and perforated fins are the three possible scenarios. The temperature distribution along the fins as well as the induction motor's surface was observed in four sub-cases of perforated fins. The results reveal that in comparison to traditional fins, the size of the perforation on the fins increases, resulting in a greater drop in temperature along the fin length. In the end, it shows that an induction motor's casing dissipates more heat.

2. Modelling of induction motor case

The induction motor is used in pumps, conveyors, winders, wind tunnels, furnaces and other industrial equipment. Table 1, which shows the specification and design information of 3-Phase, Totally Enclosed Fan Cooled (TEFC) AC Induction motor.

2.1. Geometry creation

ANSYS design modeler is used to model the casing with fins using the aforementioned data. Because of the symmetry of the motor and for simplicity of numerical simulation and analysis, just half of the full housing (ignoring the connecting box and base component) is modeled. Fig. 1 shows the actual photograph of Induction motor and Fig. 2 represents the simplified model of Induction motor.

2.2. Mesh generation

Structured mesh and unstructured mesh are the two most common mesh generating methods. For basic geometry, structured mesh is used, whereas for complicated geometry, unstructured mesh is used. Fig. 3 shows a meshed model with 15,895 mesh elements. An adequate fine mesh was created for analysis, and the influence of mesh quality on outcomes was also examined. The details are shown in Table 2. Because differing mesh element counts produce slight differences in findings, Table 2, Sr. No. 1 meshing method is used for all finned models for further analysis.

2.3. Boundary conditions

Heat is transferred from the core to the outer surface of an induction motor, where it is then convected. As a result, the temperature around the motor's core is quite high, and it will gradually drop to near the housing temperature. Around the core, the temperature is roughly 142°C(451K), while the temperature near the housing is around 130°C (403K). As a result, 130°C (403K) is applied to the inside surface of the housing [24]. The convective heat transfer between the motor case surface and the environment is assumed to be h = 20 W/m2 K, with a temperature of 28°C (301K) in the surrounding atmosphere. The motor body, which is built of cast iron, has a thermal conductivity of 53 W/m K.[24] Fig. 4 shows the applied boundary conditions.

TABLE 1. Induction Motor-design data

	Motor Specifications			*Fins Dimensions*	
Frame	90L	INS CL	B	Fin Length	164 mm
Frequency	50Hz	CONN	Star	Fin Height	14 mm
P.F.	0.723	Speed	1380 RPM	Fin Thickness	4 mm
Efficiency	76%	Voltage	415 V		
Power	2.0 HP/1.5 kW	Current	3.80 Amp	Total no. of Fins	24
Bearing	6205/6204	Weight	24.7 Kg		

Fig. 1 Induction Motor

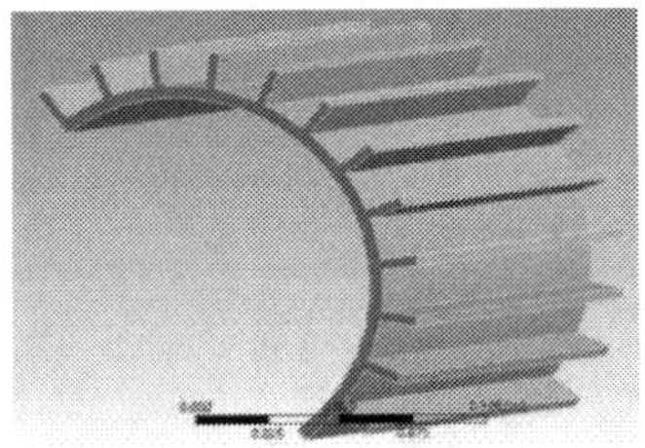

Fig. 2 Simplified model of Induction Motor housing

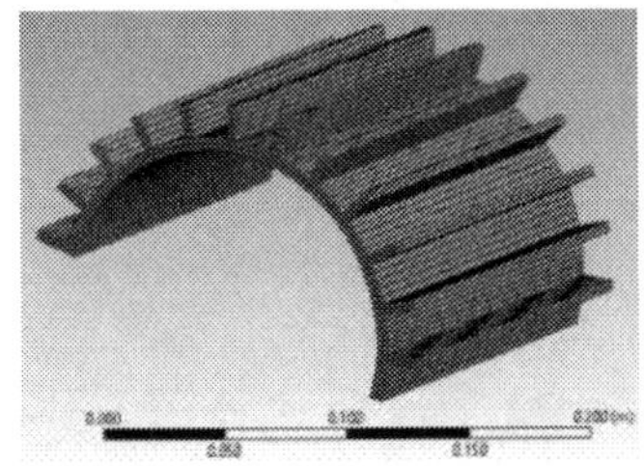

Fig. 3 Meshed Model

TABLE 2. Mesh Statistics

Sr. No.	*Number of Mesh Elements*	*Size of Element (m)*	*Max Temp (C)*	*Min Temp (C)*
1	15895	0.003	130	123.01
2	12079	0.035	130	123.02
3	10620	0.037	130	123.02
4	5617	0.004	130	123.03

3. Results and discussions

With the aid of ANSYS software, a one-dimensional steady state thermal study of the Induction motor's housing and fins on the exterior surface is performed. Only the first example was considered an unfinned housing out of a total of six cases analysed. The inside surface of the casing was kept at a constant temperature of 130°C (403K), while the rest of the outside surface, including the fins, was subjected to convection with a heat transfer coefficient of 20 W/m2K. All of the examples have been run to account for forced convection between the surface of the housing and the air surrounding it. Fig. 5 depicts the results of heat transmission from the inner surface to the outside surface, as well as convection at the induction motor's unfinned housing surface. The minimum temperature was 402.5K, with surface temperatures ranging from 402.72K to 402.78K, indicating a fairly little temperature variation. As previously stated, the total number of fins on the finned surface is 24, hence only half of the casing is used for analysis; 12 solid rectangular fins with 4 mm thickness and 14 mm height are connected. Heat is transferred from the inner surface of the housing to the outer surface, as well as from the base of the fin to the tip of the fin, and convection occurs between the outer surface and the air. The temperature distribution of solid finned housing is shown in Fig. 6, with a temperature difference of 4.01K between the maximum and minimum temperatures. The housing was next tested with 20 holes of 4 mm diameter on fins and the same boundary conditions were applied, resulting in a minimum temperature of 398.19 K near the tip of the fin (Fig. 7).

In this example, the perforation diameter was raised by 2 mm over the prior case, and 15 holes of 6 mm diameter were accommodated on the same fins. The outcome of this run is shown in Figure 8, where the minimum temperature at the fin tip was 397.5 K. The temperature difference is 5.5K. The perforation was enlarged to 8 mm for the following instance, which included a total of 13 holes on 8 mm diameter fins. The findings are shown in Fig. 9, where the highest temperature is 403K and the minimum temperature at the tip of the fins is 397.3K, a difference of 5.7K. Finally, the perforation size is set to 10 mm, and the maximum number of perforations is set at 11. As shown in Fig. 10, the minimum temperature at the tip of the fin is 396.58K, and the difference between the maximum and minimum temperatures is 6.4K, which is the greatest among the examples studied. Because the height of the fin of planned work is restricted to 14 mm, the perforation size cannot exceed 10 mm.The experimental investigation is being carried out in order to validate the results of the planned study.

To compare different geometry models, the fin tip temperature is used as an output. With reference to the Table 3, by increasing the size of the holes on the fins, the volume of the fin is reduced. As the perforation size increases from 4 mm to 10 mm, heat transfer improves while the temperature at the fin tip decreases.

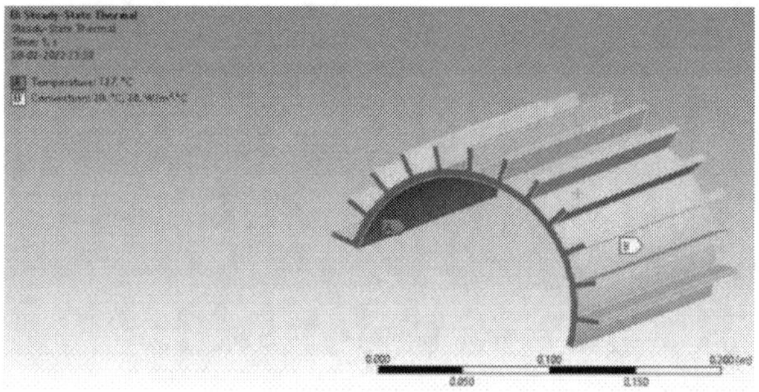

Fig.4 Boundary Conditions Applied to the Model

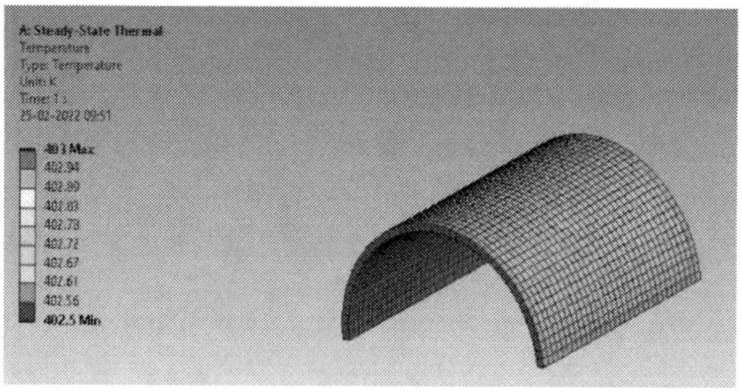

Fig.5 Unfinned Housing Surface of Induction Motor

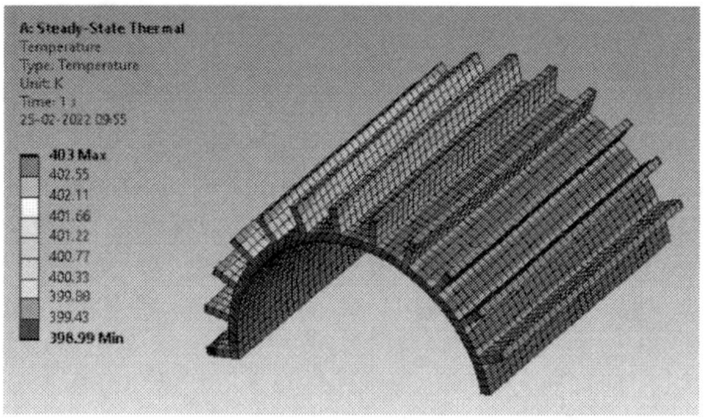

Fig. 6 Motor Housing with Solid Fins

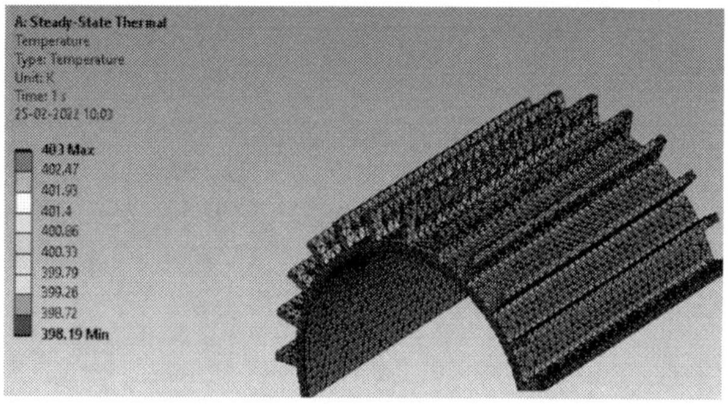

Fig.7 Fins of 20 perforations 4 mm diameter

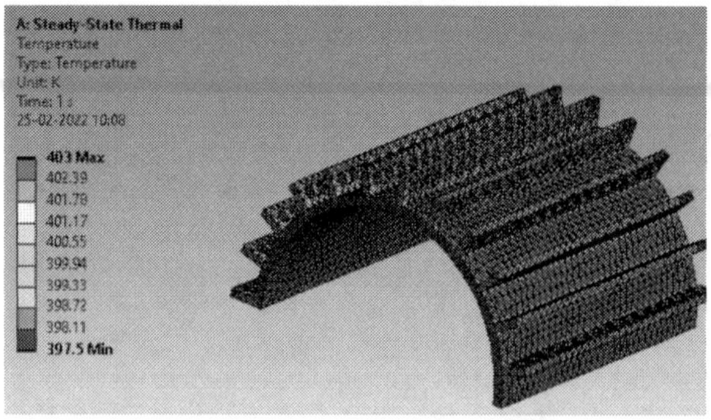

Fig. 8 Fins of 15 perforations 6 mm diameter

Fig. 9 Fins of 13 perforations 8 mm diameter

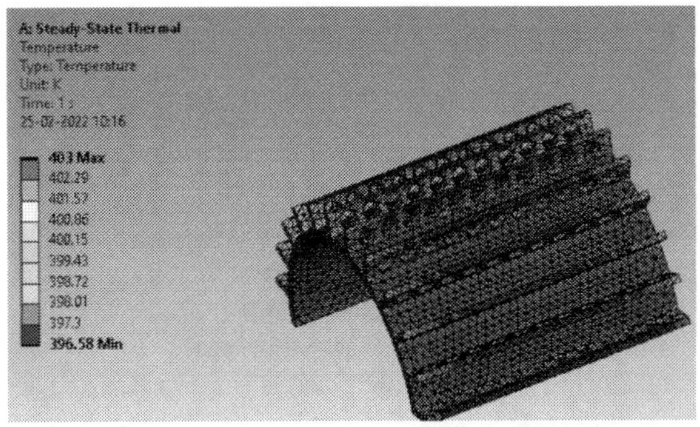

Fig. 10 Fins of 11 perforations 10 mm diameter

TABLE 3 All cases with results

Sr. No.	*Cases*	*Max. Temp.(K)*	*Min. Temp. (K)*	*Temp. difference (K)*
1	Unfinned Surface	403	402.5	0.5
2	Solid Fins (12 Fins)	403	398.99	4.01
3	20 Perforation of 4 mm diameter	403	398.19	4.81
4	15 Perforation of 6 mm diameter	403	397.5	5.5
5	13 Perforation of 8 mm diameter	403	397.3	5.7
6	11 Perforation of 10 mm diameter	403	396.6	6.4

4. Conclusions

The steady state thermal analysis of Induction motor's housing fins is performed using ANSYS software package. Conductive and convective heat dissipation from the motor surface to the atmosphere is investigated utilizing various designs such as un-finned surface housing, solid fins, and perforated fins housing. Perforation sizes ranged from 4 mm to 10 mm in diameter, with a 2 mm diameter increase. The temperature distribution over the surface of the induction motor is determined by the investigation. As a consequence of the findings, it can be concluded that increasing the size of perforation on the fins enhances heat transfer from the housing of an Induction motor. The numerical study of the presented instance indicates that the 11 number of holes of 10 mm diameter resulted in the highest temperature drop when compared to the other options.

The experimental research of heat transfer analysis on the housing of an induction motor may be performed by supplying alternative conceivable perforation on existing fins as a future scope of the suggested study. Fin material (steel/copper/aluminium) can also vary, and good material with best thermal design can offer optimal fins.

References

[1] Szogyen, John RM. "Cooling of electric motors." IEE Journal on Electric Power Applications 2, no. 2 (1979): 59-67.

[2] Farsane, K., P. Desevaux, and P. K. Panday. "Experimental study of the cooling of a closed type electric motor." Applied Thermal Engineering 20, no. 14 (2000): 1321-1334.

[3] Chen, Ya-Chi, Chung-Lung Chen, Qimin Dong, and R. W. Stephenson. "Thermal management for motor." In ITherm 2002. Eighth Intersociety Conference on Thermal and Thermomechanical Phenomena in Electronic Systems (Cat. No. 02CH37258), pp. 545-551. IEEE, 2002.

[4] Staton, Dave, Aldo Boglietti, and Andrea Cavagnino. "Solving the more difficult aspects of electric motor thermal analysis in small and medium size industrial induction motors." IEEE

Transactions on Energy conversion 20, no. 3 (2005): 620-628.
[5] Chen, Ya-Chi, Bing-Chung Chen, Chung-Lung Chen, and Jimmy Q. Dong. "CFD thermal analysis and optimization of motor cooling fin design." In Heat Transfer Summer Conference, vol. 47330, pp. 625-629. 2005.
[6] Staton, David A., and Andrea Cavagnino. "Convection heat transfer and flow calculations suitable for electric machines thermal models." IEEE transactions on industrial electronics 55, no. 10 (2008): 3509-3516.
[7] Boglietti, Aldo et al. "Evolution and modern approaches for thermal analysis of electrical machines." IEEE Transactions on industrial electronics 56, no. 3 (2009): 871-882.
[8] Jeon, C. S., M. K. Yoon, and S. Ken Kauh. "Heat transfer enhancement in the unfinned frame of an externally cooled induction motor." Heat transfer engineering 21, no. 1 (2000): 25-34.
[9] Kuria, James, and Pyung Hwang. "Investigation of thermal performance of electric vehicle BLDC motor." International Journal of Mechanical Engineering 1, no. 1 (2012): 1-17.
[10] Semidey, S. A. 2012. Thermal Design And Optimization Of High Torque density Electric Machines, PhD diss., Georgia Institute Of Technology.
[11] Badran, Omar, Hussain Sarhan, and Bilal Alomour. "Thermal performance analysis of induction motor." International Journal of Heat and Technology 30, no. 1 (2012): 75-88.
[12] Abdul Karim, et al. "Cooling system for electric motor of an electric vehicle propulsion." In Advanced Materials Research, vol. 903, pp. 209-214. Trans Tech Publications Ltd, 2014.
[13] Karlsson A. 2014. Cooling Methods for Electrical Machines. M.Tech. diss. Uppasala Univ.
[14] Almaghrabi, A. 2016. Forced Convection Cooling Of Electric Motors Using Enhanced Surfaces, M.Tech. diss. Univ. of Central Florida.
[15] Kulkarni, Pavankumar V., and Dr SR Basavaraddi. "Design optimization of Electric Motor Frame and its Study for Thermal Dynamic and Model Stability." International Research Journal of Engineering and Technology 3, no. 05 (2016): 2915-2923.
[16] Grabowski Mirosław et al. "Numerical simulation and experimental verification of heat transfer from a finned housing of an electric motor." Energy conversion and management 125 (2016): 91-96.
[17] Putra, Nandy, and Bambang Ariantara. "Electric motor thermal management system using L-shaped flat heat pipes." Applied Thermal Engineering 126 (2017): 1156-1163.
[18] Wernik, Jacek. "Investigation of heat loss from the finned housing of the electric motor of a vacuum pump." Applied Sciences 7, no. 12 (2017): 1214.
[19] Ulbrich, Stefan, Jeanette Kopte, and Jens Proske. "Cooling fin optimization on a TEFC electrical machine housing using a 2-D conjugate heat transfer model." IEEE Transactions on Industrial Electronics 65, no. 2 (2017): 1711-1718.
[20] Ronnberg, Kristian, and M. E. Beniakar. "Thermal modelling of totally enclosed fan cooled motors." In 2018 XIII International Conference on Electrical Machines (ICEM), pp. 2619-2625. IEEE, 2018.
[21] Peng, Huan-Sen, and Feng-Hsiang Lai. "Investigation of Parameters Affecting Heat Transfer and Fluid Flow of a TEFC Electric Motor by Using Taguchi Method." In IOP Conference Series: Materials Science and Engineering, vol. 491, no. 1, p. 012021. IOP Publishing, 2019.
[22] Gundabattini, Edison et al. "A review on methods of finding losses and cooling methods to increase efficiency of electric machines." Ain Shams Engineering Journal 12, no. 1 (2021): 497-505.
[23] Ramesh C. et al. "Design and Material Optimization of Cooling Fins in Electric Motors." Annals of the RSCB (2021): 1991-2004.
[24] Appadurai, M., E. Fantin Irudaya Raj, and K. Venkadeshwaran. "Finite element design and thermal analysis of an induction motor used for a hydraulic pumping system." Materials Today: Proceedings 45 (2021): 7100-7106.

Numerical Simulation of Nucleate Boiling Heat Transfer Performance using SiO_2/Water Nanofluid

Akshay Kumar Singh, Sudhir Kumar Singh, Deepak Sharma*

Mechanical Engineering Department, National Institute of Technology Hamirpur, Hamirpur 177005, H.P., India.

*Corresponding Author: dsharma@nith.ac.in

Abstract. Nucleate boiling regime is the most efficient regime that gives the maximum heat flux at the wall superheat range of 5 °C to 30 °C. In this current study, the nucleate pool boiling heat transfer performance of SiO_2/water nanofluid are studied numerically. The effect of volume fraction (0.01 – 0.06 vol. %) of nanofluid on heat transfer coefficient and wall superheat are investigated. The variation in heat transfer coefficient (HTC) & wall superheat with different heat flux are investigated. Eulerian multiphase approach with RPI boiling model is used to carry out the simulation. The pool boiling heat transfer performance increased with rise in volume concentration of nanofluid up to an optimum concentration of 0.05 % and then it starts to decrease. A reduction in wall superheat of 6.7 %, 8.43 %, 11.27 % and 9.3 % are obtained for SiO_2/water nanofluid with volumetric concentration of 0.01, 0.02, 0.05 and 0.06 vol. % respectively as compared to base fluid. Heat transfer coefficient enhanced by 3.55 %, 7.2 %, 23.87 % and 19.67 % for SiO_2/water nanofluid with volumetric concentration of 0.01, 0.02, 0.05 and 0.06 vol. % respectively as compared to base fluid.

Keywords. Nanofluids, Nucleate boiling, Heat transfer, Eulerian model, CFD

Nomenclature

Symbols		*Greek symbol*	
C_p	Specific heat [J/kg-K]	ϕ	Volume fraction
h	Heat transfer coefficient [kW/m²-K]	ρ	Density [kg/m³]
K	Thermal conductivity [kW/m-K]	μ	Dynamic viscosity [Pa-s]
Nu	Nusselt number		
P	Pressure [Pa]	*Subscript*	
Re	Reynolds number		
T	Temperature [K]	p	Nanoparticles
ΔT	Wall superheat [K]	f	Base fluid
q''	Heat flux [kW/m²]	nf	Nanofluid
Abbreviations		sat	Saturation
CFD	Computational fluid dynamics	w	Wall
HTC	Heat transfer coefficient	atm	atmospheric
CHF	Critical heat flux		
PBHT	Pool boiling heat transfer		

1. Introduction

Boiling is a phase transition from a liquid to a vapour state and occurs at the solid-liquid interface when heat is supplied to the heater surface above the saturation temperature. The application of pool boiling process are cooling purpose as well as for power generation in several fields such as industrial application, engineering application, medical fields etc. In pool boiling, nucleate boiling regime is the most-efficient regime that gives the maximum heat flux (also called as CHF) at the wall superheat range of 5 °C to 30 °C, which was discussed by Marie et al. [1]. Yuan et al. [2] described the heat transfer behaviour. The heat transfer can be increased by modification in boiling surface material or enhancing the thermo-physical properties of working fluid which further enhances the heating surface properties such as wettability, nucleation site density, roughness and porosity. Singh and Sharma [3] reviewed the pool boiling and concluded that nanofluids have the novel properties that make them conceivably huge in various applications in heat transfer. Various researchers are working in the field of heat transfer enhancement. Nanofluids has a unique acoustical property.

Das et al. [4] performed a simulation on pool boiling for silicon oxide coated heating surface using water as a base fluid. Electron beam physical vapor deposition approach has been used for the coating of silicon oxide nanoparticles on the heating surface. Maximum 58 % of enhancement in HTC has been found for the silicon oxide coated heating surface at higher thickness of coating. After three run of experimentation, small variation has been observed in the HTC. Gupta et al. [5] analysed heat transfer characteristics for water as working fluid on the copper boiling surface coated with Cu-Al_2O_3 nanocomposite material. In case of coated boiling surface, maximum HTC and CHF has been found equal to 199 kW/m^2-K, 1852 kW/m^2 respectively. These results are 273% and 72.5% respectively higher than the results of non-coated surface. Kamel et al. [6] examined the numerical simulation of nucleate PBHT inside 2-D rectangular boiling vessel using silica-water based nanofluid. This investigation used the heat flux partitioning model by improving the surface characteristics parameters like nucleation site, porosity, surface wettability then it gives better accurate and high results. Manetti et al. [7] performed experiment on the pool boiling for distilled water and Al_2O_3-water nanofluid at the low and high concentrations of nanoparticles 0.0007 vol. % and 0.007 vol. %. A boiling surface of copper and roughness values of 0.05 μm for smooth surface and 0.23 μm for rough surface was taken in their investigation. This HTC was increased for both smooth and rough surfaces up to 74 % and 16 % respectively in comparison to distilled water.

Yu et al. [8] focused on the preparation techniques of nanofluid, stability factors and their applications in several fields. They demonstrated the mechanism of nanofluid's stability and other parameters involved in the preparation of nanofluid. Hama et al. [9] experimentally analysed the nucleate PBHT characteristics for Al_2O_3-water nanofluid. The variation of different volumetric concentration of nanofluid, nanoparticle size and contact angle were investigated. The maximum CHF of 1515 kW/m^2 was obtained at 0.025 vol. % concentration and 50 nm size of Al_2O_3 nanoparticles. Sedaghatkish et al. [10] numerically studied the film boiling on a magnetic heated cylindrical boiling surface. Due to increase in magnetic force on phase, interface vapor layer thickness at the heated surface decreased and Nusselt number increased. Kang et al. [11] analysed the combined variation of heat flux and inclination angle of stainless-steel heated tubes of 19 mm dia. & 400 mm in length on PBHT. A reduction in inclination angle and upsurge in heat flux of inner tube enhanced the boiling performance of outer tube. This was due to the liquid stimulation and convective flow of fluid on outer tube caused by inner tube. Pourfattah et al. [12] studied the boiling process of CuO-water nanofluid. The Nu increased with increase in Re due to reduction in vapor phase. Kangude and Srivastava [13] investigated the performance of silica nanofluid on ITO-coated boiling surface. They concluded that, an increase in concentration of nanofluid enhances the bubble growth and bubble departure frequency, but also decrease in bubble departure diameter. Shen et al. [14] performed investigation on influence of ultrasonic waves during the pool boiling. Under the subcooled boiling conditions, the heat transfer rate of nanofluid enhanced with ultrasonic technique while for saturated boiling condition ultrasonic was not much effective at high heat flux conditions. Gobinath et al. [15] performed the thermophoresis in the nanofluid to solve the mathematical model conditions. Thermophoretic parameters like thermophoretic mobility and thermophoretic velocity was observed at different temperature value of boiling surface. Guo et al. [16] studied the magnetic field condition for flow boiling of nanofluid. VOSET interface technique was utilized to observe the movement of liquid-vapor phase interaction. The film boiling convective heat transfer increased with volumetric concentration. Dadhich and Prajapati [17] reviewed the parameter affecting the heat transfer characteristics of pool boiling. HTC and CHF can be improved by increase in surface roughness and bubble growth and their departure. Gupta et al. [18] developed the simulation of pool boiling of nanofluid using Eulerian multiphase technique.

In this current study, nucleate boiling heat transfer characteristics of SiO_2/water nanofluid are studied numerically. The variation of volume fraction of nanofluid on heat transfer coefficient & wall superheat are investigated. The change in heat transfer coefficient and incipience of degree of superheat with different heat flux are investigated.

2. Description of Computational Domain

A 2D rectangular geometrical domain is selected for computational study. A two-dimensional representation of computational domain for present simulation can be seen from Fig. 1. A three-dimensional representation of computational domain is shown in Fig. 2. A stainless-steel heater surface of 40 mm diameter is mounted centrally at the bottom of boiling vessel. The dimension of boiling vessel is 300 mm $\times$150 mm. A Eulerian multiphase approach is considered to carry out the simulation of two-dimensional geometrical model with RPI heat partitioning boiling model. For pressure-velocity coupling, phase coupled SIMPLE scheme and second order upwind scheme has been considered. Gradient for all flow conditions has been assumed to solve with least-square cell method. In case of vapor fraction, modified HRIC scheme has been considered instead of upwind scheme. Convergence criteria for the solution of all flow equations are assumed up to 10^{-6}.

Pressure based solver technique is employed in the current simulation and various forces like virtual mass, wall lubrication, lift and drag force are introduced at the liquid-vapor interface. Ranz-Marshall model is opted to obtain the HTC between liquid-vapor phase interactions. Finite volume method is used to carry out the simulation to discretise the computational geometry. In RPI boiling model, total heat flux from the boiling surface is categorized in three form which are the heat flux due to evaporation, the heat flux due to quenching and heat flux due to natural convection. Laminar and turbulent model are introduced for the viscous model selection and κ - ε (two equation) model is considered for optimistic results outcomes. In this turbulent flow simulation, effect of gravity is considered for the lift force, which is interacted between liquid and vapor phase.

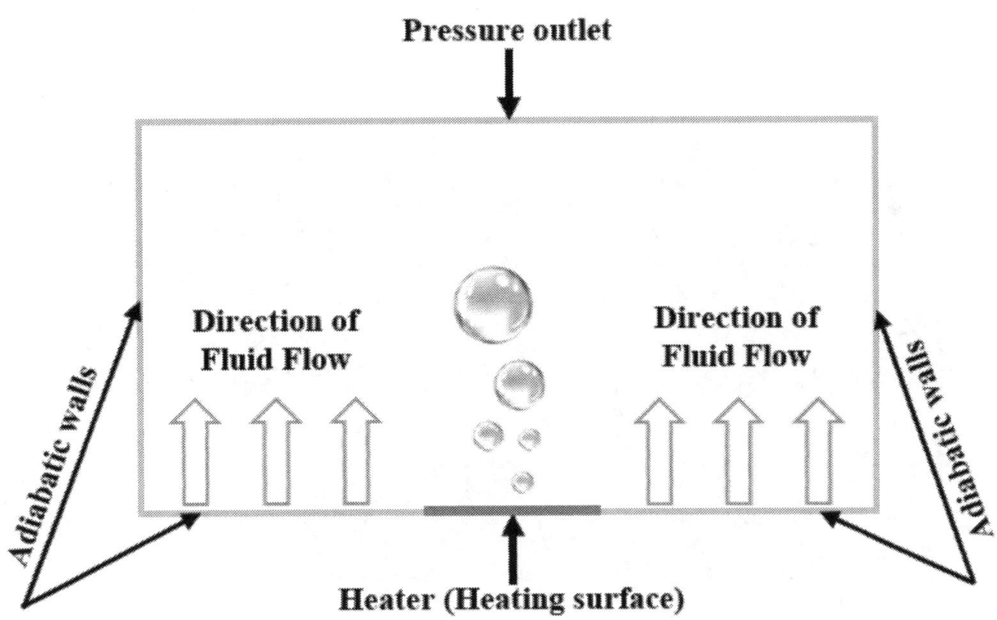

Figure 1. Computational Domain in 2D form

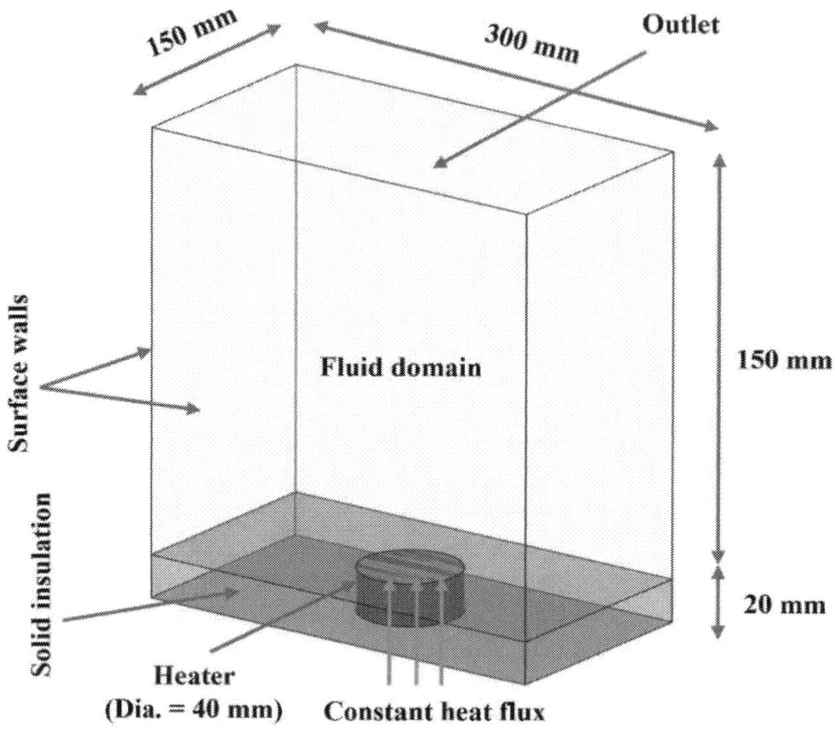

Figure 2. Computational Domain in 3D form

2.1 Mesh generation
In this simulation, a structured uniform mesh is created for two-dimensional geometrical model with the help of ANSYS fluent tool. Fig. 3 shows the mesh generation of computational domain. At the bottom portion of boiling vessel, a fine mesh is considered at the heater surface to obtain optimistic results.

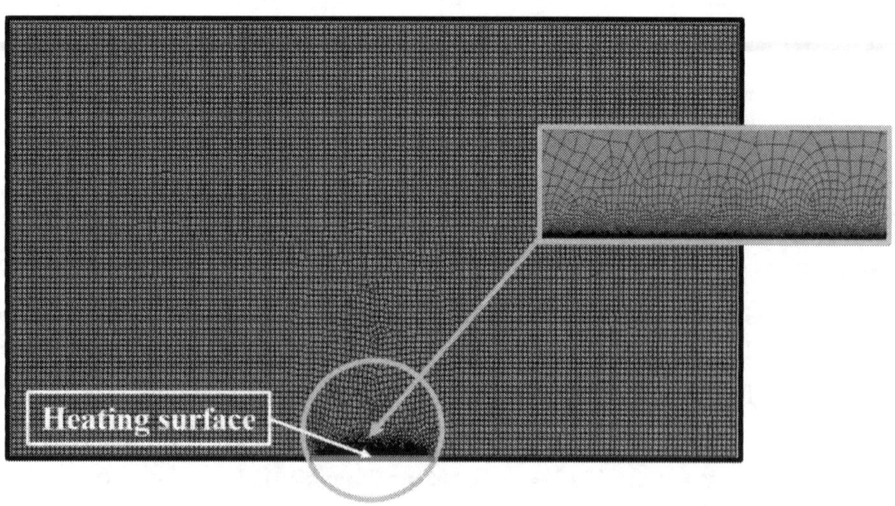

Figure 3. Mesh generation of computation domain

2.2 Grid Independent Test

Grid independence test is the most important step of any computational simulation to check the mesh sensitivity. In the present simulation, different element size of heater surface and mesh size of computational domain are considered. Table 1 shows the variation in HTC with different mesh size of heater surface. Total mesh size of computational domain increased with decrease in mesh size of heater surface at a constant heat flux of 300 kW/m^2. The variation in HTC becomes very less after the mesh size of 25456 which indicate towards the accuracy of observations. Therefore, 25456 as mesh size are considered for further simulation process and calculations.

Table 1. Variation in Heat Transfer Coefficient with different mesh size of heater surface

S. No.	Mesh size (Number of elements)	Heater element size (mm)	Heat transfer coefficient (kW/m^2-K)
1	14825	0.07	16.4562
2	16256	0.06	16.8656
3	19236	0.05	16.9961
4	25456	0.04	17.0135
5	30164	0.03	17.0269
6	37566	0.02	17.1986

2.3 Boundary conditions

In the present simulation, transient and turbulent fluid flow condition are considered while thermos-physical properties of working fluid are assumed constant at operating temperature and pressure conditions. For calculating the solutions,1 m-s time step size is considered. The boundary conditions for present simulation are as follows.

- A constant heat flux with constant temperature has been considered at the heating surface.
$$q'' = Constant \ , \ T = T_w$$
- Surface walls has been assumed adiabatic at constant saturation temperature.
$$q'' = 0 \ , \ T = T_{sat}$$
- No-slip condition at the solid-fluid interface.
$$u = 0$$
- Top portion of the boiling vessel has been assumed at atmospheric pressure conditions.
$$P = P_{atm}$$

2.4 Governing Equation

The governing equations such as continuity, momentum & energy for present simulation work are as follows [19].

$$\frac{\partial u}{\partial x} + \frac{\partial v}{\partial y} = 0 \tag{1}$$

$$\frac{\partial}{\partial t}(\alpha.\rho.u_i) + \frac{\partial}{\partial x_i}(\alpha.\rho.u_i.u_j) = -\alpha\frac{\partial P}{\partial x_i} + \frac{\partial}{\partial x_j}\left(\alpha.\rho.v.\frac{\partial u_i}{\partial x_j}\right)\alpha.\rho.g_i \tag{2}$$

$$\frac{\partial}{\partial t}(\alpha.\rho.T) + \frac{\partial}{\partial x_i}(\alpha.\rho.u_i.T) = \frac{\partial}{\partial x_i}\left(\frac{K}{C_p}\frac{\partial T}{\partial x_i}\right) \tag{3}$$

IOP Conf. Series: Materials Science and Engineering 1259 (2022) 012012 doi:10.1088/1757-899X/1259/1/012012

3. Properties of Nanofluid

The correlation used to obtain the thermo-physical properties of SiO_2/water nanofluid at varying volume fraction of 0.01%, 0.02%, 0.05% and 0.06% are given below [20]. The Thermo-Physical properties of nanofluid can be seen from Table 2.

$$\rho_{\text{nf}} = \phi\rho_{np} + (1-\phi)\rho_w \tag{4}$$

$$C_{p,nf} = \frac{\phi.\rho_{np}C_{p,np} + (1-\phi)\rho_w C_{p,w}}{\phi\rho_{np} + (1-\phi)\rho_w} \tag{5}$$

$$K_{nf} = \left(\frac{2K_w + K_{np} - 2\phi(K_w - K_{np})}{2K_w + K_{np} + \phi(K_w - K_{np})}\right)K_w \tag{6}$$

$$\mu_{nf} = \frac{\mu_f}{(1-\varphi)^{-2.3}} \tag{7}$$

Table 2. Properties of nanofluid at varying concentration

S. No.	Fluid	ρ (kg/m^3)	C_p (J/kg-K)	K (W/m-K)	μ (N-s/m^2)
1	Water	958.35	4215.72	0.67909	0.000281
2	Vapor	0.59817	2080.00	0.02509	0.000012
3	Silica/water (φ = 0.01%)	970.96	4136.34	0.68622	0.000274
4	Silica/water (φ = 0.02%)	983.58	4059.02	0.69343	0.000268
5	Silica/water (φ = 0.05%)	1021.43	3838.53	0.71554	0.000249
6	Silica/water (φ = 0.06%)	1034.04	3768.62	0.72306	0.000243

4. Results and discussion

4.1. Effect of nanofluid's concentration on wall superheat

In the pool boiling case, a reduction in wall superheat for a particular heat flux is advantageous. Fig. 4 shows the variation in wall superheat against heat flux for various concentration of nanofluid (0.01, 0.02, 0.05 and 0.06 vol. %). A reduction in wall superheat with increase in volumetric concentration of nanofluid are observed and attains minimum value at 0.05 vol. % concentration of SiO_2/water nanofluid and increased after an optimum concentration of nanofluid. A reduction in wall superheat of 6.7 %, 8.43 %, 11.27 % and 9.3 % are obtained for SiO_2/water nanofluid with volumetric fraction of 0.01, 0.02, 0.05 and 0.06 vol. % respectively as related to water. Heat transfer performance depends upon the heating surface properties such as roughness, wettability, porosity, and nucleation site density. Increase in concentration of nanofluid enhances the boiling surface properties which resulted in enhanced heat transfer performance of Nucleate pool boiling. As wall superheat is the difference in heater surface temperature and saturation temperature of working fluid. Since saturation temperature of working fluid will be same throughout (100°C for water). Therefore, reduction in wall temperature implies that, reduced temperature of heater surface for the given heat flux. A reduction in wall superheat is advantageous in the case of pool boiling which results in high heat transfer rate at relatively lower heat flux. In the present case, the application of nanofluid resulted in reduction in wall superheat. The present research work is validated with Kamel et al. [6] at similar operating condition and same volumetric concentration of SiO_2/water nanofluid. Fig. 5 shows the variation in wall superheat against heat flux for

IOP Conf. Series: Materials Science and Engineering 1259 (2022) 012012 doi:10.1088/1757-899X/1259/1/012012

SiO_2/water nanofluid at 0.01 vol. % concentration. Validation results shows a worth agreement between each other with negligible error.

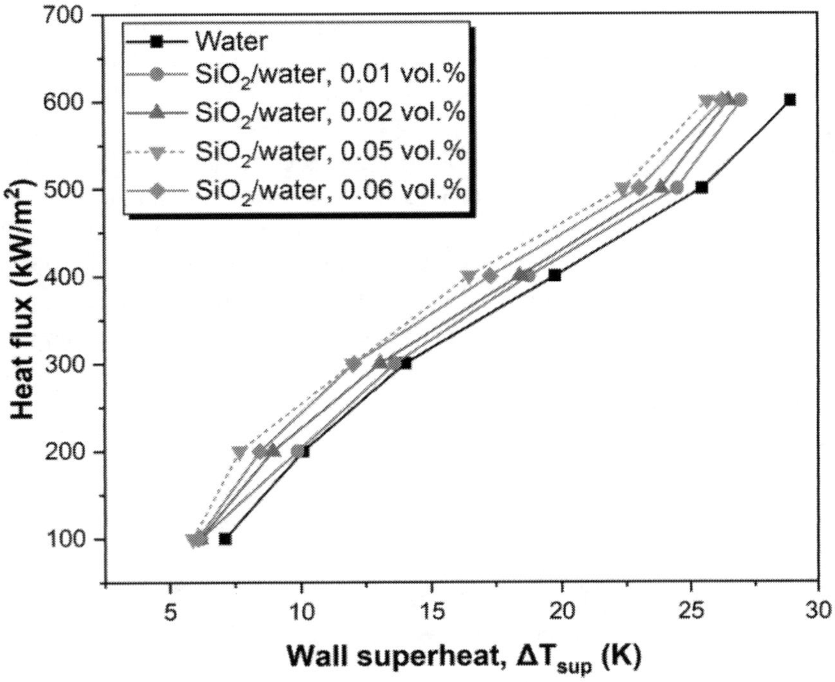

Figure 4. Variation in wall superheat with different concentration of SiO_2/water nanofluid

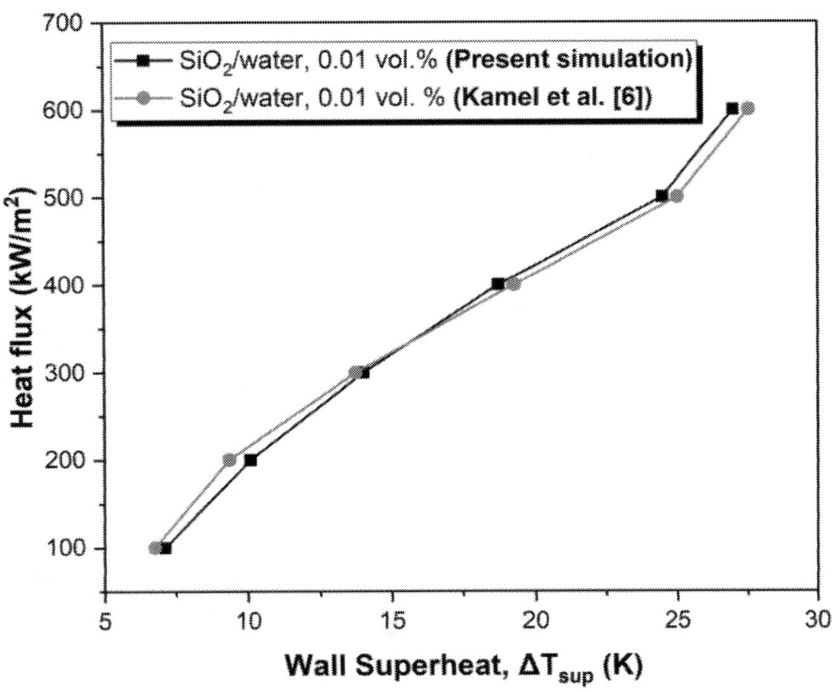

Figure 5. Validation of present simulation work with Kamel et al. [6]

4.2. Effect of nanofluid concentration on convective heat transfer coefficient

HTC depends on the different parameters of fluid flow. As we all know, the heat transfer rate is directly proportional to the heat transfer coefficient. An increase in HTC is advantageous for pool boiling. An increase in HTC increased the heat transfer rate. The cooling of high heat flux producing equipment's are operated under Nucleate boiling region of pool boiling curve. The onset and departure of nucleate boiling occurs in this region only. In Nucleate boiling region, HTC upsurges and becomes maximum (CHF) and then reduces in transition region of pool boiling curve. Fig. 6 shows the change in HTC against heat flux at different concentrations of SiO_2/water nanofluid. Nanofluid enhanced the PBHT performance as related to water. The HTC enhanced with rise in volume fraction of nanofluid up to an optimal concentration of 0.05 vol. %, after that it decreased. Heat transfer coefficient enhanced by 3.55 %, 7.2 %, 23.87 % and 19.67 % for SiO_2/water nanofluid with volumetric fraction of 0.01, 0.02, 0.05 and 0.06 vol. % respectively as related to the base fluid. In present research, the pool boiling of SiO_2/water nanofluid showed enhanced heat transfer rate, HTC, reduced wall superheat. The optimum volume fraction of 0.05 % for both wall superheat and heat transfer coefficient were achieved, and heat transfer reduced after this optimum volume fraction.

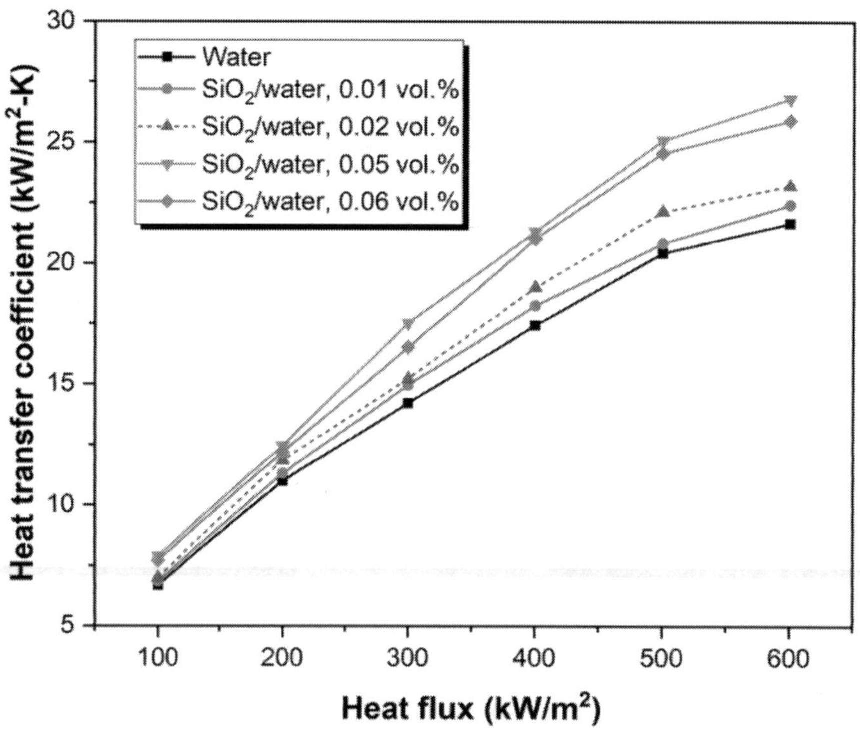

Figure 6. Deviation in HTC for varying volume fraction of SiO_2/water nanofluid

5. Conclusions

The nucleate pool boiling heat transfer performance of SiO_2/water nanofluid at varying volume concentration of 0.01 – 0.06 vol. % are studied numerically. A Eulerian multiphase approach with RPI boiling model is used to carry out the simulation. The following conclusion based the obtained results are as follows.

- Wall superheat decreased and convective heat transfer coefficient (HTC) increased with use of SiO_2/water nanofluid.

- Nucleate boiling heat transfer performance improved with rise in volume concentration of nanofluid up to an optimum concentration of 0.05 % and then it starts to decrease.
- A reduction in wall superheat (ΔT_{excess}) & enhancement in heat transfer coefficient (HTC) are obtained with rise in volumetric concentration of nanofluid.
- A drop in wall superheat of 6.7 %, 8.43 %, 11.27 % and 9.3 % are obtained for SiO_2/water nanofluid with volumetric concentration of 0.01, 0.02, 0.05 and 0.06 vol. % respectively as related to water.
- Heat transfer coefficient enhanced by 3.55 %, 7.2 %, 23.87 % and 19.67 % for SiO_2/water nanofluid with volumetric concentration of 0.01, 0.02, 0.05 and 0.06 vol. % respectively as compared to base fluid.

6. References

[1] Marie, A., Cioulachtjian, S., Lips, S. and Sartre, V., 2022. Thermal interactions between nucleation sites and the solid wall during pool boiling of a pure fluid: A review. *International Journal of Thermal Sciences*, *174*, p.107388.

[2] Yuan, X., Du, Y. and Su, J., 2022. Approaches and potentials for pool boiling enhancement with superhigh heat flux on responsive smart surfaces: A critical review. *Renewable and Sustainable Energy Reviews*, *156*, p.111974.

[3] Singh, S.K. and Sharma, D., 2021. Review of pool and flow boiling heat transfer enhancement through surface modification. *International Journal of Heat and Mass Transfer*, *181*, p.122020.

[4] Das, S., Kumar, D.S. and Bhaumik, S., 2016. Experimental study of nucleate pool boiling heat transfer of water on silicon oxide nanoparticle coated copper heating surface. *Applied Thermal Engineering*, *96*, pp.555-567.

[5] Gupta, S.K. and Misra, R.D., 2018. Experimental study of pool boiling heat transfer on copper surfaces with Cu-Al2O3 nanocomposite coatings. *International Communications in Heat and Mass Transfer*, *97*, pp.47-55.

[6] Kamel, M.S., Al-agha, M.S., Lezsovits, F. and Mahian, O., 2020. Simulation of pool boiling of nanofluids by using Eulerian multiphase model. *Journal of Thermal Analysis and Calorimetry*, *142*(1), pp.493-505.

[7] Manetti, L.L., Stephen, M.T., Beck, P.A. and Cardoso, E.M., 2017. Evaluation of the heat transfer enhancement during pool boiling using low concentrations of Al2O3-water based nanofluid. *Experimental Thermal and Fluid Science*, *87*, pp.191-200.

[8] Yu, W. and Xie, H., 2012. A review on nanofluids: preparation, stability mechanisms, and applications. *Journal of nanomaterials*, *2012*.

[9] Ham, J. and Cho, H., 2016. Theoretical analysis of pool boiling characteristics of Al2O3 nanofluid according to volume concentration and nanoparticle size. *Applied Thermal Engineering*, *108*, pp.158-171.

[10] Sedaghatkish, A., Sadeghiseraji, J. and Jamalabadi, M.Y.A., 2020. Numerical simulation of magnetic nanofluid (MNF) film boiling on cylindrical heated magnet using phase field method. *International Journal of Heat and Mass Transfer*, *152*, p.119546.

[11] Kang, M.G., 2016. Pool boiling heat transfer from an inclined tube bundle. *International Journal of Heat and Mass Transfer*, *101*, pp.445-451.

[12] Pourfattah, F., Yousefi, S., Akbari, O.A., Adhampour, M., Toghraie, D. and Hekmatifar, M., 2019. Numerical simulation of the effect of using nanofluid in phase change process of cooling fluid. *International Journal of Numerical Methods for Heat & Fluid Flow*.

[13] Kangude, P. and Srivastava, A., 2019. Performance of SiO2-water nanofluids for single bubble-based nucleate pool boiling heat transfer. *International Journal of Thermal Sciences*, *138*, pp.612-625.

[14] Shen, G., Ma, L., Zhang, S., Zhang, S. and An, L., 2019. Effect of ultrasonic waves on heat transfer in Al2O3 nanofluid under natural convection and pool boiling. *International Journal of Heat and Mass Transfer*, *138*, pp.516-523.

[15] Gobinath, N., Venugopal, T., Palani, K. and Samuel, A.A., 2018. Numerical modelling of thermophoresis in water-alumina nanofluid under pool boiling conditions. *International Journal of Thermal Sciences*, *129*, pp.1-13.

[16] Guo, K., Li, H., Feng, Y., Wang, T. and Zhao, J., 2019. Numerical simulation of magnetic nanofluid (MNF) film boiling using the VOSET method in presence of a uniform magnetic field. *International Journal of Heat and Mass Transfer*, *134*, pp.17-29.

[17] Dadhich, M. and Prajapati, O.S., 2019. A brief review on factors affecting flow and pool boiling. *Renewable and Sustainable Energy Reviews*, *112*, pp.607-625.

[18] Gupta, R.R., Bhambi, S. and Agarwal, V.K., 2019. CFD modeling for nucleate pool boiling of nanofluids. *Numerical Heat Transfer, Part A: Applications*, *75*(6), pp.402-412.

[19] Wang, J., Diao, M. and Liu, X., 2019. Numerical simulation of pool boiling with special heated surfaces. *International Journal of Heat and Mass Transfer*, *130*, pp.460-468.

[20] Mortezazadeh, R., Aminfar, H. and Mohammadpourfard, M., 2017. Eulerian–Eulerian simulation of non-uniform magnetic field effects on the ferrofluid nucleate pool boiling. *Journal of Engineering Thermophysics*, *26*(4), pp.580-597.

Optimization and fabrication of Load carrying device design using ANSYS DO Module

Mr. Ansh Gangwar[1], Mr. Sambhrant Srivastava[2], Mr.Adesh Kumar[3] , Mr. Md Umar[4]

[1,3,4]UG Student, Mechanical Engineering Department, Rajkiya Engineering College Azamgarh
[2]Assistant Professor, Mechanical Engineering Department, Rajkiya Engineering College

Abstract. Simulation technique is one of the advanced techniques in Finite Element Method. It is used as one of the tools for designing and analysis of any structure. Load carrying device is important for labors in construction works, cargo carrying etc. because the most efficient and cheap design can help them from hazardous and poisonous environment. Different load carrying device are available but their optimization regarding the load is still a challenge. In this work we have taken reference of female labors carrying load on their head like, bricks, cement, concrete on construction sites. To achieve this, a parametric design is done in ANSYS software and optimize the shape and mass of load carrying device using DO module. Then using guadua bamboo a load carrying device is made and its weight is measured. Load carrying device suitability is checked by carrying bricks on the construction site.

Keywords- Load Carrying Device, ANSYS DOModule, Bamboo, Female Labour.

1. Introduction

Safety is real issue of laborers on construction sites or loading and unloading sites of cargo. No one shows any interest for the safety to the poor people who carry the load of 30-40 kg on their head or on the back. The injuries due to heavy load was mostly seen in the areas, where labors are forced to carry heavy loads [1]. India is one of the developing countries, where many women labors get injured on the construction sites while performing the tasks of material handling. Labors on construction sites usually does the lifting of material from the ground to the specified area, loading carrying from one place to other [2]. For females, National Institute for Occupational Safety and Health,1991 recommended maximum load carrying limit to be 23 kg. However International Labor Organization (ILO 1988) reported safe load for women's to be 15 kg in Austria and 25 kg in Pakistan. India has its own safety rules for manual lifting operations of labors. In the Factory Act, 1948 each state of India has to has to make rules regarding minimum weight carrying of labors which will also include no harm or injury to the person. For precaution of female members from injury while carrying the load, a safe load of up to 30 kg is determined as limit for a female labor by Maharashtra Factory Rules, 1975 [3]. Occupational sector of India is very much prone to accidents, as they have to carry loads of about 100kg. Almost 20% of the reported accidents discontinue the jobs due to physical inability [4]. Many construction sites could be observed with the people having blood, cuts etc. On their body while performing the task. Many researchers have worked to design and develop a load carrying device, which will make the task easy and convenient for the people. A golf backpack is developed for carrying the load on the back, which is made up of polyethylene material without any pin joints [5]. Vikram Dinubhai Panchal of The National Institute of Design (NID), developed a model made up of cane with plastic fitting to carry the load on construction sites of just 2 kg (Figure 1).

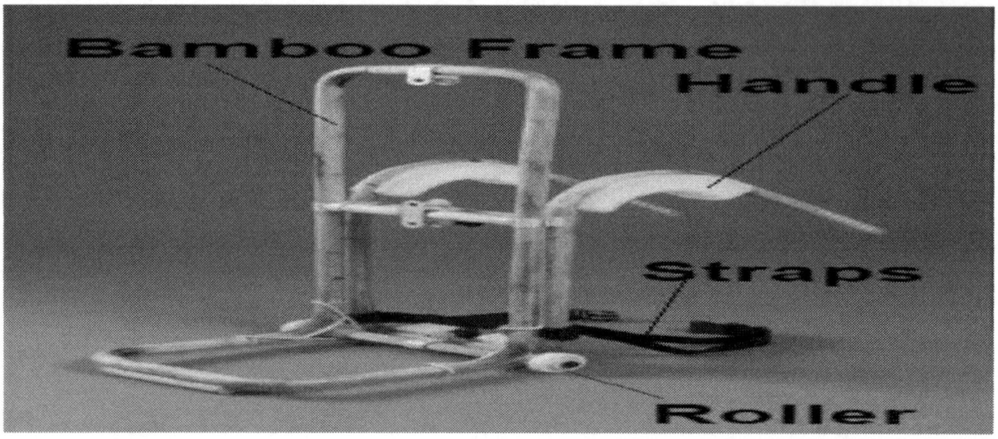

Figure 1. Description of Load carrier device for Load capacity of 2kg.

To develop a load carrier device for Indian female Labors who carry masonry loads like, cement, bricks etc. (Figure 2) on construction sites. A load carrier device with minimum weight and can carry maximum load without comfortability is the challenge. In this work we will design a frame of load carrier device using SOLIDWORKS 2019 software and a Finite Element Analysis is done in ANSYS 2021. Then by integrating with approximate model and numerical optimization algorithm for creating an automated design tool. The shape of design of load carrier device is formulated in the form of an optimization problem which is solved by numerical optimization algorithm. The constraint and objective functions are approximated by static structural load FEM analysis and get replaced by the approximations and an optimum design is obtained after the optimization problem is solved.

Figure 2. Female labor's carrying concrete bricks on their head

To carry maximum load of 30 kg, present study is focused to optimize the size and shape of a light weight, load carrying device for 5 feet to 6 feet height Labors using design module of ANSYS.

2. Materials and Methodology

Formulating a minimization problem to optimize the shape of load carrying device [6].

To Minimize: $Z_o(x)$ (1)

Subjected to:

$Zj(x) \leq 0$ (j=1,2.......nc) (2)

IOP Conf. Series: Materials Science and Engineering 1259 (2022) 012013 doi:10.1088/1757-899X/1259/1/012013

With the design space of:

$y_{i1} \leq y_i \geq y_{iu}$ (i=1,2.....N) (3)

Were Zo(x) is the objective function and Zj(x) are the constraint function and y = [y1,y2, . . .,yN] is the vector of design variables. y_{il} and y_{iu} describe the physical upper and lower bounds on design variables. nc and N are the number of constraints and number of design variables, respectively. The constraint and objective functions may correspond to weight, length, diameter of the device. Solution from equation (1-3) for optimization of dimension of load carrying device is done by replacing the objective function and constraint function with their Response Surface (RS) approximations. This approximation optimization method is implemented in ANSYS Design Optimization (DO) Module, the process used for this is shown in Fig.3. DO module generates a polynomial Of Response surface approximation for objective function and constraint function of load carrying device [7].

$$\acute{Z}(x)=a_o+\sum^{N}{}_{n} a_n y_{n,linear}+\sum^{N}{}_{n=1} b_n y_n{}^{2}{}_{,quadratic}+\sum^{N}{}_{m}{}^{=}{}_{=}{}^{1}{}_{1}\sum^{N}{}_{n=m+1} C_{mn} y_m y_n \quad (4)$$

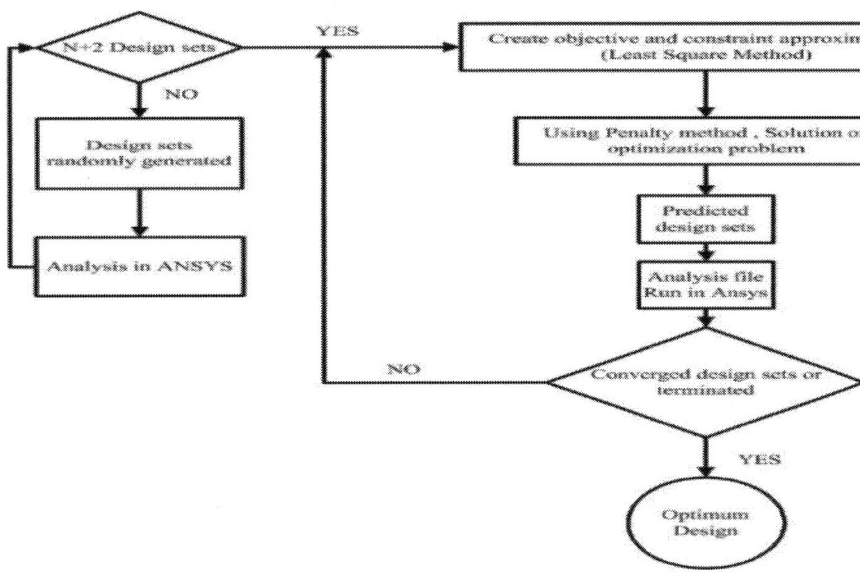

Figure 3. Approximation Design process in Design optimization module of ANSYS 2021

In this optimization process, DO module firstly create N+2 design sets, which can be randomly generated or from the existing database. Dimensions of load carrying device is optimized and so the shape from the available design sets. Analysis from FEM are used to create an approximation model of the objective and constraint functions. When there are enough design sets available RS approximation is created using Least Square Method. Based on penalty function an optimum design of load carrying device is predicted. Thus, the predicted design is verified by FEM analysis result of ANSYS and if the results of objective and constraint function matched from the ANSYS an optimum design is obtained otherwise a new approximation are created using Equation (1) – (3) and the loop continues.

2.1. Parametric and Geometric Modelling

Geometric shapes have big influence to carry load by labours. To consider the space above the head and holding the carrier by their hands a load carrier device is modelled in SOLIDWORKS 2019 and parametrized using ANSYS parametric design Language (APDL) [7]. The parametric model is shown in Fig.4 and the shape of Load Carrier device is to be formed by optimizing the parameters. Figure 4 shows the frame of the device with design parameter P4. Where, P1, P2, P3 and P4 describe the geometrical and physical properties of the frame of load carrying device. P1, indicates the young modulus of the material used in device. P2, indicates the poison's ratio of the material used. P3, Density of the materials and P4, represents the inner diameter of the rollers on the top of the frame. For the best load carrying device of minimum mass can be automated through a numerical optimization algorithm if the desired goal is expressed in the form of the objective problem definition as following: find design parameter's:

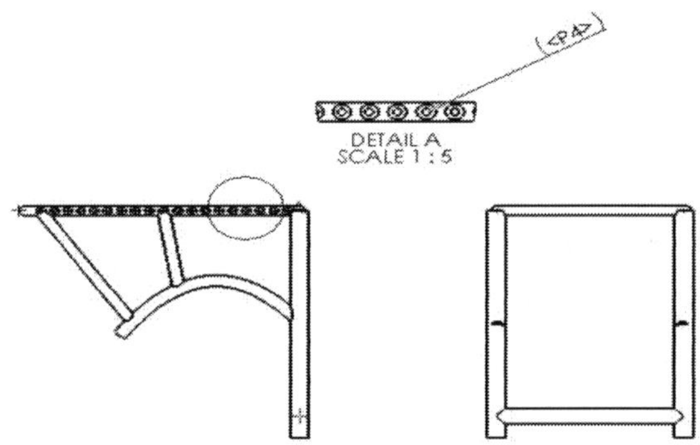

Figure 4. The load carrying device with design parameter P4

P_1, P_2, P_3 and P_4 (5)
To minimize design objective:
Maximum stress (Von Mises stress in MPa)
subjected to design constraints:
$\sigma_{\text{safe}} \leq \sigma_{\text{maximum}}$ (6)
$\text{Mass}_{\text{lighter}} \leq \text{Mass}_{\text{Maximum}}$ (7)

within the design space:
$15480 \leq P_1 \leq 18920$ (8)
$0.135 \leq P_2 \leq 0.165$ (9)

From the Equation (2)-(7) which is objective and constraint function which can only be known by applying least-square fitting to FEM analysis results for certain randomly selected design parameter values that corresponds to different type of load carrying device.

2.2. Materials

Bamboos is a type of evergreen long lived flowering plant [8]. As the middle portion of the bamboo is hollow. It is one of the fastest growing plants. It is widely used in building materials, food products and its is one of the raw material products has many faces. It is used like wood with high strength to weight ratio [9]. Mostly major material for Heavy Load Carrier is Stainless Steel. But bamboo is also an alternative of steel structure. In trials of tensile strength, bamboo outperforms most other materials, reinforcement steel included. Bamboo achieves this strength through its hollow, tubular structure and resist wind forces in natural habitat. Bamboo is stronger than steel in regards to the tensile strength per unit weight of bamboo is 3 to 4 times greater than steel. steel has a tensile strength of 23,000 pound per square inch but Bamboo surpasses steel with a noticeable lead at 28,000 pounds. bamboo compression better than steel and concreate .and a light weight, tough and cheap material. Gaudua Bamboo is available in many regions of the world, some mechanical testing is done for proving its usefulness in construction work. To construct a load carrying device we have to make sure that the weight of device shouldn't exceed. If it, does it will increase the weight of the device and thus the concept fails. Bamboo is chosen to be the cheap and lighter material. Bamboo has fast growth rate when compared to other trees, thus using it in our construction is best suited. For this a range of young modulus, poisons ratio is chosen from different bamboo test [11-12] and its range is chosen in Equation (8)-(9).

2.3. Finite Element Analysis

The FEM models are finely meshed using default mesh size in straight forward manner and a convergence study has been carried out (Figure 5).

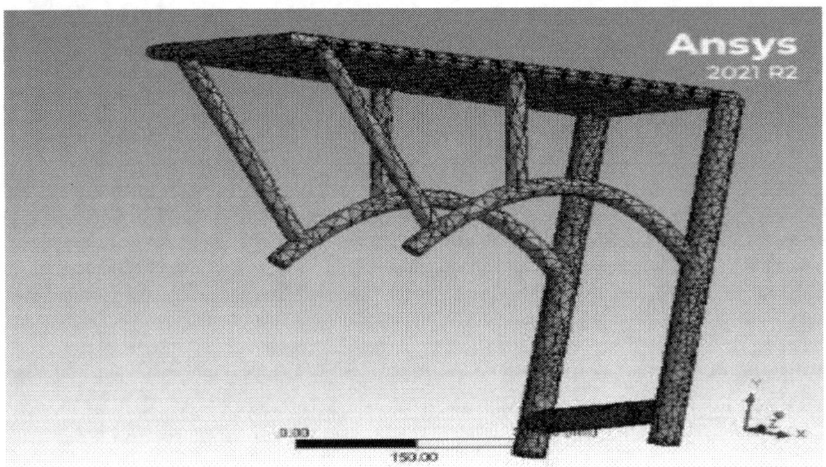

Figure 5. A converged meshed Load carrying device

Different responses and the value of Minimum mass and Von-misses stress is shown in the Table.1 is shown which is generated by RS approximation.

Table 1. RS approximated for different constraints

S.No.	P_1	P_2	P_3	P_4	Mass $_{Maximum}$	σ $_{maximum}$
1	17200	0.15	330	8.8	1.428135	2.067765
2	17200	0.15	330	7.92	1.456575	2.13632
3	17200	0.15	330	9.68	1.397044	2.567758
4	15480	0.15	330	8.8	1.428135	2.067765
5	18920	0.15	330	8.8	1.428135	2.067765
6	17200	0.135	330	8.8	1.428135	2.063044
7	17200	0.165	330	8.8	1.428135	2.072566
8	17200	0.15	297	8.8	1.285322	2.067765
9	17200	0.15	363	8.8	1.570949	2.067765

10	15988.77	0.139437	306.7612	8.180299	1.346394	2.106099
11	15988.77	0.139437	306.7612	9.419701	1.307499	2.252793
12	18411.23	0.139437	306.7612	8.180299	1.346394	2.110046
13	18411.23	0.139437	306.7612	9.419701	1.307499	2.278936
14	15988.77	0.160563	306.7612	8.180299	1.346394	2.07317
15	15988.77	0.160563	306.7612	9.419701	1.307499	2.078686
16	18411.23	0.160563	306.7612	8.180299	1.346394	2.104596
17	18411.23	0.160563	306.7612	9.419701	1.307499	2.075151
18	15988.77	0.139437	353.2388	8.180299	1.550387	2.107492
19	15988.77	0.139437	353.2388	9.419701	1.505599	2.850905
20	18411.23	0.139437	353.2388	8.180299	1.550387	2.11055
21	18411.23	0.139437	353.2388	9.419701	1.505599	2.112001
22	15988.77	0.160563	353.2388	8.180299	1.550387	2.071365
23	15988.77	0.160563	353.2388	9.419701	1.505599	2.382575
24	18411.23	0.160563	353.2388	8.180299	1.550387	2.066972
25	18411.23	0.160563	353.2388	9.419701	1.505599	2.774735

3. Results and Discussion

After performing the ANSYS DO module in collaboration with RSM (Fig.3) an optimum design is calculated. For these 4000 samples are generated initially, 800 samples per iteration and 3 candidates were found in a maximum of 20 iterations (Table. 2).The sensitivity of Mass and Von-misses stress can be easily compared in Figure 6.

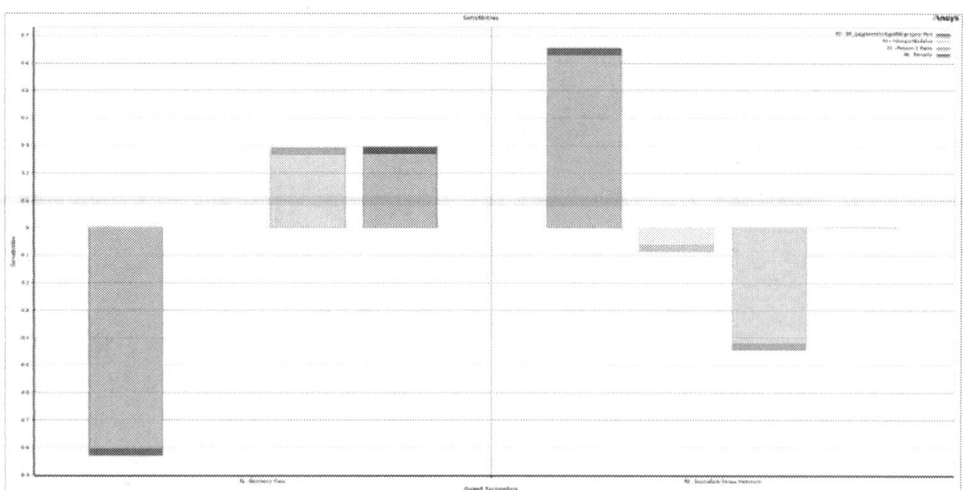

Figure 6. The sensitivity of Mass(kg) and Von-misses stress (MPa)

Table 2. Three candidates were found in a maximum of 20 iterations

	Candidate 1	Candidate 2	Candidate 3
P1-Young Modulus of the material (MPa)	17796.78	18253.21	18089.68
P2-Poisons ratio	0.163822	0.164276	0.162794
P3-Density (Kg/mm^3)	297.0315	297.0255	297.1484
P4-Inner Diameter(mm)	9.143556	9.258271	9.130015
P1 - Geometry Mass (kg)	1.27521	1.271851	1.276091
P3 - Equivalent Stress Maximum (MPa)	1.981447	1.987035	1.981075

The three possible solutions are seen in table.2 shows the minimum weight of the frame made up of bamboo can be up to 1.27 kg and minimum von misses stress is 1.98 MPa. Using the results, a new load carrying device is modelled with given optimized values the results obtained are shown in the Figure 7-9.

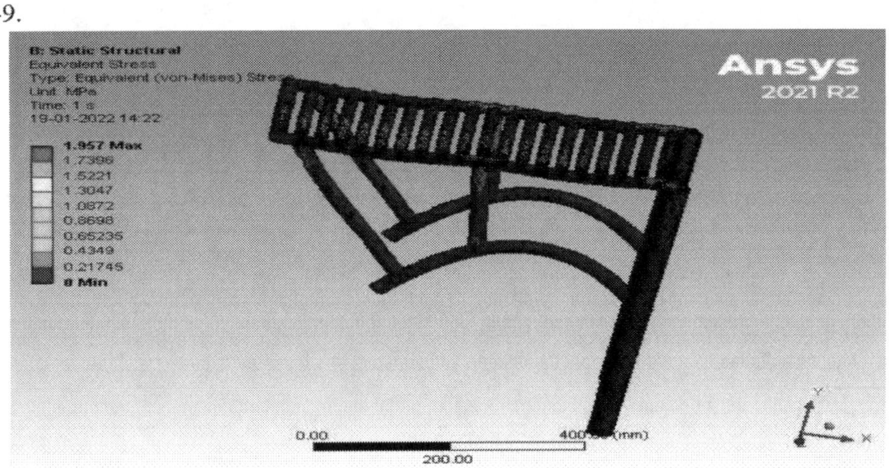

Figure 7. Predicted Von-misses stress with respect to Candidate point 1

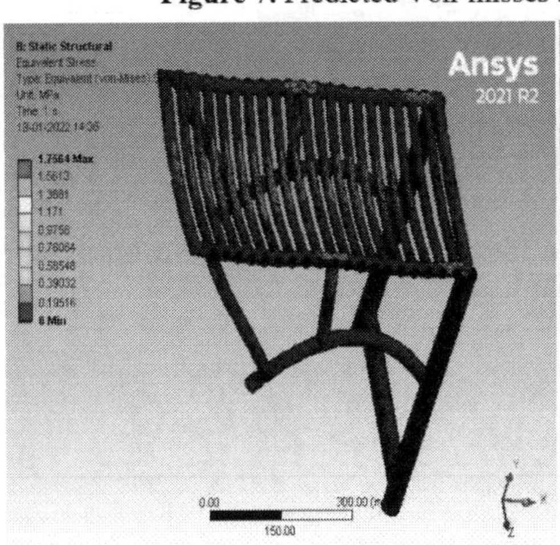

Figure 8. Predicted Von-misses stress with respect to Candidate point 2

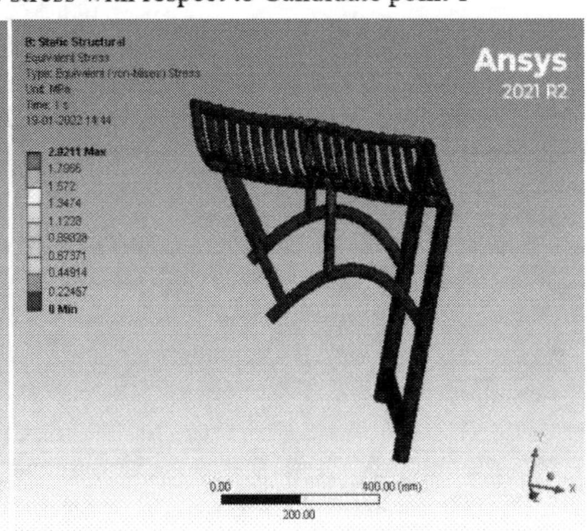

Figure 9. Predicted Von-misses stress with respect to Candidate point 3

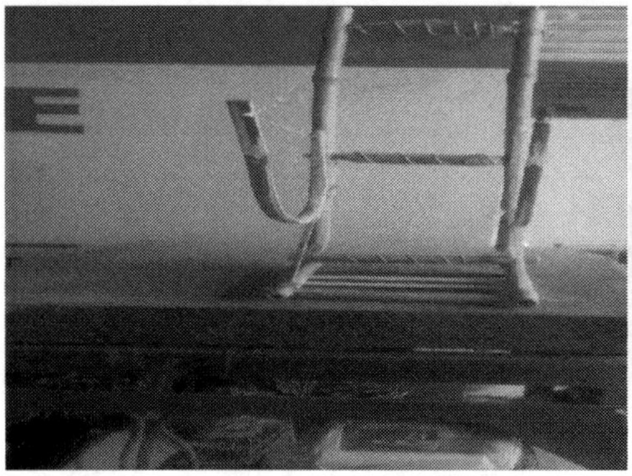

Figure 10. A 2.3 kg of Load carrying device is constructed using Gaudua Bamboo

Figure11. A Load carrying device is load with 9 bricks efficiently by a male labor

4. Conclusion

In this work, a load carrying device for the female workers of 5-6 feet tall in height are designed to secure them from hazardous substance to carry on their head. A frame is designed, to find out vonmisses stress and mass in MPa and Kg respectively, according to the density, inner diameter of the roller on the top of load carrying frame using ANSYS. Three candidate point of DOE optimization are again verified by modelling in SOLIDWORKS and analysis in ANSYS. After this a gaudua bamboo is used to construct a load carrying device to carry 15 kg(approx..) efficiently. This research is just a contribution towards the poor labors in India who get themselves hurt and unable to work in near future. So, using this designed frame one can carry up to a load of 15kg in which only 2.3 kg is its self-weight. A very nominal stress of maximum 2 MPa makes proves this frame a good design. A future scope of constructing the actual device in bulk and provide in market in cheapest price.

5. References

[1] Singh R. 2000, Indian labour statistics: *Ministry of Labour Government of Maharashtra, Industries Energy and Labour Department Mantralaya, Mumbai, India*

[2] Sahu, S., Sett, M., Basu, K., Chattopadhyay, S., and Paul, G.2008 Ergonomic evaluation of Manual Material Handling tasks performed by building construction laborers at different work sites in unorganized sectors in West Bengal: *Journal of Environmental Physiolog:* 1,2, 48-59,

[3] Dwivedi, S.L. 2000 The Factory Act, 1948 with the Maharashtra Factories Rules: *Labour Law Agency Mumbai* 1963.

[4] SINGH R. 2000 Indian Labour Statistic Ministry of Labour Government of Maharashtra, Industries: *Energy & Labour Department, Mantralaya, Mumbai, India,*.

[5] Thomas E. Wood and Calvin T. Wood 1978: Internal Frame backpack, US4074839A.

[6] Katoozian H. 1993 Three-dimensional design optimization of femoral components of total hip endoprostheses. Ph.D. dissertation. Cleveland (OH): *Case Western Reserve University.*

[7] ANSYS. 2003 Theory Reference Manual: *Ansys Inc Ansys Release 8.0.*

[8] Farrelly and David 1984 The Book of Bamboo: *Sierra Club Books*. ISBN 978-0-87156-825-0.

[9] Lakkad and Patel 1981, Mechanical properties of bamboo, a natural composite: *Fibre Science and Technology* 14 319–322.

[10] Juan Francisco Correal D, Juliana Arbeláez C. 2010,: Influence of age and height position on colombian guadua angustifolia bamboo mechanical properties: *Maderas. Cienciay tecnología* 12 105-113.

[11] DinieAwalluddin, MohdAzreenMohdAriffin, MohdHanim Osman, MohdWaridHussin, Mohamed A 2017 Ismail, Han-Seung Lee, and NorHasanah Abdul Shukor Lim. Mechanical properties of different bamboo species: *MATEC Web of Conferences* 138.

[12] Caori P. Takeuchi, Martin Estrada, Dorian L. Linero 2016,: The Elastic Modulus and Poisson's Ratio of Laminated Bamboo Guadua angustifolia: *Key Engineering Materials 668.*

[13] J R Soc Interface 2014 *The* Royal Society: *Articles from Journal of the Royal Society Interface are provided here courtesy* 11 99.

IOP Conf. Series: Materials Science and Engineering 1259 (2022) 012014 doi:10.1088/1757-899X/1259/1/012014

Effects of processes parameters on spring back, wrinkles and cracks in different flanging processes: A Review

Surendra Kumar[1], Pradeep Singh[2], Manoj Soni[1, 3], J.P. Shakya[2], Sanjay Kumar Panthi[1], Meraj Ahmed[1]

[1]Council of Scientific and Industrial Research (CSIR)-Advanced Materials and Processes Research Institute, Bhopal-462026, India
[2]Department of Mechanical Engineering, Samrat Ashok Technological Institute, Vidisha – M. P
[3]Academy of Scientific and Innovative Research (AcSIR), Ghaziabad-201 002, Uttar Pradesh, India.
E-mail address: erpradeep3408@gmail.com, surendra.javapatel@gmail.com

Abstract. Flanging is an important type of sheet metal bending process. In present review, an effort was made to explore the contributions of various researchers to drawn out the cause of defects in the flanging process. Several parameters such as geometrical, material and process parameters affect the formability of the flanging process. In straight flanging process, spring back (due to elastic recovery) is a major problem that affects dimensional and geometrical accuracy of a bent component. Several factors are responsible for springback such as – profile radius of punch and die, bend angle, thickness of sheet, clearance between punch and die, blank-holder force, Young's modulus, n-value, yield stress, plastic anisotropy of materials, velocity of punch etc. In shrink flanging process, wrinkling is the prime concern that occurs due to the excessive compressive stress in the flange portion. It is mostly affected by mechanical properties of material (R-value, n–value etc.), geometry of the tools and blank, process parameters. In stretch flanging process, the effect of material parameters is found to be less as compared to geometrical parameters for crack initiation and thinning.

1. Introduction

Flanging process is one of the important manufacturing process of sheet metal forming operation. This process is used for mass production of the products in automotive, aerospace industries and household appliances. Flanging is the process of secondary sheet metal forming which broadly used to provide smooth rounded edge of the components for aesthetics, to enhance the strength of the components, and to provide hidden joints during assembling of automotive parts [1, 2]. Fig. 1 illustrates the schematic diagram of flanging process. Table-1 shows the several symbols which is used in this study.

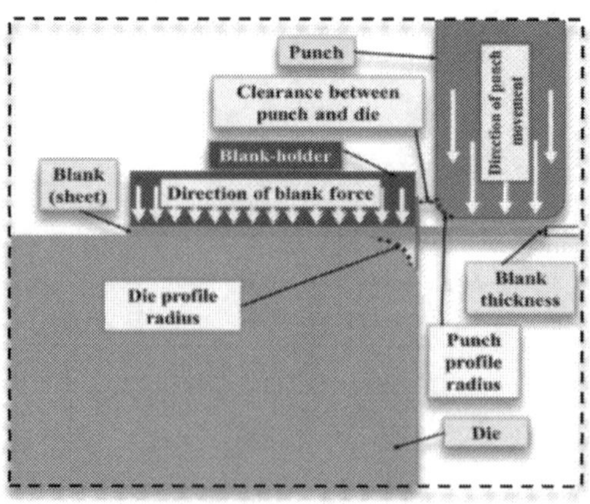

Fig. 1. Schematic diagram (description) of flanging process

IOP Conf. Series: Materials Science and Engineering 1259 (2022) 012014 doi:10.1088/1757-899X/1259/1/012014

Table 1. General terminology/symbols for flanging process

Symbol	General terminology
t	Flange thickness
r	Bend radius
W	Width of sheet metal
L	Initial flange length
R_d	Radius of die corner
R_p	Radius of punch profile
C	Clearance between the die and punch
E	Young's modulus
K	Material strength coefficient
n	Exponent of strain hardening
R	Normal anisotropy parameter

2. Types of Flanging

There are mainly three types of flanging process, known as straight, stretch and shrink flanging. These flanging processes are shown in Fig. 2. In straight flanging, edge of the sheet bends to 900 with respect to the strain deformation plane. Concave shaped curvature of bending line is found in stretch flanging, and sheet metal is circumferentially stretched. While, thinning and cracks occurs in stretch flanging process, which are the major problems. In shrink flanging, the curvature of the bending line is convex, and bending line deformation is not in plane strain condition. During the shrink flanging process, sheet metal is circumferential compressed. The dominant deformation parameters in the shrink flanging process are hoop strain compression. Wrinkles occur at the edge of the flange due to excessive compressive stress.

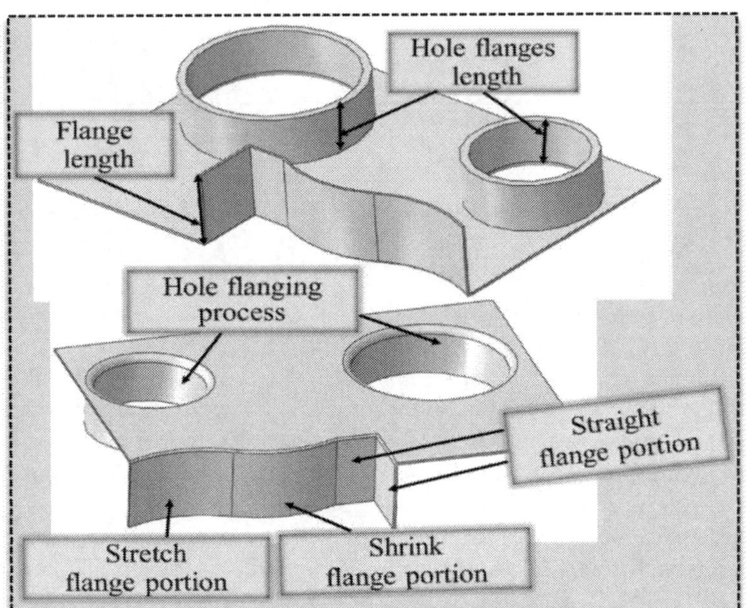

Figure 2. Different flanging processes

A new type of flanging is the group of straight, stretch and shrink flanges, known as mixed flanging in which bending is different for the different portions. The one by one explanation of all the flanging processes are following.

2.1. *Straight flanging (L-shape or wiping die bending) process*

In this process, flange is formed without concave and convex curvature and sheet edge is perpendicularly bent along a straight line, and deformation occurs in the plane strain condition. (Fig. 3). The basic problem of this type flanging process is the springback. Straight flanging process depends on the different parameters like clearance between punch and die, thickness of the blank, bend radius, types of material of the blank, geometrical parameters of tools and blank etc. R. G. Davies studied the effect of punch-die distance, ratio of thickness (t) to bend radius, types of steel and flow stress in straight flanging (double bend technique) process [3, 4]. He noted that the limiting value of the springback was independent of the flow stress for any grade of steels. The curl decreases with decrement in tensile strength and increases in thickness, and it can be eliminated by removal of non-uniform distribution of residual stresses. The relation between the t/r ratio (t = thickness of the blank, r = bend radius) and springback of different materials was established and it was found that springback reduces with the increment of t/r ratio up to a certain limit [5]. N. Song et.al. used the analytical model and numerical simulation method (finite element method and mesh free method) for prediction of springback on aluminium alloy (AA5182-O) sheet in straight flanging process using the reproducing kernel particle methods (RKPM) [6]. It was found that the kinematic hardening law provide better prediction in terms of springback than the isotropic hardening law.

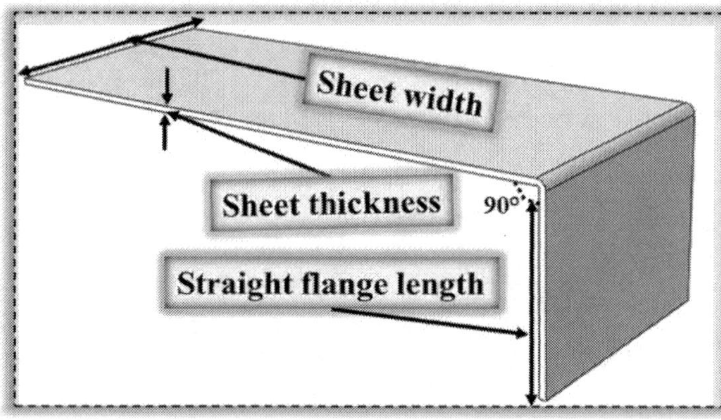

Figure 3. Straight flanging process

The springback angle goes to increase as the clearance (C) increases or the flange length (L) decreases [7]. Livatyali et al. investigated the influence of die radius, clearance (C) to thickness (t) ratio, blank holding force, and punch nose radius on springback using commercial available FEM programs of ABAQUS TM and DEFORM TM for the better dimension control [8, 9]. It was concluded that these parameters have significant effect on springback. The springback decreases with increases in blank holding force for a certain limit. The springback decreases with increment in punch nose radius or die corner radius up to a certain value [7, 8, 10, 11]. Taguchi and variance analysis methods were also applied by various researchers for optimization of process parameters (clearance (C), punch fillet radius etc.) on springback for AZ31 magnesium alloy sheets at different temperatures [12]. It was observed that at 100°C, 150°C, and 200°C bending temperatures, the respective values of springback were found as 91.06°, 90.63° and 89.84°. It explores that spring back reduces with the increment of bending temperature. It was also found that the angle of springback decreases with enhancing the blank thickness [13, 14], but increment was found with increasing the die radius [10]. Jafari et al. optimized and numerical controlled of springback effect on magnesium alloy using artificial intelligence and finite element analysis by considering different process parameters (i.e. die temperature, die clearance, fillet radius of punch, step distance etc.) [15]. In this research, it was again found that the springback effect decreases with temperature increment. Phanitwong and Thipprakmas developed the new spring-back factor for L-bending process [16]. It was concluded that the spring back enhance with increment in die radius and bend angle, while it decreases with the increment in ratio of die radius to blank thickness. Han et al. used different hardening model and yield function for the prediction of better accuracy of geometrical profile and force in straight flanging by Incremental Sheet Forming (ISF) process [17].

2.2. Shrink flanging process

The bending line curvature is convex and deformation is not in plane strain condition in shrink flanging. Also, flange is circumferentially compressed as shown in Fig. 4. Fig. 4 (a) & (b) represents the various nomenclature used in shrink flanging process. The lowest compression stress occurs at the place of along the die profile radius and the highest compression stress occurs at the free edge (flange radius). The major deformation parameter in the shrink flanging process is circumferential strain (hoop strain compression). If excessive compressive stress reaches the critical value, then it leads to wrinkling of the free edge of the flange (lower portion of the flange), which is the basic problem in shrink flanging process (Fig. 4 (c)). The material undergoes thickening in shrink flanging at the free edge.

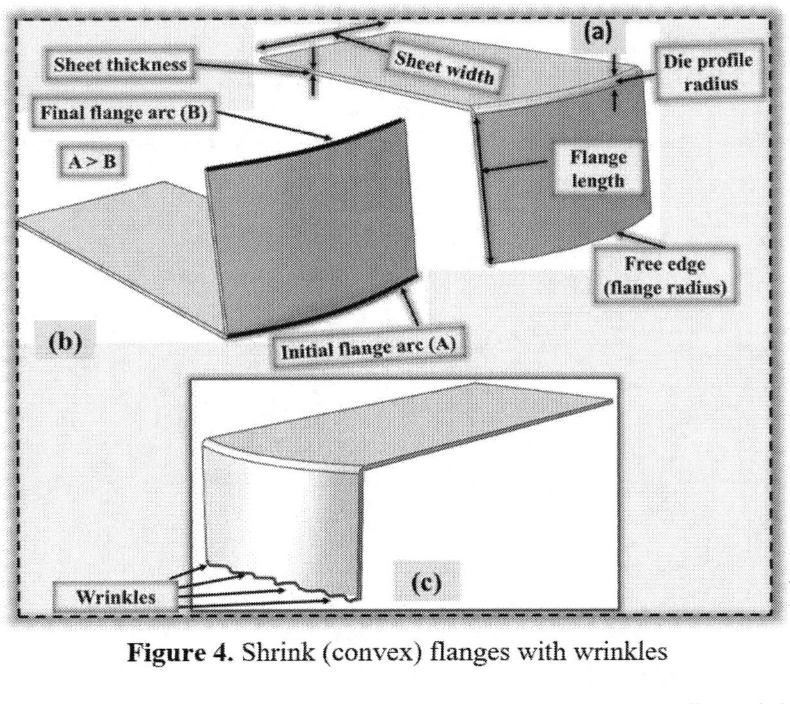

Figure 4. Shrink (convex) flanges with wrinkles

Wang et al. developed the approximate theory and mathematical model to predict wrinkling criteria in convex flanging process that was presented the results in terms of wrinkling and strain distribution [18]. Cao and Li predicted the onset of wrinkling in convex flanging process and obtained the critical condition for buckling in terms of stress and strain, geometry and material properties [19]. It was found that critical flange length is more prominent in plan view radius (PVR), strain hardening exponent (n) and sheet thickness, whereas it is less sensitive in other material properties such as R-value and material strength coefficient (K). An analytical model was developed in order to predict the hoop strain distribution and semi-empirical wrinkling criteria in shrink flanging with surface contour process [20]. The hoop strain model was based on the assumption of uniaxial stress to flange axisymmetric analysis. It was found that the ratio of blank thickness (t) to flange length (L) was the most important factor for determination of onset of wrinkling. These models may be beneficial for the designing of tools in shrink flanging process. Lei et al. used the elasto-plastic FEM model in typical convex flanging part in rubber forming technique for the analysis of springback control that was based on the updated Lagrangian formulation [21]. Finite element simulation results were verified by experimental results and found that the high pressure is useful to avoid wrinkling. Chen et al. analysed the influences of parameters such as flange length (L), fillet radius of die and forming pressure on wrinkling in shrink flanging of different aluminium alloys (i.e. AA-2024-O, AA-2024-T3 and AA7075-O) sheets by using rubber forming process [22]. Wrinkling of AA-2024-T3 was more than that of AA-2024-O and AA-7075-O. Three steps forming was the best method for controlling of wrinkling in convex flanging process by using rubber forming technique. It was found that die radius affects more as compared to other parameters on the wrinkling in shrink flanging process. Zhang et al. studied the failure behaviour and deformation mechanism in convex flanging by the use of double-sided incremental forming technique [23]. It was found that the formability limit of convex flanging is increased due to localized deformation mode and meridional tensile strain.

2.3. Stretch (concave) flanging process

In this process, the curvature of flange is concave shape and bending line deformation is not in plane strain condition, while the flange is circumferentially stretched. Thinning and cracks occurs, which are the major problem in this flanging process. Generally, two types tool set-up are used for the stretch flanging process like - non-axisymmetric stretch flanging (without hole flanging) and axisymmetric stretch flanging (with hole flanging) process. Fig. 5 represents the various nomenclature used in stretch flanging (without hole and with hole flanging) process. In this process, arc lengths of original elements are stretched in the circumferential direction (Fig. 5 (b)). In this case, the flange portion gets thinner. Hoop strain is the dominating deformation parameters in stretch flanging process. If tensile stresses is excessive from the critical limit, then it leads to start the crack in the flange (Fig. 5 (c)) [24].

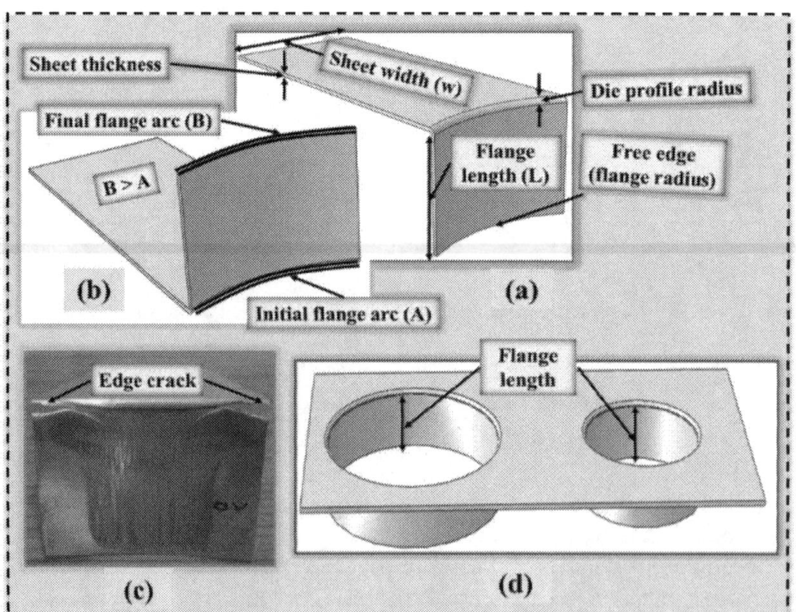

Figure 5. (a, b and c) Non- axisymmetric stretch flanging (without hole flanging) process with cracks and (d) axisymmetric stretch flanging (with hole flanging) process [25]

Table. 2 shows the constitutive material models that was applied by past researchers in stretch flanging process for finite element analysis. Chen et al. used the Gurson-Tvergaard- Needleman (GTN) based material model for finite element analysis of Al-Mg sheet AA5182 in stretch flanging process [26]. They examined the effect of strain due to void nucleation for the prediction of onset of ductile failure, and determined the level of critical porosity by the analysis of quantitative metallurgy.

Table 2. Constitutive models applied in stretch flanging process by the various researchers for finite element analysis

S.N	Reference	Use constitutive relationship and parameters
1	[27]	$\varepsilon = \frac{\sigma}{E} + \frac{(\sigma)^{1/2}}{K}$
2	[28]	$\sigma_y = \mathrm{K}\,(\varepsilon^E + \varepsilon^p)^n$
3	[29]	$\sigma = 565.32(0.007117 + \bar{\epsilon}_p)^{0.2589}$
4	[30]	$\sigma = 516.54(0.00639 + \bar{\epsilon}_p)^{0.232}$
5	[31]	$\sigma = K \; \bar{\epsilon}^n$
6	[32]	$(\sigma) = 549.30(0.0013 + \bar{\epsilon}_p)^{0.22}$
7	[33]	$(\sigma) = 597.5 \; \bar{\epsilon}^{-0.279}$
8	[34]	$\sigma = (165 + 968.6\,\epsilon - 0.206)\,(1 + 0.0145 + \mathrm{In}\,\varepsilon)\,(1 - \left(\frac{T - 20}{T\,melt - 20}\right))^{0.868})$

2.3.1. Effect of geometrical and process parameters on stretch flanging processes

There are various geometrical parameters influencing the hoop and radial strain. These parameters are initial flange length (L), sheet width, sheet thickness (t), die profile radius (Rd), curvature radius of punch, flange angle, the punch profiles, punch corner radius on the flange shape height. Sriram and Chintamani analysed the influence of different parameters (i.e. flange angle, flange width and flange radius) on major strains at the flange free edge of advanced high strength steels in stretch flanging process [35]. This study reveals that flange angle was the more dominating parameters than the other parameters. Yeh et al. determined the accurate size of sheet in stretch flanging process that was consist on a convex–concave prediction scheme by using true strain method, explicit dynamic finite element method, and fuzzy inference system based on adaptive network [30]. Li et al. studied the influence of different parameters (i.e. flange angle, flange length, etc.) in V-shaped stretch flanging process and found that the flange angle is more susceptible for the crack initiation in the blank as compared to another geometrical parameter [31]. Abe et al. used the gradually contacting type punch shape for improvement of formability on ultra-high strength steel sheet and found that tensile stress reduces in the sheet metal blank in stretch flanging process [36]. Dewang et al. studied the spreading of edge crack on AA-5052 sheet by using numerical simulation and it is verified with the experimental results [37]. It was noted that the major strain increases with increase in flange length (L) and decrease with clearance (C). Effect of binder force and thickness distribution in stretch flange process using FEM simulation was also investigated and validated with experimental one [38]. Circumferential strain, crack extension and thinning were increasing with increasing the binder force for same process. The maximum hoop strain and crack length increases with increment of initial flange length, flange height [29, 31, 39, 40]. While, it reduces with increase in punch curvature radius and flange angle, while some increment is observed with the flange width [29, 31, 35]. Circumferential and radial strain increases with decrease in punch die clearance [39]. Kumar et al. investigated the influence of different punch profiles (i.e. cylindrical, conical, hemispherical, two step, three step and six step punch) on deformation behaviour of AA5052 sheet in concave flanging [25]. It was found that the hemispherical punch profile gives better results as compared to other profile in terms of edge crack length and strain distribution. In the stretch flanging process, circumferential strain (hoop strain) is more important as compared to other two strain, namely, radial and thickness strain that are also responsible for thinning (failure) [40].

2.3.2. *Effect of material parameters on stretch flanging processes*

Material parameters influence the formability of sheet metal forming processes. Normal anisotropy (R), strain hardening exponent (n) etc. are known as the material parameters. Formability is also influenced by the orientation of the sheet in sheet metal forming process. Directional properties are used for improving and optimising the formability in sheet metal forming operations. Orientation based material properties are yield stress, Young's modulus, R-value, n value etc. Whereas, plastic strain ratio is more affected than others properties. Major strain along free edge decreases with increase in n value in stretch flanging of V-shaped sheet metal [27] and [31]. Asnafi, determined the fracture limit on the basis of parameters such as R-value, n-value and uniform strain of Al alloys by fluid forming method in stretch flanging process [28]. If value of these parameters increase, fracture limit will be higher.

2.4. *Hole flanging (axisymmetric stretch flanging) process*

To generate a flange around a hole is known as hole flanging process that is shown in Fig. 2 and Fig. 5 (d). It is also known as axisymmetric stretch flanging process. Hyun et al. estimated the formability of flange for high strength steel plates in hole flanging process [41]. During this process, it was observed that the fracture of this material was mainly caused by localized necking due to circumferential strain. Better hole flangeability was obtained in Ferrite–Bainite duplex steel as compared to TRIP steel, while better formability was found in case of lip height and lip shape accuracy in TRIP steel than the Ferrite–Bainite duplex steel. Thipprakmas et al. studied the effect of flanged shape in fine-blanked hole (FB-hole) flanging process and conventional-hole flanging process using FEM simulation [42]. Flange shapes and flangeability were obtained better by FB-hole flanging process as other hole flanging processes. Li et al. developed the model and methodology for the analysis of forming limit in hole flanging process by using direct penetration and re-penetration process [43]. Punch load was more in case of direct penetration as compared to that of re-penetration. It was found that re-penetration provides better formability as compared to direct penetration at the same condition, and formability ratio was also increased at the same case. Uthaisangsuk et al. investigated the stretch-flangeability of dual phase steels DP800 and DP600 by the use of microstructure based failure model [44]. The expansion ratio of the hole for multiphase steel was strongly influenced by phase present in the material and its distribution. Xu et al. determined the effect of stretch-flangeability on Advanced High Strength Steel (AHSS) sheet (TWIP1000) through the hole expansion test using FEM simulation of ABAQUS software [45]. They used the different types of failure criteria such as Hill 1948, yield criteria, YlD20002d and Von Mises for this study. Narayanasamy et al. studied the influences of mechanical properties and fractography behaviours of various automobile steel sheets in hole expansion test. It was noted that the hole expansion strongly affected by the microstructure of the workpieces [46]. Lee et at. studied the influence of micro-alloying on the stretch flangeability of a conventional DP (dual phase) steel in hole expansion tests [47]. It was analysed that the strength of the DP (dual phase) steel improves due to finer grain size because of suppression of formation of new ferrite from the inter-cortical austenite. W. Fracz et al. examined the influence of thickness distribution in hole flanging by using three types of punches i.e. conical, flat bottomed (cylindrical) and hemispherical, and found the uniform thickness and less thinning in the sheet metal in the case of flat bottomed punch [48]. A. Kacem et al. predicted the failure criteria (necking and cracking) of aluminium alloys AA 1050-H14 and AA 6061-O in the hole-flanging through FEM analysis using criterion of ductile fracture and damage evolution, and validated with experimental results [49]. C. Yang et al. used the bar tool and dieless tool for the analysis of thin-walled tube for the creation of branched tube through ISF process [50]. It was discussed the forming defects such as localized buckling, cracking, circularity distortion, etc., and these defects were influenced by different factors like – tool shape, processing parameters, properties of blank and pre-cut hole. Jallouli et al. used the elasto-plastic FEM computer program by the use of updated Lagrangian formulation of Aluminum alloys sheet in hole flanging [51]. The punch load, blank thickness and blank-holding force were studied in hole flanging and compared the FE simulation results with experiment one. Krawczyk et al. performed the hole expansion test using three different punch profiles (conical, cylindrical and sperical) on four different types of steels workpiece [52]. Starting hole were created by different cutting technologies in the workpiece. Hole diameter enhancement was found minimum for the cylindrical punch, while it was maximum for the spherical punch. Wen Tong et al. used the simple bar tool (tapered shoulder) in forward / backward flanging as well as hole flanging of aluminium sheet AA 6061 by using a bi-directional dieless single-point ISF process [53]. Recommended theta value for hole flanging is 200 - 300 degree. Tingting Cao et al. proposed the new hole flanging tool which consists of some geometrical feature such as hyperbolic and cylindrical sections of the tool arm that was used in ISF process [54]. It was developed and validated on the basis of comparative studied among them a numerical simulation, analytical modelling and experimentation. New hole flanging tool is the best in terms of increment of hole flanging expansion ratio, more uniform thickness distribution in the parts, because the avoid of extreme thinning, and the obtained contact area between the sheet and tool was more.

3. Conclusions

Concise review of the contributions made by the several past researchers in the area of straight, shrink, hole and stretch flanging process has been presented. Springback is the material properties, and it is impossible to eliminate completely. However, it can be mitigated by optimization of geometrical, material and process parameters. Optimization of different parameters using various methods i.e. Taguchi method, analysis of variance (ANOVA) method and artificial neural network method were used by the researchers. Several researchers have been developed the mathematical/analytical model or used the energy based approaches for predicting the onset of wrinkles by considering several parameters (R-value, n-value, flange length, die profile radius, sheet thickness etc.) in shrink flanging process. The geometrical, materials and process parameters considerably affect the formability, flangeability, failure criterion (localized necking, non-uniform thickness distribution, cracking, wrinkling, etc.) and hole expansion limit (HEL) in hole flanging process. In stretch flanging process, geometrical parameters are more prone as compare to material parameters. If magnitude of R-value, n-value and uniform strain parameters are higher, fracture limit will also be higher in the stretch flanging process.

Conflict of Interest: None

Reference

[1] Kalpakjian S and Schmid S R 2014 Manufacturing engineering and technology (p. 913). Upper Saddle River, NJ, USA: Pearson.

[2] Groover M P 2007 Fundamentals of modern manufacturing: materials processes, and systems, John Wiley & Sons.

[3] Davies R G and Liu Y C 1984 Control of springback in a flanging operation. Journal of applied Metalworking, 3(2), 142-147.

[4] Davies R G 1984 "Side-wall curl" in high-strength steels. Journal of Applied Metalworking, 3(2), 120-126.

[5] Huang M and Gerdeen J C 1994 Springback of doubly curved developable sheet metal surface-an overview. SAE transactions, 718-731.

[6] Song N, Qian D, Cao J, Liu W K, and Li S 2001 Effective models for prediction of springback in flanging. J. Eng. Mater. Technol., 123(4), 456-461.

[7] Farsi M A, Arezoo B, Alizadeh V and Mirzaee S 2011 The study of spring-back in wipe-bending processes for perforated components. Proceedings of the Institution of Mechanical Engineers, Part B: Journal of Engineering Manufacture, 225(11), 2007-2014.

[8] Livatyali H and Altan T 2001 Prediction and elimination of springback in straight flanging using computer aided design methods: Part 1. Experimental investigations. Journal of Materials Processing Technology, 117(1-2), 262-268.

[9] Livatyali H, Wu H C, and Altan T 2002 Prediction and elimination of springback in straight flanging using computer-aided design methods: Part 2: Fem predictions and tool design. Journal of Materials Processing Technology, 120(1-3), 348-354.

[10] Buranathiti T and Cao J 2004 An effective analytical model for springback prediction in straight flanging processes. International Journal of Materials and Product Technology, 21(1-3), 137-153.

[11] Gassara F, Hambli R, Bouraoui T, Halouani F E and Soulat D 2009 Optimization of springback in L-bending process using a coupled Abaqus/Python algorithm. The International Journal of Advanced Manufacturing Technology, 44(1), 61-67.

[12] Kuo C C, and Lin B T 2012 Optimization of springback for AZ31 magnesium alloy sheets in the L-bending process based on the Taguchi method. The International Journal of Advanced Manufacturing Technology, 58(1), 161-173.

[13] Kumar K D, Appukuttan K K, Neelakantha V L and Naik P S 2014 Experimental determination of spring back and thinning effect of aluminum sheet metal during L-bending operation. Materials & Design (1980-2015), 56, 613-619.

[14] Ling Y E, Lee H P and Cheok B T 2005 Finite element analysis of springback in L-bending of sheet metal. Journal of Materials Processing Technology, 168(2), 296-302.

[15] Jafari M, Lotfi M, Ghaseminejad P, Roodi M, and Teimouri R 2015 Numerical control and optimization of springback in L-bending of magnesium alloy through FE analysis and artificial intelligence. Transactions of the Indian Institute of Metals, 68(5), 969-979.

[16] Phanitwong W, and Thipprakmas S 2016 Development of anew spring-back factor for a wiping die bending process. Materials & Design, 89, 749-758.

[17] Han K, Li X, Peng X, Wang H, Li D, Li Y and Li Q 2019 Experimental and numerical study on the deformation mechanism of straight flanging by incremental sheet forming. International Journal of Mechanical Sciences, 160, 75-89.

[18] Wang C T, Kinzel G, and Altan T 1995 Failure and wrinkling criteria and mathematical modeling of shrink and stretch flanging operations in sheet-metal forming. Journal of materials processing technology, 53(3-4), 759-780.

[19] Wang X, Cao J and Li M 2001 Wrinkling analysis in shrink flanging. J. Manuf. Sci. Eng., 123(3),

426-432.
[20] Zhang G E, Yao J, Hu S J and Wu X 2003 Shrink flanging with surface contours. Journal of manufacturing processes, 5(2), 143-153.
[21] Lei C 2011 Numerical Simulation and Die Compensation on Springback of Shrink Flanging in Rubber Forming. In 2011 Third International Conference on Measuring Technology and Mechatronics Automation (Vol. 2, pp. 705-708). IEEE.
[22] Chen L, Chen H, Wang Q, and Li Z 2015 Studies on wrinkling and control method in rubber forming using aluminium sheet shrink flanging process. Materials & Design (1980-2015), 65, 505-510.
[23] Zhang H, Zhang Z, Ren H, Cao J and Chen J 2018 Deformation mechanics and failure mode in stretch and shrink flanging by double-sided incremental forming. International Journal of Mechanical Sciences, 144, 216-222.
[24] Hosford W F and Caddell R M 2011 Metal forming: mechanics and metallurgy. Cambridge university press.
[25] Kumar S, Ahmed M and Panthi S K 2020 Effect of punch profile on deformation behaviour of AA5052 sheet in stretch flanging process. Archives of Civil and Mechanical Engineering, 20(1), 1-17.
[26] Chen Z T, Worswick M J and Lloyd D J 2006 Damage-based finite element modeling of stretch flange forming of aluminium-magnesium alloy. In Materials science forum (Vol. 519, pp. 815-820). Trans Tech Publications Ltd.
[27] Wang N M, Johnson L K and Tang S C 1984 Stretch flanging of "V"-shaped sheet metal blanks. Journal of Applied Metalworking, 3(3), 281-291.
[28] Asnafi N 1999 On stretch and shrink flanging of sheet aluminium by fluid forming. Journal of Materials Processing Technology, 96(1-3), 198-214.
[29] Feng X, Zhongqin L, Shuhui L and Weili X 2004 Study on the influences of geometrical parameters on the formability of stretch curved flanging by numerical simulation. Journal of materials processing technology, 145(1), 93-98.
[30] Yeh F H, Wu M T and Li C L 2007 Accurate optimization of blank design in stretch flange based on a forward–inverse prediction scheme. International Journal of Machine Tools and Manufacture, 47(12-13), 1854-1863.
[31] Li D, Luo Y, Peng Y and Hu P 2007 The numerical and analytical study on stretch flanging of V-shaped sheet metal. Journal of materials processing technology, 189(1-3), 262-267.
[32] Bao Y D and Huh H 2007 Optimum design of trimming line by one-step analysis for auto body parts. Journal of materials processing technology, 187, 108-112.
[33] Huang Y M 2007 An elasto-plastic finite element analysis of the sheet metal stretch flanging process. The International Journal of Advanced Manufacturing Technology, 34(7), 641-648.
[34] Sartkulvanich P, Kroenauer B, Golle R, Konieczny A and Altan T 2010 Finite element analysis of the effect of blanked edge quality upon stretch flanging of AHSS. CIRP annals, 59(1), 279-282.
[35] Sriram S and Chintamani J 2005 Guidelines for stretch flanging advanced high strength steels. In AIP Conference Proceedings (Vol. 778, No. 1, pp. 681-686). American Institute of Physics.
[36] Abe Y, Mori K I and Norita K 2013 Gradually contacting punch for improving stretch flangeability of ultra-high strength steel sheets. CIRP Annals, 62(1), 263-266.
[37] Dewang Y, Hora M S and Panthi S K 2015 Prediction of crack location and propagation in stretch flanging process of aluminum alloy AA-5052 sheet using FEM simulation. Transactions of Nonferrous Metals Society of China, 25(7), 2308-2320.
[38] Dewang Y, Panthi S K and Hora M S 2019 Binder force effect on stretch flange forming of aluminum alloy. Materials and Manufacturing Processes, 34(13), 1516-1527.
[39] Dewang Y, Hora M S and Panthi S K 2014 Finite element analysis of non-axisymmetric stretch flanging process for prediction of location of failure. Procedia Materials Science, 5, 2054-2062.
[40] Wang C T, Kinzel G and Altan T 1995 Failure and wrinkling criteria and mathematical modeling of shrink and stretch flanging operations in sheet-metal forming. Journal of materials processing technology, 53(3-4), 759-780.

[41] Hyun D I, Oak S M, Kang S S and Moon Y H 2002 Estimation of hole flangeability for high strength steel plates. Journal of Materials Processing Technology, 130, 9-13.

[42] Thipprakmas S, Jin M and Murakawa M 2007 Study on flanged shapes in fineblanked-hole flanging process (FB-hole flanging process) using finite element method (FEM). Journal of Materials Processing Technology, 192, 128-133.

[43] Li C L, Huang Y M and Tsai Y W 2008 The analysis of forming limit in re-penetration process of the hole-flanging of sheet metal. Journal of Materials Processing Technology, 201(1-3), 256-260.

[44] Uthaisangsuk V, Prahl U and Bleck W 2009 Stretch-flangeability characterisation of multiphase steel using a microstructure based failure modelling. Computational Materials Science, 45(3), 617-623.

[45] Xu L, Chen L, De Cooman B C, Steglich D and Barlat F 2010 Hole expansion of advanced high strenth steel sheet sample. International Journal of Material Forming, 3(1), 247-250.

[46] Narayanasamy R, Narayanan C S, Padmanabhan P and Venugopalan T 2010 Effect of mechanical and fractographic properties on hole expandability of various automobile steels during hole expansion test. The International Journal of Advanced Manufacturing Technology, 47(1), 365-380.

[47] Lee J, Lee S J and De Cooman B C 2012 Effect of micro-alloying elements on the stretch-flangeability of dual phase steel. Materials Science and Engineering: A, 536, 231-238.

[48] Frącz W, Stachowicz F and Trzepieciński T 2012 Investigations of thickness distribution in hole expanding of thin steel sheets. archives of civil and mechanical engineering, 12(3), 279-283.

[49] Kacem A, Krichen A, Manach P Y, Thuillier S and Yoon J W 2013 Failure prediction in the hole-flanging process of aluminium alloys. Engineering Fracture Mechanics, 99, 251-265.

[50] Yang C, Wen T, Liu L T, Zhang S and Wang H 2014 Dieless incremental hole-flanging of thin-walled tube for producing branched tubing. Journal of Materials Processing Technology, 214(11), 2461-2467.

[51] Jallouli I, Rouabeh K and Attyaoui S Numerical Simulation of Hole-flanging Process of Aluminum Sheet. International Journal of Engineering and Technical Research (IJETR) ISSN, 2321-0869.

[52] Krawczyk J, Gronostajski Z, Polak S, Jaśkiewicz K, Chorzępa W and Pęcak I 2016 The influence of the punch shape and the cutting method on the limit strain in the hole expansion test. In Key Engineering Materials (Vol. 716, pp. 129-137). Trans Tech Publications Ltd.

[53] Wen T, Zhang S, Zheng J, Huang Q and Liu Q 2016 Bi-directional dieless incremental flanging of sheet metals using a bar tool with tapered shoulders. Journal of Materials Processing Technology, 229, 795-803.

[54] Cao T, Lu B, Ou H, Long H and Chen J 2016 Investigation on a new hole-flanging approach by incremental sheet forming through a featured tool. International journal of machine tools and manufacture, 110, 1-17.

IOP Conf. Series: Materials Science and Engineering 1259 (2022) 012015 doi:10.1088/1757-899X/1259/1/012015

Application of Drone Systems for Spraying Pesticides in Advanced Agriculture: A Review

Ganesh P. Borikar[1], Chaitanya Gharat[2], Sachin R. Deshmukh[3]

[1] School of Mechanical Engineering, Dr. Vishwanath Karad MIT World Peace University, Pune, India
[2] School of Mechanical Engineering, Dr. Vishwanath Karad MIT World Peace University, Pune, India
[3] School of Mechanical Engineering, Dr. Vishwanath Karad MIT World Peace University, Pune, India

ganesh.borikar@mitwpu.edu.in , chaiatanyagharat10@gmail.com , sachin.deshmukh@mitwpu.edu.in

Abstract. The farmers in agriculture fields face health problems due to diseases caused by pesticides and insecticides. Thousands of cases result in adverse health effects when spraying pesticides manually onto crop fields. The use of drones for spraying pesticides is a promising alternative to manual spraying. It represents a major emerging field in India. Therefore, this article is an attempt to explore the latest technological advances and applications in Hardware, Flight Controller (FC) & Electronic Speed Controller (ESC), Smart Agricultural Sensors, and Spraying Systems. To make a mention, a few key advancements in drone systems are: The crops can be monitored by the use of a multispectral camera, mounted on a drone. The drone spraying system makes use of GPS coordinates to auto-navigate GPS coordinates to spray the pesticides on the infected areas in real-time as soon as the camera takes a picture of the spraying area. Different types of nozzles are applied to lead to specific sprinkling speeds and adaption of Bluetooth Low Energy (BLE) via smartphone. Auto control of the quantity of pesticide as per the speed of the drone and use of Artificial Intelligence (AI) for smart drone path control. Along with challenges in spraying technology, the possible direction of future research is highlighted. These technology upgrades lead to precision spraying, along with uniformity and less time. It could attract more and more farmers to spray pesticides by drone and increase crop yields in the coming years.

1. Introduction

In today's transforming world, the need for food resources has increased with the increasing population. With this increase in the food requirement, the demands from the agriculture sectors have increased tremendously. To fulfill this food demand and increase the food production rate, the agricultural sector needs to adopt automation that combines information and communication technologies (ICT), robotics, artificial intelligence (AI), big data, and the internet of things. This use of technology for increasing the food production rate to meet is called Smart Agriculture. Drones are one of the major technologies used in smart agriculture. Drones are unmanned aircraft, also called unmanned aerial vehicles (UAVs) in technical terms. In other terms, the drone is a robot that can be controlled remotely by a human or can fly autonomously, data acquired from sensors and other peripheral devices as well as commands from the mission planner. The drone is used for applications like search and rescue, monitoring, firefighting, surveillance, and agriculture, and the most common application is ariel photography [1]. Due to their great speed, accuracy, and effectiveness, drones are becoming more and more popular for carrying out different tasks [2]. Even the use of multiple drones for carrying out a similar task is possible which could reduce the time of completing the task [3].

In the history of drones, the United States, the United Kingdom, Russia, Germany, and Israel were the first countries to conduct research on drones. The first time an unmanned flying vehicle was used by the Austrians was in August of 1849. Charles Kettering, one of the first creators of drones in collaboration with Elmer Sperrym, Orville Wright, and Robert Milikanem created in 1915. They named the aircraft "Kettering Bug" which was automatic aircraft, which used to fly based on sensors barometer for height, the amount of engine spins for the calculating distance traveled, and the position [4].

Currently, drone technology is used to monitor real-time activities, manage disasters, provide security, and observe activities in the agriculture sector, mining sector, energy sector, and construction sector, among others [5]. With new upgrades coming up in drone technology it has stepped into performing different operations for various agriculture tasks. Talking about the use of drones in the agriculture sector it has a lot of applications which range from mapping and surveying to crop dusting and spraying [6]. So, in this article, reviews different types of drones, their working, different components used in drones especially concentrating on the crop spraying application using drones.

2. UAV Platforms

Different types of drones have been used based on their application of use. Although they were used primarily for military functions, in the beginning, drones are expanding in different sectors of the industry.

A drone is classified according to its size, range, ability, and weight. Drones can be classified into two main types: fixed-wing drones and rotary-wing drones [7].

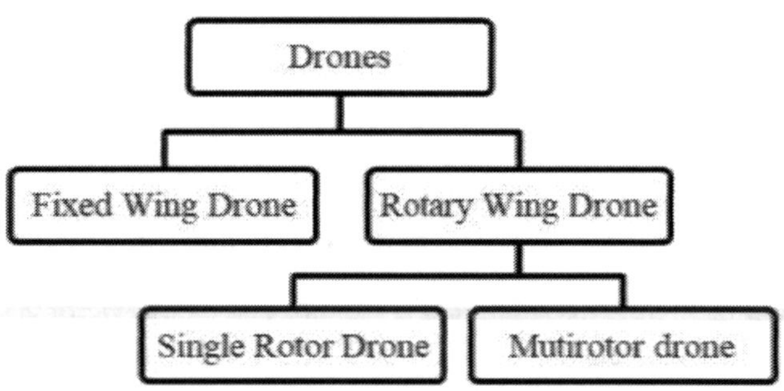

Figure 1. Type of Drones

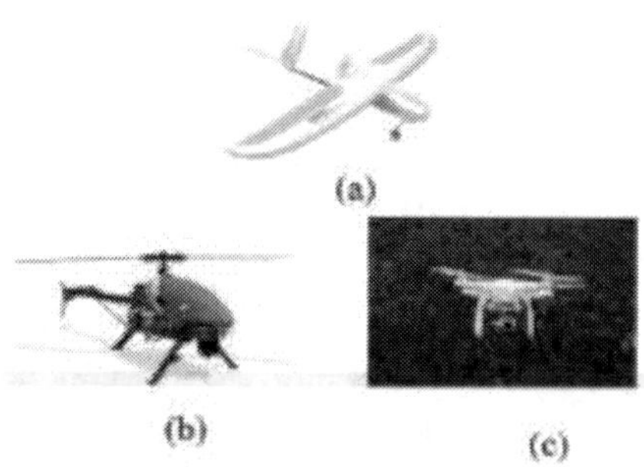

Figure 2. Type of drones (a) Fixed Wing Drone, (b) Single Rotor drone, (c) Multi-Rotor

2.1. Fixed-wing drone

The rigid streamlined wing structure of a fixed-wing drone (Fig. 2 (a)) induces lift as it moves [8]. It completely looks like a normal aircraft but without a pilot on board. It consists of one or two I.C. engines or an electric motor attached with a propellor that generates the forward thrust. The drone wings are made with a specific airfoil which generated an air pressure difference from the front surface of the airfoil to the end surface of the airfoil. The ailerons, elevators, and rudders installed on the drone's wings are used to regulate its movement. Drones with fixed wings are advantageous

IOP Conf. Series: Materials Science and Engineering 1259 (2022) 012015 doi:10.1088/1757-899X/1259/1/012015

because they can fly long distances, reach high altitudes, and operate quietly. They are used mostly for surveillance.

2.2. *Rotary-wing drone*

Rotary wing drones are those that generate lift by rotating blades called propellers [8]. Single Rotor Drones and Multi-Rotor Drones are the two types of Rotary Wing Drones. The Single Rotor Drones (Fig 2 (b)) are similar to helicopters in that they have a single rotor. They are mostly used in the construction industry for surveillance. Multi-rotor drones, on the other hand, are the most common form of drone in the market. The multirotor drone generates propulsion and manoeuvres by simply changing the speed of individual rotors with the help of a flight controller. The number of rotors on a Multi- Rotor Drone is categorised as follows: Tricopter (fig 3 (a)), Quadcopter (fig 3 (b)), Hexacopter (fig 3 (c)), and Octocopter (fig 3 (d)). The quadcopter is by far the most common type of multi-copter. They are commonly used in real-time monitoring, surveillance, mapping, etc. but the main application of multirotor drones is seen in aerial photography for beautiful cinematic videos.

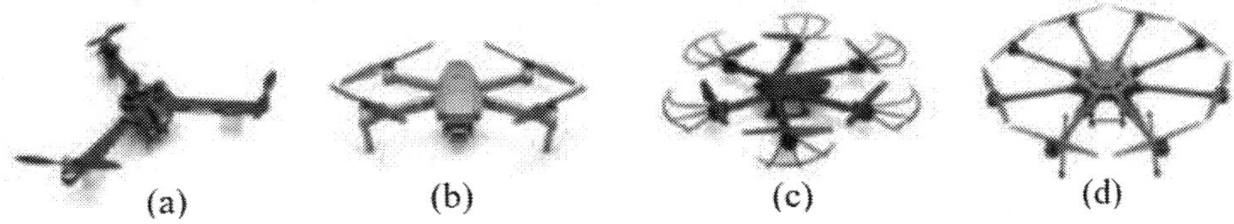

Figure 3. Multi rotor Drone (a) Tricopter, (b) Quadcopter, (c) Hexacopter,(d) Octacopter [38]

3. Working Principle

Each type of drone discussed above have their working principles but for this article, we will be concentrating mainly on the working principle of the Quadcopter. A quadcopter has 4 rotors that are equally spaced from each other with a definite circular wheelbase on the drone frame.

Drone use BLDC motors and the propellor blades are attached to the cover of the motor. The propellor blades are the most important part of the drone because they help the drone to generate proper amount of thrust which is the lift force for the drone. The blades are so designed that the lift force generated will be in the same direction on both parts of the blades. A flight controller along with the ESC controls the motor speed. The greater the motor speed (Blade speed) the greater will be the lift force.

Now comes the question of how the drone will take off. The lift force increases as the rotor speed increases, and when the lift force overcomes the weight of the drone, it is referred to as climbing. When the drone reaches a sufficient altitude, we reduce the rotor speed until the lift force equals the drone's weight, and the drone begins to hover in the air, which is known as drone hovering in technical terms. All four rotors rotate at the same pace while hovering and climbing.

Now the other question arises how does the drone move forward, sideways, and spin? Talking about moving forward, forward pitch (fig 4 (a)) is achieved when front propellors are spun at lower speed and the back propellor at higher speed. The drone's weight is balanced by the vertical component of the lift force, while the imbalanced horizontal force causes the drone to fly forward, causing a drag force on the body. The horizontal force must be increased until it matched the drag force for attaining constant speed. The same is the case for pitching backward. The same goes for rolling (fig 4 (b)) but here the side pair is spun faster than the opposite side pair. To yaw (fig 4 (c)) means to spin the drone the diagonal rotor spun at a different speed than the other diagonal rotors.

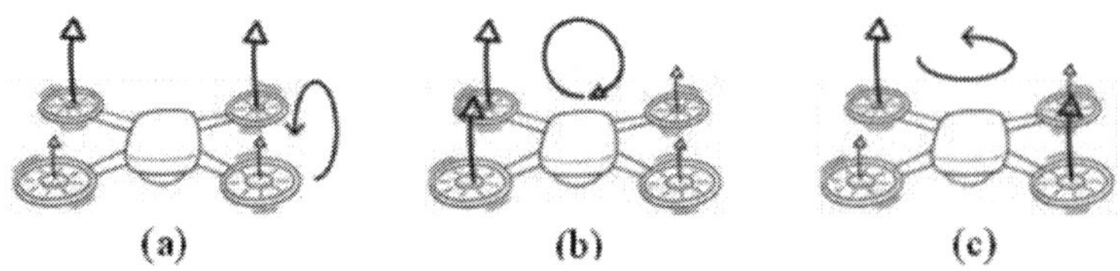

Figure 4. Movement of drone (a) Pitch, (b) Roll, (c) Yaw

4. Components of the Drone

A drone is made up of many different components which are interconnected to each other. But the key components necessary for a drone to work are a flight controller (Fig 6 (a)), BLDC Motors (Fig 6 (b)), ESC's (Fig 6 (c)), propellors (Fig 6 (d)), battery (Fig 6 (e)), and a radio controller (Fig 6 (f)). The drone's brain is referred to as a flight controller. The flight controller receives data from the human-controlled radio transmitter as well as sensors onboard. The controller then processes the information and sends commands to the ESCs, which control the motor's speed. The ESCs are attached to the power distribution board (the battery), and the flight controller receives signals from the ESCs, which govern the power delivered to each of the motors. The primary BLDC motors are used in drones. These motors are responsible for rotational motion. The propellers are connected with the motors which convert the rotational motion of the motors into thrust force which is required for lifting the drone in the air. For all these components to work we need an energy source which is the battery. The battery supplies the electric energy required to each component of the drone. The battery used in the drone is mostly LIPO batteries. Additionally, the drone also consists of two more components which are the accelerometer and the gyroscope. Because the accelerometer monitors both acceleration and force, the descending gravity will be detected as well. We can determine the device's orientation using the accelerometer's three-axis sensors. A gyroscope is a device that monitors angular velocity, or rotational speed around the three axes. A gyroscope is a gadget that helps determine orientation by using Earth's gravity. A freely revolving disc known as a rotor is attached to a spinning axis in the middle of a bigger and more stable rotor [9].

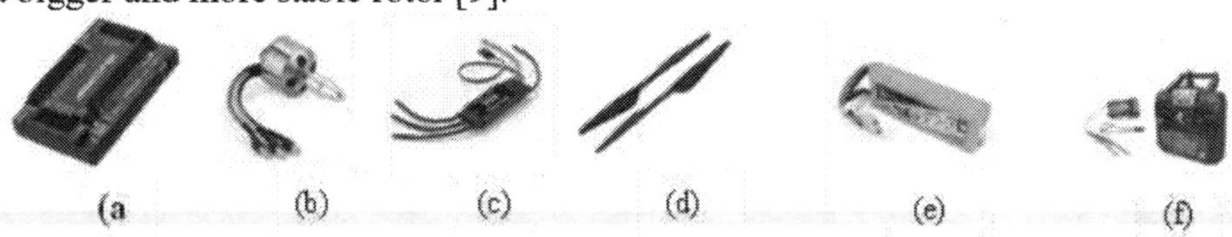

Figure 5. Drone Components (a) Flight controller, (b) BLDC Motors, (c) ESC, (d) Propellors, (e) Battery, (f) Radio Transmitter and Receiver [40]

5. Drone Control Approach

Drones are controlled with two different approaches either manually or autonomously. The manual approach relies on the commands given by humans in real-time by the use of transmitters like radio transmitters [10], Bluetooth, or Wi-Fi. The signal from the transmitter is received by the drone's receiver, which then delivers it to the flight controller for additional processing.

The Autonomous approach does not depend on the real-time control commands from the human but it follows the pre-specified instructions given by the human in the mission planning software. The commands consist of path plan, speed limit, altitude range, etc. The drone operates autonomously without human intervention and does the given task by following the instructions. The different flight controllers with their controlling approaches are given in the table below (Table 1).

IOP Conf. Series: Materials Science and Engineering 1259 (2022) 012015 doi:10.1088/1757-899X/1259/1/012015

Table 1. Flight controller with control approach

Flight controller	Control Approach	References
NAZA M-lite flight controller	Radio Transmitter	[1]
ATmega328	Radio Transmitter	[3][18][28]
Arduino mega	Wi-Fi	[2]
APM Ardupilot controller	Radio Transmitter	[10]
Pixhawk controller	Autonomous	[22][23][32]
KK 2.1.5	Radio Transmitter	[25][26][31]
Raspberry Pi Model 3 B	Autonomous	[36]

6. Recent Trends and Challenges

The drone is increasingly being used in the agricultural sector with technological advancement. These help in reducing the efforts of farmers and also on other hand providing more accuracy and fast operation. The drone is nowadays used in various fields of agriculture works like crop monitoring, seeding, spraying of insecticides and pesticides, and irrigation [11-13]. The predominant use of drone today in agricultural sector is spraying of pesticides. Details of the drones used by various users in recent time for pesticides spraying along with their technical elements are listed below in Table 2.

Table 2. Existing drones for spraying pesticides in recent time

Exiting Drones	Volume of pesticide	Max. Flight time full load	Max. Speed	Discharge rate	No. of nozzles	Reference
DJI Agrus MG-1S	10 L	10 min	12 m/s	0.379 L/min	4	[38]
3WQF120-12	12 L	30 min	5 m/s	0.8 L/min	2	[39]
3CD-15	15 L	20 min	6 m/s	0.54 L/min	4	[39]
WSZ-0610	10 L	20 min	4 m/s	0.72 L/min	2	[39]
HY-B-15L	15 L	15 min	4.5 m/s	0.38 L/min	5	[39]
N-3 UAV	25 L	-	4 m/s	0.85 L/min	2	[40]
Knapsack-type electric fog sprayer 3WBD	20 L	-	1m/s	1 L/min	1	[40]

Along with the benefits of drones, there are some challenges in agricultural drones. They are like less flight time, balancing of drone with the pesticide tank, harsh weather conditions, etc. The most important challenge is the flight time of the drone, due to the relatively higher payload. The flight time of the drones used in agriculture is short, which ranges from 10-30 minutes which leads to less coverage of land with every charge. The other challenge is that for large area of spraying, drones are less efficient and they are costlier than the other equipment used for the same purpose.

7. Drone for Crop Spraying

The drone has a wide range of applications in various sectors for different industries and hence it also has some appearance in the agriculture sector. The drone can serve the agriculture sector right from surveying and monitoring the field and crop to spraying of pesticides and seeds on the field. The main problem faced while growing crops are pests and weeds. They tend to damage the crop, which results in loss of money, time, and efforts in a short loss in productivity [14]. To kill pest and weeds, herbicides and fertilizers must be used. However, the WHO (World Health Organization) estimates that over 1 million pesticide cases occur each year. More than one lakh people die each year as a result of pesticides sprayed by humans and pesticide handling, primarily in developing countries [20]. As a result, hand fertilizer and pesticide spraying might result in lethal diseases such as cancer, asthma, hypersensitivity, and other disorders [14-20].

Here's where we'll be able to use the drones. Crop monitoring, as well as the need to spray pesticides and fertilizers at precise plant locations, is a crucial component for increasing crop output. The Drones can complete the task fast and accurately without giving any health hazards to humans. The drones used in agriculture sectors are majorly autonomous drones, which decides by themselves by sensing the surrounding and considering the task allotted to them by the human. The Pesticide spraying drones typically have a storage tank in the lower section of the drone that holds the liquid pesticide, and nozzles are connected to the storage tank through pipes and an electric pressure pump to spray the pesticides down on the crops. The storage tank, nozzle, and pressure pump work in one of two ways: manually with a controller or automatically with previously programmed commands for spraying pesticides at precise spots. The following table lists the spraying systems details /components used in pesticide spraying drones (Table 3).

Table 3. Various types of spraying systems used in drones

Tank capacity (liter)	Nozzle type	Pump discharge	Reference
5 litres	Flat fan	2.5 L/min	[17]
5.7 L	Micronair nozzles	100 mL/min	[19]
-	The Universal nozzle	1 L/min	[20]
10 L	XR11001	0.43 L/min	[21]
13.2 L	flat fan nozzles	46.8 L/ha	[24]
6 L	flat fan nozzles	2.5 L/min	[25]
250ml	-	250 mL/min	[31]
5 L	-	2.4 L/min	[35]

8. Conclusions

There is a vast scope for using drones in agricultural applications. They can be used from soil monitoring and surveying to sprinkling of seeds and pesticides. However, there are some limitations and challenges to overcome. This article briefly summarize pesticide spraying drone systems /various components. It specified various drone platforms, control approaches, and agricultural UAV applications that have been established or are currently being researched. The use of different controlling approaches has been discussed according to their application. It highlights the need to minimize pesticide spraying drone weight, maximize drone flight time and more precise autonomous control of drone.

References

[1] B. Y. Suprapto, M. A. Heryanto, H. Suprijono, J. Muliadi, and B. Kusumoputro, *Design and development of heavy-lift hexacopter for heavy payload*, 2017 International Seminar on Application for Technology of Information and Communication: Empowering Technology for a Better Human Life, iSemantic, Volume 2018-Jan.

[2] Spoorthi, S., Shadaksharappa, B., Suraj, S., Manasa, V.K., *Freyr drone: Pesticide/fertilizers spraying drone-an agricultural approach.* IEEE 2nd International Conference on In Computing and Communications Technologies (ICCCT - 2017), pp. 252-255.

[3] Prof. Swati D Kale, Swati V khandagale, Shweta S Gaikwad, Sayali S Narve, Purva V Gangal, *Agriculture Drone for Spraying Fertilizer and Pesticides*, International Journal of Advanced Research in Computer Science and Software Engineering Volume 5, Issue 12, December 2015.

[4] Piotr Kardasz, Jacek Doskocz, Mateusz Hejduk, Paweł Wiejkut, Hubert Zarzycki, *Drones and Possibilities of Their Using, Journal of Civil & Environmental Engineering*, Volume 6 Issue 3, January 2016.

[5] S. Sunderaraj, K. Dharsan , J. Ganeshraman , D. Rajarajeswari, *Structural and modal analysis of hybrid low altitude self-sustainable surveillance drone technology frame*, International Conference on Newer Trends and Innovation in Mechanical Engineering: Materials Science

[6] U. R. Mogili and B. B. V. L. Deepak, *Review on application ofdrone systems in precision agriculture,* Procedia ComputerScience, Volume. 133, pp. 502–509, 2018

[7] Jerrin Bright, R Suryaprakash, S Akash, A Giridharan, *Optimization of quadcopter frame using generative design and comparison with DJI F450 drone frame*, IOP Conference Series Materials Science and Engineering · December 2020 1012 (2021) 012019

[8] G.D. Goh, S. Agarwala, G.L. Goh, V. Dikshit, W.Y. Yeong, *Additive manufacturing in unmanned aerial vehicles (UAVs): Challenges and potential*, Aerospace Science and Technology Volume 63, April 2017, pp. 140-151.

[9] Vanitha, N., Vinodhini, V., & Rekha, S. *A Study on Agriculture UAV for Identifying the Plant Damage after Plantation.,* International Journal of Engineering and Management Research (IJEMR), Volume. 6 Issue 6, 2016, pp.310-313.

[10] MD. Faiyaz Ahmed, Mohd. Nayab Zafar, J. C. Mohanta, *Modeling and Analysis of Quadcopter F450 Frame*, International Conference on Contemporary Computing and Applications, 2020, pp. 196-201.

[11] Carlo Ferro, Roberto Grassia, Carlo Seclìb, Paolo Maggiorea, *Additive Manufacturing Offers New Opportunities in UAV Research*, 48th CIRP Conference on MANUFACTURING SYSTEMS - CIRP CMS 2015 pp. 1004-1010.

[12] Pavol Pecho,Viliam Ažaltovič, Branislav Kandera, Martin Bugaj, *Introduction study of design and layout of UAVs 3D printed wings in relation to optimal lightweight and load distribution*, 13th International Scientific Conference on Sustainable, Modern and Safe Transport (TRANSCOM 2019), May 2019 ,pp. 861-868.

[13] S. Ahirwar, R. Swarnkar, S. Bhukya and G. Namwade, *Application of Drone in Agriculture*, International Journal of Current Microbiology and Applied Sciences ISSN: 2319-7706 Volume 8 Number 01, 2019, pp. 2500-2505.

[14] N. Shahrubudin, T.C. Lee, R. Ramlan, *An Overview on 3D Printing Technology: Technological, Materials, and Applications*, 2nd International Conference on Sustainable Materials Processing and Manufacturing, SMPM 2019 pp. 1286-1296.

[15] Ebubekir KOÇ, Cemal Irfan ÇALIŞKAN, Mert COŞKUN, Hamaid Mahmood KHAN, *Unmanned Aerial Vehicle Production with Additive Manufacturing*, Journal of Aviation Volume 4 Issue (1), 2020, pp. 22-30.

[16] V. Kangunde, R. S. Jamisola, E. K. Theophilus, "A review on drones controlled in real-time," , International Journal of Dynamics and Control, vol. 2021, , 2021, pp. 1–15

[17] Yallappa, D., Veerangouda, M., Maski D., Palled V., Bheemanna M., *Development and evaluation of drone mounted sprayer for pesticide applications to crops*, IEEE Global

Humanitarian Technology Conference (GHTC) 2017 IEEE , pp. 1-7.

[18] Patrick Di Justo, Make: DIY Drone and Quadcopter Projects

[19] Huang, Y., Hoffman, W. C., Lan, Y., Fritz, B. K., & Thomson, S. J. *Development of a low-volume sprayer for an unmanned helicopter.* Journal of Agricultural Science, Volume 7 Issue 1, , 2014 pp. 148-153.

[20] Meivel S., Maguteeswarn R., Gandhiraj N., Srinivasan G. *Quadcopter UAV based Fertilizer and Pesticide Spraying System.* International Academic Research Journal of Engineering Sciences, Vol 1 issue 1, February 2016, pp.8-12.

[21] Sarghini, F.; Visacki, V.; Sedlar, A.; Crimaldi, M.; Cristiano, V.; De Vivo, A. *First measurements of spray deposition obtained from UAV spray application technique*, Proceedings of the IEEE International Workshop on Metrology for Agriculture and Forestry (MetroAgriFor), Portici, Italy, 24–26 October 2019; pp. 58–61.

[22] Budiharto, W.; Chowanda, A.; Gunawan, A.A.S.; Irwansyah, E.; Suroso, J.S. A Review and Progress of Research on Autonomous Drone in Agriculture, Delivering Items and Geographical Information Systems (GIS). 2019 2nd World Symposium on Communication Engineering (WSCE), Nagoya, Japan, 20–23 December 2019; pp. 205–209

[23] Rao, V.S.; Gorantla, S.R. *Design and Modelling of an Affordable UAV Based Pesticide Sprayer in Agriculture Applications.*, 2019 Fifth International Conference on Electrical Energy Systems (ICEES), Chennai, India, ; Volume 360, 21–22 February 2019, pp. 1–4.

[24] Li, X.; Giles, D.K.; Niederholzer, F.J.; Andaloro, J.T.; Lang, E.B.; Watson, L.J. *Evaluation of an unmanned aerial vehicle as a new method of pesticide application for almond crop protection.*, Pest Manag. Sci. 2021, 77, 527–537.

[25] Shaw, K.K., Vimalkumar, R.: *Design and development of a drone for spraying pesticides, fertilizers and disinfectants.*, International Journal of Engineering Research & Technology (IJERT), 2020, 1181–1185.

[26] Rahul Desale, Ashwin Chougule, Mahesh Choudhari, Vikrant Borhade, S.N. Teli, *Unmanned Aerial Vehicle for Pesticides Spraying*, April 2019, IJSART, ISSN: 2395-1052.

[27] Shilpa Kedari, Pramod Lohagaonkar, Monika Nimbokar, Gangaram Palve, Prof. Pallavi Yevale *Quadcopter - A Smarter Way of Pesticide Spraying*, Imperial Journal of Interdisciplinary Research (IJIR) Vol-2, Issue-6, 2016.

[28] Sadhana, B., Naik, G., Mythri, R. J., Hedge, P. G., & Shyama, K. S. B., *Development of quad copper-based pesticide spraying mechanism for agricultural applications*., International Journal of Innovation Research Electrical Electronics Instrumentation Control Engineering, Volume .5, 2017, pp.121-123.

[29] Gayathri Devi K, Sowmiya N, Yasoda K, Muthulakshmi K, Kishore B. Review on application of drones for crop health monitoring and spraying pesticides and fertilizer., 2020, Volume 7, pp.667–72.

[30] Tripicchio, P.; Satler, M.; Dabisias, G.; Ruffaldi, E.; Avizzano, C.A. *Towards smart farming and sustainable agriculture with drones.*, 2015 International Conference on Intelligent Environments, Prague, Czech Republic, July 2015; pp. 140–143.

[31] M. M. Vihari, U. R. Nelakuditi, *IoT based Unmanned Aerial Vehicle System for Agriculture Applications,* International Conference on Smart Systems and Inventive Technology, 2018, pp. 26-28,.

[32] P. Garre and A. Harish, *Autonomous agricultural pesticide spraying uav*, IOP Conference Series: Materials Science and Engineering, Volume. 455, IOP Publishing, 2018, pp. 12-30.

[33] Praveen Kumar Reddy Maddikunta, Saqib Hakak, Mamoun Alazab, Sweta Bhattacharya, Thippa Reddy Gadekallu, Wazir Zada Khan, Quoc-Viet Pham, *Unmanned Aerial Vehicles in Smart Agriculture: Applications, Requirements and Challenges*, January 2021IEEE Sensors Journal Volume 1 Issue1, 2021, pp.99.

[34] Tanha Talaviya, Dhara Shah, Nivedita Patel, Hiteshri Yagnik, Manan Shah, *Implementation of artificial intelligence in agriculture for optimisation of irrigation and application of pesticides*

and herbicides, Artificial Intelligence in Agriculture Volume 4, 2020, pp.58–73.

[35] Meivel S, Dinakaran K, Gandhiraj N and Srinivasan M, *Remote sensing for UREA Spraying Agricultural (UAV) system*, 2016 3rd International Conference on Advanced Computing and Communication Systems (ICACCS) Volume 1, 2016, pp 1–6.

[36] Saha, A. K., Saha, J., Ray, R., Sircar, S., Dutta, S., Chattopadhyay, S. P., & Saha, H. N., *IOT-based drone for improvement of crop quality in agricultural field.*, 2018 IEEE 8th Annual Computing and Communication Workshop and Conference (CCWC) ,2018, pp. 612-615.

[37] S. Bhandari, S. Pathak, R. Poudel, R. K. Maskey, P. L. Shrestha, Binaya Baidar, *Design and Development of Hexa-copter for Environmental Research*, 11th International Conference on ASIAN Community Knowledge Networks for the Economy, Society, Culture and Environmental Stability, 2016.()

[38] M. F. F. Rahman, S. Fan, Y. Zhang, and L. Chen, *A Comparative Study on Application of Unmanned Aerial Vehicle Systems in Agriculture*, Agric. 2021, Volume. 11, Jan. 2021,Page 22.

[39] Shilin, W.; Jianli, S.; Xiongkui, H.; Le, S.; Xiaonan, W.C.W.; Zhichong, W.; Yun, L.; Changling, W. *Performances evaluation of four typical unmanned aerial vehicles used for pesticide application in China.* Int. J. Agric. Biol. Eng. 2017, 10, 22–31.

[40] Qin, W.; Xue, X.; Zhang, S.; Gu, W.; Wang, B. *Droplet deposition and efficiency of fungicides sprayed with small UAV against wheat powdery mildew.* Int. J. Agric. Biol. Eng. 2018, 11, 27–32.

IOP Conf. Series: Materials Science and Engineering 1259 (2022) 012016 doi:10.1088/1757-899X/1259/1/012016

Structural and modal analysis of gas turbine blade using ansys

Hrushikesh Paricharak[1] and Dr. S .K. Tiwari

Department of mechanical enginrring, Dr. B. R. Ambedkar NIT Jalandhar, Punjab

Abstract. Gas turbine technology has advanced significantly during the last twenty years. The advancement in materials technology, new coatings, and novel cooling techniques are driving the expansion. Material selection is critical since turbomachinery design is complicated, and material performance is directly connected to efficiency. The materials should be able to withstand high vibrations, high temperature, and pressure. In this work, the maximum stress and the deformation induced due to pressure were determined to study the vibrational response, mode shapes, and natural frequencies. Four different materials have been considered, namely Inconel 617, Titanium alloy, Rhenium alloy, and Nimonic 80A. A suitable material for a gas turbine blade was suggested after analyzing these materials. The 3D CAD model is modeled in CATIA V5 and analysis is carried out using FEM-based software ANSYS

1.Introduction

Gas turbine generally used in the power plant sector produces a large amount of energy which depends on their size and weight. The use of gas turbines has experienced phenomenal growth in the power sector, both among utilities and the petrochemical industry, as well as in the merchant plants and utilities all over the world, over the last four decades. Material selection is critical since turbomachinery design is complicated and material performance is directly connected to efficiency. Similar problems exist in both gas and steam turbines, although the magnitudes of these problems vary. Components in turbines must be able to tolerate a broad variety of loads, temperatures, and erosion [1]. Whether in commercial aircraft, fighter jets, aviation sector gas turbines are crucial [2]. During the engine testing and in the service of aircraft, it is observed that the failure occurs mostly in stator and rotor blades. Hence, the analysis of the blade is required.A static structural study detects the stresses, displacements, strains, and pressures caused by loads with negligible inertia and damping in structures or components. The response and loading conditions are assumed to be steady.The process of determining fundamental dynamic features of the system, such as damping factor, natural frequencies, and mode shapes, and utilizing them to construct a mathematical model for its dynamic response is known as modal analysis. The essential premise of modal analysis is that, simple elements are used, such as natural vibration modes to describe the complicated response in structural dynamics[3].In[2] the authors studied the stress and vibrational characteristics of gas turbine blades made of four different materials. In[4] the authors replaced a turbine blade with two different materials to have a better understanding of the characteristics of the turbine blade and to reduce the risks of failure. The natural frequency of vibration of turbine blades made of IN 738 and U500 alloys was determined. In[5] the authors studied the behavior of the LPT gas turbine blade made of Nickel-based superalloys, IN 738, U500, and Mar-M246. Parameters such as maximum amplitude, natural frequency, and blade mass were determined for each turbine blade material. In the present work, gas turbine blade is examined for vibration and stressed conditions. The effect pressure acting on the blade of the turbine on the maximum stress and the maximum deformation for the selected materials have been observed. With different materials, the effect on the natural frequency of gas turbine was also observed, and hence, mode shapes were obtained for six modes of vibrations. Materials considered for

this study were Titanium alloy, Inconel 617, Nimonic 80A, and Rhenium-based alloy [6]. Among them, the well-suited material was suggested based on the results of static and modal analysis.
The study is organized into the following sections:
The materials utilized and their properties are explained in Section 2. Modeling of the gas turbine blade is presented in Section 3. Section 4 presents the simulation results. The study is concluded in Section 5.

2. Material Properties

Significant vibrations, strong loads, and high heat effects are the principal causes of turbine blade failure. A turbine blade should be made of a material with good mechanical properties including shear strength, fatigue strength, and Young's modulus. The material used should be durable and ductile at the same time.Gas turbine blades were formerly composed of pure metals such as steel and cast iron. However,with the advancement of technology pure metals are being replaced by metal alloys and composites because of their superior strength-to-weight ratio. Titanium alloy, Inconel 617, Nimonic 80A and Rhenium alloy are the materials employed in this investigation.

2.1 Material Properties

Titanium Alloy: When compared to competing materials such as steel or nickel-based superalloys, Titanium alloys have become popular for the manufacturing of gas turbine components due to improvement in ductility strength, microstructural stability, and oxidation/corrosion resistance[7].
Inconel 617: Inconel 617 is a high-performance nickel-based solid solution superalloy with outstanding engineering properties. Because of its outstanding qualities such as strong creep resistance and thermal strength, this alloy is widely used in aerospace, power plants, nuclear sectors, and chemical industries. Furthermore, due to its improved hot corrosion behavior derived from the simultaneous presence of chromium, aluminium, and molybdenum in the alloy composition, this alloy is often utilized in a variety of reducing and oxidizing circumstances[8].
Nimonic 80A: Nimonic alloys are mostly composed of chromium and Nickel. These alloys can withstand incredibly high temperatures and still work well. They are commonly used in power plants and aircraft components [9].
Rhenium Alloy: The addition of rhenium increases the mechanical properties of the material. The addition of rhenium not only enhances thermo-mechanical fatigue strength, creep resistance, but also improves environmental properties [10].

The characteristics of the materials used in study are listed in Table 1

Table 1.Material Properties

Material	Density (Kg/m^3)	Young's Modulus (GPa)	Thermal Conductivity (W/mk)	Poisson's Ratio
Titanium Alloy	7850	110	7.1	0.33
Inconel 617	8360	211	13.6	0.30
Nimonic 80A	8190	222	11.2	0.35

Material	Density (Kg/m^3)	Young's Modulus (GPa)	Thermal Conductivity (W/mk)	Poisson's Ratio
Rhenium Alloy	21030	469	39.6	0.29

3.CAD Modelling

CATIA V5.0 is used to model a three-dimensional CAD model. Aerofoils of the blade have been plotted using coordinates provided by Siemens Power Generation, Inc. [11]. The length of the blade is 165 mm along the z-axis. The blade has 12 sections along the z-axis, every section contains an aerofoil having different shapes. These aerofoils are then used to generate a three-dimensional model by using a multi-section solid command. Figure 1 shows the wireframe view and three-dimensional view.
The developed three-dimensional CAD model is then imported for further analysis in ANSYS 2021. After importing the model, different FEA steps were applied. First, the model is discretized i.e. meshed using tetrahedral elements. The mesh is converged approximately at 5 mm element size (1,20,000 elements). The graph for mesh convergence and three-dimensional view of the meshed model is presented in Figure 2 and 3.Boundary conditions are then applied for Static structural and Modal analysis. For static structural analysis boundary conditions are: the tip of the blade is free and the root of the blade is fixed. A pressure of 1.5 MPa is applied on the upper surface of the blade under the thermal condition of 850°C. For Modal analysis, boundary conditions applied are: the tip of the blade is free and the root of the blade is fixed.

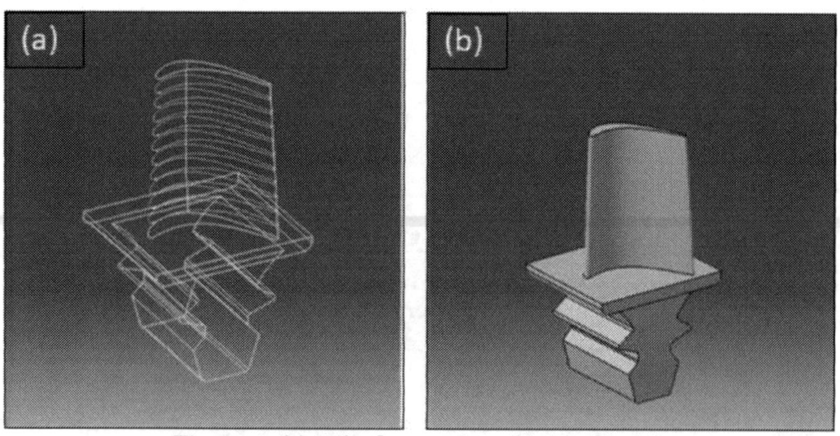

Figure 1.(a) Wireframe view (b) 3D Model

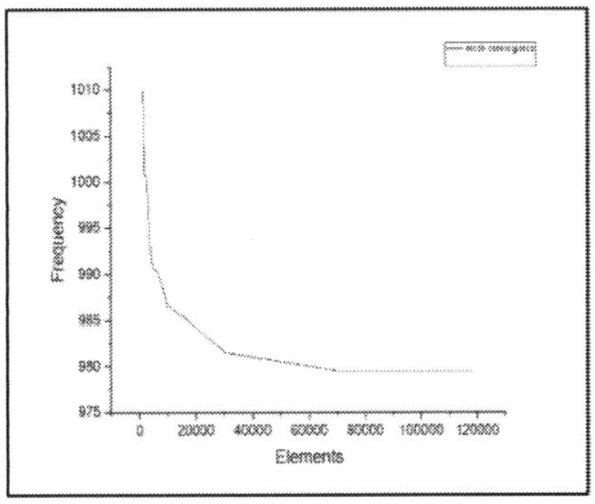

Figure 2.Mesh Convergence

Figure 3. Meshed Model

4. Results and Discussions

4.1 Static analysis

First, static analysis is performed to calculate the maximum stresses and deflections in a gas turbine blade. As per the boundary conditions, the tip of the blade was free and the blade root was fixed, and the temperature applied was 850°C. Figure 4 shows maximum deformation and stress induced in the Titanium alloy. The maximum deformation and the maximum stress generated for Inconel 617, Nimonic 80A, and Rhenium-based alloy are shown in Figure 5 to 7.The maximum deflection and maximum stress for each of the four materials have been obtained. For Titanium Alloy, maximum deflection and maximum stress were 1.9199 mm and 6439.6 MPa respectively. Similarly, for Inconel 617, 1.6242 mm and 5678 MPa, For Nimonic 80A, 1.6245 mm and 7263.5 MPa, for Rhenium alloy, were 1.9785 mm and 15533 MPa, respectively.All four materials have been evaluated based on the maximum stresses caused in the gas turbine blade. The lower the value of maximum stress induced, higher are the chances for the material of the gas turbine blade to sustain under the specified boundary conditions. From the obtained results, it can be said that the descending order of the material for the maximum stress condition is given by: Rhenium alloy>Nimonic 80A>Titanium alloy> Inconel 617. Therefore, Inconel 617 is well-suited material from the results of static structural analysis.

4.2 Modal analysis

The root of the turbine blade is fixed while the tip of the blade is free for modal analysis. Modal analysis is used to find the natural frequencies of the first six modes. Figure 8 displays the vibration data for the first six modes of a Titanium alloy turbine blade. The range of the natural frequency for Titanium alloy was from 979.36 Hz to 4507.4 Hz. For Inconel 617, it varies from 1078 Hz to 4994.4 Hz. Similarly, for Nimonic 80, 1118.2 Hz to 5152.2 Hz, and for Rhenium alloy, 1013.2 Hz to 4699.5 Hz. Six Mode shapes of Titanium alloy, Inconel 617, Nimonic 80 A, and Rhenium Alloy are shown from Figure 8-Figure 11 respectively.

The natural frequencies for all the materials for six mode shapes are enlisted in Table 2.

IOP Conf. Series: Materials Science and Engineering 1259 (2022) 012016 doi:10.1088/1757-899X/1259/1/012016

Table 2. Natural frequencies for different materials on each mode

Mode	Frequency (Hz) Titanium Alloy	Inconel 617	Nimonic 80A	Rhenium Alloy
1	979.36	1078	1118.2	1013.2
2	1356.9	1510.4	1552.1	1422.3
3	2527.5	2796	2888.2	2630.2
4	2801	3114.1	3203.3	2932.1
5	3768.3	4137.4	4300.5	3887.6
6	4507.4	4994.4	5152.2	4699.5

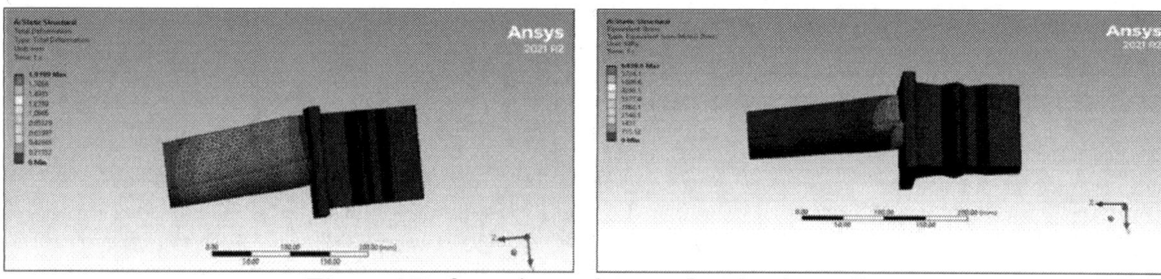

Figure 4.Deformation and Stress for Titanium Alloy

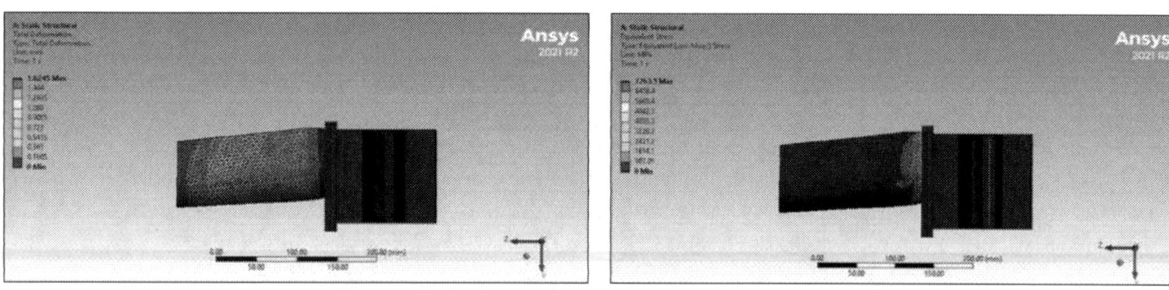

Figure 5. Deformation and stress for Inconel 617

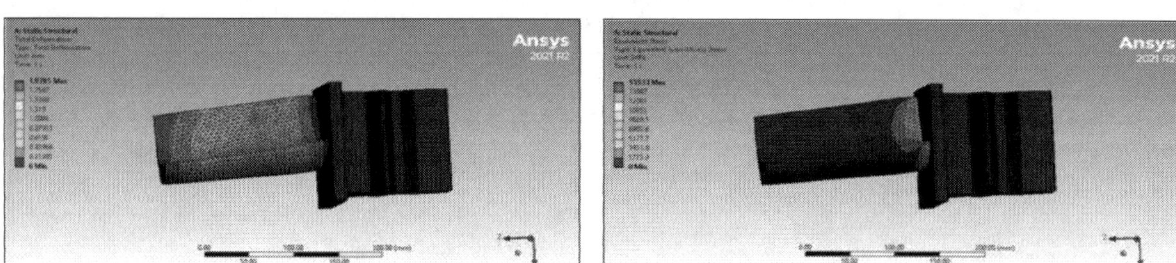

Figure 6. Deformation and stress for Nimonic 80A

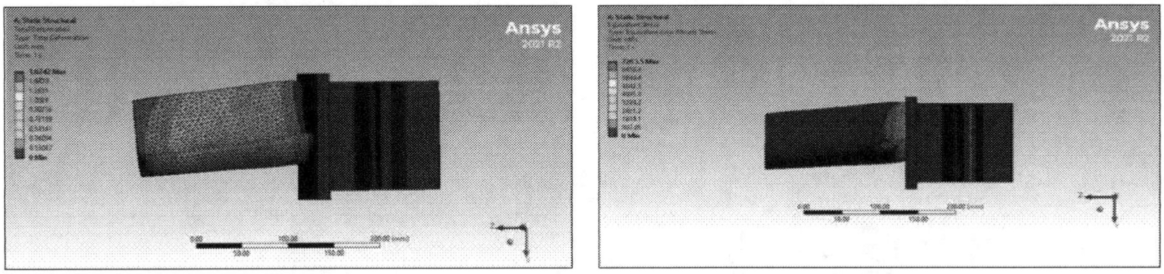

Figure 7. Deformation and stress for Rhenium Alloy

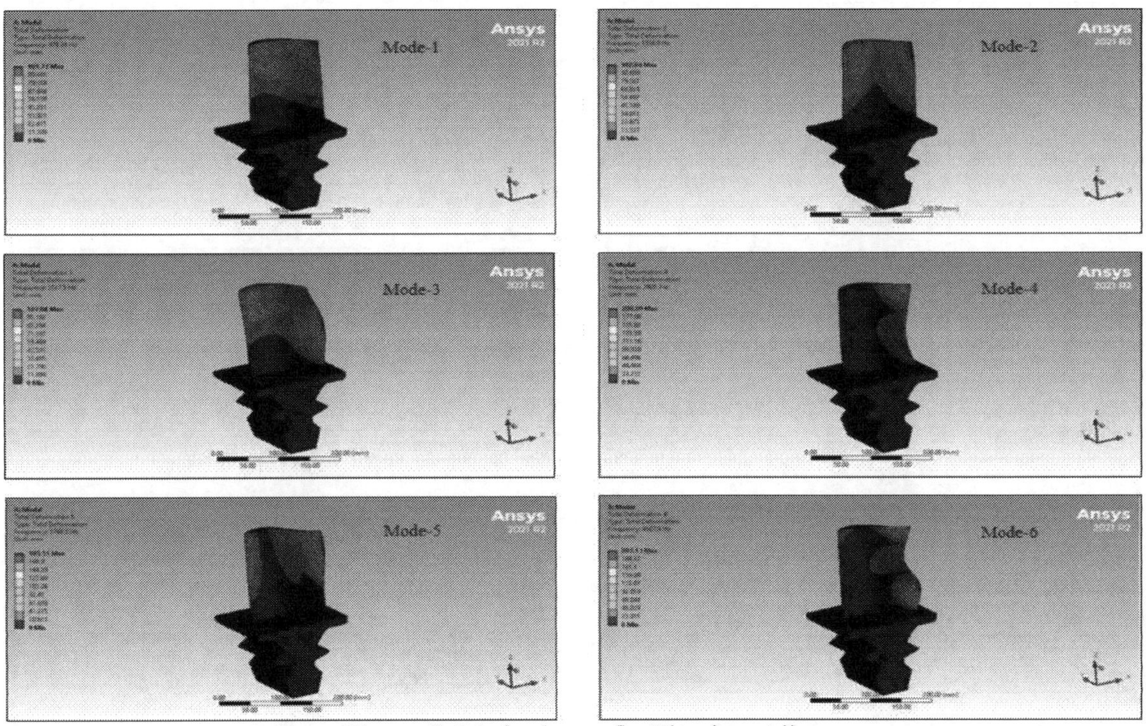

Figure 8. Mode shapes for Titanium Alloy

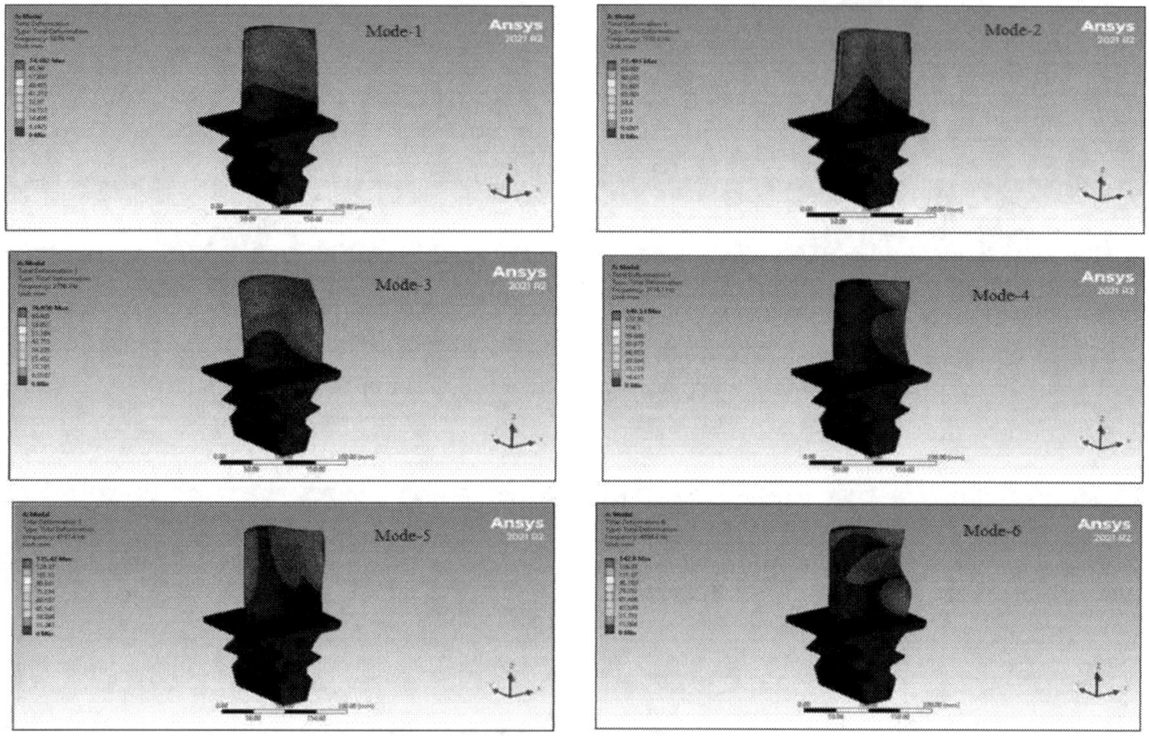

Figure 9. Mode shapes for Inconel 617

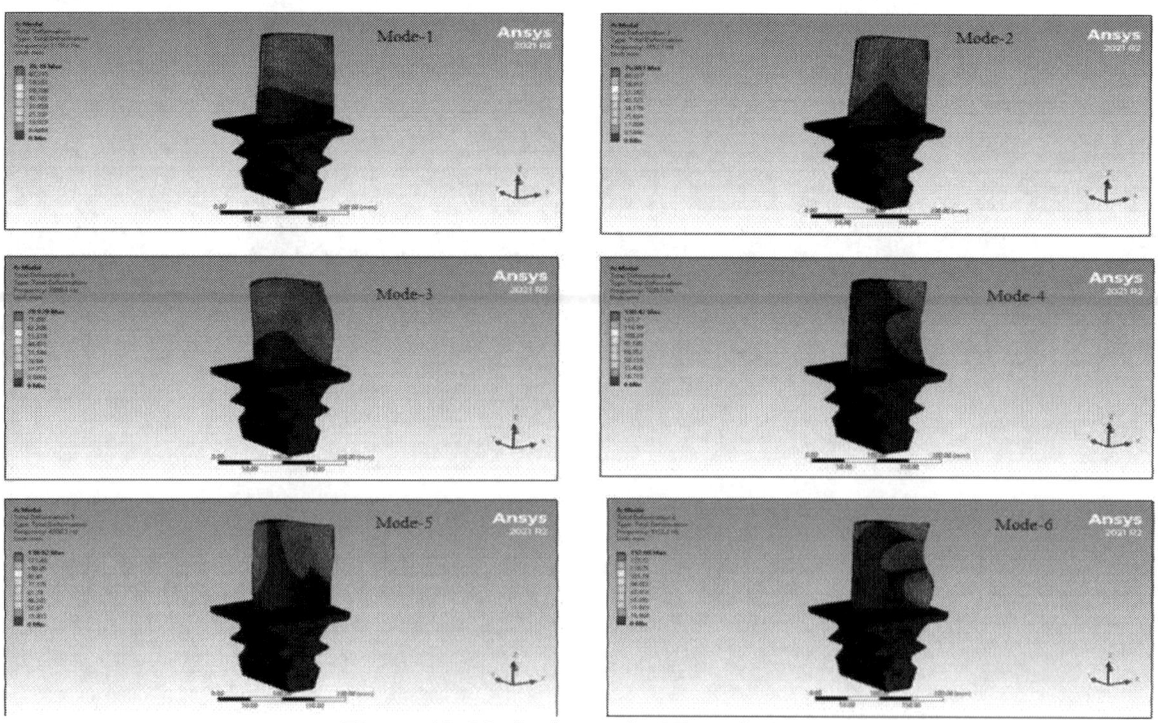

Figure 10. Mode shapes for Nimonic 80A

IOP Conf. Series: Materials Science and Engineering 1259 (2022) 012016 doi:10.1088/1757-899X/1259/1/012016

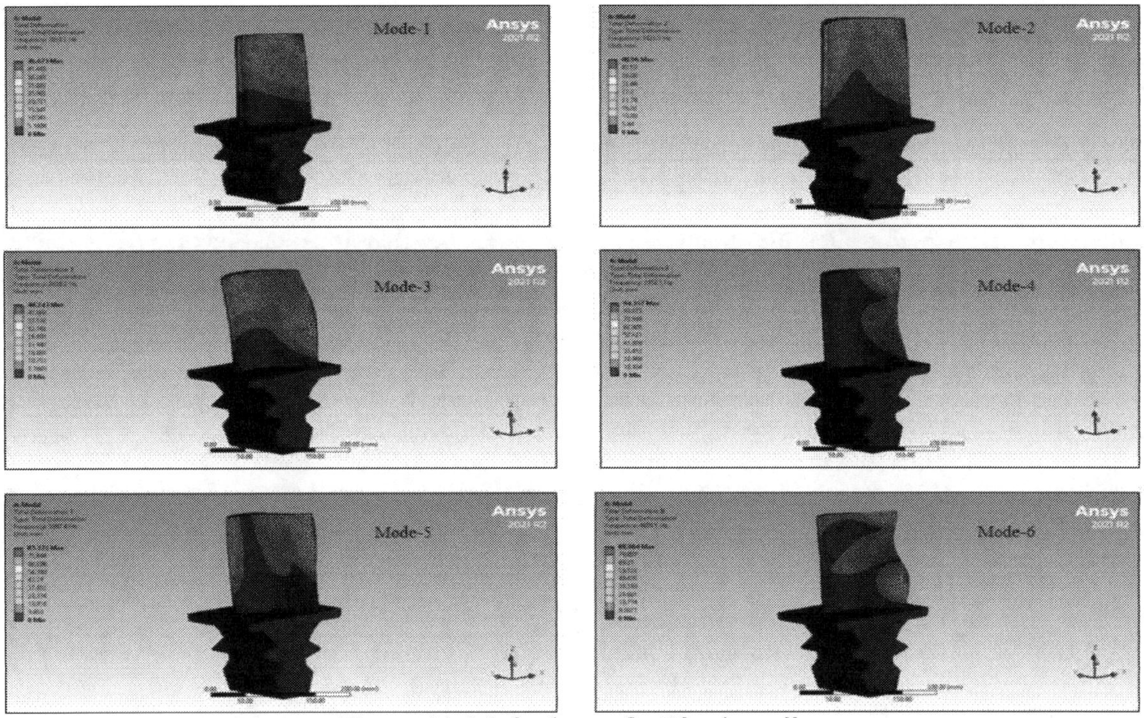

Figure 11. Mode shapes for Rhenium alloy

5.Validation

The results of the study was validated and was found to be within permissible limit with the existing results

Table 3. Validation for static analysis

	Kumar and Pamdey [12]		Results		Error	
Material	Stress (Mpa)	Deformation (mm)	Stress (Mpa)	Deformation (mm)	Stress (%)	Deformation (%)
Nimonic 80A	26287	3.3881	25958	3.587	1.26	5.54

Table 4. Validation for modal analysis

Mode	Kumar and Pamdey[12]		Results		Error	
	Frequency (Hz)	Deformation (mm)	Frequency (Hz)	Deformation (mm)	Frequency (%)	Deformation (%)
1	511.8	88.521	519.62	86.799	1.5	1.94
2	1370.7	166.32	1359.1	162.5	0.84	2.29

6. Conclusion

Four different materials, i.e., Titanium alloy, Inconel 617, Nimonic 80A, and Rhenium alloy were examined to study the characteristics of the gas turbine blade, due to the effect of pressure, temperature, and vibrations. The resulting stress and deformation due to pressure acting on gas turbine blades were determined using static structural analysis. It has been observed that Inconel 617 exhibits the least value of induced maximum stress.In addition, modal analysis of gas turbine blades was conducted to observe the vibrational behaviour of gas turbine blade on six modes of vibrations. The frequency range of vibration for different materials were: (979.36 Hz to 4507.4 Hz) Titanium alloy, (1078 Hz to 4994.4 Hz) Inconel 617, (1118.2 Hz to 5152.2 Hz) Nimonic 80A, and (1013.2 Hz to 4699.5 Hz) Rhenium alloy.From the results, it has been observed that the frequency range of Nimonic 80A was comparatively higher than the Inconel 617, but the induced stress in Nimonic 80A was considerably higher than the stress induced in Inconel 617, hence, it can be said that Inconel 617 is the best suitable material for gas turbine blade.

7. References

[1] P. Bruun, *Port engineering. Third edition.* 1981.

[2] H. P. Singh, A. Rawat, A. R. Manral, and P. Kumar, “Computational analysis of a gas turbine blade with different materials,” *Mater. Today Proc.*, vol. 44, pp. 63–69, 2021, doi: 10.1016/j.matpr.2020.06.486.

[3] M. Analysis, *Modal Analysis*. 2001.

[4] O. E. Efe-Ononeme, A. E. Ikpe, and G. O. Arravve, “Modal Analysıs of Conventıonal Gas Turbıne Blade Materıals (Udımet 500 and In738) for Industrıal Applıcatıons,” *J. Enggneerrng Technol. Applled Sccences*, vol. 3, no. 2, pp. 119–133, 2018.

[5] S. A. Khan, I. Arif, and M. U. Safdar, “Determination and Comparison of Modal Parameters of a High Lifting Low Pressure Turbine Blade under Ni-Based Superalloys Using Computational Modal Analysis,” *Proc. 18th Int. Bhurban Conf. Appl. Sci. Technol. IBCAST 2021*, pp. 169–176, 2021, doi: 10.1109/IBCAST51254.2021.9393274.

[6] R. Schafrik*, “ADVANCED MATERIALS & PROCESSES For Gas Turbine,” *Adv. Mater. Process. MAY 2004*, pp. 165–209, 2004.

[7] I. Gurrappa, “An oxidation model for predicting the life of titanium alloy components in gas turbine engines,” *J. Alloys Compd.*, vol. 389, no. 1–2, pp. 190–197, 2005, doi: 10.1016/j.jallcom.2004.05.079.

[8] H. Shah Hosseini, M. Shamanian, and A. Kermanpur, “Characterization of microstructures and mechanical properties of Inconel 617/310 stainless steel dissimilar welds,” *Mater. Charact.*, vol. 62, no. 4, pp. 425–431, 2011, doi: 10.1016/j.matchar.2011.02.003.

[9] Y. Xu, C. Yang, X. Xiao, X. Cao, G. Jia, and Z. Shen, “Evolution of microstructure and mechanical properties of Ti modified superalloy Nimonic 80A,” *Mater. Sci. Eng. A*, vol. 530, no. 1, pp. 315–326, 2011, doi: 10.1016/j.msea.2011.09.091.

[10] R. W. Broomfield *et al.*, “Development and turbine engine performance of three advanced rhenium containing superalloys for single crystal and directionally solidified blades and vanes,” *J. Eng. Gas Turbines Power*, vol. 120, no. 3, pp. 595–608, 1998, doi: 10.1115/1.2818188.

[11] R. Osann and M. M. Fasching, “(12) Patent Application Publication (10) Pub . No .: US 2008 / 0225123 A1 Patent Application Publication,” *Priv. Point Contain. Sm Card*, vol. 1, no. 19, pp. 11–14, 2008.

[12] R. R. Kumar and K. M. Pandey, “Static structural and modal analysis of gas turbine blade,” *IOP Conf. Ser. Mater. Sci. Eng.*, vol. 225, no. 1, 2017, doi: 10.1088/1757-899X/225/1/012102.

Analysis of CNC Turning Process of AA6082-T6 using Ansys Software

Mohd Aseerullah[1], C. Bhushan[2] and R. Vishwakarma[2]

[1]Research Scholar, Department of Mechanical Engineering, Babu Banarsi Das University, Lucknow, UP, India
[2]Assistant Professor, Department of Mechanical Engineering, Babu Banarsi Das University, Lucknow, UP, India

Abstract. In this new technological world, CNC and manual turning operations engage in a significant role in different types of designing and manufacturing industries. Before initiating the manufacturing processes, analysis is a crucial part of the manufacturing process. This study involves modelling and analysis of the CNC machining process of a cylindrical AA6082-T6 workpiece with a Tungsten Carbide tip tool. A 3D model is first modelled on CATIA V5 and the analysis is then carried on in Ansys R19.2's explicit module. The experimental and analytical results are carried out to justify the work. The cutting parameters used in CNC machining are Spindle rotational speed, feed rate, and depth of cut. The Normal Stress's percentage error value varied too much compared to Shear stress and Shear Strain. This shows that the aluminium is suitable for the machining processes and can also be used more efficiently.

1. Introduction

In manufacturing industries, modeling and analysis play a very significant role in determining the quality of the product. In this study, we are using AA6082-T6 as a workpiece because aluminum is one of the most machined materials available today. Aluminum CNC machining processes come second after steel in terms of frequency of execution because of their excellent machinability. Quality and Consumer are the two primary objectives of any machinery or manufacturing industry that every one of the manufacturers in the world tries to accomplish. In many manufacturing industries, machining is the main metal processing operation that is done daily. Many processing parameters decide the quality of metal cutting obtained and these process parameters affect the output response and performance characteristics of the final product that is to be made. Hence, the selection of the right process and the right parameter is the main step in the process of a product's life cycle. Poor selection of processes and parameters results in non-optimal machine running which can affect the tool sharpness which further goes to the wear and tear of the tool and in some cases workpiece too. [1].

In the process of metal cutting, a cutting tool is used to process and remove the excess material of the workpiece to make the workpiece into a desirable part to be used in many functions later. Producing a high-quality final product at a low cost is an essential part of a product life cycle which can be obtained by selecting the proper tool material, cutting parameters, machine tools, and tool geometry [2]. The world and the processes are changing very rapidly in manufacturing industries, so to cope with the changes in the modern world the optimization of the metal cutting processes is done on different levels to respond to the new demands in the manufacturing unit [3].

1.1. CNC Turning

The removal of the material from the outer diameter of a rotating cylindrical workpiece with the help of a single-point cutting tool is called turning. CNC turning or turning operation is used when we have to reduce the diameter of the workpiece, especially for a specific diameter, and also to produce a smooth surface. When this process is done with the help of a computer-based lathe then the process can be called a CNC turning operation. To carry out the CNC turning process, the part program is given to the computer by which the operation is completed without the interference of external labor.

1.2. Software used

The software used in this paper is Catia V5 and Ansys R19.2. Catia (an acronym for computer-aided three-dimensional interactive application) is software developed by the French company Dassault Systèmes which is used for designing, modeling, etc. In this paper, Catia V5 is used to create the 3D model of the workpiece and the tool to carry out the analysis. Ansys is software for product design, testing, and operation. Ansys is used to simulate and assess a product's performance with various functionalities and specifications without the need for a dummy product or crash tests. In this paper, Ansys R19.2 is used to analyze the CNC turning process based on the input parameters and get the FEA results for the same.

2. Literature Review

Mayur Verma et al. (2019). carried out experimental, analytical, and simulated investigations of CNC turning using various tool inputs. There are numerous alloys in the industry that are extremely difficult to process, necessitating the use of a precise tool, optimal machining settings, tool geometry, and cutting conditions. The simulated environment's anticipated characteristics, such as heat, forces, and strain, are confirmed using CNC turning operations utilizing L9 Taguchi orthogonal arrays on Stainless steel AISI 304 with various inserts. Cutting Speed, Feed, and Cutting depth is the CNC turning process parameters used. Finite element simulations of CNC turning operations on Stainless steel AISI 304 with various inserts such as Cubic Boron Nitride and Tungsten Carbide inserts are performed in this numerical and experimental study, and the effect of CNC turning process parameters is studied in light of experimental data. Carbide tool inserts and cubic boron nitride tool inserts are the two types of tools employed. Forces are estimated using an analytical study with a workpiece made of aluminium 6061-T6 alloy and an AISI 1045 steel tip tool. The analysis program ANSYS is used to perform numerical research on the orthogonal turning of Aluminium 6061 T6 Alloy with AISI 1045 steel. In the analytical investigation, a workpiece of Aluminium 6061-T6 Alloy is utilized with an AISI 1045 steel tooltip, and forces are determined. To verify the simulated environment, simulated studies for orthogonal, oblique, and cylindrical turning processes are done by using two different combinations, namely aluminium 6061-T6 alloy with AISI 1045 steel tooltip and aluminium 6061-T6 alloy with Tungsten Carbide tooltip to anticipate forces and temperatures. In the experiment, the workpiece is stainless steel AISI 304, and the tool elements are Tungsten Carbide and CBN.

R.K. Bharilya et al. (2015). published a paper in which the trials are conducted using three distinct materials: carburized mild steel, aluminium alloys, and brass. As input, three parameters are provided: spindle speed, depth of cut, and feed rate. When machining hard materials, the depth of cut and feed should be reduced to a minimum, whereas when machining light materials, the depth of cut and feed should then be enhanced since it gives a superior surface finish in the turning process. The purpose of this research is to investigate the optimization of machining parameters for the turning operation of a specific workpiece made of Carburized Mild Steel. Aluminium alloys and brass were manufactured on a CNC machine and tested using a cutting force dynamometer. A dynamometer is a particularly effective piece of machinery for measuring cutting force, and the rotating tool used for precision cutting is constructed of tungsten carbide. The aim is to minimize cutting forces and boost cutting speed in turning operations by utilizing a force dynamometer to optimize process parameters and determine the best surface finish of a specific workpiece. According to the findings, improved surface condition, i.e., surface polish and homogeneity, has been discovered which results in lower production costs for elevated CNC machining. The result of this article is that the optimization process for the turning operation with surface quality checks is highly efficient in this technique, with the parameters supplied as T, f, and S. The test results reveal that the optimization report proves the traditionally regarded three control factors (process parameters) such as spindle speed(S), depth of cut(T), and feed rate(f) that impact material finishing and a wide range of tests for Carburized Mild Steel, Aluminium Alloys, and Brass for surface quality/finishing are optimization.

Magdum Vikas B. et al. (2013). In this study, materials for tool and process parameters for forces in turning are chosen for a lot of different aspects. This paper shows us a method of optimization of cutting forces and other factors. ANOVA and Taguchi orthogonal array are employed for optimizing, and experimentation are carried out to get optimized results, with the outcomes of the trials providing the minimized thrust force. Experiments are carried out in this work to investigate the influence of cutting parameters such as cutting speed, feed rate, and depth of cut on surface roughness during dry turning of 40C8.The goal of this research is to develop multiple regression analyses to gain a better knowledge of the impacts of rotation speed, feed, and cutting depth on surface roughness. For the research setup, a factorial of experiments matching to sessions was used. The contribution of each element to the outcome is determined using an analysis of variance. The feed rate is shown to be the most influential characteristic impacting surface roughness, followed by cutting speed and depth of cut. A number of studies have been carried out in order to begin characterizing the elements influencing surface roughness during the turning process. The influence of rotation speed, flow rate, and cutting depth on 40C8 surface roughness was investigated.

Dr. Vijay Kumar et al. (2017). The influence of lubricating, feed rate, depth of cut, and spindle speed upon EN 19 steel has indeed been explored in this study using L18 Taguchi's orthogonal array, and the relevance of the process variables is assessed using ANOVA. The goal of this study is to cut EN 19 stainless steel material using a CNC turning process and to examine the influencing characteristics while cutting materials, which include surface roughness and material removal rate. The CNC turning process parameters of feed rate, depth of cut, spinning speed/rotational speeds, lubrication, and MRR and surface roughness have already been examined. For the trials, a carbide tip tool was employed as a cutting tool. Taguchi's L18 mixture type orthogonal array experimental design was chosen for inquiry, and optimization is accomplished using Taguchi's technique, as well as analysis of variance (ANOVA) to determine the relevance of process variables on dependent variables. The results show both feed rate and spindle speed have a considerable impact on both material removal rate and surface roughness. The CNC turning study on material removal rates and surface roughness was conducted for EN 19 steel, and the influence of lubricant, feed rate, depth of cut, and speed of rotation on the reaction was investigated under L18. ANOVA is used to examine Taguchi's orthogonal array and the importance of process variables.

3. Properties of the materials used

The materials used are AA6082-T6 as a workpiece and a tungsten carbide-tipped tool. AA6082-T6 is chosen as a workpiece because aluminum is one of the most machined materials available today. After steel, there comes aluminum in machining processes in terms of frequency of execution due to its excellent machinability. Tungsten carbide is chosen as a tool because it maintains a sharp cutting edge better than steel tools and also allows faster machining.

3.1. Tool Properties

The type of tool we are using in this paper is a single-point cutting tool and the material of the tool is Tungsten Carbide. The Mechanical properties of the tool are given in table 1. Some properties in the table are derived from other properties like Shear modulus and bulk modulus are derived from young's modulus and poison's ratio.

Table 1. Mechanical Properties of the Tool

Property	Value (Unit)
Density	20 (kg/m^3)
Young's Modulus	6e^{11}(Pa)
Poisson's Ratio	0.2
Bulk Modulus	3.33e^{11}(Pa)
Shear Modulus	2.5e^{11}(Pa)
Specific Heat	1.84e^{-7}(J/gK)

3.2. Workpiece Properties

The workpiece chosen for this paper is a cylindrical AA6082-T6 bar of length 150mm. AA6082-T6 is an aluminium alloy that is mostly used for machining because of its excellent corrosion resistance. It has strength higher than other alloys and because of that, it is replacing other aluminium alloys in the industry. The chemical composition of AA6082-T6 is given in table 2. For the failure model, the Johnson-Cook failure model is used with a linear relationship between the shock velocity and the particle velocity. The Mechanical Properties of the workpiece are given in table 3.

Table 2. Chemical Composition (wt.%) of AA6082-T6

Aluminium	Silicon	Iron	Copper	Manganese	Magnesium	Chromium	Zinc	Titanium	Others
95.2-98.3	0.7-1.3	0.5	0.1	0.4-0.1	0.6-0.1	0.25	0.2	0.1	0.15

Table 3. Mechanical Properties of the Workpiece

Property	Value (Unit)
Density	2710 (kg/m^3)
Specific Heat	0.9(J/gK)
Shear Modulus	2.69e^{10}(Pa)
Melting Temperature	555(°C)
Reference Strain Rate	1(/s)
Shock EOS Linear	
Gruneisen Coefficient	2.1
Parameter C1	5380(m/s)
Parameter S1	1.337
Johnson Cook Failure	
D1	0.0164
D2	2.245
D3	-2.798
D4	0.007
D5	3.65

IOP Conf. Series: Materials Science and Engineering 1259 (2022) 012017 doi:10.1088/1757-899X/1259/1/012017

4. Methodology

A 3D model is first created on the software Catia V5 and then the same model is imported into Ansys R19.2 explicit module for the analysis purpose. In explicit dynamics, the material properties are given to the system after which the connections are given to the tool and the workpiece. The tool is taken as the contact system and the workpiece is taken as the target system for the frictional contact between them. Orthogonal quality of fine meshing is used for the mesh of the system. The boundary conditions related to the system are then given in which a displacement of 360 is given to the workpiece to rotate it and a displacement of -5 is given to the tool on the x-axis for the tool to penetrate the workpiece. The analysis settings are given to the system like end time, automatic mass scaling, erosion controls, etc. Then the results are taken concerning the end time of the simulation. This result is justified by the analytical results calculated based on the experimental results we get by experimenting on a CNC machine and the forces are calculated.

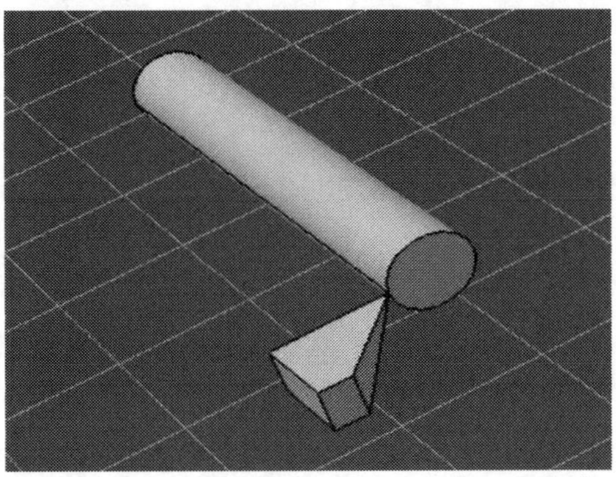

Figure 1. 3D model of the tool and the workpiece

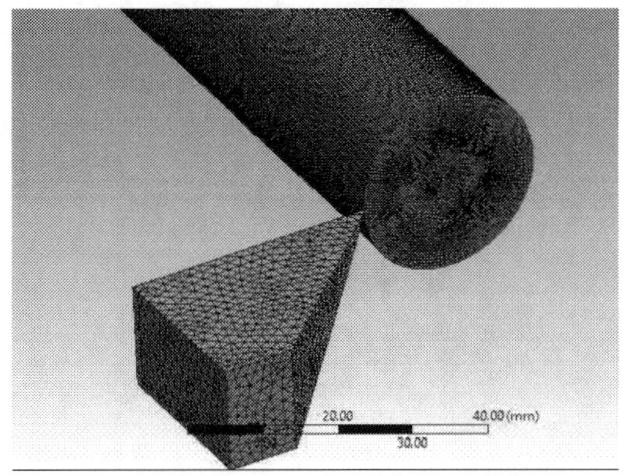

Figure 2. Fine mesh of the model

5. Results

5.1. FEA Results

After the analysis of the CNC turning operation of AA6082-T6 with the tool as Tungsten Carbide is completed, we got the result for Normal Stress, Shear Stress, and Shear Elastic Strain. The result shows us the final stage of the process of the CNC turning process with the desired results to be calculated in the explicit dynamics of the Ansys.

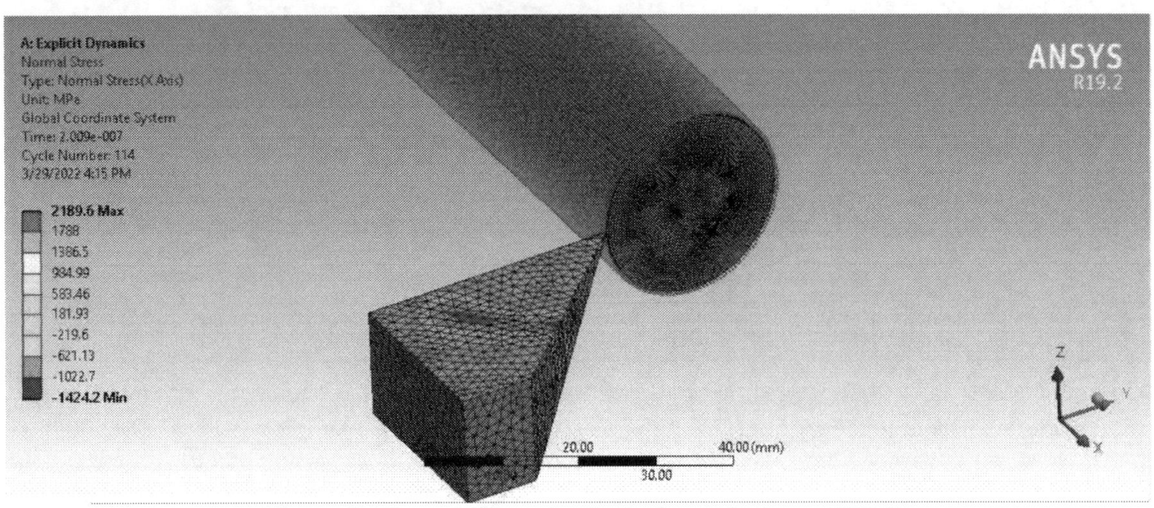

Figure 3. Normal Stress Analysis shown in Ansys R19.2

In figure 3 the analysis results of normal stress are shown. The Maximum Normal Stress in the system is 2189.6 and the minimum Normal stress is -1424.2. The red part on the workpiece shows the part which will break and flow away from it as a chip. The graph between the Normal stress and the time is plotted and shown in figure 4. The graph is plotted by using 100 Normal stress points. The graph shows the trendline for the Normal stress concerning time.

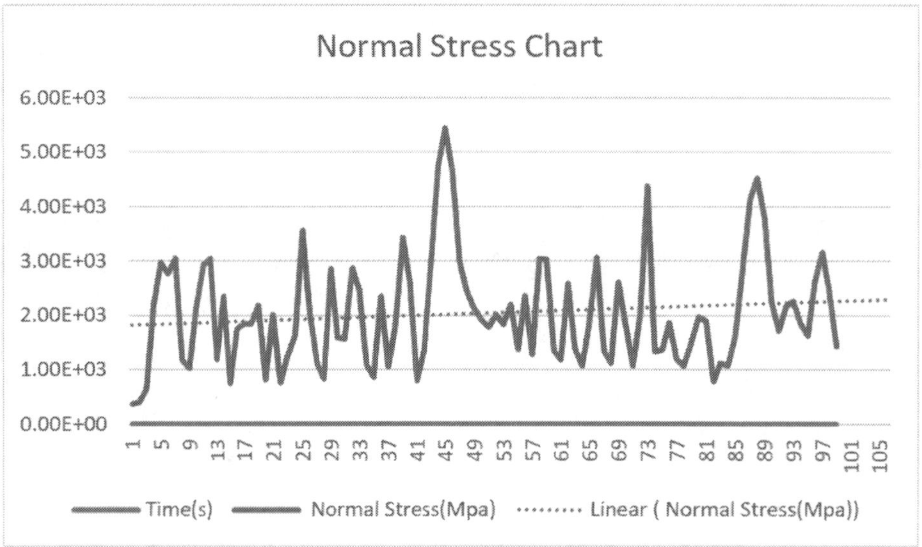

Figure 4. Line Graph of Normal Stress Analysis (Time vs Normal Stress)

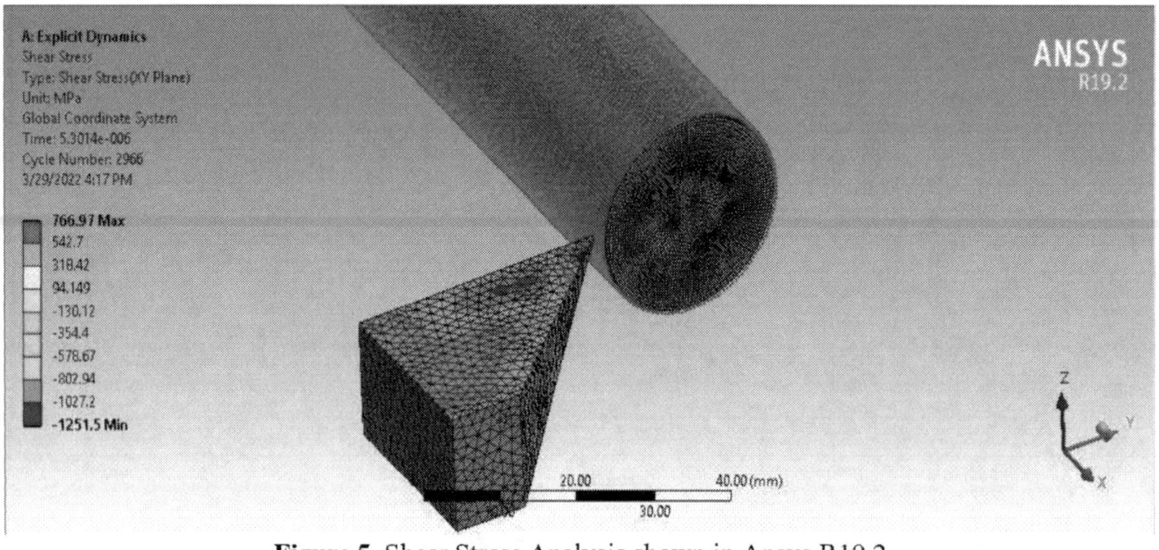

Figure 5. Shear Stress Analysis shown in Ansys R19.2

In figure 4 the analysis results of shear stress are shown. The Maximum Shear Stress in the system is 766.97 and the Minimum Shear Stress is -1251.5. The red part on the workpiece shows the part which will break and flow away from it as a chip. The graph between the Shear Stress and the time is plotted and shown in figure 6. The graph is plotted by using 100 Shear Stress points. The graph shows the trendline for the Shear Stress concerning time.

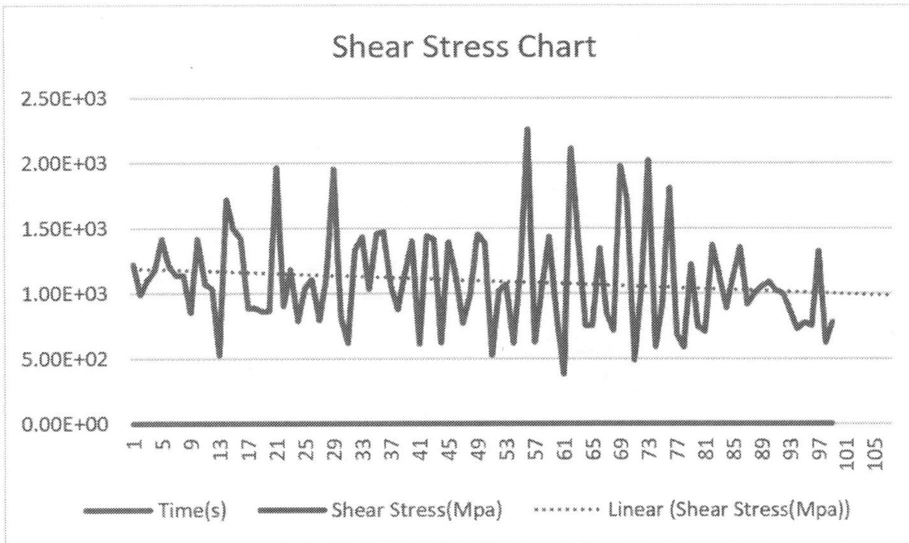

Figure 6. Line Graph of Shear Stress Analysis (Time vs Shear Stress)

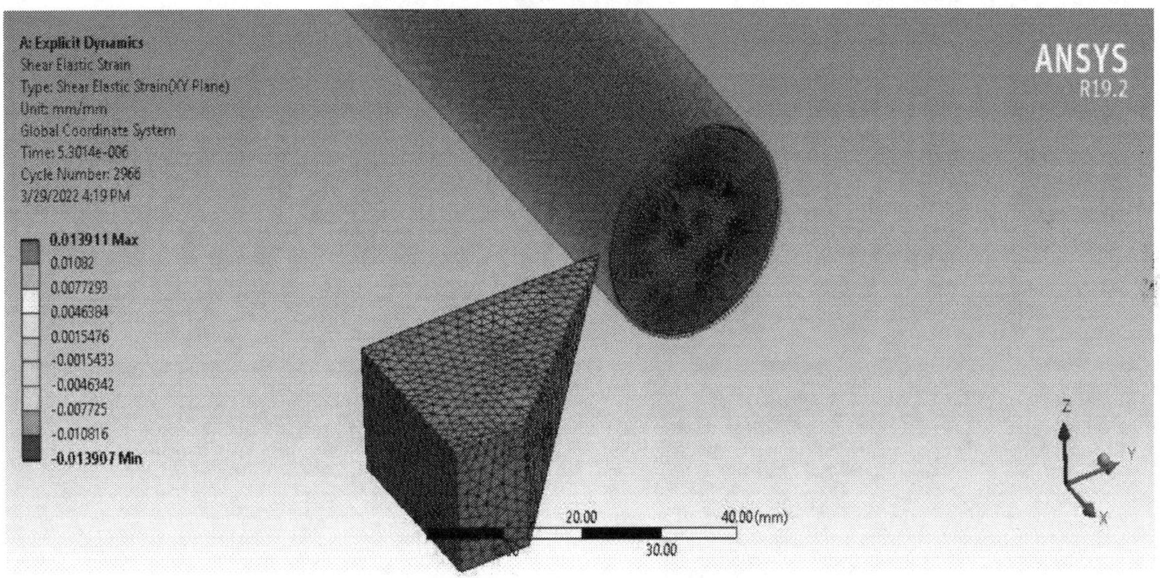

Figure 7. Shear Elastic Strain Analysis shown in Ansys R19.2

In figure 7 the analysis results of shear Elastic Strain are shown. The Maximum Shear Elastic strain in the system is 0.013911and the minimum elastic strain is -0.013907. The red part on the workpiece shows the part which will break and flow away from it as a chip. The graph between the Shear Stress and the time is plotted and shown in figure 6. The graph is plotted by using 100 Shear Stress points. The graph shows the trendline for the Shear Elastic Strain concerning time.

IOP Conf. Series: Materials Science and Engineering 1259 (2022) 012017 doi:10.1088/1757-899X/1259/1/012017

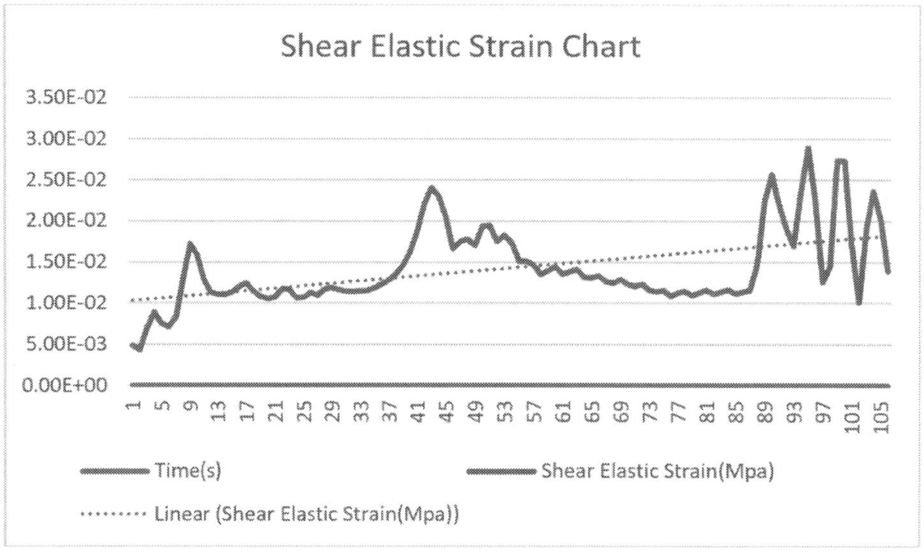

Figure 8. Line Graph of Shear Elastic Strain Analysis (Time vs Shear Elastic Strain)

5.2. Experimental Results

The Experiment is carried out on a Siemens Sinumerik 828D CNC machine and the forces are calculated with the help of the Kistler dynamometer. The diameter of the workpiece was 25.5 mm and the length of the workpiece was 150 mm. After the Turning process is carried out the diameter of the workpiece was minimized to 20mm. The cutting parameters used in CNC machining are Spindle rotational speed, feed rate, and depth of cut. In this experiment, the spindle speed taken chosen is 200 rev/min, Feed 0.1 mm/rev, and the depth of cut is 0.2 mm. The total time taken by the CNC machine to finish the turning process is 5:17 minutes.

Figure 9. Performing the Turning Process

Figure 10. Continuous Chip formed in turning process

Figure 9 shows the turning process carried out in the Siemens Sinumerik CNC machine and the setup of the dynamometer is also done so that the components of the force can be measured for further results. Figure 10 shows the chip formed in the process of cutting AA6082-T6 and It can easily be seen that the type of chip formed is a continuous chip. For the further calculation of normal stress, shear stress, and shear strain the chip thickness and the chip width is measured by a digital vernier calliper. The dimensions of the chip are as follows: chip thickness of 0.2 mm and chip width of 0.75 mm.

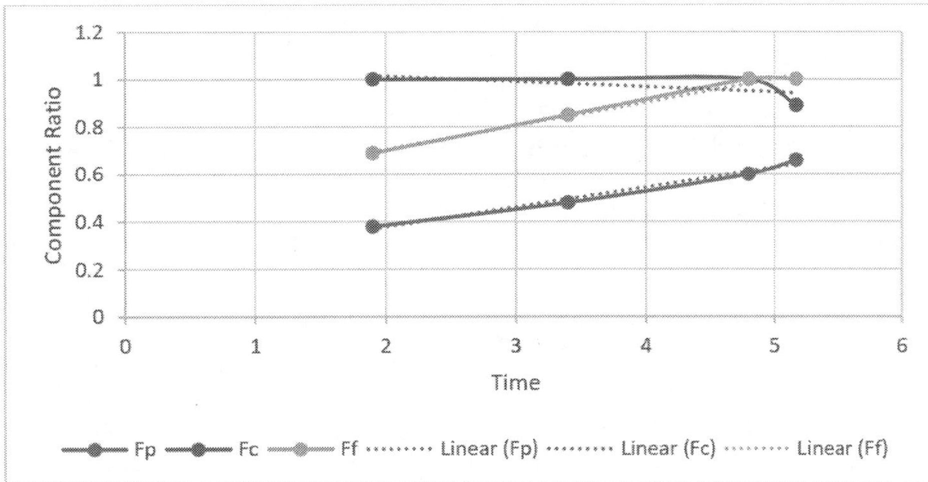

Figure 11. Component ratio vs Time Graph of the experimental Results

The component ratio vs Time graph is shown in Figure 9 which shows the results graph of the result taken from the dynamometer. In the graph, F_p is the radial component of the cutting force, F_c is the Cutting component of the cutting force, and F_f is the feed component of the cutting force. Linear F_p, Linear F_c and Linear F_f show the linearity in the cutting component of the system.

5.3. Analytical Results

With the help of the experimental results, we can find the Normal Stress, Shear Stress, and Shear Strain. For that, we first find the cutting ratio which is the ratio of uncut chip thickness to cut chip thickness.

$$\text{Cutting Ratio, } r = \frac{t}{tc} = 0.909 \quad (1)$$

$$\text{Shear Angle, } \phi = tan^{-1}[\frac{rcos\alpha}{1-rsin\alpha}] = 44.51° \quad (2)$$

$$\text{Normal Stress, } \sigma = \frac{[Fcsin\,\phi + Ftcos\,\phi]}{bt} Sin\phi = 716.14 \text{ MPa} \quad (3)$$

$$\text{Shear Stress, } \tau_s = \frac{[Fccos\,\phi - Ftsin\,\phi]}{bt} Sin\phi = 803.81 \text{ MPa} \quad (4)$$

$$\text{Shear Strain, } \gamma = cot\,\phi + tan\,(\phi - \alpha) = 0.0184. \quad (5)$$

6. Conclusions

Three parameters are calculated in this paper those are Normal Stress, Shear Stress, and Shear Strain. In the analysis done in Ansys R19.2 the value of Normal Stress is 2189.6 MPa whereas the calculated value is 716.14 MPa. For Shear Stress, the Ansys calculated value is 766.97MPa whereas the calculated value is 803.81 MPa. For Shear Elastic Strain, the Ansys calculated value is 0.013911 whereas the calculated value is 0.0184.

In Normal Stress, the Percentage error is calculated as -67.3 %.
In Shear Stress, the Percentage error is calculated as 4.80 %.
In Shear Strain, the Percentage error is calculated as 32.27 %.

The Percentage error is maximum in Normal Stress and minimum in Shear Stress which concludes that in this analysis shear stress is the most common factor that is affecting the turning process or the cutting of the workpiece. Percentage error shows that the normal stress value is very far from the expected value. The negative mark depicts that the FEA result value is very far then the analytical value that we calculated. The percentage error for the value of Normal stress is too high related to the other results because the software shows us the stress from the ending time up to the start point given for the analysis and the Normal stress, we got is the value from the starting point up to a certain point with respect to time in the relation with the force components.

7. References

[1] Dr. Vijay Kumar M, Kiran Kumar B.J, Rudresha N 2018 Optimization of Machining Parameters in CNC Turning of Stainless Steel(EN19) by TAGUCHI'S Orthogonal Array Experiments. ICMMM-2017, Materials Today: Proceedings 5 11395–11407.

[2] R.K.Bharilya, Ritesh Malgaya, Lakhan Patidar, R.k. Gurjar, Dr.A.K.Jha 2015 Study of Optimised process parameters in turning operation through Force Dynamometer on CNC Machine. Materials Today: Proceedings 2 2300-2305.

[3] Magdum Vikas B. 2013 Evolution and Optimization Parameter for Turning-En8 steel. International Journal Engineering Trends and Technology Volume 4 Issue 5 May pp 1564-1568.

[4] Mayur Verma,Sharad K. Pradhan 2019 Experimental and Numerical investigations in CNC turning for different combinations of tool inserts and workpiece material. Materials Today: Proceedings, https://doi.org/10.1016/j.matpr.2019.12.193.

[5] S.P. Palaniappan, K. Muthukumar, R.V. Sabariraj, S.Dinesh Kumar, T.Sathish 2019 CNC turning process parameters optimization on Aluminium 6082 alloy by using Taguchi and ANOVA. Materials Today: Proceedings, https://doi.org/10.1016/j.matpr.2019.10.053.

[6] N. Satheesh Kumar, Ajay Shetty, Ashay Shetty, Ananth K, Harsha Shetty 2012 Effect of Spindle Speed and feed rate on surface roughness of Carbon Steels in CNC turning. Procedia Engineering 38 (2012) 691-697.

[7] A. Saravankumar, S.C. Karthikeyan, B. Dhamotharan, V. Gokul kumar 2018 Optimization of CNC Turning Parameters on Aluminum Alloy 6063 using Taguchi Robust Design. materials today: Proceedings volume 5, issue 2 part 2, pages 8290-8298.

[8] S.K. Saini, S.K. Pradhan 2014 Optimization of multi-objective response during CNC turning using Taguchi-fuzy application. Proc. Eng.97 141-149.

[9] P. Gupta, Bhagat Singh 2021 Exploration of tool chatter in CNC turning using a new ensemble approach. Materials today: Proceedings volume 43, part 1, pages 640-645.

[10] D. Umbrello, R. M'Saoubi, J.C. Outerio 2007 The influence of Johnson-Cook material constants on finite element simulation of machining of AISI 316L steel. Int. J.Mach. Tools Manuf. 47(3-4) 462-470.

[11] Hussani T.M.EL 2010 Cutting Parameter optimization when machining different materials. Material and Manufacturing pp 1101-1114.

[12] M.Sadílek, J.Dubsk´y, Z.Sadílková, Z.Poruba 2015 Cutting forces during turning with variable depth of cut. Perspectives in Science http://dx.doi.org/10.1016/j.pisc.2015.11.055.

[13] Cep, R., et al. 2013 Experimental testing of cutting inserts cutting ability. Teh. Vjesn./Tech. Gaz. 20 (1), 21—26, 1/2013, ISSN 1330-3651.

[14] Hatala, M. 2007 Simulation of Technological Process. 1st ed. FVT TU, Preˇsov,ISBN 978-80-8073-756-6, pp. 85.

[15] Mohanraj Murugesan, Dong Won Jung 2019 Johnson Cook Material and Failure Model Parameters Estimation of AISI-1045 Medium Carbon Steel for Metal Forming Applications. Materials, 12, 609; doi:10.3390/ma12040609.
[16] M.Hanief, M.F.Wani, M.S.Charoo 2016 Modeling and prediction of cutting forces during the turning of red brass (C23000) using ANN and regression analysis. Eng. Sci. Tech., Int. J., http://dx.doi.org/10.1016/j.jestch.2016.10.019.
[17] T. Dorlin, F. Guillaume, J.P. Costes 2015 Analysis and modeling of the contact radius effect on the cutting forces in cylindrical and face turning of Ti6Al4V titanium alloy. Procedia CIRP 31 185–190.
[18] J. Xie, M.J. Luo, K.K. Wu, L.F. Yang, D.H. Li 2013 Experimental study on cutting temperature and cutting force in dry turning of titanium alloy using micro-grooved tool. Int. J. Mach. Tools Manuf. 73 25–36.
[19] A. Pramanik, L.C. Zhang, J. Arsecularatne 2006 Prediction of cutting forces in machining of metal matrix composites. Int. J. Mach. Tools Manuf. 46 (14) 1795–1803.

CAD/CAM Engineering And Artificial Intelligence In Dentistry

Vyankatesh Naidu[1] **and Santosh Jaju**[2]

[1]Researcher Department of mechanical engineering, G H Raisoni College of Engineering, Nagpur, 440016, Maharashtra, India
[2]Lecturer, Department of mechanical engineering, G H Raisoni College of Engineering, Nagpur, 440016, Maharashtra, India

Abstract. Artificial Intelligence is an instantly growing technology that has mesmerized the intellect of experimenters and researchers around the world. Since its commencement dentistry has glimpsed many incredible successes. As the fate of dentistry is adjoining the execution of AI applications, therefore, this condition needs each dentist to get educated and trained. The prelude of robotics to the healthcare sector has boosted the exactness and accurateness in a surgical procedure. However, the position of the dentist cannot be replaced by AI, but AI can be the tool for enhancing the process of dentistry.

1. Introduction

Nowadays, the information of patients is documented in a detailed form. For storing and processing this large amount of data, the software is essential. Computerized treatment is becoming popular because it can point out errors that may unnoticed by human eyes. Biomanufacturing allows for multi-disciplinary cooperation in both academia and industry, and can lead to advancements in a variety of engineering domains. The accuracy of established methods are satisfactory for normal cases but not for complex ones.[1] Impediments of traditional methods entice the need for a modern technique. The continual search for this modern technology is the reason behind for rising of Artificial Intelligence. Artificial Intelligence pertains to a system that can simulate the functioning of the human brain. The function of artificial intelligence in dentistry and modern theories can be clarified after reading this review article. In the late 50s, the word "artificial intelligence" is stamped. A machine that can conduct, execute and accomplish tasks similar to humans was referred to as artificial intelligence. The technology which is used to predict the unseen data by applying algorithms and reading structural patterns is called machine learning. Machine learning is a branch of AI. European polymath expert, Alan Turing conduct a Turing test that suggests machine can utilize accessible knowledge and purpose to unravel difficulties like humans. Due to inadequate repair outcomes and different complications, functional reconstruction of large-size mandibular continuity defects remains a serious problem in oral and maxillofacial surgery.[2] In the year 1956, John McCarthy established the term artificial intelligence. John McCarthy stated that artificial intelligence is a branch of science and technology which involved computational awareness and the generation of artifacts that display very behavior.

1.1 Popular Terms in Dentistry Artificial Intelligence

1.1.1 Clinical decision support system

A system bridge between the medical logic modules executes an inferencing product device that is a set of algorithms emanated from fact-based medical exercise and dynamic medical proficiency is called a clinical decision support system. For helping the health care experts voice command intuitive interface is developed which helps to save time, cost and also helps in more efficient dental methods.

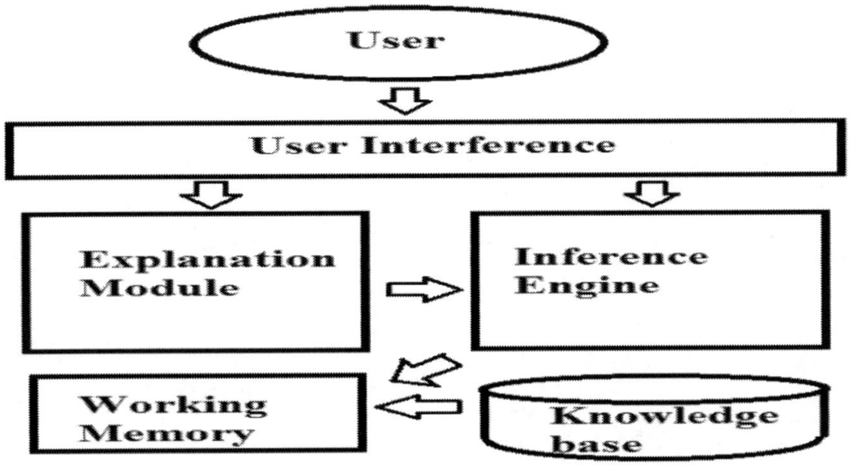

Figure 1. The basic components of CDSS

1.1.2 Augmented reality
Augmented reality is a technology which accordingly gives a rational impression of the user's perspective of the actual realm by superimposed computer-generated image.

1.1.3 Machine learning
The technology which builds algorithms by directed data is called machine learning which is a branch of computer science.

1.1.4 Deep learning
Defined as a distinctive aspect of learning based on algorithms of neural hierarchy.

1.1.5 Representation learning
For classifying the given data computer algorithm that acquires the characteristics needed is called representation learning which is a sub-branch of machine learning.

1.1.6 Artificial neural networks
In the medicinal sector, the tool which can evaluate non-linear data has the potential to learn from preceding representations, handle the imprecise message and localized facilitating the use of the model to sovereign data making it incredibly desirable. By keeping the arrangement of the brain as it is motive that can emulate the human brains this is how artificial neural networks were invented. The first development of the neural network is considered as an analogue reinforcement calculator created by Minsky and Dean Edmunds in 1951. For the very first time, the programs of AI is generated by Allen Newell and Herbert Simon in 1951. In maxillofacial surgery, new technologies are being employed to control and replace complicated bone deformities that affect both function and attractiveness. The capacity of solving complicated problems by using traditional ways is the enormous benefit of these systems.[3] They are worthwhile in several regions of medical science for eg treatment of disease, biomedical image identifier and analytical data. In the dental sector, the clinical support system also is actively advancing.

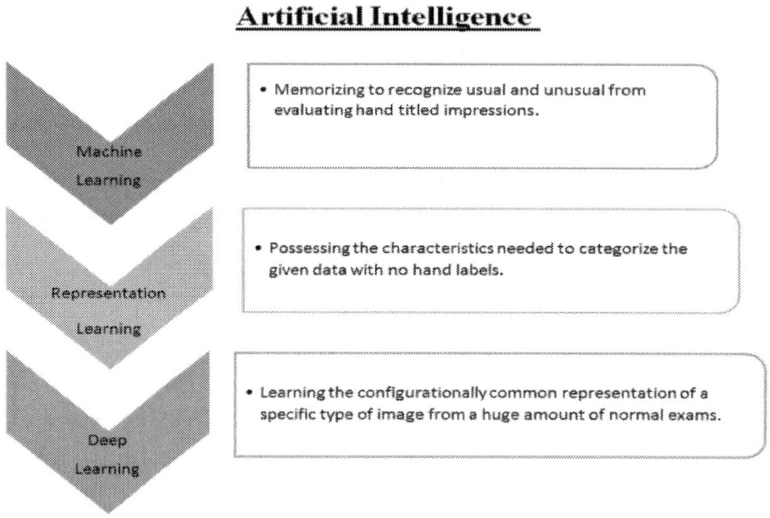

Figure 2. Configuration of Artificial intelligence fields

2. Dentistry Applications

Successful clinical exercise is dependent on accurate diagnosis. However, sufficiently acquainted neural networks are a blessing for dentists, exceptionally in circumstance having multifactorial etiology. Additive manufacturing (AM) allows for a large deal of design freedom, allowing for the fabrication of items with unique and complicated geometry and topology in a quick and cost-effective manner. The function of artificial intelligence in dentistry is relatively crucial as the emergence of data analysis and accessibility of a huge amount of patient information.[4] In radiology, a specified kind of steps created which helps further in the diagnosis and treatment of diseases.

3. Dentistry E- Learning

Chairside dentistry is extremely valuable for understanding the techniques and improving patient care. Theoretical speculative education and practical learning was only part of the preclinical conventional training. Virtual reality and augmented reality are incorporated for schooling intelligent education policy and practice in dental. Clinicians have designed and successfully employed mini-implants as a guide for orthodontic tooth movement. [5] This technology helps students and practitioners to simulate empirical procedures into 3-D models and surgical methods. The practice can be done many times until students learn perfectly without involving patient. A unified medical language system also known as UMLS is a system that helps dental students to learn about patients anatomy and diagnosis with clearer concepts.

4. Artificial Intelligence And Diagnostic Dentistry

The accuracy and time saving of treatment is increased when AI is used. In the case of aphthous ulcer, precise etiology is not known, surgeons have to perform operations based on experience. In one more case, the arrangement of temporomandibular joints was distorted. In both cases, surgeons have to perform operations without knowing the outcome. On the other hand, artificial neural networks precisely found disruptions and performed the more precise operation. Results show that ANN is more sensitive and accurate than the tasks performed by surgeons without ANN. The state-of-the-art in this rapidly evolving manufacturing field is described and analysed, with a focus on additive electrical, chemical, and physical techniques currently used to manufacture synthetic and biological parts.[6] AI can be used for solving more complex oral diseases and reducing the chances of human error. In radiography also the data came from intraoral per apical X-ray, cone-beam computer tomography and

3D scans that can be computed quickly with the help of AI. After that AI can also help to suggest treatment of oral disease. ANN can detect chances of hazardous disease like oral cancer. AI can also satisfy the need of curing oral cancer. AI can predict the chances of bond loss during the operation and can help in choosing tooth crown.

5. AI In Patient Management

AI can support the dentist by implying appointments and checkups details according to the patient's genetics data. AI can provide thorough information on patients lifestyle like chewing tobacco and taking sweet in big amount. Medical history can be provided by AI which may show allergies and other information about the patient.[7] AI can also provide tele-guidance diagnostics if the patient is unable to contact the doctor.

6. Artificial Intelligence In Prosthetic Dentistry

Anthropological data, facial dimensions, ethnicity and patients intention are the factors for aesthetic prosthetics. From the locations of placing implants to suggesting crown type and surgical guide AI plays a vital role. Computer-aided design and computer-aided manufacturing replaced the time expending and arduous work. Using a low-powered personal computer setup, design and compare the outcomes of commercial (CS) and open source (OS) software-based 3D prosthetic templates for rehabilitation of craniofacial abnormalities.[8] Virtual reality simulation technology provides post-surgical facial results before surgery. The dentist can design proper guides that can help in aesthetics.

7. AI In Peridontics

The exactness of evaluating the molars and premolars in periodontics is very high when AI is used. Rana et al. Conducted an experiment that results in favor of AI. The healthy and unhealthy gingival was identified by AI. AI can classify the patients according to their immune system.

8. AI In Orthodontics

AI is a revolutionary change in the field of orthodontics treatment. From the diagnosis of the patient to the final surgery, AI is used. AI is used in medication planning. According to the bone "Mechanostat" theory, the porous implant had a strong mechanical stimulating effect on bone tissue growth and repair. [9] 3-D scans are converted into images and 3-models which helps the dentist to perform the following tasks. Aligners can be printed with the help of 3D scans. AI suggests a suitable procedure of treatment. Information like pressure points to an exact operating tooth can be suggested by AI.

9. AI In Pathology

The traditional process of pathology is determined by the quality of microscopic morphology. AI helps to simplify the process by regulating timely inspection and accuracy of the device adjustments. The pathologist gives the diagnosis report by examining the sample. Computerized tomography (CT) has become the most widely utilised imaging modality for obtaining accurate pre-operative implant planning in the previous decade.[10] The only constraints in the pathologist report are the accuracy of the result and may vary from pathologist to pathologist. A computer-based AI system is highly accurate and more precise than the traditional way. Nonetheless, the AI in dentistry is the future of more sophisticated clinical procedure.

10. CAD-CAM Engineering Software In Dentistry

CAD/CAM software gives the most accurate and precise results while manufacturing surgical guide. Resins are commonly used to make surgical guides. Intraoral scans are also advanced technology available for dentistry. Which is able to easily give all the information of the patient's mouth. There is a lot of CAD/CAM software developed by different companies that helps with implant placement. [11] Along with implant placement, all the complications are also understood by the software. From implant placement to the size of the implant screw and the specifications of the abutment are also determined by the software. From fig.3 to fig. 6 , Materialise MIMICS Innovation suite 24.0 software and 3-Matic 16.0 is used to design surgical guide.

10.1 Advantages over Conventional Method:

10.1.1 The success rate is very high as compared to the conventional method of operation.
10.1.2 The patient can be better explained about his problem.
10.1.3 Send it to the impression lab, do the operation on it, put the implant and evaluate, it wastes a lot of time. While this time can be saved by using CAD/CAM software.
10.1.4 The doctor does not need to do much headwork for the operation. Only they have to give the case to the software operator. After this, the surgical guide becomes ready.
10.1.5 It has been seen that doctors do not use surgical guides but do implant placement work with direct hands because they have experience. But even newly educated doctors can successfully perform the operation using surgical guides.
10.1.6 The accuracy of implant placement is very high as compared to the conventional method. The operation done with the help of a surgical guide is more successful than the doctors who do the operation directly by hand. The surgical guide which is created with the help of CAD/CAM software is also capable of solving complex cases. In the case where the bone loss is more, in that case the doctor avoids implanting and goes towards other types of treatment. But using CAD/CAM software, treatment can be done by placing an implant in the right place.
10.1.7 If the doctor does the treatment for a long time with the help of CAD/CAM software, then it proves useful in saving money.
10.1.8 Usually doctors to treat one tooth side to another tooth i.e. if one implant is already there or two implants are to be placed on side, then doctors refuse to do treatment for fear of failure. In such cases also other types of treatment are suggested for the patient. Whereas the surgical guide which is built on CAD/CAM software also makes it easy to solve such cases. the sequential software processing of a computed tomography (CT) dataset for reconstructing a finite element analysis (FEA) mandibular model with a custom-made plate, as well as to provide a theoretical foundation for clinical application of this reconstruction method.[12]

10.2 Disadvantage of CAD/CAM surgical guide

10.2.1 The internal cost is high. The cost of purchasing the software and the cost of its training are expensive at the initial stage. But with prolonged use of the software, the process becomes highly productive and cost-effective.
10.2.2 Due to less research in this field, doctors are afraid of use. Incorrect case information can seriously ruin the operation. Therefore it is necessary to have complete information. Incorrect information means the patient has an old CBCT scan and an intraoral scan.

11. Treatment options for Maxilla And Mandible

- *Removable Denture*
- *Overdenture*
- *All On Four*
- *All On Six*
- *Parallel Implants*

IOP Conf. Series: Materials Science and Engineering 1259 (2022) 012018 doi:10.1088/1757-899X/1259/1/012018

12. Materials Used For Dental Implants And Surgical Guides

Materials used for dental implants and surgical guides play a vital role in successful surgeries. In recent years this field grows tremendously because biomaterials which are introduced by biotechnology are very advanced and efficient for surgeries. With the help of biomaterial, native tissue can also be reconstructed both in functional form. Multi-material printing research is needed to improve the haptic input of skeletal models and make them more human bone-like in order to better surgical planning and clinical training.[13] The exact mimicking of tissue can be produced by biomaterials. However, synthetic tissues are commonly used because it eliminates the distress of taking tissue from donor side. Many biocompatible materials are available for producing prosthesis and parts. The acquisition of imaging data is the first step in virtual surgical planning. [14] Metals are also widely used for surgery. This review article is an effort for studying different materials available for the fabrication of surgical guides. Ideally surgical guide should be lightweight and robust. But fabricating this type of surgical guides requires lots of research work. The surgical guide which can withstand drilling, osteotomy and makes no discomfort to the patient is most suitable. For completing both criteria material of the surgical guide should be optimised. The concept of "prosthetic-driven implantology" could be seen as a watershed moment in modern dental implantology. [15] It is required that the surgical guide should not leave any glimmers at the surgical site and also it should be of biocompatible material. Surgical guides are thin enough to fit into the mouth and tough enough to do not break during operation. Fulfilling every criterion mentioned above is dependent upon the selection of guide material. In recent years this sector grown-up extremely well, with remarkable advancement in biotechnology gives lots of biocompatible materials for surgical use. Surgical guidelines aid in the precise placement of dental implants for a predictable restorative result.[16] Autogenous graft, allograft and xenograft are the naturally occurring material which is used for dental implant surgery. For the fabrication of implants synthetic materials, metals are used. In this article, we will review materials for dental implants and surgical guides.

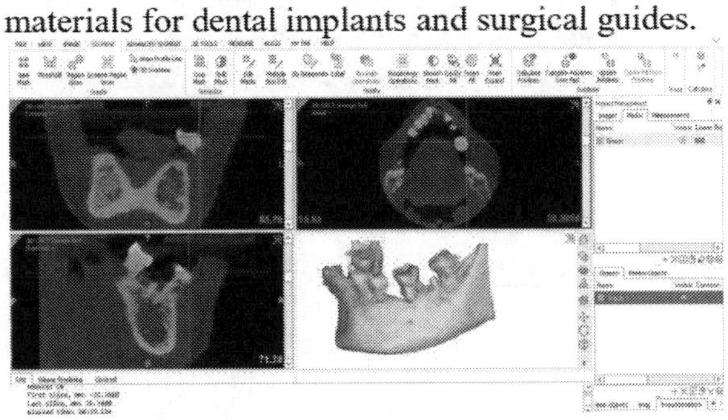

Figure 3. Creating 3-D Model from DICOM CBCT files

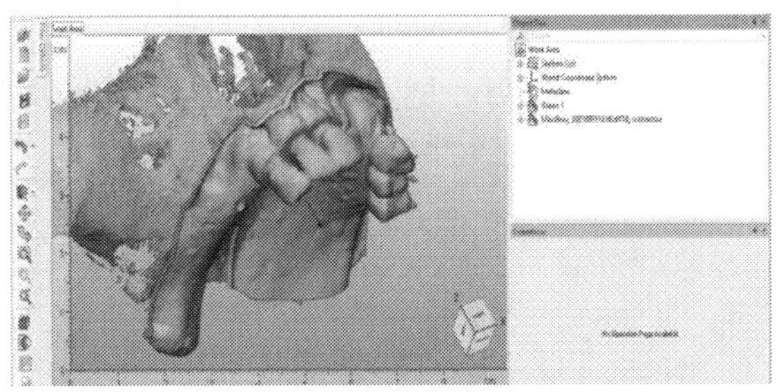

Figure 5. Validation of Surgical guide with 3D model alignment

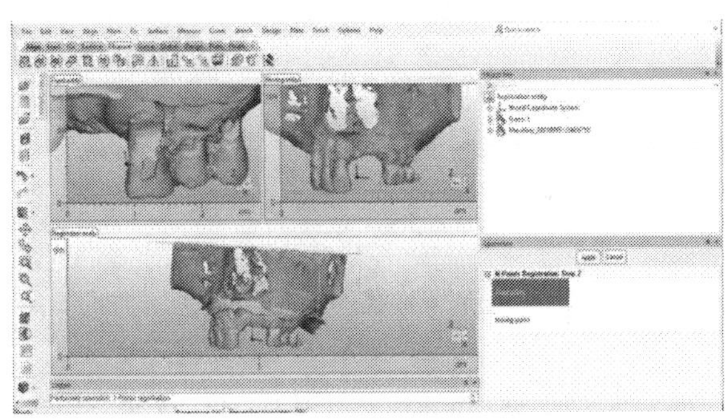

Figure 4. Positioning of Intraoral scan and CBCT 3D model

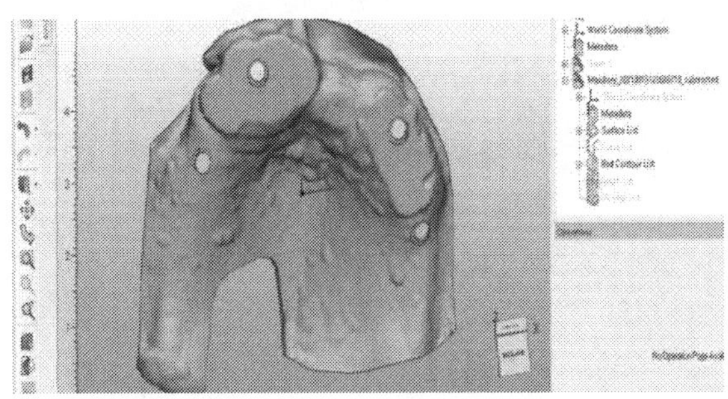

Figure. 6 Final Surgical Guide

IOP Conf. Series: Materials Science and Engineering 1259 (2022) 012018 doi:10.1088/1757-899X/1259/1/012018

13. Metals Commonly used For Dental Surgeries

13.1 The characteristics required for metals used for dental surgeries are as follows:
13.1.1 Tensile strength: tensile strength to break the material by measuring a force.
13.1.2 Shear stress: Shear stress to break the material in floating vector by measuring a force
13.1.3 Young's modulus: Young's modulus distort the material in a reversible pattern for measuring a force.
13.1.4 Yield stress: Yield stress distorts the material in an irreversible pattern for measuring force.
Gold, stainless steel and Vitallium are one of the metals which are first introduced. Vitallium is an alloy combination of cobalt, molybdenum and chromium. Due to corrosion and poor handling properties of these metals were proved troublesome. Therefore in the late '60s in the twentieth century titanium metal was introduced. The use of titanium metal increases the expectancy of surgeries.

13.2 Vitallium:

In the year 1929, Vitallium were firstly used for dental surgery. This metal is widely used for orthopaedic surgeries and maxillofacial surgeries. The first case in the late 90s shows that the construction of midface bone was performed considerably well and difficulties free. At the start, its uses were limited up to the fractured bones. Autotransplantation is a surgical procedure that involves repositioning a donor tooth from the same person into a medically created socket or the site of a previous tooth extraction.[17] Due to its less capability, it is only referred to for rigid fixation. Few studies show that in animal bodies this metal cannot compete with titanium metal. Although Vitallium metal is not infectious so it can be used as a biomaterial as long as clinicians and patient want. Even though data available for both titanium and Vitallium suggests optimum material but still one should have compared these at all factor at the time of implant placement. Surgical guides spent very little time in the mouth of the patient, therefore, Vitallium can be considered to be a suitable material.

13.3 Stainless steel (Iron-Chromium-Nickel Based Alloys):

Mechanical properties stainless steel have like flexibility, strength, ductility and biocompatibility make this material suitable for a surgical guide and dental implants. For genioplasty, the use of stereolithographic guiding, which is becoming more common in orthognathic surgery, has not been fully validated.[18] On a positive note, it is easily fabricated and relatively less expensive. Stainless steel is used for orthopaedic applications from very time. Fracture bones of the mandibular region can be treated with stainless steel. The major drawback of stainless steel is it is corrosive. This material causes scatters in the radiograph of MRI. Iron-based alloys are allergic for some patients. With all these advantages and disadvantages, stainless steel is still suitable for dental implants and surgical guides.

13.4 Gold:

In reconstructive surgeries gold implants are used for a very long time. Gold has a wide history of dental prosthesis involving inlays, inlays, crowns, bridges etc. Adequate strength and corrosion resistivity increase the credibility of gold in biocompatible materials. In the past few years, the cost and allergies to specific patients decrease the use of gold in the prosthesis. Therefore dental implants are switched to cheaper biomaterials. Although, clinical reports suggest that if the surface of the gold is encased with allopathic material then it can be used as the best alternative for a surgical guide material.

13.5 Titanium:

Remarkable mechanical properties of titanium makes this material most suitable for every dental implant and surgical guide. The biocompatibility of titanium is also way better than other materials. Computer-assisted implantology, which uses a single fixed stereolithographic surgical guide, entails a series of diagnostic and therapeutic events, with errors occurring at various stages. [19] The use of Titanium helps to overcome the defects generated by other metals previously used. Alloplastic layer to gold used to increase its property in similar pattern stabilise layer of oxide increases the success rate of titanium implants. Pure titanium is classified into four grades, grade 1 has the least amount of oxygen in it that is 0.18% whereas grade 4 has the highest percentage of oxygen in it that is 0.4%. The characteristics differ from grade to grade of titanium-based on pollutants. Titanium is the first choice for dental implants and surgical guides. When titanium mixed with polyethylene it gives freedom to shape retention and tissue ingrowth. The mechanical property of titanium is very much similar to bone tissue which helps in bone growth. Therefore titanium seems to be the only option for the material for surgical guide and dental in plants. Orthognathic surgery based on three-dimensional (3D) simulation is becoming more prevalent. As a result, consistent procedures for assessing efficacy and dependability are necessary.[20] Titanium is an inert, less corrosive and malleable metal. As disadvantage titanium is a very expensive metal and aesthetically it is not appealing. The grey colour looks very much similar to stainless steel metal. Sometimes titanium implants can be seen through soft tissues. Overall titanium is a preferable material for dental implant and surgical guide.

14. Conclusion

In conclusion, we can say that AI is the future of dentistry. The application of AI is growing day by day in dentistry. The dentist cannot be replaced by AI but AI can help the dentist to see new standards in health care. The correlation between dentist and AI is need to established quickly to get accuracy in treatment. The only limitations currently AI have the insufficient data and expensive installation. Therefore dentist, clinicians, CAD-CAM engineers and Computer science engineers must perform tasks together for providing quality treatment to society.

15. Reference

[1] Mitsuishi, M., Cao, J., Bártolo, P., Friedrich, D., Shih, A., Rajurkar, K., Sugita, N. and Harada, K., 2021.Biomanufacturing. CIRP Annals, 62(2), 585-606

[2] Kang, J., Zhang, J., Zheng, J., Wang, L., Li, D., & Liu, S. (2021). 3D-printed peek implant for mandibular defects repair - a new method. Journal of the Mechanical Behavior of Biomedical Materials, 116, 104335.

[3] Farajpour, H., Bastami, F., Bohlouli, M., &Khojasteh, A. (2021). Reconstruction of bilateral ramus-condyle unit defect using custom titanium prosthesis with preservation of both condyles. Journal of the Mechanical Behavior of Biomedical Materials, 124, 104765

[4] Oldhoff, M., Mirzaali, M., Tümer, N., Zhou, J., &Zadpoor, A. (2021). Comparison in clinical performance of surgical guides for mandibular surgery and temporomandibular joint implants fabricated by additive manufacturing techniques. Journal of the Mechanical Behavior of Biomedical Materials, 119, 104512.

[5] Benaissa, A., Merdji, A., Bendjaballah, M. Z., Ngan, P., & Mukdadi, O. M. (2020). Stress influence on orthodontic system components under simulated treatment loadings. Computer Methods and Programs in Biomedicine, 195, 105569

[6] Bartolo, P., Kruth, J., Silva, J., Levy, G., Malshe, A., Rajurkar, K., Leu, M. (2012). Biomedical production of implants by additive electro-chemical and physical processes. CIRP Annals, 61(2), 635-655.

[7] S.Khanna Artificial intelligence: contemporary applications and future compass. Int Dent J. 2010;60:269–72.

[8] Farook, T. H., Jamayet, N. B., Abdullah, J. Y., Asif, J. A., Rajion, Z. A., &Alam, M. K. (2020). Designing 3D prosthetic templates for maxillofacial defect rehabilitation: A comparative analysis of different virtual workflows. Computers in Biology and Medicine, 118, 103646.

[9] Peng, W., Cheng, K., Liu, Y., Nizza, M., Baur, D. A., Jiang, X., & Dong, X. (2021). Biomechanical and Mechanostat analysis of a titanium layered porous implant for mandibular reconstruction: The effect of the topology optimization design. Materials Science and Engineering: C, 124, 112056.

[10] Cucchiara, R., Lamma, E., &Sansoni, T. (2004). An image analysis approach for automatically re-orienteering CT images for dental implants. Computerized Medical Imaging and Graphics, 28(4), 185-201.

[11] Pacifici, L., 2016. METALS USED IN MAXILLOFACIAL SURGERY. Oral & Implantology, 9(Suppl. 1), p.107.Li, P., Tang, Y., Li, J., Shen, L., Tian, W., & Tang, W. (2013).

[12] Establishment of sequential software processing for a biomechanical model of mandibular reconstruction with custom-made plate. Computer Methods and Programs in Biomedicine, 111(3), 642-649.

[13] Shujaat, S., Da Costa Senior, O., Shaheen, E., Politis, C., & Jacobs, R. (2021). Visual and haptic perceptibility of 3D printed skeletal models in orthognathic surgery. Journal of Dentistry, 109, 103660.

[14] Hua, J., Aziz, S., & Shum, J. W. (2019). Virtual surgical planning in oral and maxillofacial surgery. Oral and Maxillofacial Surgery Clinics of North America, 31(4), 519-530.

[15] Nikzad, S., Azari, A., &Ghassemzadeh, A. (2010). Modified Flapless dental implant surgery for planning treatment in a maxilla including sinus lift augmentation through use of virtual surgical planning and a 3-Dimensional model. Journal of Oral and Maxillofacial Surgery, 68(9), 2291-2298.

[16] Reyes, A., Turkyilmaz, I., &Prihoda, T. J. (2015). Accuracy of surgical guides made from conventional and a combination of digital scanning and rapid prototyping techniques. The Journal of Prosthetic Dentistry, 113(4), 295-303.

[17] Tong, J., Rahmel, B., Hsieh, J., & Findlay, G. (2021). Use of computer-aided three-dimensional prototyping to surgically assist in tooth autotransplantation. British Journal of Oral and Maxillofacial Surgery.

[18] Kang, S., Lee, J., Lim, S., Kim, Y., & Kim, M. (2014). Validation of mandibular Genioplasty using a Stereolithographic surgical guide: In vitro comparison with a manual measurement method based on preoperative surgical simulation. Journal of Oral and Maxillofacial Surgery, 72(10), 2032-2042.

[19] Cassetta, M., Di Mambro, A., Giansanti, M., Stefanelli, L., &Barbato, E. (2014). How does an error in positioning the template affect the accuracy of implants inserted using a single fixed mucosa-supported stereolithographic surgical guide? International Journal of Oral and Maxillofacial Surgery, 43(1), 85-92.

[20] Fawzy, H. H., & Choi, J. (2019). Evaluation of virtual surgical plan applicability in 3D simulation-guided two-jaw surgery. Journal of Cranio-Maxillofacial Surgery, 47(6), 860-866.

IOP Conf. Series: Materials Science and Engineering 1259 (2022) 012019 doi:10.1088/1757-899X/1259/1/012019

Design and Development of Process Equipments for Extraction of Coconut Shell's Activated Carbon, Bio Oil and Syngas

T Parshiwanikar[1], **C Handa**[2]

[1, 2] K D K College of Engineering, Department of Mechanical Engineering, Nagpur, Maharashtra, India

Abstract. There has been an increasing demand for safe water which can be produced by conversion of agricultural waste to usable energy forms which has attracted many researchers to develop a system, because of chip & ready availability of stuff, increase in energy cost furthermore as greater pressure on the environment by the employment of fossil fuels. Thermal pyrolysis is one amongst the strategy which will be applied to effectively utilize agricultural wastes as source of energy production. Within the present work, the assorted process equipment's are design and developed for extraction of activated carbon, bio oil and syngas from waste of coconut shells. During this process, the thermal pyrolysis i.e. carbonization and activation are performed. This paper attempts to confer the generalized design consideration for pyrolysis reactor, condenser & boiler.

1. Introduction

Activated carbon, Bio oil and Syngas has been widely employed in many fields and may be produced from kind of carbonaceous raw materials. It's a very important approach to convert waste in to some value added products that would solve environmental issues like water & air pollution, large collection of agricultural waste. Also, rather than using other fuel sources like coal, using coconut shells biomass based activated carbon will mitigate the effects of global warming. As coconut shells based activated carbon, bio oil & syn gas is derived from agricultural waste and is widely available, so it'll be more economical.

Activated carbon is a type of carbon that is not graphite and can be made from any carbonaceous material, Because of its enormous adsorptive capacity, micro porous adsorbents are the most important. Various researchers have published a number of studies on the various applications of activated carbon. Pollutant removal is a common application for activated carbon prepared from coconut shell. Activated carbon is used in a variety of industrial processes, including waste water treatment, water & gas purification, desulphurization and mercury removal, refining and bleaching of vegetable oils, volatile organic compounds (VOC) adsorption, gold recovery, solvents as well as other vapours are contained in gas masks & in filters to safeguard against warm & toxic and harmful gases etc. The main characteristics of activated carbon are that it is made up of developed carbon micro porous structure with the prepared material has a large surface area and a small pore size.

Another byproduct is bio oil which is extracted from coconut and consists of over 80% saturated fat. It's typically employed in cosmetics in addition as in baking and cooking. The very first step in preparing activated carbon & bio oil is preparing biochar from raw coconut shell with pyrolysis process viz. carbonization. The moisture content and volatile matter in the coconut shells are removed during this step. From this coconut shell charcoal, activated charcoal & bio oil is produced by using steam activation of shell charcoal in a reactor. Bio oil, sometimes also referred to as bio crude may be a synthetic fuel being researched as a petroleum substitute. Bio oil has the potential to be tar-like and typically contains levels of oxygen that are too high to be classified as a hydrocarbon. Methane, which is abundant in natural gas, biogas, and mine gases, is present in the syngas product.

IOP Conf. Series: Materials Science and Engineering 1259 (2022) 012019 doi:10.1088/1757-899X/1259/1/012019

The present work is focused on the design and development of various process equipment's that are used for extraction coconut shells, activated carbon, bio oil, and syngas. The standard design process is followed for the design of reactor, condenser and boiler for capacity of 40 kg of raw coconut shells.

2. Extraction Methodology

Figure 2.1 represents the production process layout for production of activated carbon, bio oil & syngas.

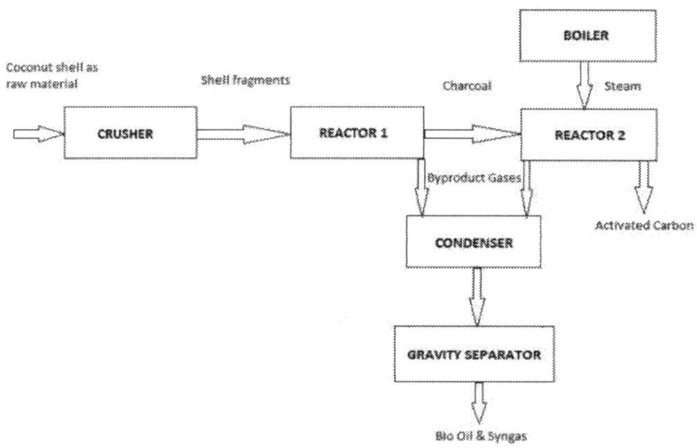

Figure 2.1: Production Process Layout

The various components used for production of activated carbon, bio oil & syngas are as follows.

Crusher

Because the shell fragments can even have rough edges as well as sharp points, coconut shells are tough and abrasive. Because it is suitable for tough and abrasive materials, a jaw crusher is used to crush the shells. The jaw crusher will have a larger crushing capacity and will be able to reduce the shell fragments to the size needed.

Reactor

In an extremely reactor, the shells are heated to 600°C at a rate of 20°C/min. A lower heating rate is required for slow pyrolysis to achieve a higher char yield. Higher gas yields result from faster heating rates, while lower char yields result from slower heating rates. However, the heating rate should not be so slow that it simply wastes time when a faster heating rate would have produced the same char yield.

Condenser

A condenser is an apparatus or equipment used to condense gases (converting a substance's physical state from gaseous to liquid). In simplest form, condenser consists of a tube through which hot gases are passed with outside water providing cooling.

Boiler

The boiler is employed to produce steam for activation process in second reactor where the porosity of char is increase and thus activated carbon is produced.

Gravity Separator

It is assumed that the vertical knock-out drum is a vertical vessel with sieve plates acting as mesh pads. Gravity separates the liquid and gas as the gas flows up through the vessel. Because the synthesis gas contains hydrogen, the vessel will be made of low alloy steel. Because the gas is corrosive, the plates are made of stainless steel.

3. Design & Development of Components

In general, two main stages are there in the production of activated carbon, bio oil & syngas from coconut shells. The first stage is started with thermal pyrolysis / carbonization of raw coconut shell biomass at a temperature of around 600^0C in the inert atmosphere inside reactor 1 for getting biochar/charcoal. The first stage of the activation process is then followed by a second stage for the enlargement of pore volume and generation of large surface area of activated carbon inside reactor 2. During both the stages, syn gases are produced which can be condensed to get bio oil. The design steps for design of reactor, condenser and boiler are as follows.

Design of Reactor:

Coconut shell is collected (40 kg) & burnt through 600-900^0 C in closed atmosphere.

1kg = 12 to 13 pieces of coconut shell

40 kg = 12×40 = 480 pieces

Volume of One hollow piece = $2/3 \times \pi \times (r_1^3 - r_2^3)$

$= 2/3 \times \pi \times (50^3 - 45^3)$

$= 70.94 \times 10^3$ mm³

Volume of 480 pieces = 34.05×10^6 mm³

$V = \pi/4 \times d^2 \times l$

$34.05 \times 10^6 = \pi/4 \times 300^2 \times l$

$\boldsymbol{l = 481.07\ mm}$

$\boldsymbol{length\ of\ Reactor \cong 500\ mm}$

Volume of Reactor (cylindrical section) = $\boldsymbol{\pi/4 \times 300^2 \times 500}$

Volume of Reactor (cylindrical section) = $\boldsymbol{35.34 \times 10^6 mm^3}$

Volume of Reactor (conical frustum section)

$= \pi/3 \times h \times (R^2 + R \times r + r^2)$

$= \pi/3 \times 200 \times (150^2 + 150 \times 50 + 50^2)$

$= \boldsymbol{6.80 \times 10^6\ mm^3}$

Volume of Reactor (with conical frustum section)

$= 35.34 \times 10^6 + 6.80 \times 10^6 = \boldsymbol{42.14 \times 10^6 mm^3}$

Considering some volume is occupied by Glass Wool

$= \pi\ (R^2 - r^2)\ h = \pi \times (200^2 - 150^2) \times 500 = \boldsymbol{27.47 \times 10^6 mm^3}$

Total Volume of Reactor = $\boldsymbol{69.61 \times 10^6 mm^3}$

Design of Shaft:

Stirrer or mixing shaft consist of 6 baffles of stroke is ensured.

It rotates at a speed of 10 – 20 rpm.

Considering 60mm × 30mm baffle/blade Subjected to 2000N of effort applied at the shaft (40×9.81=392.4 N)

(Taking Five times load as the force applied)

Torque required: - $T = F \times r$

$T = 2000 \times 60 = \boldsymbol{120 \times 10^3 N\text{-}mm}$

Power required: -

$P = 2\pi NT/60$

$= 2 \times \pi \times 20 \times 120/60$

$P = 251.32\ watts$

$\boldsymbol{P = 0.251\ kw}$

$P = 0.251/0.746$

$= 0.336\ HP$

$\boldsymbol{P \approx 0.5HP}$ (Power of motor)

Diameter of Shaft:

Considering stainless steel.......... (Design Data Book Table II-8 Cr 18%, Ni 8% Page No. 42)

Hot rolled $S_{ys} = 135\ MPa$

$\tau = 135/2$

$= 67.5\ N/mm$

$T = (\pi/16) \times d^3 \times \tau_{max}$

$120\times10^3 = (\pi/16)\times d^3 \times67.5$
$\boldsymbol{d = 20.84\ mm}$
Increasing diameter by 50% due to bending effect and torque
$= 20.84\times1.5$
$d = 31.26 \approx 32\ mm$......... (T, XI-4) ($diameter\ of\ shaft$)

Condenser Design :
(Length of tube required)
Tube diameter = 20mm
Gas temperature = 500°C Water outlet temperature = 50°C
the property of water at 40°C, Density $\rho = 1000\ kg/m^3$
$C_{pw} = 4200\ kJ/kg\text{-}K$ .... (Specific heat of water)
$K = 0.630\ W/m-K$ .... (Thermal conductivity of water)
$\mu = 0.0065\ kg/ms$ .... (Dynamic viscosity of water)
Velocity of water, $v = 0.2\ m/s$ ...... (Assume)
Diameter of tube $d_t = 0.020m$
Heat gained by water:
$Q = A_t \times v \times\rho \times C_{pw} \times(T_{wo} - T_{wi})$
$= 3.14\times10^{-4} \times0.2\times1000\times4200\times(50-40)$
$Q = 25326\ W$
The heat flow, $Q = A_s \times h \times LMTD$
$LMTD = (T_i - T_o)/(\log T_i/T_o)$
$=(40-50)/(\log 40/ 50) = 103.189$°C
$h = (N_u\times k)/d$
(h = average convective heat transfer coefficient)
$Nu = 0.332\times Re_x^{0.5} \times Pr^{0.333}$
$Re = \rho\times v \times(d_t/\mu)$
$= 1000\times0.2\times(0.02/0.0065384)$
$Re = 6153.85$
$Pr = \mu \times C_{pw}/k$
$=0.0065\times4200/0.630$
$= 4.333$
Therefore
$Nu = 0.332\times6153.85^{0.5} \times4.33^{0.333}$
$Nu = 42.42$
$h =42.42\times0.630/0.02$
Therefore $h = 1336.23\ W/m^2 -k$

$Q = h \times\pi \times d_t \times L \times LMTD$
$25326 = 1336.23\times\pi \times0.02\times L \times103.90$
$\boldsymbol{L = 2.90\ m \approx 4m}$

Design of Boiler:
Design of boiler as pressure vessel (Reference: Design Data Book Pg. No. 100)
Input Data: - At 270°C temperature the pressure of steam = 55.08 Bar (from steam table) = 5.505 N/mm^2
$\boldsymbol{D_i = 300mm}$
Where, Considering SAE1020 steel as the material of boiler
$S_d = 0.8\times S_{yt}$............ (For ductile material)
$= 0.8\times246$
$S_d = 196.8\ N/mm^2$
$n =0.95$.......... (For ductile material)
$t_b = (PD_i / 2S_d) \times n$

IOP Conf. Series: Materials Science and Engineering 1259 (2022) 012019 doi:10.1088/1757-899X/1259/1/012019

$= (5.505\times300\ /2\times196.08)\times0.95$

$\boldsymbol{t_b = 4.433\ mm \approx 5mm}$

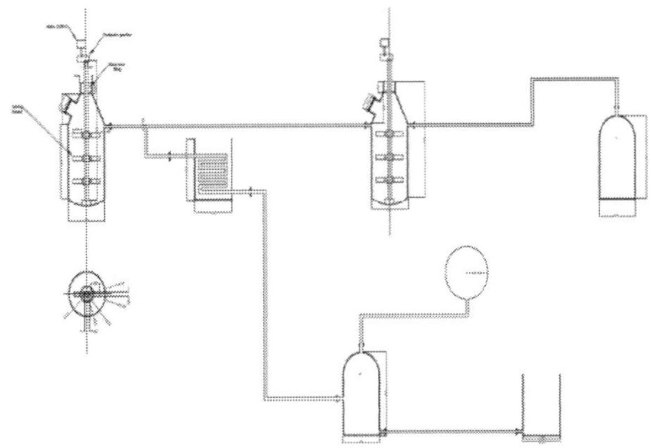

Figure 3.1: Two Dimensional View

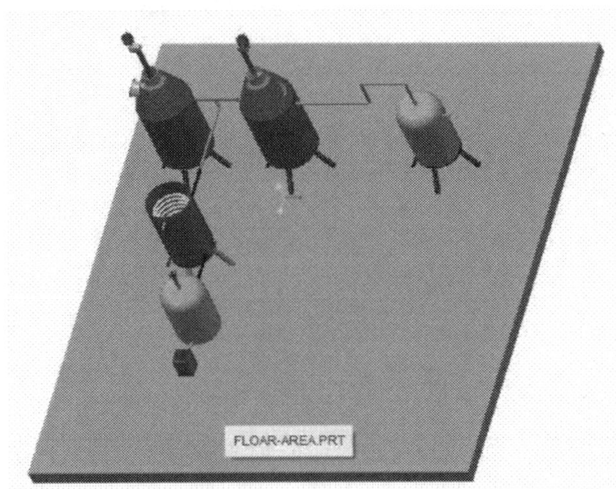

Figure 3.2: Schematic CAD Model

Figure 3.3: Developed Model (Fabricated Set up)

4. Applications of Product

Water and air pollution have emerged as major issues caused by substantial population growth and rapid development in industrial activities. As a result, controlling water and air pollution is critical to achieving long-term development. Currently, the adsorption method of scrubbing pollutants is widely used. The products are for all the concern agencies related to filtration of water and industries for use of bio oil & syngas. The activated carbon is used for absorption, while the other products are being used as fuel.

5. Result and Discussion

The extraction of activated carbon, bio oil and synthesis gas from coconut shells would improve the conversion of waste into value-added products. The various process equipment's are design and developed which supports the theoretical study. Coconut activated carbon has a high absorption capacity is critical to its successful application in water and air pollution control. Activated carbon must be modified in this way, either by adjusting the activation conditions or by applying surface treatments after activation.

6. Conclusion

This study confirmed that, for extraction of the activated carbon, bio oil along with syngas, two optimal processes thermal pyrolysis and activation are carried out by using various equipment's like reactors, condenser, boiler, oil & gas separator etc. The process will be more effectively employed for extraction of all three value added products through single set up in an economical way that reduces the environmental contamination by applied technology.

7. References

[1] Ahmad A and Idris A 2013 Preparation and characterization of activated carbons derived from bio-solid: a review *Desalination and Water Treatment*. www.deswater.com, doi: 10.1080/19443994.2013.808797.

[2] Suhas P, Carrott M and Carott R 2007 Lignin – from natural adsorbent to activated carbon: a review *Bioresource Technology*, 98, 2301–2312.

[3] Ioannidou O and Zabaniotou A 2007 Agricultural residues as precursors for activated carbon production – a review *Renewable and Sustainable Energy Reviews*, 11, 1966–2005.

[4] Gratuito M, Panyathanmaporn T, Chumnanklang R, Sirinuntawittaya N and Dutta A 2008 Production of activated carbon from coconut shell: Optimization using response surface methodology *Bioresource Technology*, 99, 4887–4895.

[5] Zhonghua H and Srinivasan M 1999 Preparation of high-surface-area activated carbons from coconut shell, Microporous and Mesoporous Materials, 27, 11–18.

[6] Laine J, Calafat A and Labady M 1989 Preparation and characterization of activated carbons from coconut shell impregnated with phosphoric acid, Carbon Vol. 27. No. 2, pp. 191-195.

[7] Li W, Peng J, Zhang L, Yang K, Xia H, Zhang S and Guo S 2009 Preparation of activated carbon from coconut shell chars in pilot-scale microwave heating equipment at 60 kW *Waste Management*, 29, 756–760.

[8] Kima J, Sohna M, Kima D, Sohnb S and Kwon Y 2001 Production of granular activated carbon from waste walnut shell and its adsorption characteristics for Cu2+ ion *Journal of Hazardous Materials*, B85, 301–315.

[9] Chandra T, Mirna M, Sudaryanto Y and Ismadji S 2009 Activated carbon from durian shell: preparation and characterization *Journal of the Taiwan Institute of Chemical Engineers*, 40, 457–462.

[10] Sun K and Jiang J 2010 Preparation and characterization of activated carbon from rubber-seed shell by physical activation with steam *Biomass and Bioenergy*, 34, 539–544.

[11] Plaza M, Pevida C, Martı´n C, Fermoso J, Pis J and Rubiera F 2010 Developing almond shell-derived activated carbons as CO_2 adsorbents *Separation and Purification Technology*, 71, 102–106.

[12] Sayan E 2006 Ultrasound-assisted preparation of activated carbon from alkaline impregnated hazelnut shell: an optimization study on removal of Cu^{2+} from aqueous solution *Chemical Engineering Journal,* 115, 213–218.

[13] Vitidsant T, Suravattanasakul T and Damronglerd S 1999 Production of activated carbon from palm-oil shell by pyrolysis and steam activation in a fixed bed reactor *Science Asia,* 25, 211–222.

[14] Daud W, Ali W and Sulaiman M 2000 The effects of carbonization temperature on pore development in palm-shell-based activated carbon *Carbon,* 38, 1925–1932.

[15] Yang T and Lua A 2003 Characteristics of activated carbons prepared from pistachio-nut shells by physical activation *Journal of Colloid and Interface Science,* 267, 408–417.

[16] Nowicki P, Wachowska H and Pietrzak R 2010 Active carbons prepared by chemical activation of plum stones and their application in removal of NO_2 *Journal of Hazardous Materials, 181, 1088–1094.*

[17] Putun A, Ozbay N, Onal E and Putun E 2005 Fixed-bed pyrolysis of cotton stalk for liquid and solid products *Fuel Process Technology,* 86, 1207–1219.

[18] Bae W, Kim J and Chung J 2014 Production of granular activated carbon from food-processing wastes (walnut shells and jujube seeds) and its adsorptive properties *Journal of the Air & Waste Management Association*, ISSN: 1096-2247.

[19] Otulana J, Oluwole O and Adeleke M 2016 A Reactor Plant for Activated Carbon Production *International Journal of Novel Research in Engineering and Science* Vol. 2, Issue 2, pp: (20-26).

[20] Mcconnachie G, Warhurst A, Pollard S, Chipofya V and Malawi V 1996 Activated carbon from Moringa husks and pods *22nd WEDC Conference, Reaching The Unreached: Challenges For The 21st Century*, New Delhi, India.

[21] Cobb A, Warms M and Maurer E 2012 Low-Tech Coconut Shell Activated Charcoal Production *International Journal for Service Learning in Engineering* Vol. 7, No. 1, pp. 93-104.

[22] Gao Y, Yang Y, Qin Z and Sun Y 2016 Factors affecting the yield of bio-oil from the pyrolysis of coconut shell *Springer Plus.*

[23] Chandra T, Maria Mirna M, Sunarso J, Sudaryanto Y and Ismadji S 2009 Activated carbon from durian shell: Preparation and characterization *Journal of the Taiwan Institute of Chemical Engineers,* 40, 457–462.

[24] Mohamad Nor N, Lee Chung L, Keat Teong L and Rahman Mohamed A 2013 Synthesis of activated carbon from lignocellulosic biomass and its applications in air pollution control- a review *Journal of Environmental Chemical Engineering,* 1, 658–666.

[25] Moreira R, Dos Reis Orsini R, Moreira Vaz J, Carlos Penteado J and Spinacé E 2017 Production of Biochar, Bio-Oil and Synthesis Gas from Cashew Nut Shell by Slow Pyrolysis *Waste and Biomass Valorization* volume 8, pages 217–224.

Acknowledgments
The authors gratefully acknowledge Rashtrasant Tukadoji Maharaj Nagpur University for providing grant-in-aid & Mr. Ajay Padole, Director, Parth Enterprizes, Nagpur in exchange for their monetary & intellectual supports.

IOP Conf. Series: Materials Science and Engineering 1259 (2022) 012020 doi:10.1088/1757-899X/1259/1/012020

Design and Development of Pneumatic Sheet Metal Cutting Machine

J. Om Prasaad[1], Krishnan A.B.[1], Parth Lokhande[1] and Biradar Mustaffa[1]

[1]Department of Mechanical Engineering, Fr. C. Rodrigues Institute of Technology, Vashi, Mumbai, India

Corresponding author: [1]omprasaadj@gmail.com

Abstract. A Pneumatic sheet metal cutting machine, is a mechanism designed to cut metal sheets at a quick pace, based on the inputs given by the operator, using a switch. The machine consists of a cutter blade system, a double acting pneumatic cylinder held over the cutter blade system, an air compressor, a pneumatic valve to regulate the air flow to the pneumatic cylinder from the compressor, and a frame to hold the machine in position. Cutting and shearing technology used nowadays, includes laser cutting, hydraulic cutters, among others. These machines are costly, with expensive maintenance. Small and medium scale industries cannot tend to afford this technology. Currently basic sheet metal cutting operations are incorporated in such industries including basic handheld sheet cutters. This is not accurate in terms of the dimensions of the final output and poses safety hazards to the operator too. Thus, the novelty of this machine is that it provides an automated system to cut aluminium sheet metal, while making sure that the system is compact and portable, which has otherwise not been implemented in industries. Pneumatic sheet metal cutters find a tremendous scope in such industries owing to their simple and easy working, low cost and ensured safety with a good degree of dimensional accuracy. The main aim of this study is to research and innovate a safe and efficient way to cut aluminium sheets at a fast and regular pace, as compared to manual cutting, using a pneumatic system.

1. Introduction

Aluminium sheet is a commonly used material, in various industries and factories, like interior and exterior design, construction of ceilings, and architecture, among many others. During its production aluminium sheet metals must undergo the shearing operation to achieve the perfect dimension, necessary for the final product. Various machines can be incorporated to carry out this shearing operation. They include manual, hydraulic, pneumatic, spring and rack and pinion operated machines.

Small and medium scale industries use aluminium sheets extensively in production as it is cheap and easily available. But when it comes to shearing operation, manual methods of cutting are incorporated to save cost. Commonly available sheet metal cutting machines are usually costly and sophisticated to use, apart from the maintenance costs, it requires periodically. This project is aimed at developing a cheap, safe, and effective system to shear aluminium sheets, using a pneumatic system for actuation.

The cutting mechanism comprises of an upper blade, constrained to rotate about an axis perpendicular to the length of the blade and a fixed lower blade. The sheet metal is placed between the blades. In shearing cutting action takes place perpendicular to the direction towards the sheet metal [1]. Upon shearing, plastic deformation occurs at the sheet, which ultimately shears the metal. The blades are actuated using a pneumatic system. In this type of machine high pressure air is used as the working fluid for the transfer of power and the motion [2]. A small clearance is present between the edges of the upper and lower tools, which facilitates the fracture of the material [3]. Solenoid Valve is used to adjust the air pressure from the source port to the desired port [4]. A compressor regulates the flow of air to the pneumatic cylinder, using a manually controlled circuit, through the solenoid valve, to carry out the cutting operation. The advantage of using a pneumatic system is that it can generate a considerable amount of force while being cost effective [5].

In this paper, the design aspects of the proposed system, have been discussed and analyzed. The proposed system is designed to cut Aluminium sheets up to a thickness of 2.5mm, using a pneumatic system.

2.Theoretical Analysis

The theoretical analysis performed asserts on the pressure necessary to cut the sheet and the best blade position for the same.

2.1. Pressure Calculations

The design calculations, to obtain the optimum pressure for cutting is performed considering the parameters from literature reviews and practical application of the project-

- The maximum width (L) of the sheet, which the machine can cut in one cycle is taken as 120 mm, based on practicality and length of the blade used.
- The strength of aluminium sheet (Tmax) is found to be 180 N/mm^2 [4].
- The maximum thickness of the sheet, which the machine can cut is taken as 2.5 mm, based on scope defined.
- The factor of safety for the calculations is assumed to be 1.2 [4].
- The mass of the cutting mechanism is taken to be 2.5 Kg, based on individual weights of the components used.
- The area of the bore was calculated to be 962.1127 mm^2, based on the blade designed on Inventor.

The Figures 1,2,3 and 4 represent the 4 different locations, the piston rod occupies in a single stroke, based on which the location of the blade changes. At these 4 positions, the force applied by the piston rod varies. At each position, each force vector, has two components. The cosine component represents the cutting force, while the sine component represents the wasted force. Thus, as per Table 1, the cosine angles of the 4 components are recorded and used for force calculations.

2.1.1. Cutting Force

$F_C = L \times t_3 \times Tmax$ [4] . . . (2.1)

The Eqn. 2.1 is implemented to calculate the force required to cut the metal sheet.

F_C – force to cut the sheet

L – width of sheet = 120 mm

t_3 – thickness of the sheet = 2.5 mm

Tmax - strength of aluminium sheet = 180 N/mm^2

IOP Conf. Series: Materials Science and Engineering 1259 (2022) 012020 doi:10.1088/1757-899X/1259/1/012020

$F_C = 120 \times 2.5 \times 180$

$\mathbf{F_C = 54000N}$

2.1.2. Stripping Force

$F_S = 0.2 \times F_C$ [4] . . . (2.2)

The Eqn. 2.2 is implemented to calculate the force required to strip the metal sheet.

F_S – force to strip the sheet

F_C – force to cut the sheet = 54000 N

$F_S = 0.2 \times 54000$

$\mathbf{F_S = 10800\ N}$

2.1.3. Static Force of the cutting mechanism

$F_{ST} = M \text{ x } g$ . . . .(2.3)

The Eqn. 2.3 is implemented to calculate the static force of the cutting mechanism

F_{ST} – static force

M – total mass of the cutting system = 2.5 Kg.

g - Gravitational moment of acceleration= 9.81 m/sec^2

$F_{ST} = 2.5 \text{ x } 9.81$

$\mathbf{F_{ST} = 24.525\ N}$

2.1.4. Total Force

$F_T = F_C + F_S + F_{ST}$ [4] . . . (2.4)

The Eqn. 2.4 is implemented to calculate the total force required to shear the sheet.

F_T – total force to shear the sheet

F_C – force to cut the sheet = 54000 N

F_S – force to strip the sheet = 10800 N

F_{ST} – static force = 24.525 N

$F_T = 54000 + 10800 + 24.525$

$\mathbf{F_T = 64824.525\ N}$

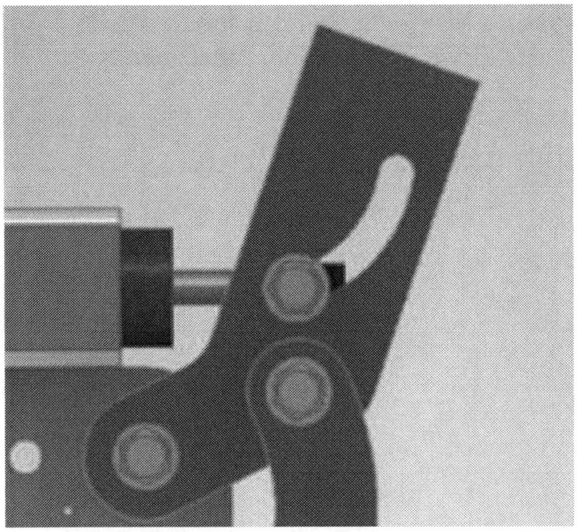

Figure 1. Position 1

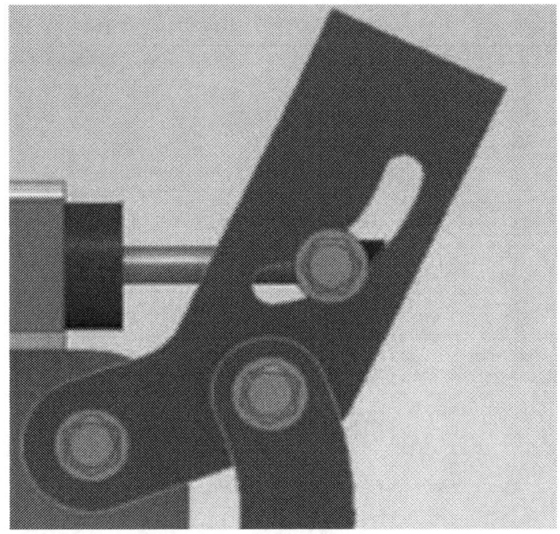

Figure 2. Position 2

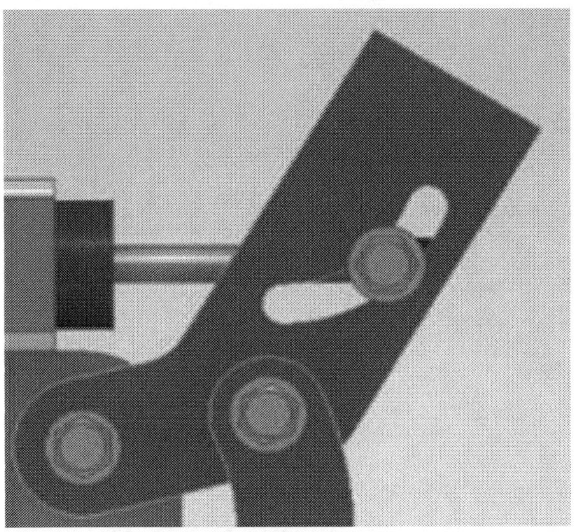

Figure 3. Position 3

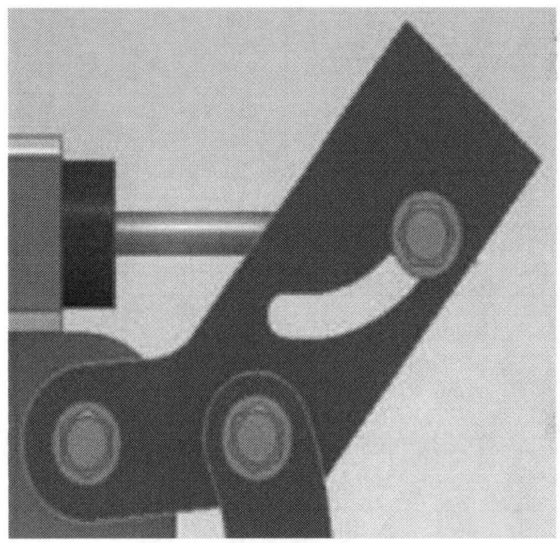

Figure 4. Position 4

Table 1. Cutting forces on 4 positions of the blade

Position	Angle of Force Component (^{0})	Cutting Force exerted (N)
1	56.88	35419.75
2	43.83	46764.26
3	32.68	54562.75
4	29.54	56398.09

IOP Conf. Series: Materials Science and Engineering 1259 (2022) 012020 doi:10.1088/1757-899X/1259/1/012020

From the force calculations at four positions of the piston, it is inferred that the maximum force required to cut the sheet is at is at position 4. Thus, the chosen force for pressure calculation is 56398.09 N.

2.1.5. Pressure

$P_C = (F/A)$ . . . (2.5)

The Eqn. 2.5 is implemented to calculate the pressure that is supplied by the pneumatic cylinder.

P_C – pressure required to cut the metal sheet

F – maximum force to shear the metal sheet = 56398.09 N

A – area of the bore = 962.1127 mm^2

$P_C = 56398.09 / 962.1127$

$P_C = 58.61 N/mm^2$

P_C = 5.86 bar

2.1.6. Total Pressure

Considering Factor of safety (FOS) to be 1.2.

$P = PC \times FOS$ [4] . . . (2.6)

The Eqn. 2.6 is implemented to calculate the total pressure required to shear the metal sheet.

P – total pressure supplied by the pneumatic cylinder

PC – pressure supplied by the pneumatic cylinder = 5.86 bar

$P = 5.86 \times 1.2$

P = 7.032 bar

Thus, as per the calculations, a pressure of 7.032 bar must be supplied to shear the aluminium metal sheet of 2.5 mm thickness.

2.2. Optimum Position of cutting mechanism for Shearing

To get the best position of the blade pivot, to shear the metal sheet completely, force calculations were carried out at 3 different positions of the rotary pivot of the blade over the frame, when the piston completes its forward stroke. Position 1 is the standard position of shearing mechanisms, while the other two positions were considered with respect to the first position. Based on the force analysis, on the three positions, the mechanism, with the maximum force output, would be selected.

As per Figures 5,6 and 7, for the simulation, the slider link is converted into a simple two link mechanism, in which the base link pivoted to the ground, is the pivot link and the second link, is the link which receives the force from the piston. In all the three cases, forces in the second link, act normal to the link, perpendicular to its axis. Thus, to find this force, the cosine component of the force applied by the piston, is found out and its product with the force, provides the cutting force of the blade.

IOP Conf. Series: Materials Science and Engineering 1259 (2022) 012020 doi:10.1088/1757-899X/1259/1/012020

2.2.1. Pivot Location 1

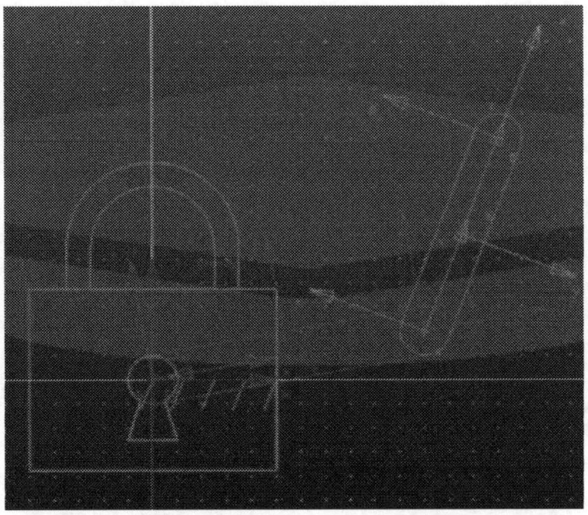

Figure 5. Position 1 of cutting mechanism

Taking cosine component,

F1 = 56398.09 cos(45.44)

F1 = 45484.5035 N.

2.2.2. Pivot Location 2

Figure 6. Position 2 of cutting mechanism

Taking cosine component,

F_2 = 56398.09 cos(46)

F_2 = 45030.89 N.

2.2.3. Pivot Location 3

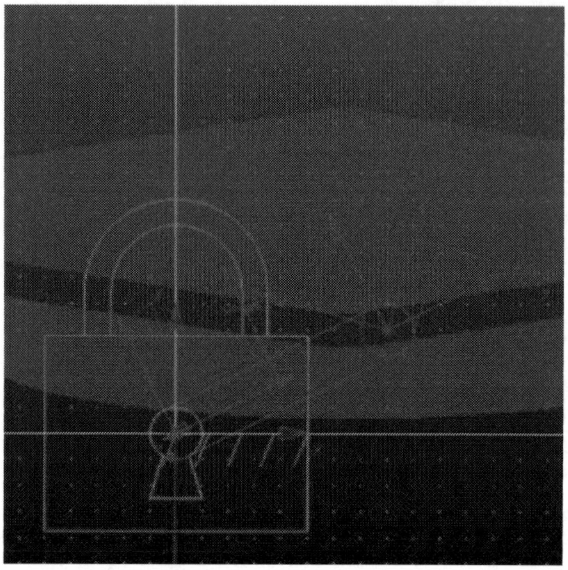

Figure 7. *Position 3 of cutting mechanism*

Taking cosine component,

$F_3 = 56398.09 \cos(47)$

$F_3 = 44210.219$ N.

Thus, based on the calculations performed, the blade exerts maximum force, when the blade is pivoted about location 1.

3. Design & Simulations

The model was made in the Inventor Modelling Environment as per Figure 8. The individual components were modelled separately and assembled in the end. So, the end model was thus achieved. The model consists of a Frame which accommodates the cutting mechanism and pneumatic cylinder over it. The pneumatic cylinder is placed over the Frame and its piston rod is attached to the cutting mechanism, held over the frame. The whole system is placed over a metal stand.

The electric circuit, for the switching operation, is designed over a PCB board. It consists of a microcontroller with the program code, which controls the other components. The circuit consists of other basic components like capacitors, LEDs, switches, and resistors.

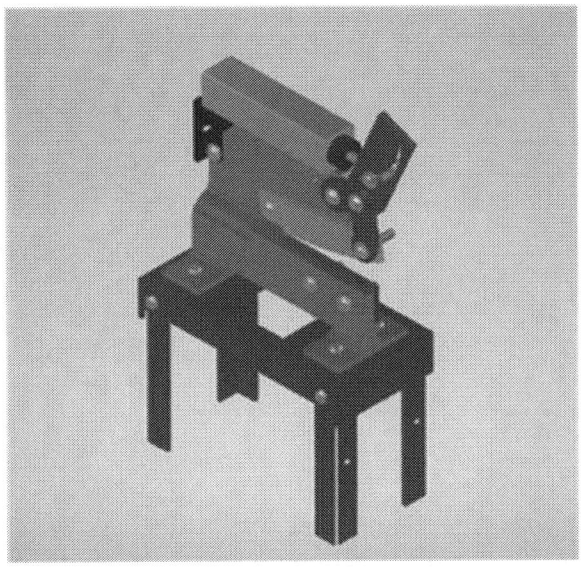

Figure 8. CAD Model of Design

3.1. Blade velocity simulation on MSC Adams

To simulate the proposed cutting operation, a primary model of the cutting mechanism, as illustrated in Figure 9, as made on MSC Adams, using simple linkages.

The mechanism consists of the blade, the two linkages attached to it, which aid in its rotary motion and an additional rotor link, which upon complete rotation, lifts the blade up and pushes it down during one cycle, through the two links attached to it.

A rotational velocity of 1 rad/s was given to the rotor as primary input. Based on this velocity, further simulations of output angular velocity of blade were obtained.

Based on the simulations performed as evident from Figure 10 according to the input angular velocity, it was observed that the output angular velocity of the blade, within the time frame of one cycle completed by the cutting mechanism ranges from 0.5 rad/sec to 0.7 rad/sec.

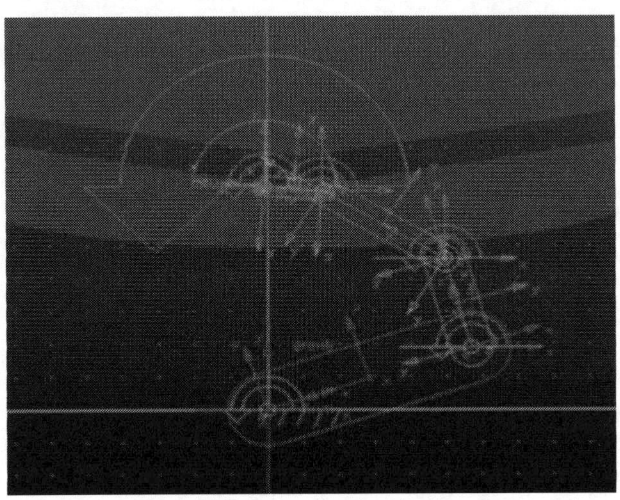

Figure 9. Primary Linkage Mechanism on MSC Adams

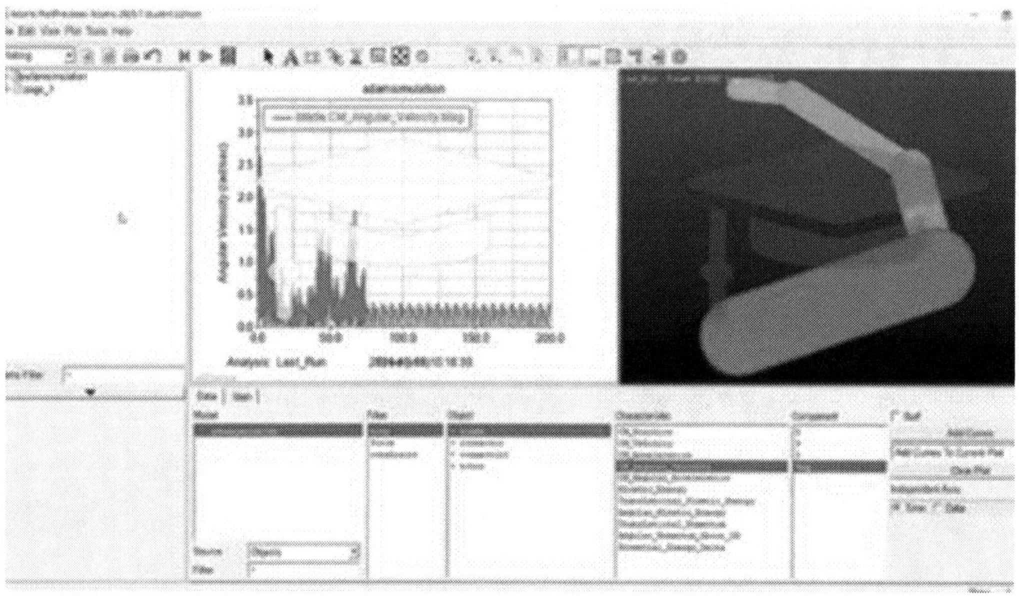

Figure 10. Output Angular Velocity of Blade

3.2. Simulation of Pneumatic System

The pneumatic system consists of the double acting cylinder, the solenoid valve, and the air compressor. A 5/2 spring return solenoid valve is incorporated in this system. The simulation of the pneumatic system was carried out on FluidSim Software. The simulation was done to test, whether a 5/2 single spring solenoid valve, was capable enough to actuate a double acting pneumatic cylinder, as shown in Figures 11 and 12.

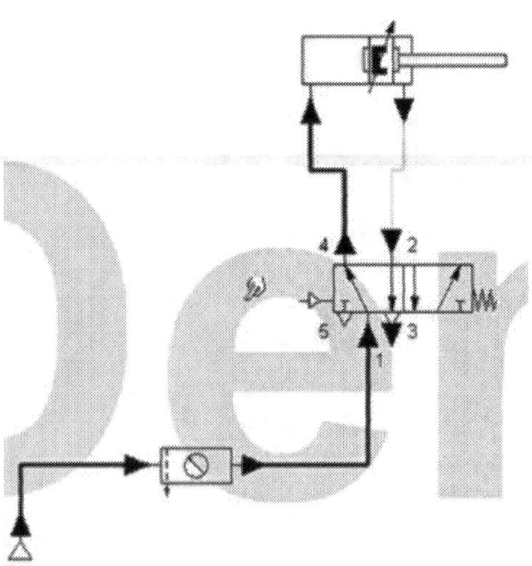

Figure 11. Expansion Stroke

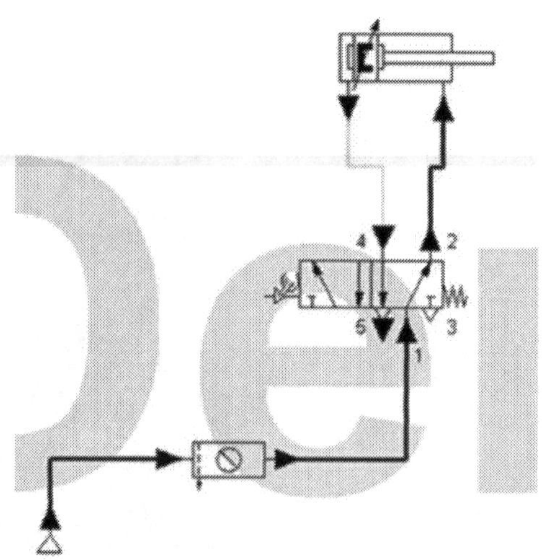

Figure 12. Retraction Stroke

4. Safety Protocols

The pneumatic sheet metal cutting machine, consists if parts which may injure the worker and can in turn affect the life of the machine too. While operating the machine workers must keep in mind-

- The ABC of safety: "Always Be Careful".
- To wear an apron or shop coat and gloves, to prevent sharp chips from causing any skin injuries.
- To protect eyes, by wearing safety glasses or an eye shield.
- To maintain safe distance from the blades, when feeding the metal sheet in.
- To tighten any loose parts before operating on the machine.
- To remove all chips caught in the working zone.
- To make sure all electrical components are at a safe distance from the work zone, to prevent any damage, from chips, during the cutting operation.
- To lubricate the pneumatic cylinder using a FRL (Filter Regulator Lubricant) system, so that the pneumatic cylinder operates effectively with less energy losses.

5. Results and Discussion

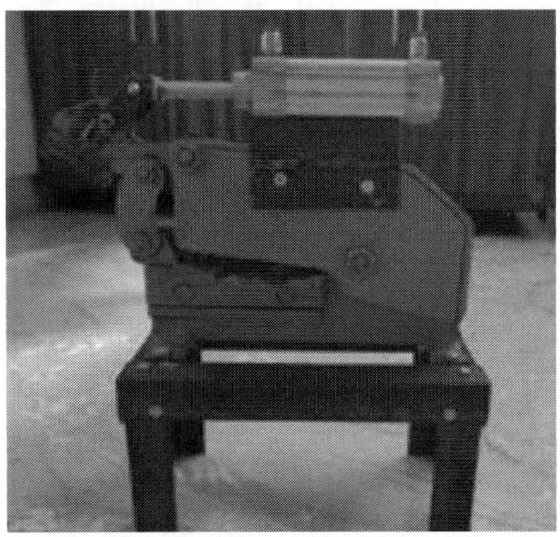

Figure 13. Basic assembly of Pneumatic Sheet Metal Cutting Machine

To check and verify the functionality of the proposed machine, the necessary components were bought based on the design calculations and assembled. A preliminary check was conducted after the assembly. The test verifies whether the pneumatic cylinder chosen, can actuate the blade down, with the necessary force to cut the blade.

The initial set up consists of the pneumatic cylinder attached to the solenoid valve, through the valves. The air compressor is attached to the other side of the solenoid valve. The cutting mechanism is held in place, by the piston of the pneumatic cylinder and the frame. The cutting mechanism consists of the upper and the lower blade.

The switching circuit, powered using an AC supply, is used for controlling the pneumatic cylinder. It is attached to the solenoid valve. It consists of 2 LED bulbs. When the circuit is powered, the first LED glows, indicating that the circuit is active. When the switch is initiated, a signal is sent to the solenoid valve, to allow the air from the air compressor, to flow into the pneumatic cylinder, to complete its forward stroke. This is the cutting stroke. After the cutting operation is performed, the switch is pressed again, to retract the piston back.

After designing and testing the proposed project it was found that the pneumatic sheet metal cutting machine, could shear metal sheets of Aluminium about 2.5 mm thickness. Before testing, it was made sure that the piston was is its retracted position, to keep the sheet, at its cutting position, over the blade. After the connections were made, the air compressor was activated, and the pressure was set to 7 bars. The testing now commenced, with a controller controlling the cutting operation of the blades, using the switch. It was observed that the machine could cut the sheet efficiently and quickly.

6. Conclusion

This paper proposes an efficient and cost-effective way to shear aluminium metal sheets, using a pneumatic actuation system which can be applicable in small and medium scale industries. The machine was also found to be accurate in its operation and safe, as opposed to manual cutting operations. Compared to its hydraulic counterparts, pneumatic systems are cheap, due to its application of air, which is abundant, inexpensive, and easy to contain and regulate. Pneumatic systems have a long operating life and require little maintenance, as gases are compressible, which makes the system less vulnerable to shock damage. Hydraulic systems on the other hand, require constant maintenance as fluids used need to be replaced over a period, which increases the cost too. Thus, pneumatic system was found to be more reliable than hydraulic system, as cost of maintenance is low apart from the total cost incurred. Thus, this project, provides a safe and effective way to cut metal aluminium sheets, with minimal cost incurred for maintenance.

7. References

[1] Aniket N S, Rahul R C, Abhishek J, Parigha B S and Ashwin P P 2018 Introduction & Overview of Pneumatic Press with Various Attachment *Int. Jour. of Inno. Sci. and Res. Tech.* **3** 450-54

[2] Suraj V B, Gajanan S G, Nikita H D and Ashish GI 2018 Design and Fabrication of Pneumatic Sheet Metal Cutting Machine-A Review *Int. Res. Jour. of Eng. and Tech.* **5** 751-53

[3] Krantikumar K, Saikiran K V S S and Jakkoju S 2016 Pneumatic Sheet Metal Cutting Machine *Int. Jour. & Mag. of Eng., Tech., Mana. and Res.* **3** 501-09

[4] Pritish C, Sachin R, Nitin M, Yogesh J, Rahul P 2018 Design and Fabrication Of Automated Pneumatic Shearing Machine to Cut Aluminium Sheet *Int. Jour. of Mech. and Prod. Eng. Res. and Dev.* **8** 499-506

[5] Aryan P, Adarsh R, Ayog K R, Aakash S and Vikas K S 2020 Design and Fabrication of Pneumatic Sheet Metal Cutting and Punching Machine *Int. Jour. of Eng. Res. & Tech.* **9** 1230-32

[6] Neeraj P, Naren K Akhshit J, Surya D S, Lakshay J, Anant K and Vishwarth S 2018 Pneumatic Sheet Cutting Machine *Int. Jour. of Scient. and Tech. Advan.* **4** 105-08

[7] Raj R 2016 Fabrication of Pneumatic Sheet Metal Cutter *Int. Jour. on Emerg. Tech.* **7** 360-65

[8] Chinmay A, Amogh G and Chandrashekar T 2014 Automatic Pneumatic Ramming Machine *Int. Jour. of Mech. And Prod. Eng.* **2** 60-62

Design, Development and Testing of Battery Electric Vehicle Sprayer

D S Karale[1], Ankita Shinde[2], N P Awate[3], and S H Thakare[4]

[1]Assistant Professor, Deptt. of Farm Power and Machinery, Dr. PDKV, Akola.
[2]Ph.D. research scholar, Deptt. of Farm Power and Machinery, Dr. PDKV, Akola.
[3]Assistant Professor, Deptt. of Mechanical Engineering, G. H. Raisoni College of Engg., Nagpur
[4]Professor & Head, Deptt. of Farm Power and Machinery, Dr. PDKV, Akola.

Abstract. To reduce the dependency on the fossil fuels a low-cost Battery Electric Vehicle (BEV) sprayer was developed in the present investigation which has a boom sprayer attachment and battery electric vehicle as a prime mover. The developed system has high uniformity of spray, the spray distribution along with the swath and easy to adjust height of boom over a target crop. The developed BEV sprayer was tested in the laboratory and on the experimental field for optimization of the operating parameters for green gram crop. It was observed that an electric motor of 1 kW, 48 V, BLDC capable to propelled the vehicle. The forward speed studied for optimization of spray nozzles were 2.0, 2.5, and 3.0 km/h and nozzles were tested at the operating pressures of 2.5, 3.0 and 3.5 kg/cm2 with orifice diameters of 0.6 mm, 0.8 mm, and 1.0 mm in the experimental field. The RSM results showed that the optimized performance of the sprayer with 0.8 mm orifice size nozzle on 3.0 kg/cm2 pressure and at a forward speed of 2.5 km/h for the green gram crop. The field tests were carried out on the optimized parameters it was observed that the effective field capacity of the sprayer was 1.09 ha/h with field efficiency of 79.81 per cent.

1. Introduction

Spraying has a very vital role to play to reduce the harvest losses and improving productivity. It was estimated that near about 52 per cent of production losses can be saved while harmful insects and diseases are eliminated by spraying [6]. The conventional manual sprayers have low field capacity and spraying of crops near about of hectare of land requires 6 to 8 hours of work. Due to this, the operator is exposed maximum time to hazardous chemicals, and attempt to consistently apply chemicals while working in a potentially hot environment without wearing a protective rubber suit with gas mask.Now a day's changing in the agricultural scenario of the country more particularly diversification in the cropping pattern and commercialization of agriculture, day by day the land fragmented in small and marginal land holdings causes more efficient equipment are required. Some of the researchers [10, 11], worked on battery electric vehicle operated automatic/robotic vehicle greenhouse sprayers but these have the limited use and high cost. The non-availability of labours in peak season, labour cost, and nonavailability of the improved spraying techniques matching the agro-technical requirements of the various crops causes a delay in the operations and ultimately there is decrease in the production. In view of the above constraints, it is proposed to develop a low-cost battery electric vehicle (BEV) sprayer which will have a boom sprayer attachment and battery electric vehicle as a prime more. The motivation behind developing a BEV sprayer for field crops application was to improve the chemical efficacy, reduce chemical and labor cost, minimize labor hazards and reduce the harmful environmental damage by avoiding the conventional fossil fuels. Adapting agricultural machinery, e.g., floating the use of agricultural purpose electric vehicles, is one way of meeting such challenges. Although interest in battery electric vehicle and sustainable farming is becoming ever stronger, in practice the usage of battery electric vehicle still remains at a relatively low level. The Indian government is gunning for its goal of making 30 per cent of Indian vehicles electric by 2030.

2. Materials and method

This section mainly deals with the detail methodology followed for design and development of battery electric sprayer and evaluation under laboratory and field conditions considering crop, machine and operational parameters.

2.1 Design and Development of Battery Electric Vehicle (BEV)

Computer Aided Design (CAD) modules were used to quickly test options for battery electric vehicle geometry along with positioning and sizing of components like spray boom, water tank, wheels, batteries and caster wheel arrangement. Many of the researcher's, different components of the machines modelled by using Computer Aided Design [1, 3, 7, 12], to save the time and repetitive fabrication work in their study. Hence to avoid the repetitive fabrication procedure the details were worked into the CAD model by using CATIA software as shown in Fig.1.

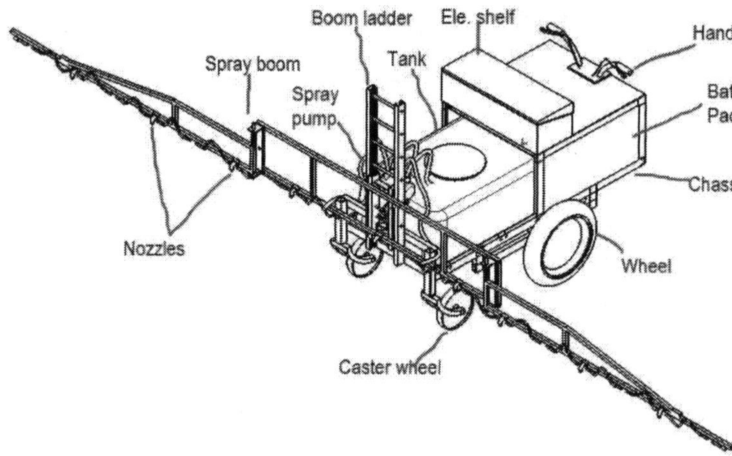

Figure 1. Concept rendering of the BEV showing the development of configuration, wheel placement, power storage etc.

A tentative BEV sprayer mass is estimated to calculate the motor power requirement to propelled the vehicle. The empirical evidence from the previous work done on self-propelled sprayer informed the mass target of the prototype BEV. From the various research reviews the mass of the vehicles in the range of 200-600 kg; so, based on this it is estimated maximum of 400 kg as the target of the BEV to minimum soil disturbance. The total mass of BEV sprayer is 3681.00 N. The row crops like greengram, blackgram, soybean and chickpea etc. were grown in the region with 45 or 60 cm row spacing's for which 33-inch differential was observed suitable to transvers the vehicle in between these the row crops. Hence a battery electric vehicle dimensions are modeled as (LBH) 1.6 x 0.9 m, whereas along with the BEV sprayer are as 1.6 m x 6 m (When boom folded width is 2 m) which enable easy to transport.

2.2. Design and Development of boom sprayer attachment suitable to Battery electric vehicle

The different components of sprayer were designed and developed on the basis of crop and machine parameters. The specifications of prototype sprayer attachment are presented in Table 1. A sprayer unit was developed by fabricating a 6 m width of spray boom with adjustable nozzle spacing and height over the row crops. The boom height and nozzle placement were kept to adjustable as per the requirement. All the nozzles on the boom could be adjusted according to the row spacing and height of the crop. To develop sufficient pressure for the sprayer, a two piston pump was used. The water dispenser pump was operated by an electric motor for which separate 12 V DC motor with power arrangement was provided from the BEV battery pack. The functional components of developed sprayer are illustrated in Fig.2.

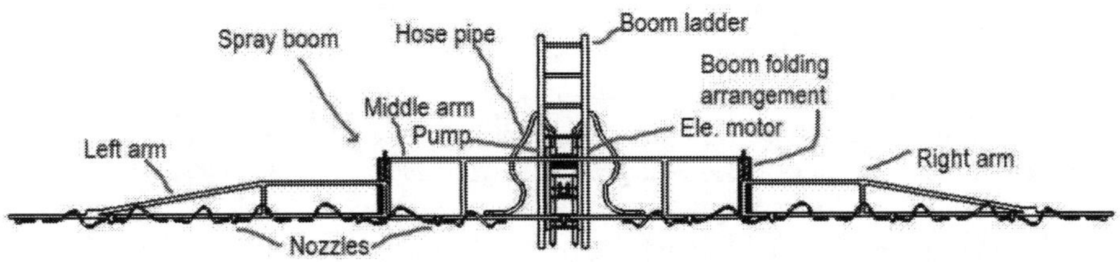

Figure 2. View of developed boom sprayer attachment.

Table 1. Specification of boom sprayer attachment for BEV.

Parameter	Value
Boom swath, m	6
Number of nozzles	9
	63
Spacing between the nozzles, cm	2 (after fold)
Folding arrangement, m	12
Max. Pump discharge, *l/m*	0-5
Operating Pressure, kg/cm^2	37
Weight, kg	

2.3. Description of the BEV sprayer

In the developed BEV sprayer, a BEV is the host vehicle and sprayer built as one unit or a multipurpose BEV based vehicle with the facility of spraying attachment (Fig. 2). The detailed specification of developed BEV sprayer is given in Table 2.

Table 2. Specifications of newly developed BEV sprayer.

Sr. No.	Particulars	Details
1	Technical descriptions	Battery electric vehicle
2	Motor	1kW W, 48 V, BLDC
3	Differential	33 inches
4	Battery pack	4x 12 V, 120 Ah
5	Speed	0 to 8.29
6	Capacity at maximum speed	50.24 km (3.84 hours)
7	Capacity on load	23.08 km (7 hours)
8	Power requirement (Max.)	443.85 Wh/km
9	Technical descriptions	Boom sprayer
10	Boom width	6 m
11	Tank capacity	200 *l*
12	Working pressure	2-5 kg/cm^2
13	Max. pump discharge	12 *l/min*
14	Max. discharge rate	7.2 *l/min*

15	Pump capacity	0.2 kW, 12lpm, 3500 rpm, Max. pressure 35 kg/cm^2
16	Weight	372 kg
17	Type and no. of nozzles used	Hollow cone nozzles, 9

2.4. Laboratory evaluation of nozzles

The uniformity of spray distribution across the boom, or within the spray swath, is essential to achieve maximum chemical effectiveness with minimal cost and reduction in non target contamination. The nozzle cone angle, nozzle discharge, nozzle spray pattern was measured by using the nozzle performance test rig at the Department of Farm Machinery and Power Engineering, Dr. PDKV, Akola, India. As prescribed in the test code pressure was set at recommended operating pressure of 3 kg/cm^2 to check the feasibility of the selected nozzles as per IS: 3652-1995 [4]. The nozzles of orifice diameter 0.6, 0.8 and 1.0 mm were tested on set pressure with three replications. The nozzle was given a preliminary run until a constant flow rate was achieved from the patternator, and the readings were subsequently taken [5].

2.5. Field performance evaluation

The performance evaluation of deigned and developed battery electric vehicle sprayers on green gram has been carried out with the optimized parameters at the Central Research Farm, Dr. PDKV, Akola. The test was conducted as per test code and followed for field testing. During the field trials, the biometric data pertaining to green gram such as row, spacing, and height of crop and stages of crop were noted. The three tests were carried out on Greengram crop to evaluate the various performance of parameters newly developed battery electric vehicle sprayer.

3. Result and Discussion

The developed BEV sprayer was tested in the laboratory and on the experimental field for optimization of the operating parameters for greengram crop. On the basis of droplet size (100 to 400 μm) the sprayer can be categorized as fine spray. The forward speed studied for spray nozzles were 2.0, 2.5, and 3.0 km/h. All the three nozzles were tested at the three operating pressures of 2.5, 3.0 and 3.5 kg/cm^2 with three orifice diameters of 0.6 mm, 0.8 mm, and 1.0 mm in the experimental field tests. The results obtained during the field tests are as given below.

3.1. Design and development of battery electric vehicle sprayer

The newly developed battery electric vehicle sprayer has the following specifications as illustrated in Table 2. During the running in test, it was noticed 443.85 Wh/km maximum power requirement of the BEV sprayer at speed of 3.5 km/h on loaded condition with a continuous work of 7 h. Whereas, the maximum speed of the design vehicle was calculated as 8.29 km/h on which the vehicle can be run up to 50.08 km with 3.84 h of continuous work on the design battery pack. As the vehicle is empty during the destination period the power requirement is calculated as 201.6 kW/h. The power consumption of the BEV sprayer was increased with the increase in the load. It was observed that an electric motor of 1kW, 12 V, 48 V, BLDC was capable to propel the vehicle as it develops a 98.84 N-m torque than the required (67.73 N-m) with the design gear ratio of 25:1. The results obtained were observed quite similar with the study of the researcher [3, 8]. During the design it was noticed that the battery weight is the major constraint. The total weight of the design BEV sprayer was observed as 372 kg and has 6 m long boom with 9 nozzles at an equal distance of 60 cm with 12 l/min pump discharge capacity.

3.2. Laboratory evaluation of nozzle

The feasibility of the spray nozzles was checked before utilization of nozzles in the newly developed battery electric vehicle sprayer for whether they conform or does not confirm with performance parameters against the prescribed procedure laid down in IS 3652:1995 [4]. Table 3 shows the average discharge and cone angle of nozzles with their orifice sizes at 3 kg/cm^2.

Table 3. Average discharge rate and cone angle of hollow cone nozzles.

Sr. No.	Orifice size, mm	Discharge rate, *ml/min*		Cone angle, °	
		Tested	Recommended	Tested	Recommended
1	0.6	255	250	55	55
2	0.8	456.66	450	60	60
3	1.0	505	500	80	80

There is no significant variation discharge is observed in three replications. (Table 3). The acceptable variation in discharge rate is ± 5 per cent and ± 3 per cent i cone angle from the performance point of view [4]. All three nozzles were found suitable to use as variation was within the prescribed limit.

3.3. Optimization of operational parameters for system performance

The input parameters such as orifice size, forward speed and pressure are required to be optimized for effective VMD, NMD, uniformity coefficient, droplet density and leaf area coverage and these parameters were found to be dependent on the machine and operational parameters. Based on filler trials, orifice size, forward speed and pressure were finalized and were kept constant throughout the complete experimentation. During the study there were 362 methylene blue @ 5 per cent concentration droplet impressed water sensitive papers were scanned for all seventeen trials of different combination as suggested by the Design Expert 11.0 software [2, 3, 7, 9].

*Volume median diameter (VMD).*From the experimental test it was observed that the treatment means of the VMD were in the range of 55.59 to 243.36 µm for all three-nozzle orifice diameter (0.6, 0.8 and 1.0 mm) with the combinations of pressure (2.5, 3.0 and 3.5 kg/cm^2) and speed (2.0, 2.5 and 3.0 km/h). Significant effect of speed and pressure was observed on VMD. Fig. 3 shows the 3D graph of the combined effect of orifice size, speed and pressure on VMD. It is cleared that the target set for the optimization for the influencing factor of orifice size (0.8 mm), speed (2.5 km/h) and pressure (3.0 kg/cm^2) the maximum VMD was found to be 124.757 µm. The response surface results showed that the orifice size and pressure have maximum effect followed by forward speed and there is the existence of quadratic terms which shows curvilinear nature of response surface graph.

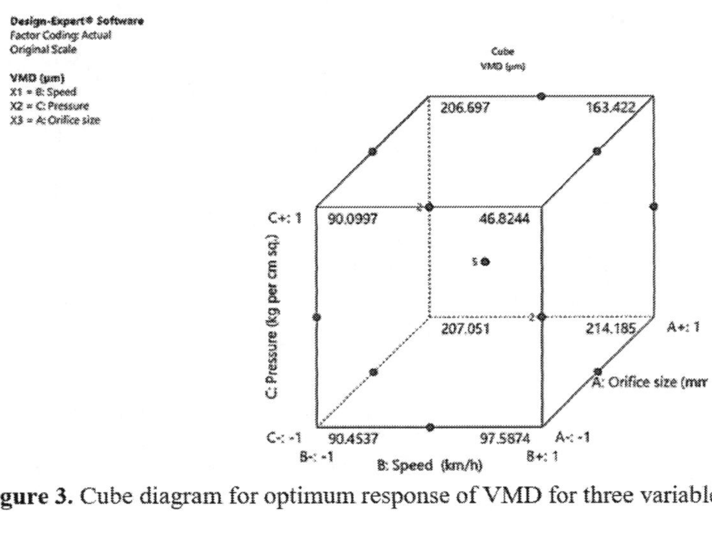

Figure 3. Cube diagram for optimum response of VMD for three variables

Number median diameter (NMD). Similarly, the treatments of the NMD were analyzed and observed in the range of 34.52 to 171.03 μm. Significant effect of speed and pressure was observed on the NMD. Fig. 4 shows the 3D graph of the combined effect of orifice size, speed and pressure on NMD. It is cleared that the target set for the optimization for the influencing factor of orifice size (0.8 mm), speed (2.5 km/h) and pressure (3.0 kg/cm^2) the NMD was found to be 105.656 μm. Increasing or decreasing the study factors effects on work of each sprayer unit and can change the amount of NMD.

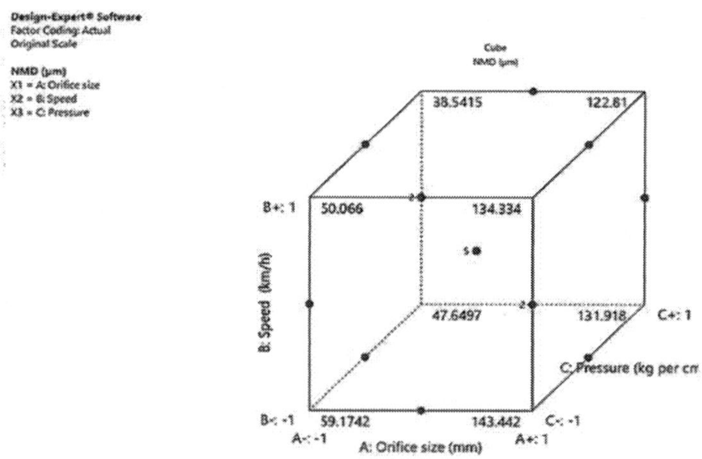

Figure 4. Cube diagram for optimum response of NMD for three variables.

Uniformity coefficient (UC). The treatment means of the UC were observed in the range of 1.12 to 1.80. The UC was calculated from the VMD and NMD values obtained for the seventeen test runs. Fig.5 shows the 3D graph of the combined effect of orifice size, speed and pressure on UC. It is cleared that the target set for the optimization for the influencing factor of orifice size (0.8 mm), speed (2.5 km/h) and pressure (3.0 kg/cm2) the UC was found to be 1.194. Increasing or decreasing the study factors effects on work of each sprayer unit and can change the amount of UC.

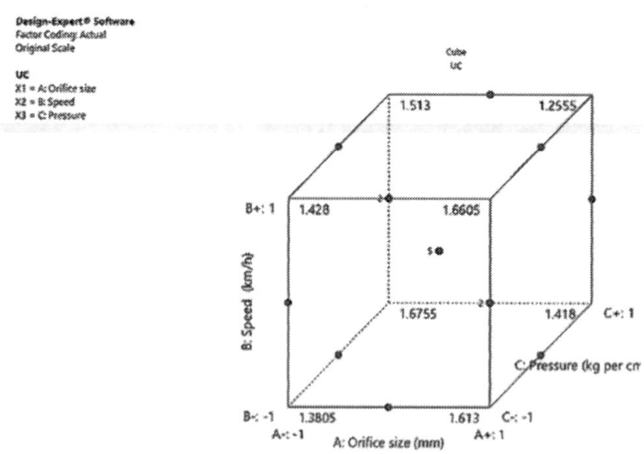

Figure 5. Cube diagram for optimum response of UC for three variables.

Droplet Density (DD). The treatments of the DD were analyzed and was observed in the range of 14.45 to 65.45 droplets/cm^2 Fig.6 shows the 3D graph of the combined effect of orifice size, speed and pressure on DD.

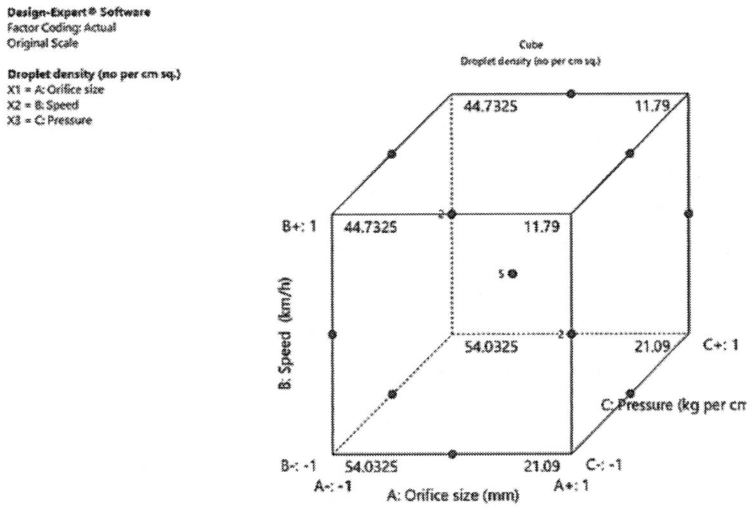

Figure 6. Cube diagram for optimum response of DD for three variables.

Increasing or decreasing the study factors effects on work of each sprayer unit and can change the amount of DD. It is cleared that the target set for the optimization for the influencing factor of orifice size (0.8 mm), speed (2.5 km/h) and pressure (3.0 kg/cm^2) the droplet density was found to be 24.952 no. of droplets/cm^2.The response surface results showed that the orifice size has maximum effect and there is the quadratic effect of orifice size shows curvilinear nature of response surface graph.

Leaf area coverage (LAC). The experimental treatment for the LAC was analyzed and observed in the range of 38.7 to 92.94 per cent. The leaf area coverage was found middle to the set value in the analysis for middle value of pressure and leaf area coverage which is bigger for the higher range of the factor. Fig. 7 shows 3D graph of the combined effect of orifice size, speed and pressure on LAC. Increasing or decreasing the study factors effect on work of each sprayer unit and can change the amount of LAC.

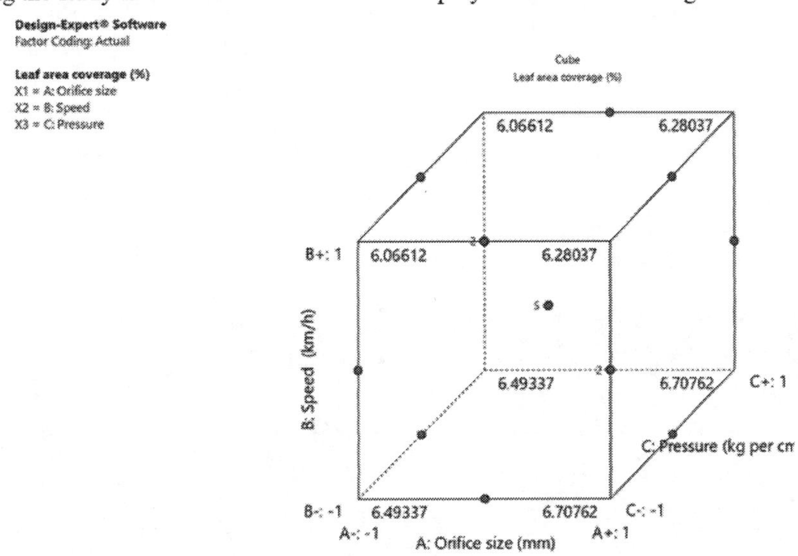

Figure 7. Cube diagram for optimum response of LAC for three variables.

3.4. Field performance of the BEV sprayer

The prototype battery electric vehicle sprayer was evaluated for optimum parameters at orifice size 0.8 mm speed 2.5 km/h and pressure 3.0 kg/cm^2. The battery electric vehicle sprayer was evaluated in greengram crop (fig.8) in accordance to the RNAM test code and test procedure. During the experiment various parameters were recorded as given below. The field performance results of the BEV sprayer are summarized in Table 4 on the optimized operating parameters.

Table 4. Field performance result of BEV sprayer.

Parameters	Test trials			
	R_1	R_2	R_3	Average
Average plat height, cm	40.12	39.69	40.01	39.94
Row to row spacing, cm	45	45	45	45
Average plant to plant spacing, cm	7.4	7.02	6.52	6.98
Average number of plants per m^2	40.54	41.56	46.01	42.70
Length of field, m	153	126	150	--
Net duration of test, h	0.97	1.06	1.18	--
Average soil moisture, %	14.24	15.16	16.04	15.15
Average battery voltage, V	48.26	47.6	48.54	48.13
Average motor current, A	37.6	37.4	37.4	37.47
Average pump motor current, A	5.10	5.20	5.08	5.13
Average pump motor voltage, V	12	12.2	12.14	12.11
Speed of operation, km/h	2.5	2.5	2.5	2.50
Average wheel slippage, %	5.08	6.74	6.93	6.25
Average swath width, m	6.09	6.14	6.10	611
Theoretical field capacity, ha/h	1.52	1.53	1.52	1.52
Effective field capacity, ha/h	1.10	1.04	1.14	1.09
Time required for one ha, h	0.90	0.96	0.87	0.91
Field efficiency, %	72.62	68.09	74.97	71.89
Water required, l/ha	152	159	155	155.33
Labour required	2	2	2	2
Cost of operation, Rs./ha	130	137	125	130.37

It was observed that the average operating cost of the sprayer was Rs. 130. 37 Rs./ha. The cost of operation may be reduced by 20-25 Rs. /ha after habitalization of the operator because it was noticed that single labor could perform the operation.

Figure 8. Field performance test on optimized input parameters.

4. Conclusion

The operating parameters of the developed BEV sprayer were optimized for application of insecticide on Greengram crop. The optimized operational parameters were *viz*., orifice size as 0.8 mm diameter, operating pressure 3.0 kg/cm^2 and forward speed as 2.5 km/h. The developed sprayer is capable to spray the liquid at top, middle and bottom position of leaves at upper and lower side satisfactorily. The travel speed and operating pressure has linear effect on uniformity coefficient of spray spectrum. The VMD observed to increase with increase in orifice size of nozzle and decreases in operating pressure and travel speed. Similar trend was observed for NMD also. The UC observed to increase with increase in orifice size pressure and forward speed. The results showed that linear effects of orifice size had significant effect on droplet density ($P<0.01$) at 1 per cent level of significance followed by the speed. Similar trend was observed for LAC also. It was observed that the average operating cost of the sprayer was Rs. 130. 37 Rs./ha. The overall performance of the newly developed BEV sprayer was found satisfactory and it concludes that present work has a wide scope in manufacturing of BEV for small and marginal land holders and can be provide the solution to farming communities and to industries from present work.

5. References

[1] Awate N. and Tidke D. J., (2016) Designing and optimization of a mobile crop residue disintegrator by using Box Behnken RSM method. *International Journal of Design and Manufacturing Technology*. Vol., **7**(2) pp. 1-13.

[2] Awate N. and Tidke D. J., (2016) Modeling and optimization of a mobile crop residue disintegrator. *International Journal of Mechanical Engineering and Technology*. Vol., **7**(3) pp. 320-328

[3] Bawden, O. J., (2015) Design of a lightweight, modular robotic vehicle for the sustainable intensification of broadacre agriculture: M. E. thesis. School of Electrical Engineering and Computer Science and Engineering Faculty, Queensland University of Technology.

[4] Bureau of Indian Standards (1995) Crop protection equipments-foot sprayer. Indian standard code IS:3652. Available at http:// www.standerdsbis.in.

[5] Bureau of Indian Standards (1980) Code of practice for operation and preventive maintenance of crop protection equipment's. Indian standard code IS: 9632. Available at http://www.standerdsbis.in

[6] Chethan M. R., Krishna S. G. G., Chennakeshava R. and Manjunath (2014) Design analysis and develop a low-cost tractor operated boom sprayer. *Engineering Research and Technology*. Vol. **2**(1) pp. 177-184.

[7] Koc Caner (2017) Structural analysis of field sprayer booms. *Journal of Agricultural Sciences*. Vol. 23, pp.147-155.

[8] Murugeshan R. (1998) Investigation on development and performance of electric powered walking tractor. Ph. D. Thesis. Department of Farm Power and Machinery. TNAU, Coimbatore.

[9] Qin W. C., Xue X. Y., Zhou Q. Q. C., Xu Z. F. and Change F. L. (2016) Optimization test for spraying parameters of cotton defoliant sprayer. *Int J. Agric and Bio. Engg.* Vol.**9** (4), pp. 63-72.

[10] Rafiq A., Kalantari D. and Mashhadimeyghani H., (2014) Construction and development of an automatic sprayer for greenhouse. *Agric Eng Int., CIGR Journal.* Vol.*16* (2), pp.36-40.

[11] Singh S., (2004) Autonomous robotic vehicle for greenhouse spraying. M. E. thesis. University of Florida, USA.

[12] Shaha V., Mulla A., Shinde G., Kumbhar M. and Mokashi S., (2019) .Design and Fabrication of Multipurpose Agricultural Vehicle. *International Journal for Scientific Research & Development.* Vol. **7** (2). pp. 7-9.

Design and Development of Portable Manure Spreading Trolley for Farmland

Dewansh Bhajane[1], Mukul Sambrey[2], Priyank Shukla[3], Lokesh Chandravanshi[4], Sangharatna M. Ramteke[5], Dr. Dheeraj S. Deshmukh[6]

[1,2,3,4]UG student at G.H. Raisoni College of Engineering Nagpur
[5] Assistant Professor, G.H. Raisoni College of Engineering Nagpur
[6] Associate Professor, Mechanical engineering, G.H. Raisoni College of Engineering Nagpur, India

bhajane_dewansh.me@ghrce.raisoni.net, sambrey_mukul.me@ghrce.raisoni.net, shukla_priyank.me@ghrce.raisoni.net, chandravanshi_lokesh.me@ghrce.raisoni.net, sangharatna.ramteke@gmail.com, dheeraj.deshmukh@raisoni.net https://orcid.org/0000-0001-6074-3111 [Orcid ID]

Abstract: As per regular requirement for farming work farm-land preparation is done by spreading manure includes cow dung. In our country India, tractors or bullock carts are preferred to carry the manure from the compost place to the farm land and manure/cow dung is stack piled in the field.[1] Farming procedures with organic farming are broadly used and adoptable for many types of lands in several regions in India.[2] Due to high increasing cost of labours, use of manpower for the purpose of organic materials are uneconomical as time requirement for this is also more.
In this work, objectives are defined, design of mechanism, development of Portable Manure Spreading Trolley is done and illustrated in this paper. The developed machine is tested and shows satisfactory performance for the specified designed parameters.

Keywords: Portable Trolley, Manure Spreading, ANSYS software

1. Introduction:

The distribution of cow dung manure is made physically by labors with spade, which will involve labor drudgery.[3] Bullock carts are used to carry the manure from the compost pit to the farm land and the manure/cow dung is stack piled in the farm field. Use of manure in farmyard is mainly being functional through manual distribution which is resulting more labors and time per unit area with poor uniformity in distribution and the extensive variation of the application rate. Solid type of manure losses nearly about 21% of its nitrogen (N_2) to the atmosphere and proper distribution and integration in the soil cause declines the loss by five percent. Therefore, world will be hopeful for this. Effectively it will not only reduce the cost but also helps to protect the environmental effect and health of farmers [4]. The distribution of the manure is achieved manually by spades, which will involve human work. Thus, Bullock used cart for manure spreading arrangement is required for adequate supply rate and equal spreading of Farmyard manure for the similar outcomes from a crop [5]. Our main objective in this research is to make a particular design for farmers who holds small to medium sized farm-lands. Also, to help them for improving crop cultivation with minimal efforts and a reduced amount of cost with easy maneuverability of manure spreading machine along the cultivated land and making it compact enough to occupy fewer space. [4] In present times, scarcity of labors and the time required for dispersal of the manure are the major problems faced by farmers. In this view for solving the problem, we proposed to

fabricate a manure spreading mechanism which is better and economic option from available methods and mechanisms available readily marketplace for the big farm owners.[6]
Thus, we have to develop initial level prototype showing suitable position for hopper and test it. In order to simulates tractor's PTO, we use an Electric motor. This project is only going to demonstrate the concept of suitable position for hopper and to fulfill the project objectives.

2. Project Objectives:

i. To study existing techniques for Manure Spreading using Trolley and design new model.
ii. To build a initial level prototype demonstrating the suitable position for hoper.
iii. To minimize dust and ensuring complete evacuation of manure.
iv. To test the developed system of Manure Spreading

3. Parameters and Methodology Adopted:

The components needed to make the design is first surveyed and the respective part design has been made in the Solidworks and according to our design parameters we then assemble them in software.

i. Farmers who are suffering from the problems including the increase in the cost of spreading of the manure, labour cost, scarcity of labours were taken into consideration by us and accordingly our main objective of our research work takes account of this. [5]
ii. Conventional methods of distribution of the compost by hands in farmland is very time consuming and inefficient process. This will require additional human energy. [7] Presently, we're used to spread the manure in traditional ways which in itself time consuming and costlier.[8]
iii. There are lot of machines available in the market but still they are very costly and not been easily acceptable for domestic farming.[9]
iv. The available machines are only fruitful for the large farm fields only.
v. The problem arises when we have to spread the manure over a small to medium range of farm field in very cost-effective manner without expending a lot into it.

So, we had considered above points to develop such a machine.

4. Steps of Experimental work:

Following steps are followed:

i. To develop prototype after design analysis of stresses using ANSYS software for demonstrating the suitable mechanism.
ii. To minimize dust and ensuring complete evacuation of manure.
iii. To test the developed system of Manure Spreading.

IOP Conf. Series: Materials Science and Engineering 1259 (2022) 012022 doi:10.1088/1757-899X/1259/1/012022

5. Operational Working Flow Diagram:

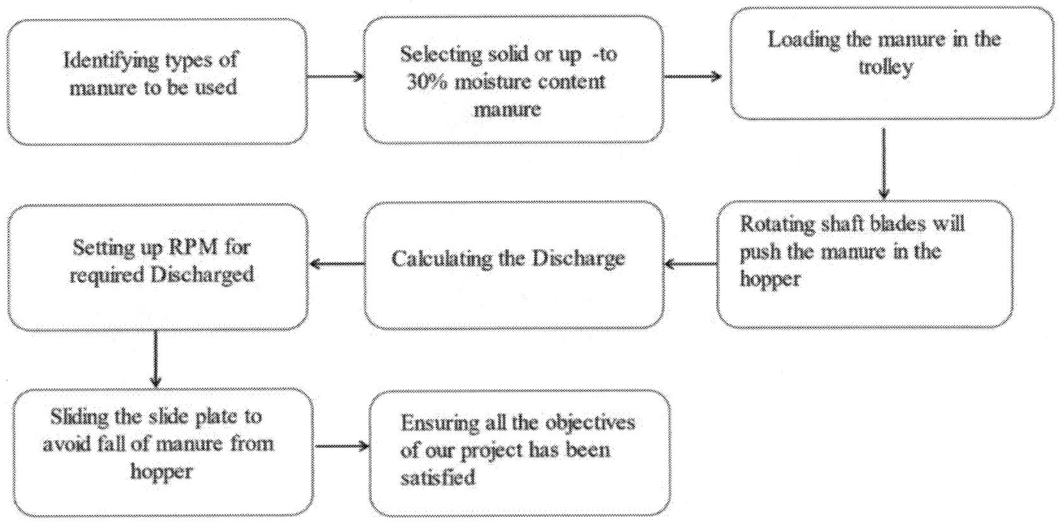

Figure 1: Operational Working Flow Diagram

6. Design and Fabrication:

During the design process, the entire model is made using Solidworks Software. For creating 2D and isometric view Solidworks Drawing software is used. The chassis design is of 1000 x 500 x 700 mm corresponding to length, width and height of project.

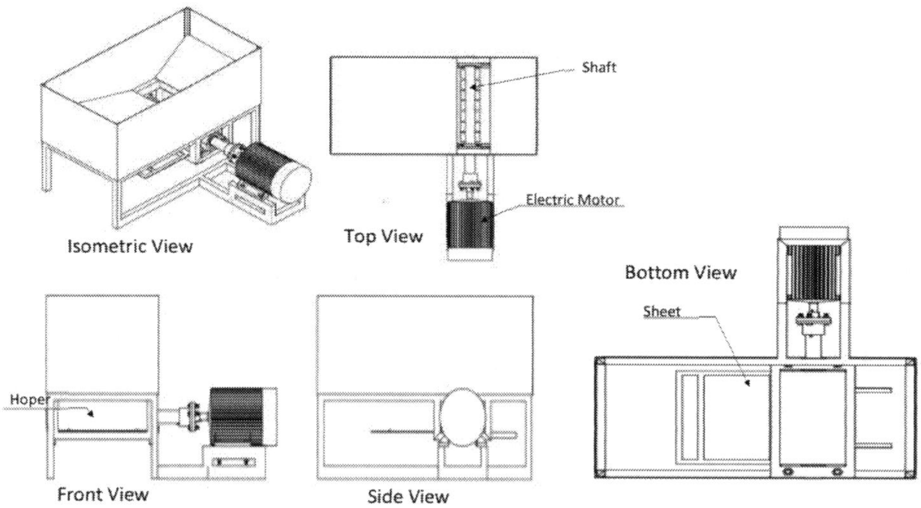

Figure 2: CAD model

6.1. CAD Design:

Figure 3:Frame

IOP Conf. Series: Materials Science and Engineering 1259 (2022) 012022 doi:10.1088/1757-899X/1259/1/012022

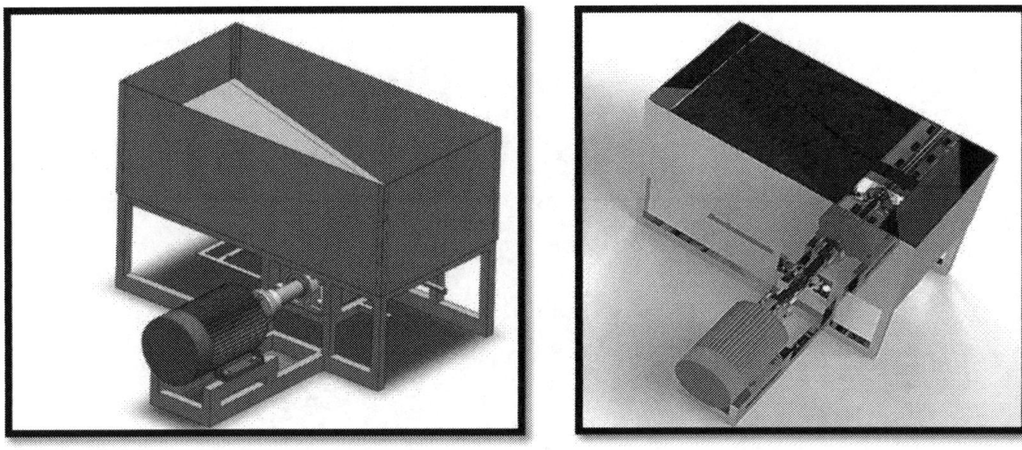

Figure 4: 3D view of Model

6.2. Ansys Analysis:

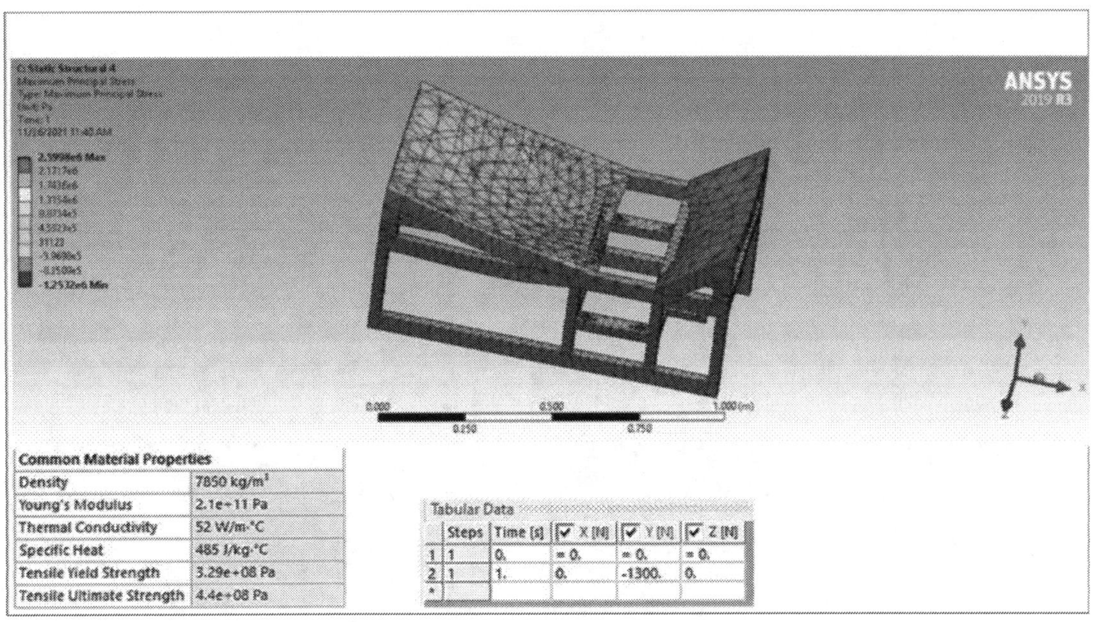

Figure 5: Maximum Principal Stress

IOP Conf. Series: Materials Science and Engineering 1259 (2022) 012022 doi:10.1088/1757-899X/1259/1/012022

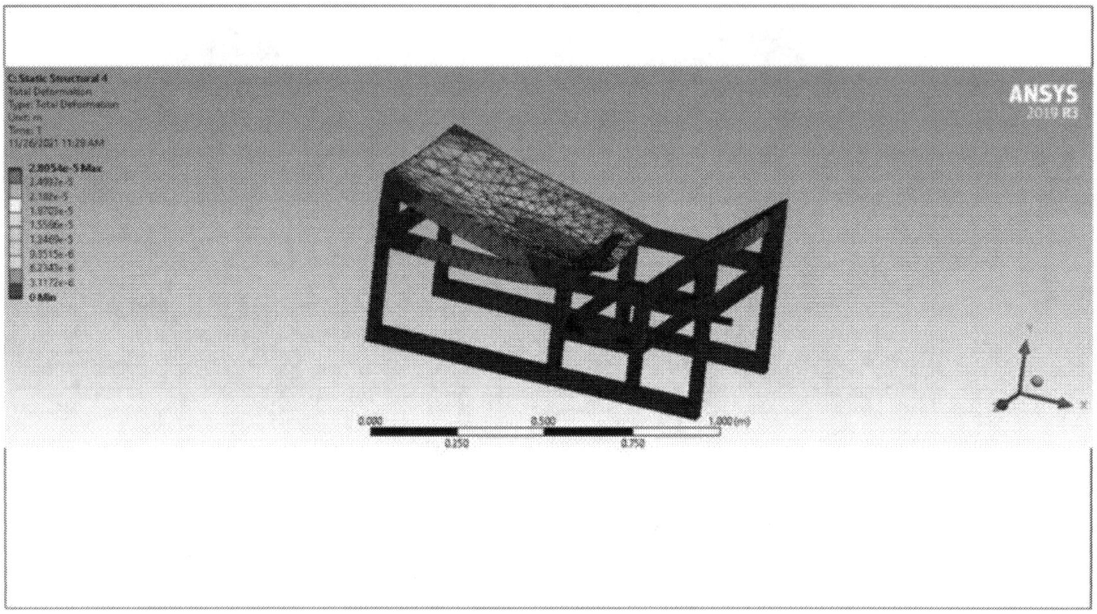

Figure 6: Total Deformation

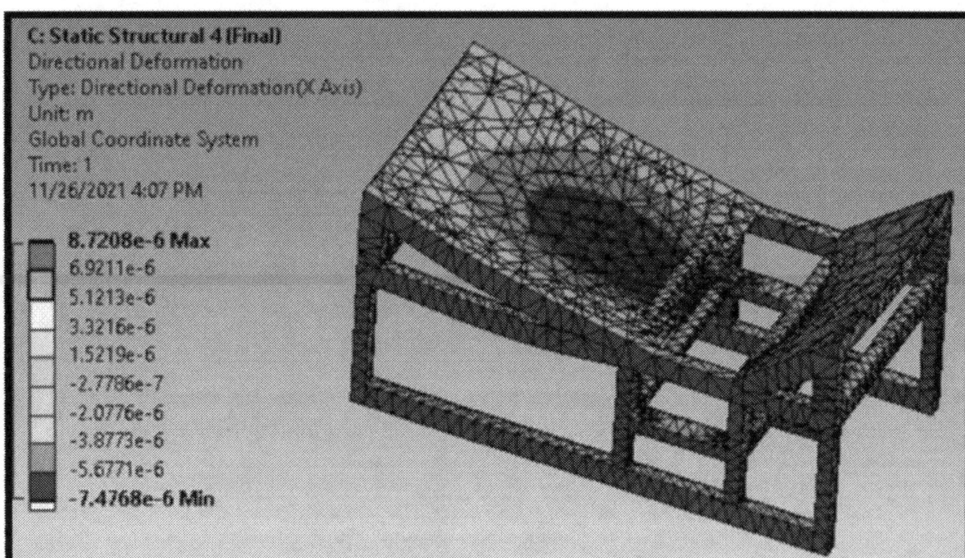

Figure 7: Directional Deformation

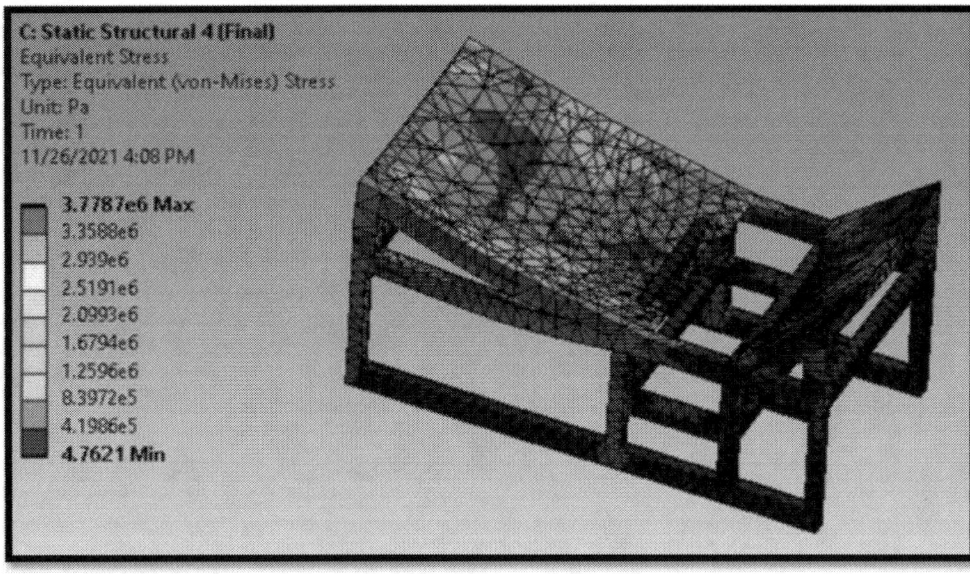

Figure 8: Equivalent Stress

Table 1: Ansys Results

Object Name	Maximum Principal Stress (MPa)	Total Deformation (meter)	Equivalent Stress (Pa)	Directional Deformation (meter)
State	Solved			
Results				
Minimum	-1.2532	0	4.7621	-7.4768e-6
Maximum	2.5998	2.8054e-5	3.7787e6	8.7208e-6
Maximum Occurs on	Near to Hopper			
Minimum on	Upper End of Trolley			

By doing Ansys analysis we've found out that the maximum stress occurs near to the hoper and the minimum stress occurs near to the upper end of the trolley.

6.3. Project Model:

Figure 9: Fabricated Project

7. Calculations:

7.1. Trolley:

Area of Trolley = 0.31649 m^3

Volume of Trolley = 0.31649 * 0.5 = 0.158 m^3

Mass Of Manure = 0.158 * 700 = 110 kg = 1079.1 N

Steel Sheet: material Cold Rolled SAE 304

Density = 8000 kg/m3

Area = Length + Width + Slope side Left + Slope side Right
= (0.31649*2) + (0.38*0.538*2) + (0.64*0.538) + (0.315*0.538)
= 1.555 m2

Volume = 1.555*0.003 = 0.004665 m^3

Weight = 0.004665 * 8000 = 37.32 kg

7.2. Shaft:

Material: Carbon Steel

T = F * R

Area Of One set of blade = (0.05*0.02) * 7
= $7 * 10^{-3}$ m^2

Force acting on set of Blades = Weight of Manure on Blades
= Volume of manure on Blades * Density Of manure
= [0.05*0.02*(0.38 + 0.6)] *7 * 700
=2.156 kg
=2.156 * 9.81
=21 N

Therefore, T = 0.525 N.m

Power, P = T*V = 0.525 * 0.523 = 0.28 hp

7.3. Rectangular Bar: material Mild Steel

Box = 1.038 + 1.038 + 0.538 + 0.538 = 3.152 m
Motor = (0.31*4) + (0.026*4) + 0.173 = 1.571 m
Stand = (0.325*4) + 1.038 = 2.338 m
Shaft = (0.47*2) + (0.15*4) + (0.213*2) = 1.966 m

Total Length = 8.973 m

A1 = 0.05*0.03 = 1.5 m^2
A2 = 0.0448*0.0248 = 1.111 m^2

Area = 1.5*10-3 -1.111*10-3
= 0.389* 10-3 m^2
= 0.000389 m^2

Volume = 0.000389*8.973 = 0.00349 m^2

7.4. L Angle: material Mild Steel

A1 = 0.02 * 0.003 = 6* 10-5 m^2
A2 = 0.017 * 0.003 = 5.1* 10-5 m^2

Area = 11.1 * 10-5 m2 = 0.000111 m^2

Length = 0.38*4 = 1.52 m

Volume = 0.00016872 m^3

Mass = 0.00016872 * 7850 = 1.32 kg

7.5. *Total Weight of Chassis*

Total Length = 8.973 m

A1 = 0.05*0.03
=1.5*10-3
A2 = 0.048*0.0248
=1.111*10-3
Area = 1.5*10-3 – 1.111*10-3
= 0.000389 m2

Volume=0.000389*8.973
=0.00349 m3

Density of Mild Steel= 7850 kg/m3

Weight of Rectangular Bar = 0.00394*7850
= 30.929 kg

Total Weight of Chassis=Weight of rectangular Bars + Weight Of L Angle
= 30.929 + 1.32
=32.249 kg

7.6. *Shaft:*

Outer Dia – 50 mm Length – 607 mm
Inner Dia – 36 mm

Cross section Area of Shaft = $\frac{\pi}{4} * (0.05)2 - \frac{\pi}{4} * (0.036)2$

= 0.000945 m^2

Volume of Shaft = Area * Length
= 9.456*10-4 *0.607
= 0.000573 m^3

Material of Shaft = MS
Density of MS = 7850 kg/ m^3

Weight Of Shaft = Volume * Density
= $5.7349*10^{-4}*7850$
= 4.5 kg

Price Of Shaft = 120 / kg
Cost Of Shaft = 4.5 * 120 = 540 Rs

7.7. Discharge Capacity:

Area of Quarter Portion $= \frac{\text{Area of 2}}{4} - \frac{\text{Area of 1}}{4}$

$= \frac{22*0.027^2}{7*4} - \frac{22*0.025^2}{7*4}$

$= 8168*10^{-5}\ m^2$

Total Length Of Blade = 0.05*7
= 0.35 m

Volume of quarter portion = 8.168*10-5*0.35
=2.8588*10-5 m^3
=0.0000285 m^3

Volume of manure in quarter portion = 0.0000285 m^3

RPM of Shaft= 200 rpm

Shaft rotation per second = $\frac{200}{60}$ =3.33 sec

Discharge of manure per Second = Volume of manure in quarter portion * No of rotation per second

= 0.00002858 * 3
=0.0000857 m^3/sec

7.8. SFD and BMD Calculations:

a) R_A and R_B

$R_A + R_B$ =1.079 * 1.038
$R_A + R_B$ = 1.12 KN 1

b) Moment about A

M_A =
R_B *1.038=1.079 * 1.038 * (1.038/2)
R_B = 0.5599 KN 2

From 1 and 2
R_A = 0.5599 KN

c) Shear Force

A_L=0
A_R = 0.5599
B_L = 0.5599 – 1.079*1.038 = -0.5599
B_R = -0.5599 + 0.5599 = 0

d) Max Bending

$M_A = M_B = 0$
M_C = 0.5599 * 0.519 – 1.079 * 0.5599 * (0.5599/2)
M_C = 0.1214 Nm

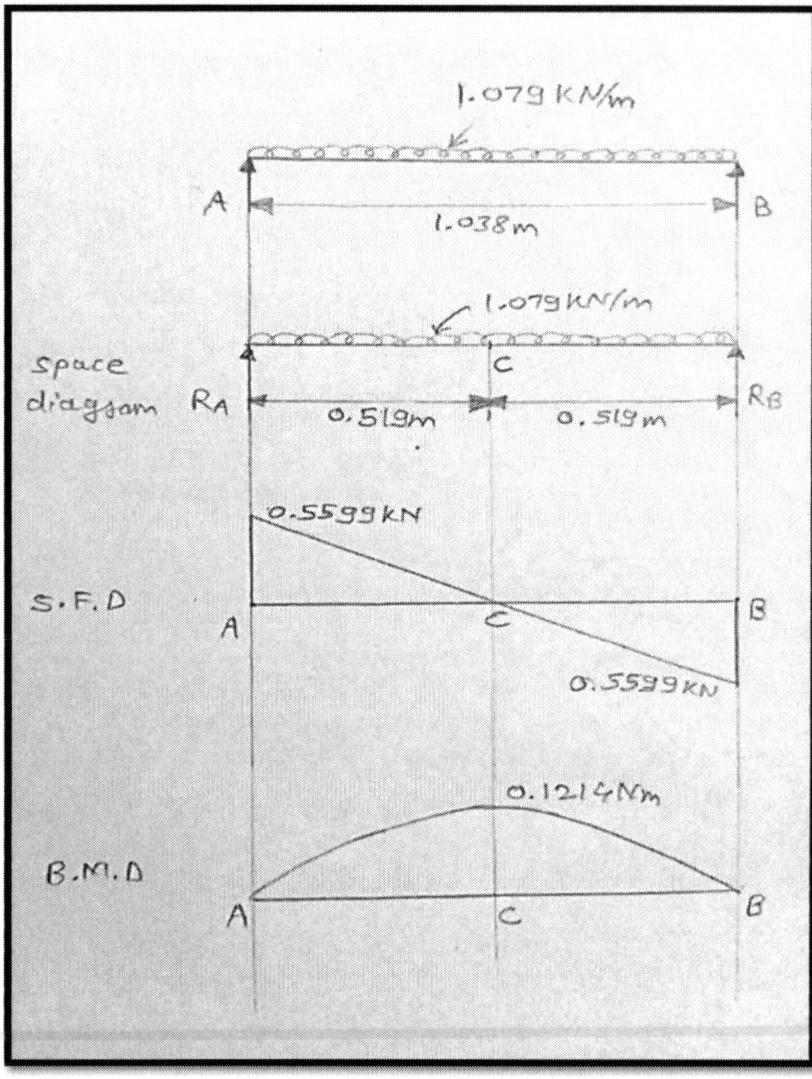

Figure 10: SFD and BMD diagrams

8. Result and Discussion:

The designed mechanism is analyzed by ANSYS software and is found safe as per the outcomes mentioned in above discussion and then developed mechanism is tested for manure spreading operation at different speed and discharge from the machine are recorded.

Table 2: Tested Results

RPM	Discharge (kg/s)
100	0.032
110	0.036
120	0.04
130	0.043
140	0.046
150	0.05
160	0.053
170	0.056
180	0.06
190	0.062
200	0.066
210	0.07
220	0.073
230	0.076
240	0.08
250	0.083

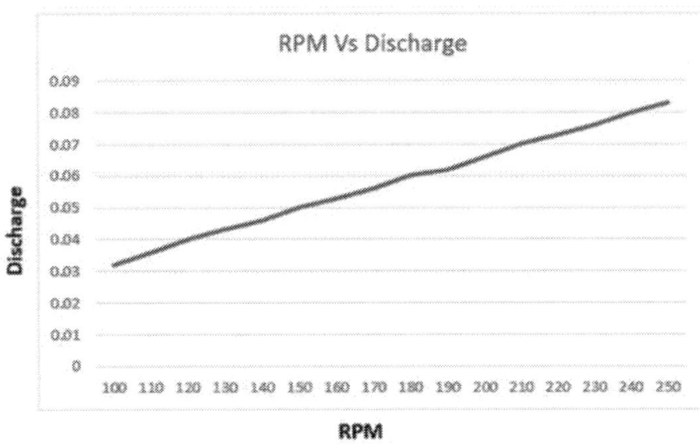

Figure 11: Graph Showing RPM vs Discharge Rate

Thus, it is clear from the testing of portable trolley developed for farmland is very useful for the targeted purpose. This concept of machine could be easily developed for any desired capacity as per the need economically.

9. Advantages:

A. Less time required to spread the manure.
B. Operating Cost is low.
C. Complete evacuation of Manure is possible from trolley.
D. No labor drudgery involved.
E. Tractor trolley can be upgraded with the hopper design.

10. Disadvantages:

The model developed is for testing of suitable position for hoper. So, the entire design is made accordingly.

11. Conclusion:

In this work the mechanism is designed and developed for Manure Spreading using Trolley with ease and quickly. Testing of mechanism gives suitable position for hoper that minimizes dust and ensuring complete evacuation of manure. Hence this work is very much useful for routine activity for manure spreading economically for small capacity farming purpose.

12. Acknowledgment:

Authors would like to thanks G H Raisoni College of Engineering, Nagpur, M. S., India for providing library facility.

13. References:

[1] S. N. Korra, E. Sudarshan, S. Venkatesulu, P. Pramod Kumar, and B. P. Yadav, "Portable manure dispenser machine," in *IOP Conference Series: Materials Science and Engineering*, 2020, vol. 981, no. 2. doi: 10.1088/1757-899X/981/2/022026.

[2] P. Sathiamurthi, A. S. Anaamalaai, M. R. Ahamed Buhari, A. Anburaja, and M. Ajmal Thahasin, "DESIGN AND FABRICATION OF MANURE SPREADER", [Online]. Available: www.ijstr.org

[3] R. C. Singh and C. D. Singh, "African Journal of Agricultural Research Design and development of an animal drawn farmyard manure spreader," vol. 9, no. 44, pp. 3245–3250, 2014, doi: 10.5897/AJAR2013.

[4] K. Sengottaian, N. Mohanrajhu, S. Palani, and C. Jayabalan, "Design of manure spreader machine for agricultural farm field," *International Journal of Innovative Technology and Exploring Engineering*, vol. 9, no. 1, pp. 4280–4284, Nov. 2019, doi: 10.35940/ijitee.A4877.119119.

[5] S. Chaudhari, M. Naeem, P. Jigar, and P. Preyash, "IJESRT INTERNATIONAL JOURNAL OF ENGINEERING SCIENCES & RESEARCH TECHNOLOGY DESIGN AND DEVELOPMENT OF FERTILIZER SPREADER MACHINE," © *International Journal of Engineering Sciences & Research Technology*, doi: 10.5281/zenodo.495879.

[6] D. P. M., "Design and Development of Manually Operated Fertilizer Spreader," *International Journal for Research in Applied Science and Engineering Technology*, vol. 6, no. 3, pp. 2402–2406, Mar. 2018, doi: 10.22214/ijraset.2018.3385.

[7] P. P. Patil, A. Subhashrao Ghodke, S. R. Kalme, M. Ashish, and R. Devshette, "UREA SPREADER MACHINE," *International Research Journal of Engineering and Technology*, p. 2345, 2008, [Online]. Available: www.irjet.net

[8] M. Jayamoorthy, A. S. Antony, P. Raj, P. Jayabharath, A. Professor, and U. G. Student, "Agriculture Manure Spreader for Natural Farming," 2018. [Online]. Available: www.ijsrd.com

[9] P. R. Sapkale, S. B. Mahalle, and T. B. Bastewad, "Performance evaluation of tractor operated manure spreader."

[10] W.H. STILITER, in july 1965 "Fertilizer spreader machine" Patent No:- US2522693

[11] J'. s CLARPER, in 1927 "MACHINE FOR SPREADING FERTILIZER".

[12] A. A. HURT, in Aug. 1960 "fertilizer mixing and spreading machine" Patent No:- US2947544

[13] Loren E. Tyler, Benson, Minn, assignor to Tyler Manufacturing Company, in Apr 1963, "fertilizer spreader" Patent No: - US3539112

[14] Carlo Bezzecchi, 3, Via dei Tigli, Campagnola,in in July 1973, "swinging tube type fertilizer spreader"

[15] Donald R. Gallogly, William E. Wheeler, in july 1971 "seed and fertilizer spreader" Patent No: - US3730395

[16] Clifford h. snyder , in march 1967 "Fertilizer spreader" Patent No:- US32236527

[17] W.H. STILITER, in july 1965 "Fertilizer spreader machine" Patent No:- US2522693

[18] P.M. Solanki, D.S. Deshmukh, V.R. Diware, M.S. Deshmukh, Experimental investigation of thermoelectric generator system, Materials Today: Proceedings, Volume 47, Part 11, 2021, Pages 3012-3017, ISSN 2214-7853, https://doi.org/10.1016/j.matpr.2021.05.478.

[19] Sheetal D Deshmukh et al 2021 J. Phys.: Conf. Ser. 1913 012093

[20] Mahesh.V. Kulkarni, D.S. Deshmukh, S.P Shekhawat, An innovative design approach of hot water storage tank for solar water heating system using artificial neural network, Materials Today: Proceedings, Volume 46, Part 11, 2021, Pages 5400-5405, ISSN 2214-7853, https://doi.org/10.1016/j.matpr.2020.09.058.

[21] Patil, P.P. and. Deshmukh, Dr.D.S and .Vaidya,, Dr.A.M and Paul, I.D., Experimental Investigation of New Innovative Flat Plate Collector for Solar Water Heater (May 17, 2019). Proceedings of the Second International Conference on Emerging Trends in Science & Technologies For Engineering Systems (ICETSE-2019), SSRN: https://ssrn.com/abstract=3509071 or http://dx.doi.org/10.2139/ssrn.3509071

[22] Shailesh Pimpale, Manish Deshmukh, Rajesh Shelke and Dheeraj Deshmukh, published a research paper titled " Experimental Investigation on Hip Implant Materials Development through Analytical and Finite Element Analysis: 3D Modelled Computed Tomography", Biointerface Research in Applied Chemistry, Platinum Open Access Journal (ISSN: 2069-5837, Volume 12, Issue 3, 2022, 4103 – 4125, online available 15 August 2021, https://biointerfaceresearch.com/, https://doi.org/10.33263/BRIAC123.41034125

[23] Dr. B. S. Bhaskar, Dr. M. S. Deshmukh, Dr. S.K. Choudhary and Dr. D. S. Deshmukh, published a research paper titled "Heat Transfer Through Porous Materials (Aluminum Foam) Empirical Optimization of a Heat Exchanger", International Journal of Mechanical Engineering, ISSN: 0974-5823, Vol. 7 No. 1, Page No. 47 – 55, January, 2022, Copyrights @Kalahari,

Design Analysis and Weight Optimization of Top Beam Up-packing 15 BPH Bale Press Machine

Ms. Bhagyashri Kandalkar[1], Mr. Vinod Bhaiswar[2], Mr. Vivek W. Khond[3]

[1]M..Tech. CAD CAM Student,G H Raisoni College of Engineering Nagpur

[2]Associate Professor, Mechanical Engineering Department, G H Raisoni College of Engineering Nagpur

[3]Assistant Professor, Mechanical Engineering Department, G H Raisoni College of Engineering Nagpur

Abstract. In the Ginning industry, the process of removing seeds from the cotton fibres and after that seedless cotton is compress to make a lint bale is carried out. The pressing of seedlesscotton is compress to make a lint bale. The pressing of seedless cotton is the last process at the ginnery. The cotton bailing press is taken under consideration, a product of Bajaj Steel Industry. It is important to modify the structure and check for suitability to improve the efficiency. Top beam modeling is done in solid edge. The static structural analysis is performed of top beam of bailing press in order to check the sustainability. ANSYS software was used to carried out the static structural analysis. The results obtained like maximum Von- mises stress and maximum deformation are used for improving design modification. Also adopting a technique of optimization it helps in provide a better outcome.

Keywords - Solid edge, ANSYS, Optimization, Cotton bailing machine.

1.Introduction

Ginning is the process of separating cotton fibres from seed in its most actual meaning. The fundamental role of a cotton ginnery is to turn a field crop into a marketable product. Current ginneriesmust clean seed cotton, divide rivers from seed, clean and position the fibres, and arrange them in a commercially acceptable container to turn seed cotton into final result. Cotton gins produce the fibres and place them in goods with a profit margin. As a result, ginning serves as a link between the production of cotton and the production of cotton. Cotton lint's bulk demands compression to make a bale in order to save money on shipping and storage. The final stage of cotton processing at theginnery is pressing (baling). Indian cotton bales are currently compressed to a high density of 500kg/m3 to 650kg/m3 (30 to 40 lb/ft3). Of course, this has a benefit in terms of volume per unit weight, given the size is lower. High-density bales, on the other hand, are more difficult to open, and a significant amount of energy is expended at the balepress and in the break to restore fluffinessacceptable for spinning.The International Cotton Advisory Committee has also underlined the necessity for a common cotton bale and processing standard. As a result, India should pursue the development of a universal density bale press, which will protect fibre quality, increase spinning capacity, and fulfill worldwide manufacturing sector demands. It is necessary to have an understandingof cotton bale standardization for enhancing efficiency in transportation, mechanical handling ofcotton bales.

1.1 Working of Cotton Uppacking Bale press machine

A framework, two hydraulic cylinders, a door less cylinder, a pusher assembly, chambers, and other components make up the cotton balepress machine. The basic function of the press machine is to compress seedless cotton into little bales. The press machine's frame structure is a critical element. It is made up of a bottom frame and a top frame that are joined by two tie bars. Cotton enters one box chamber thanks to the pusher assembly. With the use of a movement disc, the box chamber is then rotated 360 degrees. As a result, the cotton-filled box chamber is inserted into the frame construction, leaving an unfilled box chamber to be filled. There are two cylinders in the bottom frame of the press machine that are hydraulically controlled in an upward orientation. Second element is a doorless cylinder that travels downward through the top frame. The pressing procedure is then done. The pressed cotton, also known as a size bale, is removed from the machine. The bale size becomes more reliable and simple for long-distance transportation and handling.

Figure 1. 15BPH Uppacking Bale press machine

2. Literature Review

The researcher of this report discusses how globalization of cotton ginning is likely if press machines are inexpensive. The process of compressing cotton lint from a bale is a complex phenomena caused by nature's basic cotton elements, such as quality, strong point, maturity, and so on. On this press machine, a wide-ranging model of machine research and testing time and motion study was carried out for squeezing bale from the fresh lint cotton for a decentralized compostie ginnery. The framework was designed to resist the compressive load applied on the bottom most and sides of the structure's box during bale compression. The method concluded that using the main conceptual cotton press machine will significantly cut the price and space needed for a composite cotton ginnery [1].

In their research, the authors had found the lot of barriers to the implementation of advanced methods of cotton handling and processing practices, especially those related to packaging. Improvements tothe ginnery are only feasible if the intial costs of the press and cotton handling equipment areparticularly in comparison to get a complete overview about the design concepts. Using CAD softwarelike CATIA and ANSYS, design and analysis is carried out [2].

3. Problem Identification

- The Bajaj Steel Industry has manufactured cotton pressing machine for that there requirement is to improve previous design.
- The cost of machine is much higher than the other industry machine
- Plant needs to reduce machines cost.
- Changing the geometric structure and material optimization.

4. Objectives

1) Design and analysis of top beam of balepress machine for reduction of weight and cost.

2) To investigate the machine performance, stability risk analysis under the specific load condition.

5. Material Specification

Different metals are used for different applications. Materials are selected scientifically for required properties. Steels are iron and carbon alloys that are widely used in construction and other systems due to their higher quality and low cost. The mild steel used for the frame was chosen because it is soft and ductile, making it easy to weld and machine for construction purpose. Following are the mechanical properties of the material used for the analysis.
Density = 7850 kg/m^3
Yield Strength = 410 x 10^6 N/mm^2
Poisons Ratio = 0.3
Young's Modulus = 2.1 x 10^5 N/mm^2
Factor of safety = 1.25

6. Methodology

Table 1. Methodology

Detailed study of problem definition
Detailed study of existing system of present cotton bailing press machine at Bajaj Steel Industry PVT.LMT
Designing of conceptual CAD model of proposed design
Taking detailed dimension from company to develop work cell of existing system in cotton bailing press machine
Design of top beam
FEA optimization process
Analysis and testing of new improved design
Result

IOP Conf. Series: Materials Science and Engineering 1259 (2022) 012023 doi:10.1088/1757-899X/1259/1/012023

7. Procedure of design and analysis

The first stage is to research the design of the existing press, it has been discovered that the product based of the present press is insufficient. The cost prohibitively expensive for the average person.

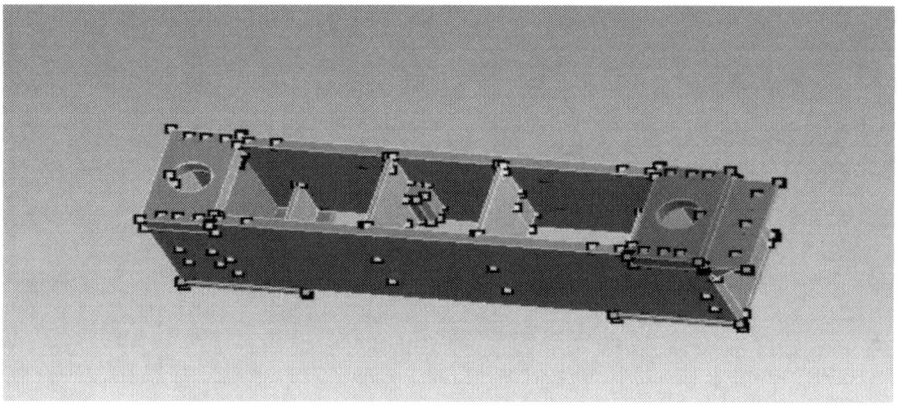

Figure 2. Frame of Top frame of machine

The outcome of the original press for maximum stress and total deformation is displayed in the diagram below.. The blue colour represents the safe region, While the red colour represents the critical region. The total deformation is 0.344 mm and the maximum stress is 539.24 Mpa.

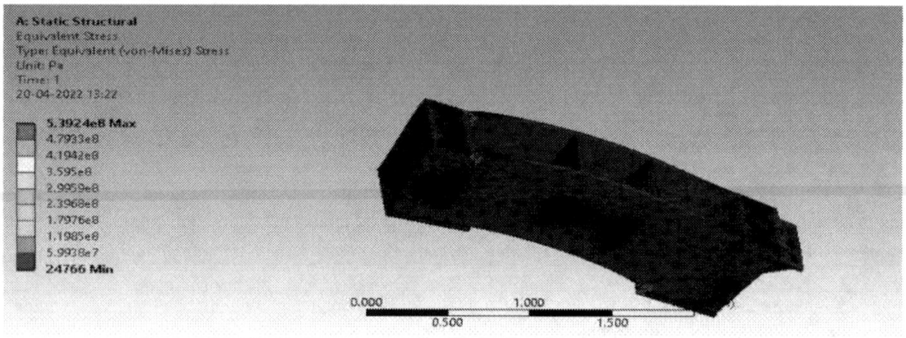

Figure 3. Maximum stress for Existing Top beam of machine

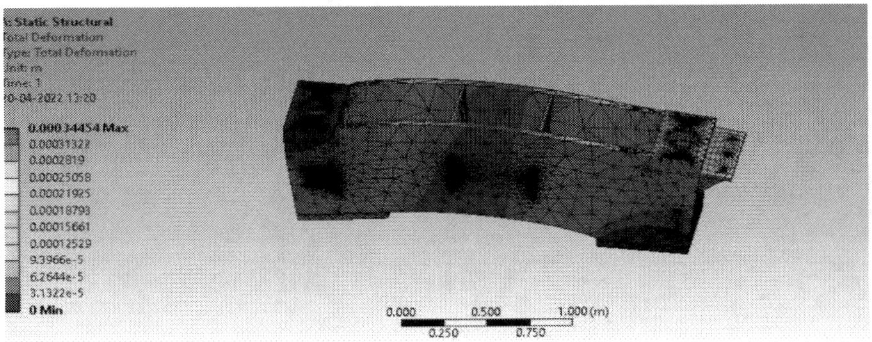

Figure 4. Total deformation of Existing Top beam of machine

The thickness of parts is reduced in the design of the balepress machine's top beam in good enough condition to optimise the weight of the top beam.

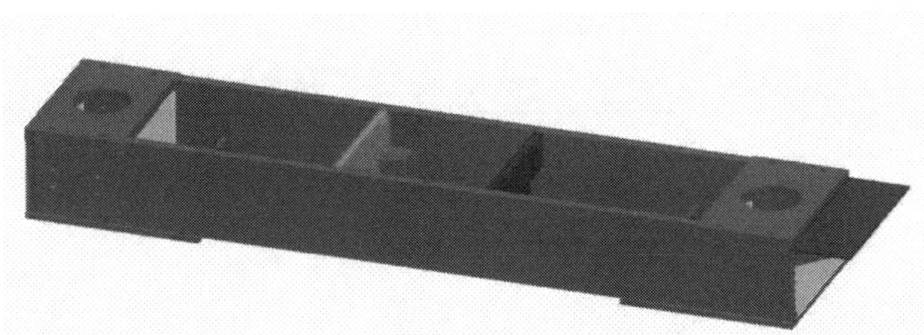

Figure 5. Frame of New top beam of machine

The analysis results for Total Deformation is 0.46mm and the Maximum stress is 776.9Mpa is obtained after doing the modification in existing top beam.

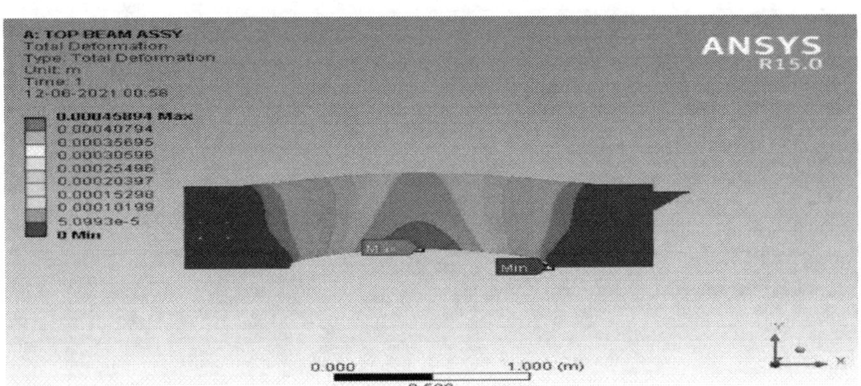

Figure 6. Total deformation for new top beam

IOP Conf. Series: Materials Science and Engineering 1259 (2022) 012023 doi:10.1088/1757-899X/1259/1/012023

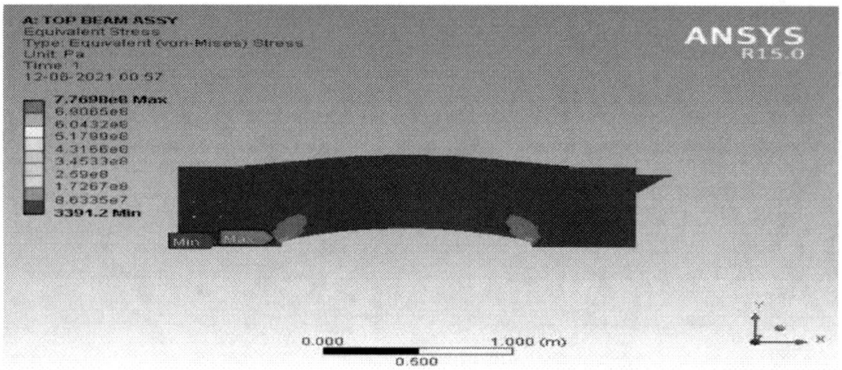

Figure 7. Maximum stress for new top beam

7.1 Result comparison

Table 2. Comparison of weight of top frame

Component	Previous weight	Optimized weight	Reduction
Top beam	2107.798 KG	1448.96KG	658.838KG

Table 3. Analysis result of existing and current top frame

Top beam	Total deformation	Maximum stress
Existing frame	0.344mm	539.24 MPa
Current frame	0.46mm	776.9MPa

8. Calculation of cost reduction for new top beam of bale press machine

Before optimization
Cost of material per kg = Rs.55
Cost of fabrication per kg = Rs.25
Cost of material 2107.798 kg = Rs.115928.8
Cost of fabrication for 2107.798 kg = Rs. 52694.9
Total Cost = Rs.168623.7

[II]
After optimization
Cost of material per kg = Rs.55
Cost of fabrication per kg = Rs.25
Cost of material1448.96kg = Rs.79692.8
Cost of fabrication for 1448.96 kg = Rs.36224
Total Cost = Rs.115916.8

[III]
Cost decreased of top beam
(Total cost before optimization) - (Total cost after optimization)/ Total cost before optimization
= (168623.7 - 115916.8)/ 168623.7
= 0.31257 x 100
= 31.25%
Reduction in cost up to 31.25%

9. Conclusion

After modification, the conclusion is that the stresses and deformation are well within the limit, implying that the change is safe to apply. The revision shows that the loading condition has no effecton the frame structure due to the decrease in thickness. Changing the thickness of the top beam provides the better answer, according to the FEA analysis. Proposed design process of analysis and optimization is successfully incorporate for minimization of weight for frame structure of hydraulic press. As a result, in attempt to correct and save the material and the machining time and cost required can be reduced. The cost reduction upto 31.25% after optimising the weight of the top beam of bale press machine.

10. References

[1] Agrawal JF, Askhedkar RD, Padole PM 2003 Design of Medium Duty Mechanical Press for Ginnery. *Journal of Industrial Engineering.*

[2] Dr. Devanand uttams 2013 Material handling in textile industry. *International Journal of Advanced Research in Engineering and Applied Science.* Vol. 2, No. 6 , June 2013, ISSN: 2278-6252.

[3] Sinha S. P. and Murarka P. D 1988 Computer-aided design of hydraulic press structures *Mathematical Computing and Modeling.* Vol 10, pp 637-645.

[4] Muni Prabaharan, V Amarnath 2011 Structural Optimization of 5Ton Hydraulic Press and Scrap Baling Press for Cost Reduction by Topology. *International Journal of Modeling and Optimization* 1.3 : 185.

[5] Le Roy Fitzwater, Richard Khalil and Ethan Hunter 2008 Topology Optimization Risk Reduction. Presented at the *American Helicopter Society 64th Annual Forum,* Montreal. Canada,April 29-May 1.

[6] Asim M. Kamate, Dr. J.S. BAGI 2006 Design Development and Analysis of A 20 Ton Hydraulic Press. *International Journal Of Innovative Technology And Research,* Volume No.4, Issue No.1, December - January, 2560 – 2563.

[7] A.G. Naik and N. K. Mandavgade 2012 Review On Analysis And Optimization Of Hydraulic Cotton Lint Bailing Press. *ARPN Journal Of Engineering And Applied Sciences,* Vol. 7, No. 5, May, ISSN 1819-6608.

[8] S. M. Bapat 2014 Design and Optimization of A 30 Ton Hydraulic Forming Press Machine *International Journal For Research In Applied Science And Engineering Technology .*

[9] H. N. Chauhan 2013 Design & Analysis of Frame of 63 Ton Power Press Machine by Using Finite Element Method. *Indian Journal of Applied Research* 3.7 (2013): 285-288. Vol. 3, Issue: 7, July, ISSN - 2249-555X.

[10] A. P. Bahale, Dr. S.S. Deshmukh 2014 Improving Material Handling Efficiency in a Ginning Machine Manufacturing Company". *International Journal of Innovative Research in Science, Engineering and Technology* (An ISO 3297: 2007 Certified Organization) Vol. 3, Issue 3, March.

[11] Badakundri, Umesh S., Santosh Kullur, and A. A. Kulkarni Finite Element Analysis of Hydraulic Press Machine. *International Journal on Recent Technologies in Mechanical and Electrical Engineering* (IJRMEE), ISSN: 2349-7947, Vol. 2, Issue: 5: 5

[12] Satish B. Mariyappagoudar, Vishal Patil, 2016 Design and analysis of hydraulic press using ANSYS. IJIRST- *International Journal for Innovative Research in Science and technology*, vol. 3, Issue 07, December, ISSN (Online): 2349-6010.

[13] Amiolemhen, P. E., and A. N. Ogie 2019 Design and manufacture of a 10-tonne hydraulic press. *J. Prod. Eng.* 22.1 (2019): 10-14.

[14] Igbinigie Osamwonyi, I. P. Okokpujie and J. O. Dirisu 2018 Development and performance analysis of horizontal waste paper bailing machine. *International Journal of Mechanical Engineering and Technology* (IJMET) Vol. 9, Issue 10, October, pp. 84-101, ISSN Print:0976-6340 and ISSN Online:0976-6359.

[15] Madhu P.S. 2014 Static Analysis, Design Modification and Modal Analysis Of Structural Chassis Frame. *J. Eng. Res. Appl. www. ijera. com* 4, no. 5 (2014): 6-10., ISSN: 2248-9622, Vol.4, Issue 5(Version3).

[16] Bapat S. M. , Desai Y.M. 2009 Design Optimization of A 30 Ton Hydraulic Press Machine. *International Journal for Research in Applied Science and Engineering Technology* (IJRASET) , Belgaum , India , p. 24-30.

[17] Malipatil S.S., Mattlikalli A.C 2014 Analysis and Structural Optimization of 5 Ton H-Frame Hydraulic Press'. *International Journal of Innovative Science Engineering and Technology*, Belgaum, India, Vol. 1, March .

[18] Agrawal, J. F, R. D. Askhedkar and P. M. Padole 2003 Optimum level of low cost automation for quality processing at ginnery. *National Conference on World Class Manufacturing, Amrita Institute of Technology and Science* Ettimadai, Coimbtorer, pp 234-238.

[19] Agrawal, J. F; R. D. Askhedkar, P. M. Padole and J. P. Modak 2003 Formulation of generalized experimental model for compressing lint cotton to form a bale with a view to generate design data for mechanical press. INCARF 2003, IIT Delhi, Paper No. 228, Technical Session IV B.

[20] Roman Schmaltz, Robert J. Wolf, and Enrich E. Salzmann, DzBaler 1993 Machine and Method of Baling. The Pennsylvania State University, University Park, USA, Vol. C201, April .

[21] Ravi, D 2014 Computer Aided Design and Analysis of Power Press. Middle-East Journal of Scientific Research 20.10 : 1239-1246.

[22] A. B. Dahake & P. G Patil Central Institute for Research on Cotton Technology, Nagpur440023, India M Padole Visvesvaraya National Institute of Technology, Nagpur, India.

[23] Piyush A. dalke, Dr. Vinod Bhaiswar 2017 Design modification and analysis of cotton lint bailing press – A review. *International journal of Innovative and Emerging research in engineering* , Vol. 4, Issue 1, e-ISSN:2394-3343.

[24] Vyankatesh Naidu, Dr. Vinod Bhaiswar 2020 Review on road marking paint machine and material. *IOP conference series materials science and engineering* , Vol :954 ISSUE: 1.

[25] Dr. Vinod bhaiswar, shilpa bhorkar 2019 Formulation of mathematical model for processing time and energy consumption by using field data based model technique. *International Journalof mechanical and production engineering research and development*, Vol 9, ISSUE: 4.

[26] Milind R. Shinde, Vinod N. Bhaiswar Jan-2016 (2008) Designing a suitable robotic arm for loading and unloading of material on lathe machine using workspace simulation software.

IOP Conf. Series: Materials Science and Engineering 1259 (2022) 012023 doi:10.1088/1757-899X/1259/1/012023

International research journal of engineering and technology, e-ISSN:2395-0056, volume:03 ISSUE:01.

[27] Shilpa P Bhorkar, Vinod N. Bhaiswar 2016 Vibration Analysis for Machinery Monitoring: A Past Review. *International journal of pollution and noise control*, vol. 1, ISSUE:2.

[28] P.M. Solanki, D.S. Deshmukh, V. R. Diware, M.S. Deshmukh 2021 Experimental investigation of thermoelectric generator system. *Materials Today: Proceedings, Vol. 47*, Part 11, Pages 3012-3017, ISSN 2214-7853, https://doi.org/10.1016/j.matpr.2021.05.478.

[29] Sheetal D Deshmukh, Shubham Gabhane and Dheeraj S. Deshmukh 2021 Osmotic dehydration of carrot strips and modelling. Published under licence by IOP publishing Ltd, Journal ofPhysics: Conference Series, Volume 1913, *International Conference on Research Frontiers in Sciences* (ICRFS 2021) 5th- 6thFebrurary, Nagpur, India, Journal of Physics: Conference Series, 1913 (2021) 012093, June 2021.

[30] Mahesh V. Kulkarni, D. S. Deshmukh, S.P Shekhawat 2021 An innovative design approach of hot water storage tank for solar water heating system using artificial neural network. *Materials Today: Proceedings,* Vol. 46, Part 11, Pages 5400-5405, ISSN 2214-7853, https://doi.org/10.1016/j.matpr.2020.09.058.

[31] Patil, P.P., Deshmukh, Dr. D.S and Vaidya, Dr.A.M and Paul, I.D., Experimental Investigation of New Innovative Flat Plate Collector for Solar Water Heater. (May 17, 2019), *Proceedings of the Second International Conference on Emerging Trends in Science & Technologies For Engineering Systems* (ICETSE-2019), SSRN: https://ssrn.com/abstract=3509071 or http://dx.doi.org/10.2139/ssrn.3509071.

[32] Shailesh Pimpale, Manish Deshmukh, Rajesh Shelke and Dheeraj Deshmukh 2021 Experimental Investigation on Hip Implant Materials Development through Analytical and Finite Element Analysis: 3D Modelled Computed Tomography. Biointerface Research in Applied Chemistry, *Platinum Open Access Journal* (ISSN: 2069-5837, Vol. 12, Issue 3, 2022, 4103 – 4125, online available 15August, https://biointerfaceresearch.com/, https://doi.org/10.33263/BRIAC123.41034125.

[33] Dr. B. S. Bhaskar, Dr. M. S. Deshmukh, Dr. S.K. Choudhary and Dr. D. S. Deshmukh 2022 Heat Transfer Through Porous Materials (Aluminum Foam) Empirical Optimization of a Heat Exchanger. *International Journal of Mechanical Engineering*, ISSN: 0974-5823, Vol. 7 No. 1, Page No. 47 – 55, January, 2022, Copyrights @Kalahari.

Acknowledgement
I would like to give special thanks to my esteemed guide Dr.V. N. Bhaiswar for his inspiring guidance and encouragement. I would like to thanks my co guide Mr.Vivek Khond for his support and guidance. And I would like to thanks to my industry guide Mr.Pravin Bhakte for his technical guidance.

Weight optimization and finite element analysis of automobile leaf spring: A design construction referred to electric vehicle

Shad Mohammad Shohel*[1], Shaptarshi Sen Gupta[1], Sabbir Hossain Riyad[1]

[1]Student, School of Mechanical Engineering, Lovely Professional University, Phagwara, Punjab, India
*Corresponding Author Email: shadshohel05@gmail.com

Abstract. In the current era, an automobile vehicle company's major focus is now on conserving natural resources, reducing consumption of fuel, and reducing the weight of vehicle components. To reduce greenhouse gas emissions and fuel consumption and mitigate global warming, an electric vehicle will be a viable option. However, the EV's main issue is the vehicle's battery longevity. More power is necessary to drive a vehicle. The best solution to the issue is to reduce the vehicle component weight. Electric Vehicle suspension systems are another sector in which these developments are performed on a regular basis. Substantial steps are being taken to improve the vehicle ride quality. In this study, static and dynamic analysis has been carried out by finite element analysis at Ansys workbench and taken structural steel material for mono leaf spring. In composite material, carbon fiber and glass fiber have been considered as reinforcement and resin epoxy has been taken as matrix for mono leaf spring. The composite material fabrication has been accomplished layer by layer at Ansys ACP. After analysis, this study has revealed that by using composite material a magnificent amount of weight reduction has been achieved without sacrificing load-bearing capability, strength, and rigidity.

Keywords. Composite material, Fiber orientation, Fabrication composite material, Ansys ACP, FEA, Modal analysis, static structural analysis, leaf spring, Electric vehicle.

1. Introduction

Environmental pollution is becoming a serious problem across the world. One of the major air pollutants is harmful emissions from combustion engines. Internal combustion engines produce toxic emissions, which are one of the major sources of pollution in the atmosphere. Internal combustion engines, such as vehicles with gas engines and diesel engines have become a severe social concern as a result of environmental pollution. Electric vehicles are being actively pushed across the world to minimize the effect of fossil fuel burning and solve environmental concerns. Electric vehicles are demonstrating to become a viable route to reducing air pollution, energy security, and economic opportunity. Electric vehicles might be a feasible alternative option for reducing carbon dioxide emissions and cutting greenhouse gas emissions [1–5] However, the electric vehicle's downside is the vehicle's battery's longevity. The greater the weight of a vehicle, the more power is necessary to drive it. To mitigate this problem in an electric vehicle, the weight of the vehicle component should be reduced. The key focus for enhancing the energy efficiency of electric vehicle components will be lightweight materials, improved manufacturing techniques, and especially the design of leaf springs with lightweight material.

A leaf spring is an elastic mechanical part that deflects when a load is applied and then returns to its original shape, depending on the size of the load. Leaf springs are manufactured from flat plates that are generally semi-elliptical in shape. It's designed to take absorb, store, and release energy. leaf spring contributes to the vehicle to the higher load-bearing capability and capacity to absorb road vibration. In addition to absorbing vibrations, it bears lateral loads, braking torque, and engine torque[6–9].To manufacture leaf springs,

Composite materials are increasingly widely used to replace the conventional materials components in the vehicle industry. Automobile manufacturers' key concerns in the current situation are power efficiency and weight reduction. Weight reduction of the leaf spring can be achieved by replacing conventional material with composite material[10–12].
A substantial amount of work has been conducted on the topic of leaf springs in terms of their materials, and substantial progress has been made in the areas of weight reduction and load-carrying capacity when using advanced material. Vijayarangan et al. [13] designed and tested a composite leaf spring with an integrated eye under static structure loading conditions. In addition, fatigue life prediction was performed to ensure a consistent number of leaf spring life cycles. but there was no mention about composite layer thickness and stacking sequence. Purusothaman Mani et al.[14] had focused on a comparative study of leaf springs with composite material to understand the static behavior, fatigue life, and natural frequency. However, there was no mention of stacking sequence, fiber orientation, and laminas thickness of composite material. Smaranika Nayak at al. [15] had reviewed on composite leaf spring. This study concluded that composite material will help to reduce the noise and vibration of leaf springs. Mahmood M. Shokrieh et al.[16] focused on the finite element analysis of leaf spring analytically and numerically. S.P. Jani et al.[17] concentrated on static structural analysis of leaf springs with carbon fiber, mild steel, and glass fiber. But there was no discussion about boundary conditions and composite layer sequence and layer thickness. V. Hariram et al. [18] carried out the analysis of leaf springs by hybrid composite and conventional material with different loading conditions. Where, they have found that hybrid composite material is inducing less stress, less deflection, and weight reduction compared to conventional steel.

2. Methodology

2.1. Design Specifications

Table 1 shows the design specifications[19] of the mono leaf spring utilized in this study.

Table 1. Design parameter

Specifications	value
Length,2L (eye to eye)	965mm
Free chamber	68mm
Width, b	50mm
Thickness, t	10mm
Number of leaves, n (master leaf)	01

2.2. Boundary Condition and Meshing

In the design modeler, mesh depicts geometry as element size and the number of an element. in this geometry, the element size is 5 mm, and the element no is 4560, which can consider a finer mesh. The front eye of the leaf spring is directly connected to the vehicle chassis/frame by a pin joint with fixed support. And the rear eye of the leaf spring is connected through a shackle, whereas the free displacement is x-direction, and it can rotate about z-direction as shown in Figure 1 & Figure 2.

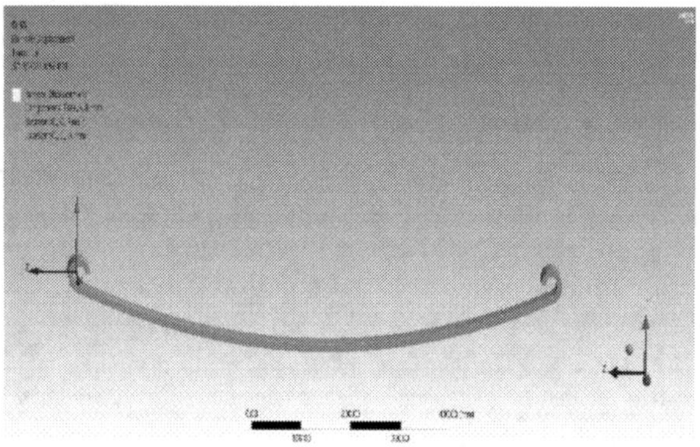

Figure 1. Remote displacement at leaf spring rear eye

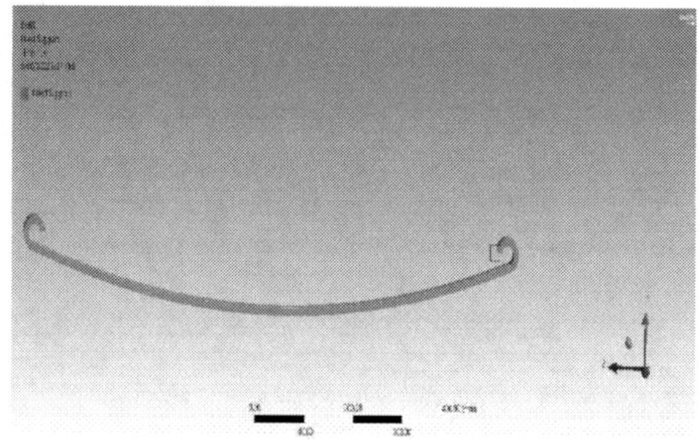

Figure 2. Fixed support at leaf spring front eye

2.3. Material Property

The mechanical property of carbon fiber and glass fiber has been depicted in Table 2 and the resin epoxy and structural steel property has been listed in Table 3 and Table 4 respectively.

Table 2. Mechanical property of carbon fiber and glass fiber[20]

Properties	Carbon Fiber	Glass Fiber
Young's Modulus X Direction, E_x (GPa)	230	90
Young's Modulus Y Direction, E_y (GPa)	23	8
Young's Modulus Z Direction, E_z (GPa)	23	8
Poisson's Ratio XY Direction, V_{xy}	0.2	0.3
Poisson's Ratio YZ Direction, V_{yz}	0.4	0.4
Poisson's Ratio ZX Direction, V_{zx}	0.2	0.3
Shear Modulus XY Direction, G_{xy} (GPa)	9	5
Shear Modulus YZ Direction, G_{yz} (GPa)	8	3
Shear Modulus ZX Direction, G_{zx} (GPa)	9	5
Density(kg/m^3)	1800	2000

Table 3. Mechanical property of resin epoxy[20]

Properties	Value
Density(kg/m^3)	1160
Poisson ratio	0.35
Young's Modulus (GPa)	3.78
Yield strength (MPa)	54
Tensile Ultimate strength (MPa)	-

Table 4. Mechanical property of Structural steel[20]

Properties	Value
Density(kg/m^3)	7850
Poisson ratio	0.3
Young's Modulus (GPa)	200
Yield strength (MPa)	250
Tensile Ultimate strength (MPa)	460

2.4. Composite Material Fabrication

In engineering material, fiber-reinforced composite is usually referred to as composite. Carbon fiber, S-glass fiber, and E-glass fiber are mostly used in the aerospace, automobile, and mechanical sector. Fiber are strong and rigid. it's very difficult to keep them together without polymer. Polymers are referred to as resin epoxy and resin polyester. And they are used as matrix.

In this study, the fabrication of composite material has been done layer by layer approach. Carbon fiber and glass fiber are used as reinforcement and resin epoxy is considered as matrix. The lamina thickness of carbon fiber and glass fiber is considered 1mm respectively and the resin epoxy thickness is 0.5mm. To make a laminated composite of carbon fiber/epoxy and glass fiber/epoxy, a unidirectional lamina is considered with zero degrees fiber orientation. In the stacking sequence process, the first layer is resin epoxy and second layer is carbon fiber, third layer is resin epoxy and fourth layer is carbon fiber. Likewise, until getting the desired thickness, this process continued to fabrication of carbon fiber/epoxy mono leaf spring. Similar way, this process had done for glass fiber/epoxy mono leaf spring. Figure 3. Shows the visual representation of laminated composite carbon/fiber epoxy.

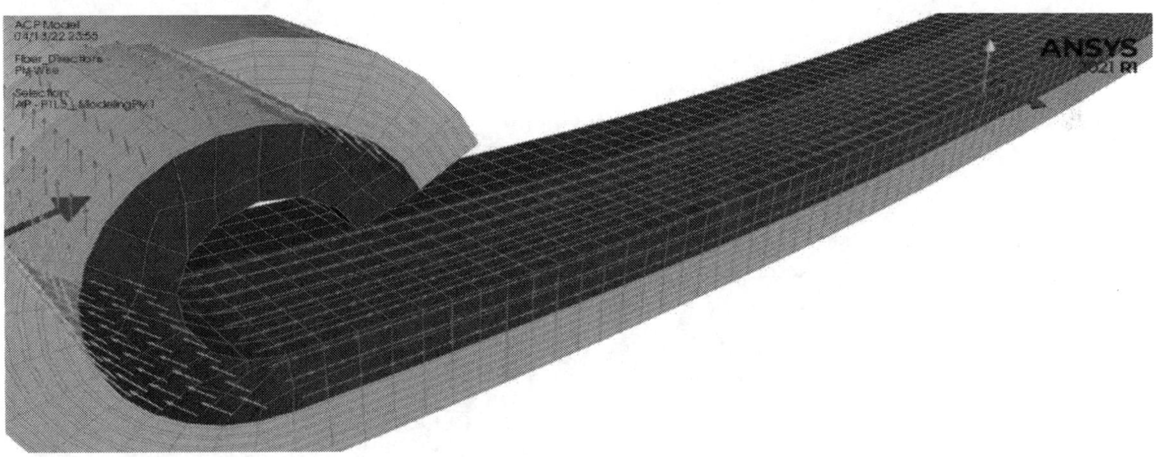

Figure 3. visual representation of fiber orientation of laminated composite carbon fiber/epoxy leaf spring.

3. Result and Discussion

3.1. Structural Steel Leaf Spring

This numerical study covered a load variation of total deformation, strain energy, and von mises stress. And each applied load has been included. Figure 4 shows the von mises stress, strain energy, and total deformation for structural steel material under a load of 2000 N, while Table 5 lists the values for different applied loads. Von mises stress is highest at the back eye portion of leaf spring and lowest in the front eye of leaf spring [Fig 4(a)]. The peak displacement takes place in the central portion of the leaf spring and gradually decreases toward the end of the leaf spring [Fig 4(b)]. The highest strain energy is also found in the rear end of the spring [Fig 4(c)].

Table 5. Parameters at various load conditions of structural steel leaf spring

Load (N)	Von mises stress (MPa)	Strain energy(mJ)	Total deformation (mm)
2000	102.55	1.247	1.011
2500	128.9	1.948	1.2641
3000	153.83	2.806	1.5169

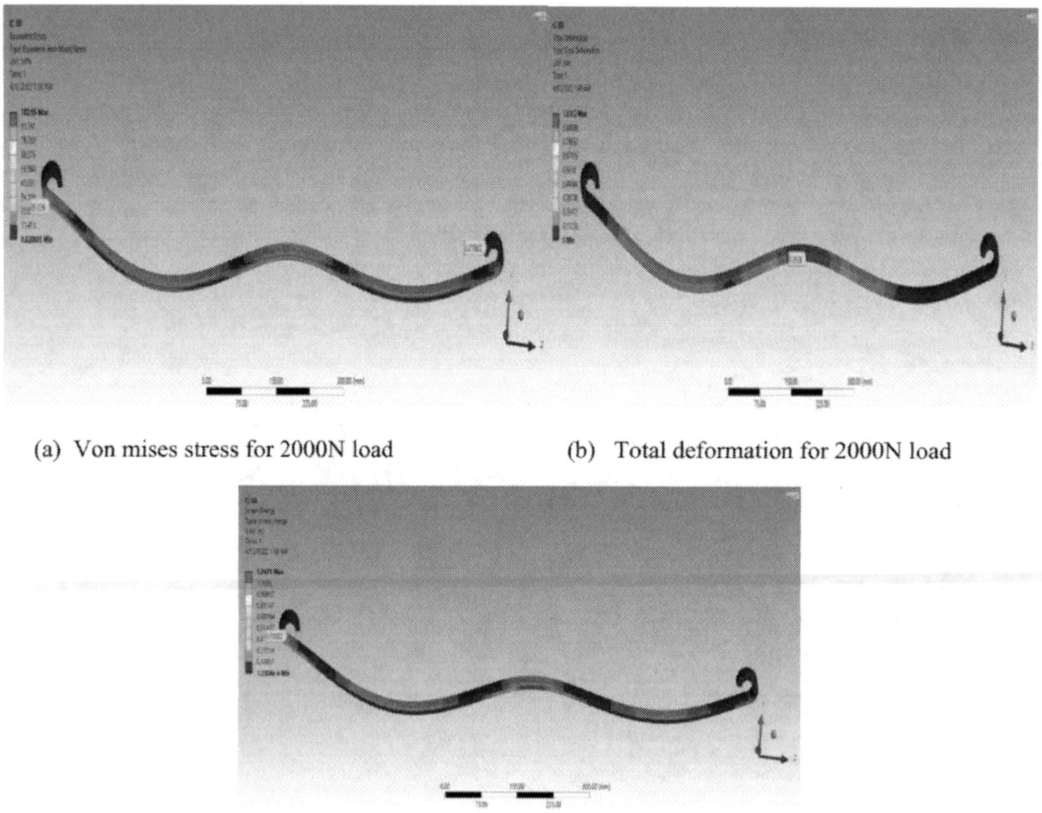

(a) Von mises stress for 2000N load

(b) Total deformation for 2000N load

(c) Strain energy for 2000N load

Figure 4. Static behavior of structural steel leaf spring

IOP Conf. Series: Materials Science and Engineering 1259 (2022) 012024 doi:10.1088/1757-899X/1259/1/012024

3.2. Carbon Fiber/Epoxy Leaf Spring

Figure 5 depicts the von misses stress, strain energy, and total deformation for the carbon fiber/epoxy material under a load of 2500 N, whereas Table 6 summarizes the results for other loading parameters.it is obvious that in Figure 5(a) the maximum von mises stress is induced at the edge portion of the composite layer. This issue should be considered During the design and fabrication of the composite layer. When the load is applied, the highest deformation takes place at the middle portion of the spring the highest strain energy is seen at the rear end of the leaf spring [Fig 5(b) & Fig 5(c)].

Table 6. Parameters at various load conditions of carbon fiber/epoxy leaf spring

Load (N)	Von mises stress (MPa)	Strain energy(mJ)	Total deformation (mm)
2000	71.648	1.4511	1.5171
2500	89.56	2.2674	1.8964
3000	107.47	3.265	2.2757

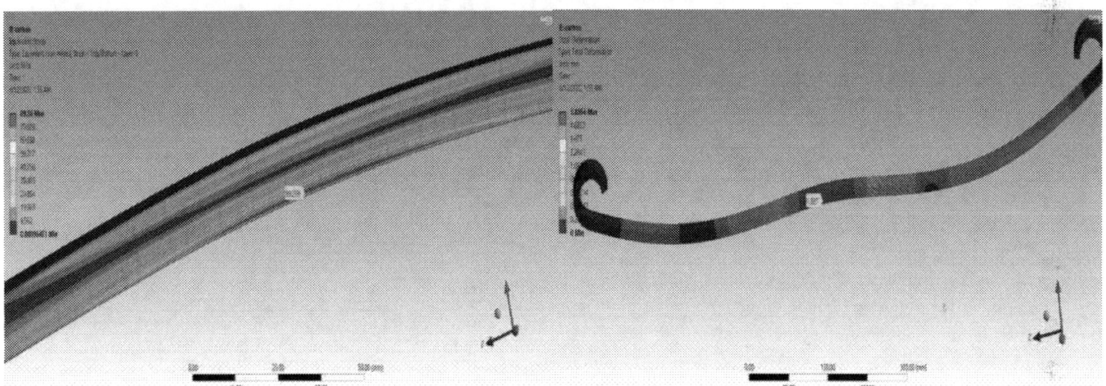

(a) Von mises stress for 2500N load (b) Total deformation for 2500N load

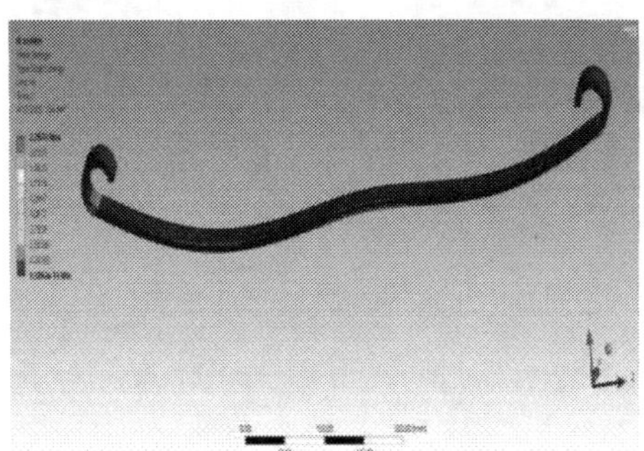

(c) Strain energy for 2500N load

Figure 5. Static behavior of carbon fiber/epoxy leaf spring

3.3. Glass Fiber/Epoxy Leaf Spring

The von mises stress, strain energy, and total deformation for glass fiber/epoxy material under a load of 3000 N has been illustrated in Figure 6, while Table 7 listed the values for different applied loads. Similarly, for glass fiber/epoxy, the maximum von mises stress is founded at the edge region of the composite layer, which has depicted in Figure 6(a). In the middle region of the leaf spring, there is a maximum deformation happened and the highest strain energy is also found at the rear end of the spring [Fig 6(b) & Fig 6(c)].

Table 7. Parameters at various load conditions of glass fiber/epoxy leaf spring

Load (N)	Von mises stress (MPa)	Strain energy(mJ)	Total deformation (mm)
2000	70.051	2.4353	3.2037
2500	87.563	3.8051	4.0046
3000	105.08	5.4793	4.8055

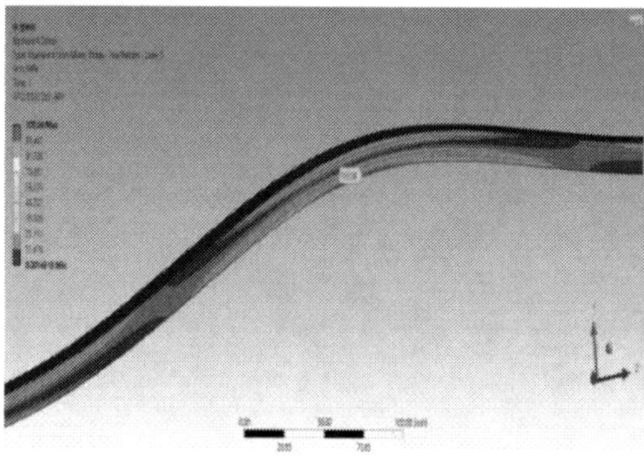

(a) Von mises stress for 3000N load

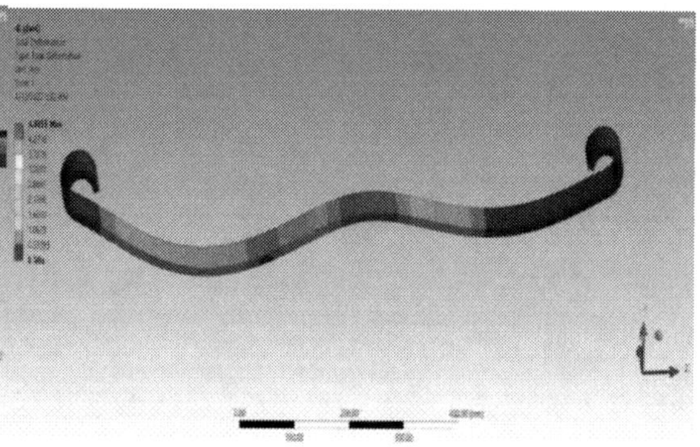

(b) Total deformation for 3000N load

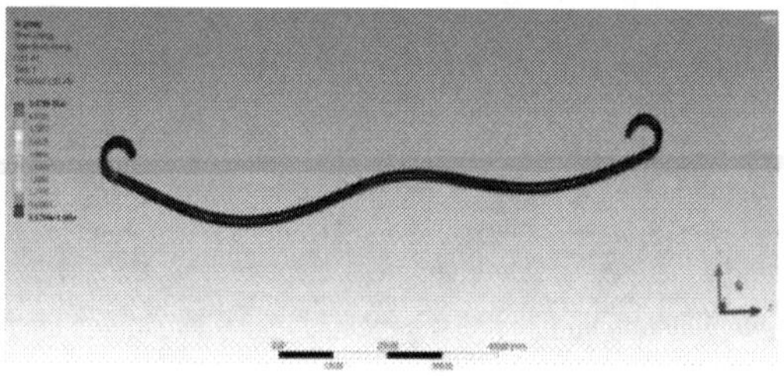

(c) Strain energy for 3000N load

Figure 6. Static behavior of glass fiber/epoxy leaf spring

3.4. Modal Analysis

Using modal analysis on the leaf spring model, the natural frequencies of a leaf spring for carbon fiber/epoxy, glass fiber/epoxy, and structural steel were determined. The natural frequency is the most important design parameter in mechanical design for a model under dynamic loads. Each material property's natural frequency must be greater than the frequency of road irregularities. TABLE 8 shows the results of a natural frequency for each material.

Table 8. Natural frequency of leaf spring

Material	Natural Frequency (Hz)					
	Mode 1	Mode 2	Mode 3	Mode 4	Mode 5	Mode 6
Carbon fiber/epoxy	239.3	281	409	686.97	776.23	787.45
Glass fiber/epoxy	151.59	242.61	257.09	460.09	460.76	670.49
Structural steel	144.24	238.37	245.46	432.53	470.42	658.14

Figures 7-10 show a graphical illustration of carbon fiber/epoxy, glass fiber/epoxy, and structural steel leaf spring. In Figure 7, the strain energy for carbon fiber/epoxy and glass fiber epoxy leaf spring is higher than the structural steel leaf spring. Figures 8 indicate that the von mises stress is induced less in carbon fiber/epoxy and glass fiber/epoxy leaf springs compared to structural steel leaf springs. Total deformation is seeming to be greater for glass fiber/epoxy leaf springs compared to carbon fiber/epoxy and structural steel leaf springs as shown in Figure 9. Carbon fiber/epoxy and glass fiber/epoxy leaf springs have a high natural frequency compared to structural steel leaf springs as shown in Figure 10.

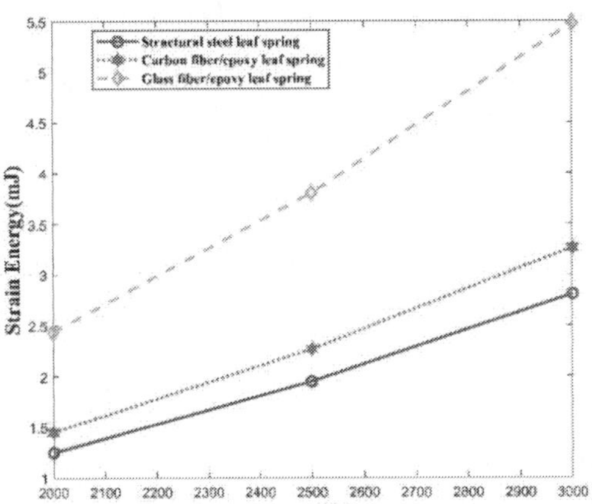

Figure 7. strain energy with different loads

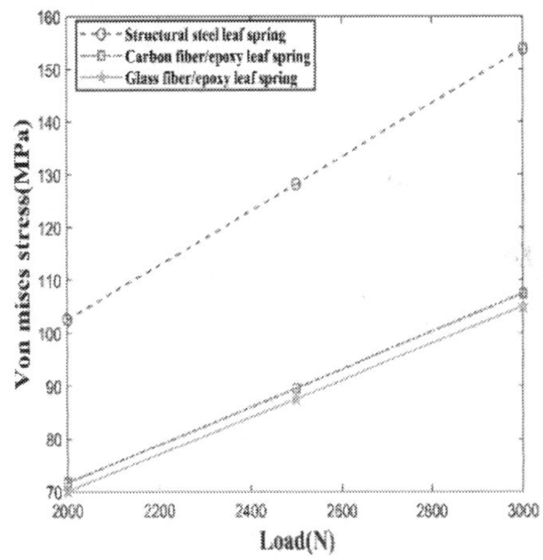

Figure 8. Von mises stress with different loads

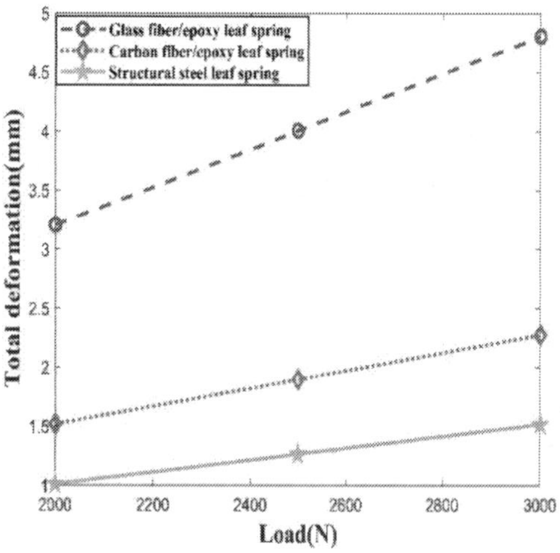

Figure 9. Total deformation with different load

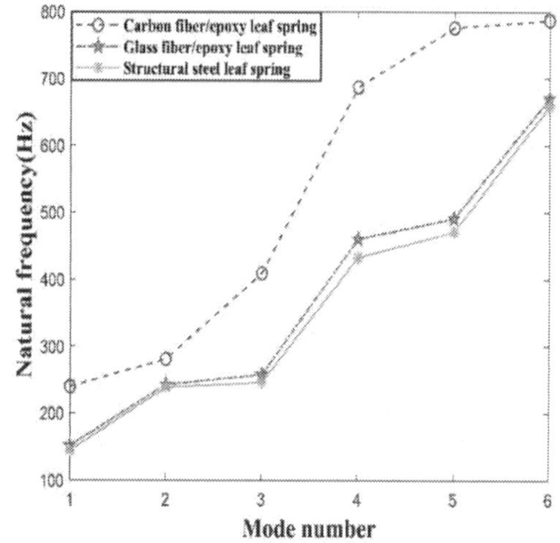

Figure 10. Natural frequency of different material

3.5. Weight Reduction

Because of pollution, the effects of combustion engine technology, like automobiles vehicle with diesel engines, have become a severe environmental concern. To alleviate this problem electric vehicles will be the most environmentally-friendly transportation. Electric vehicles are driven by electric motors, with electricity supplied by a rechargeable battery and any other portable energy storage system. These EVs are much more environmentally friendly and energy-efficient, emitting less greenhouse gas emissions and producing less noise. But the electric vehicle's weakness is the vehicle's battery's durability. More power is necessary to operate the vehicle. The best solution to the problem is to reduce the weight of the vehicle components. A magnificent amount of weight reduction has been revealed in this research by using composite material compared to structural steel. Figure 11 Demonstrated that 73.50% and 65.83% weight reduction has been achieved by using carbon fiber/epoxy and glass fiber/epoxy respectively compared to structural steel.

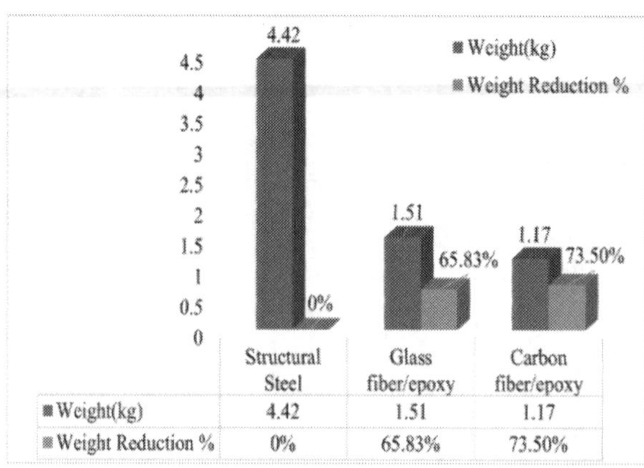

Figure 11. weight comparison of different materials

4. Conclusion

Under static and dynamic load condition, the von mises stress, total deformation, strain energy, and natural frequency has been investigated. composite material fabrication layer by layer is accomplished with the help of Ansys ACP. The following conclusion was reached from the study of the composite material and conventional material utilized in the design of the leaf spring.

- Through the static analysis, it's obvious that composite material leaf springs contain a large amount of strain energy compared to structural steel leaf springs while the load is applied, and release is slow. As a result, the suspension system is enhanced and the ride quality and handling are enhanced for composite material leaf springs compared to structural steel material leaf springs. It is because composite materials have low mass density and superior mechanical properties.
- During static structure analysis, when a load is applied, von mises stress has induced. The maximum von mises stress has been induced at the surface of the eye section for structural steel leaf spring. But, for composite material leaf spring, the maximum von mises stress has induced at the edge portion of the composite layer. During fabrication of composite layer for the manufacture of leaf spring, this factor must be considered. Finally, due to the superior mechanical property of composite material, the stress has induced lesser compared to structural steel leaf spring. For that, composite material leaf spring can handle a heavier load
- After static structural analysis, deformation seems to be greater in glass fiber /epoxy spring compared to structural steel spring. It's because glass fiber has a lower modulus of elasticity. But due to lesser stress-induced in composite leaf spring, compared to structural steel leaf spring, composite leaf spring can withstand a heavier load.
- Modal study reveals that glass fiber/epoxy and carbon fiber/epoxy leaf springs have a higher natural frequency compared to structural steel leaf springs. For that, a composite material leaf spring can withstand more amount of resonance compared to a structural steel leaf spring.

Moreover, a magnificent amount of weight reduction has been achieved, where 73.50% and 65.83% weight reduction is obtained for carbon fiber/epoxy and glass fiber/epoxy respectively compare to structural steel. Also, composite material leads to significantly lighter weight without compromising rigidity, strength, and load-bearing capability.
Finally, this study leads to conclude that, to increase the electric power efficiency and battery longevity, composites material will play an important role in electric vehicles automobile industries in the future.

References

[1] Goel S, Sharma R and Rathore A K 2021 A review on barrier and challenges of electric vehicle in India and vehicle to grid optimisation *Transportation Engineering* **4**
[2] Muratori M, Alexander M, Arent D, Bazilian M, Cazzola P, Dede E M, Farrell J, Gearhart C, Greene D, Jenn A, Keyser M, Lipman T, Narumanchi S, Pesaran A, Sioshansi R, Suomalainen E, Tal G, Walkowicz K and Ward J 2021 The rise of electric vehicles—2020 status and future expectations *Progress in Energy* **3** 022002
[3] Khurana A, Kumar V V R and Sidhpuria M 2020 A Study on the Adoption of Electric Vehicles in India: The Mediating Role of Attitude *Vision* **24** 23–34
[4] Hsu H-T, Coker C and Huang H *Optimization of an Electric Vehicle Suspension System Using CAE*
[5] WADA M 2009 Research and development of electric vehicles for clean transportation *Journal of Environmental Sciences* **21** 745–9

[6] Noronha B, Yesudasan S and Chacko S 2020 Static and Dynamic Analysis of Automotive Leaf Spring: A Comparative Study of Various Materials Using ANSYS *Journal of Failure Analysis and Prevention* **20** 804–18
[7] Rizvana S, Sri M, Murthy R, Satyanarayana D and Ganapathi A S 2017 MODELING AND ANALYSIS OF A TRUCK MULTI-LEAF TANDEM LEAF SPRING USING FEA *International Journal of Scientific and Research Publications* **7** 373
[8] Krishnamurthy K, Ravichandran P, Shahid Naufal A, Pradeep R and Sai HarishAdithiya K M 2020 Modeling and structural analysis of leaf spring using composite materials *Materials Today: Proceedings* vol 33 (Elsevier Ltd) pp 4228–32
[9] Varma N, Ahuja R, Vijayakumar T and Kannan C 2021 Design and analysis of composite mono leaf spring for passenger cars *Materials Today: Proceedings* vol 46 (Elsevier Ltd) pp 7090–8
[10] Becker F, Hopmann C, Italiano F and Girelli A 2019 Fatigue testing of GFRP materials for the application in automotive leaf springs *Procedia Structural Integrity* vol 19 (Elsevier B.V.) pp 645–54
[11] Gaylo R, Farahani S, Schmueser D and Pilla S 2020 Optimization of a mono-composite leaf spring using a hybrid fiber-layup approach *International Journal on Interactive Design and Manufacturing* **14** 407–21
[12] Sureshkumar M, Tamilselvam P, Kumaravelan R and Dharmalingam R 2014 *DESIGN, FABRICATION, AND ANALYSIS OF A HYBRID FIBER COMPOSITE MONOLEAF SPRING USING CARBON AND E-GLASS FIBERS FOR AUTOMOTIVE SUSPENSION APPLICATIONS* vol 50
[13] Siddaramanna G, Shankar S and Vijayarangan S 2006 *Mono Composite Leaf Spring for Light Weight Vehicle-Design, End Joint Analysis and Testing* vol 12
[14] Mani P, Kumar R, Kumar R, Gnanamani S, Janarthanam H and Ponnappan V S 2020 Comparative analysis of properties on leaf springs with different composite materials *AIP Conference Proceedings* vol 2311 (American Institute of Physics Inc.)
[15] Nayak S, Panigrahi I and Nayak R K 2020 Design and Analysis of Hybrid Composite Leaf Spring for Automotive Applications: An Extensive Study pp 127–32
[16] Shokrieh M M and Rezaei D 2003 Analysis and optimization of a composite leaf spring *Composite Structures* **60** 317–25
[17] Mallesh B, Gupta B, Kumar S K and Jani S P 2021 Modeling and analysis of leaf spring with different type of materials *Materials Today: Proceedings* vol 45 (Elsevier Ltd) pp 1945–9
[18] Seralathan S, Prasanna H, Arun J, Guhanesh B, Prasanth D, Hariram V and Micha Premkumar T 2020 Finite element analysis of hybrid composite material based leaf spring at various load conditions *Materials Today: Proceedings* vol 33 (Elsevier Ltd) pp 3540–8
[19] Soliman E S M M 2019 Static and Vibration Analysis of CFRP Composite Mono Leaf Spring *Journal of Failure Analysis and Prevention* **19** 5–14
[20] Anon Ansys Granta Material Properties Database for Simulation

Exploration of the factors affecting the performance of the ceiling fan: A brief review

Rubal Wandre[1], Faizan A. Gaddi[1], Vaibhav Dongre[1], Sahaj Hasija[1], Inayat Ullah*[1]

[1]Department of Mechanical Engineering G H Raisoni College of Engineering Nagpur

Abstract Ceiling fans are one of the most common appliances that are used for commercial as well as domestic purposes. This paper provides a thorough review of the literature on ceiling fan design, aerodynamics, construction, and other features. The work intends to foster a fundamental understanding of the factors associated with ceiling fans such as the shape of the ceiling fan, investigation of the flow field of a ceiling fan, aerodynamics and structural analysis of ceiling fans for different blade materials, and noise reduction techniques. This paper will assist the researchers and manufacturers by providing the crucial information related to ceiling fan in a comprehensive manner so that future research could advance in appropriate direction.

Keywords: Ceiling fan, Investigation of flow field, Aerodynamics, Blade material, Noise reduction.

1. Introduction

Ceiling fans are common devices to ensure proper ventilation in a room, particularly in humid areas. Ceiling fans are thought to be the most effective ventilation methods for distributing air in a room, even if some homes use air conditioning as their cooling system. By evaporating sweat from the human body, fans provide a cooling effect. Ceiling fans account for approximately 6% of domestic energy consumption in India [1]. When compared to air conditioning, ceiling fans consume significantly less energy and are also less expensive. A ceiling fan is an electric device that uses a hub to rotate blades and circulate air. Ceiling fans have several components, including an electric motor, blades, and blade iron that connect the blades to the motor, as well as a flywheel. Types of ceiling fansinclude energy star ceiling fan, standard ceiling fans, low profile ceiling fan, dual motor ceiling fan, wet and damp ceiling fans, remote controlled ceiling fan, commercial ceiling fans and industrial ceiling fans. These fans are further classified into three groups based on the purpose and function. Each type have some distinctive dominance from the others and therefore is appropriate for a specific approach. It includes industrial, house and large-diameter fans. Ceiling fan do not lower the room's temperature; instead, they just distribute air around our body and room, making us feel cooler. The ceiling fan generates a large amount of air flow, which disperses dust throughout the room. The fan blades are hit by a large volume of circulating air that collects on them. The accumulation of dust on ceiling fan blades is not to be underestimated [2].

Owing to the importance of the ceiling fan, the identification of the factors affecting the performance of the fan is necessary. Among others, these factors include shape of the ceiling fan, investigation of the flow field of ceiling fan, structural analysis, aerodynamics of ceiling fan, ceiling fan noise reduction. The shape of ceiling fans has a significant impact on their performance, so designing a well-balanced ceiling fan is crucial. The flow field must be investigated because it is largely responsible for the amount of air circulated in a space.

To ensure maximum efficiency, a structural analysis of the ceiling fan is performed. The length and width of ceiling fans, the design of blade tips, the number of blades, and the aerofoil of the cross-section of the blades are all aspects of ceiling fan aerodynamics. The motor of the ceiling fan has to be lubricated and the blades of the ceiling fan should be balancedto reduce noise. Parker numerically investigated various low-speed ceiling fans and suggested a new design with double the air flow, less trembling, less sound, and greater energy efficiency [3]. The induced flow pattern feature in space is determined by ceiling fan speed, diameter, no of blades, blade angle and location. He investigated the characteristics of air flow caused by a ceiling fan. The outcome demonstrated that the flow pattern has distinct characteristics as blades get detaches and progress towards the floor [4]. Another factor is construction, reducing the weight of the fan blade reduces the power required to operate the ceiling fan, using PVC for the design of the ceilingfan blade is advantageous because of light weight, has good mechanical toughness and consumes lesser power, thus saving electricity consumption, and other features affecting the performances is shape of the ceiling fan. Keeping in mind the foregoing, knowledge gained from ceiling fan size, aerodynamics, and other factors is very useful for manufacturers and researchers in designing the mostefficient ceiling fan and calculating the factors affecting ceiling fan efficiency [5]. Despite being a very common and useful device, the information regarding the different aspects of ceiling fan is rather fragmented which calls for a study to aggregate the available information in a comprehensive and absolute manner. Thus, this paper aims to fill the gap by systematically reviewing the available literature related to ceiling fan and providing the information to prompt and progress the research in the aforementioned areas [6].

The remainder of this paper is designed as follows. Section2 outlines the methodology for conducting the literature review. Section 3 discusses the different aspects of ceiling fan under five subsections, namely shape of the ceiling fan blade, investigation of flow field, structural analysis of different blade material, aerodynamics, and noise reduction of the ceiling fan. Finally, in Section 4 discussion and concluding remarks have been provided.

2 Methodology

The methodology adopted for this research is shown in figure 2, The research started with the identification of suitable literature on the design and performance of ceiling fan. The authors used 5 keywords in the search criteria, related to factors affecting the performance of the ceiling fan, including ceiling fan blade, aerofoil shape, noise reduction, optimization, blade materials. The articles were explored for a period of 25 years commencing from 1996 to 2021from various data warehouses such as Google scholar, science direct, web of science, Scopus to collect the papers of interest.

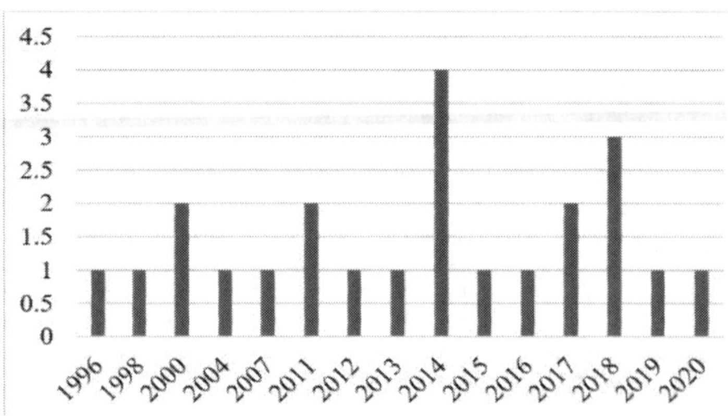

Figure.1 Research trends over the years

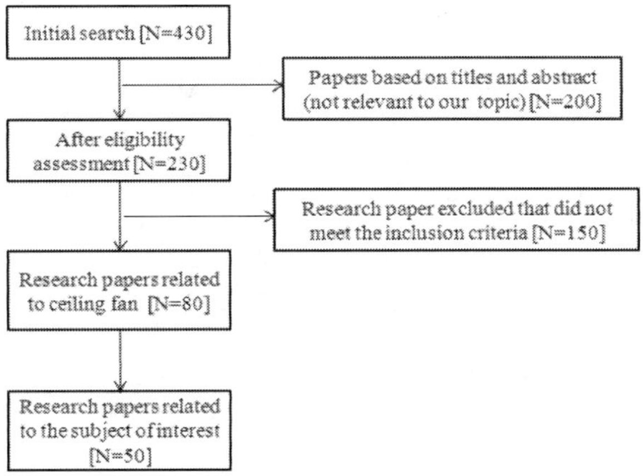

Figure.2 Flow chart

3. Design of the Ceiling Fan

3.1 Shape of the ceiling fan blade

Ceiling fan designs are described in a different perspective by various researchers in the literature. In India fans are used largely and the design of fans is in such a way so that it can generate maximum cooling. The design must be done meticulously in order for the ceiling fan blades to function properly. The cost is determined by the materials used, and the manufacturing process chosen. The design of blades is driven mostly by aesthetic considerations, while manufacturing is driven by efficiency [7]. In the majority of ceiling fans, the use of three blades is preferred instead of using more. This is because ceiling fans having three blades provide better air movement and are more efficient than others. More blades on the ceiling fans mean more aerodynamics drag on the motor and thus hampering the performance of the ceiling fan. It is obvious that the blade shape, number of blades, and weight used in ceiling fan blades all have an impact on the performance of the ceiling fans [8]. As a result, the design and functionality of the components must be in order to improve performance and save energy. As for the new design, continuous effort to improve the interface between the rotor and the blade, making it easier to attach and detach the fan blades from the rotor. It will also make the cleaning of ceiling fan blades much easier. The ceiling fan blades are also important because they determine how much air is circulated in the room, to maximize the air coming from the ceiling fan. Design changes in the shape of the ceiling fan are required [9]. Various kinds of grids and boundary conditions are used in ceiling fans. By adding grid and boundaries to the ceiling fan there will be improvement in design and shape of blades of the ceiling fan. The grid that has the finest outcome according to the flow field and calculated efficiency has been chosen for a greater extent of inspection of rake angle. The grid that is used in the ceiling fans to increase the stability and reduce the noise. Grids are also used to protect someone from putting hand or any other object in the fans accidently[10].

Fig. 3A [11]

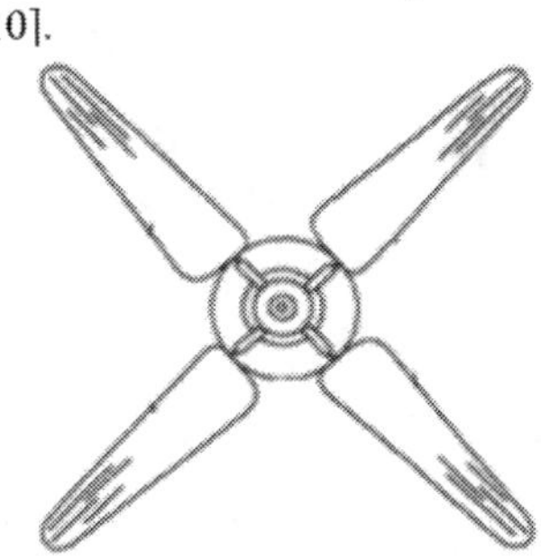

Fig. 3B [11]

To assess the field of flow variables, ceiling fans are placed at various rake angles. The configuration of the rake angles ranges from 0 to 10. The blade that is kept at higher angles create too much resistance, causing the motor to burn out. The blades of a ceiling fan are extremely important because they are a major determining factor in how much air is circulated in a room. As a result, in Figure 3A, the angle at which the ceiling fan blades are kept is a critical consideration. If the fan blades' weights aren't balanced, the fan will not work properly, causes wobbling [12]. The wobble effect, which is directly related to the blades of the ceiling fan, has a major role in flow field of the ceiling fan [13]. The bending of the blades, or the blades being of different sizes and weights, causes the wobble effect.As a result, stability of the ceiling fan is affected, and the flow field is affected as well. As a result, in Figure 3B, it is necessary to change the form and shape of ceiling fans on a regular basis in order to make them better and more efficient. Diverse academics in the literature describe ceiling fan design from various angles. In India, fans are designed in such a way that they can produce maximum cooling. Because the fans are only used as a medium to generate cooling, only three blades are used in the ceiling fan instead of more. This is because the ceiling fans with three blades have superior air circulation and are more efficient than others. More blades on the ceiling fans means more aerodynamics drag on the motor and thus hampering the work rate [14].

3.2 Investigation of flow field of ceiling fan

It is critical to understand how to properly install a ceiling fan for maximum efficiency. The intricate geometries of ceiling fans, directly modelling a rotating ceiling fan with the moving mesh method is generally prohibitively expensive [16]. To convert the air flow into a steady flow, the fan blades should be arranged at a specific angle, and the fan should vibrate less. It may be possible to improve fan performance further by adjusting geometrical parameters such as winglet height and length, or spike length and thickness [17]. Geometrical parameters affect the flow field of the ceiling fan, so improving these parameters will also improve performance. When installing the fan, it is critical that the installer adheres to the manufacturer's instructions regarding the use of proper mounting screws [18]. The ceiling fan's stability is determined in large part by the mount and hub. It is also crucial for safety reasons, as the fan should not fall from the ceiling. The flow field of a ceiling fan is required because there is a high demand for ceiling fans in humid areas, and the flow field of a ceiling fan becomes extremely important. The design of the ceiling fan, as well as the design of the room, has been observed to play an important role in the flow field over time. ASHRAE TC 216 is currently working on a standard for ceiling fans that will provide guidance on how to position them within rooms to maximize comfort [19]. This type of research is carried out to examine the air flow pattern, which may aid in the better positioning of the fan for more effective cooling results. Several similar experiments have been carried out in this field in order to determine the best way to use fans [20]. Only the blade size and rpm determine the overall quantity of air flow or displacement, and is unaffected by any other factor. Using a brushless direct current motor (BLDC) motor-based ceiling fan, the quick fix is to maintain the same amount of air movement or displacement while using lesser energy. The BLDC motor is more efficient and effective because it operates at a constant revolutions per minute (RPM) and runs on a fixed direct current (DC) voltage. There are some disadvantages to this motor as well, such as its high cost and limited utility. As a result of its scarcity of supply, finding these motors in the commercial market is difficult [21].

3.3 Structural analysis fan for a different blade material

The adjustable fan's blade design must adhere to a number of characteristics. This is reflected in varying manufacturing estimates based on material, design, manufacturing, and finishing processes. Blade manufacturing entails the creation of energy-efficient blades with minimal performance [22]. The plate at the base, the rotor, and the blades are the three basic parts of a ceiling fan. The blades commonly comprise of steel, aluminum, wood, and other materials. Researchers are working on the manufacturing of a complex blades of fan are made of glass and fiber forced polymer, and their working differs from that of traditional fan blades [23]. In table 1, the mechanical properties of different blade materials have been explained.

Ceiling fan blades have been made from a variety of materials in a variety of shapes and sizes. In case of fans designed solely for indoor use, rigid materials such as wood or plastic are commonly used. Wooden blades are used because they are less expensive, more accessible, and meet the requirements for light weight, high aesthetic value and rust resistance [24]. Timber blades produce very little noise as they move through the air, which is why they are used. Ceiling fans with metal blades are among the most durable, because metal blades are heavier, a powerful motor is required to push them to their maximum capacity [25]. There are numerous advantages of using plastic fan blades. Plastic blades are lightweight, resulting in fewer installation constraints and lower motor capacity requirements. Plastic blades are safer than wooden or metal blades because they do not require a shroud. Injection molding is used to create wooden and plastic blades and it is a time-consuming process [26].

Table 1: Mechanical properties of different blade material

Blade materials	Density (Kg/m³)	Young's modulus (GPa)	Poisson's ratio
Aluminum alloy 7075T6	2700	70000	0.33
Carbon steel	7858	20500	0.29
Rigid PVC	1800	1.5	0.42
GFRP laminate composite	1400	26	0.28

3.4 Aerodynamics of ceiling fan

The aerodynamic load can be calculated using the fan blade's design load and profile. The blade is shaped in an airfoil shape, which is a streamline cross section. The aero-elasticity study is a completely coupled performance analysis. To lodge the structural movement, the fluid mesh is in motion at each, of the time adopting a system of springs, for-which extension and compression are advised by the mode shape on the surface of the blade, with the becoming values zero at the far field. The normal prevalence of the associated mode determines the motion's frequency [27]. In table 2, different comparison of single-phase induction motor and universal motor has been depicted. The value of aerodynamic damping is calculated from the history of the slope related with time of the modal deportation to determine the stability of each mode. Ceiling fans do nothing more than boost airflow in a room. The term "airflow" refers to a current of air moving through an area. Ceiling fans improve airflow by sucking air from the ground and pushing it down from the ceiling [28].

Table 2: Comparison of single-phase induction motor and universal motor [15]:

Parameters	Sinlge phase induction motor	Universal motor
Motor power	55-78W	80-110W
Operating voltage	230V AC	230V AC
Speed Min- Max-	50 320	80 320
Speed controller	Smooth triac	Resistive
Weight	3-6 Kg	7-10 Kg
Noise	70 dB	130 dB
Efficiency	30%	22%
Air flow	270 CFM	270 CFM
Service factor	1.15	1.0

Ceiling fan consists of a single-phase induction motor. There are two windings on it. The first is initiating winding, while the second is running winding. One of the windings is linked to the capacitor to create a phase difference, which causes the electric field to rotate. Several streamlined blades are installed in a spinning housing of a ceiling fan. The cross section of the blade is aerodynamically developed. As a result, this design ensures that air flows freely between the blades without recirculation, keeping the rotor steady [29]. A suitable ceiling fan can manage large quantities of air at the highest pressures required for ventilation. For the left-hand side, a high-quality constructional finite element formulation is applied, the unsteady viscous shear stresses are not included on the right- hand side, to allow for structural movement [30].

The aerodynamic load was calculated using the fan blade's design load and profile. The blade is shaped in an airfoil shape, which is a streamline cross section. The aero-elasticity study is a completely coupled performance analysis. To accommodate structural motion, the fluid mesh is moved at each time step by employing a network of springs, whoseextension and compression are dictated by the mode shape at the blade's surface, with values zero atthe far field. The normal frequency of the associated mode determines the motion's frequency. Table 3,represents basic parameters used in MATLAB models for BLDC motor. The value of aerodynamic damping is calculated from the history of the slope related with time of the modal displacement to determine the stability of each mode [31].

Table 3: Basic parameters used in MATLAB models for BLDC motor [15]

Phase resistance	2 Ohm
Phase inductance	3 mH
Number of slots	18
Voltage constant	40 KV/rpm
Number of poles	16
Number of phases	3
Voltage input	24 V
Power input	40 W

3.5 Ceiling fan noise reduction.

Three types of noise that are related to the ceiling fans [32]:

- **Aerodynamic noise**: The noise generated by the interaction of the airflow with the impeller and the fixed component of fan.
- **Electromagnetic noise**: The electric motor noise that drives the fan. The major noise source of the motor is due to mechanical excitation of the stator by the magnetic field generated in the air gap between the rotor and the stator.
- **Mechanical noise**: Noise of the transmission and bearing system.

Aerodynamic noise characteristics depends on parameters like type and geometry of the fan, rotational speed, fan operating point, impeller diameter, connection of the fan to the ductworks. The unsteady force exerted on the blades and the effect on the noise of the steady load is negligible when the tip Mach number is lower than 0.3 [33]. Mechanical and electromagnetic noise given to the fan noise levelmay become non negligible only at low rotational speed (below 500 to 1000 rpm). A typical fan noise frequency spectrum is made up of tones at multiples of the blade passage frequency BPF (BPF= B.N where B is the number of blades and N is the rotational speed in Hz), and of a broadband [34]. The progress of the overall sound power level with volume flow is close to that of forward-curved centrifugal fans, that is the continuously increases level with flow without reaching a minimum at the BEP. The generally flow pattern within the whole fan in excess of then the fine structure of the flow on the blade side [36].

The noise generated from the blades is split into two definite mechanisms: First one is noise from the trailing-edge related with the turbulent boundary with layer on the blade and the second one is vortex shedding noise. These mechanisms may sometimes be superimposed, mainly when the property of the blade boundary layers are dissimilar on the pressure and suction sides [37]. Noise of the brushless direct current Motor (BLDC) used in a ceiling fan is evaluated in the acoustic noise testing facility andthe frequency spectrum ranging from 50Hz to 8 kHz is captured. It is noticed that there are peaks at certain frequencies in the noise spectrum [38].

4 Discussion and conclusions

Through this work relevant literature has been located, examined and presented in several subsections such as the shape of the ceiling fan blade, investigation of flow field, structural analysis of fan for different blade material and aerodynamics and noise reduction.

The shape of a ceiling fan affects the performance in a number of ways, including the irregularity inthe shape of blades, weight of the blades, and ceiling fan's stability. The current designs are quite rudimentary, and there is plenty of room for improvement in this area [39]. The fan's geometry is also essential because it is linked to the ceiling fan's flow field. The amount of air circulated in the room is mostly determined by the angle at which the fan is held. The flow field is generally gets affected from the size of the blade and the types of motor used. The flow field is also directly related with the weight and height at which the fan is placed. If there is any kind of bend or dent is on the blade of the ceiling fan the performance and efficiency get affected. If the blade surface is not flat and smooth, it will create turbulence during rotation, decreases the fan speed. If the blade spacing is too large, then it will cause insufficient wind pressure which indirectly affects the work rate [40]. Apart from the blades, the mount and hub also should be aligned properly; otherwise, the speed of the fan would be affected. The material of the blades plays a significant impact in determining the blade's weight. When steel is used, the cost is low but the weight is high, and when aluminum is used, the weight is light but the cost is high. As a result, different materials such as wood and plastics are, nowadays, becoming popular among the manufacturer of ceiling fans [41]. There are many components used in the ceiling fan and a focus on reducing the number of components by following the 'Design for manufacturing and assembly' principles will reduce the weight and enhance the performance and efficiency of the ceiling fan. The fans having geometrically symmetrical and properly balanced blades produce less noise. The literature review identified three types of ceiling fan noise, which are aerodynamic noise, electromagnetic noise and mechanical noise. It has been observed that only a few studies have been conducted to examine the aerodynamic forces acting on a fan blade made of various materials. The noise is a serious issue, thus, future studies must to look forward to identify the different sources of noise and attempt to eliminate them [42].

Another crucial observation is that there is a lack of research focusing on the optimizing the interface between the blade and rotor. The existing interface between blade and rotor needs multiple bolts for attachment which affects the ease of cleaning, repairing, and installation of ceiling fan. Ceiling fans are tough to clean because they are frequently mounted to the house's roof, which is roughly 11-12 feet high. Furthermore, the blades are fastened to the rotor with screws and washers, which takes a long time and effort to attach and detach [43]. To reach the ceiling while attaching or removing, a ladder or many tables are required. As a result, if the ladder or table slips and falls, the person doing the job mayget hurt. The above issues could be solved, if the rotor-blade interface is designed in such as a manner that it facilitates easy assembly/disassembly of the blades with rotor. The literature review revealed that to effectively control the efficiency and performance, the design modifications in the ceiling fan is imperative. Aiming to contribute to the above, presently the work has allocated an effort to highlight the important areas for improvement related to the ceiling fan. Hopefully, findings of the study would be helpful in guiding the future research in a proper direction.

Acknowledgments

This research was carried out at the Center of Excellence of Flexible Manufacturing System/Robotic Design, G H Raisoni College of Engineering, Nagpur, India. The authors are thankful for the help and support provided by Mr. Asmit Karadbhajane and Mr. Jeevan Gaidhane, Engineering students at GHRCE Nagpur.

References

[1] Adeeb, E., Maqsood, A., Musthaq, A., & Sohn, C. H 2016 Parametric study and optimization ceiling fan blades for improved aerodynamic performance. *Journal of Applied Fluid Mechanics*, 9(6), 2905-2916.

[2] MP, S., Paul Raphy, T., Menon, V., & Balachandran, V.2017. Optimisation of Fan Blade Angle.Int. *Journal of Engineering Research and Application. Vol.* 7(1), 69-71.

[3] Rajan, R.2013. Inverter Design for Improving Output Voltage of Fuel Cell Inverter with Filters. *Research Journal of Engineering & Applied Sciences*, 1(4), 01-07.

[4] Adeeb, E., Maqsood, A., &Mushtaq, 2015 Effect of number of blades on performance of ceiling fans. *In MATEC Web of Conferences*, 28(2), 216-226. EDP Sciences.

[5] Dhurvey, Priyanka; Kumar Pathak, Neeraj; Soni, Sanjay; Pandey, Ajay 2018. Aerodynamic and Structural Analysis of Ceiling Fan for Different Blade Materials. Materials, 5(9), 19604–19613.

[6] Lin, S. C., & Hsieh, M. Y 2014 An integrated numerical and experimental analysis for enhancing the performance of the hidden ceiling fan. Advances in Mechanical Engineering, 6(4), 214-254.

[7] Amazing Comfortson, S., AhamedFaiyaz, R., Mohamed Yousuf Afzal, M. R., &MohapilashaNishath, M 2014 Glass Fibre Reinforced Polymer Composite Ceiling Fan Blade. *International Journal of Emerging Technology and Advanced Engineering*, 2250-2459.

[8] Bassiouny, R., &Korah, N. 2011 Studying the features of air flow induced by a room ceiling- fan. Energy and Buildings, 43(8), 1913-1918.

[9] Tauqeer, T., Ansari, M. A., & Hasan, A 2017 Realization for low cost and energy efficient ceiling fans in the developing countries. *Renewable and Sustainable Energy Reviews*, 76(3), 193-201.

[10] Raftery, P., Fizer, J., Chen, W., He, Y., Zhang, H., Arens, E., & Palagi, G 2019 Ceiling fans: Predicting indoor air speeds based on full scale laboratory measurements. Building and Environment, 155(14), 210-223.

[11] Parker, D. S., Callahan, M.P., Sonne, J.K., Su, G.H., & Hibbs, B.D 2000 Development of a high efficiency ceiling fan, 122(2), 33-43.

[12] Dhurvey, P., Pathak, N. K., Soni, S., & Pandey, A 2018 Aerodynamic and Structural Analysis of Ceiling Fan for Different Blade Materials. Materials Today: Proceedings, 5(9), 19604-19613.

[13] Teruna, C., Avallone, F., Ragni, D., &Casalino, D 2021 On the noise reduction of a porous trailing edge applied to an airfoil at lifting condition. *Physics of Fluids*, 33(5), 55-65.

[14] Dassen, T., Parchen, R., Bruggeman, J., & Hagg, F. 1996 Results of a wind tunnel study onthe reduction of airfoil self-noise by the application of serrated blade trailing edges. *In Proceeding of the European Union wind energy conference and exhibition*, 12(2), 800-803.

[15] Bhasme, N. R., & Durge, M. P 2015 Comparative study of electrical motors for ceiling fan application. *Journal Impact Factor*, *6*(5), 08-13.

[16] Oerlemans, S., Fisher, M., Maeder, T., &Kögler, K. 2009 Reduction of wind turbine noise using optimized airfoils and trailing-edge serrations. *AIAA journal*, 47(6), 1470-1481.

[17] Sandberg, R. D., & Jones, L. E. 2011 Direct numerical simulations of low Reynolds number flow over airfoils with trailing-edge serrations. *Journal of Sound and Vibration*, 330(16), 3818-3831.

[18] Jones, L. E., & Sandberg, R. D. 2012 Acoustic and hydrodynamic analysis of the flow around an aerofoil with trailing-edge serrations. *Journal of Fluid Mechanics*, 706(13), 295-322.

[19] Chong, T. P., Joseph, P. F., & Gruber, M. 2013 Airfoil self-noise reduction by non-flat plate type trailing edge serrations. *Applied Acoustics*, 74(4), 60.

[20] Chong, T. P., & Joseph, P. F 2013 An experimental study of airfoil instability tonal noise with trailing edge serrations. *Journal of Sound and Vibration,* 332(24), 6335-6358.

[21] Chong, T. P., Vathylakis, A., Joseph, P. F., & Gruber, M 2013 Self-noise produced by an airfoil with nonflat plate trailing-edge serrations. *AIAA journal*, 51(11), 2665-2677.

[22] Xu, K., Qiao, W., Ji, L., & Tong, F 2014 An experimental investigation on the near-field turbulence for an airfoil with trailing-edge serrations at different angles of attack. In *Proceedings of Meetings on Acoustics* 167ASA, 21(1).

[23] Alawadhi, H. A., Alex, A. G., & Kim, Y. H 2014 CFD analysis of wing trailing edge vortex generator using serrations. *In EPJ Web of Conferences*, 67(3), 12-24.

[24] Arce León, C., Merino-Martínez, R., Ragni, D., Avallone, F., & Snellen, M 2016 Boundary layer characterization and acoustic measurements of flow-aligned trailing edge serrations. *Experiments in fluids*, 57(12), 1-22.

[25] Arce León, C., Ragni, D., Pröbsting, S., Scarano, F., & Madsen, J 2016 Flow topology and acoustic emissions of trailing edge serrations at incidence. *Experiments in Fluids*, 57(5), 1-17.

[26] Zhu, W. J., & Shen, W. Z 2016 LES tests on airfoil trailing edge serration. In *Journal of Physics: Conference Series*, 753(2), 13-21.

[27] Moreau, D., &Doolan, C. J 2013 Tonal noise from trailing edge serrations at low Reynolds number. *In 19th AIAA/CEAS aeroacoustics conference,* 23(2), 43-49.

[28] Avallone, F., Pröbsting, S., &Ragni, D 2016 Three-dimensional flow field over a trailing-edge serration and implications on broadband noise. *Physics of Fluids*, 28(11), 15-24.

[29] Avallone, F., Van der Ve ., Avallone, F., Scarano, F., Pröbsting, Madsen, J 2017 Effect of trailing edge serration-flow misalignment on airfoil noise emissions. *Journal of Sound Vibration*, 405, 19-33.

[30] Ryi, J., & Choi, J. S 2017 Noise reduction effect of airfoil and small-scale rotor using serration trailing edge in a wind tunnel test. *Science China Technological Sciences*, 60(2), 325- 332.

[31] Rao, C., Ikeda, T., Nakata, T., & Liu, H 2017 Owl-inspired leading-edge serrations play a crucial role in aerodynamic force production and sound suppression. Bioinspiration & biomimetics, 12(4), 22-34.

[32] Jaron, R., Moreau, A., Guérin, S., & Schnell, R 2018 Optimization of trailing-edge serrations to reduce open-rotor tonal interaction noise. *Journal of Fluids Engineering*, 140(2), 41-56.

[33] Lin, H. H 2019 Improvement of human thermal comfort by optimizing the airflow induced by a ceiling fan. Sustainability, 11(4), 44-67.

[34] D'Souza, S. 2018 Improving Performance of Ceiling Fan through EAC and Energy Saving by Smart Automation: A review. *In 2018 4th International Conference for Convergence in Technology,* 66(12), 1-7.

[35] Babich, F., Cook, M., Loveday, D., Rawal, R., & Shukla, Y. 2017 Transient three-dimensional CFD modelling of ceiling fans. *Building and Environment*, 123(2), 37-49.

[36] Adeeb, E., Maqsood, A., Musthaq, A., &Sohn, C. H. 2016 Parametric study and optimization of ceiling fan blades for improved aerodynamic performance. *Journal of Applied Fluid Mechanics*, 9(6), 2905-2916.

[37] Ho, S. H., Rosario, L., & Rahman, M. M. 2009 Thermal comfort enhancement by using a ceiling fan. *Applied Thermal Engineering*, *29*(8-9), 1648-1656.

[38] Hsiao, S. W., Lin, H. H., & Lo, C. H 2016 A study of thermal comfort enhancement by the optimization of airflow induced by a ceiling fan. *Journal of Interdisciplinary Mathematics*, 19(4), 859-891.

[39] Fazil, M., &Rajagopal, K. R 2010 Development of external rotor single-phase PM BLDC motor-blade drive for ceiling fan. In 2010 *Joint International Conference on Power Electronics, Drives and Energy Systems & 2010 Power India,* 23(2), 1-4.

[40] Jain, A., Upadhyay, R. R., Chandra, S., Saini, M., & Kale, S 2004 Experimental investigation of the flow field of a ceiling fan. In Heat Transfer Summer Conference, 469(2), 93-99.

[41] Jain, A., Upadhyay, R. R., Chandra, S., Saini, M., & Kale, S 2004 Experimental investigation of the flow field of a ceiling fan. In Heat Transfer Summer Conference, 469(2), 23-49.

[42] Shao, J 2003 Direct back EMF detection method for sensor less brushless DC (BLDC) motor drives (Doctoral dissertation, Virginia Tech), 65(3), 234-321.

[43] Dahare, S. K., &Rajvaidya, D. R 2013 Design and Analysis of Nylon66 Ceiling Fan Blade Using Finite Element Method. *International Research Journal of Engineering & Applied Sciences* Rukmini, K. V., Kumar, K. S., Manikanta, V. S., Sai, K. S. S., Vinay, K., &Satyanarayana, M. Flow Analysis over a Ceiling Fan Blade, 341(12), 112-213.

[44] Agrawal, P., Narain, R. and Ullah, I. 2020 Analysis of barriers in implementation of digital transformation of supply chain using interpretive structural modelling approach, *Journal of Modelling in Management,* 15(1), 297-317.

[45] Ullah, I. and Narain, R. 2021 Analyzing the barriers to implementation of mass customization in Indian SMEs using integrated ISM-MICMAC and SEM, *Journal of Advances in Management Research,* 18(2), 323-349.

[46] Ullah, I. and Narain, R. 2018 Analysis of interactions among the enablers of mass customization:An interpretive structural modelling approach, *Journal of Modelling in Management,* 13(3), 626-64.

[47] Ullah, I. and Narain, R.2021 Achieving mass customization capability: the roles of flexible manufacturing competence and workforce management practices, *Journal of Advances in Management Research,* 18(2), 273-296.

[48] Ullah, I. and Narain, R. 2021 Linking supplier selection and management strategies with mass customization capability, *Journal of Business & Industrial Marketing*, 36(7), 1213-1228.

[49] Shukla, S., Patil, A.P., Kawale, A., Kotta, A.B. and Ullah, I 2021, "Effect of recrystallization on degree of sensitization in nickel free austenitic stainless steel", *Anti-Corrosion Methods and Materials*, 68(6), 530-537.

[50] Bansod, A. V., Patil, A. P., & Shukla, S 2020 Electrochemical evaluation of cr-mn austenitic stainless steel in aqueous sulphuric acid and influence of thiocyanate ions. *Anti-Corrosion Methods and Materials*, 67(3), 281–293.

[51] Shukla, S., Patil, A. P., Kawale, A. P., Haldkar, R. K., Dahiwale, A., & Bansod, A. 2021 Effect of grain refinement on sensitization of high manganese austenitic stainless steel. *Materials Today: Proceedings*, 44(2), 2802–2807.

[52] Rambhad, K., Kumbhalkar, M., Kalbande, V., Jibhakate, R., Jumde, N., & Walke, P. 2022 Performance Investigation of Desiccant Wheels for Air Dehumidification. In Smart Technologies for Energy, Environment and Sustainable Development, 22(3), 257-264. Springer, Singapore.

[53] Gaidhane, J., Ullah, I., & Khalatkar, A 2022 Tyre remanufacturing: A brief review. Materials Today: Proceedings. https://doi.org/10.1016/j.matpr.2022.04.142

IOP Conf. Series: Materials Science and Engineering 1259 (2022) 012026 doi:10.1088/1757-899X/1259/1/012026

Development of Coil Spring Suspension System with Air Bellows

Abhijit Ratekkar[1], Sahil Gulhane[1], Suraj Meshram[1], Mahesh Borker[1], A.R.Khan[1], R.S. Shelke[1]

[1] Dept. Of Mechanical Engineering G.H. Raisoni College Engineering, Nagpur Maharashtra, India.

abdulrahman.khan@raisoni.net

Abstract. Most of the mass production cars use only coil springs in their suspension systems. We have developed a combination of coil spring suspension with air bellows to reduce the vibrations of the vehicle in mass production cars. Our design proposes and successfully implemented the use of coil springs and the air bellows that will reduce the vehicle's vibrations and provide some amount of cushioning effect to the passengers and Driver. The coil spring is placed on the top of air bellows only in the rear suspension. And calculated the vibration using I Dynamics app.

1. Introduction

The types of suspension systems used in today's cars are Coil spring, Leaf spring, & Air suspension system, but the owners of cars with the coil spring setup in their cars are not that satisfied. Only high-end luxury cars get the air suspension system in order to provide seamless comfort and customized handling characteristics of the car. But the big issue in the air suspension system is its maintenance and repair cost, which is not economical. Our survey found that 4 out of 10 cars stand there in most big garages due to air suspension failure issues, and the average repairing cost is approximately around one lakh INR. Hence, we are developing the suspension system, a combination of Air bellows and Coil spring with minimum maintenance and low cost. The functions of the suspension system are:

1. Reduce the overall vibration of the vehicle
2. Reduce the stress coming on the chassis due to uneven roads.
3. Reduces the chances of rattling noise of the vehicle.

The present work aims to minimize the transfer of road vibrations to the vehicle chassis and, at the same time, enhance the overall ride comfort and provide a smoother ride like in the air suspension system. The scope is to provide a better ride and improved vehicle handling at a relatively low cost and minimal maintenance compared to the air suspension system.

2. Problem Statement

It is observed from the figure 1 that in the mass production of cars, only coil springs are used in their suspension systems, but we found that coil springs alone can't restrict the overall road vibrations. Hence, in the present work, the combination of coil spring and air bellows is used to provide the advantages of both types of suspension systems. Therefore, the vibration level will be an average of pure coil spring and pure air suspension system [1].

Figure 1: Traditional Rear suspension system

3. Components

The components of the traditional coil spring suspension system are shown in figure 1. Various components of the present suspension system are as follows:

3.1. Coil Spring

Figure 2 shows the coil spring. The coil spring is located above the air bellow and mounted on the chassis at rear on both the sides. A coil spring is used to store energy and subsequently release it, absorb shock, or maintain a force between contacting surfaces [2].

3.2. Air Bellow

As shown in figure 3, the air bellow consists of an air chamber with flexible sides or ends, a means of compressing it, an inlet valve, and a constricted outlet. It is located below the coil spring and mounted on the mounting above the rear axle. They are used to absorb the road vibrations and to reduce the noise of the vehicle [3].

3.3. Shock Absorber

Figure 4 shows the shock absorber, a mechanical or hydraulic device designed to absorb the shock and damping effects of the component. We used the shock absorber as a mechanical linkage to make a stable mechanism of the suspension system. It is a link between the chassis and rear axle mounting.

Figure 2: Coil Spring

Figure 3: Air Bellow

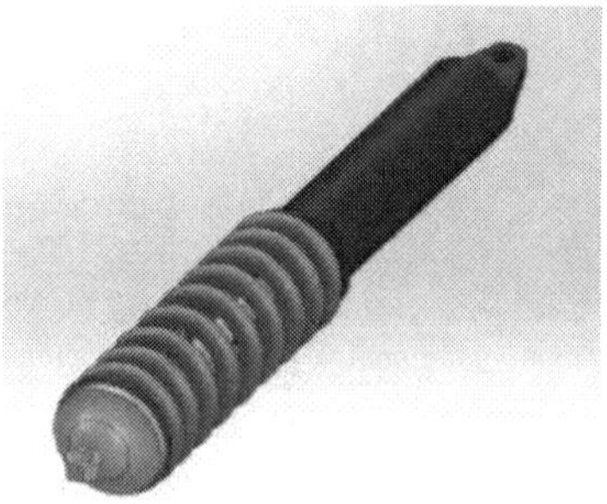

Figure 4: Shock Absorber

3.4. Chassis

Figure 5 shows the chassis of the present system. It is a primary structure upon which all major automobile components necessary to propel the vehicle are assembled. Chassis comprises everything of a vehicle except the body, cabin, and equipment. Basically, it consists of a frame, Engine, Transmission system, Suspension system, Electrical Controls etc. [4].

3.5. *Rear Axle*
Figure 6 shows the rear axle. The axle is a weight-carrying portion, whether it may be front, or rear loaded at an intermediate point. It takes vertical load due to the body's weight, shock load during braking and driving, a torque reaction due to driving and braking, and a side thrust during cornering.

3.6. *Wheel and Tyre*
As shown in figure 7, the Wheels and tire take the vehicle load, provide a cushioning effect, and cope with the steering control.

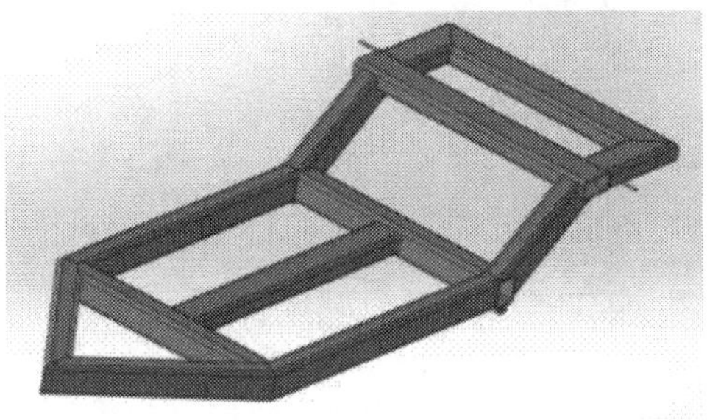

Figure 5: Chassis

Figure 6: Rear Axle

Figure 7: Wheel and Tyre

4. Mechanism
As shown in figure 8, we have developed a suspension system that combines the coil spring and air bellow. In the present system, the coil spring is mounted on the air bellow and attached to the axle. The air bellows are filled with a constant pressure value which is constant and can be changed manually without the use of an automatic compression system. When the vehicle runs on the road, the upcoming shocks and vibrations are first absorbed by the air bellow, and these reduced or damped vibrations are transferred through the coil spring to the chassis.

Figure 8: Mechanism

5. Methodology

1. •Recognization of need
2. •Research about coil spring suspension and air suspension.
3. •Selecting the best solution: research about coil spring suspension and air suspension.
4. •Planning
5. •Detailed drawing the designing process in Solidworks
6. •Design of coil spring as per requirement for prototype
7. •Simulation of coil spring
8. •Fabrication of the prototype
9. •Testing of prototype and observation of the vibration and shocks using iDynamic app.

6. Designed Part

This section shows the actual design calculations used for selecting various parameters and components of the present system.

6.1. Coil Spring Design

Table 1: Spring Dimensions

Material	Stainless steel ASTM-A313
Total number of Coils, n'	9
Number of Active Coil, n	7
Mean Diameter, D	55mm
Wire Diameter, d	9.5mm
Inner Diameter, Di	45.5mm
Outer Diameter, D0	64.5mm
Free length, LF	152mm
Solid Length, Ls	85.5mm
Pitch, p	19mm
Spring Index, C	6
Chassis Weight	170kg
Attachment Weight	50kg
Gross Weight	220kg
Factor of Safety(FOS)	1.8

Table 2: Spring Material

Material	Stainless steel ASTM-A313
Modulus of Rigidity, G	75842.33MPa
Density, ρ	7805.73kg/m^3
Young's Modulus	203.4GPa
Poisson's Ratio	0.27
Yield Strength	767.39MPa
Ultimate Strength	2192.53MPa

Load applied on the suspension system,

$W = 0.5 \times 2158.2 = 1079.1$ N

$S_{ut} = 2192.53$ Mpa

$S_{sy} = 35\%$ of ultimate strength $= 767.39$ Mpa

Torsional shear strength,

$$\tau = \frac{S_{sy}}{FOS} = 426.33 \text{ Mpa}$$

$$\tau = \frac{8WD}{\pi d^3}$$

Spring index for the oil spring $C = D/d = 6$

Torsional shear stress, $\tau_1 = \frac{8W.D}{\pi d^3} = 182.49$ Mpa

Direct shear stress, $\tau_2 = \frac{4W}{\pi d^2} = 15.22$ Mpa

The resultant shear stress $\tau = \tau_1 + \tau_2 = 197.71$ Mpa

To consider the effect of curvature on maximum shear stress

$$\tau = K_W \frac{8F.D}{\pi d^3}$$

Where,

$$K_w = \text{Wahl's factor} = \frac{4C-1}{4C-4} + \frac{0.615}{C} = 1.25$$

Therefore, $\tau = 220.34$ Mpa

Deflection of spring, $y = \frac{8FC^3 n}{dG} = 23.29$ mm

The spring rate k = F/y = 46.33 N/mm
Spring Slenderness ratio = $\frac{L_f}{D}$ = 2.76

Spring slenderness ≥ 4: spring will likely buckle or bend
Spring slenderness ≤ 4: spring will not likely buckle or bend
Pitch of coil spring wire, P = $\frac{L_f}{n'-1}$ = 19 mm
Pitch angle, $\lambda = \tan^{-1}\frac{P}{\pi D} = 6.27^0$
Maximum allowable stress

τ_{allow} = 426.33 Mpa
τ_{max} = 220.34 Mpa
$\tau_{max} \leq \tau_{allow}$

So, Design is satisfied [5][6].

6.2. Air Bellow Selection

As per the coil spring's calculated dimensions, we have selected standard-sized air bellow available in the market.

Table 3: Air Bellows Specification

Parameter	Value
Cylinder Dia.	70 mm
Weight	0.350 kg
Max. Stroke	50 mm
H. Min	65 mm
H. Max	115 mm
Max. Pressure	116 psi
Body Material	Rubber
Cup Material	Cast Iron

7. Construction and Working

Figure 9: Front view of the Prototype

Figure 10: Rear view of the Prototype

Figure 9 and figure 10 show the front and rear views of the coil spring suspension with air bellow, respectively. In which the air bellow fitted on the rear axle. The coil spring is mounted on the air bellow, and the chassis is fitted on the coil spring. The shock absorber and trailing arm are fixed between the axle and chassis to restrict the unwanted movements. When the prototype starts moving on the road, the upcoming shocks and vibrations from the uneven road surface are transferred to the chassis. Most of the vibrations and shocks are absorbed by the air bellow and remaining by the coil spring, which helps to minimize the vibration transfer to the chassis [7].

IOP Conf. Series: Materials Science and Engineering 1259 (2022) 012026 doi:10.1088/1757-899X/1259/1/012026

8. Result and Discussion

The prototype designed and fabricated for the present work is tested on actual driving conditions on the road. The vibrations transferred in the cabin from the uneven road surfaces are measured using the I-Dynamics application for android phones freely available from the Google play store. The I-Dynamics gives the displacement, velocity, and acceleration of the surface on which the mobile phone is mounted. The Z-direction is the most crucial direction from the vibration point of view that shows the direct transfer of the shocks from the tire to the vehicle cabin

The first trial is done with air bellows in the suspension system. Figure 10 shows the screenshot of the graph for acceleration values in all the individual directions generated for the trial. The prototype vehicle is operated at a 35 to 40 km/hr speed. The air bellows are inflated with a pressure of 14 psi. From figure 11 – which is the screenshot of exact values – it is observed that the maximum acceleration is recorded in the z-direction, and its value is 26.446 m/s2. Similarly, the maximum displacement is also observed in the direction with a numerical value of 0.052 m.

In the second trial, the initial operating condition is the same, and the only change in this trail is we use the suspension system without air bellows.

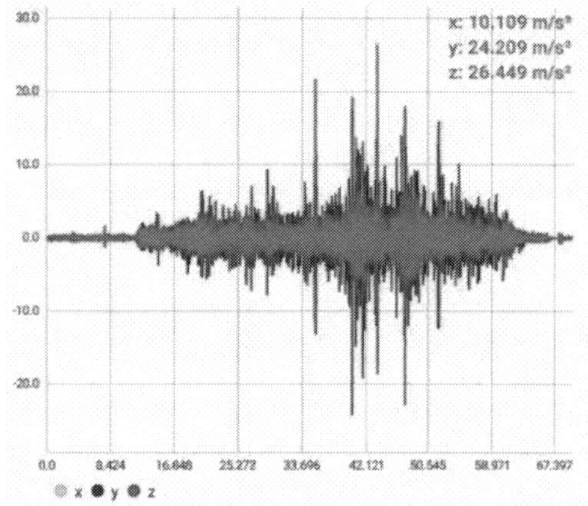

Figure 11: Graph trial 1

max. Acceleration		
X 10.11963 m/s²	Y 24.20884 m/s²	Z 26.44603 m/s²
max. Velocity		
X 0.42549 m/s	Y 0.22169 m/s	Z 0.39130 m/s
max. Displacement		
X 0.07089 m	Y 0.04460 m	Z 0.05196 m

Figure 12: Result trail 1

From figure 12, it is observed that the trend for the acceleration in various directions is similar to figure 10, but the absolute value of the maximum acceleration is 60.638 m/s2. Similarly, the displacement in the z-direction is 0.080 m which is larger than values observed in figure 11. The above observations help to distinguish the vibration level between two trial. It is evident that the suspension system with air bellows has much lower vibrations than that without air bellows. The absolute value of Z-direction acceleration with bellows is less than half of the values obtained without air bellows [8].

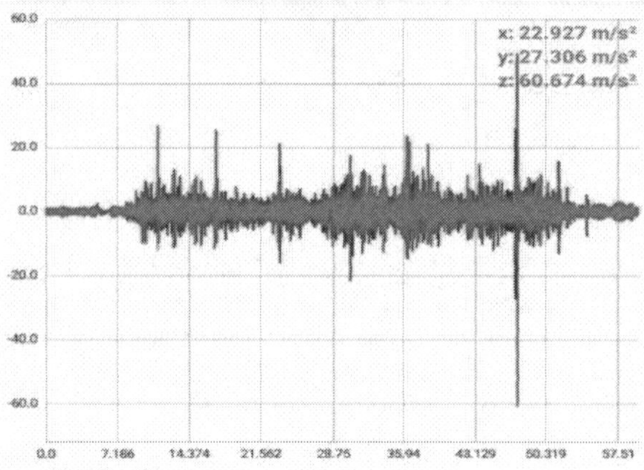

Figure 13: Graph trail 2

max. Acceleration		
X 22.95224 m/s²	Y 27.30106 m/s²	Z 60.63780 m/s²
max. Velocity		
X 0.68271 m/s	Y 0.26668 m/s	Z 0.60328 m/s
max. Displacement		
X 0.14859 m	Y 0.03858 m	Z 0.08032 m

Figure 14: Result trail 2

IOP Conf. Series: Materials Science and Engineering 1259 (2022) 012026 doi:10.1088/1757-899X/1259/1/012026

9. Conclusion

In the present work, we have successfully designed and fabricated the vehicle suspension system with air bellows that is cost-effective and capable of reducing the vibration level effectively. From the results and discussion section it is concluded that the reduction in the vibration in terms of acceleration reduction with air bellows is more than 50%.

References

[1] Gokul Parasad S. and Malar Mohan K. 2019 *ISA Transactions* **93** 244.

[2] Dr. Htay Htay Win , Dr. Nwe Ni Tun , Maung Yone Kyin Thang 2019 *IRE Journals* **2** 152.

[3] N.Lavanya , P.Sampath Rao M.Pramod Reddy *Int. Journal of Engineering Research and Applications* **4** 151.

[4] Thomas D. Gillespi 2021 *Fundamentals of Vehicle Dynamics, Revised Edition R-506* (China: SAE International)

[5] Vijayeshwar BV1 , Preetham B M2 , Bhaskar U 2017 *International Advanced Research Journal in Science, Engineering and Technology* **4** 92.

[6] C.H.Chin, S.Abdullah, S.S.K.Singh, A.K.Ariffin, D.Schramm 2020 *Measurement* **158** 107697.

[7] V.B.Bhandari 2010 *Design of Machine Elements* (India: Tata McGraw-Hill Education).

[8] A.Feldbusch, H.Sadegh-Azar, P.Agne 2017 *Procedia Engineering* **199** 2790.

Design and Analysis of tool for fractured abutment screw removal from the dental implant body

Swapnil Shedmake[1], Nilesh Awate[1], Pramod Walke[1], Barkha Bhoneja[2]

[1] G H Raisoni College of Engineering Nagpur, Digdoh Hills, CRPF Gate No 3, Hingna Road, Nagpur, 440016, India
[2] Advance Dental Clinic, Ramdaspeth, Nagpur

Abstract. Implant-based restoration is currently seen to be a more reliable option with higher success rates. Biological and technological problems, on the other hand, are common. An implant insertion complication well known to many is the loosening of the abutment screw. Fractures of the implant-abutment may cause the implant to fail to function efficiently as they may leave a fragment inside the implant. This study aimed to introduce the modified design of a tool to extract the stripped abutment screw from the implant body and to analyses the deformation, stress and strain acting on the tool using the finite element method.

1. Introduction

Dental implantology is the most recent advancement in the area of dentistry that deals with replacing lost teeth and supporting oral tissues. To support the dental prosthesis, dental implants are put into the jawbones. This therapy has progressed to the point where it can significantly improve one's quality of life. In the past 30 years, dental implants have become a popular substitute for missing teeth [1]. In the past, the only options for replacing teeth were dentures and bridges [28]. Recent decades have seen dental implants become extremely popular due to their high success rates, predictability, and relative lack of problems. [1,2,3,23,26]. A dental implant is made up of three parts: the crown, the abutment, and the implant or fixture. The crown is the component of the bridge that resembles your natural teeth. They are frequently referred to as caps. An abutment is a metal component that acts as a basis for the crown on a dental implant. It just functions as a connection, with one end linked to your jawbone and the other end fitted with the crown. Titanium is commonly used for abutments. The hardware that secures the abutment to the implant is the dental abutment screw [4,29]. The threaded portion of this fastener is tightened until it reaches its ultimate seating position. Osseointegration, which is the joining of the living jawbone with a load-bearing implant, is a crucial step in the successful healing of dental implants since it provides a strong anchor between the implant and the jawbone [24]. Even though dental implant therapy is the safest method, several issues cause dental abutment screws to break. The abutment screw becomes shattered due to biological and mechanical causes, and stays in the patient's jaw, making it extremely difficult to remove. Biological problems are divided into two categories: those that arise within the first six months following implantation and those that occur after that time. A variety of factors, including surgical stress, infection, and implant micromovements, lead to early 6-month failure because of implant micromovements related to premature loading. Later failure happens as a result of peri-implantitis, or the gradual loss of bone support caused by infection or inflammation. [5,7,8,9,10,11] Mechanical difficulties arise primarily as a result of the mastication load, which causes tension. The kind of restoration provided by the implants can impact both the amplitude and type of stresses transmitted to the implant, as Berglundh et al demonstrated [5]. The amount of the occlusal stresses is a key factor in implant loading. Parafunctional behavior's like as clenching and bruxism, according to De Boever et al., may increase prosthesis stress, resulting in mechanical issues early in the process. [6]. The screw fragments are often simpler to remove from the implant if it emerges from the implant, and they can be removed with a hemostat [12,13]. The abutment screw can be replaced with a shorter one without removing the broken screw remnant, as long as it engages the internal threads of the dental implant fixture [14]. A fracture of the dental abutment screw, on the other hand, can be a major issue since the part that remains inside the implant may prevent it from operating properly. As a result, several methods for extracting screw pieces within the implant have been documented. [18,19,20,21,22]. The goal of this research was to present a new design and analysis of a tool for retrieving fractured abutment screws from the implant body. This tool functions as a screw extractor, inserting itself into the drilled hole on the abutment surface and retrieving the cracked abutment screw.

Despite the fact that we designed the new shape, we ran a finite element analysis on the tool to confirm total deformation, equivalent elastic strain, and equivalent (von-mises) stress, and the results are reported in this study.

2. Materials and methods

*2.1. C*ad modeling of screw extractor using Mechanical Cad software.

The tool is being designed to have tapered spiral flutes. The threads are left-handed, so you can easily turn them in an anticlockwise direction. The diameter of the abutment screw when broken is generally 1 mm, therefore, the designed tool has a diameter less than 1 mm. The tang part of tool is design in such a way that it will fix with the dental ratchet easily Figure 1 illustrates the cad model of the designed tool

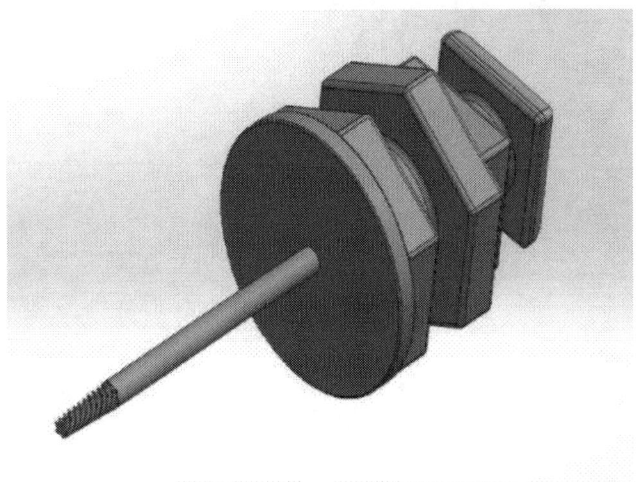

Figure 1: CAD model of Retrieval Tool.

2.2. Finite element analysis of the designed tool

2.2.1 Meshing.

The model's components were divided into smaller elements. Based on the data supplied into the software, each component of the model was separated into multiple little parts with the identical properties of the relevant material. The meshing in this project is retained with nodes 26693 and elements 8024. In a finite element model, utilizing smaller elements can improve computation accuracy. Boundary criteria were also established. The motions of nodes and their correlations are defined by boundary conditions. Figure 2 depicts the tool and body meshing

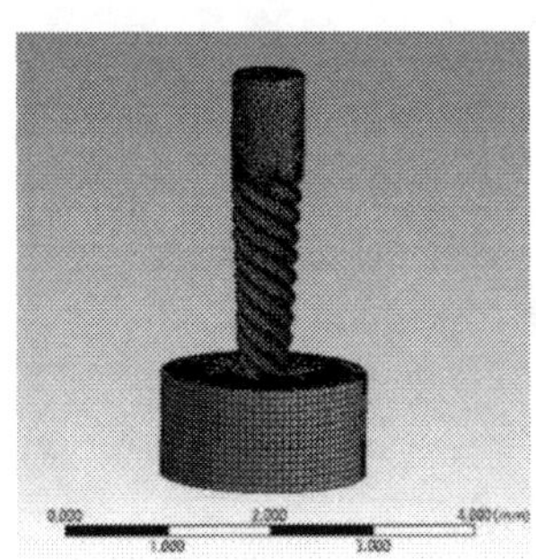

Figure 2: Meshing of parts

IOP Conf. Series: Materials Science and Engineering 1259 (2022) 012027 doi:10.1088/1757-899X/1259/1/012027

2.2.2 Material Data and Analysis

The material of the abutment body is kept TITANIUM having the shear modulus of 43400 MPa and maximum yield stress (Ymax) is 1450 MPa. The material of the tool is kept TUNGSTEN having the shear modulus of 1.6e+005 MPa and Maximum Yield Stress (Ymax) MPa is 4000 Mpa

2.2.3 Total Deformation

Total deformation is the deformation option that you can see all the deformation results related to your model, in three coordinates (X, Y, and Z). In this Analysis report total deformation is applied on the body and the total deformation applied on the tool and graphs of results has been generated. The below figure showing the total deformation and graph of total deformation

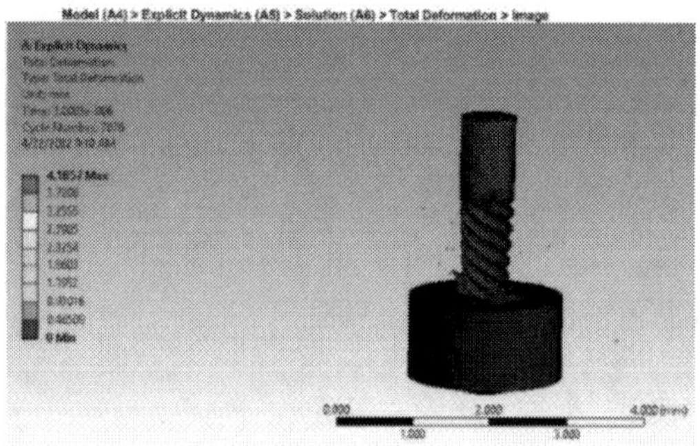

Figure 3. Total deformation in body

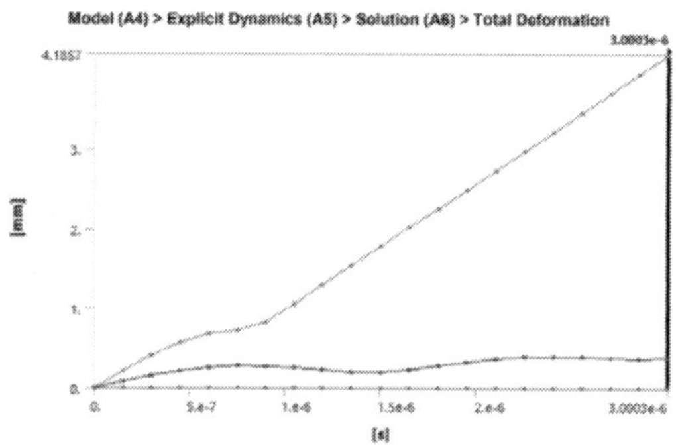

Figure 4. Graph of Total deformation in body

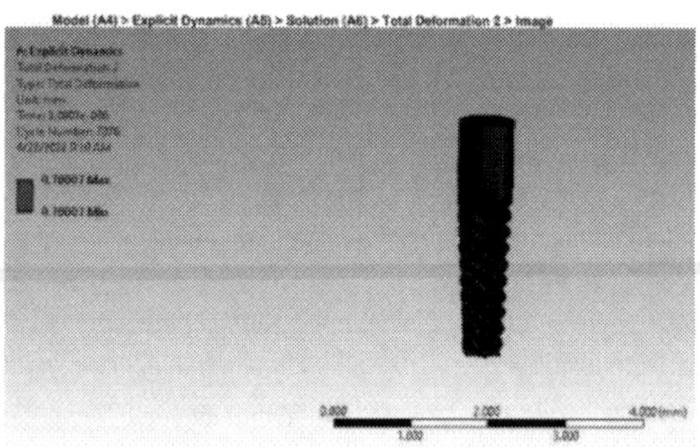

Figure 5. Total deformation in tool

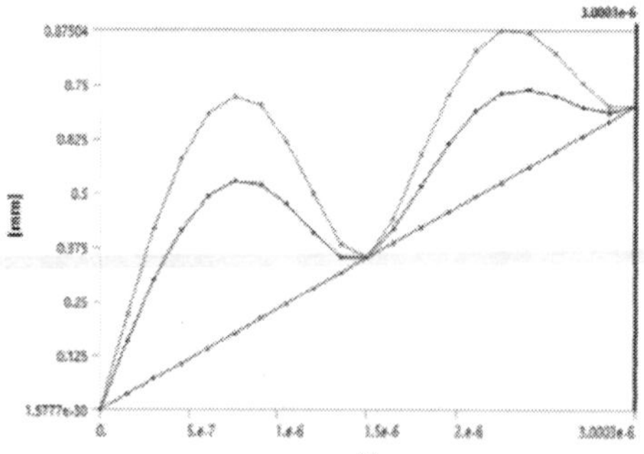

Figure 6. Graph of Total deformation on the tool

IOP Conf. Series: Materials Science and Engineering 1259 (2022) 012027 doi:10.1088/1757-899X/1259/1/012027

2.2.4 Directional Deformation

In directional deformation, you can select a coordinate (X, Y, or Z) to see the deformation result of your physical model in this direction. This analysis report shows the directional deformation on the tool in X, Y and Z direction and the graphs of results of directional deformation generated by analysis software.

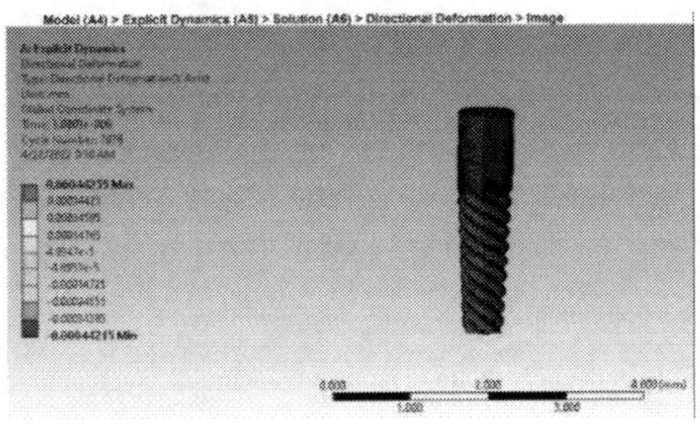

Figure 7. Directional deformation on tool (X direction)

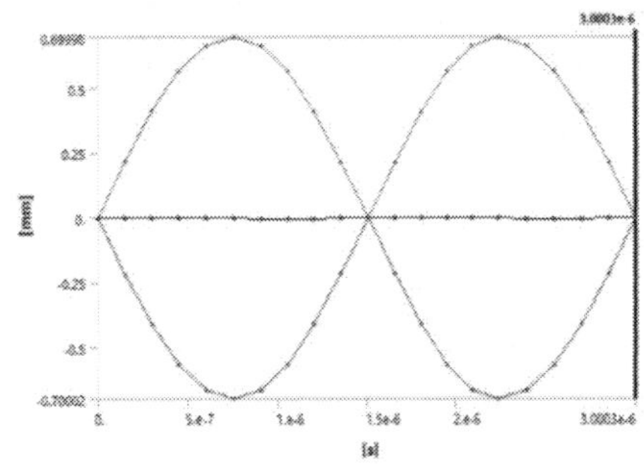

Figure 8. Graph of directional deformation (X direction)

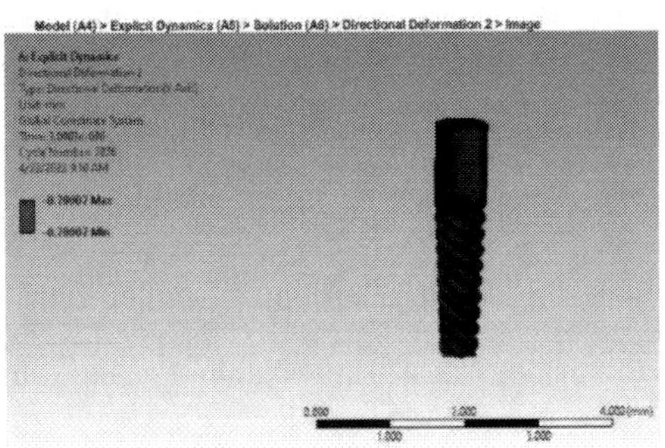

Figure 9. Directional deformation on tool (Y direction)

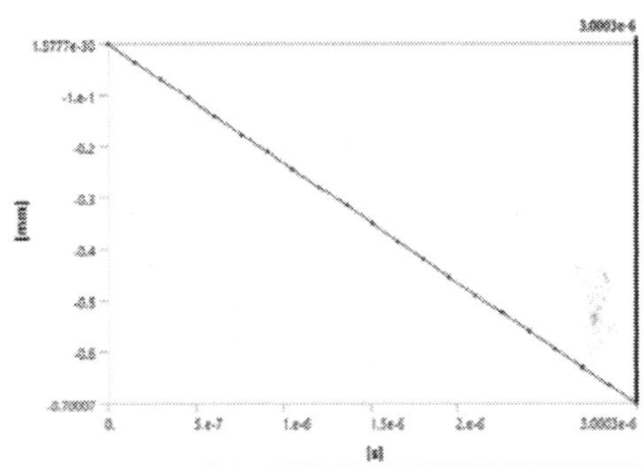

Figure 10. Graph of directional deformation Y direction)

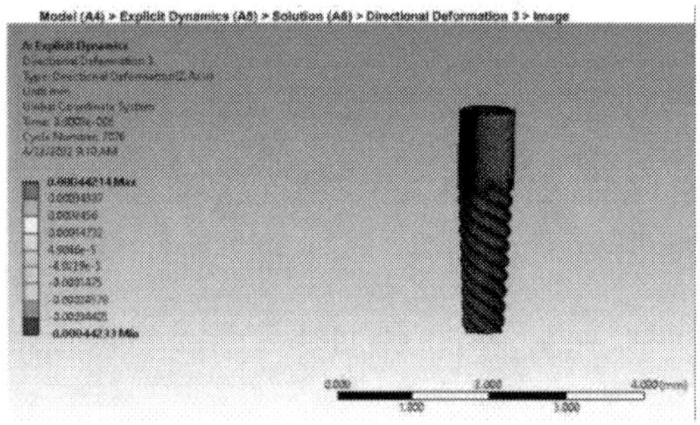

Figure 11. Directional deformation on tool (Z direction)

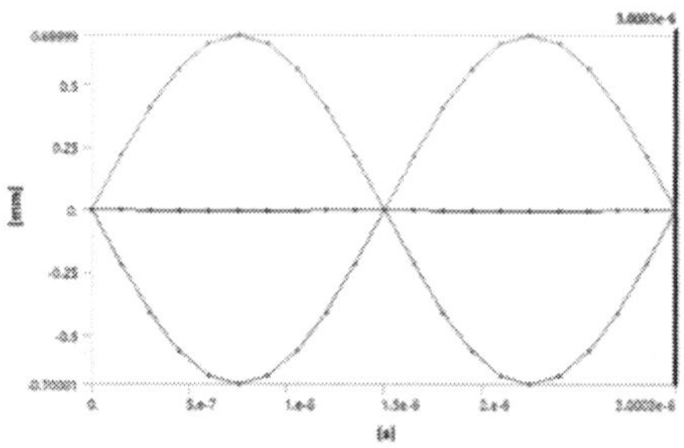

Figure 12. Graph of directional deformation on tool (Z direction)

2.2.5 Equivalent elastic strain

The equivalent elastic strain is defined as the limit for the values of strain up to which the object will rebound and come back to the original shape upon the removal of the load. Elastic limit is defined as the point on the stress-strain curve where the object changes its elastic behavior to plastic behavior. The below fig shows the equivalent elastic strain acting on the body and graph generated by analysis software.

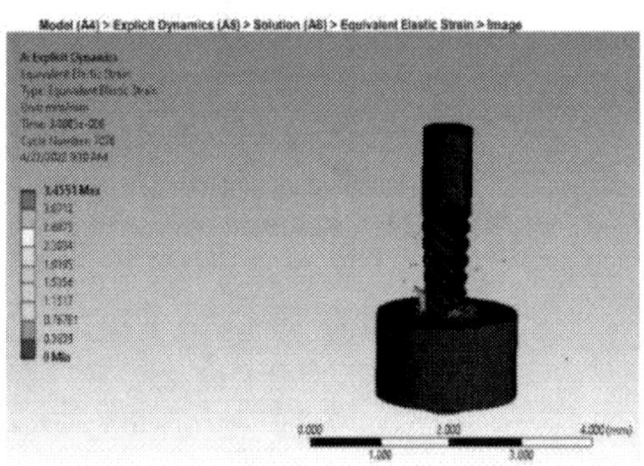

Figure 13. Equivalent elastic strain

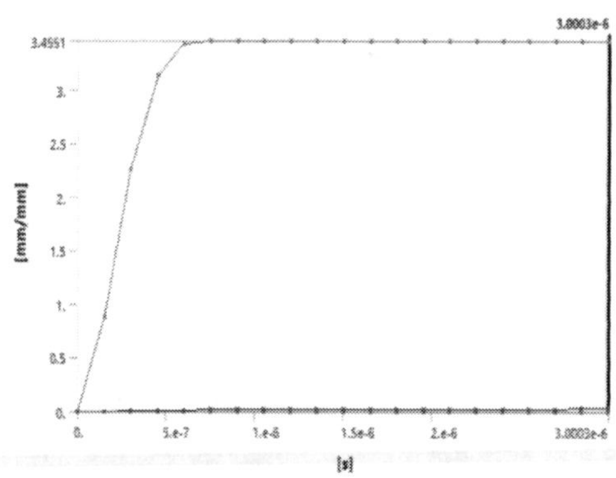

Figure 14. Graph of Equivalent elastic strain

IOP Conf. Series: Materials Science and Engineering 1259 (2022) 012027 doi:10.1088/1757-899X/1259/1/012027

2.2.6 Equivalent von mises stress

The Von Mises criteria is a formula for combining the three principal stresses into an equivalent stress, and the equivalent stress is then compared to the yield stress of the material to judge the failure condition of the material. The equivalent stress is often called the Von Mises Stress, as a shorthand description. The below figure shows the von mises stress acting on the whole body and result graph generated by analysis software.

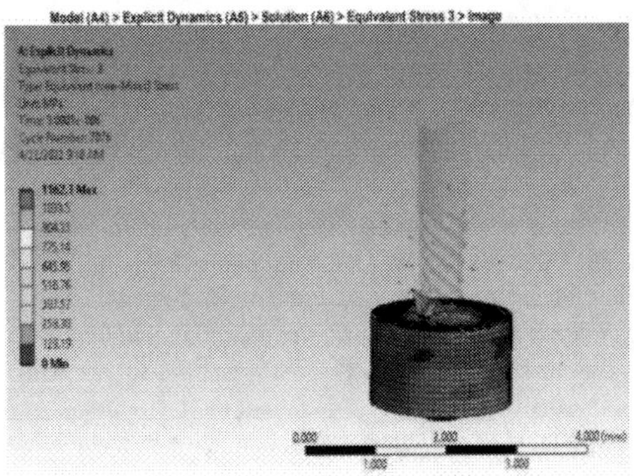

Figure 15. Equivalent von mises stress

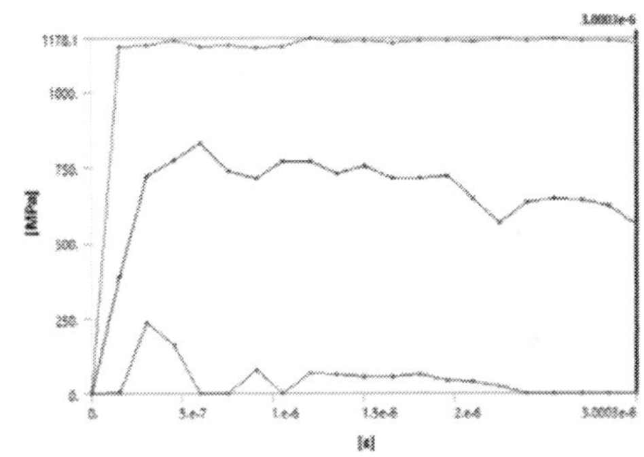

Figure 16. Graph of equivalent von mises stress

3. Results and discussion

Table 1. Results

Sr no.	Type	Minimum	Maximum
1	Total deformation on body	0 mm	4.1857 mm
2	Total deformation on tool	0.70007 mm	0.70007 mm
3	Directional deformation on tool (X Direction)	$-4.4215*10^{-4}$ mm	$-4.4255*10^{-4}$ mm
4	Directional deformation on tool (Y Direction)	-0.70007 mm	-0.70007 mm
5	Directional deformation on the tool (Z Direction)	$-4.423*10^{-4}$ mm	$-4.4214*10^{-4}$ mm
6	Equivalent elastic strain	0 MPa	3.4551 MPa
7	Equivalent (von mises) stress on the body	0 MPa	1162.7 MPa

Various author has tried various different techniques to remove broken abutment screw. By joining the revolving motion of the bur with the uneven surface of the screw, Sneha H. Gaonkar et al. (2019) determine the easiest and harmless approach for extracting a broken dental abutment screw. Using a high-speed air rotor and a fine tapered diamond bur to form a groove on the surface of the abutment. Once the bur contacts an rough surface, the shattered screw detaches and spins out of the implant body [27,28]. Yohsuke Taira and Takashi Sawase (2012) used a tungsten carbide bur and customized guide tube and to remove a faulty screw. Guide tubes were constructed using stainless steel tubes The tube was extended to a length of 4mm so that it can be reached to the fractured abutment screw. Gripping the guide-tube the tungsten-carbide bur was guided in the screw hole until it touches the surface of the failed abutment screw. Once the bur touched the surface, the hole was created on the center of the failed abutment screw using the air rotor. Following the instruction of the manufacturer the threads were restored on the screw hole with the help of retrieval instruments [27,30].
The following screw extraction tool was created to effortlessly remove a broken abutment screw by locking it into a previously drilled hole on the damaged abutment screw. Following the investigation and discussion, a cad model of the tool was created using cad software, and an analysis was done to determine the test deformation and various stress outcomes on the tool. The highest equivalent elastic strain is 3.45 MPa, while the maximum corresponding stress is 1162.7 MPa. Tungsten is used as the tool material in this study model. The Young's modulus of tungsten is 530–700 GPa, while the bulk modulus is 630–655 GPa. Tungsten has a tensile strength of 1725 MPa. As a result of the examination, the tool is safe and can withstand deformation.

4. Conclusions

Although providing an acceptable restoration plan for shattered abutment screws might be difficult, proper supervision is essential for a successful conclusion. The proposed tool for retrieving the stripped screw is safe and can tolerate deformation, according to the analysis report. The left-handed thread will aid in the extraction of the shattered screw, and the threaded profile will act as a good anchor between the previously drilled holes. The retrieved broken screw will help the dentist utilize the same implant for new restoration. However, because the physical performance (in-vitro) research on the prostatic denture is critical, subsequent investigations in the real implant system must test and validate the design tool's performance.

5. References

[1] Zohrabian VM, Sonick M, Hwang D, Abrahams JJ (2015) Dental implants. Semin Ultrasound CT MR 36:415–426

[2] Jenny G, Jauernik J, Bierbaum S, Bigler M, Gratz KW, Rucker M, Stadlinger B (2016) A systematic review and meta-analysis on the influence of biological implant surface coatings on periimplant bone formation. J Biomed Mater Res A 104:2898–2910

[3] Shemtov-Yona K, Rittel D (2015) An overview of the mechanical integrity of dental implants. Biomed Res Int 2015:547384

[4] C. E. Misch, "Generic root form components terminology," in Contemporary Implant Dentistry, pp. 26–38, Mosby, St. Louis, Mo, USA, 3rd edition, 2008.

[5] Dahibhate, R.V., Jaju, S.B., Sarode, R.I., Design Improvements in Conventional Drilling Machine to Control Thermal Necrosis during Orthopaedic Surgeries, IOP Conference Series: Materials Science and Engineering, 2020, 1004(1), 012015

[6] T. Berglundh, L. Persson, and B. Klinge, "A systematic review of the incidence of biological and technical complications in implant dentistry reported in prospective longitudinal studies of at least 5 years," Journal of Clinical Periodontology, vol. 29, supplement 3, pp. 197–212, 2002.

[7] L. De Boever, K. Keersmaekers, G. Vanmaele, T. Kerschbaum, G. Theuniers, and J. A. De Boever, "Prosthetic complications in fixed endosseous implant-borne reconstructions after an observations period of at least 40 months," Journal of Oral Rehabilitation, vol. 33, no. 11, pp. 833–839, 2006.

[8] M. S. Tonetti and J. Schmid, "Pathogenesis of implant failures," Periodontology 2000, vol. 4, no. 1, pp. 127–138, 1994.

[9] C. J. Goodacre, G. Bernal, K. Rungcharassaeng, and J. Y. K. Kan, "Clinical complications with implants and implant prostheses," Journal of Prosthetic Dentistry, vol. 90, no. 2, pp. 121–132, 2003.
[10] Dahibhate, Rajesh, V; Jaju, Santosh B, Effect of irrigation mode, volume, medium and drill on heat generation during implant surgery: A review, Published: 2022 in Materials Today: Proceedings, DOI: 10.1016/J.MATPR.2021.09.206
[11] B. E. Pjetursson, D. Thoma, R. Jung, M. Zwahlen, and A. Zembic, "A systematic review of the survival and complication rates of implant-supported fixed dental prostheses (FDPs) after a mean observation period of at least 5 years," Clinical Oral Implants Researc, vol. 23, supplement 6, pp. 22–38, 2012.
[12] Y. Manor, S. Oubaid, O. Mardinger, G. Chaushu, and J. Nissan, "Characteristics of early versus late implant failure: a retrospective study," Journal of Oral and Maxillofacial Surgery, vol. 67, no. 12, pp. 2649–2652, 2009.
[13] Shemtov-Yona, Keren, and Daniel Rittel. "An overview of the mechanical integrity of dental implants." *BioMed research international* 2015 (2015).
[14] Yilmaz, Burak, and Edwin McGlumphy. "A technique to retrieve fractured implant screws." *The Journal of prosthetic dentistry* 105.2 (2011): 137-138.
[15] Williamson, Russell T., and Fonda G. Robinson. "Retrieval technique for fractured implant screws." *The Journal of prosthetic dentistry* 86.5 (2001): 549-550.
[16] Dahibhate, R.V., Jaju, S.B., Sarode, R.I., Development of mathematical model for prediction of bone drilling temperature, Materials Today: Proceedings, 2020, 38, pp. 2732–2736
[17] Kim BJ, Yeo IS, Lee JH, Kim SK, Heo SJ, et al. (2012) The effect of screw length on fracture load and abutment strain in dental implants with external abutment connections. Int J Oral Maxillofac Implants 27: 820-823.
[18] Gooty JR, Palakuru SK, Guntakalla VR, Nera M (2014) Noninvasive method for retrieval of broken dental implant abutment screw. Contemp Clin Dent 5: 264-267.
[19] Yoon JH, Lee H, Kim MY (2015) Safe removal of a broken abutment screw with customized drill guide and rotary instrument: A clinical report. J Prosthodont 25: 170-173.
[20] Canpolat C, Ozkurt-Kayahan Z, Kazazoglu E (2014) Management of a fractured implant abutment screw: A clinical report. J Prosthodont 23: 402-405.
[21] Nergiz I, Schmage P, Shahin R (2004) Removal of a fractured implant abutment screw: a clinical report. J Prosthet Dent 91: 513-517.
[22] Walia MS, Arora S, Luthra R, Walia PK (2012) Removal of fractured dental implant screw using a new technique: a case report. J Oral Implantol 38: 747-750.
[23] Kurt M, Güler AU, Duran İ (2013) A technique for removal of a fractured implant abutment screw. J Oral Im-plantol 39: 723-725.
[24] Shah K, Lee DJ (2016) An alternative approach for the management of fractured implant abutment screws on a mandibular implant-retained overdenture: A clinical report. J Prosthet Dent 115: 402-405.
[25] Nasr HMAE (2018) The use of Dental Operating Microscope for Retrieval of different types of Fractured Implant Abutment Screws: Case Reports. Dentistry 8: 507. doi:10.4172/2161-1122.1000507
[26] Oh, Ji-hyeon. "Recent advances in dental implants." *Maxillofacial plastic and reconstructive surgery* 39.1 (2017): 1-10.
[27] Shedmake, Swapnil, Nilesh Awate, and Pramod Walke. "The dental abutment screw and new methods to remove fractured abutment screw from jawbone: A review." *Materials Today: Proceedings* (2022).
[28] SnehaH.Gaonkar et al. (2019), Retrieval of a Fractured Abutment Screw from an Implant: A Case Report. Int J Dent & Oral Heal. 5:3, 27-29

[29] Walia, Manpreet S., et al. "Removal of fractured dental implant screw using a new technique: a case report." *Journal of Oral Implantology* 38.6 (2012).

[30] Taira, Yohsuke, and Takashi Sawase. "A modified technique for removing a failed abutment screw from an implant with a custom guide tube." *Journal of Oral Implantology* 38.2 (2012): 165-169.

[31] Hong, Do Gia Khang, and Ji-hyeon Oh. "Recent advances in dental implants." *Maxillofacial plastic and reconstructive surgery* 39.1 (2017): 1-10.

[32] Shafie, Hamid R., and Scott Martyna. "Retaining abutment screws." *Clinical and Laboratory Manual of Dental Implant Abutments* (2014): 23-32.

[33] Dahibhate, R.V., Jaju, S.B., Study of influence of tool geometry and temperature on bone substructure to reduce bone drilling injury, Journal of Physics: Conference Series, 2021, 1913(1), 012114

IOP Conf. Series: Materials Science and Engineering 1259 (2022) 012028 doi:10.1088/1757-899X/1259/1/012028

Performance Investigation of Cylindrical Cavity Receiver Using Roughened Surfaces

Kamuju Naveen Kumar, Akanksha Maurya and Deepak Sharma*

Mechanical Engineering Department, National Institute of Technology Hamirpur, Hamirpur 177005, H.P., India.

*Corresponding Author: dsharma@nith.ac.in

Abstract. Solar cavity receiver is an integral part of the solar parabolic dish collectors used for absorbing concentrated solar irradiation. The performance of the receivers can be enhanced utilizing different types of heat transfer fluids (HTF) and modified geometries. The thermal performance of a redesigned cavity receiver with solid semi-circular tube attached to its surface was investigated using CFD three-dimensional simulation. The influence of operating and geometric parameters (i.e., inlet mass flow rate and inlet temperature) were investigated. The results obtained shows that, redesigned solar cavity receiver has thermal efficiency of 8.49-10.27% higher than a normal cavity receiver with an inlet flow rate of 0.01-0.04 kg/s under the same environmental and operating conditions. The temperature of outlet of HTF increases with increases in inlet temperature for Therminol66 oil. The comparison of thermal performance using different HTF (i.e., Therminol66, Ethylene Glycol, Water) is also conducted and it is found that Therminol66 oil gives 9% and 17% higher outlet temperature than ethylene glycol and water respectively at intake flow rate of 0.02 kg/s.

Keywords. Therminol 66, Cylindrical cavity receiver, helical tubes, variable heat flux, thermal performance.

Nomenclature

D	Diameter of the cavity receiver (mm)	T_s	Temperature of receiver's exterior wall, (K)
H	Height of the cavity receiver (mm)	T_∞	ambient temperature, (K)
OD	Outer diameter of the helical tube (mm)	T_{in}	Inlet temperature of HTF, (K)
ID	Inner diameter of the helical tube (mm)	T_{out}	Outlet temperature of HTF, (K)
N	No. of turns in helical coil tube	C_p	Specific heat of fluid
Q_{ab}	Heat energy absorbed by HTF (W/m^2)	K	Thermal conductivity of the HTF, [W/(m K)]
Q_{net}	Net heat supplied to cavity receiver (W/m^2)	$\dot{m}$	Inlet flow rate (kg/s)
η	Thermal Efficiency	A	Area of the cavity receiver

1. Introduction

It is well known that that recent energy issues such as global warming, rising energy consumption, fossil fuel depletion, and higher electricity prices have sparked increased interest in renewable energy research. In the domain of renewable energy, solar energy plays a vital role since it is abundantly available. It can be transformed immediately into electrical energy and also into thermal energy directly. Solar parabolic dish concentrators are among the most efficient methods of converting solar energy to thermal energy. In this system, solar radiation falls on the surface of a solar parabolic dish collector and is reflected to a cavity receiver located at the collector's focal point, where working fluid flowing through the solar cavity receiver absorbs the energy from concentrated solar rays falling on the receiver and this led to an increase in thermal energy. Solar collector and Cavity receiver play a crucial role in this process.

Furthermore, various investigations on cavity receiver optimization have been conducted. The receiver must be designed in a way that it has a low heat loss and high efficiency. Various cavity receiver forms have been studied in the past, including rectangular, conical, spherical, and cylindrical. The most prevalent cavity receiver shape in them is cylindrical, and several studies have been conducted into this configuration. R. Loni et al. [1] studied a solar-powered organic Rankine cycle, heat delivery system with a tubular cavity receiver. The major objective of this study was to improve the cavity receiver's efficiency. Zou chongzhe et al. [2] has investigation explored solar thermal conversion utilising surface partition technique for the cylindrical cavity receiver. The optimum absorption efficiency, according to this study is 87.44%. It also discusses about the methods to prevent hot melt phenomena. K.S. Reddy et al. [3] created a 2D and 3D model of a modified cavity receiver utilised with a solar dish concentrator and investigated its natural convection loss. M. Prakash [4] studied natural convection loss in cylindrical cavity receiver and discovered a new correlation that would be utilised to determine natural convection in cavity receiver systems. K. Kalidasa Murugavel et al. [5] carried out research on different shapes of solar cavity receivers to find their convection loss and found that spherical shape cavity receiver is the most efficient among them because the air movement inside cavity is stagnant, and the temperature impacted region is also less than other shape's cavity receivers. N. Sendhil Kumar et al. [6] investigated several forms of cavity receivers and created a modified cavity with lower convection loss than previous receivers. Djelloul Azzouzi et al. [7] studied a cylindrical cavity receiver and did an experimental and numerical thermal study to assess the impact of various parameters on the overall performance of the cavity receiver, as well as the total heat lost by it. R. Loni et al. [8] has experimentally studied the average thermal efficiency of the cylindrical and cubical cavity receivers with dish concentrator in steady-state phase and found as 56.44% and 65.14% respectively when receivers were employed with thermal oil as heat transfer fluid. R. Loni et al. [9] has conducted experimental test on different shape of cavity receiver with different nanofluids and found that using nanofluids as working fluid has enhanced thermal efficiency by 1.44% for cylindrical and by 12.90% for hemispherical cavity receiver. Based on CFD 3D analysis and Monte Carlo Ray Trace (MCRT) principles. Qianjun Mao et al. [10] has examined the cavity receiver with solar parabolic concentrator system to determine the radiation flux on it. The MCRT method approach was used to simulate the distribution of variable radiation heat flux on cavity receiver's wall, and the influence of incident solar irradiation on aspect ratio are analysed. It is found that at constant incident solar irradiance, the best receiver aspect ratio is found to be 1.5. Ding Wang et al. [11] examined the Conical cavity receiver's thermal performance with supercritical CO_2 as working fluid and determined how geometric and operating parameters affected it. Fuqiang Wang et al. [12] has examined conical cavity receiver's and calculated its optical efficiency using numerical analysis. It was revealed that a cylindrical receiver has a greater optical efficiency than of a conical receiver. Lan Xiao et al. [13] examined the optical and thermal conversion parameters of a helical coiled tube cavity receiver. He discovered that when the tube's diameter rises, the optical and thermal efficiency decreases. Under no-wind conditions, the ideal aspect ratio for maximising solar radiation utilisation and producing the highest $\eta_{opt-the}$ is 1.8. The best or optimal value will, however, vary openly as the wind speed increases. Walid Chakroun et al. [14] simulated semi-cylindrical solar cavity receiver for smooth and rough surfaces at various tilt angles, and the heat transfer in the receiver was investigated. Roughness was discovered to increase turbulence in the flow which results in a rise of heat transfer. Santosh Bopche et al. [15] designed a modified cavity receiver with conical copper fins of various pitches and conducted experiments in medium and high heat conditions. As a result, on smooth cavity receiver in high heat circumstances, performance was improved by 31.35% percent and 23.56% in medium heat circumstances.

The review of literature indicates, a lot of research has been done on using different shapes of cavity receiver, optimising their performance, and determining the convective losses. There is also research done using different types of Heat transfer fluid (HTF) but there is less research done on modification of cavity receiver shape to improve its performance. In this research paper, a unique design has made by surface modification which has never been tested in previous studies and its thermal performance is tested. In this research, study is conducted using a modified cylindrical cavity receiver with semi-circular tubes affixed to its walls to increase the surface area and internal reflections in the receiver and to capture more solar light which increases the thermal efficiency. The simulation is then run at varied intake mass flow rates, using different HTFs, and at different HTF inlet temperatures.

2. Modelling of Computational Domain

In this paper, Ansys software is used as platform for designing and simulation of the model using FLUENT simulation tool. A cylindrical cavity receiver with helically tubes is designed using an Ansys Design Modeller section. The material used for cavity receiver is copper.

Table 1. Parameters for the computational domain

S. No.	Parameters	Values
1.	Outer diameter of the helical tubes	0.012 m
2.	Inner diameter of the helical tubes	0.011 m
3.	Height of the receiver	0.18 m
4.	Inner diameter of cavity receiver	0.14 m
5.	Outer diameter of cavity receiver	0.164 m
6.	Number of turns	13

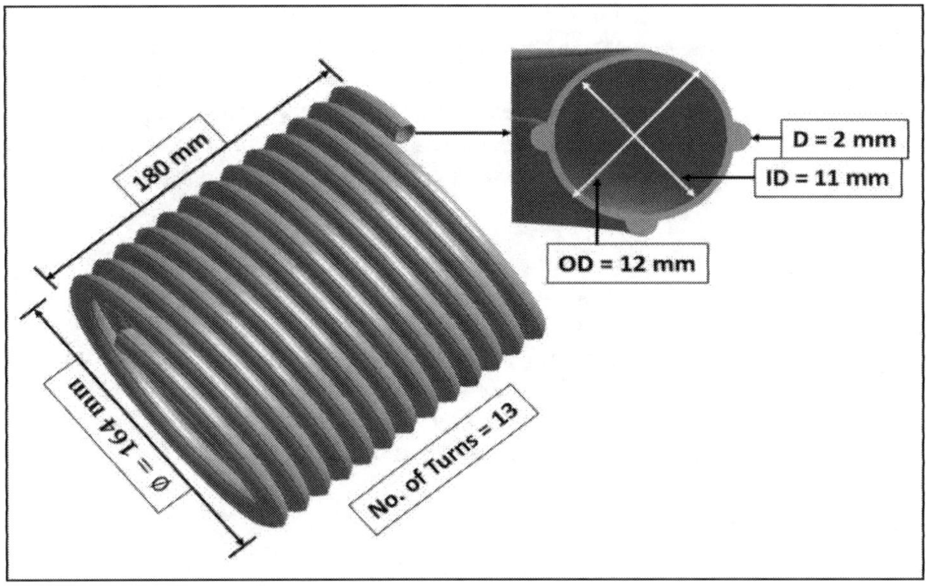

Figure 1. Computational Domain of Modified Cavity Receiver

IOP Conf. Series: Materials Science and Engineering 1259 (2022) 012028 doi:10.1088/1757-899X/1259/1/012028

In this simulation analysis, the effect of gravity is considered and effect on natural convection loss are not considered. Sun rays reflected by the parabolic dish concentrator enters cavity receiver, which is located near the solar concentrator's focal point, through the aperture opening. These rays are subsequently scattered within the cavity receiver, resulting in internal reflections. This phenomenon is responsible for the fluctuating radiative heat flux on solar cavity receiver's exterior (Helical tube) wall. To produce accurate simulation results, fluctuating heat flux is applied to the exterior wall of the receiver in this simulation., and the values of this variable heat flux are taken from Lan Xiao et al. [9].

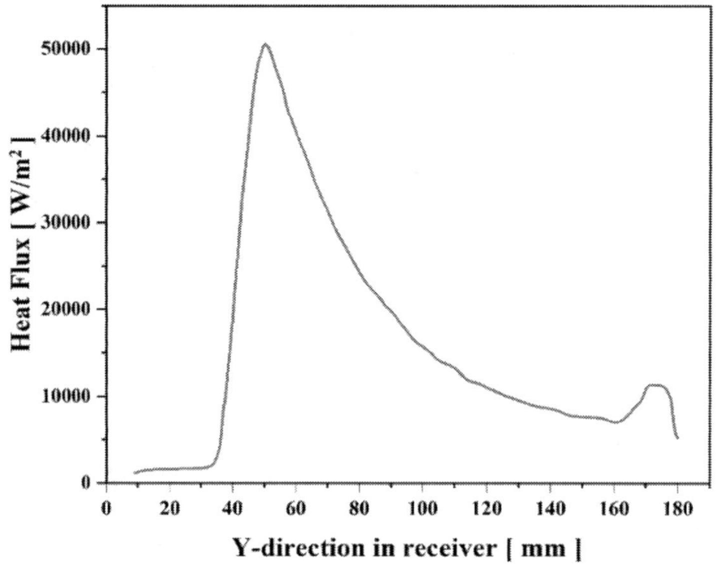

Figure 2. Variable heat flux distribution on receiver's exterior wall

The fluctuating variable heat flux distribution over the cavity receiver's length (Y-direction) with DNI of 1000 W/m^2 is illustrated in map of the variable heat flux distribution put on the exterior wall of modified solar cavity receiver is as shown in Figure.2. It is found from Lan Xiao et al. [9] that the heat flux density is very low when $Y < 35$ mm since there is no direct light irradiation hitting that region. The heat flux density progressively increases from 35mm to 50mm, and at $Y = 50$ mm, the peak heat flux density is created because direct light irradiation reflected from the solar concentrator strikes the receiver here. From $Y > 50$ mm, the heat flux density steadily decreases again.

2.1 Model Validation

For the model validation, the model same as Lan Xiao et al. [9] simulation is designed and under the same conditions, using the same materials and HTF the simulation is conducted. Figure.3 shows the comparison of the results obtained by Lan Xiao et al. research and the model that is designed. It is found that variation between the results from the model designed and the Lan Xiao et al. work is less than 4% so this model can be considered as valid. The same model is taken and further modified to create a new design with solid tubes attached to it along the length of cylindrical cavity receiver on helical coiled tube. The parameter of a modified design is as shown in table.1.

IOP Conf. Series: Materials Science and Engineering 1259 (2022) 012028 doi:10.1088/1757-899X/1259/1/012028

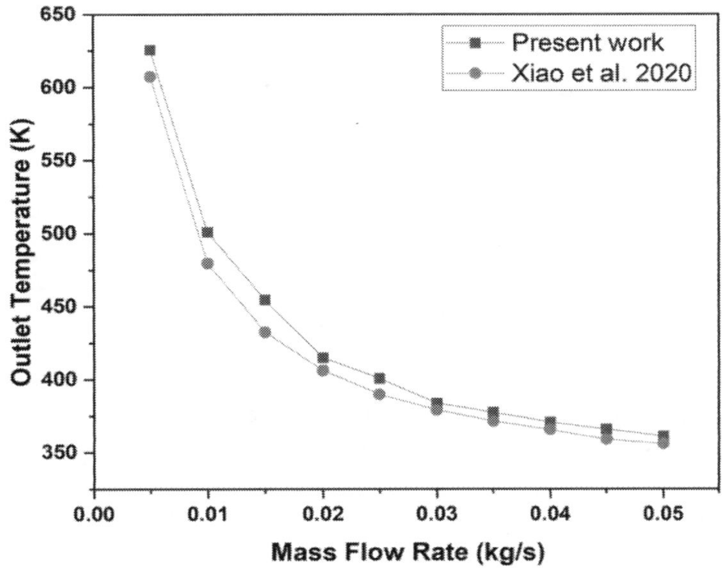

Figure 3. Inlet flow rate Vs Outlet temperature of HTF

2.2 Grid Independent Test

The model is grid independent tested to acquire the best quality mesh for simulation, which balances model efficiency and reduced computing time for the simulation. The model is tested using different mesh element counts in this test.

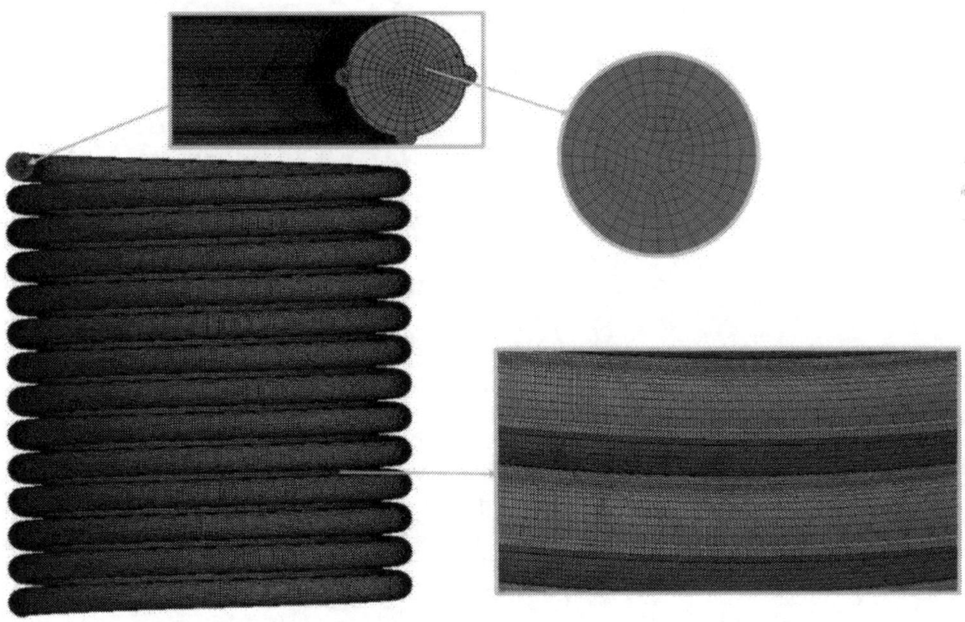

Figure. 4. Meshed Computational Domain of Cavity Receiver

IOP Conf. Series: Materials Science and Engineering 1259 (2022) 012028 doi:10.1088/1757-899X/1259/1/012028

Table 2. Grid Independent Test

Mesh type	No. of elements	Outlet temperature of HTF
Mesh-1	1531641	359.05
Mesh-2	2215709	429.99
Mesh-3	2511673	429.78
Mesh-4	2802736	429.99

From Table.2. It is found that the difference in the output (Outlet temperature of HTF) is significantly less and negligible after mesh-2. Hence, we can consider taking the mesh-2 type for further simulation with number of mesh elements 2215709. The meshing of this type is shown in Figure.4. At this point as the mesh quality is high and even increase in number of elements in mesh has negligible effect on output.

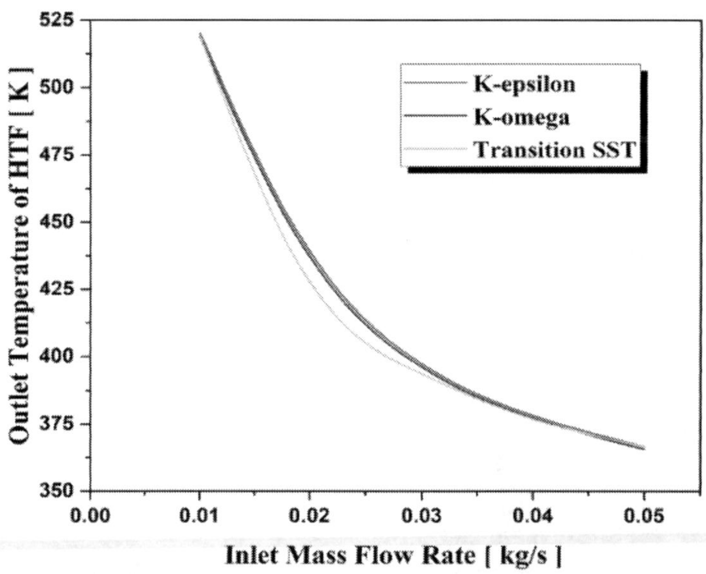

Figure 5. Inlet flow rate Vs HTF's Outlet temperature at different type of turbulence flow

Further the test is conducted on model for obtaining the best type of turbulence flow for simulation. Here, three types of turbulence flow models are taken and tested at different flow rate. The Figure. 5 is plotted using the results of this simulations. From Figure.5 it is found that all types of turbulence flow give similar results, but K-epsilon gives better results than other two type of flow. So, in this paper for further simulation K-epsilon turbulence flow model is selected.

3. Results and Discussion

The model will be now operated under various conditions to obtain the information about different parameters effecting it. Table.3 lists the operating parameters.

IOP Conf. Series: Materials Science and Engineering 1259 (2022) 012028 doi:10.1088/1757-899X/1259/1/012028

Table 3. Operating parameters setting

Parameters	Values
HTF's inlet temperature (K)	320
HTF's inlet flow rate (kg/s)	0.01–0.05
Environment temperature (K)	300
Variable Heat flux (W/m^2)	1050–50960

3.1. Effect of inlet mass flow rate on the heat transfer fluid's outlet temperature

The change in HTF's outlet temperature as a result of change in inlet flow rate is depicted in Figure.6. The heat acquired by HTF diminishes as the intake mass flow rate rises, and change in its output temperature after 0.035 kg/s is very small. It also shows the comparison of a standard cavity receiver to a modified cavity receiver with semi-circular tubes attached along its length. It is found that at any given inlet mass flow rate, the modified cavity receiver gives higher outlet temperature than normal cavity receiver. The more time the HTF element takes to pass through cavity receiver, more the heat is gained by HTF from the cavity receiver. Hence the lower flow rate helps to gain more energy from cavity receiver.

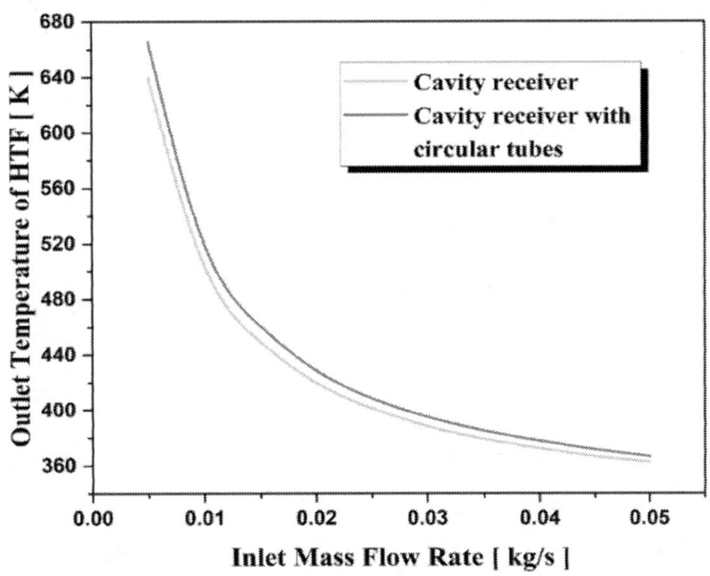

Figure 6. Comparison of outlet temperature of HTF between normal and modified cavity receiver

The relationship between the mass flow rate ($\dot{m}$) and outlet temperature (T_{out}) can be derived using polynomial regression method of order 4 using the results obtained above. The equation formed (1) can give the outlet temperature of modified cavity receiver at any inlet mass flow rate.

$$T_{out} = 4 \times 10^8 \dot{m}^4 - 5 \times 10^7 \dot{m}^3 + 3 \times 10^6 \dot{m}^2 - 56287\dot{m} + 886.72 \quad (1)$$

IOP Conf. Series: Materials Science and Engineering 1259 (2022) 012028 doi:10.1088/1757-899X/1259/1/012028

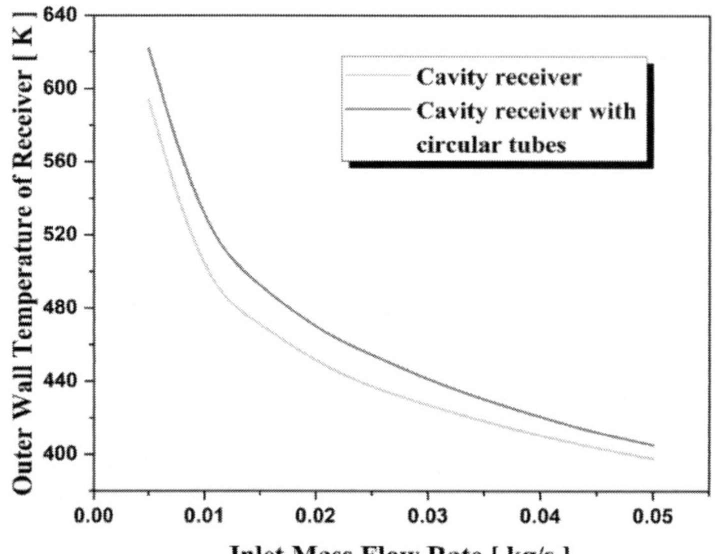

Figure 7. Comparison of outer wall temperature of receiver between normal and modified cavity receiver

The trend of this graph is similar to that of Figure.6. As depicts in Figure.7, the temperature of exterior wall of the cavity receivers decreases as the inlet mass flow rate increases. Because at lower flow rates, the HTF captures more heat from the receiver's wall, and the receiver absorbs more radiation, and at any intake flow rate, the modified receiver's exterior wall temperature is always greater than that of temperature of normal cavity receiver. Figure.8.a is a contour depicting the variation in HTF outlet temperature along its length (Y-direction). As the HTF flows from inlet to outlet, it absorbs heat from the receiver's wall, and at the outlet plane the HTF's maximum temperature can be determined. Figure.8.b depicts the temperature fluctuation on the exterior wall on the helical tube of receiver along its length (in Y-direction). It is discovered that the maximum heat flux zone at 50mm in the Y-direction is also the location where the maximum temperature on the outer wall of cavity receiver is formed.

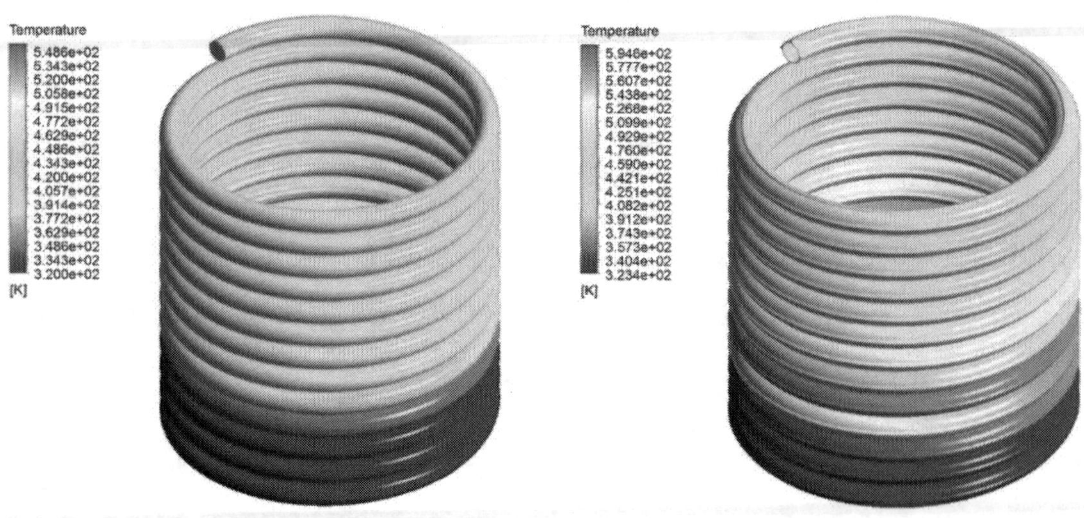

Figure 8. (a) Contour of Fluid domain **(b)** Contour of Cavity receiver with circular tubes

3.2. *Effect of Intake flow rate on Thermal Efficiency*

At varied intake flow rates, thermal efficiency of a standard solar cavity receiver and modified receiver with semi-circular tube attached along its length is calculated in this simulation.

Heat energy gained by HTF, Q_{ab}

$$Q_{ab} = \dot{m} \times C_p \times (T_{out} - T_{in}) \tag{2}$$

Weighted average of variable heat flux applied on cavity receiver, q_s

Net heat energy supplied to Cavity receiver, Q_{net}

$$Q_{net} = A \times q_s \tag{3}$$

Thermal efficiency, η

$$\eta = \frac{Q_{ab}}{Q_{net}} \tag{4}$$

The efficiencies at varied inlet flow rates for both the cavity receivers are calculated and depicted in Figure.9, and it was revealed that the cavity receiver's efficiency improves as the incoming mass flow rate rises. After 0.035 kg/s intake flow rate, the improvement in thermal efficiency of both the cavity receivers is quite minimal as the flow rate rises further. The modified receiver with a semi-circular tube attached has greater efficiency than that of standard cavity receiver with the same dimensions at any given intake mass flow rate.

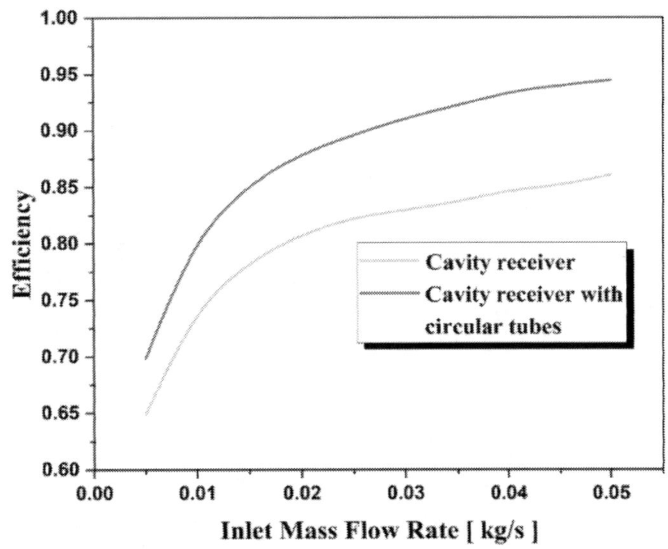

Figure 9. Inlet mass flow rate Vs Efficiency of cavity receiver

3.3. *Effect of heat transfer fluid's inlet temperature on its outlet temperature*

Because the HTF's inlet temperature isn't always consistent, the relationship between the HTF's inlet temperature supplied and its outlet temperature collected from the cavity receiver is shown in Fig.10. Therminol-66 oil is used as the HTF in this test with an inlet flow rate 0.02 kg/s. A graph is plotted using different intake temperatures of HTF, and it is discovered that outlet temperature of the HTF increases as HTF's inlet temperature rises.

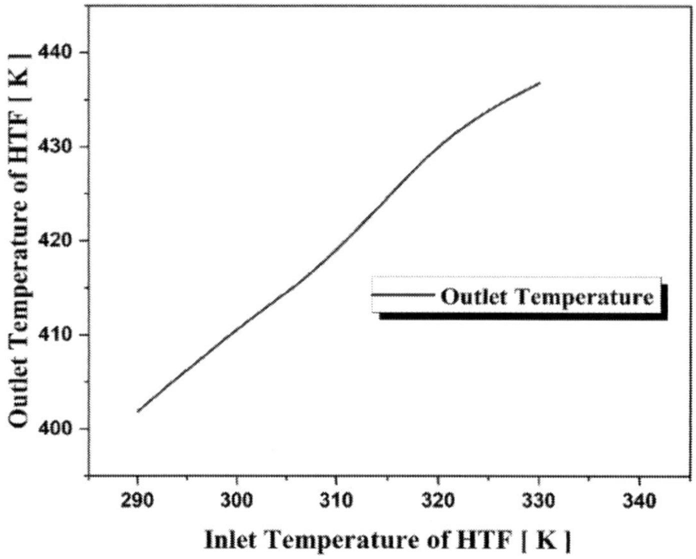

Figure 10. Inlet Temperature Vs Outlet Temperature of HTF

3.4. Comparison effect on outlet temperature by using various heat transfer fluids

In this simulation, three different HTF (i.e., Therminol 66, Ethylene Glycol, Water) were used in the modified cavity receiver at varied intake flow rates and HTF's outlet temperature was found at these conditions and shown in Figure.11. It is observed that Therminol 66 gives high outlet temperature at any given inlet mass flow rate as it is highly thermal stable and can absorb more heat as compared to water and ethylene glycol. The change in outlet temperature of all HTF gradually decreases with rise in intake mass flow rate and after 0.035kg/s the change in outlet temperature becomes even smaller.

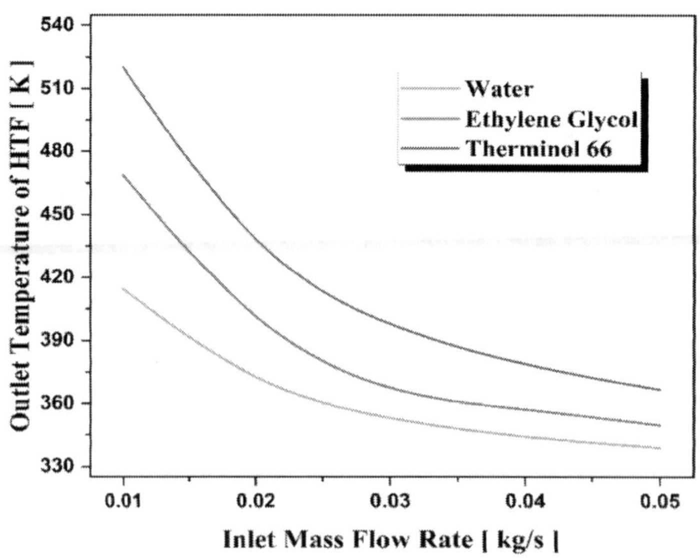

Figure 11. Inlet flow rate Vs Outlet temperature for different HTF

4. Conclusions

An extensive thermal analysis is conducted on modified cylindrical cavity receiver with semi-circular solid tubes attached to it. CFD methods were used for the simulation. The effect of surface modification and two operating factors on thermal performance, namely intake flow rate and the HTF's inlet temperature in cavity receiver, has been examined. The most important findings are summarised below as follows.

- The modified cavity receiver with semi-circular tubes attached outperforms the standard cavity receiver of the same shape. At 0.01 kg/s, 0.02 kg/s, 0.03kg/s, 0.04kg/s, and 0.05 kg/s intake flow rate, the outlet temperature of HTF of modified cavity receiver is 3.1 %, 2.1%, 1.72%, 1.44%, and 1.14% higher than the outlet temperature of standard cavity receiver.
- At any intake flow rate, the thermal efficiency of a modified cavity receiver is always greater than a standard cavity receiver, and thermal efficiency in both cavity receivers improves as HTF's inlet mass flow rate increases. In the comparison to rise in HTF's inlet flow rate, the rise in its thermal efficiency is quite small after 0.035 kg/s.
- As the inlet temperature of HTF (Therminol 66) in the cavity receiver rises, the HTF's outlet temperature rises as well.
- In comparison of outlet temperature between different HTF's. The Therminol 66 oil is found out to be best HTF with high thermal stability. The outlet temperature of Therminol 66 oil is 17% and 9% higher than the outlet temperature of Water and Ethylene Glycol respectively, at 0.02 kg/s inlet mass flow rate.

5. References

[1] Loni, R., Kasaeian, A.B., Asli-Ardeh, E.A. and Ghobadian, B., 2016. Optimizing the efficiency of a solar receiver with tubular cylindrical cavity for a solar-powered organic Rankine cycle. *Energy*, *112*, pp.1259-1272.

[2] Chongzhe, Z., Yanping, Z., Falcoz, Q. and Neveu, P., 2022. Solar-thermal conversion investigation using surface partition method for a cavity receiver with helical pipe. *Energy*, *242*, p.122943.

[3] Reddy, K.S. and Kumar, N.S., 2009. An improved model for natural convection heat loss from modified cavity receiver of solar dish concentrator. *Solar Energy*, *83*(10), pp.1884-1892.

[4] Prakash, M., 2014. Numerical study of natural convection heat loss from cylindrical solar cavity receivers. *International Scholarly Research Notices*.

[5] KalidasaMurugavel, K., Vivekanandan, M., Kumaran, E.T., Raj, R.S., Shriram, V. and Pandian, M.S., 2021. Comparative study on convection loss of different shape solar dish collector cavity receivers at different positions using CFD. *Materials Today: Proceedings*, *37*, pp.3091-3095.

[6] Kumar, N.S. and Reddy, K.S., 2008. Comparison of receivers for solar dish collector system. *Energy Conversion and Management*, *49*(4), pp.812-819.

[7] Azzouzi, D., Boumeddane, B. and Abene, A., 2017. Experimental and analytical thermal analysis of cylindrical cavity receiver for solar dish. *Renewable Energy*, *106*, pp.111-121.

[8] Loni, R., Kasaeian, A.B., Asli-Ardeh, E.A., Ghobadian, B. and Gorjian, S., 2018. Experimental and numerical study on dish concentrator with cubical and cylindrical cavity receivers using thermal oil. *Energy*, *154*, pp.168-181.

[9] Loni, R., Asli-Areh, E.A., Ghobadian, B., Kasaeian, A.B., Gorjian, S., Najafi, G. and Bellos, E., 2020. Research and review study of solar dish concentrators with different nanofluids and different shapes of cavity receiver: Experimental tests. *Renewable Energy*, *145*, pp.783-804.

[10] Mao, Q., Shuai, Y. and Yuan, Y., 2014. Study on radiation flux of the receiver with a parabolic solar concentrator system. *Energy Conversion and Management*, *84*, pp.1-6.

[11] Wang, D., Chen, Y., Xiao, H. and Zhang, Y., 2022. Effects of geometric and operating parameters on thermal performance of conical cavity receivers using supercritical CO2 as heat transfer fluid. *Renewable Energy*, *185*, pp.804-819.

[12] Wang, F.Q., Tan, H.P. and Shuai, Y., 2012. The numerical investigation on optical efficiency and heat flux distribution of conical cavity receiver. In *Advanced Materials Research* (Vol. 347, pp. 1530-1536). Trans Tech Publications Ltd.

[13] Xiao, L., He, S., Wu, S.Y. and Chen, Z.L., 2020. Optical-thermal conversion characteristics of cylindrical receiver with built-in helically coiled tubes. *Sustainable Energy Technologies and Assessments*, *37*, p.100626.

[14] Chakroun, W. and Quadri, M.M.A., 2002. Heat transfer measurements for smooth and rough tilted semi-cylindrical cavities. *International journal of thermal sciences*, *41*(2), pp.163-172.

[15] Bopche, S., Rana, K. and Kumar, V., 2020. Performance improvement of a modified cavity receiver for parabolic dish concentrator at medium and high heat concentration. *Solar Energy*, *209*, pp.57-78.

[16] Yan, J., Peng, Y.D. and Cheng, Z.R., 2018. Optimization of a discrete dish concentrator for uniform flux distribution on the cavity receiver of solar concentrator system. *Renewable energy*, *129*, pp.431-445.

[17] Xiao, L., Guo, F.W., Wu, S.Y. and Chen, Z.L., 2020. A comprehensive simulation on optical and thermal performance of a cylindrical cavity receiver in a parabolic dish collector system. *Renewable Energy*, *145*, pp.878-892.

[18] Harris, J.A. and Lenz, T.G., 1985. Thermal performance of solar concentrator/cavity receiver systems. *Solar energy*, *34*(2), pp.135-142.

[19] Hassan, A., Quanfang, C., Abbas, S., Lu, W. and Youming, L., 2021. An experimental investigation on thermal and optical analysis of cylindrical and conical cavity copper tube receivers design for solar dish concentrator. *Renewable Energy*, *179*, pp.1849-1864.

[20] Barbero, R., Rovira, A., Montes, M.J. and Val, J.M.M., 2016. A new approach for the prediction of thermal efficiency in solar receivers. *Energy Conversion and Management*, *123*, pp.498-511.

[21] Li, H., Huang, W., Huang, F., Hu, P. and Chen, Z., 2013. Optical analysis and optimization of parabolic dish solar concentrator with a cavity receiver. *Solar energy*, *92*, pp.288-297.

[22] Forristall, R., 2003. *Heat transfer analysis and modeling of a parabolic trough solar receiver implemented in engineering equation solver* (No. NREL/TP-550-34169). National Renewable Energy Lab., Golden, CO.(US).

[23] Li, S., Xu, G., Luo, X., Quan, Y. and Ge, Y., 2016. Optical performance of a solar dish concentrator/receiver system: Influence of geometrical and surface properties of cavity receiver. *Energy*, *113*, pp.95-107.

[24] Qiu, Y., Li, M.J., He, Y.L. and Tao, W.Q., 2017. Thermal performance analysis of a parabolic trough solar collector using supercritical CO2 as heat transfer fluid under non-uniform solar flux. *Applied Thermal Engineering*, *115*, pp.1255-1265.

[25] Zou, C., Zhang, Y., Feng, H., Falcoz, Q., Neveu, P., Gao, W. and Zhang, C., 2017. Effects of geometric parameters on thermal performance for a cylindrical solar receiver using a 3D numerical model. *Energy Conversion and Management*, *149*, pp.293-302.

Welding defect detection using artificial neural network and support vector machine

[1]**Swapnil Gundewar**0000-0002-9510-8625, [1]**Prasad Kane**0000-0002-1250 9846, [1]**Santosh Behara**0000-0003-3595-3699, **and** [1]**Uttam Kumar**

[1]Department of Mechanical Engineering, Visvesvaraya National Institute of Technology, Nagpur, India
Email: swapnilfundewar32@gmail.com

Abstract. Welding is an important operation in manufacturing that finds wide applications while joining components. Many destructive and non-destructive techniques are applied to ensure the quality of welded joints. In this paper, an attempt is made to apply the vibration-based technique along with the Artificial Neural Network (ANN) and Support Vector Machine (SVM) to classify the defects of welded joints. Features datasets extracted from the acquired vibration signals during experimentation on the test's samples fabricated with and without defects are applied to pattern recognition techniques for fault identification. The accuracy of detection of defects using ANN is found to be 90.1% while for SVM it is found to be 92.85% for test datasets. The accuracy of classification obtained for the detection and classification of defects is found to be encouraging demonstrating the suitability of the proposed vibration-based approach to the development of a decision support system for non-destructive testing for defect identification.

Keywords: Welding joint defects, Vibration signal, Artificial neural network, Support vector machine

1. Introduction

Welding is an important operation in manufacturing and found a wide application in the fabrication and assembling process. It is necessary to ensure the quality of the welded joints. Many destructive and non-destructive techniques are applied to ensure the quality of welded joints. There are many faults like cracks, blowholes, porosity, and slag inclusion that may get developed during the welding process [1, 2]. The monitoring of welded joints is crucial to ensure their safety and reliability of welded joints.

This conventional NDT is used to assess the level of damage, the most popular is an ultrasonic technique or radiographic technique [3, 4]. The major limitation with these techniques is that it needs the expertise of a human operator with experience to interpret the results of these techniques. The results of these tests are therefore dependent on the skill and experience of the operator. The test becomes monotonous and time-consuming and it is based on an individual and there is subjectivity which may lead to biased results. Hence to overcome these efforts are made to apply ANN and SVM in this work.

Vibration-based techniques are also getting popularity due to their ability to capture the dynamic behaviour of components. The vibration is the result of the dynamic behaviour of the welded joint that can be captured defects if any and hence it is selected in this work for fault identification [5]. An attempt is made to apply vibration-based techniques along with artificial intelligence techniques to classify the faults of welded joints. The ability of different pattern recognition techniques is compared to the vibration features datasets that are acquired during experimentation on the test's samples.

The review of literature is carried out to study the methodology to be followed. Suyama F. M. et al. [6] applied Deep Neural Network (DNN) models along with the radiographic image features as input for fault identification in welding joints. The visual Recognition Challenge model (VGG-VD-16) is proposed by authors who achieved the highest average F-score of 96.00% with extended inputs. Amiri, N et al. [7] reported the application of the machine learning method to predict the fatigue behaviour of the spotwelded joints using ultrasonic techniques. Kumar, M.G. et al. [8] applied the decision tree algorithm and Naive Bayes algorithm for fault diagnosis of welded joints where the classifier gave an accuracy of 89.38% for fault identification. Luo, H et al. [9] and Maurya M et al. [10] applied an artificial neural network to identify faults in the welding process. A combination of these with some of the most effective and optimized techniques such as neural networks and fuzzy inference systems in collaboration with intelligent artificial systems is widely used. Riaz, S et al. [11] elaborated on the application of pattern rection techniques in the rotating machine where the vibration features are the input to the classifier. With this review, vibration-based techniques along with pattern recognition techniques are applied to identify defects in the welding joint. The objective of the proposed technique is to provide a technique which can identify the defects present in the welding using vibration monitoring and supervised learning techniques.

2. Experimental setup and procedure

The experimental setup used to simulate various faults is shown in Fig 1. It consists of a Mounting table for clamping the welded joint specimen with fault and without fault; also, an impact ball is used for giving the impact on the welded joint specimen. An accelerometer is connected with a data acquisition system (DAQ) system for acquiring the vibration signal. The DAQ card is connected with a laptop where they give output through LabVIEW as interfacing software for data acquisition. With the help of software, the program with a block diagram is developed to obtain vibration and acoustic signals to extract signal statistical features and psychoacoustic features in healthy and faulty conditions. The test piece used in this work is a corner joint structure that is fabricated. It was fabricated with mild steel plates having measurements of 100 mm × 50 mm × 6 mm. Welding faults (lack of penetration and porosity) were produced during the welding progression as shown in Fig. 3. Also, in the weld area of the corner joint structure, notches having depths 10%, 5%, and 2.5% of wall thickness were made at the outer or inside of the corner joint

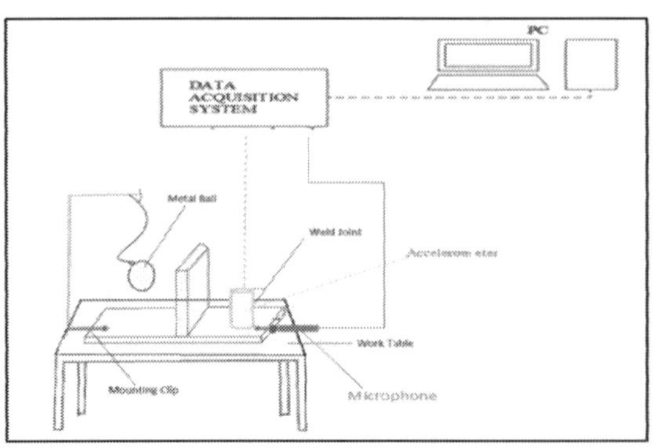

Figure 1. The layout of the Project setup

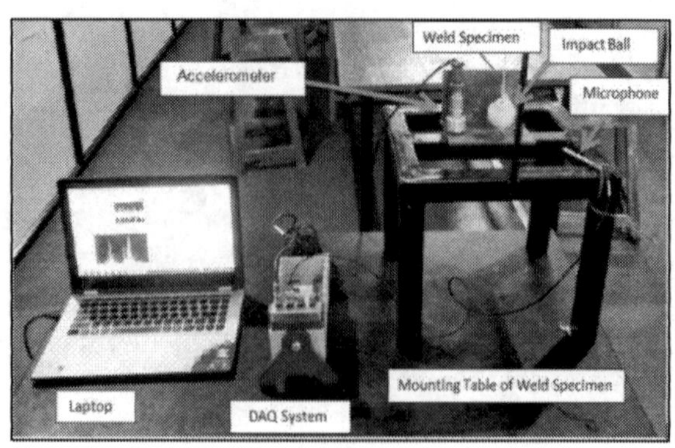

Figure 2. Experimental setup

Three welded joint samples were fabricated as shown in Fig. 3. The first welded joint specimen is in a healthy condition without defects welded in a controlled environment. The second specimen (fault level 1) was welded with lack of penetration and porosity defect. The third specimen (fault level 2) was welded with the inclusion of foreign particles and water in the joint. The specimen was prepared from mild steel with standard dimensions for all three specimens. The vibration signals were recorded for healthy and simulated faulty conditions of welded joints.

The experiments were performed using a table which was used for the mounting purpose of welding plate's joints. A metal ball was hung with the table of a certain height for giving impact force with it. Samples of vibration signals were acquired from healthy conditions and for each faulty condition. Signal for 2 sec time interval was acquired when impact ball hits specimen which is shown in Fig 2. Signals are acquired in the time domain for healthy welded joints and other faulty welded joints.

Figure 3. Welded Specimen in healthy and faulty conditions

3. Methodology

The methodology followed in this work is as shown in Fig. 4. The specimens used in this experimental setup NI DAQ card are used along with LabView software to acquire vibration signals. The sampling frequency of 20 kHz was maintained. The program is written to extract the statistical features such as root mean square (RMS), kurtosis, skewness, mean, maximum, minimum, etc. Different methods are used to acquire the data. The tools available that are used are vibration analysis as seen in the 3.6 flow map. The most widely used method for diagnosing faults is vibration analysis. The procedure consists of three main steps: acquisition of data, processing of data, and making decisions. Several methods for identifying the meaningful data can be applied to the vibration data. These methods are vibration measurements in the period, frequency, and time-frequency domain.

In this work, the impact is made on the welding specimen using the impact ball. After the impact, vibration signal data is acquired using the vibration sensor. i.e., accelerometer which is connected to the DAQ card. The statistical features are extracted from the acquired vibration signal. These statistical features are used for the training and testing of the supervised learning techniques such as ANN and SVM. Finally, the classification performance for both the classifiers (ANN and SVM) are analysed as shown in the Fig. 4

IOP Conf. Series: Materials Science and Engineering 1259 (2022) 012029 doi:10.1088/1757-899X/1259/1/012029

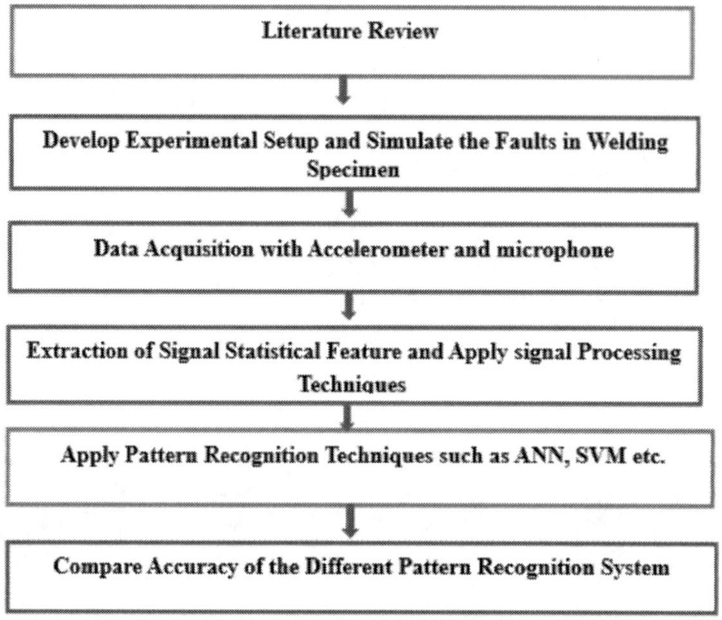

Figure 4. Flow-Chart Fault Identification System

4. Pattern Recognition Techniques

The vibration time-domain signal statistical features are extracted are input to the ANN and SVM which are shown in Fig. 5. In this work, ANN and SVM are used as pattern recognition techniques using MATLAB. The architecture of the selected ANN is shown in Fig. 6. The activation function used in the hidden layer is the tan-sigmoid and log-sigmoid function respectively.

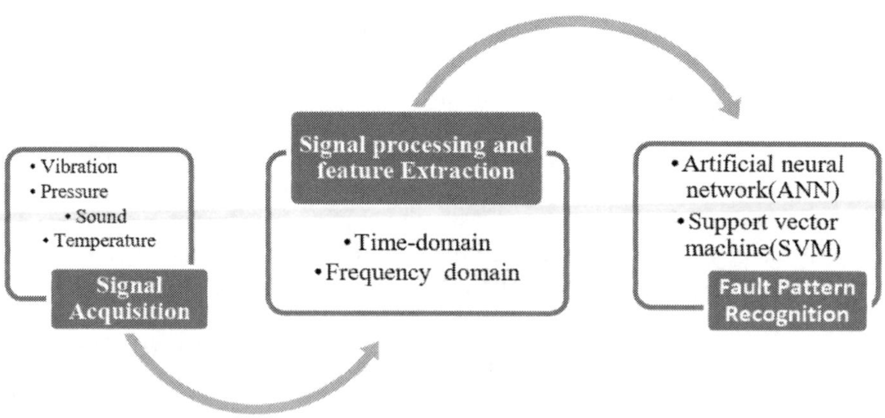

Figure 5. Pattern Recognition Technique- ANN and SVM

The training is found to be successfully obtained from the performance and regression plots. The regression plot shown in Fig.7 shows a satisfactory value of 0.95 which indicates that the output of the trained ANN is in close agreement with the target value.

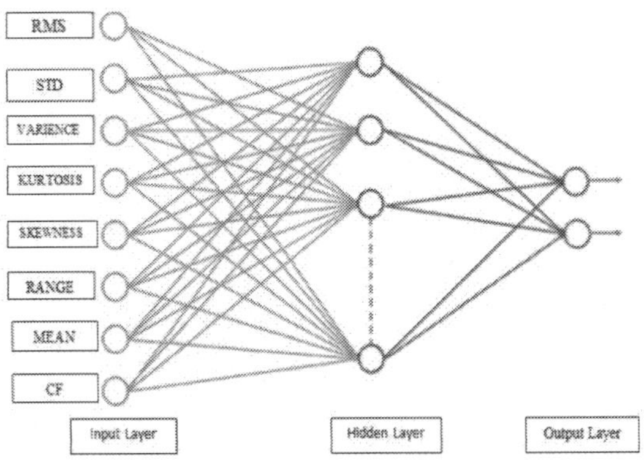

Figure 6. The architecture of the Artificial Neural Network

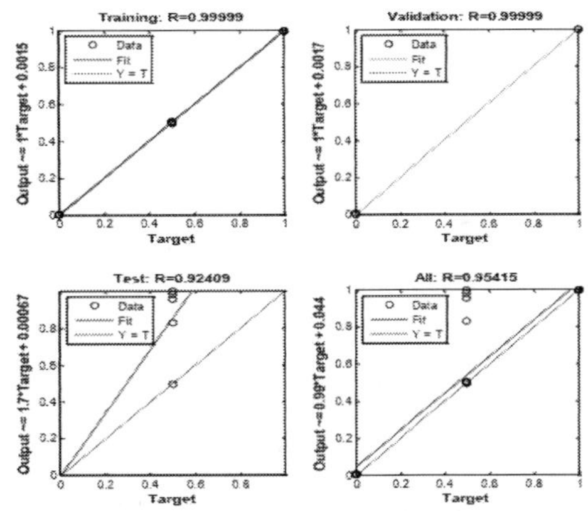

Figure 7. Regression plot of ANN training

Similarly, the support vector machine which is based on the statistical learning framework is also applied to classify the features obtained from the healthy and faulty conditions. In this case, 70% of the dataset is used for training while 30% is used for the testing. The training accuracy of 99% and the testing accuracy of 90.1% are found with the ANN, while SVM has shown training accuracy of 100% and testing accuracy of 92.85% to classify the healthy and faulty conditions of the welded joints.

5. Validation of results with the other testing techniques

The FFT of the acquired signal was plotted as shown in Fig 8. Where the blue signal indicates the FFT of the vibration signal acquired from the healthy specimen and the red FFT plot superimposed on blue indicates the FFT of the vibration signal obtained from the faulty specimen. It is found that FFT is not able to identify faults by looking at the FFT plot. Hence ultrasonic testing of the component was carried out which reported the healthy and faulty condition of the specimen. The test reports obtained were found to be in close agreement with the result of the pattern recognition technique.

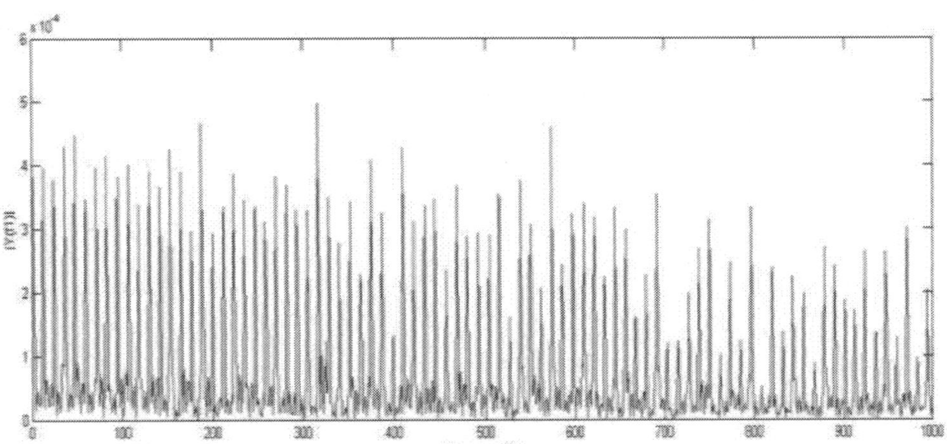

Figure 8. FFT Plot of Vibration Signal of Healthy and Faulty Specimen

6. Conclusions

The vibration-based technique along with the pattern recognition technique is studied for ensuring the quality of the welding joint in this work. The conventional technique found application to inspect the welding joint is ultrasonic testing or radiographic technique. The interpretation of the results obtained from these techniques needs human intervention to identify the presence of a fault. The application of the proposed vibration technique would be advantageous to overcome this issue of human technique when it is applied with a pattern recognition technique. The accuracy of 90.1% is obtained with ANN and 92.85% accuracy is obtained with SVM. The proposed methodology would help to ensure the quality of the welding joint which will enhance the quality and reliability of the product.

References:

[1] Boaretto, N. and T.M. Centeno, Automated detection of welding defects in pipelines from radiographic images DWDI. Ndt & E International, 2017. **86**: p. 7-13.

[2] Wang, B., et al., Intelligent welding system technologies: State-of-the-art review and perspectives. Journal of Manufacturing Systems, 2020. **56**: p. 373-391.

[3] Gong, Y., et al., A deep transfer learning model for inclusion defect detection of aeronautics composite materials. Composite structures, 2020. **252**: p. 112681.

[4] Hou, W., et al., Review on computer aided weld defect detection from radiography images. Applied Sciences, 2020. **10**(5): p. 1878.

[5] Razvarz, S., R. Jafari, and A. Gegov, A review on different pipeline defect detection techniques, in Flow Modelling and Control in Pipeline Systems. 2021, Springer. p. 25-57.

[6] Suyama, F.M., et al., Deep neural networks based approach for welded joint detection of oil pipelines in radiographic images with Double Wall Double Image exposure. Ndt & E International, 2019. **105**: p. 46-55.

[7] Amiri, N., et al., Applications of ultrasonic testing and machine learning methods to predict the static & fatigue behavior of spot-welded joints. Journal of Manufacturing Processes, 2020. **52**: p. 26-34.

[8] Kumar, M.G., et al., Fault diagnosis of welded joints through vibration signals using Naïve Bayes algorithm. Procedia Materials Science, 2014. **5**: p. 1922-1928.

[9] Luo, H., et al., Application of artificial neural network in laser welding defect diagnosis. Journal of Materials Processing Technology, 2005. **170**(1-2): p. 403- 411.

[10] Maurya, M., J. Sadarang, and I. Panigrahi, Detection of crack in structure using dynamic analysis and artificial neural network. Engineering Solid Mechanics, 2020. **8**(3): p. 285-300.

[11] Riaz, S., et al., Vibration feature extraction and analysis for fault diagnosis of rotating machinery-a literature survey. Asia Pacific Journal of Multidisciplinary Research, 2017. **5**(1): p. 103-110.

IOP Conf. Series: Materials Science and Engineering 1259 (2022) 012030 doi:10.1088/1757-899X/1259/1/012030

Enhancement in the teeth cutting operation on gear hobbing machine

Mr. Gaurav Bore[1], Dr. Vinod Bhaiswar[1], Dr. Rupesh Shelke[1], Mr. Navin Verma[2]

[1]Department of Mechanical Engineering, G H Raisoni College of Engineering Nagpur, Digdoh Hills CRPF Gate No 3, Hingna Road, Nagpur, 440016, India.

[2]Production Manager , Mahesh Casting Works, Hingna MIDC Nagpur, 440016.

Corresponding author E-Mail: gauravbore1234@gmail.com

Abstract. Gear is mechanical component mounted on the shaft to transmit the power from one shaft to another. Gear is mostly manufactured by using the hobbing process. In hobbing process the tooth are cut on the gear blank with the help of the indexed hobs. The hobbing process the gear blank and the hob are rotated with each other simultaneously with certain speed. The rotating movement of the hob and the blank creates the cuts and forms the tooth on the gear blank. In the process of the gear hobbing the rotating speed of the hob and rotating speed of the gear blank plays an important role in teeth cutting operation and the formation of the teeth on the gear blank. The rotating speed of the hob also determines the tool life of the hob. As the hob rotates with high speed the wear occurs in the hob due to variation of the cutting force with respect to the cutting speed. This paper will conclude how the variation of the cutting speed of the hob and gear blank affects the cutting of tooth on the gear blank.

Keywords: hob, gear blank, cutting speed.

1. Introduction

Gear hobbing is a process of cutting gears during the production process by simultaneously rotating a gear blank and a cutter called a cutter with a fixed gear ratio between the cutter and the gear blank [18]. During this process, the gear blank is fed towards the rotating cutter until the required depth is reached [4][26]. The cutter is fed along the face of the work piece until all teeth are complete. Gear hobbing is a process of cutting gears during the production process by simultaneously rotating a gear blank and a cutter called a hob with a fixed gear ratio between the cutter and the gear blank [2][3][6][7][25]. Hob tool is multipoint cutting tool on which the multiple cutting teeth are designed. During this process, the gear blank is fed towards the rotating cutter until the required depth is reached [12][13]. The cutter is fed along the face of the work piece until all tooth are complete [11]. No intermediate links are used by the gears to transmit the motion by direct contact. In gear drive two bodies make tangential contact. Gear drive is used when the distance between driver and driven shaft is very less [16][17].
The hob and the gear blank rotates at the certain RPM (revolution per minute) which tends to the cutting operation of the teeth on the blank. The hobbing operation is being carried out for the various types of the gear such as spur gear, helical gear, worm gear, etc [14][15][27]. The angle of the cutting edge of the hob plays an important role in the teeth cutting operation on the gear blank. The angle of the teeth of the hob decides the pitching angle on the gear blank [4][20]. The parameters which are considered while the performing the gear hobbing operation are indexing, angle between the axis of

IOP Conf. Series: Materials Science and Engineering 1259 (2022) 012030 doi:10.1088/1757-899X/1259/1/012030

gear blank and the hob, feed rate [19]. The hobbing process is very widely used for the manufacturing of gear [8][9][10]. It is very conventional process. Gear hobbing process is considered as very economical process because of its continuous action[24][25][26]. In gear hobbing process the teeth are cut properly by indexing so that the accurate module is obtained [5]. In gear hobbing process the fast production rate is obtain because of lower production cycle time[28][29][30]. The various new innovations are carried out in the gear hobbing machine such as dry hobbing process are introduced to reduced the costing of the lubricants for cooling [1][31][32]. In this process various parameters such as tool wear, selection of material used for manufacturing of gear and tool, tool life are considerd [4][11].

2. Problem Statement

The gear hobbing machine cuts only the certain amount of teeth on the gear blank. If two gear blanks of same diameter are mounted on the machine the machine will cut equal number of teeth on the gear blank. If machine operator wants to cut different number of teeth on the gear blank of same diameter and same size then it is not possible on machine.

3. Methodology

In the specific gear hobbing machine in the industry the machine configuration has been set in such a way that the machine cuts only certain number of teeth on the gear blank if the two gear blank were cut on the gear hobbing machine it will cut exact same number of teeth on the gear blank. the main aim to get the variation in the number of teeth on gear blank of the same size. First the specification of the machines has been studied.

Table 1. Table for feed sequence of hob at first position of lever.

Position of Lever			Rolling action of table (mm/rev)		To divide (mm/min)		Tangential (mm/rev) of table at division in		
3C	3A	3B	Vertical	Horizontal	Vertical	Horizontal	Inch	mm	module
			Feed Sequence 1						
		1	3.99	2.49	235.644	147.279	3.62	3.8	4.48
		2	3.16	1.97	186.892	116.808	2.87	3.01	3.55
		3	2.51	1.56	148.224	97.641	2.28	2.39	2.81
		4	1.98	1.23	116.807	73.005	1.79	1.88	2.22
	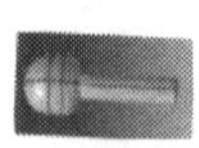	1	1.62	1.01	95.532	59.708	1.47	1.54	1.81
		2	1.28	0.80	75.766	47.354	1.16	1.22	1.44
		3	1.02	0.63	60.091	37.557	0.92	0.97	1.14
		4	0.80	0.80	47.354	29.597	0.73	0.76	0.90
		1	0.64	0.40	37.710	23.569	0.58	0.51	0.72
		2	0.51	0.31	29.908	18.693	0.46	0.48	0.57
		3	0.40	0.25	23.72	14.825	0.36	0.38	0.45
		4	0.32	0.20	18.692	11.683	0.29	0.30	0.36

Table 2. Table for feed sequence of hob at second position of lever

Position of Lever			Rolling action of table (mm/rev)		To divide (mm/min)		Tangential (mm/rev) of table at division in		
3C	3A	3B	Vertical	Horizontal	Vertical	Horizontal	Inch	mm	module
Feed Sequence 2									
		1	0.637	0.384	37.289	23.303	0.572	0.501	0.709
		2	0.499	0.313	29.571	18.482	0.454	0.436	0.562
		3	0.397	0.248	23.452	14.658	0.361	0.378	0.446
		4	0.313	0.195	18.432	11.551	0.285	0.297	0.351
	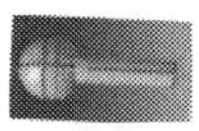	1	0.256	0.160	15.115	9.303	0.232	0.243	0.285
		2	0.202	0.127	11.988	7.493	0.183	0.193	0.229
		3	0.161	0.101	9.307	5.942	0.145	0.15	0.180
		4	0.126	0.073	7.493	4.683	0.116	0.120	0.152
	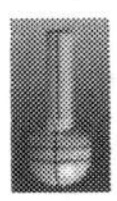	1	0.101	0.063	5.967	3.729	0.092	0.096	0.114
		2	0.080	0.050	4.732	2.958	0.073	0.076	0.090
		3	0.063	0.040	3.753	2.346	0.057	0.060	0.071
		4	0.051	0.031	2.958	1.848	0.048	0.047	0.055

The maximum number of teeth that can be cut on the gear blank as per the module which can be set in the machine is

$$number\ of\ teeth = pitch\ diameter/module \quad (1)$$
$$= 120 / 0.055$$
$$= 2{,}181.8$$

The maximum number of teeth that can be cut on the gear blank of diameter 120 mm is 2,181.8 as per the theoretical calculation. Here the gear pitch diameter is considered as the 120 mm for the experiment. The variation in the number of teeth can be achieved by the carrying out the changes in the indexing process[20][21].

The changes are made in the indexing process in such a way that the variation in number of the teeth cut in the gear blank are to carry out the changes in the gear blank and the hob as per the design conditions of them respectively. The hob in which pitch of the hob is 4mm the distance between the two successive teeth of the hob is 4mm. The gear blank should revolve at the angle of the 4 degree for the proper pitching as well as the for smooth gear cutting operation.

IOP Conf. Series: Materials Science and Engineering 1259 (2022) 012030 doi:10.1088/1757-899X/1259/1/012030

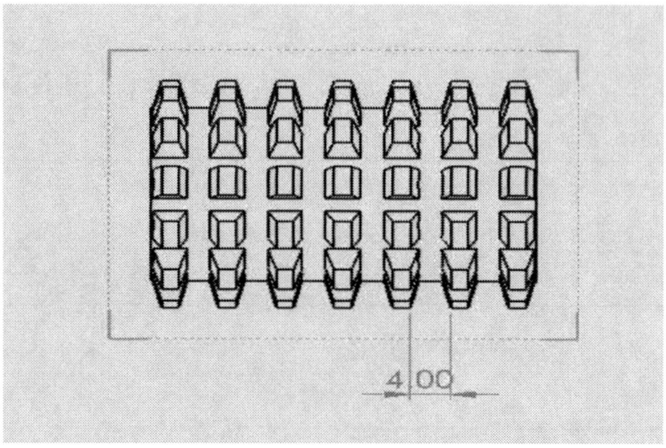

Figure 1. Diagram of the pitch of hob.

The calculations for the RPM of the gear blank which are carried out are as follows
Diameter of the gear = 120mm

$$Circumference\ of\ the\ circular\ gear\ blank\ = \pi * Diameter\ (Here\ \pi = 3.14) \quad (2)$$

$$= 3.14 * 120$$

$$= 376.8 \text{ mm}$$

For covering circumference 376.8mm the total angle occupied by the circular gear blank is 360°

$$Therefore\ to\ cover\ 4mm\ of\ the\ angle\ is\ = (360 * 4)/376.8 \quad (3)$$

$$= 3.80°.$$

The 4mm part of the circumference of the gear blank lies at the 3.8° angle of the gear blank.

When the gear blank will complete rotate at the 3.8 degree then at that time the hob will complete its one rotation.
As the hob complete its one rotation when gear hob rotates at the 3.8° therefore the total revolution of the hob when gear blank will completes its one rotation will be

$$The\ total\ RPM\ of\ the\ hobber\ = (360 * 1)/3.8 \quad (4)$$

$$= 94.73$$

$$\approx 95 \text{ RPM}$$

The ratio obtain from the above calculations is 1:95. When the gear blank will completes its one rotation the hob will completes its 95 rotation for the smooth cutting operation of the gear.

Gear ratio calculation for the gear train of the gear blank

T1= Number of teeth on gear1

T2= Number of teeth on gear2

T3= Number of teeth on gear3

T4= Number of teeth on gear4

N1= RPM of gear1

N2= RPM of gear2

N3= RPM of gear3

N4= RPM of gear4

Gear ratio calculation for gear1 and gear2

$$N1/N2 = T2/T1 \tag{5}$$

1/N2=47/59

N2=59/47

N2=1.25 RPM

Gear Ratio calculation for the gear3 and gear4

$$N3/N4 = T4/T3 \tag{6}$$

1.25/1.5=T4/54

T4 = (1.25*54)/1.4

T4= 48 number of teeth.

Gear ratio calculation for the gear train of the hob

Gear ratio calculation for the gear1 and gear2

T1 = 60

T2 = 36

N1 = 95

$$N1/N2 = T2/T1 \tag{7}$$

95/N2 = 36/60

N2 ≈ 158 RPM

Gear ratio calculations for gear3 and gear4

$$N3/N4 = T4/T3 \tag{8}$$

158/200 = T4/43

T4 = (158*43)/ 200

T4 = 33 number of teeth.

Figure 2. CAD model of the gear train of the gear blank

Figure 3. CAD model of the gear train of the gear hob

Figure 4. Analysis of total deformation carried out in the teeth of the hob while gear cutting operation.

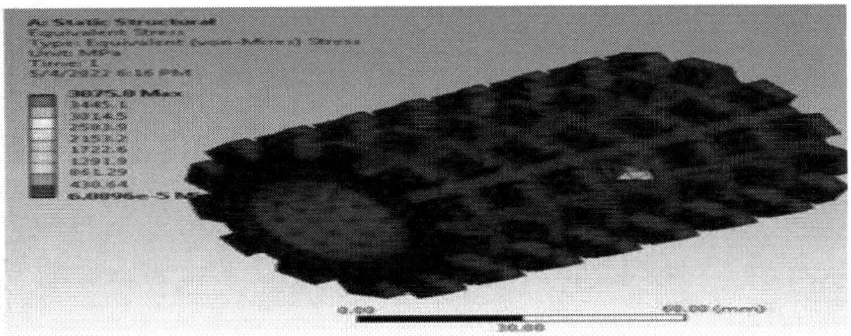

Figure 5. Analysis of stresses induce out in the teeth of the hob while gear cutting operation.

4. Results and Discussion

Various changes are carried out in the indexing process in gear hobbing operation by performing the certain calculations for the RPM of the hob and the RPM of the gear blank to carry out the smooth gear cutting operation to get variations in number of teeth. The reduction in the rotating speed of the gear blank and increase in the speed of the hob results in the reduction of the distance of the two successive cuts on the gear blank which results to increase in the number of teeth on the gear blank. In this particular way by changing the RPM of hob and gear blank the changes in the number of teeth can be done. As the gear hobbing operation takes place due to the cutting force the various changes occur in the hob. The teeth of the hob came in contact with gear blank the maximum deformation due to cutting force occurs in the teeth. The maximum stresses are induced near the teeth of the hob. To resolve this problem in future the changes in the material of the hob should be done so that the less wear should occur in the hob and the hob life will get increase. To achieve the required RPM of the hob and the gear blank the initial configuration of machine the changes in the machine are mentioned below in the table.

Table 3. Initial and final configuration for the gear train of the gear blank

Gears in the gear train of the blank	Initial RPM of the gears	Final RPM of the gear	Initial number of teeth on the gear	Final number of teeth on the gear
Gear 1 (Driven gear)	2	1	47	47
Gear 2 (Driver Gear)	2.5	1.25	59	59
Gear 3 (Driven Gear)	2.5	1.25	54	54
Gear 4 (Driver Gear)	3	1.5	56	48

Table 4. Initial and final configuration for the gear train of hob

Gears in the gear train of the hob.	Initial RPM of the gears	Final RPM of the gear	Initial number of teeth on the gear	Final number of teeth on the gear
Gear 1 (Driven gear)	87	95	60	60
Gear 2 (Driver Gear)	85	158	43	43
Gear 3 (Driven gear)	85	158	36	36
Gear 4 (Driver Gear)	93	200	24	33

5. Conclusion

The changes carried out in the gear hobbing machine so that the smooth gear cutting operation should be carried out. The changes are done in the indexing process of the gear hobbing by carrying out the calculations for the RPM (revolutions per minute). The study states that as variation will be the done in the RPM of the hob and the gear blank there will be variations in the number of teeth which will be cut on the gear blank. To achieve the variation in the RPM gear blank and the hob the changes in the machine configuration are carried out. The parameters of the gear in the gear train such as number of teeth on the gear RPM of the gear train are varied this will tends to the variation in the number of teeth cut on the gear.

6. References

[1] Claudin, C., & Rech, J. Effects of the edge preparation on the tool life in gear hobbing. In *Proceedings of the 3rd International Conference on Manufacturing Engineering (ICMEN)* (pp. 57-70).

[2] Li, D. Z., Xia, L., Liu, Y. Y., & Han, J. (2014). Research on non-circular gear hobbing simulation based on piecewise cubic spline fitting. In *Key Engineering Materials* **Vol. 620**, pp. 357-362.

[3] Sohn, J., & Park, N. (2017). Study on the influence of gear hobbing and shaft misalignments on the geometric interference of cylindrical worm gear set. *Proceedings of the Institution of Mechanical Engineers, Part C: Journal of Mechanical Engineering Science*, ***231*(24)**, 4646-4654.

[4] Bouzakis, K-D., et al. (2002) "Gear hobbing cutting process simulation and tool wear prediction models." *J. Manuf. Sci. Eng.* **124.1**: 42-51.

[5] Azvar, Milad, et al. (2021) "Chip geometry and cutting force prediction in gear hobbing." *CIRP Annals* .

[6] Guba, Nikolai, Tobias Hüsemann, and Bernhard Karpuschewski. (2020) "Influence of gear hobbing feed marks on the resulting gear quality after discontinuous profile grinding." *CIRP Journal of Manufacturing Science and Technology* **31**: 314-321.

[7] Radzevich, Stephen P. (2002) "About hob idle distance in gear hobbing operation." *J. Mech. Des.* **124.4** 772-786.

[8] Innocenti, Carlo. (2008) "Optimal choice of the shaft angle for involute gear hobbing." *Journal of Mechanical Design* **130(4)**.

[9] Wang, S. L., Yang, Y., Zhou, J., Li, Q., Yang, S., & Kang, L. (2011). Effect of machining precision caused by NC gear hobbing deformation. In *Applied Mechanics and Materials* (Vol. **86**, pp. 692-695). Trans Tech Publications Ltd.

[10] Liu, W., et al. (2013) "A gear cutting predictive model using the finite element method." Procedia CIRP **8**: 51-56.

[11] Hirogaki, Toshiki, et al. (2004) "Design systems for gear elements made of cotton fiber-reinforced plastics." Composite structures **66.1-4** 47-52.

[12] Vasilis, Dimitriou, Vidakis Nectarios, and Antoniadis Aristomenis. (2007) "Advanced computer aided design simulation of gear hobbing by means of three-dimensional kinematics modeling." 911-918.

[13] Simon, Vilmos. (2007) "The influence of gear hobbing on worm gear characteristics." 919-925.

[14] Bhattacharyya, Amitabha, and Satya Ranjan Deb. (1970) "Mechanics of gear hobbing.": 103-108.

[15] Lümkemann, A., Beutner, M., Morstein, M., Köchig, M., Cselle, M. W. T., & Karpuschewski, B. (2014, October). A New Generation of PVD Coatings for High-Performance Gear Hobbing. In *Coatings Conference. Thessaloniki, Greece.*

[16] Suri, Amar Raj Singh, and A. P. S. Sethi. 2012 "Development of Gear Hobbing Fixture Design for Reduction in Machine Setting Time." International Journal of Scientific and Research Publications **2.10** 417-419.

[17] Simon, Vilmos. (2007) "The influence of gear hobbing on worm gear characteristics." 919-925.

[18] Colbourne, J. R. (2012). The geometry of involute gears. Springer Science & Business Media.

[19] Jain, N. K., & Chaubey, S. K. (2017). Review of miniature gear manufacturing. Elsevier Inc.

[20] Khond, V. W., Taiwade, R. V., Butaley, S., & Bhaiswar, V. (2021). Experimental analysis, modelling of evaporation and ignition characteristics of different biodiesels using RSM. Materials Today: Proceedings, **47**, 2801-2807.

[21] Naidu, V., & Bhaiswar, V. (2020, October). Review on road marking paint machine and material. In IOP Conference Series: Materials Science and Engineering **Vol. 954, No. 1**, p. 012016. IOP Publishing.

[22] Dalke, P. A., & Bhaiswar, V. N. Design Modification And Analysis Of Cotton Lint Bailing Press-A Review.

[23] Bhorkar, S. P., & Bhaiswar, V. N. (2016). Vibration Analysis for Machinery Monitoring: A Past Review. International Journal of Pollution and Noise Control, **1(2)**, 24-29.

[24] Chirag K. Patil, Manoj D. Chorghade and Dr. Vinod Bhaiswar. (2017) "Review Paper (Solar Pond/Solar Thermal Energy Collector)." International Journal for Scientific Research and Development **5.9**: 618-619.

[25] P.M. Solanki, D.S. Deshmukh, V.R. Diware, M.S. Deshmukh, 2021 Experimental investigation of thermoelectric generator system, Materials Today: Proceedings, Volume **47**, Part 11, , Pages 3012-3017, ISSN 2214-7853, https://doi.org/10.1016/j.matpr.2021.05.478.

[26] Deshmukh, S. D., Gabhane, S., & Deshmukh, D. S. (2021, May). Osmotic dehydration of carrot strips and modelling. In *Journal of Physics: Conference Series* (Vol. **1913**, No. 1, p. 012093). IOP Publishing.

[27] Kulkarni, M. V., Deshmukh, D. S., & Shekhawat, S. P. (2021). An innovative design approach of hot water storage tank for solar water heating system using artificial neural network. *Materials Today: Proceedings*, ***46***, 5400-5405.

[28] Patil, P. P., Deshmukh, D. D., Vaidya, D. A., & Paul, I. D. (2019, May). Experimental Investigation of New Innovative Flat Plate Collector for Solar Water Heater. In *Proceedings of the Second International Conference on Emerging Trends in Science & Technologies For Engineering Systems (ICETSE-2019)*.

[29] Pundlik N. Patil, Dheeraj S. Dehmukh and Vilas S. Patil, July 2019, "Analysis of the effect of LPG on the Performance and Frictional Power Loss for SI Engine", International Journal of Innovative Technology and Exploring Engineering (IJITEE), ISSN: 2278-3075, Volume-**8**, Issue-**9S3**, Page No. 1406 -1409..

[30] Shailesh Pimpale, Manish Deshmukh, Rajesh Shelke and Dheeraj Deshmukh, 2022, "Experimental Investigation on Hip Implant Materials Development through Analytical and Finite Element Analysis: 3D Modelled Computed Tomography", Biointerface Research in Applied Chemistry, Platinum Open Access Journal (ISSN: 2069-5837, Volume **12**, Issue 3, 4103 – 4125, online available 15 August 2021, https://biointerfaceresearch.com/, https://doi.org/10.33263/BRIAC123.41034125

[31] Pimpale, S. S., Deshmukh, M. S., Shelke, R. T., & Deshmukh, D. S. (2021). Biomaterial Properties of Femur Implant on Acetabulum Erosion: A Review. *Journal of Biomimetics, Biomaterials and Biomedical Engineering*, ***51***, 39-62.

[32] Dr. B. S. Bhaskar, Dr. M. S. Deshmukh, Dr. S.K. Choudhary and Dr. D. S. Deshmukh, January 2022, "Heat Transfer Through Porous Materials (Aluminum Foam) Empirical Optimization of a Heat Exchanger", International Journal of Mechanical Engineering, ISSN: 0974-5823, Vol. **7** No. 1, Page No. 47 – 55.

Acknowledgement

I would like to thank my guide and co-guide Dr. Vinod Bhaiswar sir and Dr. Rupesh Shelke sir for their esteemed guidance support and encouragement. I would like to thank my industry guide Mr. Navin Verma for his guidance to carry out my research study under his supervision in industry.

Experimental investigation and optimization of WEDM process for AISI 420 stainless steel

Mukulanand Jha[1], Rajiv Kumar garg[1], Shekhar Shrivastava[1]

[1]Department of Industrial and Production Engineering, Dr. B. R. Ambedkar National Institute of Technology, Jalandhar-144011(India)

Mukulanandj.mt.20@nitj.ac.in https://orcid.org/0000-0002-4796-8823

Abstract. AISI 420 Stainless steel is martensitic steel having sufficient corrosive resistance and outstanding hardness used for making knife blades, shear blades, hand tools, and surgical instruments for the medical industry. For the machining of this kind of high-strength material, WEDM is a popular advanced machining technique for creating items with complex forms and profiles. WEDM of AISI420 stainless steel is accomplished utilizing Zn coated copper wire as an electrode in this study. Machine input control parameters include pulse on time (Ton), pulse off time (Toff), peak current (Ip), and wire feed rate (Wf), whereas measured output response parameters include material removal rate (MRR), Kerf width (KW), and surface roughness (SR). Experiments were constructed using Taguchi L16 orthogonal arrays, and control parameters were optimized using Taguchi-based grey relational analysis (TGRA). Parametric analyses are done for multi-response optimization by obtaining grey relational grade (GRG). It has been found from the ANOVA table that Ton, Toff, Ip, and Wf are having % contributions of 85.67%, 4.38%, 8.47%, and 0.26% respectively. The optimal level of input parameters obtained by Taguchi-based GRA is Ton 125μs, Toff 40μs, Ip 220A, and Wf 8m/min. To validate the results confirmation tests are performed on optimal input parameters.

Keywords: AISI420 stainless steel; Wire electric discharge machining; Taguchi-based grey relational analysis; ANOVA; MRR; KW; SR.

1. Introduction

The presence of a minimum chromium concentration of 12% gives AISI 420 stainless steel exceptional corrosion resistance qualities. It is made of high carbon steel and has the maximum hardness of any stainless steel at 50HRC. Under hardened circumstances, AISI 420 is more resistant to alkalis, freshwater, mild acids, and food. It performs well on smooth surfaces. It is utilized in a variety of applications because of its excellent corrosive resistance and hardness, including molds, hand machines, medical tools, pump axles, steel balls, knife blades, scissor blades, and so on. The production process is constantly evolving as the need for dependable, high-quality, and better components grow. Traditional machining procedures make it impossible to manufacture items with inherent forms and geometries from materials such as Nimonics, Stainless steels, Titanium, Ceramics, and so on. For this purpose, advanced machining technologies are an excellent option[1]. WEDM is a cutting-edge machining technique that produces products with high surface qualities and dimensional precision. The WEDM process is dependent on a variety of input factors, and selecting the needed input parameters for desired response characteristics is critical. Major input factors such as pulse on duration (Ton), pulse off duration (Toff), gap voltage (Ig), wire tension, current, dielectric type, working time, wire speed, flushing pressure, density, and electrode material, among others, have the greatest influence on response characteristics such as material removal rate (MRR), surface finish (Ra), electrode wear rate (EWR), wear percentage, and surface integrity. A superior surface polish, a higher MRR, and a reduced tool abrasion rate and kerf breadth may come from the selection of proper settings[2]. WEDM material erosion is achieved by firing a cascade of discontinuous sparks between the work material and the tool wire. The molten waste is flushed away by the dielectric medium (generally deionized water).The tool electrode used by WEDM is a wire with a diameter of 0.05 to 0.25 mm. Instabilities and chattering, which can introduce defects into machined goods, are avoided since this is a non-contact process[3]. Every machining method is concerned with single and/or multi-objective characteristics, both of which could be equally essential in terms of product features and affordability. Several approaches are available to optimize single and multi-objective features, such as the Taguchi robust design approach, Factorial design techniques, Statistical regression analysis, GRA, Principal component analysis, GA, RSM, utility theory, and so on. Several writers have used various optimization strategies to increase process performance with success[4].

The remaining part of this manuscript is as follows
The various selected input parameters, response parameters, materials, and optimization techniques by former researchers are discussed in section 2. The selection of material, a machine tool for performing experiments, selection of machine control parameters, design of experiment, and methods for obtaining responses are explained in section 3. The analysis and optimization of response parameters and machine control parameters are further shown in section 4. The optimized results are discussed in result and discussion section 5. In section 6 the conclusion obtained from the optimization of experimental work is discussed.

2. Literature review

2.1. On the basis of work piece and electrode material used

Shivade et al.[4]used a 0.18mm diameter molybdenum wire electrode to perform WEDM on D3 steel. To identify the optimal value of four performance characteristics MRR, dimensional variation, machining duration, and gap current. Pramanik et al.[5]used 2205 duplex stainless steel as a work material and brass wire as an electrode material to investigate the effect of WEDM input parameters. Muniappan et al.[6] investigated WEDM of Al6061/SiC/graphite (aluminium hybrid composite) using Zn coated brass wire as an electrode. Kumar et al.[7] used Inconel 825 as a work material and 0.25mm plain brass wire as electrode on WEDM for optimization of performance parameters using GRA methodology. Ravi Kumar et al.[8] utilized WEDM of Al (6082)/tungsten carbide composites using diffused brass wire as electrode material for optimization of MRR and SR using RSM methodology. Kavimani et al. [9] optimized control parameters of WEDM for AZ31 Mg alloy(magnesium metal matrix composite) using Cu wire electrode.

2.2. On the basis of parameters used

Gajjar et al.[10]investigated and optimized KW, SR, and MRR for wire EDM of EN-31 steel utilizing 0.25mm molybdenum wire as an electrode using GRA methodology. Kumar et al.[11] used the RSM approach to study and improve three response parameters in WEDM of the nickel-based alloy Nimonic-90: CS, SR, and radial overcut (RoC), using Ip, Ton, Toff, and SV as input machine variables.It was found that wire failure occurs when the pulse on time is extended and the WT is increased. Camposeco-Negrete [12] studied the impact of input parameters Plate thickness (plate A -15.8 mm plate B- 25.4mm), Ton, servo voltage and voltage on response parameters machining time(MT), electric power used in cutting(P), total energy consumption during machining(EC), SR, and MRR. Jebarose Juliyana et al.[13] investigated WEDM of AMCs (LM5/ZrO2) using brass wire for optimizing control variables Pulse on time, Pulse off time, Gap voltage, Wire feed, and Reinforcement (wt. %) for responses MRR, SR and KW. Ugrasen et al.[14]studied SS304 WEDM with molybdenum wire as an electrode material. According to the findings, the Ton has a stronger influence on surface roughness, while bed speed has a bigger influence on MRR.

2.3. On the basis of types of wire used

Tahir et al.[15] used a wire electric discharge machine to study the effects of cold treated(CT) and non-cold treated(NCT) brass wire electrodes on machining hardened (HSLA) steel.(CT) brass wire generates a smoother surface finish than (NCT) brass wire, and(NCT) brass wire creates a lower kerf width, according to the research. Saini et al.[16]employed Diffused & Zn-coated wire to examine the WEDM of 16MnCr5 alloy steel and applied the multi-response optimization strategy based on Taguchi's design approach to optimize the WEDM. Zn-coated wire has a greater optimal response parameter. The research demonstrated that Ton was the most important factor for MRR and surface roughness. Reolon et al.[17]researched wire, electric discharge machining with Zn coated copper wire and brass wire as the tool electrode. Overall performance was superior in all studies where Zn-coated copper wire outperformed raw brass wire.

2.4. On the basis of different techniques used

Alduroobi et al.[18]investigated response parameters MRR and SR for work material AISI 1045 for the WEDM process using the artificial neural network optimization technique. According to the findings, an increase in Ton correlates to an increase in MRR. Harish et al.[19]did a comparative study of WEDM of D2 steel utilizing three distinct electrodes: uncoated, coated, and annealed copper wire. For optimization grey fuzzy logic approach combined with Grey relational analysis is used. The tool wear rate of annealed copper wire is reduced when compared to other wires. Furthermore, the surface smoothness and kerf breadth are approximately comparable for all wires. Vishwasrao[20] using response surface methods, optimized WEDM control settings for H11 tool steel with molybdenum wire over a wide range of responses. Goyal et al. [21] used Grey wolf optimization for optimization of the WEDM process parameter for AZ31 Mg-alloy using Zn coated brass wire.

In present literature review shows a selection of important work done over the last few years by many scholars, which provides the direction of our study presented in this article. This will help in determining the most significant process variables to be optimized for various response parameters using various optimization techniques.

Research gap

- Only a few research publications on AISI 420 stainless steel have been discovered.
- Few studies have employed multi-response optimization strategies for WEDM of AISI420 stainless steel, according to the literature review.

The Taguchi method generates a one-of-a-kind set of optimum answers for a variety of replies. When one component increases, another may decrease.[13]. So, the Taguchi-based grey relational analysis (TGRA) methodology is utilized in this research work for multi-response optimization. It has also been cleared that Zn coated copper wire gives a better response compared with other kinds of wire[19].

The objectives of the present experimental work are as follow

- To determine which machine control factors have the greatest impact on response parameters.
- TGRA Multi response optimization approach is used to identify the ideal level of machine control parameters for maximized MRR, lesser KW and low SR.

3. Material and method

The work material in this study is an AISI420 stainless steel plate with dimensions of 100mm*120mm*4mm, and the electrode material is 0.25mm Zn coated copper wire. The experiments are performed on WEDM (MODEL ELPLUS 40A DLX SPRINTCUT) machine situated at tooling solution Okhla Delhi shown in fig. 1 below. For the design of the experiment, Taguchi L16 OA is used using four machine control parameters (Ton), (Toff),(Ip,) and (Wf) each at four levels to record the response of (MRR), (Kw) and (SR). The machine control parameters and their levels are shown in table 1 below.

IOP Conf. Series: Materials Science and Engineering 1259 (2022) 012031 doi:10.1088/1757-899X/1259/1/012031

Table 1. Selected parameters and levels.

	Machining control parameters			
	pulse on time(μs)	pulse off time(μs)	peak current(A)	wire feed rate(m/s)
Levels	T_{on}(110-125)	T_{off}(30-60)	I_p(130-220)	W_f(4-10)
1	110	30	130	4
2	115	40	160	6
3	120	50	190	8
4	125	60	220	10

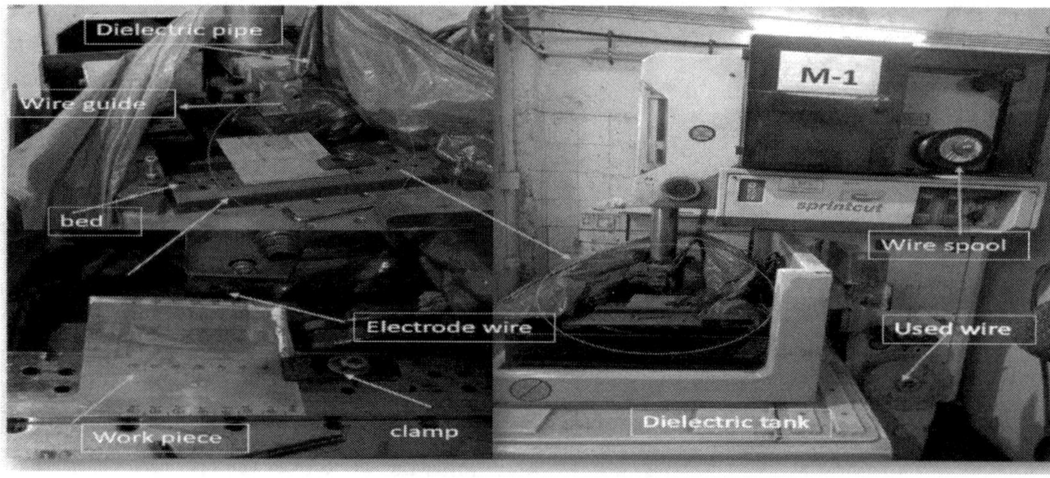

Figure 1. Experimental setup.

A straight cut of 50mm length is created in each pass, and the machining time is recorded for further material removal rate computation using equation 1 below.

$$MRR\ (mm3/min) = \frac{(kerf\ width\ (Kw) * length\ of\ cut * thickness\ of\ plate)}{(machining\ time\ (Mt))} \tag{1}$$

For kerf width (Kw) optical microscope is used and surface roughness (SR) is measured using the SJ-210 surface roughness tester of samples prepared from 50mm cut samples by cutting from a distance of 30mm from the end. The samples used for the measurement of KW and SR are shown in figure 2 below.

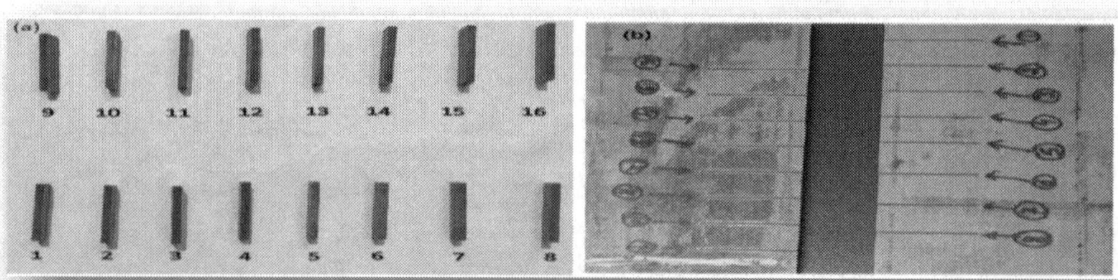

Figure 2. Sample for response measurement (a) Samples for surface roughness (b) Samples for kerf width.

4. Methodology

There is a heretical method for carrying out the experiment that must be followed. The heretical approach consisted of steps from finding the work piece material to optimization of machine control parameters. The experiment runs are created with L16 OA and MINITAB19 software. Table 2 shows the experimental runs and response outcomes.

Table 2. Experimental runs and Responses.

	MACHINE CONTROL PARAMETERS				RESPONSE PARAMETERS		
Experiment No.	Ton(μs)	Toff(μs)	Ip(A)	Wf(m/s)	MRR (mm3/min)	KW (microns)	SR (microns)
1	110	30	130	4	3.181	312.843	2.41
2	110	40	160	6	3.178	319.135	2.46
3	110	50	190	8	3.288	321.930	2.53
4	110	60	220	10	3.280	318.516	2.56
5	115	30	160	8	3.469	323.875	2.61
6	115	40	130	10	3.252	324.015	2.44
7	115	50	220	4	3.847	328.103	2.7
8	115	60	190	6	3.461	328.210	2.53
9	120	30	190	10	5.259	339.221	2.86
10	120	40	220	8	5.466	358.063	2.91
11	120	50	130	6	3.732	347.062	2.45
12	120	60	160	4	3.765	352.076	2.48
13	125	30	220	6	6.041	352.410	3.03
14	125	40	190	4	6.099	363.403	2.85
15	125	50	160	10	6.114	362.750	2.58
16	125	60	130	8	5.821	358.982	2.42

Furthermore, Taguchi-based grey relational analysis (TGRA) is used to optimize machine control settings. The stages used in TGRA are depicted in the flow chart in figure 3 below.

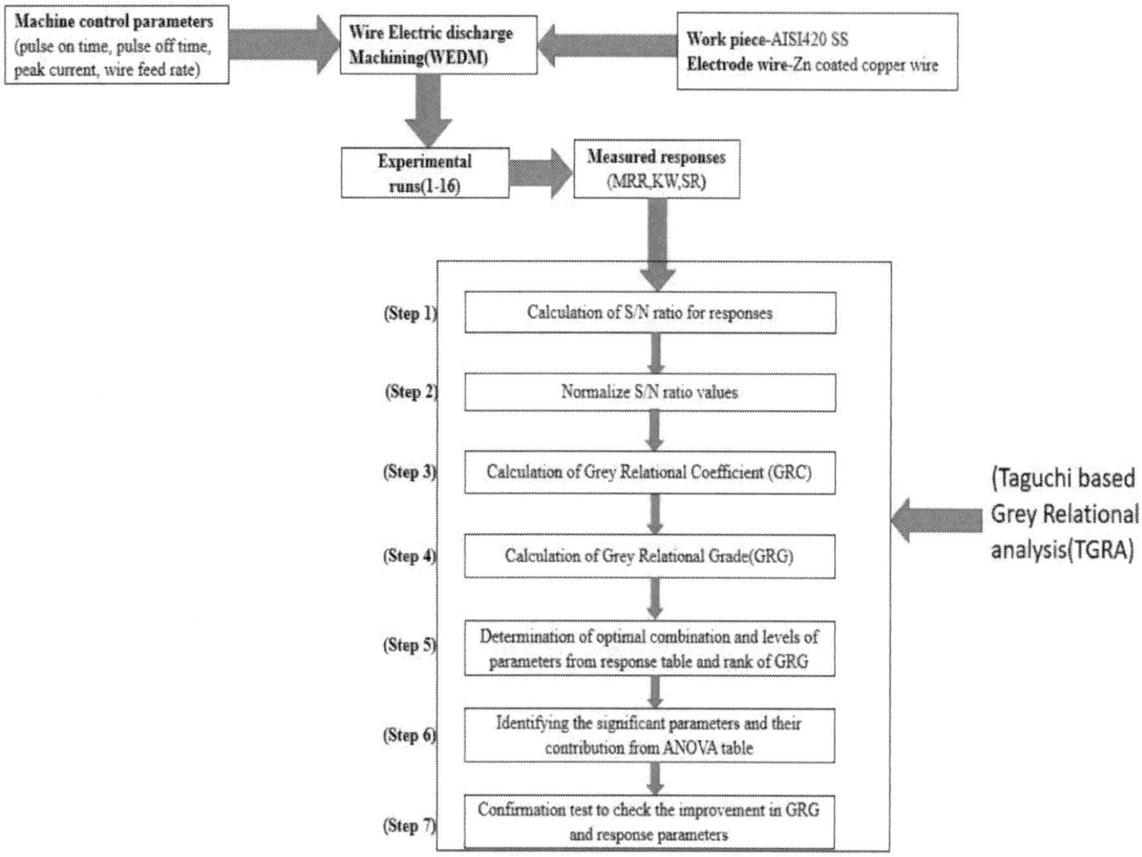

Figure 3. Flow chart of steps in Taguchi based grey relational analysis.

For MRR larger the better criteria are utilized because the objective is to remove the material at a higher rate for quicker production. S/N ratio for MRR is determined using equation 2.

$$S/_N\, ratio = -10 \times log\left(\Sigma\left(\frac{1}{y^2}\right)\Big/_n\right) \tag{2}$$

Where y is the response for the particular experimental run and n represents the total number of repetition of experimental run.

For KW and SR lower the better criteria is opted for because lower surface roughness is required for better surface quality and lower kerf width is for dimensional accuracy. So, S/N ratio for KW and SR is determined using equation 3.

$$S/_N\, ratio = -10 \times log\left(\frac{\Sigma y^2}{n}\right) \tag{3}$$

Where y is the response for the particular experimental run and n represents the total number of repetition of experimental run.

Normalizing of S/N ratio is utilized to convert the original sequence of values into a decimal sequence between 0.00 to1.00 for comparison.[9] For normalization of MRR equation 4 and for KW and SR equation 5 is used.
For larger the better

$$x_i(r) = \frac{y_i(r) - miny_i(r)}{maxy_i(r) - miny_i(r)}$$

(4)

For smaller the better

$$x_i(r) = \frac{maxy_i(r) - y_i(r)}{maxy_i(r) - miny_i(r)}$$

(5)

Where $x_i(r)$the normalized reference sequence for the ith experiment is, $y_i(r)$ is the S/N ratio of the ith response. Table 3 shows the determined S/N ratios and normalized values for each response.

Table 3. Normalization of S/N ratio of responses.

	S/N RATIO of Responses			**NORMALIZED S/N RATIO**		
SR. NO.	SR	MRR	KW	SR	MRR	KW
1	-7.640	10.051	-49.907	0.000	0.002	0.000
2	-7.819	10.042	-50.079	0.090	0.000	0.133
3	-8.062	10.338	-50.155	0.212	0.052	0.191
4	-8.165	10.318	-50.063	0.264	0.049	0.120
5	-8.333	10.805	-50.208	0.348	0.134	0.231
6	-7.748	10.242	-50.211	0.054	0.035	0.234
7	-8.627	11.703	-50.320	0.496	0.292	0.318
8	-8.062	10.783	-50.323	0.212	0.130	0.320
9	-9.127	14.418	-50.610	0.748	0.770	0.540
10	-9.278	14.754	-51.079	0.823	0.829	0.901
11	-7.783	11.438	-50.808	0.072	0.246	0.693
12	-7.889	11.516	-50.933	0.125	0.259	0.789
13	-9.629	15.622	-50.941	1.000	0.982	0.795
14	-9.097	15.704	-51.208	0.732	0.996	1.000
15	-8.232	15.726	-51.192	0.298	1.000	0.988
16	-7.676	15.300	-51.101	0.018	0.925	0.918

Following the normalization of the S/N ratio values of replies, we must determine the GRC for each response in order to calculate the deviation sequence using equation 6.

Deviation sequence,

$$(\Delta_{0i}) = |x_0(r) - x_i(r)| \tag{6}$$

Where (Δ_{0i}), $x_0(r)$, and $x_i(r)$ are the corresponding deviation, reference, and comparability sequences. Equation 7 below calculates the GRC using the values of the deviation sequence for each response.

$$(GRC)\xi_i(r) = \frac{\Delta_{min} + \zeta\Delta_{max}}{\Delta_{0i}(r) + \zeta\Delta_{max}} \tag{7}$$

Where, $\xi_i(r)$ denotes the GRC of individual responses, Δ_{min} is the minimum value of deviation and Δ_{max} is maximum value of deviation the minimum and maximum deviation response. ζ is characteristics or identification coefficient, specified in the range [0,1], is generally allocated equal weights to every response. In this research, three responses are measured so 0.33 weightage is given to each response.

Grey relational grade (GRG)

Following the computation of GRC values for each response, the next step is to compute GRG by averaging the GRC of each response for each run of the experiment using equation 8 below.

$$(GRG)Y_i = \frac{1}{n}\sum_{i=1}^{n} \xi_i(r) \tag{8}$$

Where Y_i the GRG value computed for the ith experiment, and n is is the total number of response characteristics. Table 4 shows the derived GRC values for each response, GRG, and their relative rankings.

Table 4. Grey Relational Analysis of responses.

	GREY RELATIONAL COEFFICIENT (GRC)			GRG	RANK
Sr. No.	SR	MRR	KW		
1	0.248	0.254	0.248	0.250	16
2	0.266	0.254	0.276	0.265	15
3	0.295	0.264	0.290	0.283	12
4	0.309	0.263	0.273	0.282	13
5	0.336	0.282	0.300	0.306	10
6	0.259	0.261	0.301	0.273	14
7	0.396	0.324	0.326	0.349	9
8	0.295	0.281	0.327	0.301	11
9	0.567	0.596	0.418	0.527	6
10	0.652	0.665	0.770	0.695	4
11	0.262	0.311	0.518	0.364	8
12	0.274	0.315	0.610	0.399	7
13	**1.000**	**0.949**	**0.617**	**0.855**	**1**
14	0.552	0.989	1.000	0.847	2
15	0.320	1.000	0.965	0.762	3
16	0.252	0.819	0.802	0.624	5

IOP Conf. Series: Materials Science and Engineering 1259 (2022) 012031 doi:10.1088/1757-899X/1259/1/012031

5. Result and discussion

The present study employs a multi objective optimization approach based on TGRA methodology. MINITAB19 software is used to analyze these responses. Figure 4 depicts the influence of machine control settings on each responses.

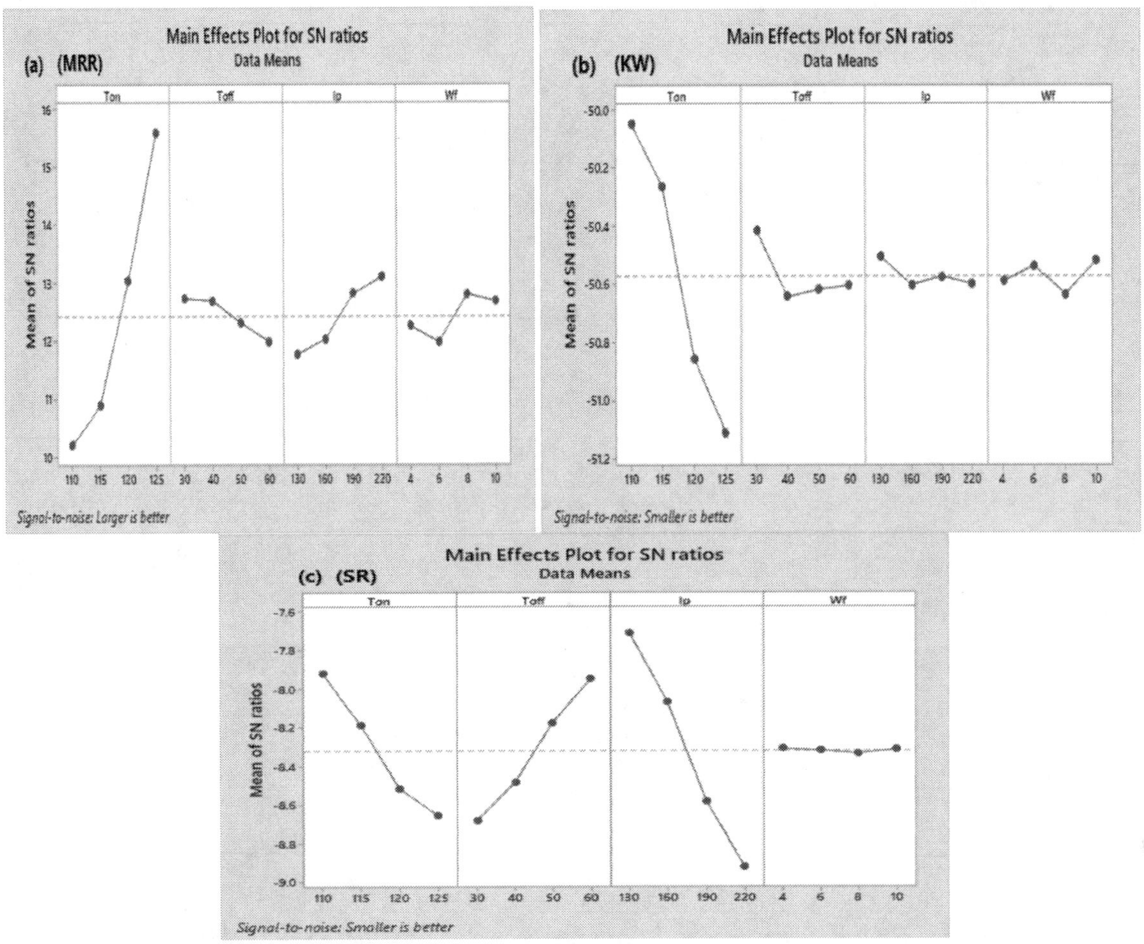

Figure 4. Main effect of machine control parameters for S/N ratio on each response (a) MRR (b) KW (c) SR.

The influence of machine control settings may be seen in the above major effect charts for the S/N ratio of each response. Ton has been discovered to be the most important parameter for MRR, kerf width, and surface roughness. Toff has an inverse relationship with MRR since MRR reaches its maximum at the lowest value of Toff. The minimal value of kerf width is at the minimum value of Toff, but beyond that, the value of KW has no influence as Toff increases. Peak current increases material rate, which is ideal, but it also increases SR, which is not. Wf has almost little influence on any of the responses. A single response grey relational grade (GRG) is computed to discover the most ideal level of parameters to match the requirements of maximum MRR, minimal kerf width, and surface roughness. Tablè 4 shows the highest value of GRG for experiment number 13. This signifies that experiment 13 has the best level of machine control settings. The Taguchi approach is used to further improve the GRG response. Figure 5 and table 5 illustrate the primary effect plot and response table for GRG.

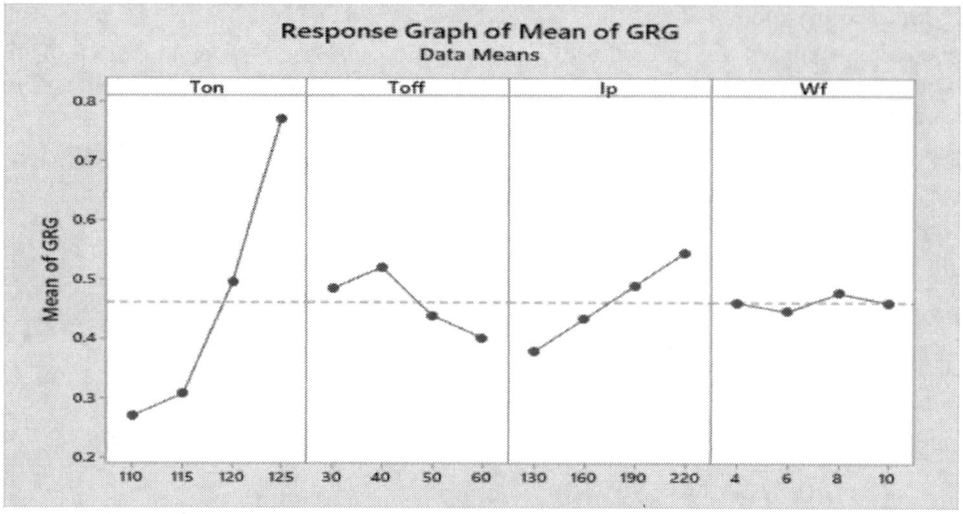

Figure 5. Effect of machine control parameters on GRG.

The value of GRG increases with Ton and peak current and decreases with Toff, according to the main effect plot for GRG. Wf has no discernible impact on GRG. A response table is developed to find the appropriate amounts of machine control parameters.

Table 5. Response Table for GRG.

Level	T_{on}	T_{off}	I_p	W_f
1	0.2700	0.4846	0.3778	0.4613
2	0.3074	**0.5203**	0.4331	0.4463
3	0.4964	0.4392	0.4895	**0.4772**
4	**0.7720**	0.4016	**0.5453**	0.4610
Delta	0.5020	0.1187	0.1675	0.0309
Rank	1	3	2	4

The mean value of GRG at each level of machine control settings is shown in the response table. The ideal level of machine control parameters is determined by GRG maximum values. Delta values are the difference between maximum and minimum values of mean GRG each machine control parameters which determines the rank. The highest delta value parameter has highest influence on response. It is found that Ton at (4), Toff at (2), Ip at (4) and Wf at (3) is optimal levels of machine control parameters. For determining the significance of machine control parameters ANOVA table is drawn using general linear model in MINITAB19 software shown in table 6 below.

Table 6. ANOVA table for GRG.

Source	DF	Seq SS	%Contribution	Adj SS	Adj MS	F-Value	P-Value
Ton	3	0.632233	85.67%	0.632233	0.210744	69.71	0.003
Toff	3	0.032296	4.38%	0.032296	0.010765	3.56	0.162
Ip	3	0.062479	8.47%	0.062479	0.020826	6.89	0.074
Wf	3	0.001916	0.26%	0.001916	0.000639	0.21	0.883
Error	3	0.009070	1.23%	0.009070	0.003023		
Total	15	0.737993	100.00%				

IOP Conf. Series: Materials Science and Engineering 1259 (2022) 012031 doi:10.1088/1757-899X/1259/1/012031

ANOVA table shows that Ton is most significant parameter and wire feed rate is not significant parameter. It also shows the contribution% of each machine control parameters. Ton has the highest contribution of 85.67% and wire feed rate has the least contribution of 0.26%. The optimum levels of parameters are obtained and the improved value of GRG at optimum level is predicted using Taguchi optimization methodology using MINITAB19 software.

Table 7. Optimized result.

	MACHINE CONTROL PARAMETERS				RESPONSE PARAMETERS			
Parameters	Ton	Toff	Ip	Wf	MRR	KW	SR	GRG
Initial value(At max GRG exp. no. 13)	125	30	220	6	6.041	352.41	3.03	0.855
optimized values	125	60	220	8	6.269	342.170	2.950	0.933
improvement					0.228	10.240	0.080	0.078
%improvement					3.774	2.906	2.640	9.064

The response values at optimum levels of parameter shown in table 7 are obtained by performing validation experiments three times at optimum levels of machine control parameters.

6. Conclusion

The following conclusions were reached based on the data acquired from the current examination of WEDM machined AISI420 Stainless Steel utilizing Zn coated copper wire tool electrode.

- Ton is most significant parameter for all three responses and its contribution% is 85.67% followed by peak current with 8.47% contribution and pulse off time and wire feed rate has negligible contribution.
- The optimal level of machine control parameters obtained from Taguchi based grey relational analysis are Ton 125μs, Toff 40μs, Ip 220A, and Wf 8m/min.
- The value of GRG is increased from 0.855 to 0.933 by using TGRA.
- The response values MRR, KW and SR are improved by 3.774%, 2.906% and 2.640% respectively by use of TGRA.

References

[1] S. Kumar, S. K. Ghoshal, P. K. Arora, and L. Nagdeve, "Multi-variable optimization in die-sinking EDM process of AISI420 stainless steel," *Mater. Manuf. Process.*, vol. 36, no. 5, pp. 572–582, 2021, doi: 10.1080/10426914.2020.1843678.

[2] S. Sharma, U. K. Vates, and A. Bansal, "Parametric optimization in wire EDM of D2 tool steel using Taguchi method," *Mater. Today Proc.*, vol. 45, pp. 757–763, 2021, doi: 10.1016/j.matpr.2020.02.802.

[3] A. Mukhopadhyay, T. K. Barman, P. Sahoo, and J. P. Davim, "Modeling and optimization of fractal dimension in wire electrical discharge machining of EN 31 steel using the ANN-GA approach," *Materials (Basel).*, vol. 12, no. 3, pp. 1–13, 2019, doi: 10.3390/ma12030454.

[4] A. S. Shivade and V. D. Shinde, "Multi-objective optimization in WEDM of D3 tool steel using integrated approach of Taguchi method & Grey relational analysis," *J. Ind. Eng. Int.*, vol. 10, no. 4, pp. 149–162, 2014, doi: 10.1007/s40092-014-0081-7.

[5] A. Pramanik, A. K. Basak, A. R. Dixit, and S. Chattopadhyaya, "Processing of duplex stainless steel by WEDM," *Mater. Manuf. Process.*, vol. 33, no. 14, pp. 1559–1567, 2018, doi: 10.1080/10426914.2018.1453165.

[6] A. Muniappan, A. P. Jaivaakheish, V. Jayakumar, A. Arunagiri, and R. Senthilkumar, "Multi objective optimization of process parameters in WEDM of aluminum hybrid composite using taguchi and topsis techniques," *IOP Conf. Ser. Mater. Sci. Eng.*, vol. 402, no. 1, 2018, doi: 10.1088/1757-899X/402/1/012002.

[7] P. Kumar, M. Meenu, and V. Kumar, "Optimization of process parameters for WEDM of Inconel 825 using grey relational analysis," *Decis. Sci. Lett.*, vol. 7, no. 4, pp. 405–416, 2018.

[8] K. Ravi Kumar and Nishasoms, "Desirability-Based Multi-objective Optimization and Analysis of WEDM Characteristics of Aluminium (6082)/Tungsten Carbide Composites," *Arab. J. Sci. Eng.*, vol. 44, no. 2, pp. 893–909, 2019, doi: 10.1007/s13369-018-3353-5.

[9] V. Kavimani, K. Soorya Prakash, and T. Thankachan, "Multi-objective optimization in WEDM process of graphene – SiC-magnesium composite through hybrid techniques," *Meas. J. Int. Meas. Confed.*, vol. 145, pp. 335–349, 2019, doi: 10.1016/j.measurement.2019.04.076.

[10] D. Gajjar and J. Desai, "Optimization of MRR, Surface Roughness and KERF Width in wire EDM Using Molybdenum Wire," *Int. J. Res. Educ.*, vol. 4, no. 2, pp. 9–17, 2015.

[11] V. Kumar, K. K. Jangra, V. Kumar, and N. Sharma, "WEDM of nickel based aerospace alloy: optimization of process parameters and modelling," *Int. J. Interact. Des. Manuf.*, vol. 11, no. 4, pp. 917–929, 2017, doi: 10.1007/s12008-016-0298-3.

[12] C. Camposeco-Negrete, "Analysis and optimization of sustainable machining of AISI O1 tool steel by the wire-EDM process," *Adv. Manuf.*, vol. 9, no. 2, pp. 304–317, 2021, doi: 10.1007/s40436-021-00353-2.

[13] S. Jebarose Juliyana, J. Udaya Prakash, A. Divya Sadhana, and C. Sarala Rubi, "Multi-objective optimization of process parameters of wire EDM for machining of AMCs (LM5/ZrO2) using grey relational analysis," *Mater. Today Proc.*, vol. 52, pp. 1494–1498, 2022, doi: 10.1016/j.matpr.2021.11.213.

[14] G. Ugrasen, M. R. Bhagawan Singh, and H. V. Ravindra, "Optimization of Process Parameters for SS304 in Wire Electrical Discharge Machining using Taguchi's Technique," *Mater. Today Proc.*, vol. 5, no. 1, pp. 2877–2883, 2018, doi: 10.1016/j.matpr.2018.01.080.

[15] W. Tahir and M. Jahanzaib, "Multi-objective optimization of WEDM using cold treated brass wire for HSLA hardened steel," *J. Brazilian Soc. Mech. Sci. Eng.*, vol. 41, no. 11, pp. 1–14, 2019, doi: 10.1007/s40430-019-2028-9.

[16] T. Saini, K. Goyal, and D. Bhandari, "Multi-response optimization of WEDM parameters on machining 16MnCr5 alloy steel using Taguchi technique," *Multiscale Multidiscip. Model. Exp. Des.*, vol. 2, no. 1, pp. 35–47, 2019, doi: 10.1007/s41939-018-0027-7.

[17] L. W. Reolon, C. A. Henning Laurindo, R. D. Torres, and F. L. Amorim, "WEDM performance and surface integrity of Inconel alloy IN718 with coated and uncoated wires," *Int.*

J. Adv. Manuf. Technol., vol. 100, no. 5–8, pp. 1981–1991, 2019, doi: 10.1007/s00170-018-2828-6.

[18] A. A. A. Alduroobi, A. M. Ubaid, M. A. Tawfiq, and R. R. Elias, "Wire EDM process optimization for machining AISI 1045 steel by use of Taguchi method, artificial neural network and analysis of variances," *Int. J. Syst. Assur. Eng. Manag.*, vol. 11, no. 6, pp. 1314–1338, 2020, doi: 10.1007/s13198-020-00990-z.

[19] I. Harish and P. S. Rao, "a Comparative Study on the Performance of Wire Edm Machining of D2 Steel Using Uncoated , Coated and Annealed Wire," vol. 7, no. 13, pp. 2941–2955, 2020.

[20] P. R. Vishwasrao, "Optimization of Process Parameter in Wire Electrical Discharge Machining of H11 Tool Steel Using Response Surface Methodology," *Int. J. Res. Appl. Sci. Eng. Technol.*, vol. 9, no. 12, pp. 2150–2157, 2021, doi: 10.22214/ijraset.2021.39693.

[21] K. K. Goyal *et al.*, "Measurement of performance characteristics of WEDM while processing AZ31 Mg-alloy using Levy flight MOGWO for orthopedic application," *Int. J. Adv. Manuf. Technol.*, no. 0123456789, 2022, doi: 10.1007/s00170-021-08358-8.

An Investigation of the Factors affecting Flexible Manufacturing Competence of Organizations

Asmit Karadbhajane[1], Jeevan Gaidhane[1], Inayat Ullah*[1], Sourabh Shukla[1], Anand Babu Kotta[1]

*Corresponding author
[1]Center of Excellence of FMS/Robotic Design, Department of Mechanical Engineering, G H Raisoni College of Engineering Nagpur, India

Email : inayatu6@gmail.com

Abstract:

Owing to the fluctuating demands of customers and intensified competition, organizations are searching ways to enhance their flexible manufacturing competence (FMC) as it has been recognized as one of the most important approaches to handle the aforementioned situations. The existing literature provides several factors that significantly affect FMC, however, the interrelationships among these factors are somewhat unexplored. The insights to the interactions among these factors could provide crucial managerial implications that could pave ways for success of organizations in today's uncertain business environment. To this end, the present work makes an effort and highlights the important factors and investigates their interactions with each other for the enhancement of FMC of organizations. In this paper, the interpretive structural modelling (ISM) methodology has been implemented to explore the interactions and develop a structural model depicting the hierarchy of the factors. Further, MICMAC analysis has been used to categorize the factors into different groups based on their driving and dependence powers.

Keywords: Flexible Manufacturing Competence, Interpretive Structural Modelling, MICMAC analysis, Manufacturing flexibility.

1. Introduction

In today's fast paced world, industries are striving to increase their efficiency and responsiveness so that they can cope with the environmental turbulence and better serve their customers. To better serve customers, companies are looking for the ways to adjust themselves swiftly to the changes brought on by customers, competitors, and technologies, and hence, flexibility has emerged as a critical weapon in this battle of survival and growth. The ability of an enterprise to meet the growing expectations of customers without sustaining excessive expenditures, time, and performance losses is referred as flexibility. More specifically, industries are focusing on strengthening their flexible manufacturing competence (FMC) which is widely recognized as a cause of competitive advantage in response to the growing environmental unpredictability. FMC, from a strategic standpoint, enhances a company's business strategy by reducing uncertainty and maintaining a smooth production flow [1-2]. Flexible manufacturing capabilities, such as volume and mix flexibility, is an extrinsic dimension of competition that customers value. Volume flexibility is described as an organization's capacity to modify volume levels profitably and with minimal interruption in response to changing socioeconomic situations [3]. Mix flexibility refers to a production system's capacity to cope with increasing product variation while maintaining system performance [4]. These external dimensions are valued by the customers as these dimensions directly benefit them. FMC, which encompasses machine, labour, material handling, capacity, and routing flexibility, is an important internal component of competitiveness that customers are unaware of [5]. When a firm focuses on improving its flexibility and succeeds in doing so, the firm becomes able to effectively handle the fluctuating demands of products in the market. Nonetheless, to achieve that level, firms need to identify and implement certain factors which could help them achieve greater FMC [6]. The existing literature offers several factors that may significantly influence the FMC of organizations however there is a lack of study that provides the information related to the crucial factors in a comprehensive manner [12]. Furthermore, the research rendering information regarding the importance of these factors and interactions among them is scant [18]. In the present work, the effort has been undertaken to provide an inclusive list of factors that has a positive effect on the FMC of organizations. In addition, ISM technique has been utilized to develop a structural model showing the hierarchy and interactions among the factors of interest. Furthermore, MICMAC analysis is used to classify the factors based on their importance into four different categories. The findings of the study provide crucial managerial implications by interpreting the hierarchy of the factors and their interactions with each other. Having a clear picture of the importance and interrelationship of the factors, managers can better devise an effective action plan and prioritize their efforts and resources in the course of enhancing their FMC and winning the competition in the market.

The remainder of the paper is designed as follows. The study begins by providing a brief overview of the crucial factors that significantly impact the FMC of organizations. Following that, research methodology is outlined, including the procedure of implementing ISM technique. Next, the step-by-step development of the structural model based on ISM and the categorization of factors into four groups depending on their driving and dependence power using MICMAC analysis have been performed [6]. This is now followed by the discussion of findings from above analyses. Finally, the conclusions of the study are presented.

2. Literature Review

The present study has identified a total of 12 factors that are particularly relevant to the improvement of FMC of organisations from the available literature (Table 1). The following factors have been recognized as FMC facilitators.

2.1 Advanced manufacturing technology (AMT)

AMT refers to a collection of tools that automatize and integrate various design, production, planning, and control processes [7]. AMT shortens set-up times, improve workflow, and improve plant integration, allowing the manufacture to offer a wide range of product varieties in different volume and mix and in small lead time [8].

2.2 Operational improvement practices (OIP)

OIPs are a collection of Just-In-Time (JIT) methods that emphasis on constant improvement in manufacturing, such as set-up time reduction, preventative maintenance, cellular layout, pull production, overall quality management, etc. and contribute to volume and mix flexibility of organizations [8]. The deployment of AMTs in combination with OIPs has a greater impact on an organization's flexibility [23].

2.3 Internal and external learning

Internal learning refers to the situation where the employees get to learn new skills and knowledge while working as a team or individuals during the execution of routine and challenging tasks. External learning refers to the acquisition of skills, knowledge, and experience that the employees gain while working with suppliers, customers and other third-party service providers [24]. The organization which facilitates learning behaviour among employees develops competitive capabilities to handle a wide variety of scenario that occur in both internal and external environments [25].

IOP Conf. Series: Materials Science and Engineering 1259 (2022) 012032 doi:10.1088/1757-899X/1259/1/012032

2.4 Adaptive human resources

Adaptability is a key trait of an individual which refers to his ability to modify himself with the changes in the environment. It has very well established that adoption of advanced technologies is conducive to the FMC and competitiveness of the organization [26]. However, even the most advanced technologies cannot handle the demand fluctuations single-handedly and they would require the help of adaptive human resources [40]. Adaptive human resources possess the ability to deal with diverse, ambiguous, and complex work environment and make effective decisions under different circumstances [27].

2.5 Absorptive Capacity

Absorptive ability is based on prior learning experiences and current knowledge, and it influences an organization's creative actions in response to the environmental disturbances, in both proactive and reactive manner [9]. In other words, the term "absorptive capacity" refers to organizations' ability to manage inside innovation as well as to acquire and absorb external ideas, data, and expertise. Organizations with absorptive capacity can better assimilate and disseminate the available information and use it to innovate and develop superior capabilities [28].

2.6 Modularity in products and processes

Product modularity is commonly cited as an aim of good design practise, although it has not received the same level of research as other current design practises like design-for-assembly. Modularity is becoming a focal point of attention as companies attempt to streamline their product lines and give even more diversity at reduced costs [29]. Improving product modularity can lead to the betterment of product flexibility [15]. Process modularity refers to the technique of standardising manufacturing sub-processes such that they may be readily re-ordered or new modules can be swiftly introduced in response to changing product needs. The three foundations of process modularity include process standardization, process re-sequencing, and process postponement [10].

2.7 *Time-based manufacturing*

Time-based manufacturing is a production approach that focuses on responding quickly to the changing customer demands. To increase responsiveness and other competitive capabilities such as FMC, time-based manufacturing relies on time reduction approaches [11]. These practices include active participation of employees in problem solving, setup redesign, cellular manufacturing, total preventive maintenance, quality management, use of reliable suppliers, and make-to-order production [22]. The implementation of time-based manufacturing practices leads to the significant reduction in lead time which, in turn, enables the organizations to exhibit quick responsiveness to customers' orders and thereby allows the organizations to show more competitiveness.

2.8 *Capacity Flexibility*

It is referred to as the capability to modify overall production volume and varieties, with the possibility of using conditional as well as fixed resources [30]. This factor can help the organizations in absorbing the short-term demand fluctuations with high degree of effectiveness [31]. Some of the practices that increases the capacity flexibility are hiring temporary staff, use of overtime/flexi-time, installation of new machinery, use of subcontracting, etc.

2.9 *Postponement Implementation*

Postponement is a strategy in which a manufacturer creates a generic product that can be reconfigured in the later stages of production before the final delivery to the customer [32]. The implementation of postponement allows the manufacturers to incorporate the diversified demands of customers into the products without affecting the existing production schedules and operational routines, thereby, renders flexibility to cope with demand fluctuations [33].

2.10 *Organizational Structure*

The system by which work flows through an organisation is stated as the organizational structure. It empowers the groups or departments to team up within their particular functions to manage different activities [16]. The organizational structure can be formal or flexible, depending on the nature of business [34]. Although the formal organizational structure are comparatively stable than the informal or flexible organizational structure, however, the formal structure is somewhat constrained in its ability to adapt to new or unforeseen circumstances. In such cases, the flexible organizational structure is much more suitable as it allows the organizations to adapt to the changing course of events [35].

2.11 *Value Chain Management*

Value chain management is the approach that offers both supplier flexibility and sourcing flexibility, allowing for quick changes in production levels, raw material procurement, and transportation capacity and schedules in response to the volatile business environment [36]. Sourcing flexibility represents the capability of a manufacturer to respond to the changing supply needs by swiftly reconfiguring the supply nodes [14]. On the other side, supplier flexibility refers to the competence of individual suppliers to conform to the variations in order in terms of design, volume, mix, and delivery schedule [37].

2.12 *Top Management Commitment*

The devotion of top management is regarded as critical to achieving the organizational objectives. Top management makes strategic decisions, and the outcomes of those decisions have an impact on the success of the business [38]. Managers hold the responsibility of deciding how the resources will be utilized, what strategies needs to be followed and what changes need to be implemented. In other words, the quest of enhancing FMC cannot be achieved until the top management shows their whole hearted commitment towards achieving the same [39].

Table 1: Factors affecting the flexible manufacturing competence of organizations

Code	Factors name	References
X1	Advanced manufacturing technologies	[7] [8]
X2	Operational improvement practices	[8] [23]
X3	Internal and external learning	[24] [25]
X4	Adaptive human resources	[26] [27] [40]
X5	Absorptive capacity	[9] [28]
X6	Modularity in products and processes	[10] [15] [29]
X7	Time-based manufacturing	[11] [22]
X8	Capacity flexibility	[30] [31]
X9	Postponement implementation	[32] [33]
X10	Organizational structure	[16] [34] [35]
X11	Value Chain Management	[14] [36] [37]
X12	Top management commitment	[38] [39]

3. Methodology:

The objective of the present work is to provide insights to the manufacturing organizations that can help them in enhancing their FMC. In the beginning, the relevant literature published in the span of past 20 years has been searched and reviewed to identify the factors of interest. The databases such as Scopus, Web of Science, Researchgate, and google scholar have been used for the above. The research papers published in the areas including manufacturing flexibility, supply chain flexibility, mass customization, and lean/agile manufacturing have been given the priority during the search. The literature review rendered a total of 20 factors related to FMC of manufacturing organizations. A group of experts from academia evaluated the relevance and mutual exclusiveness of these factors. Based on the experts' suggestions, the factors were screened and a total of 12 factors has been chosen for the study. Following this, the relationship between the factors has been established with the assistance of experts and ISM technique was employed to construct the structural model of these factors.

3.1 Interpretive structural modelling

Individuals or groups, it is often assumed, face difficulty coping with complicated situations or systems having large number of variables as the interactions among these variables contribute to the complexity of the issues or systems [13]. The inclusion of variables that are directly or indirectly connected complicates the system's structure, which may or may not be stated clearly [17]. Dealing with a system whose structure isn't well defined becomes tough. As a result, it is necessary to employ a technique that assists in revealing the structure inside a system, ISM is an example of this kind of techniques [2]. The ISM method converts poorly, vaguely outlined models of systems into comprehensible structural models [5].

The steps in the ISM methodology are as follows [18]:

(i) Identify enablers that need to be examined and modelled.

(ii) Create a SSIM by associating the variables pair-wise and allocate contextual relationships between them.

(iii) From the SSIM, create a reachability matrix (RM) and test it for transitivity.

(iv) Divide the final RM into various levels.

(v) Using final RM, create a digraph and eliminate transitive links.

(vi) Transform the digraph into the structural model by adding factors in place of nodes.

(vii) Examine the ISM model for intangible inconsistencies and, if required, introduce necessary corrections.

3.2 MICMAC analysis

The concept of MICMAC analysis, asserts that if the variable 'a' directly influences the variable 'b' and variable 'b' directly influences variable 'c', then any act on variable 'a' will have implications on variable 'c', is formulated on the multiplication characteristics of matrices [18]. The analysis is employed to find out the power of the factors to affect others and being affected by others, i.e. driving and dependence power. To compute the driving power of a variable, all the entries of '1' in the respective row of the final RM are added, and its dependence power is calculated by adding the number of '1' in respective column of the final RM. Based on their powers, the variables can be classified to following four categories [19-21].

(i) Autonomous factors (I^{st} quadrant): These factors contain low driving and low dependence powers and have very strong linkages, however, the number of linkages are very few and therefore they remain isolated from the network.

(ii) Dependent factors (IInd quadrant): These factors display weak driving power and high dependence power, Thus, create difficulty in decision making as any action on any variable can greatly affect these factors.

(iii) Linkage factors (IIIrd quadrant): These factors have high driving and high dependence powers which means that the action taken over these factors will impact others, as well as themselves.

(iv) Independent factors (IVth quadrant): They have low dependence power but high driving power. They are referred to as the system's primary factors.

4. Analysis and results

4.1 Development of structural model using ISM

The work began with the identification of the relevant factors through literature review. A total of 12 factors are selected for the study and the description of these factors are provided in section 2. A panel of experts containing five academicians has been approached for establishing the interrelationship among the selected factors. Through a rigorous brainstorming session with these experts, the interconnection between each pair of the factors which is contextually meaningful is established. Following this, a structural self-interaction matrix (SSIM) is created by using VAXO analysis (Table 2). In VAXO analysis, the term 'V' represents that the factor 'a' will help in reaching factor 'b', the term 'A' represents that the factor 'b' will help in reaching factor 'a', the term 'X' represents that both the factors will help in reaching each other, and the term 'O' represents that both the factors do not have any relationship with each other [19]. The SSIM is then converted into final RM by replacing the entries of SSIM with binary digits and employing the concept of transitivity where few of the cells are filled by inference (Table 3).

Table 2: Structural self-interaction matrix

	(i)	(ii)	(iii)	(iv)	(v)	(vi)	(vii)	(viii)	(ix)	(x)	(xi)	(xii)
X1	1	V	O	O	A	V	V	V	O	O	A	O
X2		1	A	A	A	A	V	A	V	O	A	A
X3			1	X	X	V	V	V	V	X	A	A
X4				1	V	V	V	V	V	V	A	V
X5					1	X	V	V	V	V	A	A
X6						1	V	V	V	A	A	O
X7							1	X	V	A	A	A
X8								1	V	A	A	O
X9									1	A	A	A
X10										1	A	O
X11											1	V
X12												1

Table 3: Final reachability matrix

	(i)	(ii)	(iii)	(iv)	(v)	(vi)	(vii)	(viii)	(ix)	(x)	(xi)	(xii)	DRP*
X1	1	1	0	0	0	1	1	1	0	0	0	0	5
X2	0	1	0	0	0	0	1	0	1	0	0	0	3
X3	0	1	1	1	1	1	1	1	1	1	0	0	9
X4	0	1	1	1	1	1	1	1	1	1	0	1	10
X5	1	1	1	0	1	1	1	1	1	1	0	0	9
X6	0	1	0	0	1	1	1	1	1	0	0	0	6
X7	0	0	0	0	0	0	1	1	1	0	0	0	3
X8	0	1	0	0	0	0	1	1	1	0	0	0	4
X9	0	0	0	0	0	0	0	0	1	0	0	0	1
X10	0	0	1	0	0	1	1	1	1	1	0	0	6
X11	1	1	0	1	1	1	1	1	1	1	1	1	11
X12	0	1	0	0	1	0	1	0	1	0	0	1	5
DPP**	3	9	4	3	6	7	11	9	11	5	1	3	72/72

***DRP: Driving power; **DPP: Dependence power**

Table 4: Assigning the factors into different levels

Factor	Reachability Set	Antecedent Set	Intersection Set	Level
1	1	1,11	1	5
2	2	1,2,3,4,5,6,11,12	2	3
3	3,4,10	3,4,10	3,4,10	5

4	3,4,10,12	3,4,11	3,4	2
5	1,3,10	3,4,11,12	3	4
6	5,6	1,3,4,5,6,10,11	5,6	4
7	7,8	1,2,3,4,5,6,7,8,10,11,12	7,8	2
8	2	1,3,4,5,6,10,11	2	3
9	9	2,3,4,5,6,7,8,9,10,11,12	9	1
10	3,10	3,4,10,11	3,10	5
11	11	11	11	6
12	12	4,11,12	12	5

The level partitioning of the factors is then established by identifying the reachability and antecedent set from the final RM [5]. The reachability set comprises of any factors it may affect, whereas the antecedent set comprises of factors that may affect it [19]. The intersection of these sets is calculated for all of the factors, and the level of each factor is computed. The top level of the ISM hierarchy is inhabited by the factors with the same reachability and intersection sets. The procedure is performed in repetition to until the levels of each of the factors are determined. The diagraph (Figure 1) and the ISM model (Figure 2) is then constructed.

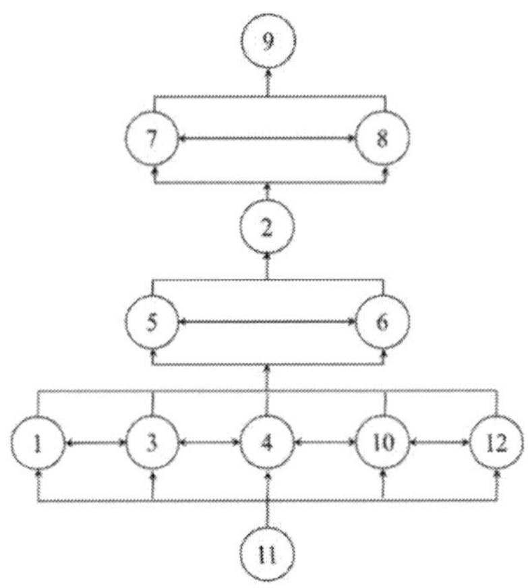

Figure 1: Final digraph

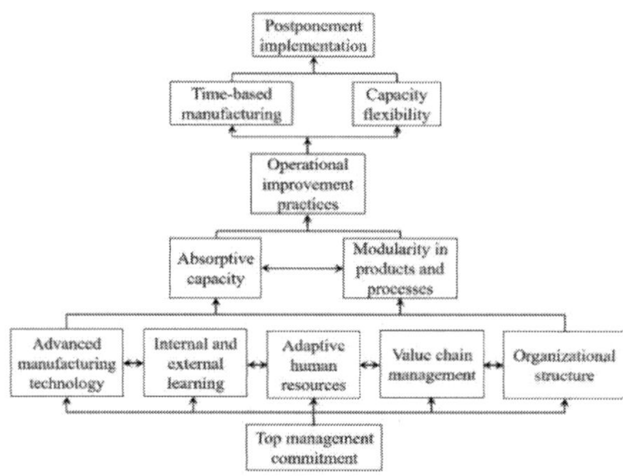

Figure 2: Final ISM model

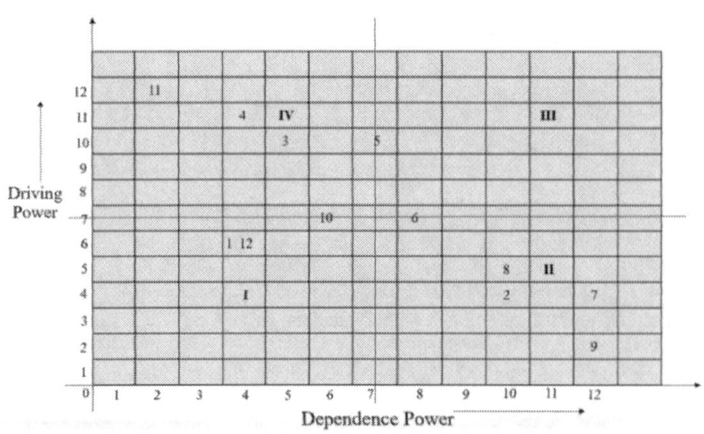

Figure 3: Diagram representing driving and dependence power

4.2 MICMAC analysis

In this analysis, the entries of final RM are used to evaluate the driving and dependence powers of the selected factors. The driving power of a factor is estimated by summing the entries of '1' into the respective row while the dependence power is computed by putting together the entries of '1' in the corresponding column. These values are presented in Table 3. Following the above, the values of driving and dependence power are plotted over a graph and separated into four parts as discussed in section 3.2. The diagram shown in Figure 3 furnishes critical insights into the relative significance and interdependence of the factors selected for the study. It can be observed from the Figure 3 that the factors X3, X4, X10, and X11 stay in the fourth quadrant and henceforth considered as independent factors. The X2, X7, X8, and X9 are categorized as dependent factors. The factors X1 and X12 are autonomous factors containing the least driving and dependence power among all factors. Furthermore, the factors X5 and X6 are found as linkage factors in the analysis having both strong driving and dependence powers.

5. Discussion

As the development and enhancement of FMC asks for significant amount of resources and efforts, organizations must be able to recognize the most important factors, how they interact with one another, and which strategy should be followed to achieve the maximum benefits in terms of FMC. The hierarchy, interactions and categorization of the factors presented in this work assists businesses in gaining pivotal insights regarding the factors that need immediate attention in order to gain better capabilities of coping with demand volatility and intense competition [19]. It can be observed from Figure 3 that 'top management commitment' is located at the lowermost position of the model. It denotes that this factor is the most crucial factor for firms seeking to improve their FMC. The finding seems logical, given that the top management is the in-charge of defining policies, guidelines, and strategic intent, as well as giving purpose and guidance for achievement of the organizational goals. The observation finds support from the MICMAC analysis as the factor lies in the fourth quadrant of the driving-dependence power diagram and can be considered as independent factor having strong driving power. This factor can be referred to as a primary factor in the quest of improving FMC. The next level of the hierarchy contains five factors, namely AMT, internal and external learning, adaptive human resources, value chain management and organizational structure. These factors hold almost equal importance since they are all in the same level of hierarchy. The driving-dependence power diagram shows that the factors including internal and external learning, value chain management and adaptive human resources are independent factors, therefore corroborates the findings from the ISM model. Another noteworthy observation from the MICMAC analysis is that the factors AMT and organizational structure are found as autonomous factors having strong links with other factors on the same level but with weak driving and dependence power. The findings could be explained by the fact that the benefits of AMT and organisational structure can only be realised in conjunction with other 0factors, not on their own. For example, AMT can be used to produce a variety of products and fulfil the variety of demands of customers only if the workers are adaptive and motivated to improve their skills and knowledge while work together with internal departments, suppliers, and customers. This kind of work culture depicts the prevailing structure in the organizations. The third level of hierarchy has two factors in it, namely absorptive capacity and modularity in products and processes. Both of these factors lie in the third quadrant of the driver-dependence diagram, indicating their dominance. These factors play pivotal role in the enhancement of FMC of organizations as they influence other factors as well as get influenced by others. The implication is that organizations should concentrate on improving their absorptive capacity by implementing appropriate information technologies and systems, establishing close ties with their suppliers and customers, and encouraging employees to place a greater emphasis on learning from others as well as from adverse circumstances. Having the required technologies, skilled and experienced workers and streamlined value chain, it would become extremely easy to incorporate modularity in the products and process design, resulting in a higher level of flexibility to fulfil the specific and diverse customers' needs. The fourth level of hierarchy has a single factor in it, namely operational improvement practices and this factor lies in the second quadrant of driving-dependence power diagram. This factor has high dependence and low driving power which means that it can lead to the continuous improvement in manufacturing only when it is implemented together with other factors such as AMT, adaptive human resources motivated to learn, absorptive capacity, modularized products and processes, etc. The factors time-based manufacturing and capacity flexibility are located at level 5 of the model. On the topmost level of the ISM model lies the factor postponement implementation. Similar observations are received from MICMAC analysis as these three factors lie in second portion and are mostly dependent on others. The interpretation of these findings is that the achievement of flexible capacity and the implementation of time-based manufacturing and postponement is possible only when the factors from the lower levels of hierarchy are successfully implemented, monitored, and managed, leading to augmented FMC of organizations.

6. Conclusions

This study adds to the existing knowledge and wisdom regarding the development and augmentation of FMC of manufacturing organizations by identifying key factors and highlighting the dynamics of interactions among them. The study categorises the factors of interest into autonomous, dependent, linkage, and independent factors, that can aid managers in formulating and implementing strategies and putting their best foot forward to develop flexible operational capabilities and convert the existing systems into an adaptive system. With environmental turbulence increasing at an alarming rate and expected to worsen, it is critical for businesses to come up with new ideas and develop unique strategies to amplify their FMC in the course of seizing opportunities to excel and addressing the current business environment. In this regard, the present work will be useful to organizations seeking a competitive advantage through the development of FMC.

7. References:

[1] Zhang Q, Vonderembse MA and Cao M 2006 Achieving flexible manufacturing competence: the roles of advanced manufacturing technology and operations improvement practices. *International Journal of Operations & Production Management*, 26(6), 580 – 599.

[2] Ibraimi S, Bexheti A, Zuferi R, Rexhepi G and Ramadani V 2016 Enhancing flexible manufacturing competence. *The Eurasia Proceedings of Educational and Social Sciences*, 5, 378-384.

[3] Raj T, Shankar R and Suhaib M 2007 An ISM approach for modelling the enablers of flexible manufacturing system: *The case for India. International Journal of Production Research,* 46(24), 1-30

[4] Ullah I and Narain R 2021 Achieving mass customization capability: the roles of flexible manufacturing competence and workforce management practices. *Journal of Advances in Management Research*, 18(2), 273-296

[5] Agrawal P, Narain R and Ullah I 2020 Analysis of barriers in implementation of digital transformation of supply chain using interpretive structural modelling approach. *Journal of Modelling in Management*, 15(1), 297-317

[6] Ullah I, Narain R and Singh A 2012 Role of buyer-supplier relationship and trust in organizational performance. *Delhi Business Review*, 13(2), 73-82

[7] Cagliano R and Spina G 2000 Advanced manufacturing technologies and strategically flexible production. *Journal of Operations Management*, 18(2), 169-190

[8] Mishra R 2020 Empirical analysis of enablers and performance outcome of manufacturing flexibility in an emerging economy. *Journal of Manufacturing Technology Management*, 31(6), 1301-1322.

[9] Wang Y, Guo B and Yin Y 2017 Open innovation search in manufacturing firms: the role of organizational slack and absorptive capacity. *Journal of Knowledge Management*, 21(3), 656-674.

[10] Makhloufi L, Azbiya Yaacob N, Laghouag AA, Ali Sahli A, and Belaid F 2021 Effect of IT capability and intangible IT resources on sustainable competitive advantage: Exploring moderating and mediating effect of IT flexibility and core competency. *Cogent Business & Management*, 8(1)
[11] Jin Y, Norbis M and Awudu I 2021 Supplier-dedicated resources and flexibility's roles in a manufacturer's superior performance. *Journal of General Management*, 47(2), 86-96
[12] Ivanov D, Das A and Choi T 2018 New flexibility drivers for manufacturing, supply chain and service operations. *International Journal of Production Research*, 56(10), 3359-3368.
[13] Kumar S, and Sharma RK 2015 An ISM based framework for structural relationship among various manufacturing flexibility dimensions. *International Journal of System Assurance Engineering and Management*, 6(4), 511-521.
[14] Agarwal A, Shankar R and Tiwari M 2007 Modelling agility of supply chain. *Industrial Marketing Management*, 36(4), 443- 457
[15] Ravi V, Shankar R and Tiwari MK 2005 Productivity improvement of a computer hardware supply chain. *International Journal of Production and Performance Measurement*, 54(4), 239-255
[16] Barve A, Kanda A and Shankar R 2007 Analysis of interaction among the barriers of third party logistics. *International Journal of Agile System and Management*, 2(1), 109-129
[17] Attri R, Dev N and Sharma V 2013 Interpretive Structural Modelling (ISM) approach: An Overview. *Research Journal of Management Sciences*, 2(2), 3-8
[18] Ullah I and Narain R 2018 Analysis of interactions among the enablers of mass customization: An interpretive structural modelling approach. *Journal of Modelling in Management*, 13(3), 626-645
[19] Ullah I and Narain R 2021 Analysing the barriers to implementation of mass customization in Indian SMEs using integrated ISM-MICMAC and SEM. *Journal of Advances in Management Research*, 18(2), 323-349.
[20] Ullah I, Narain R, and Singh A 2015 Supply chain management practices in SMEs of India: Some managerial lessons from large enterprises. *International Research Journal of Engineering and Technology (IRJET)*, 2(5), 256-269.
[21] Ullah I and Narain R 2020 The Impact of Customer Relationship Management and Organizational Culture on Mass Customization Capability and Firm Performance. *International Journal of Customer Relationship Marketing and Management*, 11(3), 60-81.
[22] Belay AM, Kekäle T and Helo P 2011 Time-to-market and concurrent engineering in product development processes. *International Journal of Innovation and Learning*, 10(1), 60-84.
[23] Da Silveira GJ 2006 Effects of simplicity and discipline on operational flexibility: an empirical reexamination of the rigid flexibility model. *Journal of Operations Management*, 24(6), 932-947.
[24] Okoro CS, Nkambule M and Kruger A 2020 The state of restroom facilities as a measure of cleaning service quality in an educational institution. *Journal of Corporate Real Estate*, 23(1), 55-68.
[25] Harryson S, Kliknaite S and Dudkowski R 2008 Flexibility in innovation through external learning: exploring two models for enhanced industry? university collaboration. *International Journal of Technology Management*, 41(2), 109-137.
[26] Sabuhari R, Sudiro A, Irawanto D and Rahayu M 2020 The effects of human resource flexibility, employee competency, organizational culture adaptation and job satisfaction on employee performance. *Management Science Letters*, 10(8), 1775-1786.
[27] Do BR, Yeh PW and Madsen J 2016 Exploring the relationship among human resource flexibility, organizational innovation and adaptability culture. *Chinese Management Studies*, Vol. 10(4), 657–674
[28] Patel PC, Terjesen S, and Li D 2012 Enhancing effects of manufacturing flexibility through operational absorptive capacity and operational ambidexterity. *Journal of Operations Management*, 30(3), 201–220.
[29] Wang Z and Zhang M 2020 Linking product modularity to supply chain integration and flexibility. *Production Planning & Control*, 31(14), 1149-1163
[30] Alp O and Tan T 2008 Tactical capacity management under capacity flexibility. *IIE Transactions*, 40(3), 221-237

[31] Holtewert P and Bauernhansl T 2016 Increase of Capacity Flexibility in Manufacturing Systems by Substitution of Product Functions. *Procedia CIRP*, 57, 92–97.

[32] Saghiri S and Barnes S 2016 Supplier flexibility and postponement implementation: An empirical analysis. *International Journal of Production Economics*, 173, 170-183

[33] Ferreira KA, Flávio LA, and Rodrigues LF 2018 Postponement: bibliometric analysis and systematic review of the literature. *International Journal of Logistics Systems and Management*, 30(1), 69-94.

[34] Sawhney R 2006 Interplay between uncertainty and flexibility across the value-chain: towards a transformation model of manufacturing flexibility. *Journal of operations management*, 24(5), 476-493.

[35] Koçyiğit Y and Akkaya B 2020 The role of organizational flexibility in organizational agility: A research on SMEs. *Business Management and Strategy*, 11(1), 110-123.

[36] Soon QH and Udin ZM 2011 Supply chain management from the perspective of value chain flexibility: an exploratory study. *Journal of Manufacturing Technology Management*, 22(4), 506-526.

[37] Siagian H, Tarigan Z and Basana S 2022 The role of top management commitment in enhancing competitive advantage: The mediating role of green innovation, supplier, and customer integration. *Uncertain Supply Chain Management*, Vol.10(2), 477-494.

[38] Ullah I and Narain R 2021 Linking supplier selection and management strategies with mass customization capability. *Journal of Business & Industrial Marketing*, 36(7), 1213-1228.

[39] Shukla AR, Khalatkar AM and Ullah I 2022 Scrap Reduction and Proper Utilization of Raw Materials During Cutting Operations in SMEs. In *Smart Technologies for Energy, Environment and Sustainable Development*, 1, 725-737.

[40] Ullah I and Narain R 2021 Linking supply network flexibility with mass customization capability. *Journal of Business & Industrial Marketing*.

An application of quality tools to improve the tyre remanufacturing process

Jeevan Gaidhane[1], Asmit Karadbhajane[1], Abhay Khalatkar[1], Inayat Ullah*[1].
*Corresponding author

[1]Department of Mechanical Engineering, G H Raisoni College of Engineering Nagpur, India.

*Corresponding Author Email: inayatu6@gmail.com

Abstract. The tyre remanufacturing industries are facing quality and performance issues of the remanufactured tyres as they have developed their own tyre retreading methods due to the lack of standardized retreading process guidelines. This study focuses on identifying the challenges of the tyre remanufacturing process, highlighting various defects in retread tyres and their causes so that the challenges could be mitigated and the quality and performance of the tyres could be improved. To achieve the objectives of the study, the quality tools including fishbone diagram and Pareto chart has been used. For the above analysis, the data has been collected through multiple field visits and semi-structured interviews. The findings of the study indicate that organizations can improve the utilization of worn-out tyres by identifying tyre retreading defects and their causes and by implementing corrective actions at different stages in the tyre remanufacturing. According to cause-and-effect analysis, the defects found are failure of tread due to lack of standard tread joint method, side bulge generation due to sidewall degrading, and failure of the buffer layer due to tread and sidewall crack. Besides, a few recommendations are made for tyre remanufacturing to increase productivity and minimize the defects generation in retreaded tyres.

Keywords: Fishbone diagram, Pareto chart, Root-cause analysis, Retreading tyres defects, Tyre remanufacturing.

1. Introduction

Tyre remanufacturing is an eco-friendly, sustainable and cost-effective method in which the worn-out tyres are repurposed by applying new precure tread. Tyre remanufacturing increases the tyre life which, in turn, reduces the environmental impacts of discarded tyres and hence, the process contributes to sustainability and circularity [1]. Earlier, the worn-out tyres were disposed of using various disposal practices such as stockpiling, burning, and illegal landfilling. These practices are still being used by the majority of people which leads to serious environmental and health hazards [2]. Owing to these hazardous impacts, academicians, policymakers, and practitioners have started looking into these issues seriously. Many organizations, mostly small businesses, have started tyre retreading practices in order to generate revenues from waste rather than minimizing environmental impacts [3]. In developing countries such as India, tyre remanufacturing sector is still unorganized and there exists a lack of standard process guidelines, thus, these organizations have developed their own methods of tyre remanufacturing and recycling processes. Due to the lack of standardised process guidelines and other challenges, the adoption of tyre remanufacturing practices is very low and the disposal rates are very high. Majority of the existing processes are ineffective which results in poor quality, low life, and poor performance of the remanufactured tyres [4,5]. To satisfy the expectations of the customers and to popularize the use of remanufactured tyres, the remanufactured tyres must be more safe and durable, have high levels of quality and capability of rendering better performance, which necessitates a clear

IOP Conf. Series: Materials Science and Engineering 1259 (2022) 012033 doi:10.1088/1757-899X/1259/1/012033

change in the existing tyre remanufacturing processes [6]. The efforts are needed to develop best operational practices, standards and quality systems to increase the process efficiency and productivity as well as to improve the tyre life in terms of safety, quality, performance and durability [7]. Besides, more research is needed to examine and identify new perspectives for increasing the use of tyres for remanufacturing rather than disposal. However, to achieve the above, there is a need to identify the prevailing challenges of tyre remanufacturing process, pinpoint the defects in retread tyres and their causes so that necessary actions can be taken to rectify the crucial issues [8]. In line with the above, the present study has been carried out to investigate the challenges of tyre remanufacturing process and the defects in remanufactured tyres as well as their probable causes [9]. The study has used following methodologies: data collection through field visits and semi-structured interviews, brainstorming, pareto chart and cause-and-effect analysis. The findings of study provides crucial managerial implications that the top management should look forward to implement for the development of better operations routines and an effective and optimized tyre remanufacturing process that could help in eliminating the defects of remanufactured tyres and increasing their quality, safety and performance.

The remainder of this study has been organized as follows. Section 2 provides a brief overview of the existing tyre remanufacturing process and the use of quality tools. The methodology adopted for the study are described in section 3. In section 4, the analyses performed to achieve the objectives of the research has been illustrated. Following this, the findings of the study are discussed in section 4. In the end, the conclusions and the directions of future research are outlined.

2. Materials and Methods

2.1 Tyre remanufacturing process

There are two methods of tyre retreading, namely hot tyre method and cold tyre method. In both of these methods, a proper bonding of tread with tyre base is achieved through the application of heat and pressure for a certain time period. Among these two methods, the cold tyre method is most popular among the organizations which are involved in retread business [11,10]. These organizations work closely with tyre manufacturing industries and keep an eye on the innovations/developments in the areas of materials and other performance parameters which are then incorporated with tyre remanufacturing process [12, 13]. The steps involved in the tyre remanufacturing process are as follows.

• **Primary Inspection:** In this step, the worn casing, cuts, punctures, and other damages are identified to decide the extent of retreading.
• **Buffing:** Buffing is performed to remove the outer layers of worn-out tyres using high-speed buffers. More specifically, buffing removes the undesired or worn tread.
• **Cleaning:** After buffing, the casing of the tyre is cleaned using high pressurized air.
• **Casing repair:** In this step, the casing is inspected for the damages that need to be corrected. These corrections are necessary to make sure that the quality of casing would remain within the acceptable limits and the retreaded tyre would exhibit similar performance as a new tyre.
• **Cementing:** A temporary bond between the tread and casing is created in this step to prevent the buffed surface from corrosion.
• **Curing or mold curing process:** After adding the layers of tread up to the required diameter, using uncured rubber compound, the casing is placed into the mold. The molds are then closed and the tyre body is filled to the right pressure. This casing is allowed to expand within the mold so that so that the uncured material achieves the desired structure of the product along with the tread pattern [14,15].

2.2 Pareto chart

The Pareto chart is an excellent quality tool for identifying significant issues in the system. It is made up of a sequence of bars, the heights of which represent the frequency or severity of issues. The bars are organized in ascending order of height from left to right. This indicates that the issues depicted by the tall bars on the left are more crucial than those represented by the short bars on the right [16]. The tool is incredibly helpful in identifying the issues that have the most significant cumulative effect on the system. Furthermore, the tool renders ability to categorize the issues according to the weight of impact so that the necessary actions can be employed in a prioritized manner. The literature states that approximately 80% of the issues arise from 20% of the causes in any system. The appropriate data must be gathered and analysed from the process or system which is under scrutiny in order to produce a Pareto Chart [17]. When the list of issues contains a mix of problems including both the issues with potentially major repercussions and issues with minor consequences, one of the following two approaches can be used, (i) the author weighting method to adjust count frequency, or (ii) frequency analysis paired with a cost or exposure Pareto chart. The various versions of the basic Pareto chart are routinely used in quality improvement approaches to identify the bulk of issues that may arise due to small reasons [18].

2.3 Brainstorming

Brainstorming is a technique which is used for generating a large number of creative solutions and innovative ideas for crucial problems that occur during the various stages of value creation and distribution. Brainstorming is not a method for analysing ideas or creating ideas; nevertheless, its output can be considered as an input to other quality improvement techniques such as the fishbone diagram [19,20,21]. In this method, a group of experts, having different opinions regarding the subject of interest, gather in a room and sit around a table to discuss and find all possible and valid solutions to the problem at hand [22]. During the session, the suggestions and queries of all group members are catalogued in a way that everyone can see them and, if necessary, modify them. To facilitate the brainstorming session, it is preferable to adhere to the fundamental rules and explain them to the members. Furthermore, to ensure the success of the meeting, the role of team leader in creating a friendly and comfortable environment for the team members is crucial. The technique can be used for a variety of purposes, including problem identification, determining causes and effects of a problem, and arriving at optimal solutions with shared understanding of all the members of the team [23,24].

2.4 Cause-and-effect Diagram

The Fishbone diagram (also known as the Ishikawa diagram or cause-and-effect diagram) is a visual representation of how different factors are sorted into main categories and subcategories. This technique can aid in the generation of ideas by encouraging a balanced approach in a brainstorming session when people list their observations about the origin (cause) of a problem (effect) [25]. Depending on the case at hand, the causes are classified differently, but it is a widely recognized approach to assume six key components that can contribute to an inappropriate characteristic, response, or impact. These components are materials, machine, man, measurement, process, and environment. Each of these items can then be explored for sub-causes through a brainstorming session among a group of experts [26]. The fishbone diagram improves the quality of the processes that are currently in use through the monitoring and control of the causes and effects of different issues. It also aids in identifying the root causes of an existing problem as well as analyzing the likelihood of future events and preventing them from occurring.

2.5 Industrial applications of Pareto chart and cause-and-effect diagram

There are several studies in the literature that have utilized Pareto chart and fishbone diagram as quality improvement tools. For example, Behnam and Alvelos (2011) discussed the application of quality tools in tyre remanufacturing with the specific goal to improve the understanding of root-cause analysis regarding the quality issues that has been discovered during the retreading process. In this work, the authors identified the critical steps in the process, determined their root cause, and evaluated all the potential causes of risks by using the high-quality tools such as Pareto chart, brainstorming, bond diagram, and metrics diagram [27]. Laddha (2021) evaluated the performance of retreaded tyre and their environmental impact. To meet the objectives of the work, a total of 60 retreaded tyres of different brands and sizes were remanufactured and their quality and performance was assessed [28]. Banerjee (2020) investigated the cold remanufactured process by studying the characteristics and composition of materials in order to perform the quality checks that are needed at various stages of retreading [29]. A brief summary of relevant literature has been provided in Table 1.

Table 1: A summary of literature on the use of quality tools and identification of defects

References no.	Research objectives	Tools used
[27]	Find out the root cause analysis of the process in the tyre retreading industry from ken tyre.	Pareto Chart, Brainstorming, Fishbone Diagram, Matrix Diagram.
[33]	Risk management analysis for improvement of performance and safety of tyres and reliability in the deployment stages.	Pie chart.
[34]	Which method used retreading in government ST. the workshop follows also finds the various retreading defects during remanufacturing.	Bar chart.
[30]	Improvement of enveloped packing process before the tyre is placed in the mold.	Bar chart.
[32]	The objective of the paper is to investigate the retread and new tyres on the basis of distance travel of 1000km.	Bar chart.
[29]	Investigate and explain the tyre retreading stepwise process of cold and hot working and material properties of the compound.	Fishbone diagram.
[28]	Investigate the various factors affecting tyre retreading brands and performance off retread tyres.	Bar chart.
[31]	Investigate the tyre retreading defect of fatigue growth crack rate generation in retreaded tyres.	Scatter diagram bar chart.

3. Methodology

The methodology adopted for this work is shown in Figure 1. The study began with the selection of the tyre retreading process, i.e., cold retreading process. Further, data has been collected over a period of four months (September 2021 to December 2021) through field visits and semi-structured interviews of industry personnel. A total of 25 industries, located in Nagpur, Maharashtra, have been visited and 50 interviews have been conducted. During the data collection phase, the information regarding various types of tyre defects and their causes and other important factors that play significant roles in the assuring the quality and performance of retread tyres have been collected. Besides, a literature review has been conducted to cross-validate the accuracy of the collected data. In the next step, the factors influencing retreading process have been selected through the analysis of the data and literature review; next, these factors are classified into the following categories, man, machine, technique, material, process, and environment (Table 2). Following the categorization of factors, a fishbone diagram is developed using all the valid tyre retreading factors (Figure 2). The various defects that occur in the retreaded tyre are identified in the next step and a Pareto chart has been prepared to depict the severity of defects (Figure 4). Using the output of the Pareto chart another fishbone diagram (Figure 5) is prepared, containing the most severe defects. Following this, the root-cause analysis using the fishbone diagram and brainstorming has been carried out to identify the major causes of the defects. Finally, an action plan is prepared to improve the quality, safety, durability, and performance of retreaded tyre (Table 3).

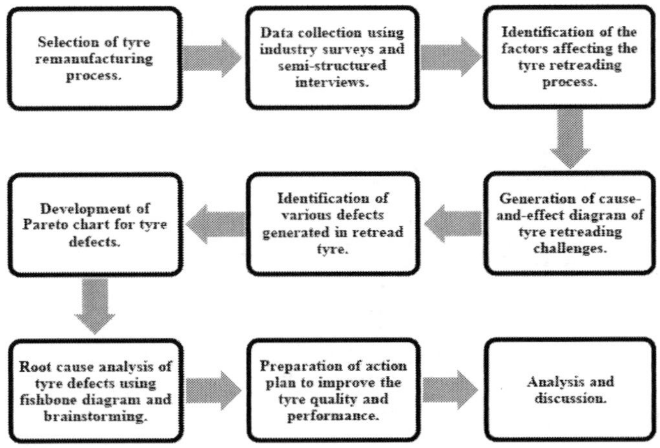

Figure 1. Research methodology.

4. Analysis and results

4.1 Challenges in tyre remanufacturing

The tyre remanufacturing industry faces several challenges during the application of the retread over the worn-out tyre. Since the number of times for which tread can be applied on casing is limited, the identification, monitoring, and control of the critical process parameters as well as challenges are necessary. Through the analysis of the data, a total of 32 challenges have been identified. These challenges are further classified into six categories. These categories and the challenges are shown in Table 2. One of the major challenges is casing preparation, which requires skilled workers having the ability to pinpoint the problems in the casing. As there is no automated system to check the casing defects, it is done manually. Based on their skills and experience, the workers decide whether to accept the casings and proceed with the steps of tyre retreading process. After the casing preparation, another challenge is to apply the pre-cure tread on the casing due to difficulty in finalizing the pre-cure tread length. To avoid the issue, the operator should measure the casing perimeter as well as tread length while assembling the tread in the tyre casing. Further, by using all the valid challenges, a fishbone diagram is prepared that depicts the different categories of crucial challenges that one could face in tyre remanufacturing. The fishbone diagram is shown in Figure 2.

Table 2. Challenges faced during the tyre retreading process.

Category	Valid and invalid tyre retreading challenges
Man	• Lack of process knowledge • Lack of awareness regarding tyre retreading defects • Uneven speed (mura, muda, muri) • Lack of training regarding inspection and quality control • Lack of classification of defects and causes • Lack of systematic work plan
Machine	• Improper mold design • Old technology-based machines • Lower tyre curing rate • Lack of calibration of tools, machines and instruments • Absence of preventive maintenance plan • High energy consumption
Material	• Lack of testing of material composition • Lack of testing of material properties

Process	• Lack of standards for manufacturing methods • Lack of standard guidelines for tyre casing preparation • Lack of proper loading and unloading methods of tyres • Lack of variations in tread pattern designs • Lack of proper guidelines for deciding the tread gauge • Lack of standard for retreading pressure • Lack of standard for retreading temperature • Lack of standard setup time
Environment	• Poor weather conditions (hot, cold, humid) • Dusty (pollution) • Shortage of natural resources. • Lack of floor area • Unspecified light intensity
Method	• Lack of standard measuring instrument • Lack of FIFO, LIFO • Lack of 5S • Lack of destructive and non-destructive testing • Lack of compound testing methods

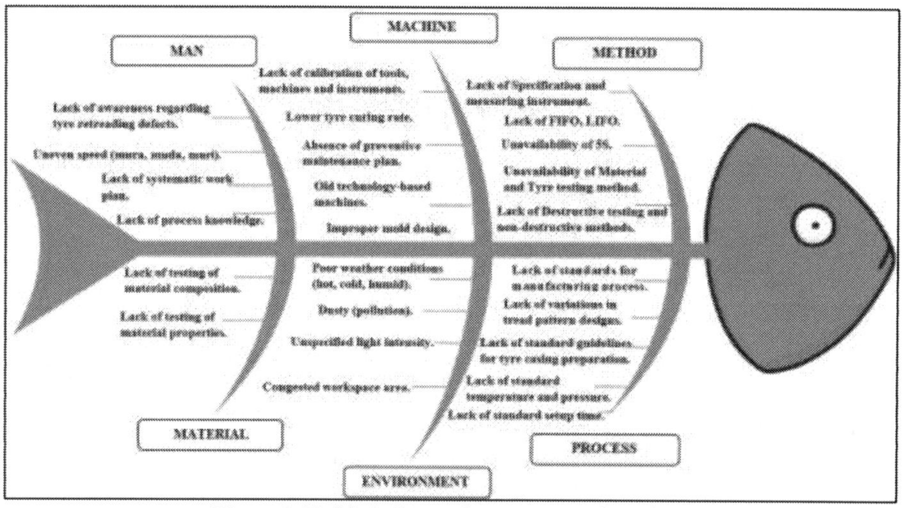

Figure 2. Valid challenges in tyre remanufacturing.

4.2 Use of Pareto chart to display the severity of retreading defects.

The most common defects in remanufactured tyres include failure of tread, failure of the sidewall, failure of the buffer layer, bead delamination, penetration damage, sidewall cut, exposed ply, and bulges. The defects are shown in Figure 3. In this study, a Pareto chart is used to organize the remanufacture tyre defects and to help the operations managers in determining the severity of defects and identifying the appropriate solution to overcome the defects. The Pareto chart depicting various defects and their severity is presented in Figure 4. The Pareto chart shows the percentage frequency of various defects which are arranged in descending order. The steps involved in drawing the Pareto chart are as follows. First, the common defects in retreaded tyres have been identified through literature review, field visits, and interviews. The identified defects are further arranged in a sequence based on their severity level as specified by the data collected from the industrial survey. Following this, using MS-Excel, the data has been plotted in the form of a bar chart showing the percentage-wise rating of defects and their cumulative percentage. It can be observed from the Pareto chart that the most severe defect of retread tyre is the failure of tread which is found about 39% of total defect generations in the retreaded tyre. The other important defects include exposed ply, failure of the sidewall, failure of the buffer layer, bulge, casing defects, and bead delamination.

4.3 Analysis of the causes of tread failure through fishbone diagram

By analysing the data, a portfolio of important causes of tread failure and their sub-causes have been identified. These factors are further used for the development of a fishbone diagram, depicting the major causes and their sub-causes. To ensure the well-defined identification of the various causes and sub-causes in each of the major categories, a team of experts, containing 6 members, has been consulted and a rigorous brainstorming session has been conducted.

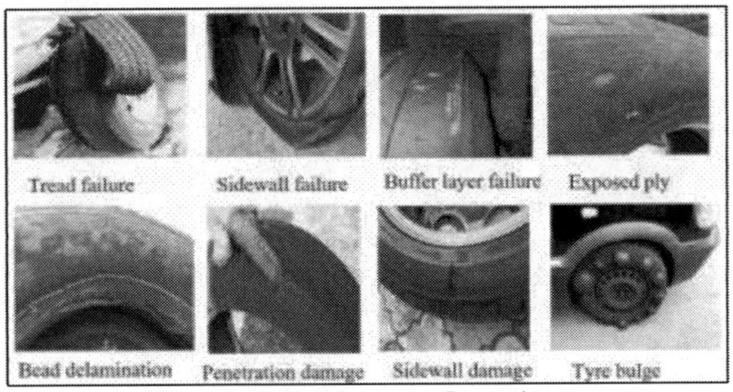

Figure 3. Defects in remanufactured tyres.

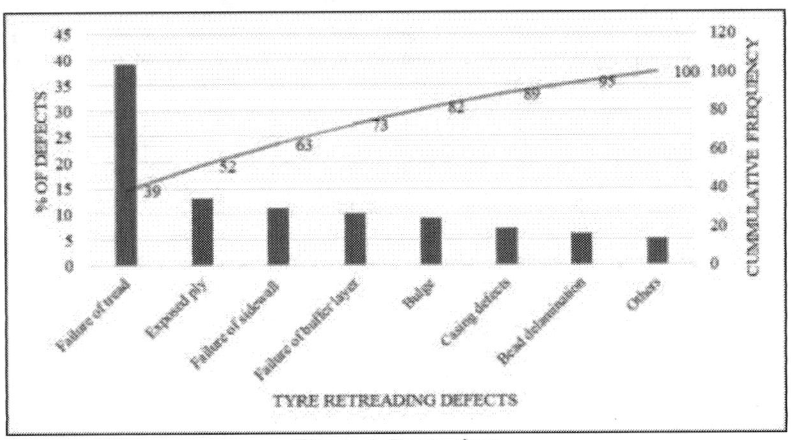

Figure 4. Pareto chart.

The experts included a remanufacturing and planning supervisor, casing preparation worker, tyre storage manager, quality worker, sales representative, and a representative of the factory manager who was also the team's leader. In the brainstorming session, the team leader, first, explained the objectives of the session. After that, the conversation regarding the essential process factors, tyre defects and their causes has been started. The activity proceeded with a discussion of the group members' ideas and suggestions, which they had written down on paper. In the final step, the conclusion, the final evaluation, and modification (some identified sub-causes were deleted while others were added) have been incorporated into the common solution. Figure 5 illustrates the fishbone diagram based on the causes and sub-causes of tread failure.

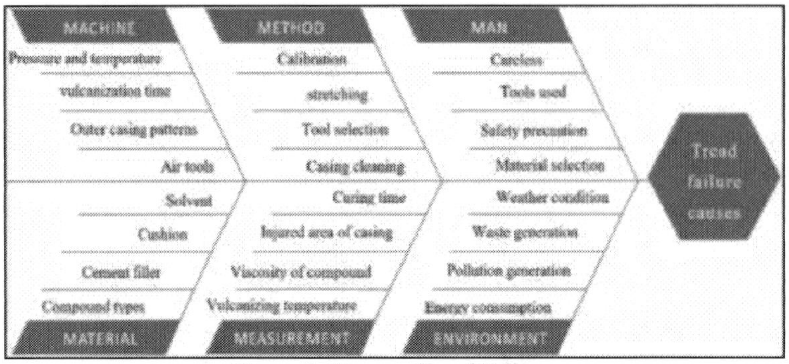

Figure 5. Causes of tread Failure.

5. Discussion

The purpose of the present study is to identify the tyre remanufacturing process challenges and the various causes of tread failure so that appropriate actions can be employed by the managers to improve the above. To this end, data has been collected through industrial visits and semi-structured interviews of industry personnel. The information collected is further used to develop fishbone diagrams and Pareto chart. The identified challenges of tyre remanufacturing are classified into five sections, namely man, machine, process, material, and environment. It has been observed that workers lack the process knowledge due to the lack of specific training and therefore, they do not perform the work systematically. The workers do not possess the knowledge regarding the different tyre defects, their causes and the approaches to prevent the defects. The machines which are being used by the majority of the industries are based on old technologies and need to be upgraded to ensure the production of high-quality remanufactured tyres. The other significant observation related to machine includes the lack planning for preventive maintenance, lack of proper mold design, poor curing rate due to the lack of standard operating procedure, high energy consumption, and lack of calibration of tools and instruments. Besides, the findings show that the firms do not focus on employing the crucial operational practices such as inspection and testing of material quality and composition, developing standard process guidelines for casing preparation, ensuring the effectiveness of process parameters such as pressure, temperature, and time, developing standard procedure for minimizing loading and unloading of tyres on machine as well as setup and changeover time, and implementing 5S methodology at the shop-floor.

While investigating the tyre defects, it has been observed that tread failure is the most prominent defect in a retread tyre which is followed by exposed ply failure of sidewalls, failure of buffer layer, bulge, casing defects, and bead delamination. Among these defects, tread failure accounts for 39% of total tyre retreading defects. The majority of the firms follow their own tread joint procedure, such as using high additive gum, bonding gum, and cushion strip in the joint area, and a few of them apply staple pins to achieve better tread joints. The application of staple pins in the joint area is extremely unsafe and leads to generation of most of the defects in retread tyres. The fishbone diagram (Figure 5) illustrates the various causes of these defects which could be beneficial for managers in deciding how to control the occurrence of these defects. Overall, the use of quality tools reveals that one of the most important reasons for the defects is the lack of standard process guidelines regarding tyre remanufacturing which have led these firms to develop their own work procedure. These procedures are mostly faulty as the firms do not have the required skilled personnel, advanced technologies as well as the innovative mindset.

In addition to the above, a list of action plan is formulated to minimize the defects and to mitigate the process challenges which is presented in Table 3. As the pollution is increasing day-by-day and the hazardous impacts of environmental deterioration are now becoming visible in the form of changing climatic conditions, weather pattern, losses of flora and fauna, and various diseases, it has now become imperative more than ever for everyone to look for ways to minimize the resource consumption, pollution, and carbon footprint. The adoption of tyre remanufacturing business approach could be one of the most effective means to achieve the above. Using tyre remanufacturing as a business strategy one can generate revenue from wastes, however, the process the not that simple as it needs

IOP Conf. Series: Materials Science and Engineering 1259 (2022) 012033 doi:10.1088/1757-899X/1259/1/012033

a lot of efforts in terms of resources and time. Hopefully, the findings of the study would help the firms to move ahead towards the implementation of an effective system for tyre remanufacturing.

Table 3. Action plan to mitigate tyre retreading challenges.

Type	Suggested action plan
Man	• The workers should be trained for the retreading process. • Experienced workers should be assigned for product and process quality checks. • Fostering problem identification and joint problem-solving approach. • Use of concurrent engineering and value stream mapping to develop standard work plan. • The development of standard operating procedure for handing tyre defects and their causes.
Machine	• Maintaining the proper standard pressure during the curing process. • Maintaining the standard temperature during the curing process. • Machine design mold patterns used the latest technology tread patterns design. • Using either technological advanced machines or developing and installing innovative attachments in old machines. • Using calibrated measuring instruments to enhance the accuracy of process parameters. • Calculation of optimum curing time in the mold. • Developing periodic preventive maintenance plan.
Method	• Development of standard process specifications in tyre retreading process. • Development of standard work procedure for hot and cold tyre retreading. • Development of standard inspection methods. • Use of FIFO approach in tyre retreading. • Use of standard material testing techniques to ensure the product quality.
Material	• Performing material composition testing. • Procuring raw material from certified suppliers • Implementing quality inspection for worn-out tyres as well as tread to be applied.
Process	• Development of standard procedure for casing preparation and providing necessary training to workers. • Development of standard operating procedure for loading and unloading of tyres • Selection of best procedure to make a perfect tread joint to avoid tread failure. • Development of standard operating procedure for the selection of specific tread pattern for specific tyre size for better performance. • Selection of optimum tread depth for a particular application and providing TWI (tread wear indicator) to all remanufactured tyres.
Environment	• Minimizing the energy consumption per kg of tyres. • Minimizing the waste generation during tyre retreading operations. • Development of guidelines for recycling and reuse of the waste generated during the process. • Investigating the effects of weather conditions (hot, cold, and humid) on the tyre retreading process and developing the control parameters.

6. Conclusions

The present work highlights the significant challenges of tyre remanufacturing and classifies them into different categories that could be used by the managers while looking to overcome the impediments. In addition, the most prominent defects and the reasons for their occurrences have been identified and a comprehensive list of plans of action has been provided which could provide necessary insights to the managers while developing a roadmap for the development of high quality and high-performance remanufactured tyres. The use of quality tools including the Pareto chart and fishbone diagram has provided a clear picture of the current situations as well as the actions that can be taken to improve them. Hopefully, the present work would prove beneficial for the firms to adopt and excel in the tyre remanufacturing business.

Acknowledgement
This work has been conducted under the Center of Excellence Flexible manufacturing system/ Robotic Design, G H Raisoni College of Engineering Nagpur India. The authors are also thankful for the support provided by the organisation Bharat Tyre Retreading Works Nagpur, India.

References

[1] Ortíz-Rodríguez, O. O., Ocampo-Duque, W., & Duque-Salazar, L. I. Environmental impact of end-of-life tires: Life cycle assessment comparison of three scenarios from a case study in Valle Del Cauca, Colombia. *Energies,* (2017). 10 (12), 2117. p.1-13.

[2] Araujo-Morera, J., Verdejo, R., López-Manchado, M. A., & Santana, M. H. Sustainable mobility: The route of tires through the circular economy model. (2021). *Waste Management,* 126, p.309-322.

[3] Mrad, M., & El-Samra, R. Waste tire management: Lebanon case study. *Journal of Waste Management and Disposal,* (2020). 3(1), 101.

[4] Gaidhane, Jeevan, Inayat Ullah, and Abhay Khalatkar. "Tyre remanufacturing: A brief review." *Materials Today: Proceedings,* (2022). 60, (3), p.2257-2261

[5] Ferronato, N., & Torretta, V. Waste mismanagement in developing countries: A review of global issues. *International journal of environmental research and public health,* (2019). 16(6), 1060.

[6] Gupta, V., Jain, R., Meena, M. L., & Dangayach, G. S. Six-sigma application in tire-manufacturing company: a case study. *Journal of Industrial Engineering International,* (2018). 14(3), p.511-520.

[7] Campbell-Johnston, K., Friant, M. C., Thapa, K., Lakerveld, D., & Vermeulen, W. J. How circular is your tyre: Experiences with extended producer responsibility from a circular economy perspective. *Journal of Cleaner Production,* (2020). 270, 122042.

[8] Qiang, W., Li, J., Yunlong, W., Xiaojie, Q., & Guotian, W. Discussion on Tyre Retreading and Reuse Technology. *In IOP Conference Series: Earth and Environmental Science,* (2020). 512(1), p. 120-146.

[9] Shukla, A. R., Khalatkar, A. M., & Ullah, I. Scrap Reduction and Proper Utilization of Raw Materials During Cutting Operations in SMEs. *In Smart Technologies for Energy, Environment and Sustainable Development,* (2022). Vol 1 p.725-737.

[10] Siegfried, G. An International View on Tyre Retreading. *Vale bridge publications,* (2000). United Kingdom.

[11] Lindquist, K.; Wendt, M. Retreaded Tire Use and Safety: Synthesis, *Proceeding of Washington State Department of Transportation,* (2009). 2–16.

[12] Ullah, I., & Narain, R. Achieving mass customization capability: the roles of flexible manufacturing competence and workforce management practices. *Journal of Advances in Management Research.* (2020). ISSN: 0972-7981.

[13] Ullah, I., & Narain, R. Linking supply network flexibility with mass customization capability. *Journal of Business & Industrial Marketing,* (2021). ISSN: 0885-8624.

[14] Sharma, A. Retreading of tyres. *International Journal of Engineering and Advanced Technology (IJEAT),* (2013). 2(6). ISSN, 2249-8958.

[15] Kumar, S. A. Tyre Retreading by Hot Retreading Process. *International Journal of Applied Science and Engineering,* (2016). 4(2), p.61-65.

[16] Karuppusami, G., & Gandhinathan, R. Pareto analysis of critical success factors of total quality management: A literature review and analysis. *The TQM Magazine*, (2006). 18, p.372–385.

[17] Jayswal, A., Li, X., Zanwar, A., Lou, H. H., & Huang, Y. A sustainability root cause analysis methodology and its application. *Computers and Chemical Engineering,* (2011). 35, p.2786–2798.

[18] Benjamin, S. J., Marathamuthu, M. S., & Murugaiah, U. The use of5-WHYs technique to eliminate OEE's speed loss in a manufacturing firm. *Journal of Quality in Maintenance Engineering,* (2015). 21, p.29–39.

[19] Rietzschel, E. F., Nijstad, B. A., & Stroebe, W. Productivity is not enough: A comparison of interactive and nominal brainstorming groups on idea generation and selection. *Journal of Experimental Social Psychology,* (2006). 42(2), p.244-251.

[20] Ullah, I., & Narain, R. Analyzing the barriers to implementation of mass customization in Indian SMEs using integrated ISM-MICMAC and SEM. *Journal of Advances in Management Research.* (2020). Vol. 18 No. 2, p.323-349.

[21] Ullah, I., & Narain, R. The impact of customer relationship management and organizational culture on mass customization capability and firm performance. *International Journal of Customer Relationship Marketing and Management (IJCRMM),* (2020). 11(3), p.60-81.

[22] Treffinger, D. J., & Isaksen, S. G. Creative problem solving: The history, development, and implications for gifted education and talent development. *Gifted Child Quarterly*, (2005). 49(4), p.342-353.

[23] Jachak, P. R., Khalatkar, A., & Narkhede, N. M. Real-Time Problem of Knob Assembly in Automobile Manufacturing Industries. In Smart Technologies for Energy, Environment and Sustainable Development, *Springer, (*2022). Vol 1, p 831-835.

[24] Sutton, R. I., & Hargadon, A. Brainstorming groups in context: Effectiveness in a product design firm. *Administrative science quarterly,* (1996). p.685-718.

[25] Raghavan, A. Root cause analysis. In Management and Leadership–A Guide for Clinical Professionals. *Springer, Cham*. (2015). p.105-121.

[26] Davis, S. Introduction to total quality: quality management for production, processing, and services. *Prentice Hall*. (1997).

[27] Behnam, B., & Alvelos, H. Exploring the potential of quality tools in tyre retreading industry: A case study. *International Journal of Engineering Science and Technology (IJEST),* (2011). Vol. 3, Issue 6, p.5337-5345.

[28] Laddha, S. Tyre Retreader's Perception About Quality & Brand. *PalArch's Journal of Archaeology of Egypt/Egyptology*, (2021). 18(7), p.2425-2431.

[29] Bireswar Banerjee, Manufacturing technology for tyre Retreading. *Resource, Uppsala University, Sweden Library Authenticated.* (2020). p. 47-74. https://doi.org/10.1515/9783110643893-003

[30] Patel HV, Kayasth BD. A review on Design and Analysis of Tyre Envelope Expander for Tyre Retreading Process. *International Journal of Engineering Innovation and Scientific Research.;* (2015). vol.1(1): p.65-68.

[31] Gudsoorkar U, Bindu R. Fatigue crack growth characterization of re-treaded tire rubber. *Materials Today: Proceedings.* (2021). Vol 29(7), p.8697-8708.

[32] Vankar Bharat, A., Gosai Dipak, C., & Shah Amit, V. AN EXPERIMENTAL ANALYSIS ON LIFE CYCLE OF TYRE RETREADING. (2015) Vol. No.4, Special Issue (01), p.30-39.

[33] Xie, M., Yao, H., & Zhao, Z. Failure analysis on tread throw of the tire of B737-800 airplane. *Engineering Failure Analysis,* (2011). 18(2), p.789-796.

[34] Mehta Divyesh S, Gohil Sachin A, Shah Amit V, Patel Vipul R. A Case Study on Tyre Retreading of ST Workshop. (2015). Vol. 3, Issue: 2, p.131-134.

Bearing fault diagnosis using signal processing and machine learning techniques: A review

Viplav Barai[1](0000-0002-1372-4150), Sangharatna M. Ramteke[1](0000-0001-7533-7167), Vismay Dhanalkotwar[1], Yatharth Nagmote[1], Suyash Shende[1], Dheeraj Deshmukh[1](0000-0001-6074-3111)

[1] Dept. Of Mechanical Engineering G H Raisoni College of Engineering, Nagpur Maharashtra, India.

Sangharatna.ramteke@gmail.com

Abstract. In the majority of machines, bearings are among the most crucial components. Bearings are so important that they have been the subject of intensive research and ongoing development throughout the years. Often, bearing fails to reach its expected service life, resulting in failures that cause economic losses. Therefore, there has been a growing interest in research on bearing failure diagnosis systems due to the availability of condition monitoring techniques. Fault feature extraction techniques with the application of signal processing methods and machine learning techniques introduce an Intelligent Fault Diagnosis system that can identify and diagnose the bearing faults. Many researchers have been interested in such techniques in recent decades, which use artificial intelligence to diagnose machine health conditions. In this article, the authors have explored certain fault diagnosis methodologies based on signal processing and machine learning. From the discussed literature review, a research gap for future work has been defined.

Keywords: Bearing, Fault diagnosis, Signal processing, Machine learning.

Nomenclature:

IFD	Intelligent fault diagnosis.
ANN	Artificial neural network.
SVM	Support vector machine
DNN	Deep neural network.
PCA	Principal Components Analysis.
DWT	Discrete wavelength transform.
WPT	Wavelet packet transform.
FFT	Fast Fourier transform.
PSDs	Power spectral density.
CWT	Continuous wavelet transform.
STFT	Short time Fourier transform.
MLP	Multilayer perceptron
TMI	Time marginal integration.
FBG	Fiber Bragg grating

R2	Coefficient of determination.
RMSE	Root mean squared error.
KFCM	Kernel-based fuzzy c-means.
EMD	Empirical mode decomposition.
VMD	Variational mode decomposition.
LMD	Local mean decomposition.
SNR	Signal to noise ratio.

1. Introduction:

Condition-based monitoring and diagnosis of faults in machine components have become major concerns in the industrial sector recently [1]. Millions of money in emergency maintenance expenditures can be saved by detecting machine faults. One of the most important components of rotating machinery is rolling element bearings [2]. which on failure cause the majority of system failures [3]. The effective operation of these elements of machines is essential in the industries to avoid long-run, expensive destructive downtimes [4]. An online system to monitor the condition of the machines can help a large number of industries to detect an impending fault in machines and prevent machine performance degradation, malfunctions, or even catastrophic failures [5]. The numerous ways to perform the condition monitoring of bearing elements is shown in Figure. 1. The processing of the acquired data from these monitoring techniques involves the use of signal processing and machine learning techniques. If human life is at stake, it is clear that the rolling element bearing condition is a topic that requires more attention from researchers.[6] As a result, advanced technologies are required to be employed efficiently and effectively to monitor the health status of bearings [7].

Sliding bearings and rolling bearings are the two basic types of bearings as shown in Figure. 2(a) and (b). Sliding bearings are those that work by generating sliding friction [8]. They are used for slippery and smooth surfaces in a large number of fields because of their lower weight and lower cost [9]. Rolling element bearings are the components in the machines that allow the shafts to rotate in machinery [10]. They are used in a large number of applications such as engines, aircraft gas turbines, various marine machinery, automotive industries, and power transmissions. Rolling bearings have numerous advantages over other types of bearings [11]. These bearings are commonly referred to as antifriction bearings as their requirement for lubrication is low. They have a low friction coefficient and are ideal for shafts with high speed that require long life[12]. They consist of four basic components: an inner ring, an outer ring, rollers, and a cage. Rollers are the elements in the bearing that divides the inner ring and outer ring. Rollers make it possible for the bearing to operate with less friction [13].

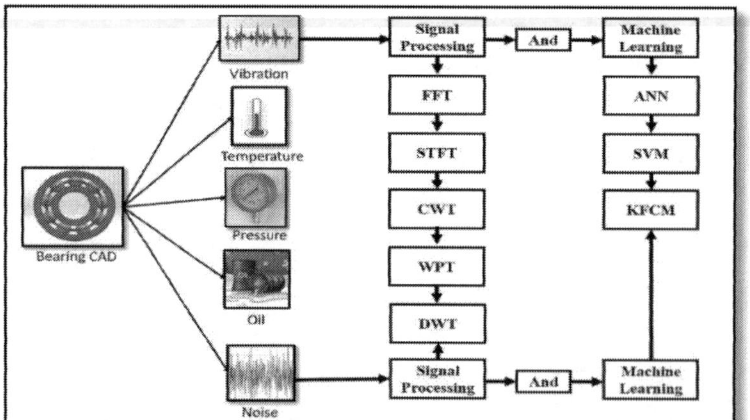

Figure.1. Various ways of condition monitoring techniques for bearing elements [14].

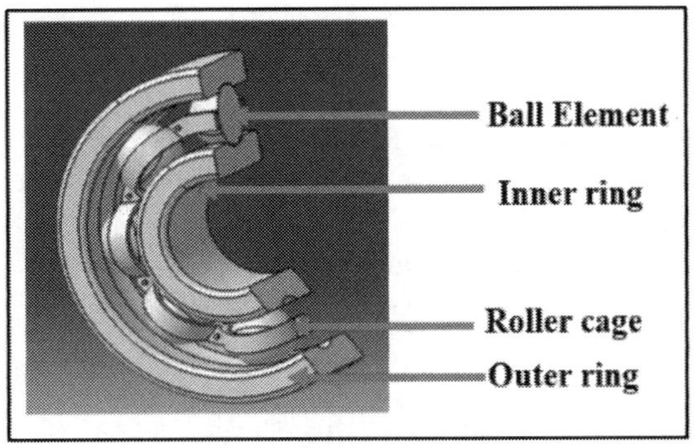

(a)

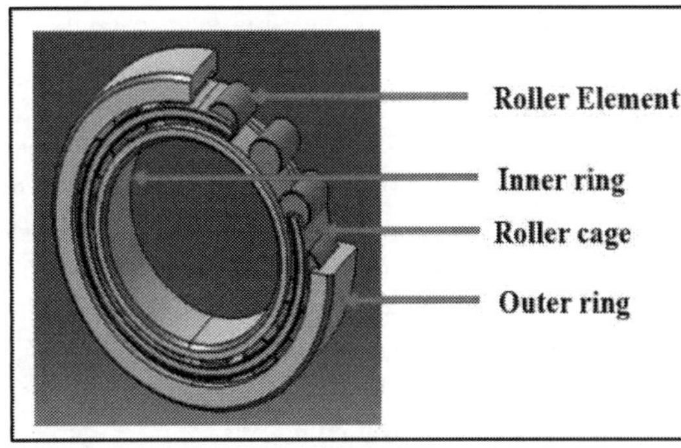

(b)

Figure.2. CAD model of bearing elements (a) Ball bearing (b) Roller bearing.

In recent years, the manufacturing community has focused on developing new ways and new methods to improve the designs of bearing. Any process that includes a lot of bearings increases the chance of risk of system failure [15]. Experimental models, various mathematical models, and simulation models based on computer-aided engineering (CAE) are some of the methodologies that are being employed in the research [16]. The primary purpose of this work is to provide a thorough explanation of the various methodologies for the measurement of bearing fault used in the recent past [17].

2. Critical Reviews:

2.1. Bearings fault diagnosis using signal processing techniques.

Priyadarshini et al. [18] used vibration analysis to conduct an experimental investigation to diagnose a rotating machine bearing failure. FFT and cepstrum together with Bearing Characteristics Frequencies were used for fault diagnosis. Cepstrum kurtosis and cepstrum skew factor were retrieved. The accuracy of the FFT-BCF, Cep-BCF, and Cep-Ks techniques was 98.11 percent, 98.11 percent, and 100 percent, respectively. The Cep-ks approach was found to outperform the FFT-BCF and Cep-BCF methods.

ZHOU et al. [19] proposed an effective and reliable signal processing method and classification mechanism for fault detection in rolling element bearing. Both healthy and faulty rolling bearing vibration signals were extracted. FFT, ANN along with Ensemble Rapid Centroid Estimation (ERCE), and Principal Components Analysis (PCA) were employed for fault detection. The simulation results showed that the classification was accurate in detecting faults and that the subsequent three approaches produced similar results. The PCA approach was determined to be the most efficient approach of all.

Wescoat et al. [20] conducted an experimental study to classify rolling bearing defects using the total energy acquired from FFT and PSD. Data was collected from a faulty lift that had a bearing fault and a healthy lift with no fault. Fault detection was validated using the distribution of energy values. The energy content concerning the associated FFT and also the time corresponding to it was used to highlight the difference between a healthy lift and a faulty lift. The results revealed that finding bearing faults from healthy data using the energy content obtained from the PSD was a valid method.

Bellini et al. [21] investigated the impact of stator current rolling element bearing faults, including the effects of load and electromagnetic torque. The capacity to detect bearing faults using vibration and current signals was evaluated. Signal processing techniques were examined and compared. The results showed that the bearing fault diagnostics were only successful for faults with a low critical frequency rate. As a result, it was determined that the current signal may be used as a suitable medium for detecting bearing faults while vibration signals were a reliable indicator of bearing failures.

Abid et al. [22] carried out an experimental investigation to diagnose bearing faults in induction motors using a combination of advanced signal processing tools. Spectral analysis using PSD and also the time-frequency analysis using wavelet techniques were used. The Welch and Multiple Signal Classification PSD estimations were used to identify signals in the current spectrum. DWT, WPT, and SWPT were also utilized and compared. Wavelet analysis and SWPT demonstrated their capacity to detect bearing faults signals more accurately and proved more efficient than other methods.

Kulkarni et al. [23] highlighted the importance of comparing some vibration measures to characterize the bearing's distributed faults. Vibration analysis with parameters like RMS, Peak, and Peak to Peak was used in bearing fault diagnosis. Fault detection ability with outside and inner race defects was successfully obtained by using statistical analysis. Peak to peak amplitude response, followed by RMS amplitude and peak amplitude, produced superior results for faulty bearings. Kurtosis had demonstrated high sensitivity in the identification of bearing faults.

Chaudhari et al. [24] conducted an experimental investigation for bearing fault detection in electric motors using vibration analysis. Vibration signals were recorded in three different conditions: good, having an outer race fault, and having a fault on both the outer race and the ball. DWT and FFT were used for the analysis. Wavelet analysis based on time frequency produced better results. As a result, DWT was viewed as a more effective tool for monitoring the condition and maintenance of electrical motors. For a faulty bearing, FFT of the 4th level approximation of the vibration signal revealed improved fault frequencies.

Hariharan et al. [25] investigated the most effective signal processing methods to detect the faults in roller bearing. Vibration monitoring was used to detect faults in the outer ring, inner ring as well as the rolling element. For both healthy bearing conditions and faulty bearing conditions, time-domain data and frequency-domain data were collected. The results showed that the frequency spectrum was found to be useful in determining the specific nature of the fault in bearing. The amplitude of vibration obtained was 2-7 times larger for the outer race fault, 3-8 times bigger for the inner race fault, and 2-8 times larger for the rolling element fault when compared to the healthy bearing.

Osman et al. [26] developed a fault detection methodology for fault detection in the outer ring of the bearing and estimated the remaining life of the bearings using vibrational analysis. Vibration signals were extracted and compared from both healthy and faulty bearing conditions in a pumping system. The result showed that the inner race and outside raceway had the earliest discrete bearing frequency and the highest probability of wear. Any change in amplitude at the bearing component frequencies was regarded as a sign of component failure in the making.

2.2. Bearing fault diagnosis using machine learning techniques.

Cardoso et al. [27] explained the difficulties of current-based condition monitoring and early bearing damage detection. Traditional methods in a supervised learning perspective that extract features in the time as well as frequency, and also the time-frequency domain provided high accuracy in a vibration-based analysis. These methods, however, were insufficient for current-based approaches due to low SNR and low magnitude harmonics. Low SNR, information loss in the magnetic field, saturation harmonics, electrical faults, interference, and indirect measures are among the main drawbacks of current-based bearing condition monitoring. As a result, in current-based analysis, traditional signal processing techniques that denoise and extract information from vibration-based signals performed poorly. There is less information available (e.g., indirect measure) and more feature extraction complexity with current-based bearing condition monitoring (e.g., a poor SNR).

Sharma et al. [28] created an ANN model and SVM model and used them for fault classification with the help of WEKA software. Vectors for training have already been created and are being used as input. In the study, ANN and SVM were

IOP Conf. Series: Materials Science and Engineering 1259 (2022) 012034 doi:10.1088/1757-899X/1259/1/012034

used for the classification of the defects. A total of 73 cases were taken under consideration for testing including balls with corrosion pitting cases, bearing cases with rough inner race surface, cases with bearing elements faults combined, the case with healthy bearings, as well as cases with outer race crack in bearing, respectively. After training the models, all these cases were correctly classified by SVM.

Amirat et al. [29] proposed a diagnosis method for bearing fault detection which was validated by using the Western Reserve University experimental dataset. The ability of the proposed diagnosis methodology to classify between different severities of a given fault has been highlighted in particular. The performance of detection and diagnosis has also been compared to the relevant modern equivalent techniques. In comparison to EMD and traditional VMD, VMD which was used as a notch filter has been shown to be the preferred solution for fault detection. Further, it was compared to MLP, RNN, and LSTM, convolution layers (1DCNN) are more useful in feature extraction for classification.

Gupta et al. [30] proposed a comparison between techniques of supervised and unsupervised learning for the diagnosis of bearing fault. Vibration signals from various fault classifications were used to determine the best scale selection of wavelet coefficients, and the permutation entropy method was applied. For both horizontal and vertical response circumstances, features were retrieved using the best scale. Permutation entropy, form factor as well as crest factor for both horizontal and vertical responses along with loaders and shaft rotation, were all taken into account. When SOM and SVM classification results are compared, it was clear that SVM produces significantly better results than SOM because of its superior generalization performance.

Patidar et al. [31] aimed to summarize the current advancement in vibration analysis approaches for detecting the faults in rolling element bearing. According to the study's findings, time-domain methods were only able to identify the presence of faults in a bearing but cannot locate its location. The location of a bearing's fault can be determined using frequency-domain techniques. At the frequencies of the bearing characteristics, vibration peaks appear in the spectrum, allowing us to determine which bearing element is defective. For detecting rolling element bearing incipient failure, envelope analysis is an important method.

Goyal et al. [32] investigated the vibration analysis based on a laser beam that was developed for measuring machine vibrations. The vibration signals were acquired and then preprocessed using the Hilbert transform. To eliminate redundant features and rank them in order of importance, PCA and SFFS were used. Furthermore, the features which were selected for classification and performance evaluation were fed into SVM and ANN. The vibration signals which were obtained with the help of a contactless sensor are comparable to accelerometer data obtained under similar conditions. The results showed that the proposed contactless sensor was used to develop a dynamic, powerful condition-based maintenance system that can prevent destructive failures and lower operating costs.

Sikora et al. [33] developed a multiple health state assessment-based RUL concept. During the health assessment, the total bearing life was separated by the framework into several health states. A separate regression model was developed for every state. A model with a combination of unsupervised and supervised elements was used. PCA and fuzzy c-means were used by the unsupervised element to extract the knowledge in the form of health state labels automatically. Further, an SVM model was used to identify the fault by using the generated labels and to calculate the RUL.

Kandukuri et al. [34] investigated the feasibility of developing a real-time bearing fault detection and recognition algorithm. As a classifier, the support vector machine algorithm was used. To train classifiers, time-frequency domain features were collected from experimental results. Experimental data was used to validate the trained classifiers, which showed that time-frequency domain features produce accurate results. A bearing fault was detected 2.5–3.5 days before failure using the run-to-failure test. As an outcome, it was concluded that the proposed method could be developed further to produce a reliable real-time fault diagnosis system.

Elforjani et al. [35] discovered that the suggested ANN model, fault diagnostic SIE, and modified regression function were successful in forecasting the parameter for speed bearings, validating Elforjani's well-known position. On the other hand, SVMR and GPR models did not include sensitive tools for analyzing the RUL of the bearing according to a comparison study. Finally, while these diagnosis techniques had been successfully implemented to specifically made test rigs, instrumentation, and particular AE tests, it was a fundamental concept to conduct additional research and analyses to evaluate their validity in real-world applications where other variables such as structural noise and other operational parameters exist.

Shaner et al. [36] used real-world vibration signals froma 2 Megawatt wind turbine to evaluate their suggested model. The regression and ANN model outcomes were compared. RMS, Kurtosis, and EI were used as condition indicators in the regression. RMSE, R2, and adjusted-R2 were the three factors that were used to evaluate the efficacy of each regression model. The exponential model performed the best, according to the study results. Using monotonicity and trend capability factors, the capacity of condition indicators to be used for diagnosis was analyzed. The results indicated that RMS provided the best overall performance, it was selected as the better data output again for the neural network. The suggested ANN model's outcome revealed that it can accurately forecast a bearing's remaining life, and this achievement may be due to the relationship formed between the linear regression and the ANN through the perfect-suited condition signal.

Zhiyi et al. [37] suggested an improved project-based auto-encoder for detecting bearing faults in a variety of machines. In their proposed model, they applied a scaled exponentially currently carrying to increase the quality of the mapping vibrational data acquired from bearings, as well as a positive integer constraint to adjust the loss function to increase the reconstructed effect. The parameter knowledge from the well model structure was then passed to the target model, and target training data with limited labeled information were utilized to perfect the target model to fit the targeted testing cases' features. The suggested model was evaluated using vibration signals from bearings installed in different machines.

3. Methodology:

This methodology for fault diagnosis of bearing using signal processing techniques and machije learning approaches includes three stages as shown in Figure. 3. Firstly, vibration signals will be acquired using accelerometer for both faulty bearing condition of cage wear and ball wear. Secondly, this extracted signals will be then analyse using the Fast fourier transform technique which is one of the most efficient signal processing tool. This signals will also be processed using Power spectrum density. Along with that, machine learning techniques like Artificial nueral network and Support vector machine will be used for the classification of these vibration signals for both cage wear and ball wear bearing condition. At the end, all the results of signal processing tools and machine learning techniques will be compared for both the bearing condition for determining the optimized fault detection methodology.

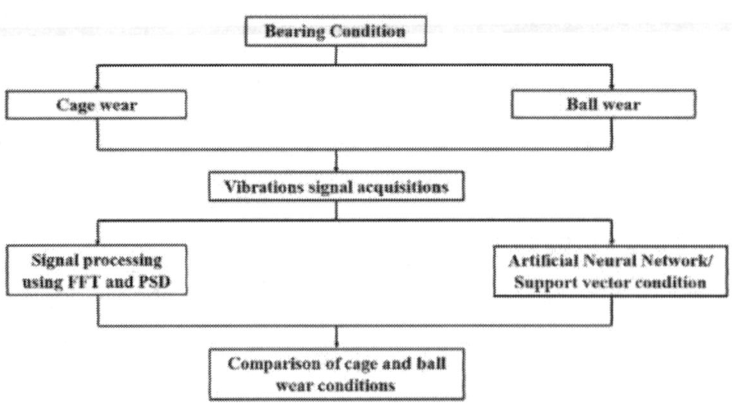

Figure. 3. Flow chart showing the condition monitoring steps will be involved in bearing fault diagnosis.

4. Conclusions

In this article, an attempt has been made to highlight recent trends in research on the diagnosis of faults in bearing. The majority of the work has been focused on the diagnosis of bearing faults using vibration analysis and machine learning approaches to analyze faults in the inner rings and outer rings. Researchers have devised several vibration measurement and other signal processing methods such as Fast Fourier transform (FFT), wavelet packet transform (WPT), Short-time Fourier transform (STFT), Principal Components Analysis (PCA), continuous wavelet transforms (CWT), Ensemble Rapid Centroid Estimation (ERCE), and many other techniques. Furthermore, many machine learning methods and artificial intelligence methods such as support vector machine (SVM) and the artificial neural network (ANN) has been utilized to detect and classify bearing faults. However, Very little research and studies have been done on identifying cage wear and balls wear faults in bearing and analyzing their impact on a machine's vibration and performance characteristics. As a result, the authors have taken this research gap into account for their study. The flowchart showing the methodology for the work taken into consideration is shown in Figure. 3. Experiments are underway, result will be published in future publications.

5. References:

[1] A. R. Bhende, G. K. Awari, and S. P. Untawale, "Comprehensive bearing condition monitoring algorithm for incipient fault detection using acoustic emission," *J. Tribol.*, vol. 2, no. July, pp. 1–30, 2014.

[2] S. M. Ramteke, H. Chelladurai, and M. Amarnath, "Diagnosis of Liner Scuffing Fault of a Diesel Engine via Vibration and Acoustic Emission Analysis," *J. Vib. Eng. Technol.*, vol. 8, no. 6, pp. 815–833, 2020, doi: 10.1007/s42417-019-00180-7.

[3] V. P. Kalbande, P. V. Walke, and K. Rambhad, "Performance of oil-based thermal storage system with parabolic trough solar collector using Al2O3 and soybean oil nanofluid," *Int. J. Energy Res.*, vol. 45, no. 10, pp. 15338–15359, 2021, doi: 10.1002/er.6808.

[4] R. V. Dahibhate, S. B. Jaju, and R. I. Sarode, "Development of mathematical model for prediction of bone drilling temperature," *Mater. Today Proc.*, vol. 38, no. xxxx, pp. 2732–2736, 2020, doi: 10.1016/j.matpr.2020.08.537.

[5] S. Ramteke, H. Chelladurai, and K. K. S. Pandian, "Simulation and analysis of magnetic field strength with magneto-resistive sensor: A future application," *2017 IEEE 3rd Int. Conf. Eng. Technol. Soc. Sci. ICETSS 2017*, vol. 2018-Janua, pp. 1–5, 2018, doi: 10.1109/ICETSS.2017.8324161.

[6] V. Khond, "Experimental investigation of desert cooler performance using four different cooling pad materials," *Am. J. Sci. Ind. Res.*, vol. 2, no. 3, pp. 418–421, 2011, doi: 10.5251/ajsir.2011.2.3.418.421.

[7] N. Tandon, G. S. Yadava, and K. M. Ramakrishna, "A comparison of some condition monitoring techniques for the detection of defect in induction motor ball bearings," *Mech. Syst. Signal Process.*, vol. 21, no. 1, pp. 244–256, 2007, doi: 10.1016/j.ymssp.2005.08.005.

[8] S. M. Ramteke and H. Chelladurai, "Effects of hexagonal boron nitride based nanofluid on the tribological and performance, emission characteristics of a diesel engine: An experimental study," *Eng. Reports*, vol. 2, no. 8, pp. 1–20, 2020, doi:

10.1002/eng2.12216.

[9] S. G. Chakrabarty, U. S. Wankhede, R. S. Shelke, and T. B. Gohil, "Investigation of temperature development in salinity gradient solar pond using a transient model of heat transfer," *Sol. Energy*, vol. 202, no. March, pp. 32–44, 2020, doi: 10.1016/j.solener.2020.03.052.

[10] S. M. Ramteke, H. Chelladurai, and M. Amarnath, "Diagnosis and Classification of Diesel Engine Components Faults Using Time–Frequency and Machine Learning Approach," *J. Vib. Eng. Technol.*, 2021, doi: 10.1007/s42417-021-00370-2.

[11] V. Srinivasan, "Analysis of dynamic load characteristics on hydrostatic bearing with variable viscosity and temperature using simulation technique," *Indian J. Sci. Technol.*, vol. 6, no. SUPPL.6, pp. 4797–4803, 2013, doi: 10.17485/ijst/2013/v6isp6.23.

[12] S. D. Deshmukh, S. Gabhane, and D. S. Deshmukh, "Osmotic dehydration of carrot strips and modelling," *J. Phys. Conf. Ser.*, vol. 1913, no. 1, 2021, doi: 10.1088/1742-6596/1913/1/012093.

[13] M. Gajbhiye, A. Chaudhary, and S. Ramteke, "EasyChair Preprint Simulation of Vision-Based Model of Robotic Work Cell (6-Axis)," 2021.

[14] A. D. Patange, R. Jegadeeshwaran, N. S. Bajaj, A. N. Khairnar, and N. A. Gavade, "Application of Machine Learning for Tool Condition Monitoring in Turning," *Sound& Vibration*, vol. 56, no. 2, pp. 127–145, 2022, doi: 10.32604/sv.2022.014910.

[15] V. W. Khond, K. Rambhad, and V. Shahadani, "New diesel-neem biodiesel blend (D75NB25) containing nano iron oxide, silicon dioxide and zinc oxide for diesel engine: An experimental investigation," *Mater. Today Proc.*, vol. 47, no. xxxx, pp. 2701–2708, 2021, doi: 10.1016/j.matpr.2021.03.004.

[16] K. S. Rambhad, V. P. Kalbande, M. A. Kumbhalkar, V. W. Khond, and R. A. Jibhakate, "Heat Transfer and Fluid Flow Analysis for Turbulent Flow in Circular Pipe with Vortex Generator," *SN Appl. Sci.*, vol. 3, no. 7, 2021, doi: 10.1007/s42452-021-04664-8.

[17] V. N. Patel, N. Tandon, and R. K. Pandey, "Defect detection in deep groove ball bearing in presence of external vibration using envelope analysis and Duffing oscillator," *Meas. J. Int. Meas. Confed.*, vol. 45, no. 5, pp. 960–970, 2012, doi: 10.1016/j.measurement.2012.01.047.

[18] S. S. Priyadharsini and N. Aiswarya, "JASC : Journal of Applied Science and Computations ISSN NO : 1076-5131 Bearing Fault Analysis using Signal Processing Techniques ISSN NO : 1076-5131," vol. VI, no. 826, pp. 826–833, 2019.

[19] W. Li, M. Qiu, Z. Zhu, B. Wu, and G. Zhou, "Bearing fault diagnosis based on spectrum images of vibration signals," *Meas. Sci. Technol.*, vol. 27, no. 3, 2016, doi: 10.1088/0957-0233/27/3/035005.

[20] E. Wescoat, L. Mears, J. Goodnough, and J. Sims, "Frequency energy analysis in detecting rolling bearing faults," *Procedia Manuf.*, vol. 48, pp. 980–991, 2020, doi:

10.1016/j.promfg.2020.05.137.

[21] F. Immovilli, A. Bellini, R. Rubini, and C. Tassoni, "Diagnosis of bearing faults in induction machines by vibration or current signals: A critical comparison," *IEEE Trans. Ind. Appl.*, vol. 46, no. 4, pp. 1350–1359, 2010, doi: 10.1109/TIA.2010.2049623.

[22] F. Ben Abid and A. Braham, "Advanced signal processing techniques for bearing fault detection in induction motors," *2018 15th Int. Multi-Conference Syst. Signals Devices, SSD 2018*, pp. 882–887, 2018, doi: 10.1109/SSD.2018.8570403.

[23] S. Kulkarni and A. Bewoor, "Vibration based condition assessment of ball bearing with distributed defects," *J. Meas. Eng.*, vol. 4, no. 2, pp. 87–94, 2016.

[24] Y. K. Chaudhari, J. A. Gaikwad, and J. V. Kulkarni, "Vibration analysis for bearing fault detection in electrical motors," *1st Int. Conf. Networks Soft Comput. ICNSC 2014 - Proc.*, pp. 146–150, 2014, doi: 10.1109/CNSC.2014.6906681.

[25] V. Hariharan, P. Thangavel, G. Rajeshkumar, and D. Deepa, "Investigations of Antifriction bearing defects using Vibration Signatures," *IOP Conf. Ser. Mater. Sci. Eng.*, vol. 1084, no. 1, p. 012126, 2021, doi: 10.1088/1757-899x/1084/1/012126.

[26] A. H. Osman, A. A. Salman, and K. M. Fawzy, "Vibration Signature of Roller Bearing's Faults," *Eur. Sci. J. ESJ*, vol. 15, no. 12, 2019, doi: 10.19044/esj.2019.v15n12p425.

[27] A. S. Barcelos and A. J. Marques Cardoso, "Current-based bearing fault diagnosis using deep learning algorithms," *Energies*, vol. 14, no. 9, 2021, doi: 10.3390/en14092509.

[28] P. K. Kankar, S. C. Sharma, and S. P. Harsha, "Fault diagnosis of ball bearings using machine learning methods," *Expert Syst. Appl.*, vol. 38, no. 3, pp. 1876–1886, 2011, doi: 10.1016/j.eswa.2010.07.119.

[29] H. Habbouche, Y. Amirat, T. Benkedjouh, and M. Benbouzid, "Bearing Fault Event-Triggered Diagnosis Using a Variational Mode Decomposition-Based Machine Learning Approach," *IEEE Trans. Energy Convers.*, vol. 37, no. 1, pp. 466–474, 2022, doi: 10.1109/TEC.2021.3085909.

[30] V. Vakharia, V. K. Gupta, and P. K. Kankar, "Ball Bearing Fault Diagnosis using Supervised and Unsupervised Machine Learning Methods," *Int. J. Acoust. Vib.*, vol. 20, no. 4, p. 482005, 2015, doi: 10.20855/ijav.2015.20.4387.

[31] S. Patidar and P. K. Soni, "An Overview on Vibration Analysis Techniques for the Diagnosis of Rolling Element Bearing Faults," *Int. J. Eng. Trends Technol.*, vol. 4, no. 5, pp. 1804–1809, 2013, [Online]. Available: http://ijettjournal.org/archive/ijett-v4i5p97.

[32] D. Goyal, S. S. Dhami, and B. S. Pabla, "Non-Contact Fault Diagnosis of Bearings in Machine Learning Environment," *IEEE Sens. J.*, vol. 20, no. 9, pp. 4816–4823, 2020, doi: 10.1109/JSEN.2020.2964633.

[33] S. Schwendemann, Z. Amjad, and A. Sikora, "A survey of machine-learning techniques for condition monitoring and predictive maintenance of bearings in grinding machines,"

Comput. Ind., vol. 125, p. 103380, 2021, doi: 10.1016/j.compind.2020.103380.

[34] R. Wang, H. Chen, C. Guan, W. Gong, and Z. Zhang, "Research on the fault monitoring method of marine diesel engines based on the manifold learning and isolation forest," *Appl. Ocean Res.*, vol. 112, no. May, p. 102681, 2021, doi: 10.1016/j.apor.2021.102681.

[35] M. Elforjani and S. Shanbr, "Prognosis of Bearing Acoustic Emission Signals Using Supervised Machine Learning," *IEEE Trans. Ind. Electron.*, vol. 65, no. 7, pp. 5864–5871, 2018, doi: 10.1109/TIE.2017.2767551.

[36] F. Elasha, S. Shanbr, X. Li, and D. Mba, "Prognosis of a wind turbine gearbox bearing using supervised machine learning," *Sensors (Switzerland)*, vol. 19, no. 14, pp. 1–17, 2019, doi: 10.3390/s19143092.

[37] P. KABIRI and H. GHADERI, "Automobile Independent Fault Detection based on Acoustic Emission Using Wavelet," *Ndt.Net*, no. November, pp. 3–4, 2011, [Online]. Available: http://www.ndt.net/article/SINCE2011/papers/10_Kabiri_Rev2.pdf.

IOP Conf. Series: Materials Science and Engineering 1259 (2022) 012035 doi:10.1088/1757-899X/1259/1/012035

Research Progress in Life Cycle Assessment for Sustainable Manufacturing Industries: A Bibliometric Analysis

Alok Yadav[1], Rajeev Agrawal[2], Rajiv Kumar Garg[1], Anish Sachdeva[1]

[1]Department of Industrial and Production Engineering, Dr. B. R. Ambedkar National Institute of Technology, Jalandhar-144011(India)

[2]Department of Mechanical Engineering, Malaviya National Institute of Technology Jaipur-302017 (India)

Aloky.ip.21@nitj.ac.in

Abstract. Abstract. Many tools for analysing the sustainability of manufacturing industries, such as life cycle assessment (LCA), have been created as the debate over sustainable manufacturing has grown. A bibliometric analysis is conducted for a scientific paper in Scopus databases on the subject of LCA, focusing on studies related to manufacturing industries from a sustainable point of view. This study compiled a portfolio of papers on LCA-based sustainable manufacturing industries. In the present study, a total of 133 publications are found from the Scopus database between the periods when LCA was added in the ISO series (1997) to 2020. The study was carried out using R studio and the Bibliometrix R package. The primary data for the present study is collected from the database of Scopus in BibTex format, after which is analysed in R studio for exploration using the command. The annual scientific production of Life cycle Sustainability Assessment papers is discovered, with a 13.39 % increase in article publication. According to the findings, developed countries produced the most publications compared to developing nations. When comparing a number of publications, India ranks fourth. According to the thematic analysis, the bulk of papers tried to examine manufacturing industries' environmental implications. However, other publications have combined LCA with additional methodologies to provide a more comprehensive analysis that considers the social and economic pillars of sustainable manufacturing. The study also emphasises the importance of LCA in manufacturing process decision-making. There are numerous novel contributions, such as categorising published data using scientific indexes and bibliometrics, describing the main research approaches, and highlighting future research opportunities to provide academics and practitioners with an overview of LCA implementation in the field of sustainable Manufacturing industries.

Keywords: Life cycle sustainability assessment, Manufacturing Industries, Sustainability, Bibliometrics, Manufacturing, tools

1. Introduction

Environmental, social, and economic factors should all be considered when making decisions, and the best Manufacturing techniques should take all of these factors into account [1]. Furthermore, applying sustainability principles throughout the manufacturing process helps to limit environmental and community impacts, minimise expenses, and effectively control human health and environmental hazards [2]. Many elements must be addressed for a technique to be considered sustainable [3]. Even in the production phase, it's not enough to just look at how various innovations affect the market and the bottom line. Sustainability encompasses a wide range of factors, from the initial collection of raw materials through the ultimate disposal of waste products, as well as the social and economic impacts of the production process as a whole. To evaluate the sustainability of manufacturing sectors over the long term, it is necessary to examine the entire process from the perspective of life-cycle assessment, making it imperative to create appropriate approaches and tools in this area. [4].

LCA has been extensively researched, with various studies published in various disciplines, including the manufacturing industry life cycle assessment [5]. In this framework, the Life cycle assessment has been used in manufacturing as a more significant and more complex environmental research instrument regarding the implications of manufacturing processes, taking the entire life cycle into account [6]. In this case, though, there was no mention of sustainability. As a result, the goal of this research is to fill a scientific gap by investigating studies on LCA, Manufacturing Industries, and sustainability, as well as established relationships. Bibliometric studies are also gaining popularity. These studies provide readers with an overview of a specific thematic analysis, which may aid future research planning. Another reason for writing this article is that this type of bibliometric technique was not used in the LCA and manufacturing industries, and sustainability was not taken into account. As a result, this work aims to stimulate further research into related papers on this topic.

The objective of this study is to perform a statistical analysis of published publications from the Scopus database. In this regard following research questions were developed in mind:

1. What is the current publication pattern, keywords, and authors working in this field?
2. What are writers, institutes, and countries involved in this study?

Rest paper in this manuscript is organised as follows: Section 2 represent the methodology for this study, section 2.1 describes the boundaries of search for bibliographic portfolio (BP), section 2.3 introduce method of obtaining BP, section 3 explains different result obtains by R-studio. Finally, section 4 explains about conclusion and future scope of this study.

2.Methodology

Bibliometric analysis is performed using a statistical method to classify and evaluate present findings on LCA and sustainability in current research. The research results provide considerable insight into ongoing and prospective research directions by summarising what is known about the topic. The evaluation of research output indexes such as the number of articles, writers, journals, and nations is a critical component of a systematic and bibliometric study [7,8]. In addition, this research technique aims to have a scientific assessment of the essential works on a particular issue. Creating a bibliography is the first step. Then, bibliometric studies are conducted on a wide range of topics, including theoretical developments, authors, countries producing the most works, periodicals, and publishing focuses.

Bibliometric is a collection of all the resources you have used your work. In other words, Bibliometric is a collection of methods for reviewing science and technical literature quantitatively (Jamwal et al., 2021b). Alan Pritchard used the word "Bibliometric" in his paper statistical Bibliography or Bibliometric. A more unambiguous definition of Bibliometric is that it is statistical analysis of publication published in a particular field [10]. Bibliometrics describes the body of literature by evaluating objects such as physical publications, bibliographic citations, etc.

2.1 Boundary of research

Scientific papers such as articles, book chapters, review and conference papers that had been subjected to a robust peer review procedure were included in this study. In this bibliography analysis, we consider the publication between the year 1997 to 2020. Sustainability is used as a key factor for this study. It is possible to validate the LCA application in the manufacturing industries using this method. This bibliography makes use of the Scopus database. Elsevier BV is the operator of the Scopus database (Netherlands Private Limited Company). Because it offers consistent and uniform records and a comprehensive scope of Top journals and publications on LCA, the Scopus database is crucial for bibliometric study in the science sector, which is why these databases are considered for analysis. Finding publications by author name and the paper title is simple with Scopus, and there's even an advanced search option.

2.2 Method of obtaining bibliographic portfolio

The Scopus databases are searched using the terms Life cycle sustainability assessment, sustainability, Sustainable, Manufacturing sustainability, Bibliometrics, tools, and manufacturing industries, as follows: "Life cycle sustainability assessment" AND "sustainability OR sustainable" AND "Manufacturing sustainability" AND "bibliometrics" AND "tools" AND "Manufacturing industries" The Boolean operators "AND" and "OR" are used to combine keywords. When the "AND" operator is used in a database search, it means that any words may show up in the outcomes. The "OR" controller specifies that the probe includes at least one of the terms.

133 publications were discovered in the Scopus database between the time LCA was added to the ISO series (1997) and 2020. By categorising the data by nations, years, journals, authors, and universities, may be used for a more bibliometric study. The bibliographic portfolio is built by combining the search parameters with the representativeness of the papers in the chosen topics. The database also excluded articles written in languages other than English. The bibliography portfolio of publications is depicted in Figure 1.

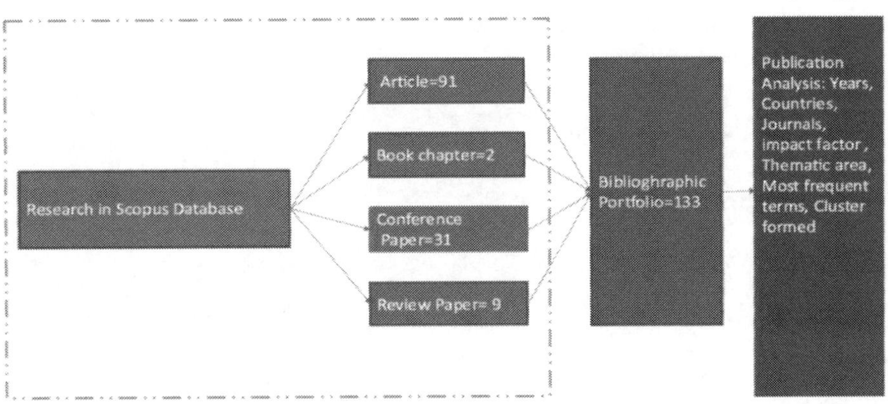

Fig1: Search portfolio Scopus database

3. Results and discussion:

The results of bibliometric-based study of articles from the Scopus database were reported in this section. R studio's Bibliometrix R software was used to conduct the analysis. The raw bibliographic data is exported from the Scopus database in BibTex format and used in R studio for research with the command.

Figure 2 below summarises the main points of the publication Life cyclic sustainability assessment in the manufacturing industry.

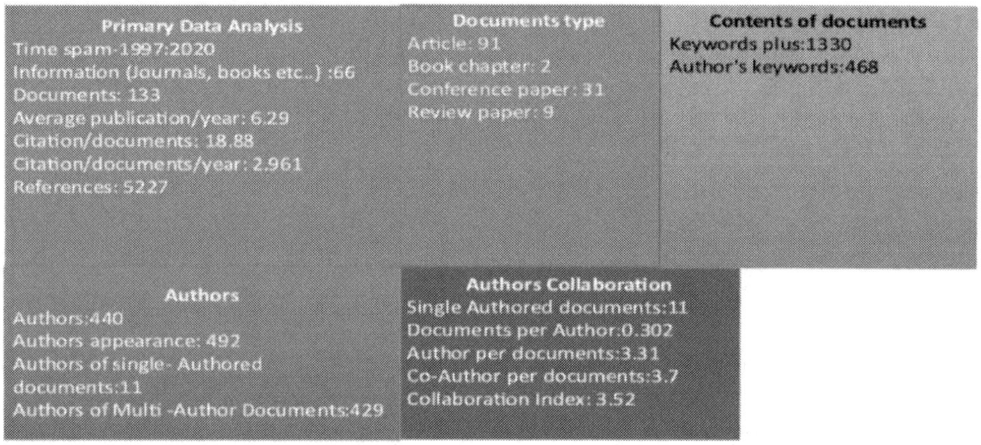

Fig2: Main information about data

3.1Annual Scientific Article Production:

The R package discovered the annual scientific production of Life cycle Sustainability assessment articles, with a 13.39 per cent yearly growth rate in article publication. Until 2013, there was very little publication in the area because, at the time, sustainability in the manufacturing sector was still in its early stages. Significantly few authors were aware of the capabilities and benefits of using sustainability. Figure 3 shows that there has been an increase in the number of publications since 2012. Many researchers are presently working in this field around the world, pushing this field and identifying new study scopes in design and production.

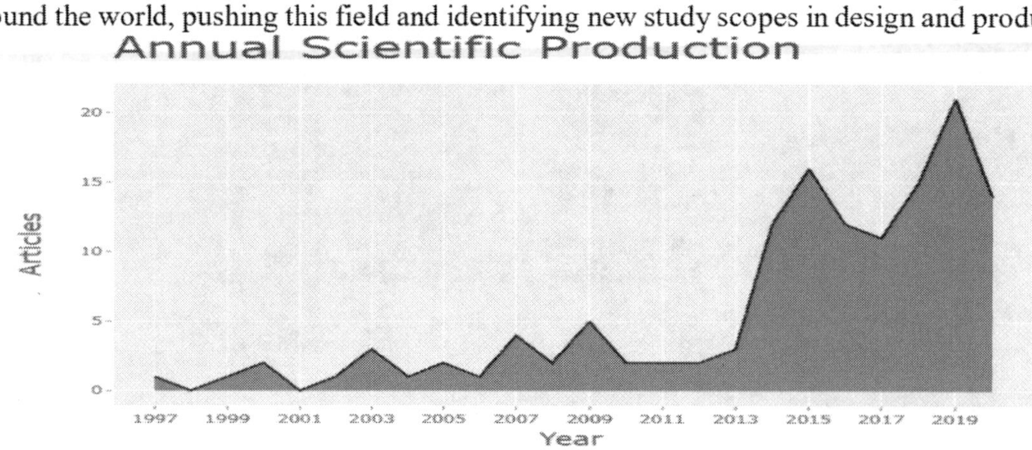

Fig3. Annual scientific production is included in the Scopus publication database.

3.2 The most productive and relevant sources

Identifying the most active and vital so vital particular study subject is crucial since this will help authors figure out which sources are producing more research articles in that sector. As we all know, developing and developed countries are increasingly embracing sustainable manufacturing. This field's research has never been more active. As a result, it's more crucial than ever to find and publish studies about sustainable manufacturing in journals and other key places. By analysing the data, we identified the most productive journals. We discovered that "Journal of Cleaner Production" has the most significant number of publications in this field, followed by "Sustainability" and "Resource conservation and recycling." These magazines are concerned with sustainability and have been publishing articles on sustainable manufacturing and other concerns for quite some time. We can also see that a few sources are freely accessible and have acceptable citation structures, indicating that these publications are easily accessible to academics and practitioners. Figure 4 depicts the most productive journals and magazines.

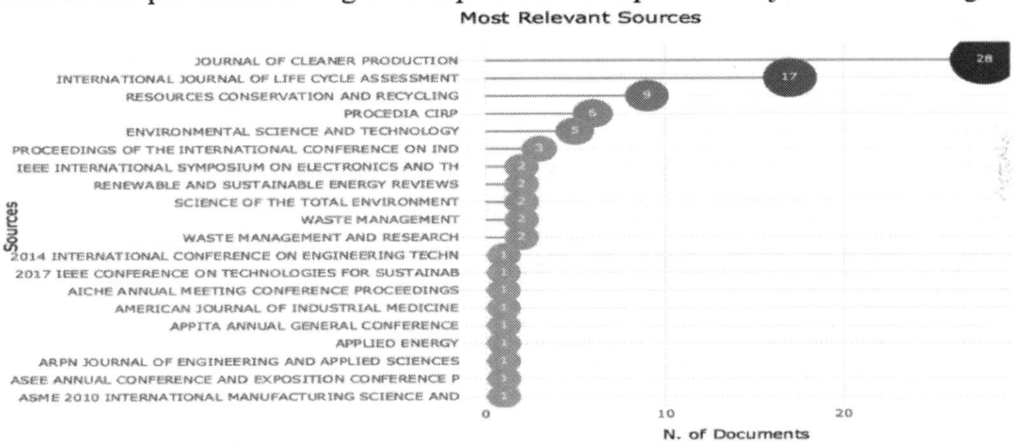

Figure 4: The most productive journals and other resources (Scopus database)

3.3 Word cloud in LCSA in the manufacturing industry

We used R studio software to build a word cloud of keywords for this study. There are two keywords in the Scopus bibliographic database:(1) Author Keywords (2) Keyword Plus .We used author keywords in this report to more precisely map our research. Figure 5 depicts a keyword cloud of author keywords.

Figure 5: a cluster of a word in LCSA and manufacturing industry (Scopus database)

Based on the analysis results, we can conclude that Life Cycle is the most frequently used keyword by authors in their articles. Also included are the terms "sustainable development," "manufacturing industry," "waste management," and "carbon footprint."

IOP Conf. Series: Materials Science and Engineering 1259 (2022) 012035 doi:10.1088/1757-899X/1259/1/012035

3.4 Annual mean citation frequency

The annual mean citation frequency between 1997 and 2020 is shown in table 1.

Table1: Average number of citations per year

Year	N	MeanTCperArt	MeanTCperYear	CitableYears
1997	1	4	0.166666667	24
1998	0	0	0	0
1999	1	26	1.181818182	22
2000	2	42.5	2.023809524	21
2001	0	0	0	0
2002	1	7	0.368421053	19
2003	3	15.33333333	0.851851852	18
2004	1	32	1.882352941	17
2005	2	10.5	0.65625	16
2006	1	26	1.733333333	15
2007	4	25	1.785714286	14
2008	2	2	0.153846154	13
2009	5	74.8	6.233333333	12
2010	2	51	4.636363636	11
2011	2	4.5	0.45	10
2012	2	1.5	0.166666667	9
2013	3	56.33333333	7.041666667	8
2014	12	27.75	3.964285714	7
2015	16	16.6875	2.78125	6
2016	12	19.66666667	3.933333333	5
2017	11	10.81818182	2.704545455	4
2018	15	16.53333333	5.511111111	3
2019	21	10.80952381	5.404761905	2
2020	14	5.214285714	5.214285714	1

3.5 Bradford's law is used to cluster sources.

By calculating the link among the most productive journals (n) and their cumulative yield C, the Bradford law aids in source clustering (n). In general, the literature yield of a publication determines its quantitative value, which may be computed as: $C(n) = p \log(n q + 1) + C(0)$, for $n >= 0$. In this case, p, q, and C (0) are constants.

Table 3 shows the source clustering using Bradford's law, with the top ten journals ranked using Bradford's law. The most productive abundant source was Journal of cleaner production, with a frequency of 28 and a cumulative frequency of 28. Similarly, International journal of life cycle assessment ranks second with a cumulative frequency of 76. The top ten journals in Table 2 in terms of rank, frequency, and cumulative frequency.

Table 2: Source clustering through Bradford's law

Source output	Rank	Freq	Cumfreq	Zone
JOURNAL OF CLEANER PRODUCTION	1	28	28	Zone 1

IOP Conf. Series: Materials Science and Engineering 1259 (2022) 012035 doi:10.1088/1757-899X/1259/1/012035

INTERNATIONAL JOURNAL OF LIFE CYCLE ASSESSMENT	2	17	45	Zone 1
INTERNATIONAL JOURNAL OF LIFE CYCLE ASSESSMENT	3	9	54	Zone 2
PROCEDIA CIRP	4	6	60	Zone 2
ENVIRONMENTAL SCIENCE AND TECHNOLOGY	5	5	65	Zone 2
PROCEEDINGS OF THE INTERNATIONAL CONFERENCE ON INDUSTRIAL ENGINEERING AND OPERATIONS MANAGEMENT	6	3	68	Zone 2
IEEE INTERNATIONAL SYMPOSIUM ON ELECTRONICS AND THE ENVIRONMENT	7	2	70	Zone 2
RENEWABLE AND SUSTAINABLE ENERGY REVIEWS	8	2	72	Zone 2
SCIENCE OF THE TOTAL ENVIRONMENT	9	2	74	Zone 2
WASTE MANAGEMENT	10	2	76	Zone 2

3.6 Lotka's law to calculate author productivity:

The Lotka law calculates how frequently authors publish in a given field. According to Lotka law, the proportion of an author's single contributions is calculated using the formula:
The general procedure is:
$X^{n}Y = C$
or
$Y = C/X^{n}$
X is the number of publications, Y is the frequency of authors linked with X publications, and C and n are constants in this equation.
The investigation revealed that 405 authors wrote only one piece, whereas 29 wrote two. Only two authors, representing 0.001% of the total number of authors, have authored more than ten publications. Table 3 shows the number of authors with the proportion of authors.

Table 3: productivity of author by Lotka's Law

Documents written	No. of Authors	Proportion of Authors
01	405	00.92
02	29	00.066
03	3	00.007
04	1	00.002
08	2	00.005

3.7 The most common structure for global and local citations

Local citations are the number of times a document or author in this collection has been cited by other documents (articles, conference papers). The following formula is used to compute the local citation score of the articles:

Local Citations (M, fast. search = FALSE, sep = ";", verbose = FALSE)

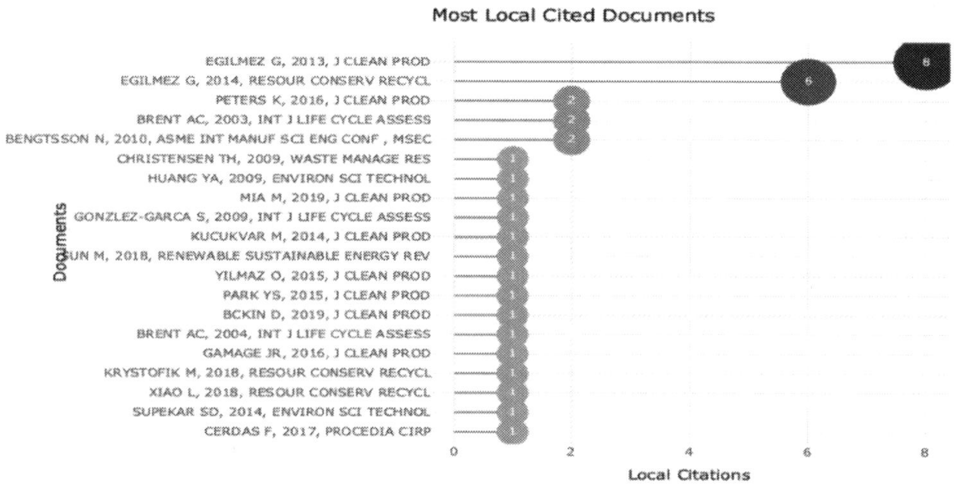

Fig6: Local citation structure (Scopus database)

The global citation refers to the number of times a certain author or document in a single collection has been cited by documents in another set. The following is how the articles' global citation score is calculated: global Citations (M, fast. Search = FALSE, sep = ";", verbose = FALSE).

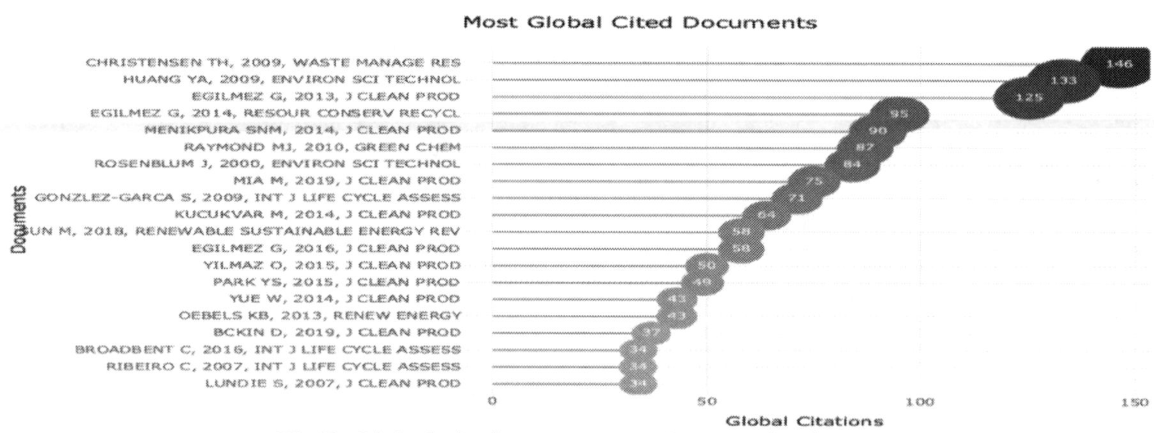

Fig7: Global citation structure (Scopus database)

3.8 Co-occurrence networks analysis

It is easier to see the logical connection between terms in bibliographic data with the help of co-occurrence network analysis. A co-occurrence analysis of 1594 terms was presented in this study. The Louvain clustering algorithm was implemented alongside a normalisation strategy for associations, and 50 nodes were used for this analysis. There are graphical considerations such as an opacity of 0.7, 50 labels, and six label sizes.

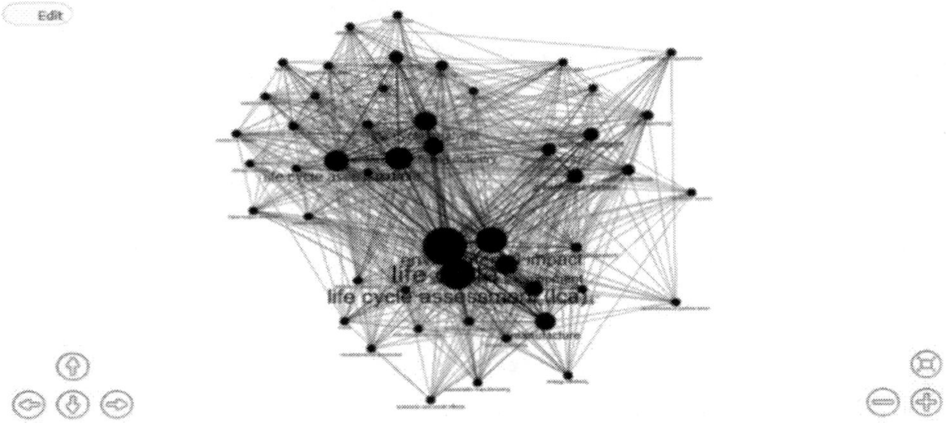

Fig.8: Co-occurrence network for keywords (Scopus database)

3.9Multiple corresponding analyses resulted in a conceptual structure map.

The conceptual structure function of the bibliometric package aids in building three map structures, one of which is a factorial map of the research papers with the highest contributions and citations. The LCSA clusters and variables are depicted in the diagram.

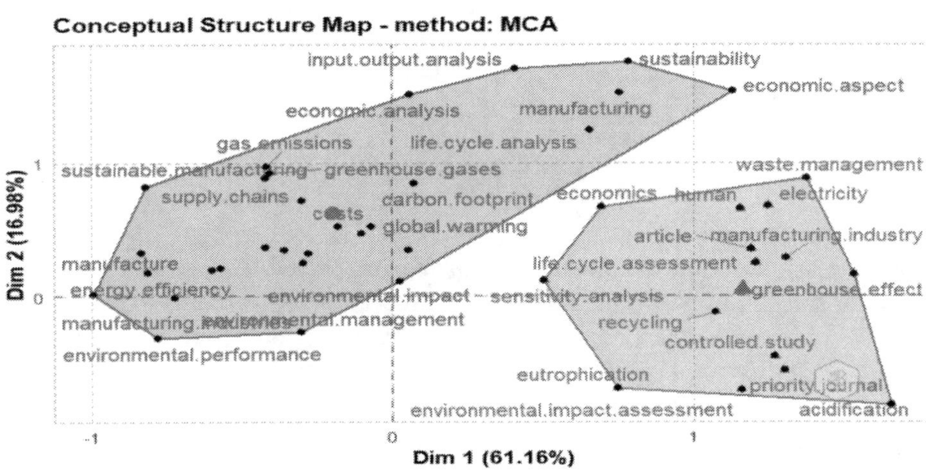

Fig.9: Conceptual structure map by multiple corresponding analyses (Scopus database)

Multiple comparison analysis of the conceptual map is conducted using the graph's vertical and horizontal dimensions. The horizontal group has a vital research area for LCSA research on the left side of the chart, and a weak one for LCSA research on the right side. The research in LCSA is represented in the graph by two separate clusters.

3.10 Top 10 keywords According to the Scopus database:

There are a few keywords that frequently appear throughout the paper. These keywords also provide information about what is currently trending or essential. The data is gathered using the Scopus database, and the top ten keywords are listed in the figure. The most commonly used keyword is "Life cycle," which represents the fundamental concept and way of thinking about LCA. Other keywords, such as environmental impact, LCA, sustainable development, and life cycle analysis, represent ecological issues that arise throughout the life cycle. Another set of keywords, such as environmental impact assessment and environmental management, refers to methods of lowering total emissions. The keyword sustainable development appears 40 times, indicating that sustainability is the primary goal, and LCA provides a path toward sustainability by offering data on total environmental emissions.

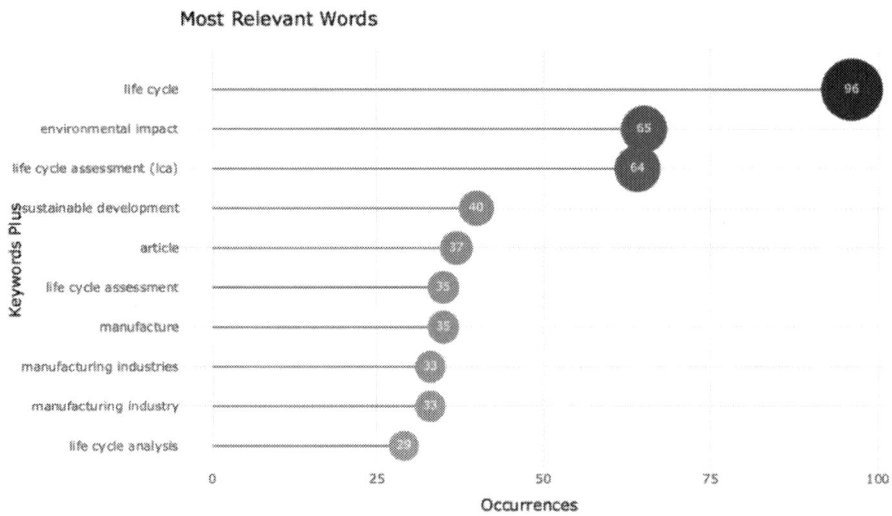

Fig10: Top 10 keywords (Scopus database)

3.11 Country Scientific Production:

It is critical to understand which country is contributing the most. In this research field, the United States ranks first with the most publications (57), assuming that the top ten countries contribute equally. If the top ten countries contribute similarly, the United States' contribution is per cent. China ranks second with 41 publications, according to the Scopus database. The contributions of all other countries are shown in the table.

Table 4: Top 10 countries with publication according to the Scopus database

Region	USA	China	Italy	India	Brazil	Spain	Germany	Japan	Sweden	Canada
Freq.	57	41	16	15	14	13	11	11	11	10

3.12 Top 10 Most Relevant Affiliations (Scopus database)

From figure 11 given below, it can be seen that North Dakota State University holds the top position globally in the field of LCSA with seven publications, according to the Scopus database.

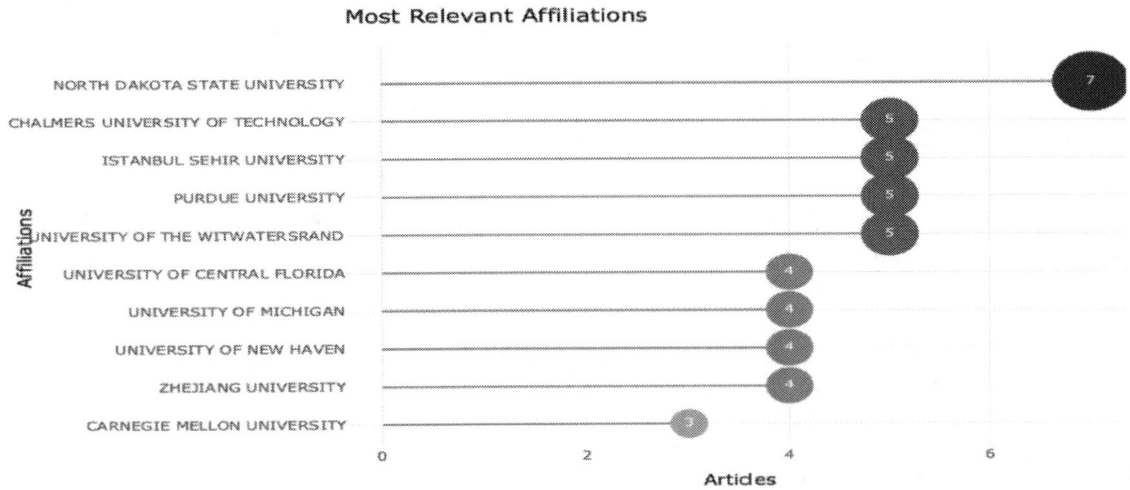

Figure11: Top 10 organisation focus on LCSA (Scopus database)

4. Conclusions and future scope:

Only a few systematic and chronological studies on LCSA in the manufacturing industries were conducted in the past. This Bibliometric paper shed some light on the current state and trend of LCSA. This Bibliometric research work provides an overview of research in LCSA by summarising the institute's contribution, top countries in the LCSA research domain, a top author working in this field, how the growth of LCSA research varies year by year, and top keywords primarily used in publications, top Indian institutes working in LCSA field, and also most cited paper of each year from 1997 to 2020. This bibliometric analysis will aid both researchers and practitioners in comprehending the current state and future direction of LCSA research in the industrial industries. This publication presents viewpoints from leading scholars and research centres in the topic. The Scopus database was used to obtain articles about "Life Cycle Sustainability Assessment in Manufacturing Industries" for this study. Although the use of LSA in manufacturing can benefit industries. Furthermore, we discussed the research progress in this area using R studio and the Scopus database. This research can be expanded by mapping out the study using Web of knowledge and another database that is relevant.

References:

[1] Amicarelli V, Lagioia G and Bux C 2021 Global warming potential of food waste through the life cycle assessment: An analytical review *Environ. Impact Assess. Rev.* **91**

[2] Benis K and Ferrão P 2017 Potential mitigation of the environmental impacts of food systems through urban and peri-urban agriculture (UPA) – a life cycle assessment approach *J. Clean. Prod.* **140** 784–95

[3] Arora L, Kumar S and Verma P 2018 The Anatomy of Sustainable Growth Rate of Indian Manufacturing Firms *Glob. Bus. Rev.* **19** 1050–71

[4] Burchart-Korol D 2011 Green manufacturing based on life cycle assessment (LCA) *Carpathian Logistics Congress CLC*

[5] Blanco C F, Cucurachi S, Dimroth F, Guineé J B, Peijnenburg W J G M and Vijver M G 2020 Environmental impacts of III-V/silicon photovoltaics: Life cycle assessment and guidance for sustainable manufacturing *Energy Environ. Sci.* **13** 4280–90

[6] De Menna F, Davis J, Östergren K, Unger N, Loubiere M and Vittuari M 2020 A combined framework for the life cycle assessment and costing of food waste prevention and valorization: an application to school canteens *Agric. Food Econ.* **8**

[7] Gundu K, Jamwal A, Yadav A, Agrawal R, Jain J K and Kumar S 2022 Circular Economy and Sustainable Manufacturing: A Bibliometric Based Review *Recent Advances in Industrial Production* (Springer) pp 137–47

[8] Upmanyu S, Upmanyu A, Jamwal A and Agrawal R 2022 Machine learning in CAD/CAM: What we think we know so far and what we don't *Recent advances in industrial production* (Springer) pp 495–507

[9] Jamwal A, Agrawal R, Sharma M, Kumar A, Luthra S and Pongsakornrungsilp S 2021 Two decades of research trends and transformations in manufacturing sustainability: a systematic literature review and future research agenda *Prod. Eng.*

[10] Donthu N, Kumar S and Pattnaik D 2020 Forty-five years of journal of business research: a bibliometric analysis *J. Bus. Res.* **109** 1–14

Maintenance scheduling under multiple resource constraints using a project scheduling technique

Amol Chaudhary *[1], **Sachin Meshram** [2]

[1] Research Scholar, G H Raisoni University, Amravati, 444701, India
[2] Mechanical Engineering Department, G H Raisoni University, Amravati, 444701, India

Abstract. The work presented in this paper deals with the scheduling problem under multiple resource constraints.A heuristic method called the current float technique of resources scheduling under multiple resource constraints is applied to solve this problem.The maintainance activities of some machines of coal handling plant of a thermal power station is considered.The activities are arranged in network diagram and a schedule is developed which will give the information of current status of activities,utilization of resources and idle resources.

Keywords: Project scheduling, resource constraints, project duration, optimum schedule, idle resources
* Corresponding Author. *E-mail:* a.choudhari@raisoni.net

1. Introduction

Maintenance shutdown management is one of the most important activity in a thermal power plant. This activity needs forecasting of sequencing of maintenance activities. Hongming Zhou et al [1] presents policy for optimization in maintenance which is preventive in nature. Paper describes method of optimization for production line operations. Heuristic algorithm used here derives optimal parameters considering maintenance cost and operating speed as objective. Joanna Józefowska et al [5] considers make span and multiple execution modes as the criteria of minimization in resource constrained project scheduling problem (RCPSP).Luca Benini et al [8] use Cell BE platform for scheduling its processors, memories and communication channels.J. Arturo Castillo-Salazar et al [10] suggest methods to solve problems related to scheduling of workforce and routing problems . Altannar Chinchuluun et al [11] discuss optimization techniques, optimality conditions and applications of multiobjective optimization (MO).Morteza Davari et al [13] studies contribution of optimal proactive and reactive policy. Ulrich Junker [21] studies preference based search to handle preferences between multiple criteria. Olivier Lambrechts et al [22] studies the effect of unexpected resources on duration of activity and develop a technique for inserting explicit idle time into the project schedule. Chakrabortty R.K. et al [27] provides a heuristic technique for optimization for non-renewable and deterministic renewable resources.

F. Le Hu´ed´e et al [2] proposes an algorithm for combinatorial multicriteria optimization problems. Zdenˇek Hanzálek et al [3] presents a scheduling technique which can be applied for vehicle routing problem to minimize the distance travelled. Catalina Beatriz et al [4] studies generalized parabolic

IOP Conf. Series: Materials Science and Engineering 1259 (2022) 012036 doi:10.1088/1757-899X/1259/1/012036

distribution (GBP) as an alternative distribution method for Project Analysis and Review Technique (PERT). J Alcaraz et al [6] carries out a series of experimentation with a specially designed genetic algorithms for allocation of resources in project scheduling. Haldun Aytug et al [7] considers strategies based on genetic algorithms. Markus Bohlin et al [9] considers preventive maintenance optimization to prove that parallel execution of maintenance activities at a reduced production level is cost effective. R.H.Clark et al [12] presents algorithm for the optimization of generalized networks. Eric V. Denardo et al [14] analyses a Markov chain decision problem called multi arm bandit. Paper presents optimum solution using column generation and elementary row operations. Gülcin Ermis et al [15] develop a meta heuristic and search algorithm to solve academic timetabling problem. Laureano F. Escudero et al [16] presents a 0-1 model applicable for a wide area of project scheduling problems. Warren B. Powell et al [17] deals with the network resource by stockastic programs with expected resource function bounds. Debasish Ghoseet al [18] applies game theory concepts in resource allocation problem in military operations. Stefano Gualandi et al [19] attempts pricing problems with column generation framework with constraint programming. Sönke Hartmann et al [20] use genetic approach to solve the scheduling of activities with multiple execution modes. Olivier Liess et al [23] presents approach of constraint programming to solve RCPSP.Filtering algorithm is used to solve the resource constraints problem producing effective numerical results when used with known filtering algorithms. Christian Artigus et al [24] proposes to solve multiple mode scheduling problem with cumulative constraints using polynomial algorithm. The method requires low computational complexities to find optimal solution.Peter Brucker et al [25] presents the branch and bound algorithm for determination of schedule with minimum makespan of RPCSP.The algorithm gives better results for capacity factor greater than 0.5.Martin Tritschler et al [26] proposes hybrid metaheuristic to solve flexible resource constrained project scheduling problem (FRCPSP).This technique is a combination of non-greedy resource allocation, delayed scheduling and genetic algorithm.

2. The maintenance scheduling problem

The present paper considers the shutdown maintenance of a coal handling plant of a thermal power plant.The machines which undergo the maintenance are Wagon tippler, Wagon tippler marshalling unit, Wagon tippler Hopper, Apron feeder, Double roll crusher to conveyor transfer chute, Impact crusher, Roller screens and Impact crusher. For sample scheduling, maintenance activities of wagon tippler are considered. Duration of activities are in normal working shifts.Duration of each shift is eight hours.

Table 1. Duration of activity and requirements of resources for Wagon Tippler machine

Activity	Description of activity	Duration (Shifts)	Resource requirements		
			Type 1 (F)	Type 2 (W)	Type 3 (H)
1-2	Removing of gear box, motor decoupling, removing of motor etc.	2	1	1	3
1-3	Removing of bottom frame	2	1	1	5
2-4	Fabrication of bottom frame	2	1	1	4
3-5	Fitting of gear box	2	1	1	4
4-6	Alignment of gear box	2	1	1	5
5-7	Fitting of motor and alignment	2	1	1	4

6-8	Lifting of one side of barrel.	2	2	1	4
7-9	J1 pin making free	2	1	1	4
8-10	J2pin making free	2	1	1	4
9-11	Releasing of barrel	1	2	1	4
11-12	Lifting of other side of barrel	1	1	1	5
12-13	J3 pin making free	2	1	1	4
13-14	J4 pin making free	1	1	1	4
14-15	Releasing of barrel from other side	1	1	0	4
15-16	Fitting of bottom frame	1	0	0	1
10-16	Hydraulic system –package oil leakage attending, Valve servicing and trials	2	2	1	3

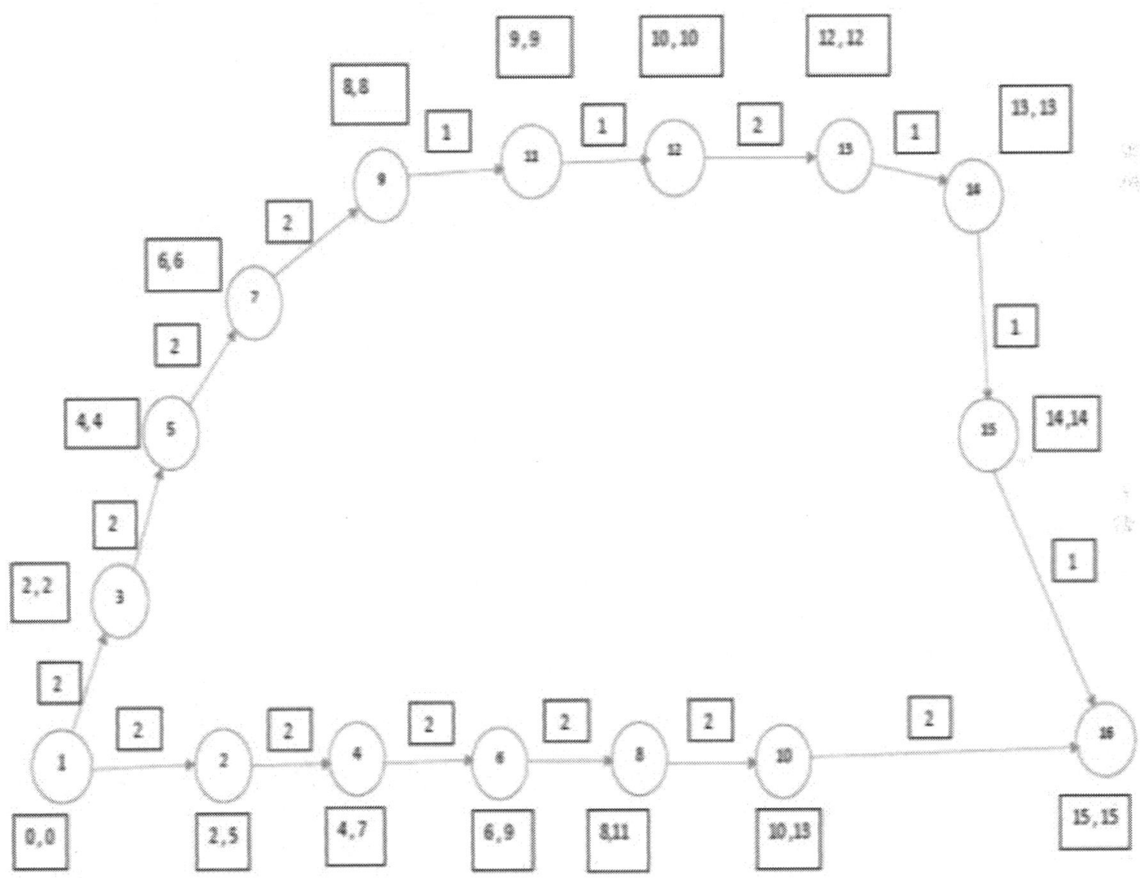

Figure 1. Maintenance activities network diagram for Wagon tippler machine.

Table 2. Availability of resources

Resource Type	Resource Description	Maximum availability of resources.
1	Fitter (F)	2
2	Welder (W)	2
3	Helper (H)	8

3. The scheduling model.

There are several heuristics and algorithms to deal with the scheduling and allocation of resources. Main problem is to give priority to an activity which is in conflict with other activities for allocation of resources under multiple resource constraints.Previous study proves that activity having minimum total float results a close optimum solution.The total float model has a disadvantage that it requires many computations based on status network .To solve the present problem, current float model is used which is first proposed by V.Shanmuganayagam [27].

4. Priority rules

The current float method assigns the priority for an activity which has minimum current float.The next priority is the activity with least earliest start time.Then the activity with longest duration. The last priority is for the activity.

The formulations used in the current float model are as follows:

Current float (CF) = Latest finish time(LFT) - duration (d) – Current time (CT)

Finish Time = LFT – CF (For the activity which are started)

CT = Minimum finish time.

Table 3. Computations for activity scheduling

CT	Activities Considered	Resource Requirements			d	LFT	CF	Decision	Finish Time
0	1-3	1	1	5	2	2	0	Start	2[b]
	1-2	1	1	3	2	5	3	Start	2
2	3-5	1	1	4	2	4	0	Start	4[b]
	2-4	1	1	4	2	7	3	Start	4
4	5-7	1	1	4	2	6	0	Start	6
	4-6	1	1	5	2	9	5	Can not start	-
6	5-7	1	1	4	2	8	0	Start	8[b]
	4-6	1	1	5	2	9	1	Can not start	-
8	4-6	1	1	5	2	9	-1	Start	10[b]
	9-11	2	1	4	1	9	0	Can not start	-
10	9-11	2	1	4	1	9	0	Start	-

	6-8	2	1	4	2	11	-1	Start	-
11	11-12	1	1	5	1	10	-2	Start	12[b]
	6-8	2	1	4	2	11	-2	Can not Start	-
12	6-8	2	1	4	2	11	-3	Start	14[b]
	12-13	1	1	4	2	12	-2	Can not Start	-
14	12-13	1	1	4	2	12	-4	Start	16[b]
	8-10	1	1	4	2	13	-3	Can not Start	16
16	13-14	1	1	4	1	13	-4	Start	17[b]
	10-16	2	1	3	2	15	-3	Can not Start	-
17	14-15	1	0	4	1	14	-4	Start	18[b]
	10-16	2	1	3	2	15	-4	Can not Start	-
18	10-16	2	1	3	2	15	-5	Start	20[b]
	15-16	1	0	0	1	15	-4	Can not Start	-
20	15-16	1	0	0	1	15	-6	Start	21

[b]Next current time

5. Results

The project duration is 15 days without considering constraints on resources.If the resource constraints are considered,the project duration increases to 21 days.

Table 4. Utilization of Resources

Resource types	Available resource shifts	Used resource shifts	Idling resource shifts
1	42	30	12
2	42	30	12
3	168	97	81

6. Conclusion

The scheduling procedure applied for maintenance scheduling gives information about utilization of resources and idle resources .It also provides the actual duration of the project under multiple resource constraints. The heuristic models will yield same number of resource which are utilized but the number of idle resources will depend upon the type of model. The model resulting into minimized idle resources will be nearer to the optimum model.

7. References

[1] Hongming Z, Sufen W, Faqun Qi, Shun G 2019 Maintenance modeling and operation parameters optimization for complex production line under reliability constraints J Annals of Operations Research

[2] F. Le Huede, M Grabisch, C Labreuche, P Saveant 2006 MCS—A new algorithm for multicriteria optimisation in constraint programming J Annals of Operations Research 147 pp143–174

[3] Zdenek H., Premysl Šucha 2016 Time symmetry of resource constrained project scheduling with general temporal constraints and take-give resources. J Annals of Operations Research

[4] Catalina Beatriz G, José Garcia P, Salvador C 2010 Proposal of a new distribution in PERT methodology JAnnals of Operations Research 4181 pp 515–538

[5] Joanna J, Marek M, Rafał R, Grzegorz W, Jan W 2001 Simulated Annealing for Multi-Mode Resource-Constrained Project Scheduling.*J Annals of Operations Research* 102 pp 137–155

[6] J. Alcaraz, C. Maroto 2001 A Robust Genetic Algorithm for Resource Allocation in Project Scheduling *J Annals of Operations Research* 102 pp 83–109.

[7] Aytug H, Bhattacharyya S, Koehler G, 1998 Genetic learning through simulation: An investigation in shop floor scheduling *J Annals of Operations Research* 78 pp 1 – 29

[8] Luca B, Michele L, Michela M, Martino R, 2011. Optimal resource allocation and scheduling for the CELL BE platform. *JAnnals of Operations Research 184* pp 51–77

[9] Markus B, Mathias W, 2012 Maintenance optimization with duration-dependent costs. *J Annals of Operations Research* 224 pp 1-23

[10] J Arturo Castillo-S , Dario Landa-S, Rong Qu 2014 Workforce scheduling and routing problems: literature survey and computational study. *J Annals of Operations Research* 239 pp 39–67

[11] Altannar C, Panos M. P 2007 A survey of recent developments in multiobjective Optimization. *J Annals of Operations Research* 154 pp 29–50

[12] Clark R.H, Mayer R.R 1990 Parallel arc-allocation algorithm for optimizing generalized networks. *J Annals of Operations Research* 22 pp129-160

[13] Morteza D, Erik D 2018 Important classes of reactions for the proactive and reactive resource-constrained project scheduling problem. *J Annals of Operations Research* 274 pp 187-210

[14] Denardo E V, Feinberg E A, Rothblum U G2013 The multi-armed bandit, with constraints. *J Annals of Operations Research* 208 pp 37–62

[15] Ermis G, Akkan C, 2019 Search algorithms for improving the pareto front in a timetabling problem with a solution network-based robustness measure. *J Annals of Operations Research* 275 pp 101–121

[16] Escudero L F, Salmeron J 2005 On a Fix-and-Relax Framework for a Class of Project Scheduling Problems *J Annals of Operations Research* 140 pp 163–188

[17] Powell W B 1996 Restricted resource strategies for bounding the expected network resource function. *JAnnals of Operations Research* 64 pp 261-287

[18] Ghose D,Speyer J L, Shamma J S2002 A Game Theoretical Multiple Resource Interaction Approach to Resource Allocation in an Air Campaign *J Annals of Operations Research* 109 pp 15–40

[19] Gualandi S, Malucelli F, 2013 Constraint Programming-based Column Generation *J Annals of Operations Research* 204 pp 11–32

[20] Hartmann S 2001 Project Scheduling with Multiple Modes: A Genetic Algorithm *J Annals of Operations Research* 102 pp 111–135

[21] Junker U 2004 Preference-Based Search and Multi-Criteria Optimization *J Annals of Operations Research* 130 pp 75–115

[22] Lambrechts O, Demeulemeester E, Herroelen W 2011 Time slack-based techniques for robust project scheduling subject to resource uncertainty *J Annals of Operations Research* 186 pp 443–464

[23] Liess O, Michelon P, 2008 A constraint programming approach for the resource-constrained project scheduling problem *J Annals of Operations Research* 157 pp 25–36

[24] Christian A, Francois R 2000 A polynomial activity insertion algorithm in a multi-resource schedule with cumulative constraints and multiple modes. *European Journal of Operational Research* 127 pp 297-316

[25] Peter B, Sigrid K, Arno Schoo, Olaf T 1998 A branch and bound algorithm for the resource-constrained project scheduling problem *European Journal of Operational Research* 107 pp 272-288

[26] Tritschler M, Naber A, Kolisch R 2017 A hybrid metaheuristic for resource-constrained project scheduling with flexible resource profiles *European Journal of Operational Research* 262 pp 262–273

[27] Shanmuganayanam V 1989 Current float techniques for resources scheduling *Journal of construction Engineering and Management* Vol 115 No 3

IOP Conf. Series: Materials Science and Engineering 1259 (2022) 012037 doi:10.1088/1757-899X/1259/1/012037

Automated Sanitizing Machine for Recording Thermal Detection and Hygiene Application

Prashant D. Kamble[1], Ketan Fulzele[2], Nikhil Shahare[3], Ritik Rane[4], Shilpa B. Sahare[5] and Yashika Gaidhani[6]

[1,2,3,4,5,6] Department of Mechanical Engineering, YCCE, Nagpur, MS India, 441110

Abstract. A multipurpose automatic disinfection booth is an automated, non-contact, alcohol-based disinfection station, which finds its use in hospitals, work, premises, offices, schools, and other relevant places. It serves multiple purposes of sanitizing not only the human body but all of the components they are carrying. By using UV lights, it sanitizes the whole object or accessories carried by a person. Along with this functioning, it also provides a fully automated hand sanitizing mechanism which is followed by temperature detection of the person entering the machine, and it is loaded with smart sensors to take appropriate decisions. It has a security mechanism in form of a barricade to restrict the potentially unwanted person from entering. Also, it stores data of the person in form of an image for security purposes along with providing free energy.

Keywords: Automated hand sanitizing dispenser, disinfection chamber, ESPN CAM, sanitization tunnel, temperature detection.

1. Introduction

Viruses such as COVID-19 are transferable through human interactions. There are WHO guidelines to clean or wash or sanitized hands every day to reduce the risk of spreading the infection. Sanitizer storage will be manual. The corona pandemic is affecting the education system very hard. Still, after 8 months schools are unable to reopen the online system of education is not appealing to students as well as teachers as the presence of the teacher is most important for the learning process. To reopen the educational institutes, it needs a strong sanitizing machine for sanitizing 1000s of the students entering the premises. No such machine or arrangement is available at the Institute which sanitizes not only the person and its equipment but also marks its attendance with a face reading application. This project proposes a new design of a disinfectant machine without touching to reduce the risk due to contact and measurement of body temperature and various uses of the tripod or fields. Which is widely seen in hospitals, workplaces, offices, schools, and other relevant places. The microprocessor has received all input in terms of motion detection or captures the radiation of the body, from the ultrasonic sensor which detects proximity. The microprocessor processes the sensed data which was collected by sensors and activates the pump. The disinfectant liquid flows through the spray nozzle. Problems find after researching all results got. The Corona pandemic has impacted the world and triggered a global social and economic crisis, and science through invention and innovation holds the key to fighting against this outbreak. No such machine or arrangement is available at the Institute which sanitizes not only the person and its equipment but also marks its attendance with a face reading application.

Due to the pandemic of COVID19, temperature measurement of persons became one of the major tasks to detect the affected. To do this, non-contact measurement and social distance increasingly become a temperature and off sound abnormal it gives danger alarm sound so that every one of us may be alert.

It believes in delivering sanitization services to the community with the safest and cost efficiency to reach hygiene Expectations that strive to provide people with preventive and curative methods keeping in mind the environmental challenges come across. Provide full body sanitation service, cell phones, wallets, bags, and other essentials. Some objectives related to sanitization are as follows:

Design or manufacture a simple and efficient room that will be portable. To sanitize your full body while going inside any institute. To replace old techniques of sanitization. To capture the temperature of the person who is passing through the chamber and able to store the data in the database, to mark the attendance of the person. To reduce manpower requirements for personal sanitization. The booth should successfully be able to sanitize students entering the college. The attendance and count of the person should be automatically marked with the help of biometrics.

2. Construction

A fully automated body sanitizing tunnel comprises the different applications and various functions. It comprises the various component as follow,

2.1 Automatic hand sanitizing dispenser:

Which will be used to sanitize the hand of the person who will use that machine. The purpose is to keep our machine also free from any bacteria or viruses.

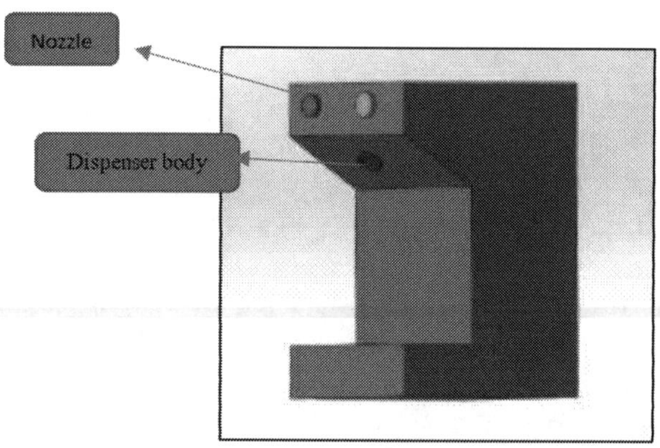

Figure 1 Hand sanitizing dispenser

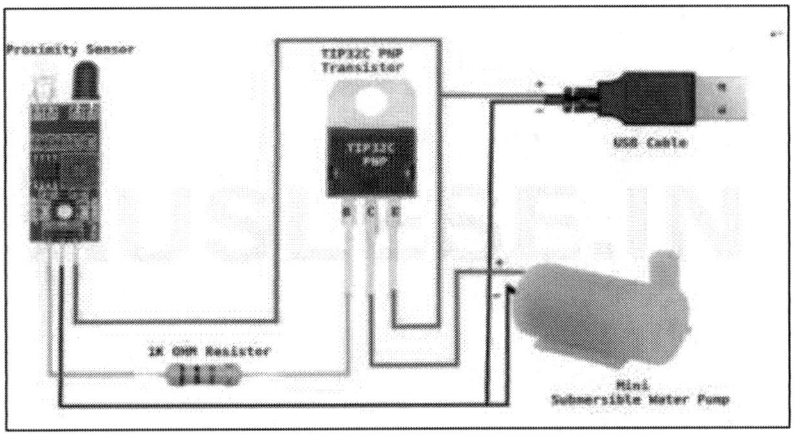

Figure 2 dispenser circuit with all components

2.2 *Automatic entry and exit gate:*

This setup consists of a barricade that is attached to the thermal detection setup. When a person enters his hand inside the detection section, a reading of the temperature is taken and the program decides whether to let the person enter or not based on the temperature level. The signal is given to the motor if allowed to enter else the gate remains closed and the person is restricted from entering the organization.

Figure 3 Automatic entry and exit gate

2.3 *Object sanitizing container:*

This sanitization chamber is an extra layer of protection provided in this system which is utilized to sanitize the objects carried by the person.it uses UV radiation to disinfect objects from any kind of viruses and bacteria. It has doors from both sides which are utilized to take out the belongings of the person from the other side once he is sanitized.

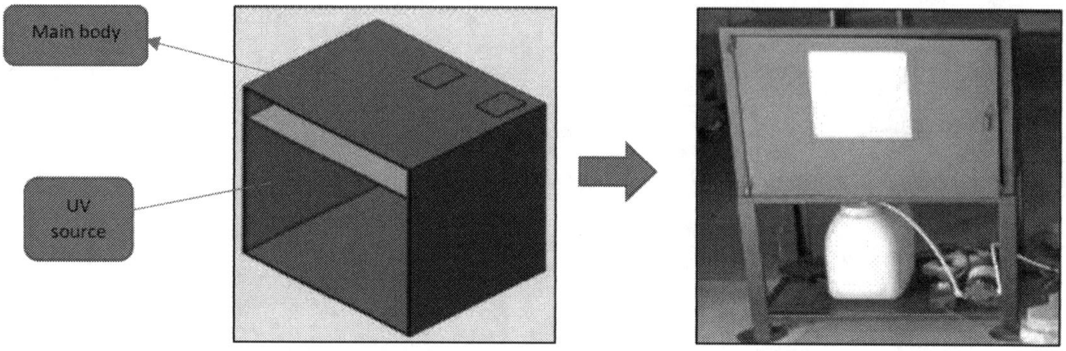

Figure 4 UV light disinfectant chamber

2.4 ESPN32 cam motion detector:

This will help to capture the picture of that person as well as store the info about the person like temperature, time, image, date, etc. stored into the database on our college server

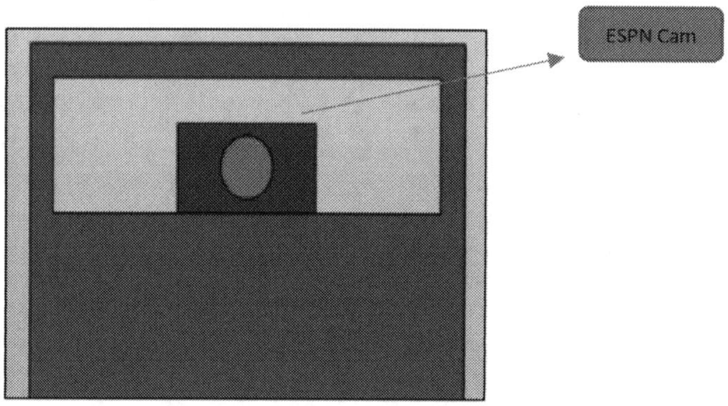

Figure 5 ESPN-32 CAM

2.5 Sanitizing tunnel:

The tunnel from which a person passes and body will get fully sanitized automatically with the help of the fine nozzle.

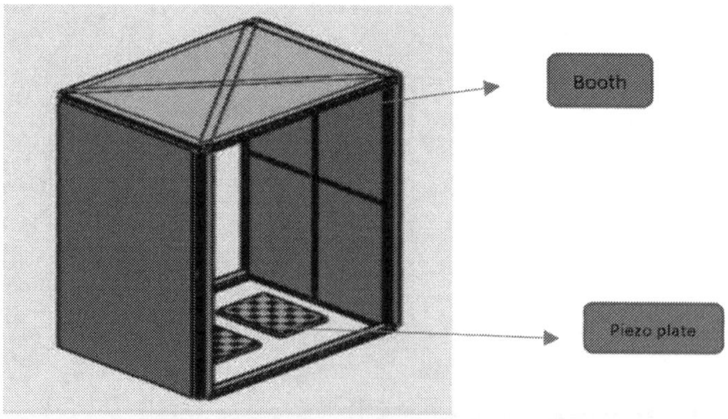

Figure 6 sanitization tunnel

2.6 Temperature detection setup:

The purpose to detect the temperature of the body and take the appropriate action against the person means to allow the person to enter our department or not.

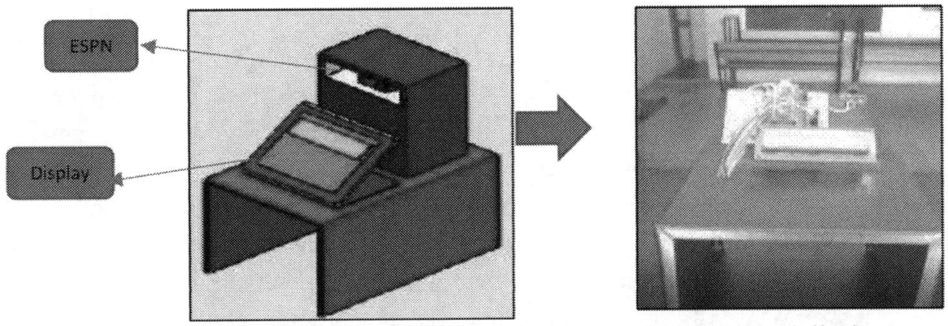

Figure 7 temperature detection setup with ESPN and display

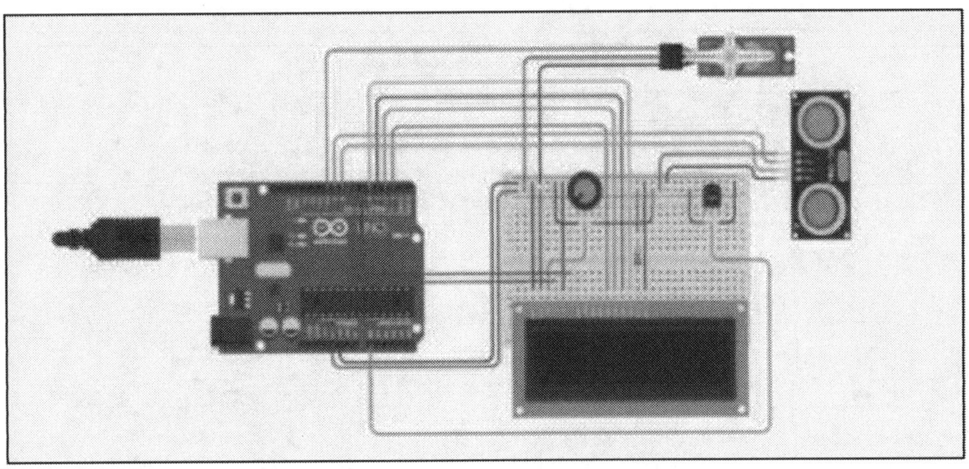

Figure 8 Thermal detection setup circuit

2.7 Pump, sensor, and sanitizer tank:

The pump is used to spray the sanitizer over the person attached with the time module proximity sensor.

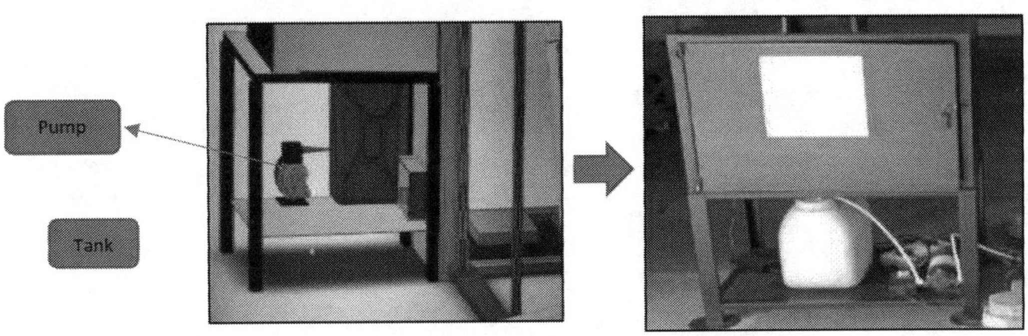

Figure 9 Pump and sanitization tank

3. Working

- As per workflow first component is an automatic hand sanitizing dispenser who use to sanitize the hand before touching or infecting our machine.
- The second step in the workflow is to keep our all accessories in the disinfectant chamber for sanitization purposes by using UV rays.
- The third step in workflow keeps the wrist inside the temperature detection setup and will detect the temperature of the body and decide according to their body temperature to allow that person to be inside our department or not along with that data like temperature, date, time, as well as image sent to the server or drive and stored it for security purpose.
- The fourth step in the workflow plan as per the temperature range as it decided how much value temperature, it's had to allow the person to enter the department if the temperature is within the range, then the gate open otherwise it will remain close.
- The fifth step in the workflow plan automatic sanitization tunnel which is used to spray the sanitizer on the body with the help of a nozzle only up to the shoulder.

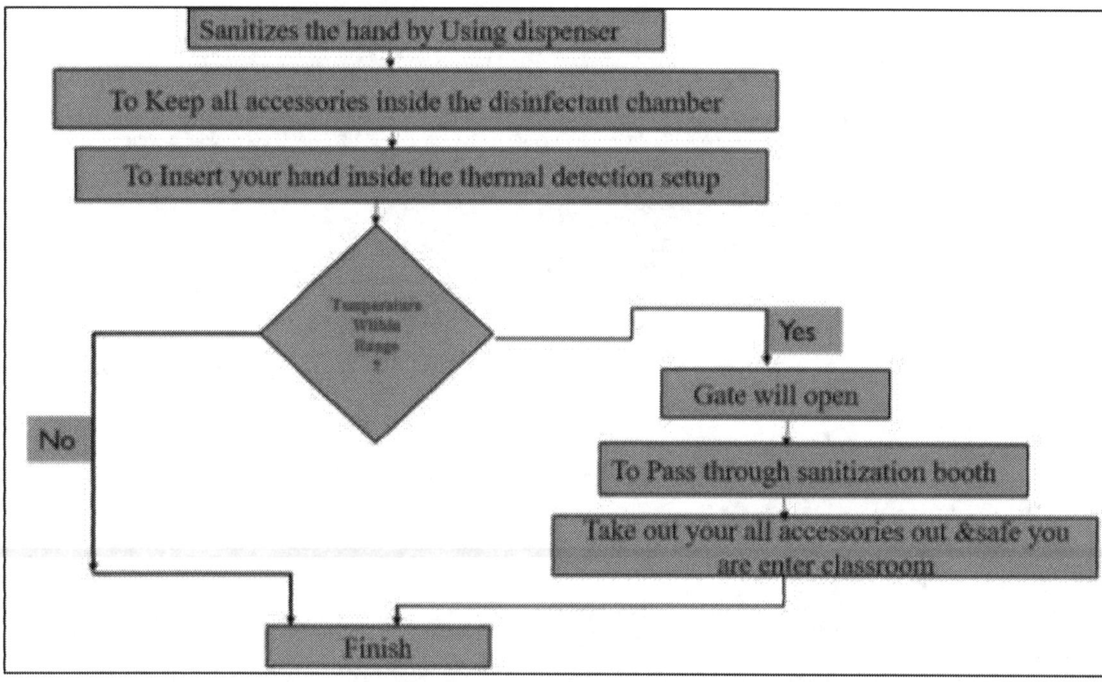

Figure 10 Workflow for the execution of a task.

4. Design Calculation

- DESIGN
- Specifications
  - Frame (CAST IRON-SQUARE ROD):2.5*3*7ft_(l*b*h)
  - UV Chamber (SHEET METAL) :2.6 * 2ft
  - Sheet Metal: Steel

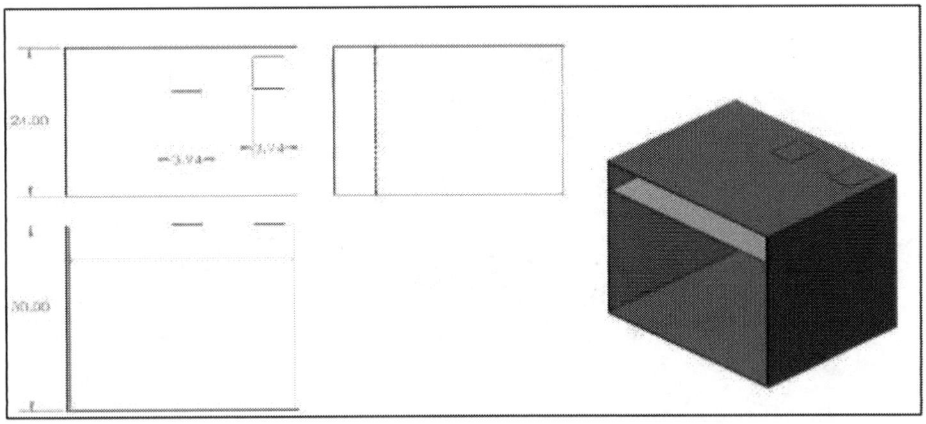

Figure 11 CAD model of disinfectant chamber

For that purpose, it considers the average Height and width of a human and by practicing all possible practical ways, it's decided to keep at least 7 feet of height.

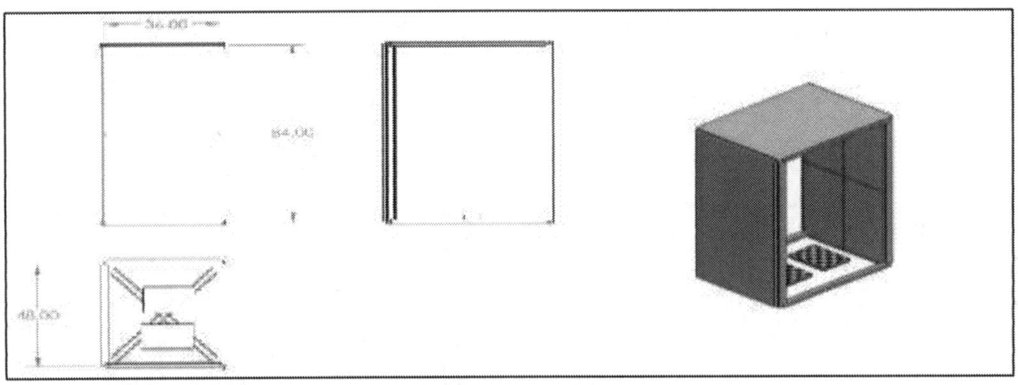

Figure 12 CAD model of sanitization boot

- To a Human sanitizing chamber and a width of 2.5 feet and a depth of 3 feet. Decided to Fixed the joints of a frame by attaching it with Nut & Bolt system.

- Rest from this weal so have different frame structure for the UV CHAMBER for sanitizing the other human accessories that he/she carries with it such as wallet, bag, mobile, etc.

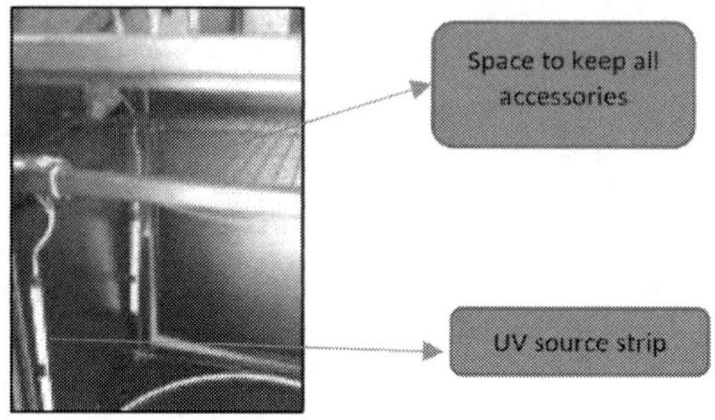

Figure 13 inside view of the disinfectant chamber

- Material of UV Chamber– Sheet Metal
  - Has high strength and corrosive resistance properties. Use of UV Chamber–Sanitization of Mobile Phones, Laptops, Wallets, Keys, Bags, etc.

- The material used for the dispenser body is corrugated shit which is lightweight weighted and cheap.

5. Final Result of Research

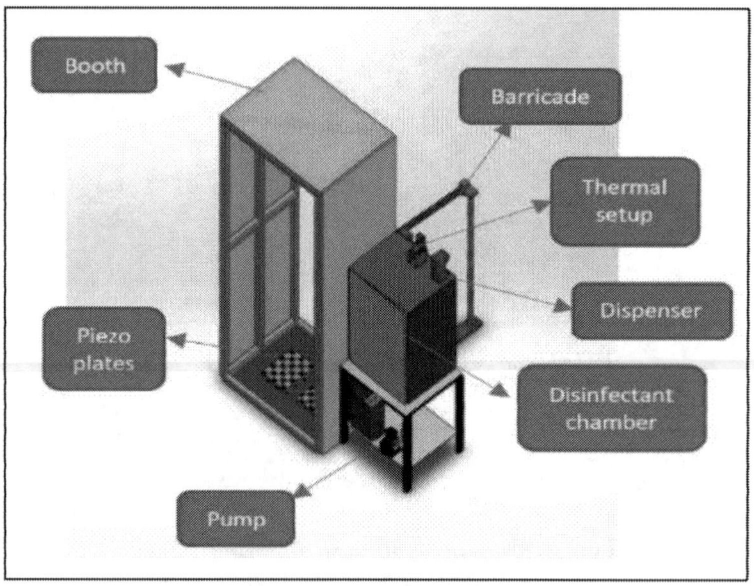

Figure 14 Complete CAD model assembly

It explained all the components individually with proper functioning so now it's time to assemble that all together to see how that work or helps in our actual life.

As a person use that setup for sanitization purpose then our sample test result comes out like that, It's stored the data of the person like temperature, image, date, and time when he or she enter our department in the cloud storage service like google cloud or google drive in the following format:

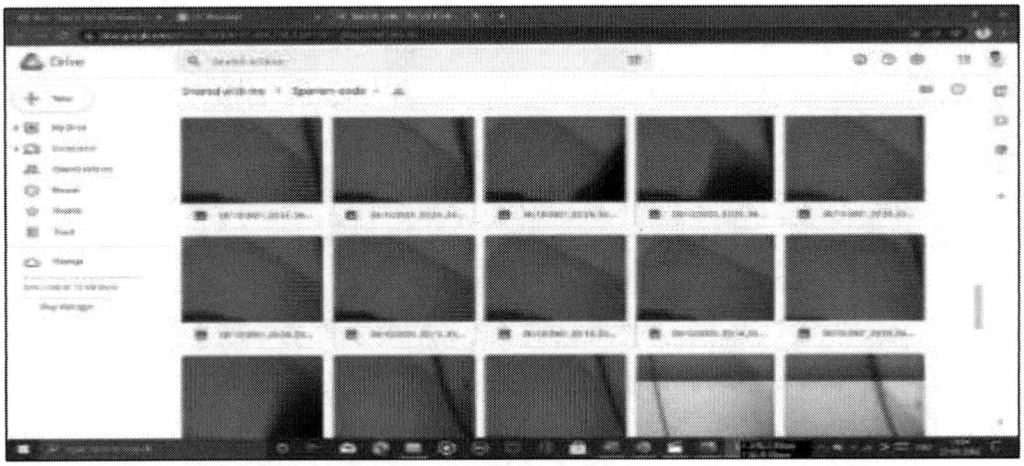

Figure 15 Format for getting our image, time, date, and temperature

Here is our main concern in doing that research to store the data in that format for future use so that it can identify all all-possible solutions.

6. Conclusion

This research paper deals with the concept of multipurpose sanitizing booths mainly for colleges and offices. The setup is successfully Fabricated and implemented. The booth is collapsible and can be used to curb the spread of the coronavirus and help in starting the school and colleges, and big offices in full fledge which are in running in a limited capacity. This strand aims to disinfect the human body and its personal effects to the maximum reduce the risk of infections. Life and health are the big concerns nowadays. This COVID19 pandemic has hit so hard that millions of people around the world have been affected. This is an alarming and eye-opening situation for the prevention of such epidemics in the coming future. Thus, our project act as a barrier between harmful and deadly viruses to keep the human race safe. Sanitize the person as well as his other accessories and also security purposes. The person is almost safe before contacting other people. If the person is not fit and healthy your sanitizing booth grasp that information immediately and alert that person about his condition. This is concept is useful for the corporate world

References

[1]. Sharma A S, July-2020 Review on automatic sanitizer dispensing machine, *IJERT* **Vol. 9 Issue 07.**

[2]. Puput Wanarti Rusimamto, Nurhayati Nurhayati, Eppy Yundra Reza Rahmadian Arif and Widodo Much Ade Dermawan, 2020 Automatic Hand Sanitizer Container to Prevent the Spread of Corona Virus Disease, *IJCSE* **volume 196**

[3]. E. minazara, D. Vasic, and F. Costa, March 2008 Piezoelectric generator harvesting bike vibrations energy to supply portable devices, *IJCSE* **Vol. 1, No.6**

[4]. Singh U. K. and Middleton R. H., October (2007) Piezoelectric power scavenging of mechanical vibration energy, *AMTC* p 111-118.

[5]. Sarkar Abhinandan, June 2020 Design of automatic hand sanitizer with temperature sensing, *IJISRT*

[6]. B Varshini, HR Yogesh, Syed Danish Pasha, Maaz Suhail, V Madhumitha and Archana Sasi, 2021 IoT-Enabled smart doors for monitoring body temperature and face mask detection, *science direct* p 246-54

[7]. Roundy S., Wright P. K., and Rabaye, 2003 A. study of low-level vibrations as a power source for wireless sensor nodes, *IJCSE* p 1131–44.

[8]. T.Deepika and Dr.P.Srinivasa Babu, January 2014 Motion Detection In Real-Time Video Surveillance with Movement Frame Capture And Auto, *IJIRSET* **Volume 3.**

[9]. Leo Louis, April 2016 working principle of Arduino and using it as a tool for study and research, *IJCACS* **Vol 1**

[10]. Jithin K, Subash, 2019 Disinfection Gateway & Facemask disposal bin to fight COVID-19, *SCTIMST*

Implementation of AHP to evaluate variables of prognosis agent technology managing manufacturing organizations

Vikram Singh[1],Somesh Sharma[1]

[1] *PhD. Research Scholar, Mechanical Engineering Department, National Institute of Technology Hamirpur, Himachal Pradesh, India, 177005, Email*: vikram_phdme@nith.ac.in

[1]*Associate Professor, Mechanical Engineering Department, National Institute of Technology, Hamirpur, Himachal Pradesh, India, 177005, Email*: somesh.sharma@gmail.com

Abstract:- This research aims to use AHP model to evaluate variables of PAT managing manufacturing organizations. Using AHP approach, an evaluation model can be constructed, capable to measure the weight and ranking of variables. Literature helps to identify twenty six variables of PAT. These variables are arranged on the basis of expert's opinion and construct a conceptual domain in which variable are divided as five major and twenty one sub variables. This domain is used to draw hierarchy framework of AHP model. Thereafter, pair wise comparison is performed on the expert's responses to measure weight of PAT variables, the responses and matrices, validated by computing consistency ratio. Finally, weight and ranking of entire variables are evaluated in AHP methodology section and concluded in last section which gives interesting suggestions to the manufacturing organizations so as to manage their manufacturing system.

Keywords: Prognosis Agent Technology (PAT), Analytical Hierarchical Process (AHP), Variables, weight, rank and maintenance

1. Introduction:

In this current scenario, most of the manufacturing organizations have faced troubles in production due to pandemics of COVID 19. Functionality of every domain have become complicated specially the manufacturing system faced many ups and downs due to unavailability of raw material, man power, obstacle in the supply of goods. To overcome these situations, the manufacturing system needs to be configured with flexible system. This flexible system may includes the combination of hardware and software and constructs a hybrid manufacturing system known as Digital manufacturing, which is operated by an agent based technology. Agents based system is quite flexible, secure, works online as well as offline and also provide facility to the manufacturing organizations so as to operate system from outside of the production plant. Therefore, researchers are working very dedicatedly to explore agent's technology in manufacturing system especially in maintenance domain. Maintenance is the most challenging domain for any manufacturing organizations. About more than 50% of the assets are used by the manufacturing organizations on its maintenance every year. Maintenance provides safety, security, durability and enhances productivity & efficiency [1-4] of manufacturing system. Maintenance can be performed online through e-maintenance [5] so as to monitor, fault detection and prognosis. The term prognosis means to estimate the remaining life of products [6-10]. Integration of agent technology with prognosis is recommended by manufacturing organizations for working efficiently in maintenance to reduce maintenance expenses and also to improve economy of manufacturing organizations. So, with the passage of time it's important to evaluate the variables of PAT in manufacturing organizations.

The current research is a small effort in this area in which AHP method is used to evaluate the variables of PAT managing manufacturing organizations.

Rest of this paper is arranged as Section-2, literature review helps to identify the variables of PAT, Section-3 contains AHP methodology, Section 4 analysis and discussion outcomes from the section 3 and at finally, Section-5 completes the research with conclusion, limitations and future scope.

2. Literature review:

Research in the area of agent's technology is going very rapidly to automate the functionality of existing traditional domains. So, the hardware components are replaced with the software packages which help to increase quality and quantity of organizations. These rapid changes in the infrastructure make the system complicated and require a flexible system to manage production and maintenance process. To achieve this, diagnostics and prognosis are integrated in a common platform to maintain manufacturing system. PAT in manufacturing performs monitoring, estimate remaining life and maintain life of equipment. In views of such attractive characteristics of PAT, we have studied literature and identified twenty six variables of PAT as listed in the table 2.1

Table 2.1 identified variables

Sr.	Variable	Symbol	Reference
1.	Manufacturing process	S^1	11
2.	Fault detection	S^2	9
3.	Merger of manufacturing system with maintenance	S^3	12
4.	Organizational productivity	S^4	13
5.	Organization control	S^5	14
6.	Production Planning.	S^6	15,11
7.	Production Scheduling.	S^7	16
8.	Manufacturing Disturbances.	S^8	17
9.	Virtual Manufacturing.	S^9	18
10.	Decision support tool.	S_{10}	19
11.	Distributed digital manufacturing.	S_{11}	20
12.	Merger of monitoring, diagnosis, and prognosis.	S_{12}	7
13.	Internet enabled predictive maintenance.	S_{13}	21
14.	Diagnosis.	S_{14}	8
15.	Prognosis.	S_{15}	10
16.	Proactive maintenance.	S_{16}	22
17.	Integration of manufacturing system	S_{17}	4
18.	Machine down time	S_{18}	23
19.	Inventory Expensive	S_{19}	24
20.	Setup time	S_{20}	25
21.	Cycle time	S_{21}	26
22.	Organization input	S_{22}	27
23.	Organization output	S_{23}	28
24.	Control System architect	S_{24}	29
25.	Networking between agent	S_{25}	3
26.	Just in time	S_{26}	30

These variables help to construct a conceptual model used in whole research. On the basis of expert's opinion, twenty six variables are arranged as, five major variables (S^1 to S^5) and twenty one sub-variables (S^6 toS^{26}). All twenty one sub- variables are grouped under five variables as shown in table 2.2.

Table 2.2 Conceptual model

Sr. no	Factors	Sub-factors
1	S_1	S_6 S_7 S_8 S_9 S_{10} S_{11}
2	S_2	S_{12} S_{13} S_{14} S_{15} S_{16}
3	S_3	S_{17} S_{18} S_{19} S_{20}
4	S_4	S_{21} S_{22} S_{23}
5	S_5	S_{24} S_{25} S_{26}

3. AHP Methodology

AHP approach was implemented by Saaty in 1980 to solve decision problems related to quality and quantity [31]. This was used to build a hierarchical structure of complex problem and also weigh and rank the factors that affect functionality of organization. This model works on the opinion collected from the experts to achieve the objective. Opinions were collected on a rating performa designed with the help of model created in the previous section; a questionnaire was prepared to collect the opinions from the selected experts. The experts belong to academics and industries. Considering the convenience of experts, the questionnaire performa was prepared in MS office and the presence of every variables of PAT in proper questions was made mandatory. The AHP approach is effectively applied on single response and provides better result. But, we have approached ten experts, out of these six respond to our communication. Experts were asked to give responses to variables of PAT, this responses help to weigh and rate the variables of PAT in manufacturing organizations. The Questionnaire was designed accordingly. This article discussed and implements AHP model step by step as discussed below-

IOP Conf. Series: Materials Science and Engineering 1259 (2022) 012038 doi:10.1088/1757-899X/1259/1/012038

Step 1: Establish the target.

Implementation of AHP method to evaluate the variables of PAT managing manufacturing organizations

Step 2: Divide the target into sub-target

The target is divided into sub-target in conceptual model table 2.2 and used to construct a hierarchy structure.

Step 3.2.3: Create a framework of research hierarchy.

AHP approach begins with hierarchy structure consisting of three levels. Level 1 holds the objective of the research, level 2 has sub-targets i.e. variables and level 3 based on the sub variables of PAT. The prepared model is shown in figure 3.1.

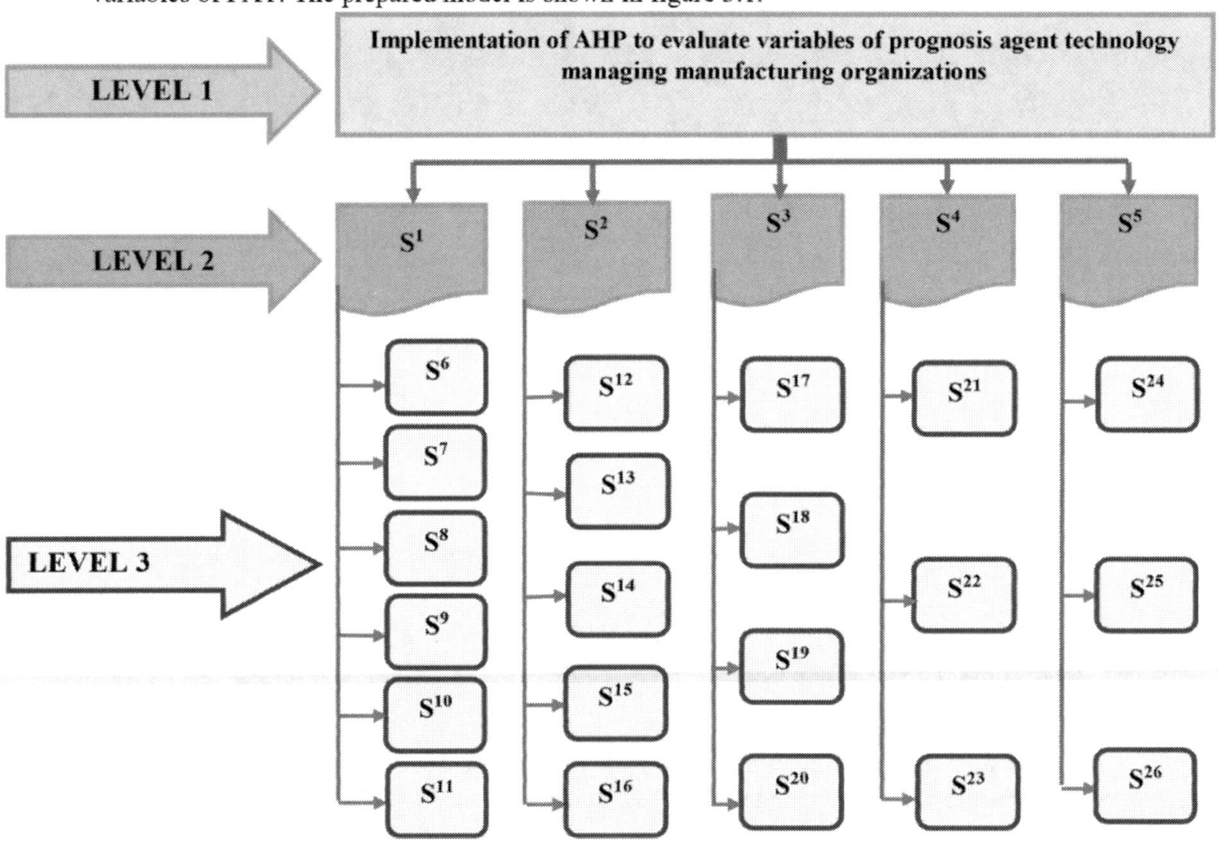

Fig. 3.1 Hierarchical structure of PAT

Step 4: Pair-Wise Comparisons of variables.

After hierarchy model, next stage considered as spine of AHP is used to evaluate the weight of variables by doing pair wise comparisons. In this comparison elements in every set of arrangements are compared with their corresponding group members. To achieve this, a Rating from one to nine, shown in table 3.1, is used to determine the relative weight of the variables.

Table 3.1 Model Rating scale

Rating scale	meaning
1	*Equally essential*
3	*Moderately essential*
5	*Strongly essential*
7	*Very strongly essential*
9	*Extremely essential*
2,4,6,8	*Intermediate values between two adjacent choice*

Questionnaire based on rating scale is send to expert and data collected from experts are used in pair wise comparisons as shown in table 3.2. Then, sum of each column is calculated and then we divide elements column wise w.r.t. calculated sum. This evaluates the normalize matrix as shown in table 3.3. Compute weight of variables as shown in figure 3.2, after computing average of each row in the matrix.

Table 3.2 Pair wise contrast of variables

Factor	S^1	S^2	S^3	S^4	S^5
S_1	1	5	3	9	9
S^2	0.20	1	2	4	7
S^3	0.33	0.50	1	1	3
S^4	0.11	0.25	1	1	3
S^5	0.11	0.14	0.33	0.33	1
Sum	1.76	6.89	7.33	15.33	23

Table 3.3 Normalized matrix

Factor	S^1	S^2	S^3	S^4	S^5	weight
S_1	0.570	0.725	0.409	0.587	0.391	0.536
S^2	0.114	0.145	0.273	0.261	0.304	0.219
S^3	0.190	0.073	0.136	0.065	0.130	0.119
S^4	0.063	0.036	0.136	0.065	0.130	0.086
S^5	0.063	0.021	0.045	0.022	0.043	0.039

Table 3.4 Matrix to measure consistency ratio

Factor	Δ	Eigen vector
S_1	3.117	5.811
S^2	1.182	5.389
S^3	0.611	5.135
S^4	0.436	5.057
S^5	0.198	5.092

IOP Conf. Series: Materials Science and Engineering 1259 (2022) 012038 doi:10.1088/1757-899X/1259/1/012038

Table 3.5 Saaty's random consistency

n	1	2	3	4	5	6	7	8	9	10
ri	0.0	0.0	0.58	0.90	1.12	1.24	1.32	1.41	1.45	1.49

Step 5: Consistency in comparison of matrices

Consistency Ratio (CR) has a great importance in AHP. CR helps to determine whether the calculated weights of elements are consistent or not. Because, data collected from the experts is not always consistent, so to overcome this discrepancy, AHP approach is applied and gives best solution. The consistency is measured by calculating CR which is the ratio of the Consistency Index (CI) calculated from table 3.4 to Random index (RI) table 3.5. Calculated value is always less than 0.1. Here, CR is 0.087 and is less than 0.1. So, the data collected from expert and calculation done is consistent with highest degree of accuracy.

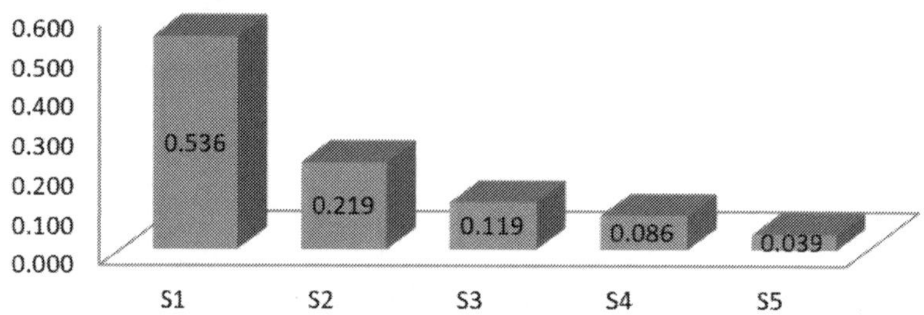

Fig 3.2 weight of variables of PAT evaluated to manage manufacturing organizations

4. Analysis and discussion

AHP is utilized in the previous section to evaluate variables and compute local & global weights with ranking of selected variables of PAT shown in table 4.1. The variables have same, local & global weights; on the other hand sub variables have separate local and global weights shown in table 4.1.

Table 4.1 Local & Global weight and ranking of the criteria and sub-criteria.

Major Variables	Local Weight	Ranking	Sub-Variables	Local Weight	Ranking	Global Weight	Ranking
			S_6	0.355	1	0.190	1
			S_7	0.274	2	0.147	2
			S_8	0.154	3	0.083	4
S_1	**0.536**	**1**	S_9	0.109	4	0.058	7
			S_{10}	0.081	5	0.043	9
			S_{11}	0.032	6	0.017	14
			S_{12}	0.477	1	0.104	3
			S_{13}	0.260	2	0.057	8
S_2	**0.219**	**2**	S_{14}	0.154	3	0.034	10
			S_{15}	0.067	4	0.015	16

IOP Conf. Series: Materials Science and Engineering 1259 (2022) 012038 doi:10.1088/1757-899X/1259/1/012038

			S_{16}	0.042	5	0.009	17
			S_{17}	0.530	1	0.063	6
			S_{18}	0.244	2	0.029	11
S_3	**0.119**	**3**	S_{19}	0.171	3	0.020	13
			S_{20}	0.056	4	0.007	19
			S_{21}	0.737	1	0.063	5
S_4	**0.086**	**4**	S_{22}	0.186	2	0.016	15
			S_{23}	0.077	3	0.007	20
			S_{24}	0.688	1	0.027	12
S_5	**0.039**	**5**	S_{25}	0.234	2	0.009	18
			S_{26}	0.078	3	0.003	21

First group variables S^1 from table 4.1 are considered on the basis of local weight & ranking, S^6 is ranked in the 1st position with 0.355 local weight value, S^7 is ranked in 2nd position, S^8 is ranked 3rd position, S^9 is ranked 4th, S^{10} is ranked 5th position and S^{10} is ranked in 6th position by obtaining least 0.032 local weight value. Figure 4.1 shows local weight of PAT sub-variables evaluated to manage manufacturing process.

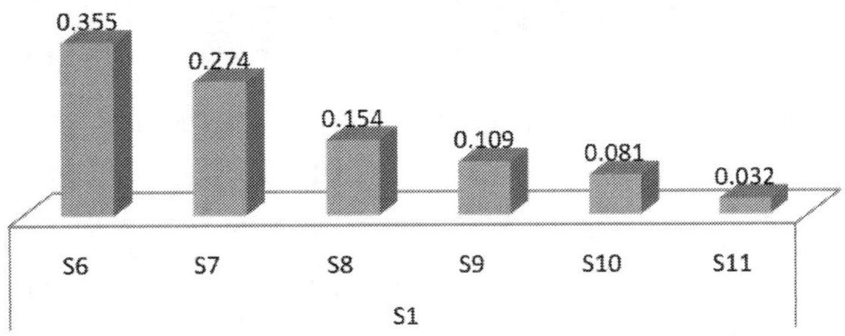

Fig 4.1 Weight of sub-variable evaluated to manage S^1

Second group variables S^2 from table 4.1 are considered on the basis of local weight & ranking, S^{12} is ranked in the 1st position with 0.477 local weight value, S^{13} is ranked in 2nd position, S^{14} is ranked 3rd position, S^{15} is ranked 4th and S^{16} is ranked 5th position obtaining least 0.042 local weight value. Figure 4.2 shows local weight of PAT sub-variables evaluated to manage fault detection.

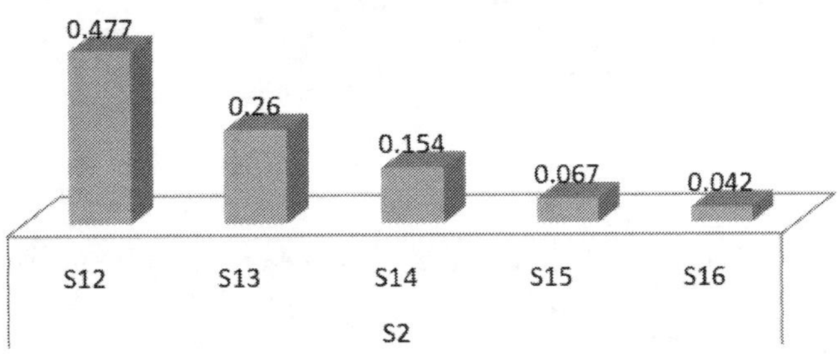

Fig 4.2 Weight of sub-variable evaluated to manage S^2

Third group variables S^3 from table 4.1 are considered on the basis of local weight & ranking, S^{17} is ranked in the 1st position with 0.530 local weight value, S^{18} is ranked in 2nd position, S^{19} is ranked 3rd position and S^{20} is ranked 4th by obtaining least 0.056 local weight value. Figure 4.3 shows local weight of PAT sub-variables evaluated to manage merger of manufacturing system with maintenance.

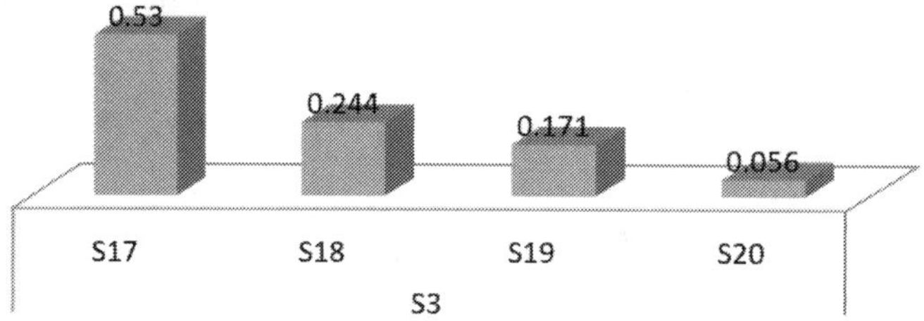

Fig 4.3 Weight of sub-variable evaluated to manage S^3

Fourth group variables S^4 from table 4.1are considered on the basis of local weight & ranking, S^{21} is ranked in the 1st position with 0.737 local weight value, S^{22} is ranked in 2nd position and S^{23} is ranked 3rd position by obtaining least 0.077 local weight value. Figure 4.4 shows local weight of PAT sub-variables evaluated to manage merger of organization productivity.

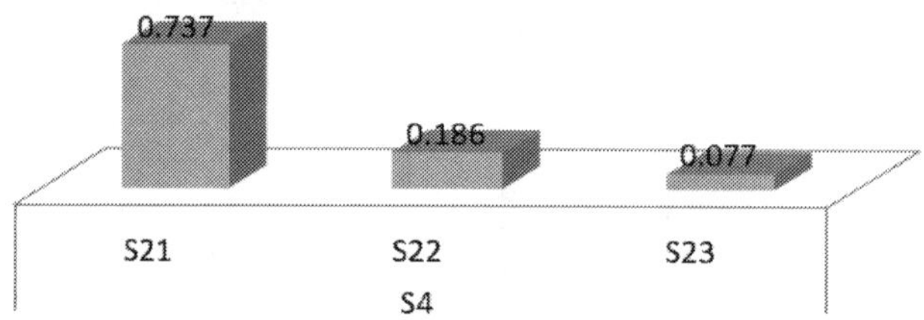

Fig 4.4 Weight of sub-variable evaluated to manage S^4

At last , we have considered S^5's group variables in our discussion on the basis of local weight & ranking, S^{21} is ranked in the 1st position with 0.530 local weight value, S^{18} is ranked in 2nd position, S^{19} is ranked 3rd position and S^{20} is ranked 4th by obtaining least 0.056 local weight value. Figure 4.5 shows local weight of PAT sub-variables evaluated to manage merger of manufacturing system with maintenance.

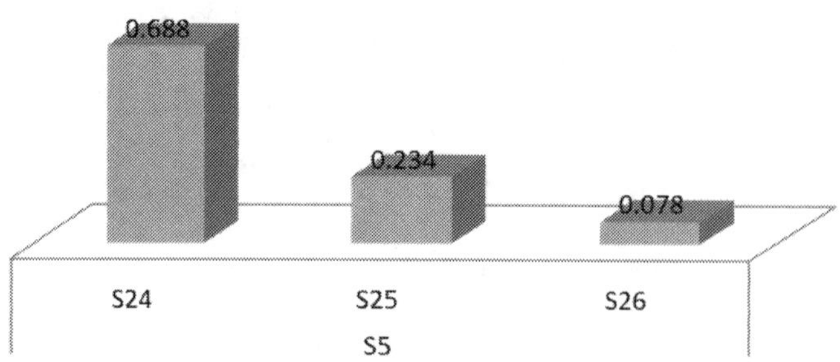

Fig 4.5 Weight of sub-variable evaluated to manage S^5

Figure 4.6 shows the global weights of PAT sub- variables. This helps to show the global ranking of all the selected sub-variables of PAT. From this figure, we can see that S^6 variable with highest 0.19 global weight value construct the longest bar and achieved 1st rank. Similarly,

S_7, S_{12}, S_8, S_{21}, S_{11}, S_9, S_{13}, S_{10}, S_{14}, S_{18}, S_{24}, S_{19}, S_{11}, S_{22}, S_{15}, S_{16}, S_{25}, S_{20}, S_{20} obtained the ranking from 2nd to 20th position and the last 21st global ranking is obtained by the variable S^{26} with least value of global weight.

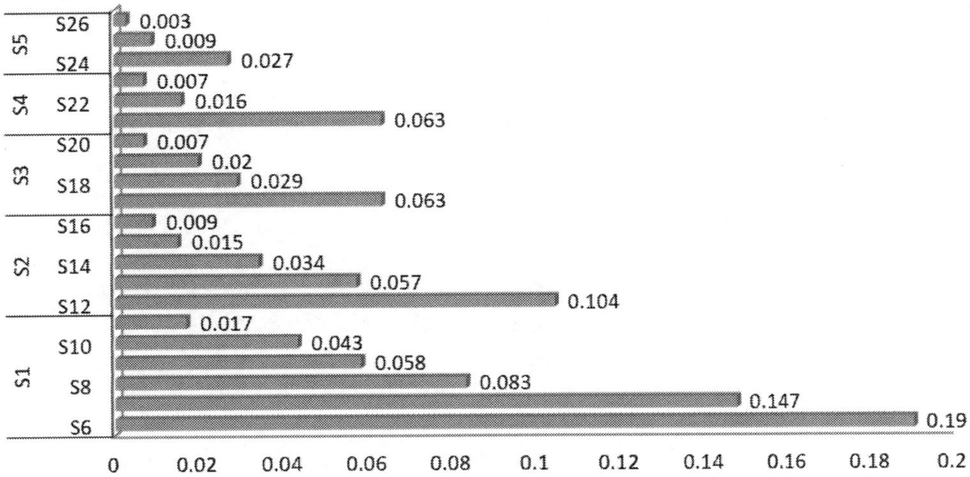

Fig 4.6 Global weight of sub variables

5. Conclusions, Limitations and future scopes.

Automation in manufacturing system increases complexity, due to this manufacturing system faces many challenges to manage process like maintenance. We have used AHP approach to construct a model. This model is designed specially to serve manufacturing organizations; also highlight the weight of variables of PAT influencing manufacturing organizations. Model was prepared on the basis of data collected from the experts of different domains. This data was utilized to construct a hierarchical structure, followed by pair wise comparisons of variables and sub variables of PAT. Thereafter, consistency of variables and sub variables of matrices was verified by calculating the consistency ratio CR. The outcome from the model shows that manufacturing process has highest 0.536 weight value and achieved 1st rank followed by fault detection, merger of manufacturing system with maintenance, organization productivity and organization control as shown in table 4.1.

Findings from this study helps to understand the most and least weighted variables of PAT. This understanding will help manufacturing organizations to enhance their vision and implement variables in manufacturing system as per their weight. This study suggests that all variables and sub-variables have similar contribution to achieve the defined objective. So, manufacturing organizations have not only to focus on the major variables, but also need to care for the sub variables of PAT influencing manufacturing system.

Finally it is concluded from this research that if this model will be followed effectively and the outcomes will be implemented properly, then manufacturing organizations will definitely overcome their challenges like maintenance and save their assets invested on the maintenance.

The model discussed in our research has some limitations; it consumed much time to collect the data from the experts and about 50% of the data collected from experts was found inconsistent. These findings and suggestions explored from this model will only be valid for a shorter period of time. So, the future study will suggests utilizing other multi criteria decision making approach to validate the results.

Reference

[1]. Sharma SK(2013) Maintenance reengineering framework: a case study. J QualMaintEng 19(2) :96-113
[2]. Chan FTS, Swarnkar R, Tiwari MK (2007) Infrastructure for coordination of multiagents in a network-based manufacturing system. Int J AdvManufTechnol 31:10281033.
[3]. Mahesh M, Ong SK, Nee AYC (2007) A web-based multi-agent system for distributed digital manufacturing. Int J Comp IntegManuf 20(1) :11-27.
[4]. Schrader PG, Lawless KA (2004) The knowledge, attitude. & behavior approach: how to evaluate performance and learning in complex environments. Perform Improv 43(9): 8-15.
[5]. Zhang Q, Cheng L, Boutaba R (2010) Cloud computing: state-of the-art and research challenges. J of Inter Serv and Applicat 1:7-18.
[6]. Yu R, Iung B, Panetto H (2003) A multi-agents based e maintenance system with case-based reasoning decision support. EngApplArtifIntell 16:321-333.
[7]. Kothamasu R, Huang SH, VerDuin WH (2009) System health monitoring and prognostics-a review of current paradigms and practices. In handbook of maintenance management and engineering, Springer, London.
[8]. Zhenyou Z, YiW, Kesheng W (2013) Intelligent fault diagnosis and prognosis approach for rotating machinery integrating wavelet transform, principal component analysis and artificial networks. The IntJAdvManufTechnol 1 68:763-773.
[9]. Sikorska JZ, Hodkiewicz M, Ma L (2011) Prognostic modeling options for remaining useful life estimation by industry. MechanSyst and Sign Process 25:18031836.

[10]. Linxia L, Jay L (2010) Design of a reconfigurable prognostics platform for machine tools. Expert SystAppl 37:240-252.
[11]. Sharma SK, Vishwakarma S, NishantJha (2017) Prognosis agent technology: Influence on manufacturing organization. Int J AdvManuf Techno 92:135-446.
[12]. Cheeseman MJ, Swann P, Hesketh GB, Barnes S(2005) Adaptive manufacturing scheduling: a flexible and configurable agent-based prototype. Prod Cont: ManagOperat 16(5):479-487.
[13]. Santiago A, Rosa MS (1997) An information system for computer integrated manufacturing systems. Rob & Comp-IntegManuf 13(3):217-228.
[14]. Ercan O, Esra KT (2009) A general framework of a reference Model for integrated manufacturing system (REMIMS). EngApplArtifInte 22:855-864.
[15]. Shaw CF, keith AS, Kevin KJ (2005) Manufacturing planning and predictive process model integration using software agents. AdvanEng Inform 19:135-142.
[16]. Cowling PI, Ouelhadj D, Petrovic S (2003) A multi agent architecture for dynamic scheduling of steel hot rolling. J Intel Manuf 14:457-470.
[17]. Brennan RW, William O (2004) Performance analysis of a mullti-agent scheduling and control system under manufacturing disturbances. Prob Plan Cont: ManagOperat 15(2):225-235.
[18]. Ottaway TA, Burns JR (2000) An adaptive production control system utilizing agent technology. Int J Prod Res 38(4):721-737.
[19]. Yam RCM, Tse PW, Li L, Tu P (2001) Intelligent predictive decision support system for condition-based maintenance. Int J AdvManufTechnol 17(5) :383-391.
[20]. Wörn H, Längle T, Albert M, Kazi A, Brighenti A, Seijo SR, Christopher Senior Bobi MAS, Collado JV (2004) DIAMOND: distributed multi-agent architechcture for monitoring and diagnosis. Prod Plan Count: ManagOperat 15(2):189-200.

[21]. Jaime CV (2009) Development in the application of ICT in Condition monitoring and maintenance. ComputInd 60:1-20.

[22]. Weerahandi S, Kurien TV, Sadrian A (1996) Estimating and optimizing the efficacy of predictive systems in proactive maintenance. Telecom Syst 6(1):315-327.

[23]. Xinhua L, Xiaolong C, Guomin S, Bihong X (2014) Development of a virtual maintenance system with virtual hand. The Int J AdvManufTechnol 70:2241-2247.

[24]. Huiskonen J (2001) Maintenance spare parts logistics: special characteristics and strategic choices. Int J Prod Econ 71(1):125-133.

[25]. Yunfei C, Fengqi Y, John MW (2014) Hybrid method integrating agent –based modeling and heuristic tree search for scheduling of complex batch processes. CompChemEng 60:277-296.

[26]. Miltenburg J, Sparling D (1996) Managing and reducing total cycle time: models and analysis. Int J Prod Econ 46:89-108.

[27]. Lin HF, Lee GG (2005) Impact of organizational learning and knowledge management factors on e-business adoption. ManagDecis 43(2):171-188.

[28]. Maurice B, Mohamad YJ (2013) Developing an input-output activity matrix (IOAM) for environmental and economic analysis of manufacturing system and logistics chains. Int J Prod Econ 143:589-597.

[29]. Changqing L, Yingguang L, Weiming S (2014) Integrated manufacturing process planning and control based on intelligent agents and multi-dimensional features. Int J AdvManufTechnol 75:1457-1471.

[30]. Yang J, Deane RH (2002) A lot size reduction model for just-in-time manufacturing systems. IntegManufSyst 13(7):471-488.

[31]. Longaray AA, Gois JDDR, Silva PRD (2015) Proposal for using AHP method to evaluate the quality of services provided by the outsource companies. Procedia Computer Science 55 (2015):715-724.

Review on study of design optimization for additive manufacturing of various mechanical/machine components

P Upare[1], A Kalatkar[2], R Shelke[3]

[1] G H Raisoni College Of Engineering Nagpur, Digdo Hills,CRPF Gate No.3, Hingna road, Nagpur,440016,India
[2] G H Raisoni College Of Engineering Nagpur, Digdoh Hills,CRPF Gate No.3, Hingna road, Nagpur,440016,India
[3] G H Raisoni College Of Engineering Nagpur, Digdoh Hills,CRPF Gate No.3, Hingna road, Nagpur,440016,India

Abstract:- In this paper we are going to know about various researchers who had done various projects on design optimization and then further they manufactured them by additive manufacturing machine which we need while performing our project. Lightweight design provides many advantages in the product development process, such as reduced material consumption, process reduction, and energy usage. This review article explains the ideas of design optimization, topology optimization, and additive manufacturing and how they may be applied to any machine or automobile component. The study of numerical optimization techniques used to the design of various components in which size, form, and structure optimization may be done is known as design optimization. Optimization of Topology is a structural optimization. It optimises the material distribution in a design part or product. Additive manufacturing often called as 3Dimensional printing, a technique which involves layering of materials & creates a 3D component. During the creation of a product, these principles are critical in determining the best size, form, and topology.

Keywords: Topology; Optimization; Additive; Manufacturing; Components; Analysis; Original; Redesigning

1. Introduction

Structural (topology) optimization is mathematical process that optimizes the distribution of a part body that has to be developed inside a certain model domain for a specific stress condition without changing its limitations. Structural optimization is used to meet certain design goals while keeping the design restrictions in mind [26]. It's utilized to get rid of parts that don't add to the part's overall performance. The topological procedures are carried out using the Finite Element Analysis (FEA) idea of Computer Aided Design (CAD).

Topology optimization uses the CAD idea to create a before optimized product model, while FEA is utilized to examine the stress and displacement distribution across the product. Topology optimization is used to increase part stiffness [22-23].Topology optimization tools produce the complicated natural shape of the product/component based on the restrictions stated in the design challenge after the unneeded material is eliminated from the part. The design acquired via the topology optimization method is then completed on CAD software in order to proceed with the production process. Finally, using FEA tools, the design is tested to see if it is of the criteria and the product performs as expected [27]. Additive manufacturing is a method of combining materials to create items using 3D model data, generally layer by layer, and is capable of producing complicated forms [24-25].In compared to conventional / traditional processes, additive manufacturing offers new options that must be considered early in the design process. Additive manufacturing allows for complete design flexibility, allowing for the production of even the most complicated forms. It's simple to put together, and it's also less expensive.

2. Methodology

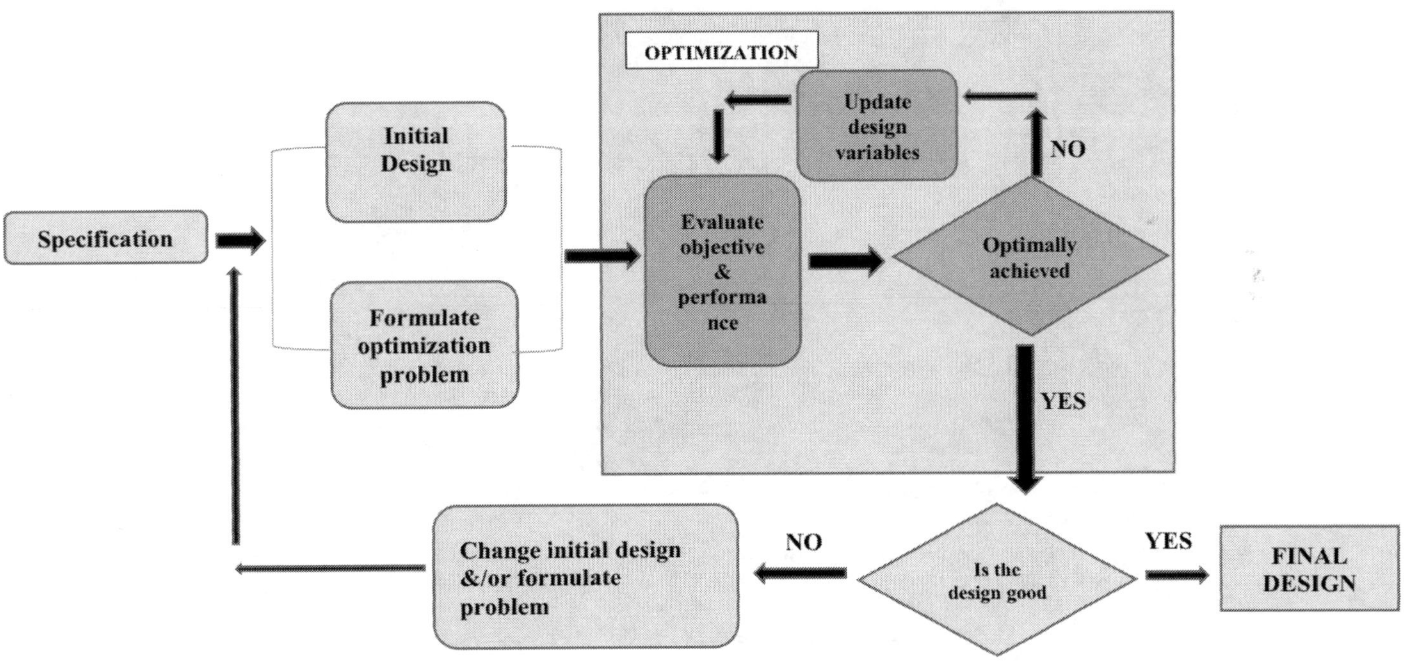

Figure 1. step by step process of design optimization

3. Literature Review

S Mantovani , SG Barberi M Giacopini A croce A sola and E Bassoli , 2020 [1] have published a paper in which there goal was to redesign a steering upright for a Formula SAE racing vehicle . Weight minimization was the goal of topology optimization, while strict structural restrictions, such as global and local stiffness performance, were met. Vehicle situations such as cornering, banging, and braking were taken into account. On a reference component, the optimization constraints were assessed using finite element analysis, using the loading circumstances received via telemetry data. The reference item was made from a solid aluminum block. The revised upright weighed 9% less than the standard component.

A W Gebisa and H G Lemu in 2017 [2] has used structural optimization for modelling of jet engine bracket. The project goal was to minimize the bracket's weight meeting all of the design criteria. The part was of Ti-6Al-4V, which has a yield strength of 904MPa (assumed) at 230^{0}C, a wall thickness of 1.13mm, an interface of 1:19 mm pin diameter (infinitely rigid), and an interface 2-5: 9.5-24^{0}C. The AS3239-26 machine bolt has a nut face with a max ID of 10.287 mm and a min OD of 14.173 mm. According to their study, the final design resulted in a 65 % weight decrease, from 2.067 kg to 0.72 kg.

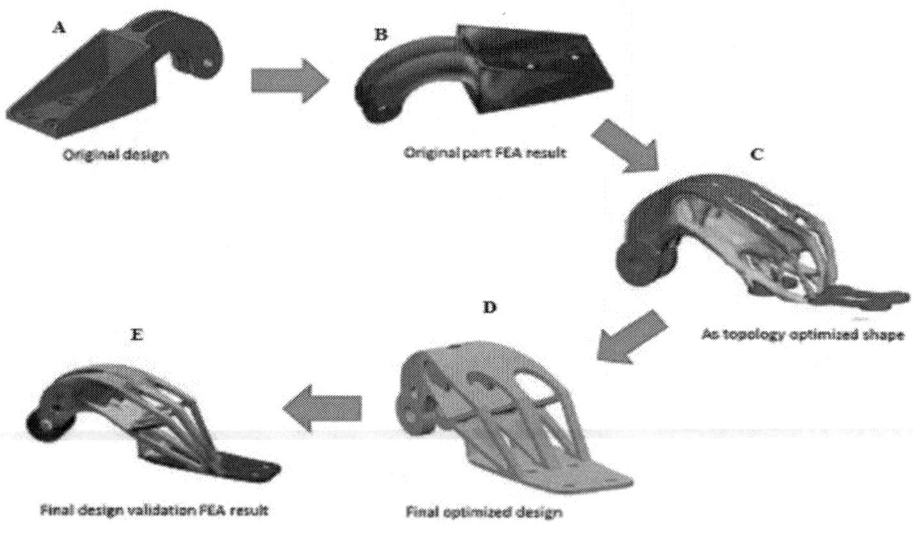

Fig 2. Topology optimization process of jet engine bracket. (A):original design (B):original part FEA result (C):As topology Optimised shape (D):Final optimized design (E):Final design validation FEA result

Meisam Abdi, Ian Ashcroft, Ricky Wildman 2015 [3] A braking pedal has been modified. The optimization was carried out using the ISO-XFEM topology optimization method. After the additive manufacturing process, the lattice structure design was also employed. Selective Laser Melting was the additive manufacturing method employed (SLM). The two designs varying volume fraction been made after optimization were compared with original design by finding values of their stiffness and equivalent von-mises stresses. After comparison the optimized design with vf=0.15 was considered , then again the redesigning was done. The main purpose of this method was to decrease internal stresses produced in SLM process. This research can be used for producing lightweight components.

Miguel Seabra, Jose Azevedo, Aurelio Araujo, Luis Reis, Elodie Pinto, Nuno Alves, Rui Santos and Joao Pedro Mortagua [4] together, they investigated aero plane bracket for topology optimization. Originally, the product was built of an aluminum alloy 7500-T7451. Its objective is minimizing mass as retaining the same stress level as titanium alloy Ti6Al4V. Then, because the material has changed, topology optimization is performed on the design space. The improved bracket is then redesigned in CAD software before being analyzed in FEA software. The amount of material was discovered to be 54 % smaller than the original component, resulting in a 28 % weight decrease, the safety factor was enhanced by two times its original amount.

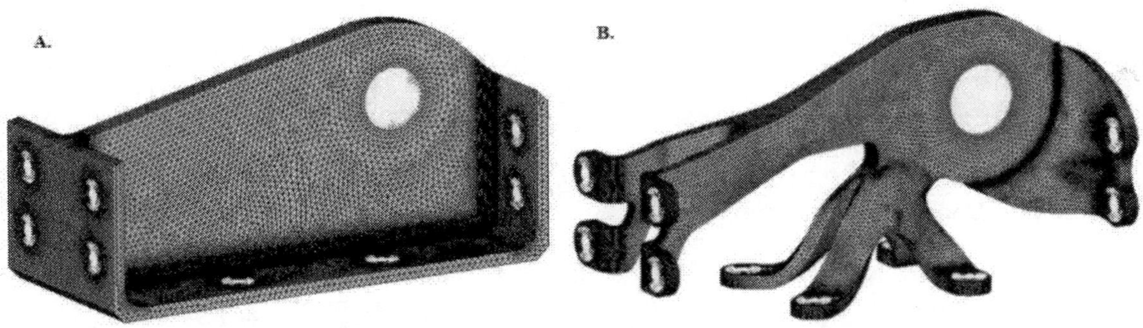

Figure 3. (A) before optimized product (B) after optimization product (right).

Mohd Nizam Sudin, Musthafah Mohd Tahir, Faiz Redza Ramli, Shamsul Anuar Shamsuddin [5] has suggested a proposal to use topology(structural) optimization on an blueprint of an automobile part at pre-development. The material is unchanged since the goal of this study was reducing weight of original brake pedal without changing the substance. CATIA V5 solid modelling software was used to create a 3D model of an original brake pedal. Under linear static stress analysis, topology optimization was accomplished by Altair Opti struct software. At last a brake pedal has been developed which is of low weight. The study found weight of an improved brake pedal was 22 percent lower than the original brake pedal without violating performance requirements.

Shen-En Chen , Navanit Sri Shanmugam ,David Boyajian , Vidya Subhash Chavan , Erika Weber ,Keith Baarsons 2021 [6] The modelling and prototype of train line tracks is described in this study. Electric MPMs have been shown to be useful for short-distance travel. The material chosen for prototyping in this work is PLA (polylactide) Contraction and bending stresses performed to establish the sustainability of the printed tracks. The breakdown shift from material failures to structure failures as the fibre content increased from 20% to 100%.Complicated deviation reacting on several peak loads are common during transitions. The transition zone occurs in flexural testing for specimens with 60 percent and 80 percent fibre content. The transition zone appears in compression testing for specimens with 20 to 40% fibre content.

Saverio Giulio Barbieria, Matteo Giacopinia, Valerio Mangerugaa, Sara Mantovania 2017 [7] A approach for designing a motorbike piston being examined in particular, with the goal of replacing the typical aluminum piston, which is generally constructed by manufacturing methods, with an apparent made of steel and built using an Additive Manufacturing technique. The smallest weight of the product is the goal function, and a strength of essential piston sections is used as a design constraint in this technique The optimization process's outcomes were described using relative density contour maps.

Sai Nithin Reddy K, Vincent Maranan, and Timothy W. Simpson Todd Palmer and Corey J. Dickman[8] By using design for additive manufacturing (DfAM) guidelines and topology optimization, they investigate the tradeoffs between component weight, support, costs, and efficiency the case study uses Direct Metal Laser Sintering to rebuild an upright on an SAE Formula student racer to reduce volume and costs (DMLS).Manufacturing the optimum design required up to 202.4 percent of the model's volume in support material if DfAM requirements were not followed. The upright is designed and produced with DfAM, which is less than 15% of the model's volume.

Nilesha.U.Patil,Akshay Sawant, Harsh Patel, Vishakha Salve , Vikas Khadk [9] The existing design of brake pedal appears to be overdesigned, thus FEA utilised one load at one and other fixed, and an AMSES used for removing material. CATIA V5 solid modelling was used to create a model of an existing brake pedal. The existing design will be put to the test utilising the strain gauge technique on a UTM machine under tensile loading. . There is a significant possibility to decrease material, risk reducing is fulfil by eliminating material in low-stress areas, hence lowering weight without impacting structural behaviour. The greatest displacement is likewise rather little. As a result, matter in the low-stressed zone removed keeping toughness, and it still without plastic deformation.

Ghasem Azamirad and Behrooz Arezoo [10] In this paper, stamping die components used for remodelling of low weight. The evolutionary structural optimization algorithm is used to accomplish this. The parts of the die are first created, examined, and relate to die that is currently in use. The results demonstrate a decrease in volume of 37% and maximum displacement of 8%, respectively.

Anton Wiberg, Johan Persson and Johan Ölvander [11].The structural aircraft component was designed to be as light as feasible while satisfying the criteria of a variety of static and fatigue structural load conditions. Another aim was to maintain production costs as low as feasible. Designs with and without post-process (centrifugal finish) tested on two various AM machines (EOS M400 and Arcam Q201).The designs developed in this investigation exhibit a significant decrease in both mass and price comparing to one AM generated &one design milled in aluminium.

Sara Mantovani , Giuseppe A Campo and Andrea Ferrari [12]This study demonstrates the application of topology optimization in a dash component from a car. In order to reduce mass, the components go through various test, are being examined. A gradient-based approach is used to tackle the problem. Steel, aluminium, and magnesium alloys are all put to the test.The new dash idea was evaluated using finite element analysis. The recommended dash weighs 721g as the reference dash, a weight reduction of 53%.

Bhavesh N. Khichadia,Dipeshkumar M. Chauhan [13] This document presents a review on design of mechanical press frame. The press frame was created by the CAD software and analysis performed by the FEA tool. The optimized design of press frame was compared to original design. The pre-tensioning of the press frame increases rigidity and stiffness as well as lowering material costs and weight.

Dr. H. K. Amarnath, Vaibhav Deshpande, Abhishek Kulkarni ,Mahesh Patil ,Omkar Ravan, Sagar Chougule [14] Industries are having a difficult time storing and processing scrap. The scrap requires a lot of room to be stored. Scraps can be compacted and kept in brick form to solve these issues. Today, all contemporary manufacturing organisations are attempting to make the most optimised reduced weight and cost effective products that functionally and reliably satisfy the original design .Topology optimization was used on several components of a 5Ton hydraulic paper baling press in this project. Shape optimization, it is assumed, resulting in better and more reliable design under the same stress circumstances, limitations, and intended design aims. The machine's cost is minimal when compared to market products; the cost is reduced by half.

Jihong ZHU, Han ZHOU, Chuang WANG, Lu ZHOU, Shangqin YUAN, Weihong ZHANG [15] in this article we can see various applications whose remodelling was done by using various remodelling methods

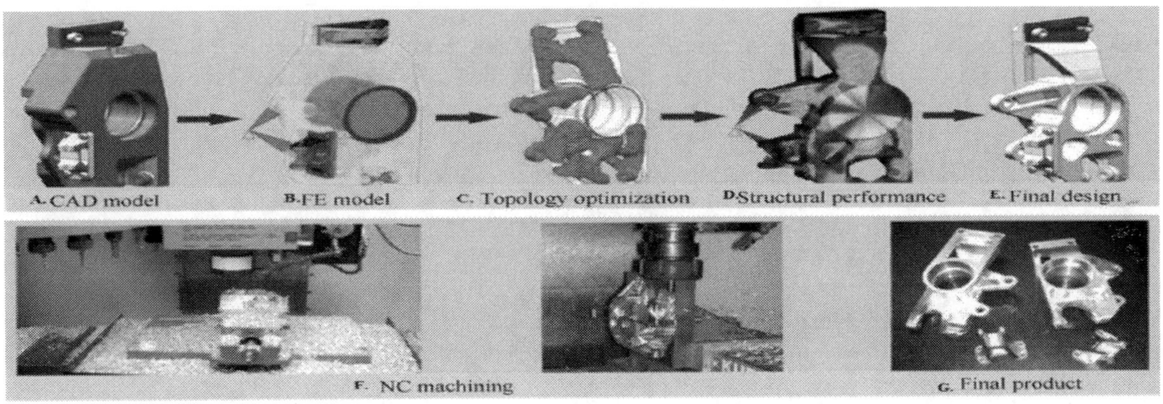

Figure 4. A topology optimization design cannot be fabricated as an integrity in NC machining, which have to be segmented then assembled (A).CAD model (B):FE Model (C):Topology Optimization (D):Structural Performance (E):Final Design (F):.NC Machining (G):.Final Product

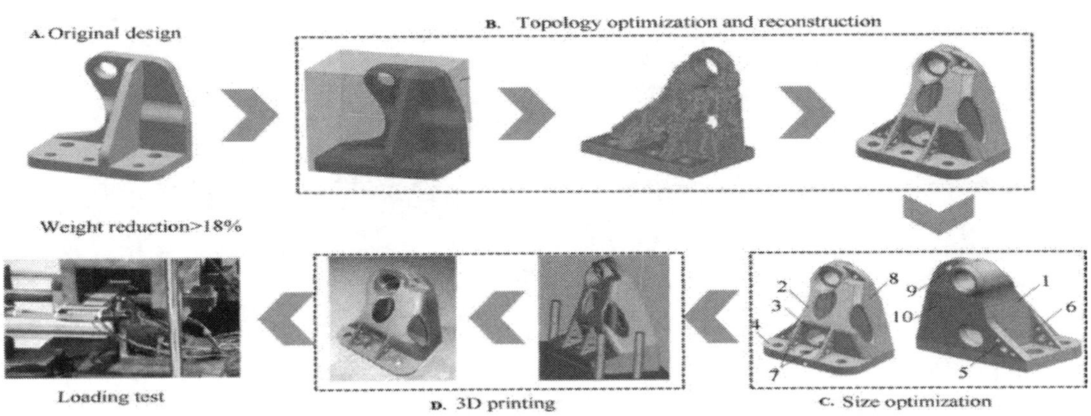

Figure 5. Aerospace bracket designed by topology optimization and manufactured by AM (A):Original Design (B):Topology Optimization and reconstruction.(C):Size Optimization (D):3D Printing

Tong Wua, Suchana A. Jahana, Yi Zhanga, Jing Zhanga, Hazim Elmounayria* and Andres Tovar [16] This study use injection tools for their experiment. They offers step by step process for its remodelling.from this paper we will study about injection moulding tool how it is prepared and what was the problem in previous model and what changes were made in new design of injection.

Stefan Junk,Benjamin Klerch,Lutz Nasdala,Ulrich Hochberg [17] The goal of this study is to optimize a part of the humanoid robot "Sweaty" ; the part is same the below abdomen area in humans. The "pelvis" module built of AlMgSc-alloy will be additively created for this purpose using laser melting. This result, light weight joints was successfully being used.

Jikai Liu & Andrew T. Gaynor & Shikui Chen & Zhan Kang & Krishnan Suresh & Akihiro Takezawa & Lei Li & Junji Kato & Jinyuan Tang & Charlie C. L. Wang & Lin Cheng & Xuan Liang & Albert. C. To [18] Production-oriented topology optimization has been intensively researched over the last two decades, particularly for traditional manufacturing processes like as machining and injection moulding or casting. Additive manufacturing (AM) has recently gained a lot of interest from both academia and business. As a result, the objective for this perspective article is to outline the most recent design optimization approaches for a wide range of AM subjects. Also this article provides writers' thoughts on the issues' . The aim make active both researchers & engineers to find innovative ideas to these problems.

Y.Tateishi, V.Parque, T.Miyashita H.Gohara, R.Kato, Y.Ikeda [19] In this paper a active model for absorbing excessive heat was remodelled means a heat absorbent was redesigned by optimizing and by using mathematical formula which was further followed by a 3D printing using the finite element approach and for better efficiency by building a new heat sink. In compared to the old one, the new one resists the heat flow by around 15%.

Yunlong Tang, Kieran Mak, Yaoyao Fiona Zhao [20] This study presents a new general technique for accurately assessing the environmental effect of an AM-fabricated product, namely the binder-jetting process. By comparing environmental effect of CNC by binder-jetting production processes, results says binder-jetting uses much less energy and emits significantly less CO_2.

CONCLUSION

From this review paper we are able to study about various researchers and the study they carried on various components and how the design optimization process is carried on various components by various processes optimizing methods and various formulaes then further modelling in cad software after redesigning checking for analysis stress deformation etc. and after successful analysis they went for additive manufacturing processes. This all the processes in details we can study from all this research papers. This will further help us to use this processes in our project.

REFERENCES

[1] S Mantovani , SG Barberi M Giacopini A croce A sola and E Bassoli . "Synergy between topology optimization and additive manufacturing in the automotive field" *J Engineering Manufacture* , 1-13,2020

[2] A W Gebisa and H G Lemu . "A case study on topology optimised design for additive manufacturing" *IOP Publishing*, 1-11, 2017

[3] Meisam Abdi, Ian Ashcroft, Ricky Wildman. "Design optimization for an additively manufactured automotive component",*engineering and physical science research council,*2015

[4] Miguel Seabra, Jose Azevedo, Aurelio Araujo, Luis Reis, Elodie Pinto, Nuno Alves, Rui Santos & Joao Pedro Mortagua, "Selective laser melting(SLM) and topology optimization for lighter aerospace components", *Procedia structural Integrity* 1, 289-296, 2016.

[5] Mohd Nizam Sudin, Musthafah Mohd Tahir, Faiz Redza Ramli, Shamsul Anuar Shamsuddin,"Topology optimization in automotive brake pedal design".*IJET*,398-402,2014.

[6] Shen-En Chen , Navanit Sri Shanmugam ,David Boyajian , Vidya Subhash Chavan , Erika Weber Keith Baarsons, "Prototyping rail track for micro-people movers using additive manufacturing" *Elsevier journal,* 1-11, 2021.

[7] Saverio Giulio Barbieria, Matteo Giacopinia, Valerio Mangerugaa, Sara Mantovania, "A design strategy based on topology optimization techniques for an additive manufactured high performance engine piston," *Elsevier*, 641-649, 2017.

[8] Sai Nithin Reddy K, Vincent Maranan, and Timothy W. Simpson Todd Palmer and Corey J. Dickman,"Application of topology optimization and design for additive manufacturing guidelines an automotive component".*ASME* ,1-10,2016.

[9] Nilesha.U.Patil,Akshay Sawant, Harsh Patel Vishakha Salve , Vikas Khadk,"Design and analysis of brake pedal", international journal of future communication and networking, 1113-1127,2020.

[10] Ghasem Azamirad and Behrooz Arezoo,"Topology optimization of stamping die components using evolutionary structural optimization method,"*J engineering manufacture* ,1-9,2015.

[11] Anton Wiberg, Johan Persson and Johan Ölvander,"An optimisation framework for designs for additive manufacturing combining design,manufacturing and post-processing ," *rapid prototyping journal*,90-105,2021.

[12] Sara Mantovani , Giuseppe A Campo and Andrea Ferrari , "Additive manufacturing and topology optimization: A design strategy for a steering column mounting bracket considering overhang constraints, *J engineering manufacture* ,1-21,2020.

[13] Bhavesh N. Khichadia,Dipeshkumar M. Chauhan, "A review on design and analysis of mechanical press frame", *International Journal of Advance Engineering and Research Development (IJAERD)* 2348 – 4470,2014.

[14] Dr. H. K. Amarnath, Vaibhav Deshpande, Abhishek Kulkarni ,Mahesh Patil ,Omkar Ravan, Sagar Chougule , "Design and Fabrication of Paper Bailing Machine" *International Research Journal of Engineering and Technology (IRJET)*,770-775,2017.

[15] Jihong ZHU, Han ZHOU, Chuang WANG, Lu ZHOU, Shangqin YUAN, Weihong ZHANG , "A review of topology optimization for additive manufacturing:status and challenges"*ELSEVIER*,1-26,2020.

[16] Tong Wua, Suchana A. Jahana, Yi Zhanga, Jing Zhanga, Hazim Elmounayria and Andres Tovar, " Design optimization of plastic injection tooling for additive manufacturing",*ELSEVIER*,923-925,2017.

[17] Stefan Junk,Benjamin Klerch,Lutz Nasdala,Ulrich Hochberg, "Topology optimization for additive manufacturing using a component of a humanoid robot"*ELSEVIER*,102-107,2018.

[18] Jikai Liu & Andrew T. Gaynor & Shikui Chen & Zhan Kang & Krishnan Suresh & Akihiro Takezawa & Lei Li & Junji Kato & Jinyuan Tang & Charlie C. L. Wang & Lin Cheng & Xuan Liang & Albert. C. To, "Current and future trends in topology optimization for additive manufacturing ," *SPRINGER*,7-27,2018.

[19] Y.Tateishi, V.Parque, T.Miyashita H.Gohara, R.Kato, Y.Ikeda, "Design Optimization of Heat Sink Using Additive Manufacturing",*IEEE*,91-94,2017.

[20] Yunlong Tang, Kieran Mak, Yaoyao Fiona Zhao, "A framework to reduce product environmental impact through design optimization for additive manufacturing",*Journal Of Cleaner Production*,1-26,2016.

[21] Khalatkar, Abhay, Vijay Kumar Gupta, and Rakeshkumar Haldkar. "Modeling and simulation of cantilever beam for optimal placement of piezoelectric actuators for maximum energy harvesting." *Smart Nano-Micro Materials and Devices. Vol. 8204. International Society for Optics and Photonics*, 2011.

[22] Mohite, Prachi Dashrath, Shweta Abhay Khalatlkar, and Abhay M. Khalatkar. "Wattages Optimization of Small Hydro Power Station to Build up Energy." Smart Technologies for Energy, Environment and Sustainable Development, Vol 2. *Springer, Singapore*, 2022. 707-713.

[23] Dahake, Shrutika, Nilesh Awate, Rupesh Shelke, and Abhay Khalatkar. "Shape Memory Alloy as a Smart Material: A Review." *Smart Technologies for Energy, Environment and Sustainable Development, Vol 2* (2022): 395-401.

[24] Jachak, Prabinkumar R., Abhay Khalatkar, and Nilesh M. Narkhede. "Real-Time Problem of Knob Assembly in Automobile Manufacturing Industries." Smart Technologies for Energy, Environment and Sustainable Development, Vol 1. *Springer, Singapore*, 2022. 831-835.

[25] Khalatkar, Abhay M., et al. "Arduino-based tuned electromagnetic shaker using relay for MEMS cantilever beam." Smart technologies for energy, environment and sustainable development. *Springer, Singapore*, 2019. 795-801.

[26] Chakrabarty, Shyamal G., Uday S. Wankhede, Rupesh S. Shelke, and Trushar B. Gohil. "Investigation of temperature development in salinity gradient solar pond using a transient model of heat transfer." *Solar Energy 202* (2020): 32-44.

[27] Chakrabarty, S. G., Wankhede, U. S., & Shelke, R. S. (2022). Thermal Analysis of Salinity Gradient Solar Pond in Nagpur City. In Smart Technologies for Energy, Environment and Sustainable Development, Vol 2 (pp. 487-492). *Springer, Singapore.*

[28] Shingare, Prashant P., Vilayatrai M. Kriplani, and Rupesh S. Shelke. "Heat Transfer Enhancement of Multi-Walled Carbon Nanotubes and Al2O3 Nanofluids in a Smooth Tube." *Heat Transfer* 6.11 (2019).

[29] Gaddamwar, Sagar S., and Rupesh S. Shelke. "experimental investigation of heat transfer characteristics of high pressure gas in an augmented heat exchanger." *Springer, Singapore* (2013).

[30] Patankar, Vaibhav A., and Rupesh S. Shelke. "A Review on Open Sorption Thermal Energy Storage System Using MgCl2. 6H2O." *Springer, Singapore* 2015

Experimentation and Optimization of Milling Machine Parameters Using Taguchi Method

Amol Vikas Joshi[1] Pranav Charkha[2] Avinash Panchal[3] Devanand R. Tayade[4]

[1,2,3,4] *Department of Mechanical Engineering, G. H. Raisoni Institute of Business Management Jalgaon, Maharashtra India*
[1]amolvjoshi10@gmail.com

Abstract. : In this research work, the problem is to reduce the surface roughness of the Jacquard body which isa casting part to optimize the effects of different process parameters of surface roughness & to reduce surface roughness for milling operation by using the Taguchi method. Milling is better than other machines in the surface finish and accuracy. Greater quality product and higher productivity is the main vision of every entrepreneur. To achieve this, the process parameters of each process need to be optimized. In a milling machine, some of the important parameters like feed rate, depth of cut, speed, etc, play an important role in achieving the required qualityproduct. To optimize these parameters, the most suitable and widely used method is the Taguchi method. The mostsignificant objective of this research is to obtain effective quality parameters. Experimentation is done on the Jacquard body by varying the milling machine parameters. The experimental results indicate that the minimum surface roughness is obtained for the process parameters feed 0.7 mm per revolution, speed 225 rpm and 1.5 mm depth of cut. This study recommended that more surface finish be obtained by increasing speed, low feed rate, and depth of cut.

Keywords: Milling Machine, Process parameters, Surface roughness, Taguchi Method, Jacquard body

1. Introduction

In mechanical machines, roughness is the responsible factor for the wear of sliding or rolling surfaces. Surface irregularities are often responsible for the foundation of cracks and corrosion. Usually, roughness is unacceptable. Butthe surface finish is directly proportional to manufacturing cost, so a smart way to achieve the required results is an optimization of existing parameters. Finishing of the surface depends on the depth of cut, feed rate & cutting speed. Milling is known for its dynamic role in maintaining accuracy at work. So to improve the surface smoothness, specialattention needs to pay to influencing parameters as well as the material removal rate (MRR). Some of the uncontrollable factors like material properties and tool geometry of tool as well as workpiece also influence badly onsurface finish.[1]

The irregularities available on the work surface observed in the form of valleys and hills varied in spacing and height.The metal removal rate (MRR) in the milling operation by the effects of machining parameters values. The factors affecting surface roughness are analyzed to determine the optimum cutting combination [2].

2. Problem Definition

A field visit for understanding real-time problems at "Manudevi Engineering Works" located in the V- 92 M.I.D.C., Jalgaon has done where surface grinding of a complete set of casted pasts of Mat sewing machine is brought to get proper finishing dimensions as well as removing casting defects. While machining mat sewing machine parts, some problems faced by operators such as surface roughness and chattering as shown in Fig. 1. So the challenge was to reduce the surface roughness of the Jacquard body which is a casting part. To address this problem, the team has observed the whole process of machining and studied the process parameters. The work has been carried out to optimize the different cutting parameters of the milling machine to machine Jacquard's body. Experimentation has been carried out to study the results of varying parameters. The objective of this study is to reduce surface roughnessby analyzing the different cutting parameter values in the face milling of a casting job.

The selection of correct machining process parameters is responsible for the determination of optimized cutting conditions [3]. The feed rate is directly proportional to the product surface roughness. Also, tool life and cutting tool performance is directly dependent on cutting parameters [8].

Figure 1. Jacquard body

3. Methodology

3.1 Taguchi Method

Taguchi is a statistical method [1]. Selection of control factors must be made such that it nullifies the effect of noise factors. This method identifies proper control factors and obtains the required optimum results of the process. To perform a set of experiments, Orthogonal Arrays (OA) are used. The quality of components produced is predicted through the results of experiments that are conducted virtually [4].

A large number of experiments requires for Full Factorial Design. The value of the loss function can be transformed into a signal-to-noise (S/N) ratio. There are normally three types of the performance characteristics are available for analyzing the S/N ratio are; Nominal-the-best, larger-the-better, and smaller-the-better [4]

3.2 Controlling Factors at various levels

The controlling factors at various levels are decided, based on a brainstorming session, carried out by team membersfor conducting the experiments. The control factors with the levels are shown in table 1.

Table 1. Controlling Factors at various levels

Controlling Factors	Levels		
	1	2	3
Speed (rpm)	125	175	225
Feed (mm/rev)	0.6	0.7	0.8
Depth of cut (mm)	1.5	2	2.5

3.3 Selection of Orthogonal Arrays

In an appropriate selection of OR for conducting the experiments, the degrees of freedom are to be computed [9].1 DOF for Mean Value, and 8= (2x4), 2 each for the remaining factors.

Total Degrees of Freedom = 9

The total number of experiments to be carried out is nine as shown in table 2[9].

Table 2. Orthogonal Array (OA) L9

Experiment No.	Control Factors		
	1	**2**	**3**
1	1	1	1
2	1	2	2
3	1	3	3
4	2	1	2
5	2	2	3
6	2	3	1
7	3	1	3
8	3	2	1
9	3	3	2

3.4 Conduction of Matrix Experiment

According to OA, conduction of experiments is done as shown in table 3. Each of the above 9 experiments was conducted 2 times to compensate for the variations which may be occurred because of the presence of noise factors. The control factors are varied as given the table 3 and the Roughness results are obtained by using a surface roughnesstester.

Table 3. Values of surface roughness obtained from different experiments

Experiment No.	Control Factors			Surface Roughness Ra (μm)	
	Speed(rpm)	Feed(mm/rev)	Depth of cut(mm)	Result 1	Result 2
1	125	0.6	1.5	1.753	3.210
2	125	0.7	2	1.998	2.120
3	125	0.8	2.5	6.883	4.176
4	175	0.6	2	1.751	2.979
5.	175	0.7	2.5	1.710	2.906
6	175	0.8	1.5	2.849	3.979
7	225	0.6	2.5	1.990	2.320
8	225	0.7	1.5	1.678	1.890
9	225	0.8	2	2.515	2.340

3.5 Design of Experiment

A structured and well-organized method called, Design of Experiment (DOE) is used to set experimental sets to determine the relevance of various parameters of inputs and the output of the process [10]. Optimal conditions can beidentified from the results of these experiment helps to analyze the most influencing factors [11].

4. Experimental Work

4.1 Experimental arrangement

The experimentation work has been carried out at "Manudevi Engineering Works" on Vertical Milling Machine. Forthis experimental work, we selected Sand Casting part-Jacquard body of Mat Machine

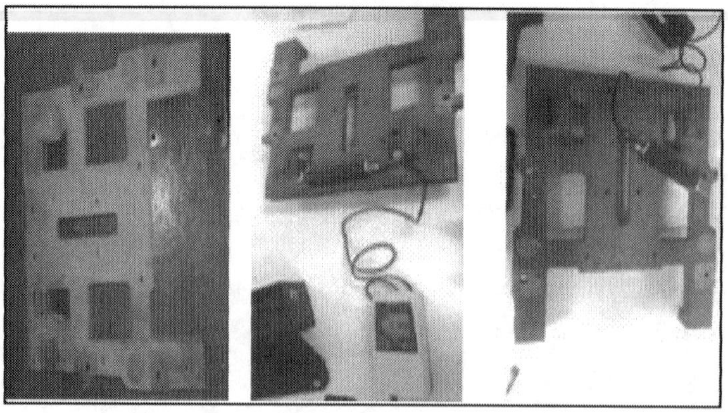

Figure 2. Jacquard body with surface roughness tester

4.2 Material Specifications used

Name of part: - Jacquard body (Mat machine part), Material of part: - Sand casting, Dimensions of part: - Width 340mm, height 400mm, thickness 30mm

We have selected vertical milling machine for machining operation. In the machining section, we have conducted 9 experiments according to the orthogonal array prepared by the Taguchi method. We have performed a face milling operation on the Jacquard body of the Mat machine by using a Conventional Milling Machine.

5. Results and Discussions

Table 4 shows the 9 experiments with their surface roughness values and S/N ratio values obtained from MINITAB software. The 8th experiment represents the lowest surface roughness value and higher S/N ratio values which is theoptimized results

Table 4. The 9 experiments with their surface roughness values and S/N ratio values

Expt No.	Control Factors			Surface Roughness Ra			S/N Ratio
	Speed (rpm)	Feed (mm/rev)	Depth of cut(mm)	Result 1	Result 2	Mean (Ra)	
1	125	0.6	1.5	1.753	3.210	2.4815	-8.2533
2	125	0.7	2	1.998	2.120	2.0590	-6.2769
3	125	0.8	2.5	6.883	4.176	5.5295	-15.1064
4	175	0.6	2	1.751	2.979	2.3650	-7.7599
5	175	0.7	2.5	1.710	2.906	2.3080	-7.5469
6	175	0.8	1.5	2.849	3.979	3.4140	-10.7826
7	225	0.6	2.5	1.990	2.320	2.1550	-6.6943
8	**225**	**0.7**	**1.5**	**1.678**	**1.890**	**1.7840**	**-5.0432**
9	225	0.8	2	2.515	2.340	2.4275	-7.7088

Table 5. Response for surface roughness

Level	Speed	Feed	Depth of cut
1	-9.789	-7.569	-7.249
2	-8.698	-6.289	-7.249
3	-6.482	-11.199	-9.783
Delta	3.397	4.910	2.534
Rank	2	1	3

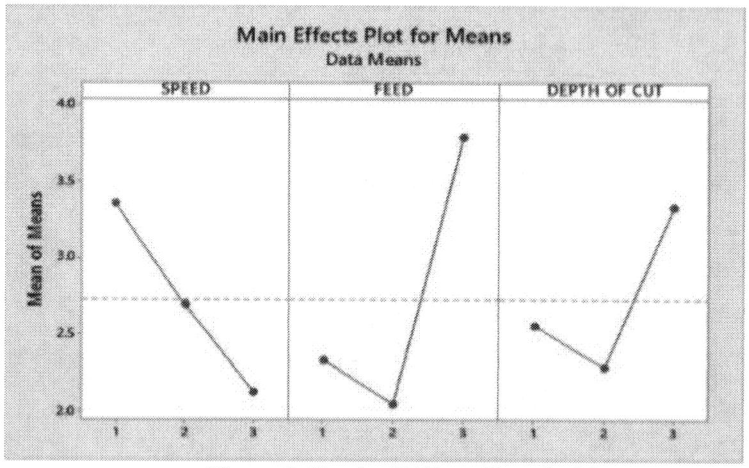

Figure 3. Graph of Effects Plots for Mean

In the above research, the smaller-the-better case says that the ideal condition of surface roughness will be the smallest.

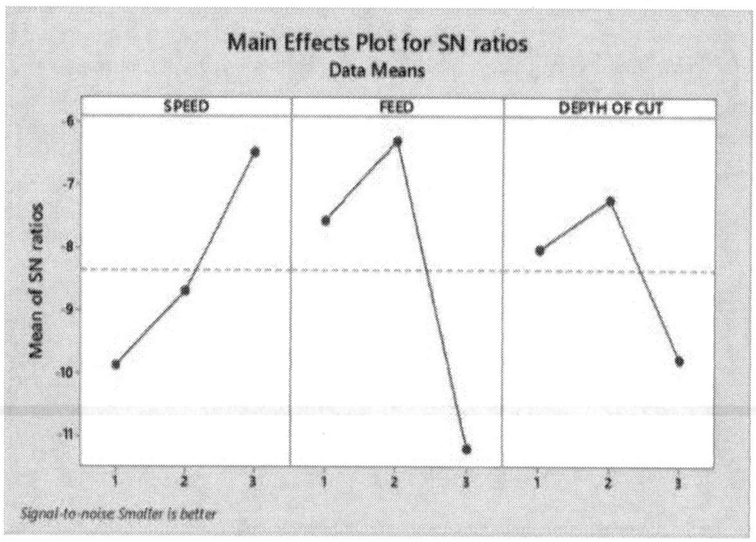

Figure 4. Graph of Effects Plot of S-N Ratios

One of the control factor, spindle speed (225 rpm) identified the best result at level 3. Second control factor is, feedrate (0.7 mm/rev) identified the best result at level 2. The third and less influencing control factor, depth of cut identified with the least surface roughness and adequate S/N ratio at level 1. So, level 1 identified the cutting depth(1.5 mm) at the optimal cutting conditions.

6. Conclusions

1. The experimental results indicate that at the speed of 225 rpm, 0.7 mm/rev of feed, and 1.5 mm with a depthof cut, the minimum surface roughness is obtained.
2. From Taguchi Method, it may be observed that a medium feed rate and low depth of cut gives more surfacefinish at high speed

7. References

[1] Kannan. S, Baskar. Modeling and Optimization of Face Milling Operation Based on Response Surface Methodology and Genetic Algorithm Kannan. S. et.al / International Journal of Engineering and Technology (IJET)

[2] Arokiasamy et. al. Optimisation of shock absorber process parameters using failure mode and effect analysis and genetic algorithm Mariajayaprakash et al. Journal of Industrial Engineering International 2013, 9:18

[3] G.M.Sayeed Ahmed et. al. Experimental Investigation of Effect of Tool Length on Surface Roughness during Turning Operation and its Optimization IOSR Journal of Mechanical and Civil Engineering (IOSR-JMCE) e- ISSN: 2278-1684,p-ISSN: 2320-334X, Volume 7, Issue 2 (May. - Jun. 2013), PP 73-80

[4] Srinivas Athreya, Dr Y.D.Venkatesh, Application Of Taguchi Method For Optimization Of Process Parameters In Improving The Surface Roughness Of Lathe Facing Operation International Refereed Journal of Engineering and Science (IRJES) ISSN (Online) 2319-183X, (Print) 2319-1821 Volume 1, Issue 3 (November 2012), PP.13- 19

[5] Neeraj Kumar, Prof. K.K. Chhabra An Overview of Optimization Techniques for CNC Milling Machine International Journal of Engineering, Management & Sciences (IJEMS) ISSN 2348 –3733, Volume-1, Issue-5, May 2014

[6] Surasit Rawangwong et. al., "An investigation of Optimum Cutting Conditions for Quality of Surface Roughness in Face Milling Mold Steel AISI P20 Using Carbide Tool",10th Eco-Energy and Materials Science and Engineering (EMSES2012), (2012).

[7] Vikas Pare et. al., "Optimization of Cutting Conditions in End Milling Process with the Approach of Particle Swarm Optimization", International Journal of Mechanical and Industrial Engineering (IJMIE)ISSN No. 2231 – 6477, Volume-1, Issue-2, (2011).

[8] D. R Kotkar, Wakchaure V. D, Vibration control of newly designed Tool and Tool-Holder for internal treading of Hydraulic Steering Gear Nut, IJMER Date uploaded on Nov 24, 2014, pp. 46-57

[9] Kashi, Sima & Souza, M. & Al-Assafi, Salwan & Varley, Russell. (2019). Understanding the Effects of In-Service Temperature and Functional Fluid on the Ageing of Silicone Rubber. Polymers. 11. 388. 10.3390/polym11030388.

[10] Mr. Dhole N.S Optimization of Milling Process Parameters of EN33 Using Taguchi Parameter Design Approach, Journal of Engineering Research and Studies E-ISSN0976-7916

[11] Mr. Prabhavalkar M.S., Mr. Waghmare S N, Optimization of Machining Parameters of M.S. in Vortex Tube Cooling, International Journal of Research in Advent Technology, Vol.2, No.6, June 2014 E-ISSN: 2321-9637

[12] Mr. Amol Vikas Joshi, Mr. Devanand Tayade, Mr. Wasim Patel, An Experimental evaluation of FEA based Helical Compression Spring to design a Shock Absorber" Published in Smart technologies for Energy, Environment and Sustainable Development,Vol 2, Springer Proceedings in Energy 2022, Page-299-308

IOP Conf. Series: Materials Science and Engineering 1259 (2022) 012041 doi:10.1088/1757-899X/1259/1/012041

Analytical study of HRM practices in industry 5.0

S D Ganer[1], S O Kediya[2], A K Suchak[3], S K Dey[4] and G. Band[5]

[1] Student, Department of Management Studies, G H Raisoni College of Engineering, Nagpur, India. shivani.ganer.mba@ghrce.raisoni.net

[2] Associate Professor, Department of Management Studies, G H Raisoni College of Engineering, Nagpur, India. Orcid ID – 0000-0002-2144-7800 shailesh.kediya@raisoni.net

[3] Associate Professor, Central Institute of Business Management Research and Development, Nagpur, India. anupsuchak@gmail.com

[4] Associate Professor, Faculty of Management Studies, ICFAI University, Jharkhand, India. subratokumardey@iujharkhand.edu.in

[5] Assistant Professor, Shri Ramdeobaba College Of Engineering and Management, Nagpur, India. bandgs@rknec.edu

Abstract. The industry 5.0 is the new revolution in which transition is going right now from industry 4.0. While industry 4.0 consisted of technological developments in the fields of data analytics, cyber physical systems and huge exposure to Artificial Intelligence, Machine Learning, IoT etc., Industry 5.0 is all about mass customisation or personalization of these existing technologies. While the world is going through this transition the current Human Resource Management of the organizations also has to go through this transition. The present research analyses certain factors affecting the HRM practises during this process. There are hypotheses formed on factors such as better performance, emotional intelligence, automation, and cross disciplinary workforce. The authors also try to propose certain possible and suitable suggestions based on analysis of these hypotheses towards smooth processing of the transition.

1. Introduction

The term Industry 5.0 was first used in 2015 in an article on LinkedIn entitled "Industry 5.0 – from virtual to physical" which highlighted the idea of Industrial Upcycling. The basic idea behind the article was to make people understand that it is necessary for the world to go towards industry 5.0 leaving 4.0 behind and the motive that machines are not "Working For Us" rather "Working With Us."

The industry 5.0 gives the personal touch of human beings to the technologies, machines and robots which will be a big part of this revolution. As these technologies or machines will be working collaboratively with human beings this will bring forward new challenges to the Human Resources Management in the organization.

The present research paper has certain hypotheses designed based on various factors which are identified through various research paper and articles published and also testing is conducted on them.

IOP Conf. Series: Materials Science and Engineering 1259 (2022) 012041 doi:10.1088/1757-899X/1259/1/012041

2. Literature Review

The literature review consists of various research papers and articles, blogs published on the concerned topic. In "Industry 5.0 and Human-Robot Co-working" [1] the authors have highlighted the meaning of co working of human beings and robots and at the same time also highlighted the pros and cons of this process which is essential for industry 5.0. The authors in research paper "Industry 4.0 and Industry 5.0—Inception, conception and perception" [2] focused on the inter-relation between industry 4.0 and 5.0 particularly telling the fact that industry 5.0 relies heavily on 4.0 for its technological advancements but at the same time adding societal needs, values and responsibilities as ultimate goal since 5.0 is value-driven in nature. "Rethinking companies' culture through knowledge management lens during Industry 5.0 transition." [4] has tested hypotheses which says that when empowerment is given in diverse culture of human beings and also considering machines a part, leads to greater profitability of the organization. The authors in "Electronic Human Resource Management (e-HRM): A New Concept for Digital Age." [6] have highlighted that the revolution of 5.0 not only remains limited to operations or other departments but also to HR department as well and tries to come up with solutions for e-HRM. In "Gamifying employer branding: An integrating framework and research propositions for a new HRM approach in the digitized economy." [7], the authors have tried to explain the fact that gamified version of processes help in better employee branding and uplifts them as a person with better working knowledge. Cosmin Tataru in "Human Resources in the Digital Age A Manager's Realities and Perspectives." [8] emphasized on reinventing HR, getting close to build more customer-focussed organization and to understand importance and welcome workforce from diversity including humans and smart machines and robots. The paper "Artificial intelligence and human workers interaction at team level: a conceptual assessment of the challenges and potential HRM strategies." [9] focuses on Performance Evaluation being a key factor for HR managers to address to as there will be humans and robots working in collaboration so special importance being given to Humans. In the paper "Influence of Emotional Intelligence on the Workforce for Industry 5.0." [14], the authors give significance to emotional intelligence of employees for better performance of the overall organization.

3. Research Objectives

- To identify various factors affecting industry 5.0 in the area of HRM
- To explore relevant factors in order to gain more insights on industry 5.0
- To come up with a model explaining holistic approach towards HRM in industry 5.0

4. Hypotheses

- H01 – Industry 5.0 has created coordination between man and technology to improve industrial performance.
- H02 – Industry 5.0 enhances the emotional intelligence of people.
- H03 – Industry 5.0 has created automation in HR processes.
- H04 – Industry 5.0 requires cross disciplinary work force for better performance.

5. Research Methodology

5.1. Research Design

In the literature review various papers, articles and blogs that are present on topic are studied. After that, certain hypotheses are proposed to examine, based on the factors identified according to the objective of the research. [19][20] Then to examine those hypotheses the responses from working professionals from different industrial background are collected. The data collected from these responses is the primary source of data for the study.[23] According to the hypotheses, certain questions based on each hypothesis were asked to which answer were to be given on Likert scale.[21][22]

IOP Conf. Series: Materials Science and Engineering 1259 (2022) 012041 doi:10.1088/1757-899X/1259/1/012041

5.2. Primary Data
The responses to the questions asked to employees working in various organizations through Google Forms.

5.3. Sample
Convenience sampling method is employed. The sample size is of 144 working and HR professionals of various organizations in India and abroad.

5.4. Statistical tool used
Anova

6. Data Analysis and Hypotheses Testing

To perform the test on hypotheses and analyze the data the statistical tool ANOVA is used and One-way Anova is employed.

To test the hypothesis "Industry 5.0 has created coordination between man and technology to improve industrial performance" one-way Anova Test has been applied taking employment of technology as fixed factor and coordination between manpower and technology as fixed factor. The results of the analysis are shown below:

Table 1. Description of H01

There is proper coordination between manpower and these technologies.

	N	Mean	Std. Deviation	Std. Error
Strongly Disagree	3	2.0000	0.00000	0.00000
Disagree	3	4.0000	0.00000	0.00000
Neutral	21	3.1429	0.85356	0.18626
Agree	75	3.3600	0.84853	0.09798
Strongly Agree	42	4.1429	0.75131	0.11593
Total	144	3.5417	0.91510	0.07626

The above descriptive table shows the mean, standard deviation and standard error of the two above mentioned variables, The mean score of more than 3 indicates proper coordination between the manpower and technology. The mean value obtained is 3.54 which is more than the central score of 3, this indicates that according to respondents there is proper co-ordination between manpower and the technologies. Following table helps to prove this further:

Table 2. Analysis of H01

There is proper coordination between manpower and these technologies.

	Sum of Squares	df	Mean Square	F	Sig.
Between Groups	28.756	4	7.189	10.982	.000
Within Groups	90.994	139	.655		
Total	119.750	143			

IOP Conf. Series: Materials Science and Engineering 1259 (2022) 012041 doi:10.1088/1757-899X/1259/1/012041

The above ANOVA table gives the F-value i.e. ANOVA in the fifth column and sig. value i.e. p-value in the sixth column. The ANOVA i.e. F-value obtained at the given degree of freedom is 10.982 and the p-value obtained is 0.00, which is less than the alpha value of 0.05 ($F = 10.982$, $p < 0.05$), this indicates that there is significant impact of Industry 5.0 in creating a coordination between man and technology to improve industrial performance. Hence, the hypothesis i.e., "Industry 5.0 has created coordination between man and technology to improve industrial performance" is accepted

To test the hypothesis "Industry 5.0 enhances the emotional intelligence of people" one-way Anova Test has been applied taking employment of technology as fixed factor and enhancement in emotional intelligence and its benefits to the organizations as dependent factor. The results of the analysis are shown below:

Table 3. Description of H02

		N	Mean	Std. Deviation	Std. Error
The Emotional Intelligence of the employees has seen enhancement with technological enhancement taking place.	Strongly Disagree	3	4.0000	0.00000	0.00000
	Disagree	3	5.0000	0.00000	0.00000
	Neutral	21	3.5714	0.74642	0.16288
	Agree	75	3.5600	0.94783	0.10945
	Strongly Agree	42	4.0714	0.71202	0.10987
	Total	144	3.7500	0.88104	0.07342
The enhanced Emotional Intelligence of people can be benefitting for the company.	Strongly Disagree	3	3.0000	0.00000	0.00000
	Disagree	3	5.0000	0.00000	0.00000
	Neutral	21	3.5714	0.74642	0.16288
	Agree	75	3.7200	0.78085	0.09017
	Strongly Agree	42	4.1429	0.75131	0.11593
	Total	144	3.8333	0.80210	0.06684

The above descriptive table shows the mean, standard deviation and standard error of the two above mentioned variables, The mean score of more than 3 indicates proper coordination between the manpower and technology. The mean value obtained in care of enhancement of emotional intelligence is 3.75 and benefits of emotional intelligence to the company is 3.83 which is more than the central score of 3, this indicates that according to respondents there is positive enhancement of emotional intelligence due to Industry 5.0. Following table helps to prove this further:

Table 4. Analysis of H02

		Sum of Squares	df	Mean Square	F	Sig.
The Emotional Intelligence of the employees has seen enhancement with	Between Groups	12.591	4	3.148	4.446	.002
	Within Groups	98.409	139	.708		

technological enhancement taking place.	Total	111.000	143			
The enhanced Emotional Intelligence of people can be benefitting for the company.	Between Groups	12.594	4	3.149	5.512	.000
	Within Groups	79.406	139	.571		
	Total	92.000	143			

The above ANOVA table gives the F-value i.e., ANOVA in the fifth column and sig. value i.e. p-value in the sixth column. The ANOVA i.e., F-value obtained at the given degree of freedom in case of enhancement of emotional intelligence is 4.446 and the p-value obtained is 0.002, which is less than the alpha value of 0.05 ($F = 4.446$, $p < 0.05$), this indicates that there is significant impact of Industry 5.0 in enhancing the emotional intelligence of the employees. The ANOVA i.e., F-value obtained at the given degree of freedom in case of benefits of emotional intelligence to the organisation is 5.512 and the p-value obtained is 0.000, which is less than the alpha value of 0.05 ($F = 5.512$, $p < 0.05$), this indicates that there is significant positive impact of Industry 5.0 in enhancing the work efficiency of the employees working in the company which ultimately benefits the company. Hence, the hypothesis i.e. "Industry 5.0 enhances the emotional intelligence of people" is accepted.

To test the hypothesis "Industry 5.0 has created automation in HR processes" one-way Anova Test has been applied taking employment of technology as fixed factor and use of collaborative robots will affect the evaluation in negative way, use of cross disciplinary workforce carries most updated knowledge as dependent factor. The results of the analysis are shown below:

Table 5. Description of H03

		N	Mean	Std. Deviation	Std. Error
When it comes to performance evaluation, if in future collaborative robots are employed then you feel your evaluation maybe be slightly affected in negative way	Strongly Disagree	3	4.0000	0.00000	0.00000
	Disagree	3	4.0000	0.00000	0.00000
	Neutral	21	3.0000	0.77460	0.16903
	Agree	75	3.3600	0.89503	0.10335
	Strongly Agree	42	3.4286	0.73726	0.11376
	Total	144	3.3542	0.83179	0.06932
Cross disciplinary workforce which consists of human beings as well robots or any smart machines, carries most updated knowledge of work as well as technology, this	Strongly Disagree	3	5.0000	0.00000	0.00000
	Disagree	3	4.0000	0.00000	0.00000
	Neutral	21	3.7143	0.46291	0.10102
	Agree	75	3.6400	0.74689	0.08624
	Strongly Agree	42	3.7857	0.68202	0.10524
	Total	144	3.7292	0.70183	0.05849

aspect can be benefitting to the Organization.

The above descriptive table shows the mean, standard deviation and standard error of the two above mentioned variables, The mean score of more than 3 indicates more automation in HR process. The mean value obtained in care of both the variable is 3.35 and 3.72 which is more than the central score of 3, this indicates that more automation has been applied effectively in an organisation with the help of Industry 5.0. Following table helps to prove this further:

Table 6. Analysis of H03

		Sum of Squares	**df**	**Mean Square**	**F**	**Sig.**
When it comes to performance evaluation, if in future collaborative robots are employed then you feel your evaluation maybe be slightly affected in negative way	Between Groups	5.372	4	1.343	1.995	.099
	Within Groups	93.566	139	.673		
	Total	98.938	143			
Cross disciplinary workforce which consists of human beings as well robots or any smart machines, carries most updated knowledge of work as well as technology, this aspect can be benefitting to the Organization.	Between Groups	5.800	4	1.450	3.118	.017
	Within Groups	64.637	139	.465		
	Total	70.438	143			

The above ANOVA table gives the F-value i.e., ANOVA in the fifth column and sig. value i.e. p-value in the sixth column. The ANOVA i.e., F-value obtained at the given degree of freedom in case of variable i.e., use of collaborative robots will affect the evaluation in negative way is 1.995 and the p-value obtained is 0.099, which is more than the alpha value of 0.05 ($F = 1.995$ $p > 0.05$), this indicates that there is no significant impact of Industry 5.0 in use of collaborative robots to affect the evaluation in negative way. The ANOVA i.e., F-value obtained at the given degree of freedom in case of benefits of cross disciplinary workforce to the organization is 3.118 and the p-value obtained is 0.017, which is less than the alpha value of 0.05 ($F = 3.118$, $p < 0.05$), this indicates that there are significant benefits of cross disciplinary workforce to the organization. Hence, the hypothesis i.e. "Industry 5.0 has created automation in HR processes." is accepted.

To test the hypothesis "Industry 5.0 requires cross disciplinary work force for better performance" one-way Anova Test has been application of Industry 5.0 as fixed factor and requirement of cross disciplinary workforce for better work performance as dependent factor. The results of the analysis are shown below:

Table 7. Description of H04

You feel Cross Disciplinary workforce is required to handle these new age technologies for better work processing.

	N	Mean	Std. Deviation	Std. Error
Disagree	9	3.6667	0.50000	0.16667
Neutral	39	3.4615	0.64262	0.10290
Agree	84	3.7500	0.69204	0.07551
Strongly Agree	12	2.7500	0.86603	0.25000
Total	144	3.5833	0.73380	0.06115

The above descriptive table shows the mean, standard deviation and standard error of the two above mentioned variables, The mean score of more than 3 indicates requirement of cross disciplinary workforce for better performance. The mean value obtained in care of both the variable is 3.58 which is more than the central score of 3, this indicates that majority of the respondents agree that industry 5.0 required cross disciplinary work force for better performance. Following table helps to prove this further:

Table 8. Analysis of H04

You feel Cross Disciplinary workforce is required to handle these new age technologies for better work processing.

	Sum of Squares	df	Mean Square	F	Sig.
Between Groups	11.308	3	3.769	8.033	.000
Within Groups	65.692	140	.469		
Total	77.000	143			

The above ANOVA table gives the F-value i.e., ANOVA in the fifth column and sig. value i.e. p-value in the sixth column. The ANOVA i.e., F-value obtained at the given degree of freedom in case of variable i.e., requirement of cross disciplinary workforces for better work processing is 8.033 and the p-value obtained is 0.000, which is less than the alpha value of 0.05 ($F = 8.033$ $p < 0.05$), this indicates that there is significant requirement of cross disciplinary work force for better work processing.

7. Results and Discussion

Following is the discussion on points identified based on analyses of hypotheses testing:

7.1. Performance – The recent technologies, smart machines or robots when work in association then the performance of organization improves to a great extent. Now for that the existing workforce need to always keep them updated and learn ways to operate in cohesion with each other

7.2. Emotional Intelligence – Many research papers prove that when an employee has better emotional intelligence the performance of the organization always improves. The industry 5.0 presents challenge to the employees to renew and revise their knowledge in this competitive environment. So, HR managers can look forward to guide their employees in a way to improve their knowledge and be ambitious which will ultimately improve their emotional intelligence. They can also conduct personalized sessions with employees for better achieving this goal.

7.3. Work-process Automation – The industry 5.0 asserts that this revolution brings upon automation. However, the HR professionals need to be cautious while working in this environment. One of the main tasks of HR being performance evaluation, the managers need to consider that while evaluating they might have diversified workforce consisting of human beings and robots so the human beings should be assessed based on his/her capabilities. Also, the gamified version of training and development activities of employees prove to be more fruitful since employees enjoy that at the same time update their knowledge.

7.4. Cross-disciplinary Workforce – It is inevitable that human beings will be working together with collaborative robots or other smart machines. The present data suggest that there is less co-ordination between humans and these technologies. Hence there is a dire need of training to existing workforce to work together in association with them. The data in research papers published earlier suggests that the performance and growth improve when such cross-disciplinary workforce is employed. So, the proper training to employees is a must in this scenario.

8. Conclusion

The basic idea behind conducting this research is to identify factors affecting HR functions while transitioning to industry 5.0. The objective of identifying and exploring the factors is covered. And last to come up model explaining holistic approach for transitioning to 5.0 is mentioned in results and discussion section. To sum up, 'The emotional intelligence of employees does need to be taken care of while cross disciplinary workforce is conducting the processes in the organization which provides automation while showing improved performances.'

The research project presents results based on the responses of the people received. This study brings some relevant factors which requires attention from managers, however there is much scope ahead to look when the horizon of the study is made broader, and aspects are differed.

References

[1] Demira K. A., Dövena G, Sezen B. (2019). Industry 5.0 and Human-Robot Co-working. *Procedia Computer Science (Elsevier)* 158:688-695

[2] Xu X, Lu Y, Vogel-Heuser B, Wang L. (2021). Industry 4.0 and Industry 5.0—Inception, conception and perception. *Journal of Manufacturing Systems* :530-535

[3] Nahavandi S. (2019). Industry 5.0—A Human-Centric Solution. *Sustainability 11*, 4371; doi:10.3390/su11164371

[4] Cillo V, Gregori G. L., Daniele L. M., Caputo F, Bitbol-Saba N. (2021). Rethinking companies' culture through knowledge management lens during Industry 5.0 transition. *Journal of Knowledge Management ahead-of-print(ahead-of-print)* DOI:10.1108/JKM-09-2021-0718

[5] Martynov V. V., Shavaleeva D. N., Zaytseva A. A. (2019). Information Technology as the Basis for Transformation into a Digital Society and Industry 5.0. *International Conference "Quality Management, Transport and Information Security, Information Technologies"* (IT&QM&IS) DOI:10.1109/ITQMIS.2019.8928305

[6] Berber N, Đorđević B, Milanović S. (2018). Electronic Human Resource Management (e-HRM): A New Concept for Digital Age. *Vol. 23 No. 2 (2018): Strategic Management*

[7] Küppera D. M., Kleinb K, Völckner F. (2019) Gamifying employer branding: An integrating framework and research propositions for a new HRM approach in the digitized economy. *Human Resource Management Review* Volume 31, Issue 1, 100686 https://doi.org/10.1016/j.hrmr.2019.04.002

[8] Tataru C. (2019) Human Resources in the Digital Age A Manager's Realities and Perspectives. *Review of International Comparative Management* Volume 20, Issue 4, October 2019

[9] Arslan A, Cooper C, Khan Z, Golgeci I, Ali I. (2021). Artificial intelligence and human workers interaction at team level: a conceptual assessment of the challenges and potential HRM strategies. *International Journal of Manpower ahead-of-print(ahead-of-print)* DOI:10.1108/IJM-01-2021-0052

[10] Meijerinka J., Boonsb M., Keeganc A., Marlerd J. (2021). Algorithmic human resource management: Synthesizing developments and cross-disciplinary insights on digital HRM. The International *Journal of Human Resource Management* 32(3):1-18 DOI:10.1080/09585192.2021.1925326

[11] Selvaraj A. (2019). Role of Humans in Industry 5.0. https://www.columbusglobal.com/en/blog/role-of-humans-in-industry-5.0

[12] Saxena A., Saxena A., Sharma R., Parashar M. (2021). Emergence of Futuristic HRM in Perspective of Human - Cobot's Collaborative Functionality. *International Journal of Engineering and Advanced Technology* 10(5):292-296 DOI:10.35940/ijeat. E2763.0610521

[13] Kleinings H. (2021). Industry 5.0: Leadership in the era of AI. https://levity.ai/blog/leadership-ai

[14] CHIN S. T. S. (2021). Influence of Emotional Intelligence on the Workforce for Industry 5.0. *Journal of Human Resources Management Research*, Vol. 2021 (2021), Article ID 882278, DOI: 10.5171/2021.882278

[15] Kumar K. (2021). How Industry 5.0 Might Deliver Human-Centric Work Cultures https://www.cmswire.com/digital-workplace/how-industry-50-might-deliver-human-centric-work-cultures/

[16] Maddikunta P. K. R., Pham Q., Prabadevi B. , Deepa N. , Dev K., Gadekallu T. R., Rukhsana Ruby R., Liyanage M. Industry 5.0: A survey on enabling technologies and potential applications. *Journal of Industrial Information Integration*, https://doi.org/10.1016/j.jii.2021.100257

[17] Strategic Knowledge Management in the Digital Age. *Journal of Business Research*, Volume 94, January 2019, Pages 223-226

[18] Halid H., Yusoff Y. M., Somu H. The Relationship Between Digital Human Resource Management and Organizational Performance. Advances in Economics, *Business and Management Research*, Series volume number 141 Proceedings of the First ASEAN Business, Environment, and Technology Symposium (ABEATS 2019) (Article)

[19] Kediya, S. O., Singh, D. K., Shukla, J., & Nagdive, A. S. (2021, November). Analytical Study of Factors Affecting IoT in SCM. In 2021 *International Conference on Computational Intelligence and Computing Applications (ICCICA)* (pp. 1-4). IEEE.

[20] Kediya, S. O., & Kumar, S. (2021, November). An Analysis of Factors Affecting IoT Adoption by Indian Retail Industry. In 2021 *International Conference on Computational Intelligence and Computing Applications (ICCICA)* (pp. 1-3). IEEE.

[21] D. K. Singh, S. Kediya, R. Mahajan and P. K. Asthana, "Management Information System in context of Food grains: An Empirical Study at Eastern Maharashtra," (2021) *International Conference on Computational Intelligence and Computing Applications (ICCICA)*, 2021, pp. 1-5, doi: 10.1109/ICCICA52458.2021.9697144D.

[22] Singh, D. K., Kediya, S., Mahajan, R., & Asthana, P. K. (2021, November). Study of non-technical factors responsible for power losses at MSEB. In 2021 *International Conference on Computational Intelligence and Computing Applications (ICCICA)* (pp. 1-3). IEEE.

[23] Kediya, S., Ganer S. (2022)An analysis of Consumer Buying Behaviour in Street food industry in Nagpur during Pandemic. *14th CIMC 2022 on PPF (Past, Present & Future) - Kick Start to Business and Financial Recovery Post Pandemic Conference (Online)* 11 March 2022

[24] Ravekar, K., & Jaju, S. (2021). Design and analysis of material discharge plate of micro surfacing machine. *Materials Today: Proceedings*, 38, 2385-2391.

IOP Conf. Series: Materials Science and Engineering 1259 (2022) 012042 doi:10.1088/1757-899X/1259/1/012042

Application of Design for Manufacturing Technique to Lubricating Oil Pump Body for Reduction of Its Machining Time

P R Kulkarni[1], T S Pathak[2] and S B Tuljapure[3]
[1] Professor, Walchand Institute of Technology, Solapur, India
[2] Student- Masters, Walchand Institute of Technology, Solapur, India
[3] Associate Professor, Walchand Institute of Technology, Solapur, India

Email: tejaspathak047@gmail.com

Abstract: Design, manufacturing (production) & quality are linked to each other. Design has large impact on manufacturing ease & time along with the quality of product. Hence design need to be modified taking into consideration the manufacturing aspect. In this work, a lubricating oil pump body is studied from manufacturing point of view. The operations requiring more processing time were identified. An effort is made to change the design of the lubricating oil pump body to reduce the manufacturing time & efforts. Use of finite element method-based software ANSYS is done to determine the stresses during the machining with reference to the present design. The lubricating oil pump body is redesigned from manufacturing point of view keeping the end product as it is, to avoid the modifications of the other components in assembly. The machining parameters are predicted for this new design. At last, comparison of the machining parameters for existing & proposed design is done. There seems to be overall improvement in processing time.

1. Introduction

Manufacturing of products is closely concerned with design of product along with selection of material. The raw material is processed & finished components are obtained. Types of manufacturing processes are Casting, deformation, material removal, joining processes.

In case of product development cycle one of the critical parts is Design for Manufacturing (DFM) or Design for Manufacturing & Assembly (DFMA). It includes optimization of the product design for the manufacturing & assembly process & consideration of production method during design.

2. Literature review

Munanga P et al. (2020) used Design for Manufacturing and Assembly methodology for an intelligent single axis solar tracking system. This research was aimed at developing an intelligent solar tracking device.

Diegel O et al. (2020) studied the process of redesign of hydraulic manifold which was conventionally manufactured to the one designed for the additive manufacturing (DFAM). During the DFMA process, removal of the material which was not contributing to any particular function was done. The effects of various print orientations, elimination of the support material were explored during the study.

Thompson M et al. (2018) performed research in the area of high-speed product development. Working with the DFMA expert, late engineering changes were suggested.

Naiju C et al. (2017) applied DFMA for minimizing the cost of a product. A pedestal fan was chosen for this as this product is an extensively used in houses and offices. Reduction in number of parts was one. The redesign resulted into reduction of assembly time of product by 15.17%.

Ngatilah Y et al. (2017) designed energy saving lamps using "Light Emitting Diode" (LED). The method used was Design for Manufacture and Assembly (DFMA), with the result of design development the percentage increase in assembly efficiency was 39.76%. The decrease in material costs was 8.54%.

Naiju C et al. (2016) analyzed two products folding chair and bunk bed using DFMA software. The products were redesigned for cost reduction. Total number of parts reduced was from 9 to 3 & 29 to 9 respectively for folding chair & bunk bed. The overall cost of the folding chair was reduced by 10% & the reduction was 20% for bunker bed.

More N et al. (2015) performed analysis using the method of Design for Manufacture and Assembly (DFMA). The object for the analysis was an automotive tool. Total 23.5% of cost reduction was achieved resulting into the reduction of 21% assembly process time. After successful DFMA analysis, the assembly design efficiency was improved by 26.5%.

Wankhade N et al. (2014) researched on the product fluid flow control valve. The Design for manufacturing and assembly (DFMA) concept applied for the re-design of the existing valve. The outcome of the research work led to the optimum design of the valve. The quality of the valve was improved & the cost was also reduced.

Barbosa, G.F and Carvalho (2013) used the method of Design for Manufacturing and Assembly (DFMA) for design and manufacturing of aircrafts. In the final assembly, not only cycle time was reduced but manpower requirement was also reduced. The other gains were possibility of segments (fuselage & half-wings) stuffing independently & separately.

Salim E et al. (2011) applied principles of industrial design and design for manufacturing and assembly (DFMA) to juicer-mixer-grinder. The cost of the proposed juicer-mixer-grinder was at least 10 to 15% lower when compared to the similar models available in the market.

Huang G and Mak K (1999) discussed about providing design for manufacture and assembly (DFMA) techniques on the Internet. It was found that web-based DFX techniques give more chances for integration with other decision-support systems for instance Computer Aided Process Planning (CAPP), Computer Aided Design (CAD) and Computer Aided Production Management (CAPM) in the process of product realization.

3. Problem identification

Lubricating oil pumps: Use of lubricating oil is done widely for the purposes of reducing friction in machineries, bearing, turbines etc. This lubricating oil also helps in cooling of the working parts. Pumps are used to take the oil from reservoir & force the same through lubrication points. Again, this oil comes back to the reservoir & process continues.

One of the major components of the lubricating oil pumps is its body (Figure 1). This lubricating oil pump body is made up of Cast iron (FG26 grade). It involves number of steps/operations in its manufacturing. These operations are performed on different machines like Centre lathe, CNC lathe & Vertical machining center.

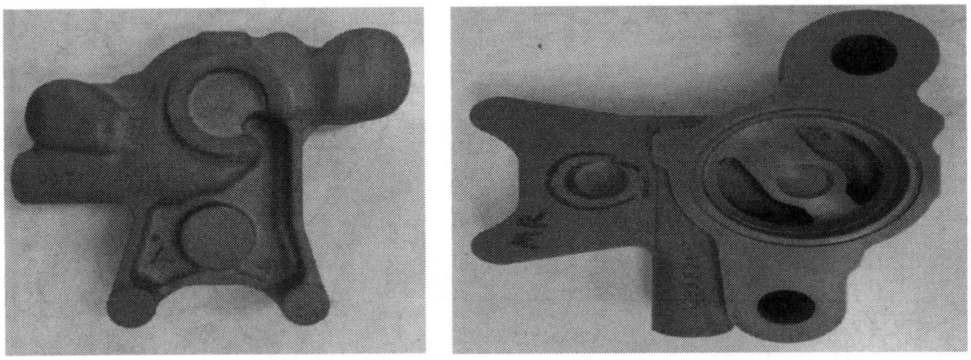

Figure 1. Lubricating oil pump body

One of the operations which are causing more processing time & difficulty is being carried out on centre lathe. This operation is involving facing, drilling & boring operations.

The drawing of the lubricating oil pump body is as below. Considering the secrecy, the details of dimensions are not included here.

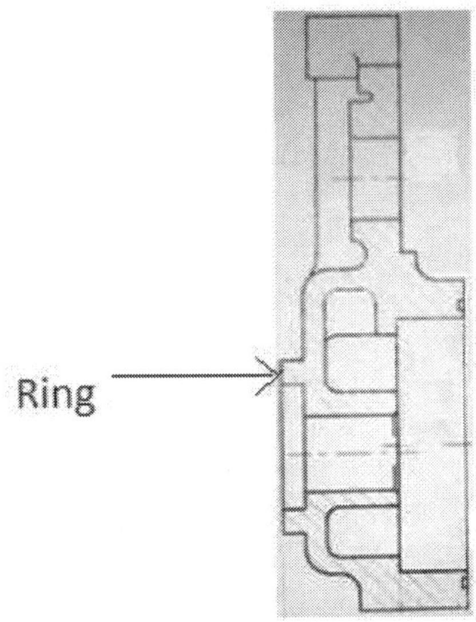

Figure 2. Drawing of Lubricating oil pump body

On the left side, a small boss (ring of 7 mm width or thickness & diameter 55 mm) is there, which is to be finished by facing operation & turning operation to get the desired dimensions of it. On the right hand side, it needs facing operation & boring operation along with one centre drilling operation for further machining on CNC & VMC.

Initially part is held from inner side of right hand part with the help of the four jaw chuck. The small boss on left side is then turned & faced to final size.

When the operation is being performed on the right side of the part, the holding can be done only on the small boss available on the left hand side. This holding/clamping on the small boss leads to lot of time spent on the truing of the part to satisfy the geometrical requirements. Even though clamping is done somehow, due to the small boss size for clamping, large cuts can't be taken & machining is requiring more time. Thus, this large processing time is the problem considered here for the possible design modifications.

IOP Conf. Series: Materials Science and Engineering 1259 (2022) 012042 doi:10.1088/1757-899X/1259/1/012042

4. Existing process study

Machining Process details taken for study:

Below are the details of the machining process details of left & right sides of the lubricating oil pump body.

4.1 First Process:

1. Rest the job (casting) on the right face & hold it with 4 jaw chuck at inner diameter.
2. Do truing operation.
3. Light Facing operation – to keep the width (Speed: 198 rpm, Feed: 0.2 mm/rev)
4. Turn the outer diameter -- to get it desired diameter (Speed: 198 rpm, Feed Hand)

Cutting Tool : 'V' tool PSDNN 2525 M12, Side tool: PCLNL 2525 M12

4.2 Second process:

1. Hold on small projected diameter in three jaw chuck and get machined bore diameter & inner face (Speed: 310 rpm, Feed: 0.1 mm/rev)
2. Drill 5 mm drill (not shown in figure 2) x 0.5 / 0.7 depth (Speed: 310 rpm, Feed: Hand)

Cutting Tool : 'V' tool PSDNN 2525 M12, Brazed boring tool & 5 mm drill

Machine : Centre Lathe machine

4.3 Issues in process:

- Holding (in three jaw chuck) of job small project diameter of 55 mm & width 7 mm & truing the surface is requiring more time.
- As the clamping is not much rigid, machining operations are requiring more time due to lower depth of cuts, feed & speed.

5. Proposed solution

As it is very difficult to hold the job on a small boss, it is proposed to increase the width (height) of the boss. Initially the diameter of the boss will be finished to the required size keeping the width as it is. Later on, when the operations of the right-hand side are completed, then the width (height) of the boss will be finished to the required size by facing operation. The increase in the boss width (height) will lead to more rigid clamping/holding. This will facilitate increase depth of cut, feed rate or speed of cutting. This will reduce the machining time. But overall new machining time needs to be calculated considering the removal of extra material added to increase the boss width (height).

6. Calculation of cutting forces for prediction of improvement (reduction in processing time)

To predict the percentage of increase in depth of cut, feed rate or speed of cutting, finite element analysis of the part is done. Considering the secrecy only few dimensions were provided by the company. These dimensions are used to prepare simplified model of the part.

6.1 Calculation of tool angles for machining of existing part:

Orthogonal cutting is assumed here, as oblique cutting is very complex for the analysis.

Designation of tool:

ASA system of the tool designation is used here.

IOP Conf. Series: Materials Science and Engineering 1259 (2022) 012042 doi:10.1088/1757-899X/1259/1/012042

Table 1 Tool designation

Angle	Symbol	Value
Back rake angle	α_b	0 °
Side rake angle	α_s	5°
End relief angle	θ_e	6 °
Side relief angle	θ_s	6 °
End cutting edge angle	C_e	10 °
Side cutting edge angle	C_s	10 °
Nose radius	R	1 mm

The tool approach angle, $\lambda = 90° - Cs = 90° - 10° = 80°$
Friction angle, $\beta = \tan^{-1}\mu$
μ = Coefficient of friction at tool work-piece interface = 0.7 (approx.)
Friction angle, $\beta = \tan^{-1}\mu = \tan^{-1} 0.7 = 34.992°$
Orthogonal rake angle is calculated as,
$\text{Tan } \alpha = \text{Tan } \alpha_s \text{ Sin } \lambda + \text{Tan } \alpha_b \text{ Cos } \lambda$
$= \text{Tan } (5°) \text{ Sin } (80°) + \text{Tan } (0°) \text{ Cos } (80°)$
$= 0.086$
$\alpha = 4.933\,°$
Using Merchant's relation, shear angle is calculated as
$\phi = 45° + \alpha/2 - \beta/2 = 45° + (4.933°)/2 - 34.992°/2 = 29.97°$

6.2 Calculation of cutting force:
[Reference: 13]
Let F_c is cutting force. The cutting force is calculated using following relation.

$$F_c = \frac{\tau_s \; x \; b \; x \; f}{Sec\,(\beta-\alpha)\;Cos\,(\varphi+\beta-\alpha) Sin\,\varphi}$$

b = width of cut = depth of cut / Sin λ = 2 / Sin (80°) = 2.03 mm
f = feed mm/ revolution = 0.2 mm/revolution……... (Process sheet of job)
Tensile strength of material is 260 MPa
Using the formula,
Ultimate Shear Strength = 1.3 x Ultimate Tensile Strength
= 1.3 x 260
= 338 MPa

$$F_c = \frac{338 \; x \; 2.03 \; x \; 0.2}{Sec\,(34.992-4.933)\;Cos\,(29.97+34.992-4.933) Sin\,29.97}$$

$$= \frac{137.228}{1.15 \; x \; 0.499 \; x \; 0.499}$$

$$= \frac{137.228}{0.286}$$

= 479.81 N

6.3 Calculation of thrust force:
The thrust force is $F_t = F_c \times \tan(\beta - \alpha) = 479.81 \times \tan(34.992° - 4.933°) = 277.67$ N
This thrust force is perpendicular to the tool-job interface. i.e., perpendicular to the principal cutting edge of the tool.

6.4 Calculation of feed force:
Feed force (along the axis of the job) = $F_f = F_t$ x Sin λ = 277.67 x Sin (80°) = 273.45 N

6.5 Calculation of radial force:
Radial force (Normal to the axis of the job) = $F_r = F_t$ x Cos λ = 277.67 x Cos (80°) = 48.21 N
$V = \frac{\pi D N}{1000 x 60} = \frac{\pi x\ 51 x 310}{1000 x 60}$ = 827.8 mm/sec = 0.8278 m/s

6.6 Calculation of cutting power:
Cutting power = F_c x V = 479.81 x 0.8278 = 397.1 watts

7. Finite element analysis
Finite element analysis is done to determine the stresses due to the forces during cutting. Here only static structural analysis is done using Finite Element Method based software ANSYS.

7.1. Model
Figure 3 shows different views of the model of the existing part.

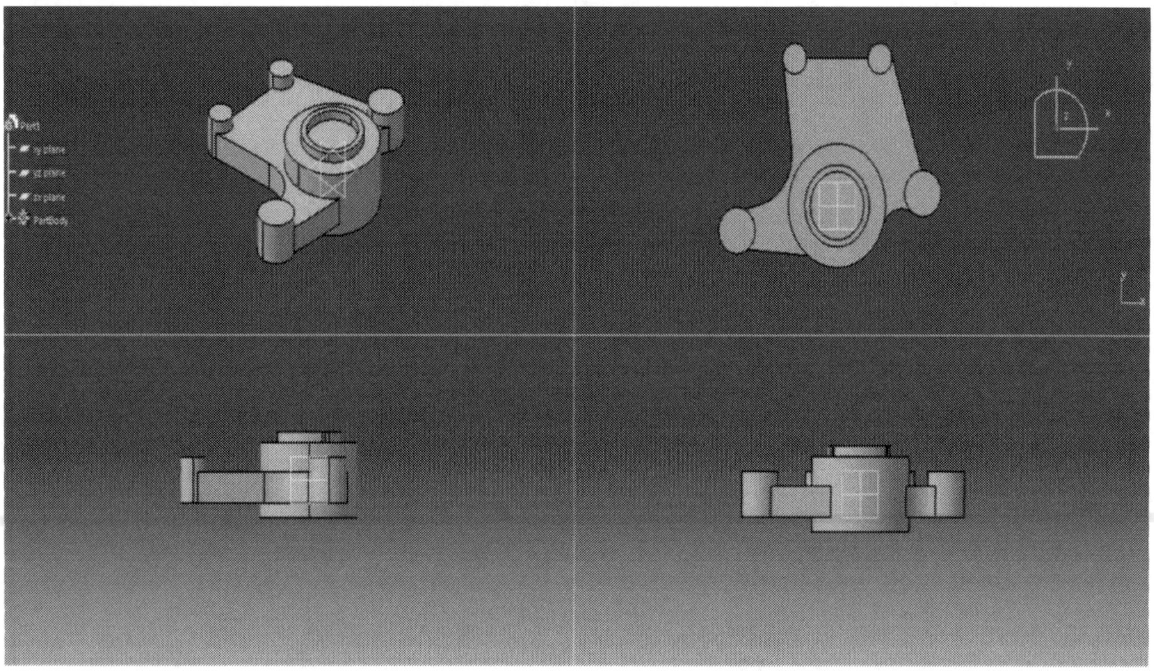

Figure 3. Different views of the model of the lubricating oil pump body

Modifications: Modification done is increase in ring width from 7 mm to 17 mm. This can be observed in Figure 4 & Figure 5. The ring with blue spots is modified portion.

IOP Conf. Series: Materials Science and Engineering 1259 (2022) 012042 doi:10.1088/1757-899X/1259/1/012042

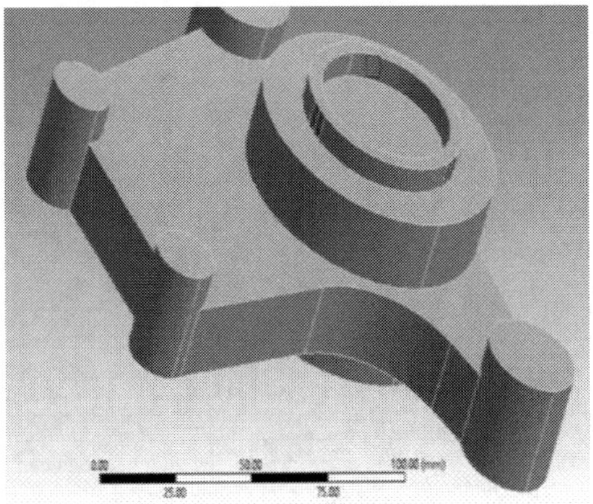

Figure 4. Existing part (ring width 7 mm)

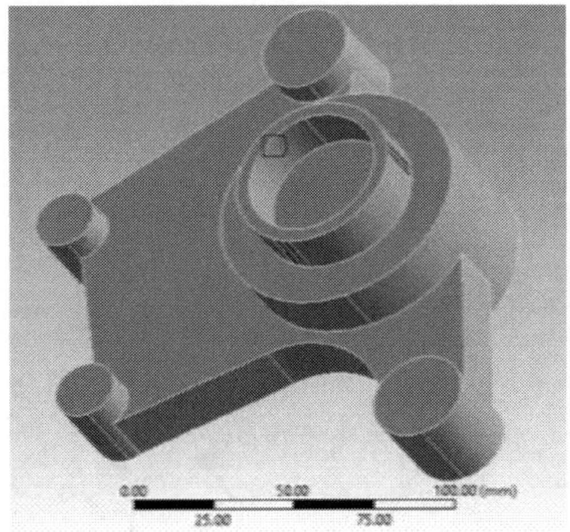

Figure 5. Modified part (ring width 17 mm)

7.2. Meshing

Initially a convergence study is done to decide the mesh size (element length). Table 2 represents the values of deformation & maximum principal stresses for different mesh sizes.

Table 2 Effect of mesh size on stresses & deformation

Mesh Size (mm)	Max. Deformation 1. (mm)	Max. Principal Stress (MPa)
10	0.0027199	6.2909
9	0.0026705	5.5754
8	0.0029033	7.1559
7	0.0027830	6.2805
6	0.0029032	6.8393
5	0.0030520	7.462
4	0.0031206	9.0422
3	0.0341150	21.554
2	0.0036051	23.327

Considering the values of the stresses for different mesh sizes, mesh size of 6 mm is finalized.

Meshed models:

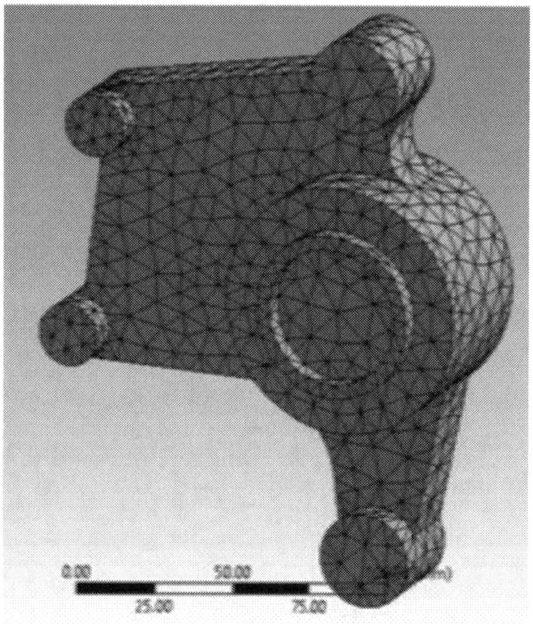

Figure 6. Meshed model of Existing part

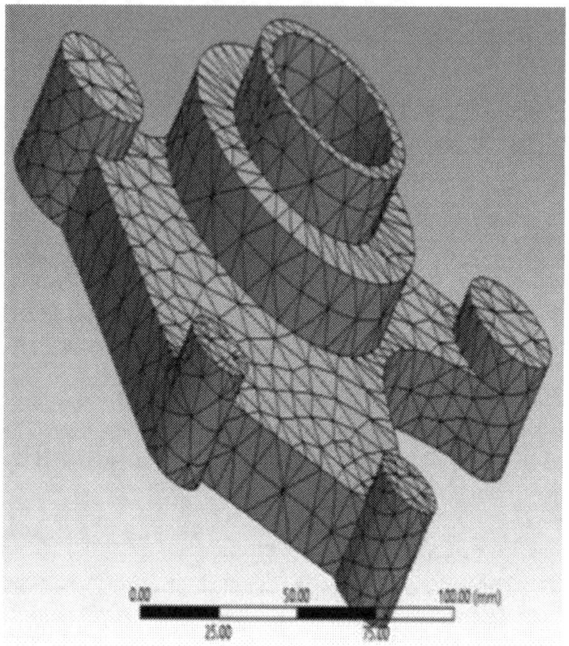

Figure 7. Meshed model of modified part

Table 3 Mesh details

	Existing part	Modified part
No. of Nodes	10465	10975
No. of Elements	5719	6056

7.3. Boundary Conditions

As the job is clamped in the three-jaw chuck at 3 positions on the ring at 120 degree to each other, the movements of the region at the clamping is restricted completely. The blue colored region shows the applied boundary conditions.

Figure 8. Clamping on small zone of the ring

Figure 9. Clamping on large zone of the ring

7.4. Load conditions

As the load conditions applied are same, only the figures corresponding to the forces applied on the existing job are shown below. Small groove is made on the job to facilitate application of forces properly.

a) Cutting force (vertical) = 479.81 N

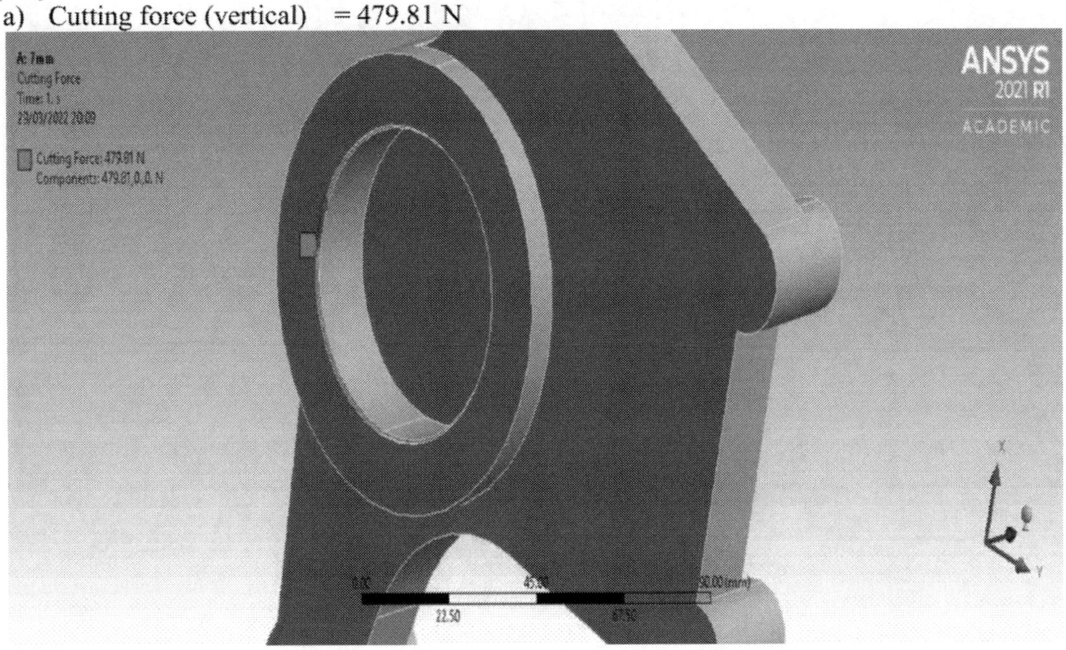

Figure 10 Load condition (Cutting force)

b) Feed force (along the axis of the job) = 273.45 N

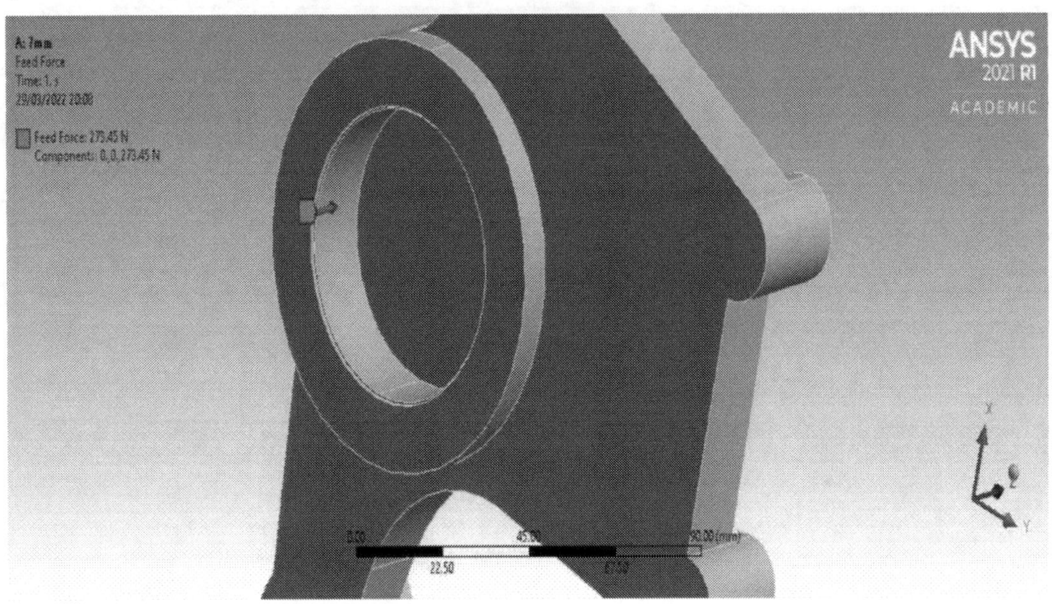

Figure 11 Load condition (Feed force)

IOP Conf. Series: Materials Science and Engineering 1259 (2022) 012042 doi:10.1088/1757-899X/1259/1/012042

c) Radial force (Normal to the axis of the job) = 48.21 N

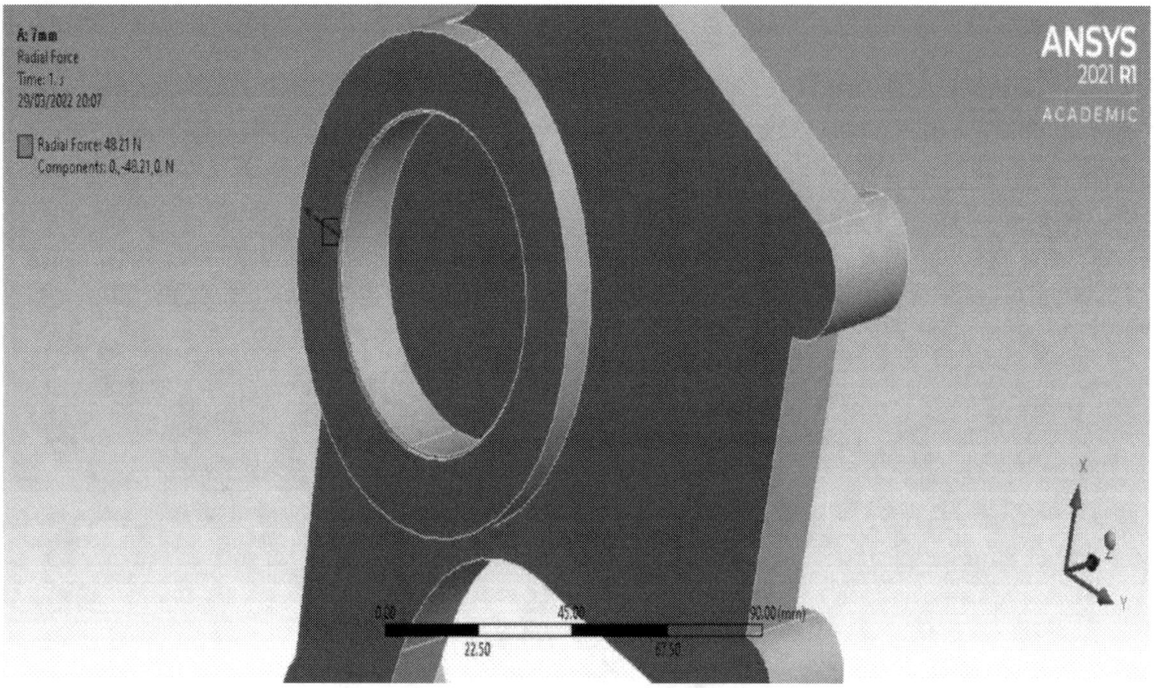

Figure 12 Load condition (Radial force)

8. Results and Discussions

The results (Maximum deformation & Maximum Principal Stresses) are shown below.

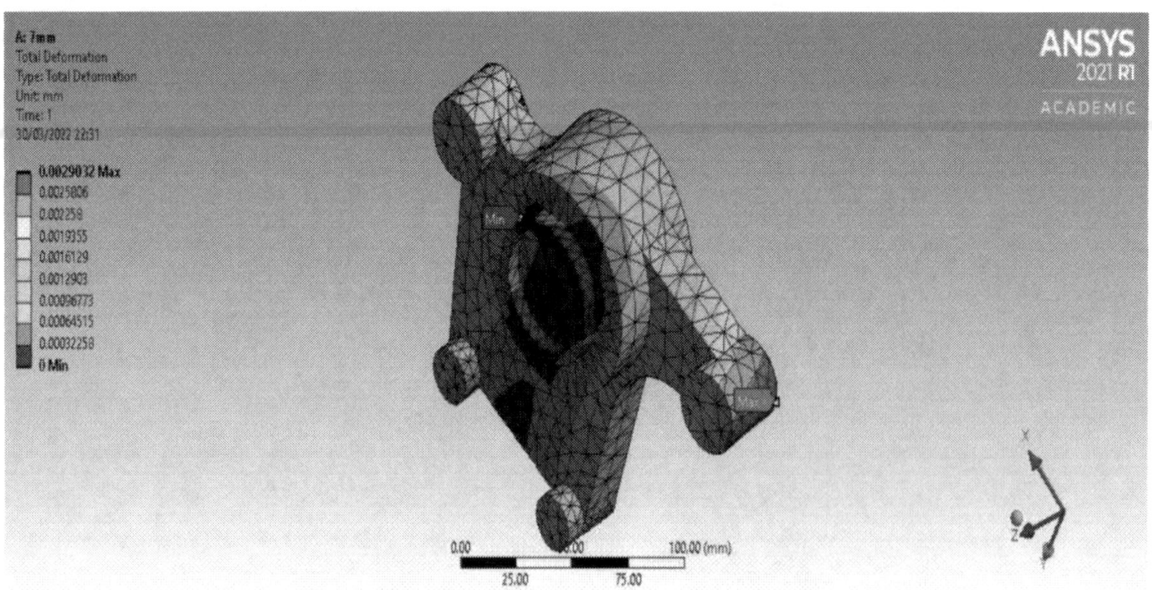

Figure 13 Total deformation –Existing part

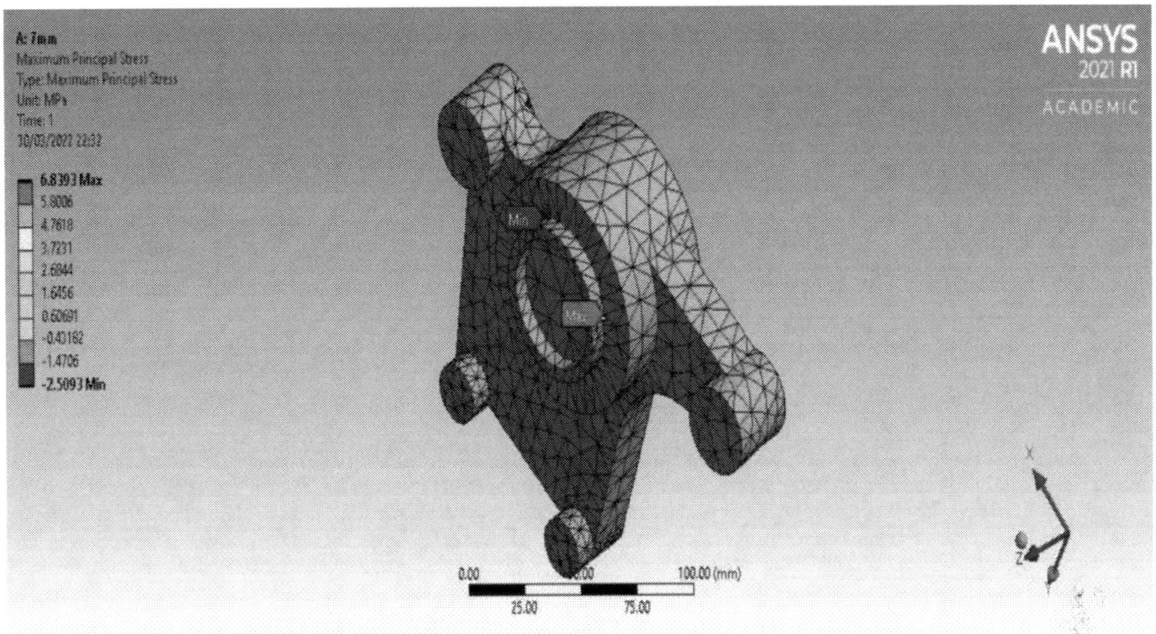

Figure 14 Maximum Principal Stress–Existing part

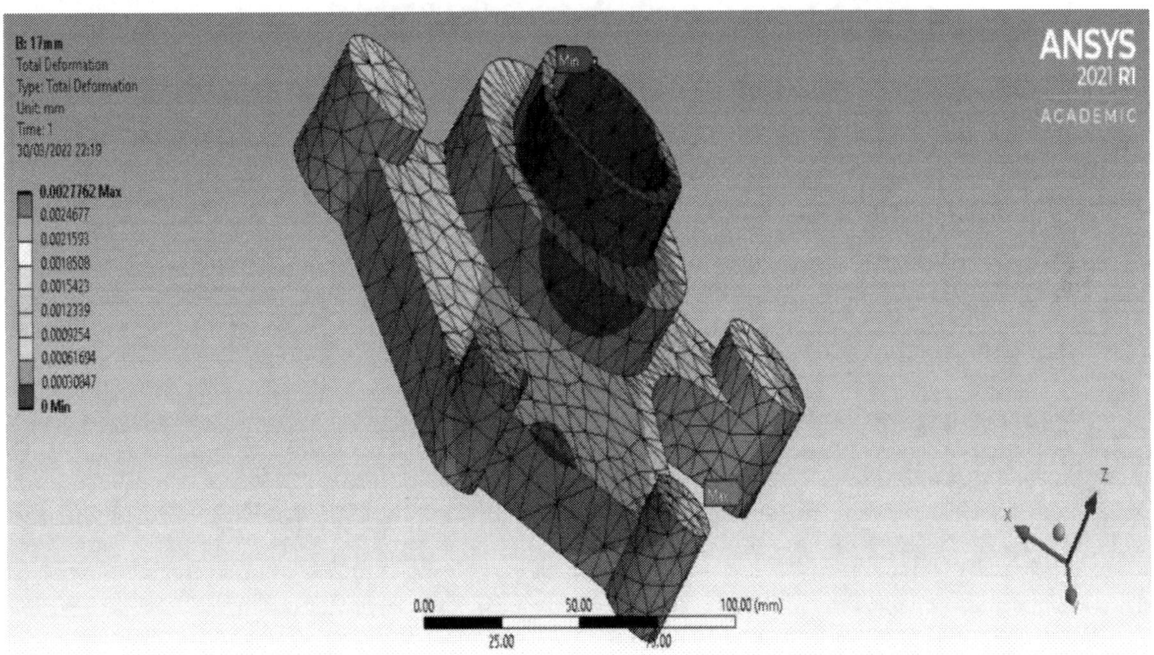

Figure 15 Total deformation –Modified part

IOP Conf. Series: Materials Science and Engineering 1259 (2022) 012042 doi:10.1088/1757-899X/1259/1/012042

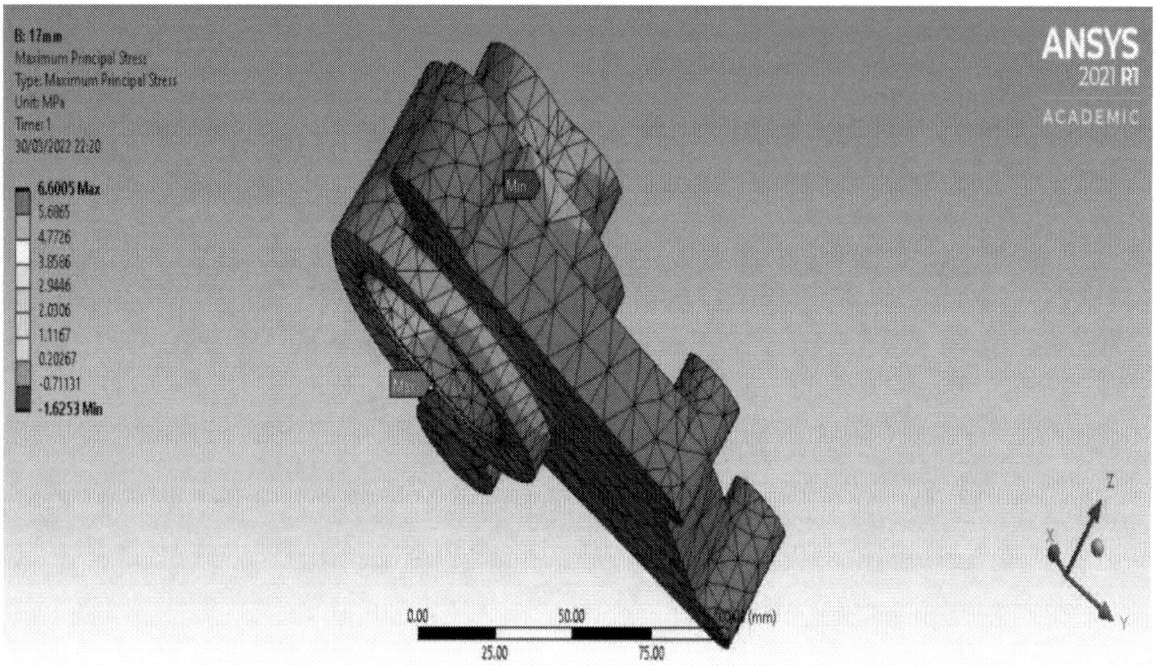

Figure 16 Maximum Principal Stress–Modified part

The output values are summarized in the Table 4 below.

Table 4 Comparison of results

Parameters for comparison	Existing part	Modified part	Reduction due to modification
Total deformation (mm)	0.0029032	0.0027762	4.37 %
Max. Principal Stress (MPa)	6.8393	6.6005	3.49%

It is observed that due to rigidity in clamping after modification, deformation is reduced. Even though it seems that we can increase the cutting parameter by small amount, the reduction in time required for truing of part & clamping of part will be very large.

Material added & removed by turning:

Increase in width of the ring = 17- 7 = 10 mm

Inner & Outer diameters of ring = 33 mm & 55 mm

Volume of material added = $\frac{\pi}{4}(0.055^2 - 0.033^2)x\ 0.01$ = 0.0000152 m^3

Weight of material added = Volume of material added x density of material

= 0.0000152 x 7850 = 0.11932 Kg = 119 gm

This weight of added material may be reduced by optimizing the ring width.

9. Conclusions

Design for manufacturing technique, if used effectively, may reduce the manufacturing time of the components largely thereby increasing productivity. In case of the machining of cast components, after redesign/ modifications, the reduction of machining time as well as non-machining time should be considered. Reduction in non-machining time (truing, clamping etc.) may be much more as compared to reduction in machining time.

References

1. Munanga P, Chinguwa S and Nyemba W 2020, Design for manufacture and assembly of an intelligent single axis solar tracking system, *30th CIRP Design Conference*
2. Diegel O, Schutte J, Ferreira A, Chan Y 2020 Design for additive manufacturing process for a lightweight hydraulic manifold, *Additive Manufacturing*, Vol. 36,
3. Thompson M, Jespersen I, Kjaergaard T 2018 , Design for manufacturing and assembly key performance indicators to support high-speed product development, *28th CIRP Design Conference*, Nantes, France
4. Naiju C, Jayakrishnan V and Warrier P 2017, DFMA for Early Cost Estimation of Pedestal Fan - A Case Study, *Journal of Industrial and Intelligent Information* Vol. 5, No. 1,
5. Ngatilah Y, Pulansari F, Ernawati D, Pujiastuti C, Parwati C, Prasetyo B 2017, Design for Manufacture and Assembly for Product Development (Case study : Emergency Lamp), *The 2nd International Joint Conference on Science and Technology* (IJCST)
6. Naiju C, Jayakrishnan V, Warrier P and Rijul B 2016, Application of Design for Manufacturing and Assembly (DFMA) Methodology in the Steel Furniture Industry, *National Conference on Technological Advancements in Engineering* –Payyanur
7. More N, Buktar R, Ali S, Samant S 2015, Design for manufacture and assembly (DFMA) analysis of burring tool assembly, *International Research Journal of Engineering and Technology* (IRJET), Volume: 02 Issue: 06
8. Wankhade N, Sridhar V and Annamalai K 2014, New product development by dfma and rapid prototyping, *ARPN Journal of Engineering and Applied Sciences*, Vol. 9, No. 3
9. Barbosa, G.F and Carvalho 2013, *11th IFAC Workshop on Intelligent Manufacturing Systems The International Federation of Automatic Control*, São Paulo, Brazil
10. Salim E, George A, Naiju C, Annamalai K 2011, Conceptualization, design for manufacture and assembly (dfma) of juicer mixer grinder, *Proceedings of National Conference on Advances in Mechanical Engineering "NCAME"/SBCEC/*
11. Huang G and Mak K 1999, Design for manufacture and assembly on the Internet, *Computers in Industry*, Vol. 38, Issue 1

Productivity enhancement by using VSM tools in farm equipment manufacturing company

Hemant Khubalkar[1], Rahul A Jibhakate[1], PV Walke[1], Rajendra Kadu[2]

[1]Department of Mechanical Engineering, G H Raisoni College of engineering, Nagpur 440016, India
[2] Mahindra & Mahindra Ltd. tractor division, Nagpur 440016, India

jibhakate.rahul3@gmail.com (https://orcid.org/ 0000-0002-3484-8830)

Abstract. Lean manufacturing improves product quality and productivity, minimizes takt time, manufacturing waste, and inventory and eradicates NVA activities. company victimization the quality methodology and it's featured the heaps of disadvantage in terms of high takt times, in correct line levelling, high transportation, and material handling, etc. the foremost objective of this analysis is to increase the productivity by victimization lean tools like, scrutiny, work standardization through time and time-and-motion study, and line levelling to attenuate the wastages that's found in company, and it helps to company to increase market demand. Among the investigation, high work-in-process, poor line levelling, high cycle time and production interval, unbalance work distribution/assignment, among the corporate area unit ascertained. once implementation of some lean tools, cycle time is reduced to twenty initial, cycle time is balanced with takt time, 12.3 min. being the standard cycle time of product, production takt time decreases by eighteen whereas production can increase by thirty third. Some new manufacturing processes area unit typically applied to urge improved product quality and involving further workers to spice up manufacturing processes, the merchandise of company. The profit of the company is in addition enlarged to spice up the quality product and minimum delivery time, and shopper satisfaction has in addition been considered in addition.

Keywords: Boring operation process, Tool study, Design of tool, process mapping, VSM, Optimization, Tool life study.

1. Introduction

The process of changing raw materials into merchandise, together with choice of raw materials, the merchandise style and therefore the order of the producing method is outlined as machine producing. As per existing market, the quality of final product to be ensured in stages of manufacturing. As a result, the demand of market for perfect manufacturing processes with low cost, best quality and to reduce the environmental considerations. In addition, company have to build the consistency of manufacturing parts to benefit of market change. Changing the tool within the time to get benefits may provide desired results. High cutting tool cost and high processing cost, despite all the advantages of the re-engineering concept [1.2,3].

1.1. Boring process

Metal cutting is very important in manufacturing process. The required product can be obtained through this process as the sharp cutting tool takes off the material from inner surface of work piece. The most conventional machining process are milling, drilling, and turning. Circular inner profiles are generated in component by removing extra material from inner diameter of the components [4,5]. This is called boring process. A boring bar have single or no of cutting edges which is rotating at recommended RPM the tool is goes inside the work piece and comes out with certain velocity. To make the specific bore size with specified dimension. Boring process is primarily used. Operation is doing in auto mode with help CNC code programming which done in CNC control while the boring process is done on horizontal machines or vertical machines [6,7].

Since tool wear deteriorates surface roughness, productivity, and cost etc, it is one of the most significant deliberations in machining operations. Selecting cutting parameters for achieving great cutting performance is a major task during boring operation. Hence, in terms of time and materials, the desired model building is very expensive. The use of Taguchi method is also suggested. The desired cutting parameters were also compared with tool life in turning process [8].

2. Experiment

In this process we have done process mapping and all related parameter is studied in term of tool cutting speed, nose radius, diameter, length, and material, cutting oil pressure and concentration and depth of cut which is put in CNC programme to run the components and also help to improve the tool life and machine efficiency to get the better quality products. Due to this we have get the benefit in cycle time to improve the productivity. Also study the casting material to improve the tool life and geometrical parameter. Also, with considering the casting material we have select the tool geometry to get lowest cycle time [9, 10]. On this base of study some variable process is find out so that NVA time is getting wastages and also production rate is low. All line balancing is done. In that case all finish operation cycle time is studied and optimisation of cycle time is done same on all machines. We must work in which is shown in layout Engine block machining cell as shown in fig. 1.

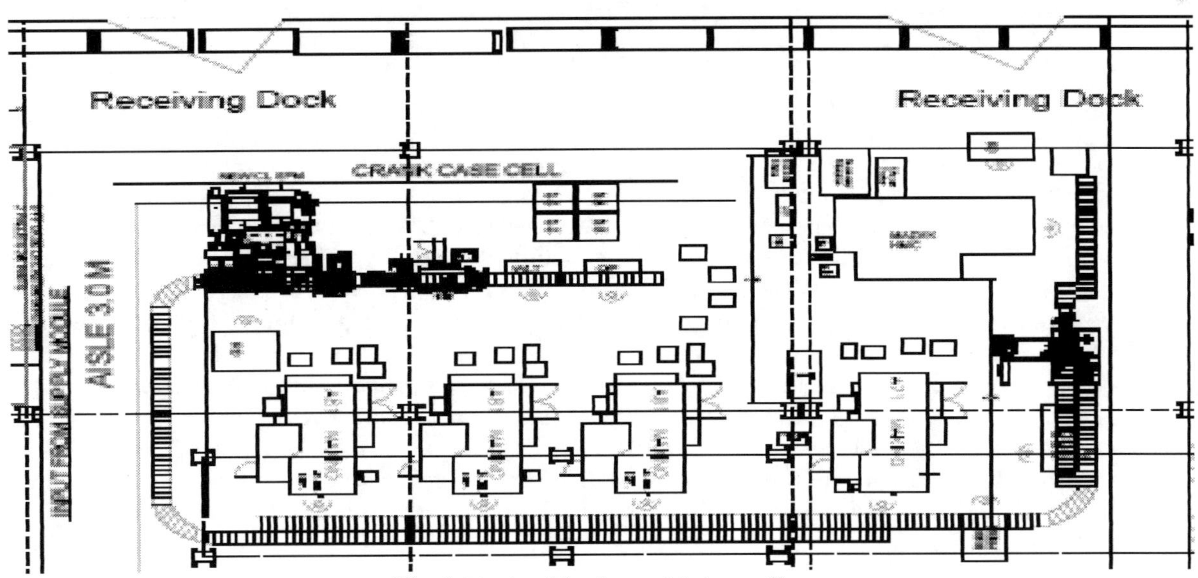

Fig.1 Engine block machining cell

IOP Conf. Series: Materials Science and Engineering 1259 (2022) 012043 doi:10.1088/1757-899X/1259/1/012043

3. Experimental setup

Hollow work items may be created victimisation boring processes, by internal diameter surface removal off the work piece. During this paper, formulating the mathematical model and its analysis to extend the tool life in casting operation is shown. 2 stages for each freelance parameter is needed for the experiment. to gather data through experimentation for formulating the mathematical model for cast-iron machining operation, the target of experiment is ready in tool life improvement method [11, 12]. Measure of varied parameters like surface end, bore size variation, tool life, operation time and spindle load is dole out throughout forged iron machining processes by meter scale, digital DIA. check plug gauges, digital timer, surface end tester and current otherwise amp. The investigation of methodology parameters like length, nose radius, diameter, cutting speed, and material of the cutter, cutting fluid pressure and concentration and depth of cut for casting machining methodology are used for experimental style. The verification of determined values for surface end, operation time, spindle load, tool life, and bore size variation is completed for formulating the mathematical model shown in Table a pair of (b) completely different instruments for measurements [13].

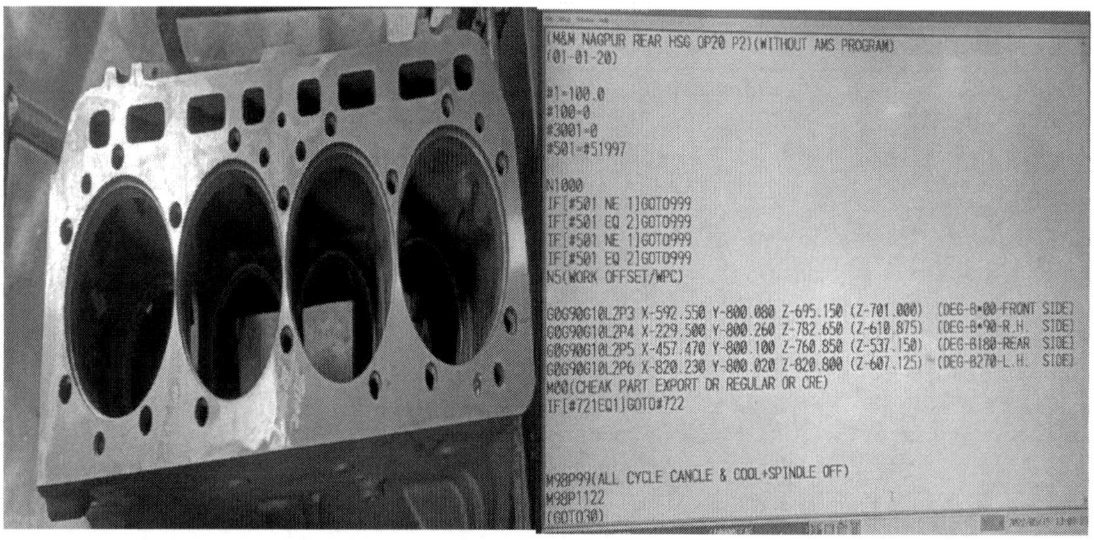

Fig.2 CNC G-code for Engine block machining

Tractor business is one in all the industries that have potential and believed in developing the economy of the country. History tells that this business has been a base for many prosperous industrial developments. Government has made public a policy where one in all the tasks recognized is quick export growth through production of high worth agricultural product and increasing support to export aligning manufacturing sectors. The rise in labour wage in developed countries has diode to a shift in attire producing a lot of towards low wage developing countries instead of the high wage developed world. In producing industries of developing countries are a lot of fixated on getting the casting and lowering delivery value than productivity value because of the innovation of tool & changes in method [14]. As a result, Tool style is lower during this sector than within the developed ones. Currently the concern is concerning Tool style. Creating production flexible; because of the market demand is a lot of in farm business and it's too unsure, such a lot so if the orders aren't consummated on time, one would possibly lose the business. In tool style following geometrical parameter to be studied altogether facet. Following are the parameter are studied to design tool for rising the housing block production to fulfil the demand. Out of all cells I even have elite the housing Block Machining Cell to extend the capability to fulfil the demand for creating tractor production as per market demand [15].

IOP Conf. Series: Materials Science and Engineering 1259 (2022) 012043 doi:10.1088/1757-899X/1259/1/012043

4. Process mapping

In this experimentation process mapping has done of entire cell. Total 9 stage in this cell for engine block machining. Total five HMC machine. Out of this one machine is used for roughing operation and four machine is used for finishing operation. Demand cannot fulfil due to high Cycle time of five machine. Total nine tool is selected to reduce the cycle time of semi finish & finish operation. All tools are studied in terms of cutting parameter, tool material, tool geometry, tool life, geometrical study [13].

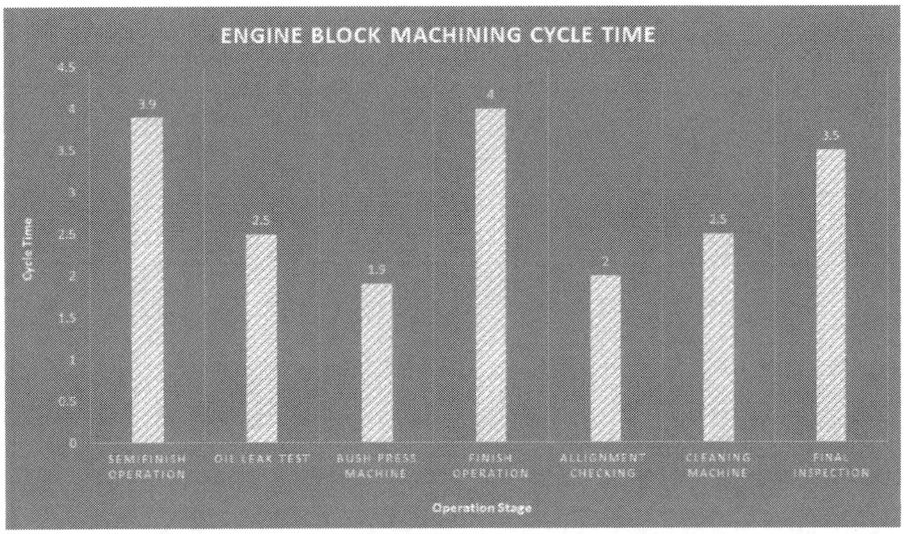

Fig. 3 Value Stream Mapping of Engine block Machining Operation Tool Mat

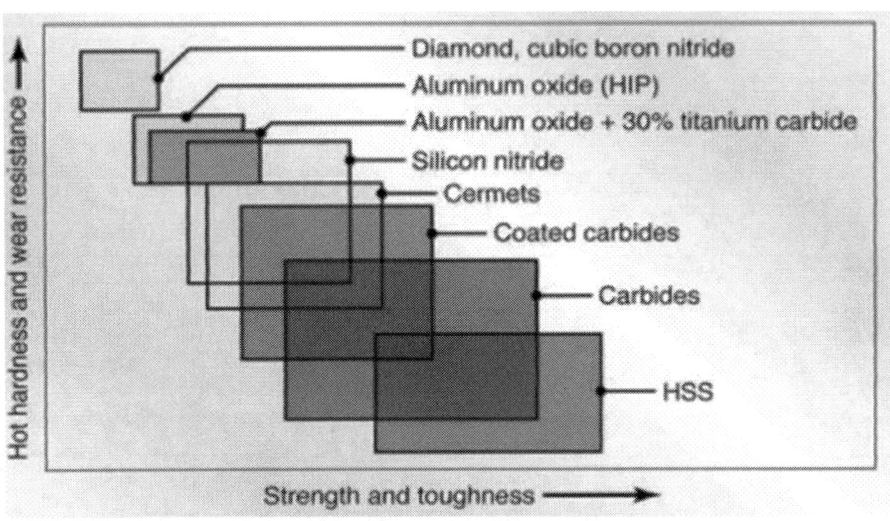

Fig. 4 Common cutting tool materials

IOP Conf. Series: Materials Science and Engineering 1259 (2022) 012043 doi:10.1088/1757-899X/1259/1/012043

Cemented Carbides Consists of tungsten carbide (WC) in a cobalt binder (2 to 20%) Can be tailored to give different combinations of abrasion resistance and toughness by varying the amount of cobalt and the WC grain size. Has high hot Hardness (up to 800 to 900ºC –87 to 92 Ra) Straight tungsten carbides are more ductile and less brittle For Cast Iron machining -ductile straight carbide material should be employed. For machining steel -wear resistant titanium –tungsten carbide material is called for machining [14].

5. Result

Table 1: Tool cycle time analysis for engine block machining

No.	Stage	Cycle Time (Min.)	Capacity (Qty.)	Target Cycle Time (Min.)	Target Capacity (Qty.)
1	Semi finish Operation	3.9	335	3.1	421
2	Oil leak Test	2.5	522	2.5	522
3	Bush Press Machine	1.9	687	1.9	687
4	Finish Operation	4	326	3.1	421
5	Alignment checking	2	653	2.0	653
6	Cleaning Machine	2.5	522	2.5	522
7	Final Inspection	3.5	373	3.0	435

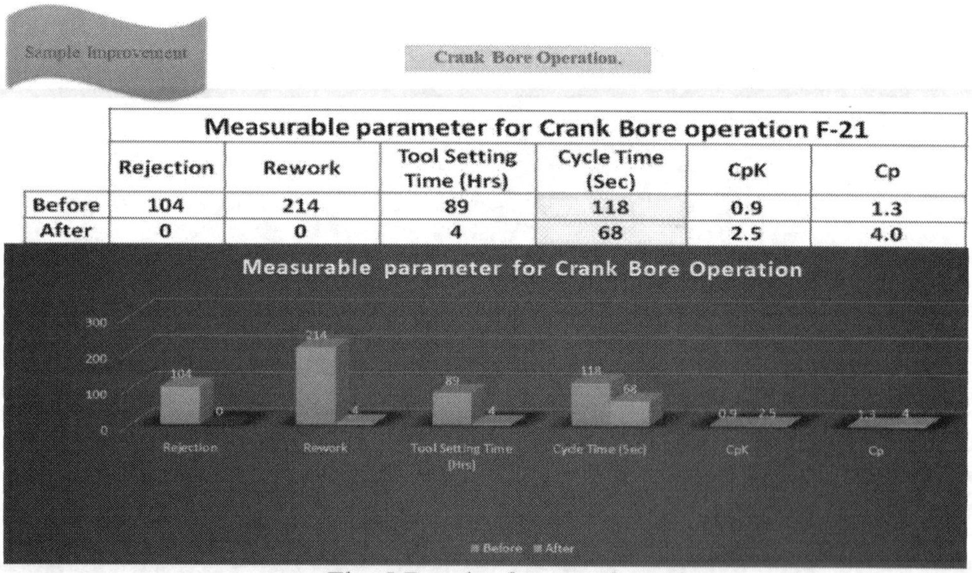

	Rejection	Rework	Tool Setting Time (Hrs)	Cycle Time (Sec)	CpK	Cp
Before	104	214	89	118	0.9	1.3
After	0	0	4	68	2.5	4.0

Fig. 5 Results for sample tool

IOP Conf. Series: Materials Science and Engineering 1259 (2022) 012043 doi:10.1088/1757-899X/1259/1/012043

Before	Bore -1	Bore -2	Bore -3	Bore -4	Bore -5
Y	103.9580	103.9514	103.9611	103.9513	103.9501
X	88.87370	88.89170	88.90320	88.86210	88.87260

After	Bore -1	Bore -2	Bore -3	Bore -4	Bore -5
Y	104.0220	104.0170	104.0264	104.0214	104.0304
X	88.89040	88.91600	88.91080	88.89010	88.91800

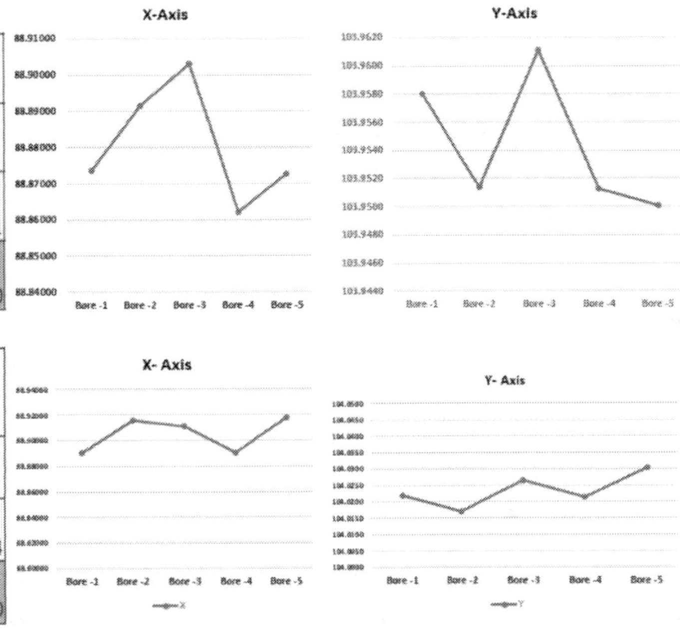

Fig. 6 Graphical representation of old and new tool machining results

6. Conclusions

In this study by mistreatment lean VSM tools and techniques utilized in case company (M&M Ltd). Consequently, the matter known during this study and study the prevailing layout and method mapping shows that there's an excessive amount of non-value-added activities, movement, and transportation of labour items between workstations. Also, as per takt time, the road isn't balanced and ends up in massive WIP is formed between processes. Employees rapt from one digital computer to a different station due poor line equalization & in correct line equalization. In order that high WIP inventory in cutting. No specific time for the merchandise that contains a high demand has been set by the corporate. Lack of time-study or work activity techniques ends up in line equalization is extremely poor because of lack of method mapping study. The bottlenecks method particularly on digital computer of casting provides the best cycle time by four.2 min. whereas different workstations square measure awaiting it to complete its target. Supernumerary quality checking points at every stage on the end operations to extend the cycle time. The purchasers complain lately delivery of finished product thus, the assembly cycle time isn't balanced consistent with takt time. Restricted talent of operators ends up in poor flexibility of style-change.

Acknowledgement

The authors would like to thank Mahindra & Mahindra Ltd. tractor division, Nagpur, India for providing their support for the conduction of experimentation work.

References

[1] Morales Mendez JD, Rodriguez RS. "Total productive maintenance (TPM) as a tool for improving productivity: a case study of application in bottleneck of an auto-parts machining line" Int J Adv Manuf Technol. 2017;92(1-4):1013-1026. doi:10.1007/s00170-017-0052-4.

[2] Arunagiri P, Gnanavelbabu A. "Identification of high impact lean production tools in automobile industries using weighted average method. Procedia Eng. 2014;97:2072-2080. doi:10.1016/j.proeng.2014.12.450.

[3] Panwar A, Jain R, Rathore APS. Journal of Manufacturing Technology Management Lan Implementation in Indian Process Industries – Some Empirical Evidence Lean Implementation in Idian Process Industries – Some Empirical Evidence. Vol 26.; 2015. http://dx.doi.org/10.1108/JMTM-05-2013

[4] Sundar R, Balaji AN, SatheeshKumar RM. A review on lean manufacturing implementation techniques. Procedia Eng. 2014;97:1875-1885. doi:10.1016/j.proeng.2014.12.341.

[5] Rahani AR, Al-Ashraf M. Production flow analysis through Value Stream Mapping: A lean manufacturing process case study. Procedia Eng. 2012;41 (Iris):1727-1734. doi:10.1016/j.proeng.2012.07.375.

[6] Singh R, Gohil AM, Shah DB, Desai S. Total productive maintenance (TPM) implementation in a machine shop: A case study. Procedia Eng. 2017;51(NUiCONE 2012): 592-599. doi:10.1016/j.proeng.2013.01.084.

[7] Ezugwu E.O, D.A. FadareBonney, R.B. Da Silva, and W.F. Sales. "Modelling the correlation between cutting and process parameters in highspeed machining of Inconel 718 alloy using an artificial neural network., International Journal of Machine Tools & Manufacture, vol. 45 (12-13), pp.1375–1385, Oct 2005.

[8] EzugwuE.O,"Key improvements in the machining of difficult-to-cut aerospace superalloys." International Journal of Machine Tools and Manufacture, vol.45 (12-13), pp. 1353-1367, Oct 2005.

[9] Kalpakjian S., Manufacturing process for engineering material, 3rd ed. Menlo Park, CA: Addison-Wesley Longman, 1997

[10] Lee K.S., Seah K.H.W., Wong Y.S., Lim L.K.S.. "In-process tool-failure detection of a coated grooved tool in turning." Journal of Materials Processing Technology, vol. 89-90, pp. 287-291, May 1999.

[11] Modak J. P., "A Specialized course on Research Methodology in Engineering and Technology", at Indira college of Engineering and Management, Pune (India), during 12th to 14th March, 2010.

[12] Rao, S.S., "Optimization Theory & Applications", Wiley Eastern Ltd., 2nd Ed., 1984.

[13] Sahin Y., "Comparison of tool life between ceramic and cubic boron nitride (CBN) cutting tools when machining hardened steels". journal of materials processing technology (2 0 0 9) 3478–3490, Elsevier.

[14] Upadhye N, Deshmukh SG, Garg S. Lean manufacturing system fr medium size manufacturing enterprises: An Indian case. Int J Manag Sci Eng Manag. 2010;5(5):362-375. doi:10.1080/17509653.2010.10671127.

[15] Rahul A. Jibhakate, Nilesh W. Nirwan, Kishor S. Rambhad, Enhancing the effectiveness of green technology in manufacturing industry, Materials Today: Proceedings,Volume 47, Part 14,2021, Pages 4298-4305, ISSN 2214-7853, https://doi.org/10.1016/j.matpr.2021.04.592.

IOP Conf. Series: Materials Science and Engineering 1259 (2022) 012044 doi:10.1088/1757-899X/1259/1/012044

An overview of recent advancement in grade cladded surface using overlay techniques and various materials

Sudarshan Butaley [1], Lalit P. Dhole[2] and Vivek W. Khond [3]

[1]Research Scholar, Mechanical Engineering Department, Government College of Engineering, Chandrapur, India
[2] Mechanical Engineering Department, Government College of Engineering, Chandrapur, India
[3] Mechanical Engineering Department, G H Raisoni College of Engineering, Nagpur, India
vivek.khond@raisoni.net

Abstract. In this review paper, previous studies on effect of cladding on surface of welding using overlay was discussed. The Research papers up to 2021 related with various cladding methods, effect on mechanical properties and overlay methods were discussed. Also, paper discussed the effect on tribological behavior. This paper reported the various metal and its alloy, recent development in advanced materials for cladding in various high temperature applications. The merits, demerits, limitations and future development and possible area of research in welding cladding methods were discussed.

1. Introduction

The process of deposition of thick layer of anti-corrosion material on steel surface for improvement in material properties like corrosion resistance, strength etc is known as cladding. The processes which used to produced cladding are roll bonding, centric cast, welding, explosive cladding, [1-2].The overlay cladding materials are used in many application like pressure vessels in nuclear reactor, gas and steam turbine, combustion chambers, ovens, welding etc. Cladding produce during arc welding process are porn to cracking due to uncontrolled heat input [3-5]. In a marine application, for pressure vessel repairing, SMAW (Shielded Metal Arc Welding) was selected for producing coating to prevent severe localized corrosion [6]. Gas tungsten arc welding (GTAW) process was used in cladding of steel to improve the strength and reduction of residual stresses for gas refinery [7]. Sedighi et al [8] investigated cladding using TIG (tungsten inert gas arc welding) technique, followed by PWHT at temperatures varying from 650-850°C. The results reveals that no crack was observed thought the specimen. The various welding methods are studied for cladding with alloy as cladding materials. The other useful method for cladding are Laser-assisted wire cladding, Electron beam welding powder plasma transferred arc welding, etc. The objectives of paper is to study various methods of cladding and its effects on various properties.

2. Effect of Cladding parameters

Ebrahimnia M, Xie Y, and Ch C. investigated [9] direct laser metal deposition method and examined the effect of intensity of laser, powder feed, and scanning speed. They have observed cracks in multitrack clad samples. Due to insufficient power, trapped are accumulated inside the molten pool. Such cracks are solidification cracks. Due to high tensile stress, another cracks which is occurs at the heat affected zones. The reason are oxide and sulfide presence called as reheating cracks. The highest power intensity, medium scanning speed and proper feed rate (PFR) of powder can avoid cracks. Qimeng Wang and Liying Yang [17] tested the effect of power intensity on cladding properties. The various parameters like Shield gas flow (12LPM), PFR (7LPM), Laser power intensity (LPI) 1000-1300 W and scan speed (SS) 400-600 LPM were selected for experimental testing. The excessive LPI causes burning of laser powder and results in yellow surface. The more effective parameter is SS, when it increased, cracks increases at constant LPI. Reason may be different melting speed of the molten pool and the cooling rate of the surface which causes the different stress at different at portion of sample, as a results of formation of cracks.

3. Various alloy Material used for cladding

Yongjian Li et al [10] investigated nodular CI due to poor weld ability and observed behaviour of cold structure during cladding process at interface region. Zhang et al [11] examined effect of niobium metal using plasma cladding on the properties like microstructure, wear resistance. He observed that Coatings layer was pores free and crack free with good strength and bonding.. The SS and PFR affects the coating thickness around 2–2.5 mm Similarly Zhang et al.[12] examined NiCrBSiW (C88) alloy coating over 316 L SS using plasma arc welding, They reported that erosion and corrosion properties using the C75 and C88 coating have increased about 7 times and 30 times than of 316 L SS. They obtained steel good combined hardness and toughness interface. Similarly, Most of studies indicates that the layers of NbC, TiC and VC were pasted on a carbon steel. He reported that the W–VC coating have the good tensile strength and toughness. Also, In W–TiC and the W–NbC coating reported same behaviour. FeCrC composite coatings on AISI 5115 was investigated steel microstructure and wear resistance properties [13-14]. They reported increased in hardness about 5times and wear resistance about 2.5times. Kılıçay et al [15] examined TiC composite coatings on 38MnVS6 for tribological behaviour in micro alloyed steel continuous with cryogenic cooling. They obtained decreased coefficient of friction (COF) and wear rate. COF was about to be 0.373, However, TiC composite layers, varying between 0.236 and 0.33. Siva et al. [16] examined the effect of PTAW Process Parameters of Nickel coating on Corrosion behaviour. They reported decreased Pitting phenomenon when welding current increased Also, Other parameter like oscillation width indicated less pitting corrosion behaviour [18-19].

4. Conclusive Summary of literatures

Literature interrelated to welding of dissimilar base metals is ample, but the comparative detailed study of dissimilar welding behavior with conventional welding and plasma cladding is scanty. Hence, the current attempt is made to correlate the behavior of joining of two dissimilar materials by above two methods which provides better understanding of metallurgical, mechanical and Tribological properties under operated service conditions which is already used in industry. It was observed that plasma cladding shows most promising technique due to its high performance efficiency, lower thermal input. This method can be widely used in wider material application.

References

[1] Senthilkumar B. and Kannan T. 2015 Effect of flux cored arc welding process parameters on bead geometry in super duplex stainless steel cladding. *Measurement*, 62, 127–136.

[2] Fu Q. and Wagner T. 2007 Interaction of nanostructured metal over layers with oxide surfaces. *Surface Science Reports,* 62, 431–498.

[3] Cao X. Dinga X. Lu Y. Zhub P. and Shoji T. 2015 Influences of Cr content and PWHT on microstructure and oxidation behavior of stainless steel weld overlay cladding materials in high temperature water. *Journal of Nuclear Materials*, 467, 32-41.

[4] Georgiopoulos I. Vourdas N. Mirza S. Andreouli C.and Stathopoulos V. 2018 LaAlO3 as over layer in conventional thermal barrier coatings. *Procidia structural Integrity*, 10, 280–287.

[5] Wahlmann B. Körner C.and Nunn M. 2021 Electron beam wire cladding of nickel alloys and stainless steel on a reactor pressure vessel steel. *Materials Science and Engineering*, 811, ,141082.

[6] Fabio P. Araujo D,Fernando B. Mainier, Brígida B. and Almeida D, 2021 Evaluation of Ni-Cr-mo alloy applied by weld overlay cladding on carbon steel for use in NaCl 3.5% mass solution, Proceedings on Engineering Sciences, 03, 3 355-364.

[7] Britto G, Valarmathi T, John A. Murugesan G. and Prabhakaran R. 2021 Characteristics study of post weld heat treatment in SA 387 grade 22 steel by cladding using gas tungsten arc welding process. *Materials Today: Proceedings*, 44, 5, 3798-3802.

[8] Sedighi, M, Shajari, Y. and Razavi, S. 2021 The Effect of Post Weld Heat Treatment (PWHT) on the Microstructure, Microhardness and Sulphide Stress Corrosion Cracking (SSCC) of Ni-Base Superalloy IN625 Hot Wire TIG Cladding on AISI 4130 Steel. *Prot Met Phys Chem Surf* **57**, 113–120.

[9] Ebrahimnia M. Xie Y., and Ch C. 2021 Effect of cladding parameters on microstructure and defects in direct laser metal deposition of 24CrNiMo steel.*Journal of Laser Applications*, 33, 012007.

[10] Li Y. Dong S. Liu X., He P. Ren X. Yan S. and Xu B, 2021 Interface phase evolution during laser cladding of Ni-Cu alloy on nodular cast iron by powder pre-placed metho.*Optics & Laser Technology*.135, 106684.

[11] Zhang L. Sun D. and Yu H. 2008 Effect of niobium on the microstructure and wear resistance of iron-based alloy coating produced by plasma cladding. *Materials Science and Engineering A*, 490 57–61.

[12] Zhang S. Wang S. Wu C. Zhang C. Guan, M. and Tan J. 2017 Cavitation erosion and erosion-corrosion resistance of austenitic stainless steel by plasma transferred arc welding. *Engineering Failure analysis* 76 , 115-124.

[13] Hou Q. Luo L. Huang Z. , Wang P. Ding T. and Wu Y. 2015 Comparison of three kinds of MC-type carbide modified thick W coatings fabricated by plasma transferred arc surfacing. *Surface & Coatings Technology*. 283 ,52–60.

[14] Ulutan M. Kilicay K. Celik O. and Er U. 2016 Microstructure and wear behaviour of plasma transferred arc(PTA)-deposited FeCrC composite coatings on AISI 5115 steel. *Journal of Materials Processing Technology* 236, 26–34.

[15] Kılıçay K.and Ulutan M. 2017 Effect of cryogenic treatment on tribological behaviour of TiC composite coatings. *Surface Engineering*.

[16] Siva K. and Murugan N. 2013 A study on the Influence of PTAW Process Parameters on Pitting Corrosion Resistance of Nickel Based Overlays. *Procedia Engineering*, 64, 1147 – 1156.

[17] Wang O. and Yang L. 2021 Effect of Laser Cladding Process Parameters on Microstructure of Titanium-aluminum Composite Coating, *IOP Conf. Series: Earth and Environmental Science* 714.

[18] Li X. Zhang C. Zhang S. Wu C. Liu Y. Zhang J. Shahzad M. 2019 Manufacturing of $Ti3SiC_2$ lubricated Co-based alloy coatings using laser cladding technology,*Optics & Laser Technology*,Volume 114, 209-215.

[19] Zuo P. Mingwei C. Liang J. Liu C., Wang M. 2019 Thermal behavior and grain evolution of 24CrNiMoY alloy steel prepared by pre-laid laser cladding technology,*Optics & Laser Technology*,119,105613.

Study on Polyether-ether-ketone (PEEK) dental implant, future alternatives to Titanium

Archi Parekh [1], **Hitakshi Doifode** [1], **Prasad Patil** [1] **Ritik Katole** [1],
Prof. Amol Pitale [2], **Dr. Nilesh Awate** [2]
Dr. Praneeta Priya [3]

[1]B.E at G.H. Raisoni College of Engineering Nagpur, India
[2]Assistant Professor at G.H. Raisoni College of Engineering Nagpur, India
[3]Advance Dental Hospital Nagpur, India

parekh_archi.me@ghrce.raisoni.net, doifode_hitakshi.me@ghrce.raisoni.net, patil_prasad.me@ghrce.raisoni.net, katole_ritik.me@ghrce.raisoni.net, amol.pitale@raisoni.net, nilesh.awate@raisoni.net, praneeta.52@gmail.com

Abstract: Dental implant can be defined as the process of inserting an artificial tooth in the place of the natural tooth. This artificial tooth consists of a) an abutment b) Custom made crown and (c) Implant. Dental implants are of two major types Endosteal and Subperiosteal Implant, endosteal is more widely used and nowadays Subperiosteal is very rare. Apart from types, it also has variety of materials usage. The widely used materials are titanium and zirconia. Both titanium and zirconium have lots of disadvantages, to overcome those we can use PEEK that is polyether-ether-ketone. PEEK has many advantages like Flexibility, Radiolucency, Aesthetic features, Thermal resistance, Cost-Efficient and its Biocompatibility. One of main factor responsible for selection of dental implant material is Osseointegration. This all advantages make us see PEEK as a reliable alternative to the present materials that is titanium and zirconia. PEEK also has long life and it is insoluble having low moisture absorption, low density (1.32 g/cm3), and low elastic modulus. This study shows mechanical analysis of PEEK abutments.

Keywords: Endosteal, Subperiosteal, Radiolucency, Biocompatibility, Osseointegration

1. Introduction

Nowadays, dental implants are used to provide support for the replacement of missing teeth. The use of gold or its alloys, with and without resin veneering, has been discarded for esthetic, economic, and

IOP Conf. Series: Materials Science and Engineering 1259 (2022) 012045 doi:10.1088/1757-899X/1259/1/012045

functional reasons. The choice of the material used for implant-supported prosthesis manufacturing plays a very important role as it is a crucial issue due to the dynamic characteristics of the stomatognathic system. Endosseous dental implants are currently used, which are inserted into the bone are of screw-shaped, and used to replace the tooth root. Dentist implant can be used to replace one tooth, multiple teeth, replace all teeth or to support removable dentures. These implants are majorly made of titanium and zirconia material. According to studies we have found that PEEK is a material which can be used a reliable alternative to titanium and zirconia as it is biocompatible and has many advantages. [1]–[3]

2. Types of Dental Implant

2.1 Endosteal Implant: The most commonly used type of implant is endosteal implant. They have screw types, cylinder types, or bladed types. Endosteal is the most stable type of implant; they are close to natural teeth. An endosteal implant is fixed by screwing the implant into the jawbone. If the person has a narrow, short, or worn-out jawbone then the endosteal implant may not be properly used.[4], [5]

2.2 Subperiosteal Implant: Subperiosteal is used in very rare cases. Subperiosteal implants are not asstable as endosteal implants; this is because the implant merely stays on the soft tissues whereas endosteal implants are skewed into jawbones. However, a subperiosteal implant has more stability than that of dentures without an implant. A subperiosteal implant requires two or more appointments forthe treatment which is shorter than the treatment required for an endosteal implant. [6], [7]

3. Types of Dental Implant materials

There are two types of most commonly used dental implants material. 1. Titanium metal. 2. Zirconia metal.

3.1 Titanium Dental Implant Material: As per studies titanium metal was one of the very first materials used for dental implants. Titanium dental implant is also the most commonly used materialsimplant. The material was discovered as an implant in the year 1950 and since then it has been used. During the discovery of titanium as an implant material, some of the discoveries showed that bone could be grown when the implant made up of titanium is used. The material is also cost-effective andlong-lasting. [8]

3.2 Zirconia Dental Implant Material: Zirconia metal as a dental implant is a recent discovery. It was discovered in the early 1990s but was not used until 2003. However, in the year 2003, corol tooth released the material-based all-in-one crown. It is also believed that zirconia metal is the future of dental science. [9]

4. Problems faced by implants currently in use

PEEK can be a solution for problems faced by Titanium and zirconia. The problems are as follows:
- Autoimmune illness, due to which it has resistance
- Galvanic toxicity in titanium implants
- There are many Allergic reactions to titanium.
- Low temperature degradation in case of zirconia before the product reaches it age
- Zirconia implant provokes inflammatory reactions. [10][11], [12]

5. PEEK as an alternative

PEEK is a semi-crystalline organic thermoplastic polymer. The outstanding properties of PEEK, such as, high durability, low moisture absorption, bone-like modulus, load bearing, thermal resistance, flexibility, toughness etc. making it a popular and standard biomaterial.

In a bone, a homogeneous distribution of stress is necessary to ensure the maintenance of implant and bone combination for a long period. By using PEEK, we can solve a variety of stress-related problems. It can be used to make abutments and prosthetics because of its high mechanical propertiesas compared to metal-supported ceramic. PEEK is more inert compared to other materials. PEEK canbe made biocompatible by coating its surface with various materials, which can help in the process ofOsseointegration and hence can be used in dentistry. It eliminates the risk of corrosion when inserted into the human mouth. PEEK material alleviates the stress generated on the teeth due to its elasticity during chewing.
PEEK has low solubility when in contact with water and low reactivity which makes it unreacted when in contact with other substances, it is also useful for patients with metal allergy or patients whohave susceptibility to metallic taste. [13]

5.1 Chemical formulae of PEEK

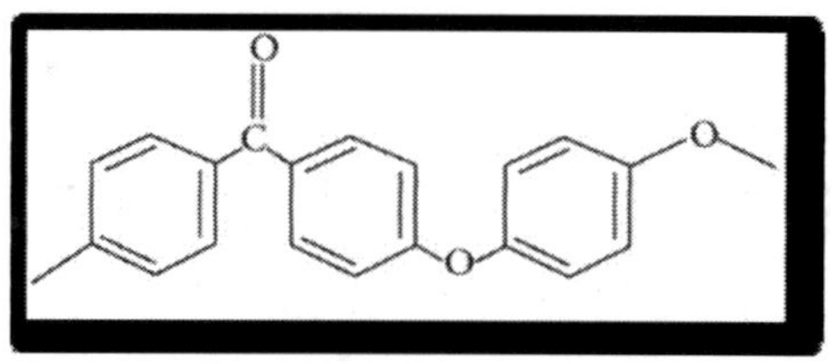

Fig.1 Chemical Formula

5.2 Biological Properties of PEEK

PEEK is a biologically inert material. As per the studies, there is no proof of the effects of the material on mutagenicity, carcinogenicity, cytotoxicity and immune system. it shows resistant to hydrolysis and shows good thermal, mechanical resistance. Resistant to chemical abrasions is seen. During various sterilization processes it is found that it is a degradation-resistant material. PEEK material has a low elastic modulus like the bone and its properties are also similar to that of cortical bone. PEEK material allows Magnetic Resonance Imaging (MRI). They do not distort images obtained by an MRI scanner. [14], [15]

5.3 Advantages of using PEEK

- Flexibility is a key advantage in PEEK Material
- Radiolucency is good in PEEK making it opaque to radio waves
- Aesthetic features of this material make it show little changes
- Thermal resistance is better in PEEK
- It is Cost-Efficient[15]

5.4 Properties of PEEK profitable to Dentistry

- Long life makes it profitable in dentistry
- Insoluble
- low density (1.32 g/cm3)

IOP Conf. Series: Materials Science and Engineering 1259 (2022) 012045 doi:10.1088/1757-899X/1259/1/012045

- low elastic modulus
- low moisture absorption[15]

6. The load-carrying capacity of teeth

Gnathodynamometer is used to determine how much load a tooth can carry. Methods to measure the maximum load capacity of an individual pair of teeth are of many types. These types are made of mechanical and electric measuring instruments. The electrical and mechanical instrument shows identical results. According to the survey conducted on thousands of human teeth, the load carrying capacity of teeth was determined. A variety of different instruments were used to conduct a gnathody-nanometric test. [16]

The load-carrying capacity varies by many factors, from gender to gender as well as person to person. A study was conducted to determine the average strength by Dr. G. E. Black, Chicago DentalUniversity, where thousands of people's jaws were tested. The results were as follows:
In males, the results were 20 kgf for Incisors and 50 kgf for molars.
And in females, the results were 10 kgf for Incisors and 40 kgf for molars.The conversion is 1 kg-f = 9.8 Newton and 1 Pa = 1 N/m^2 [17]

7. CAD model of abutments

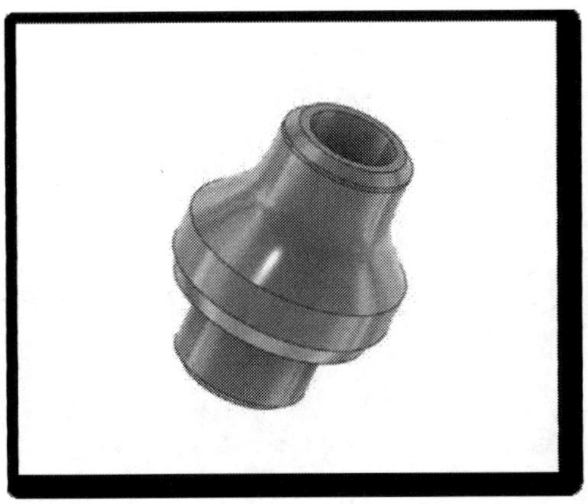

Fig.2 CAD model (3D Projection View)

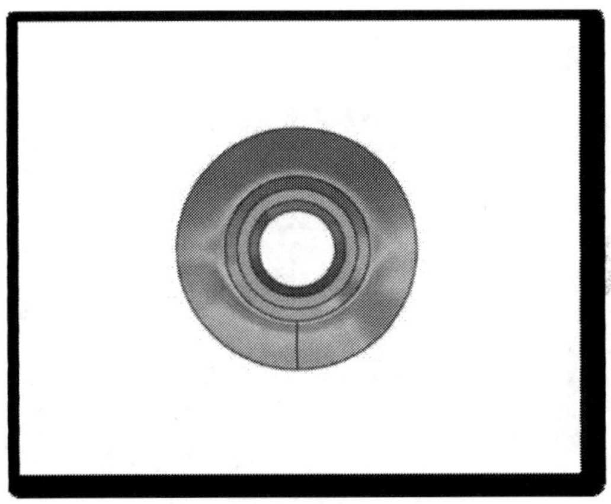

Fig.3 CAD model (Top View)

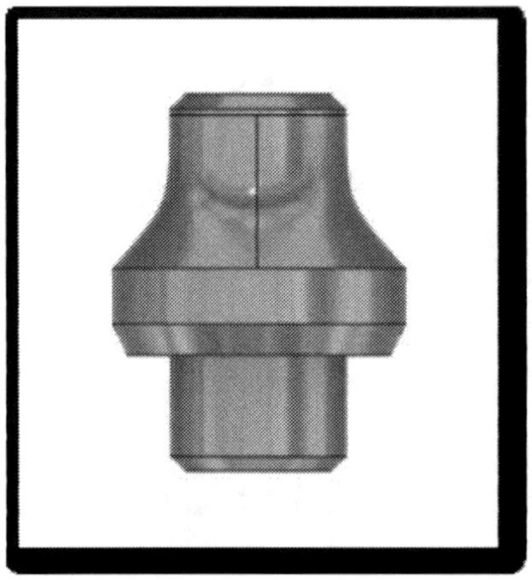

Fig.4 CAD model (Front View)

8. Analysis of PEEK abutment

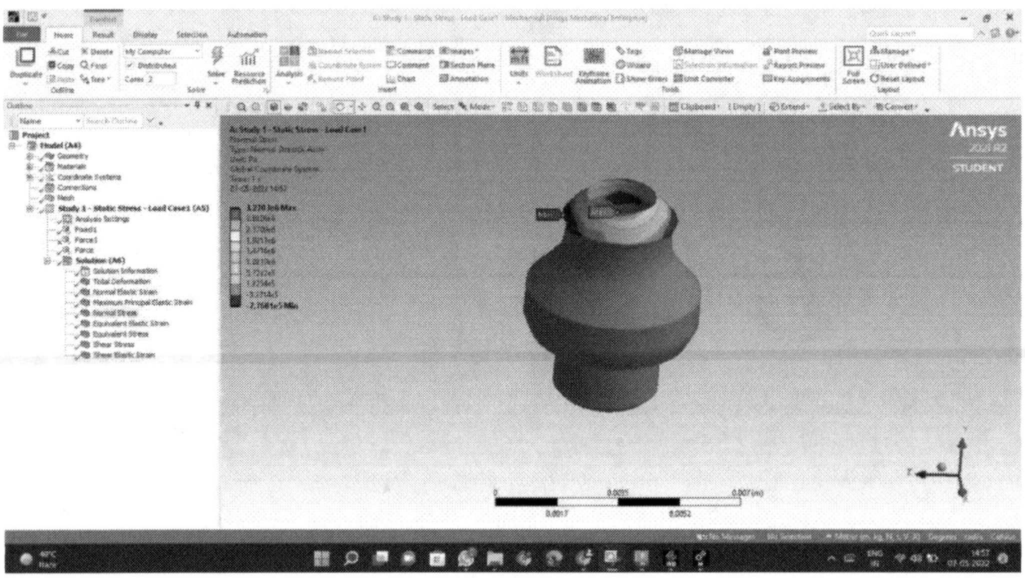

Fig.5 Normal Stress

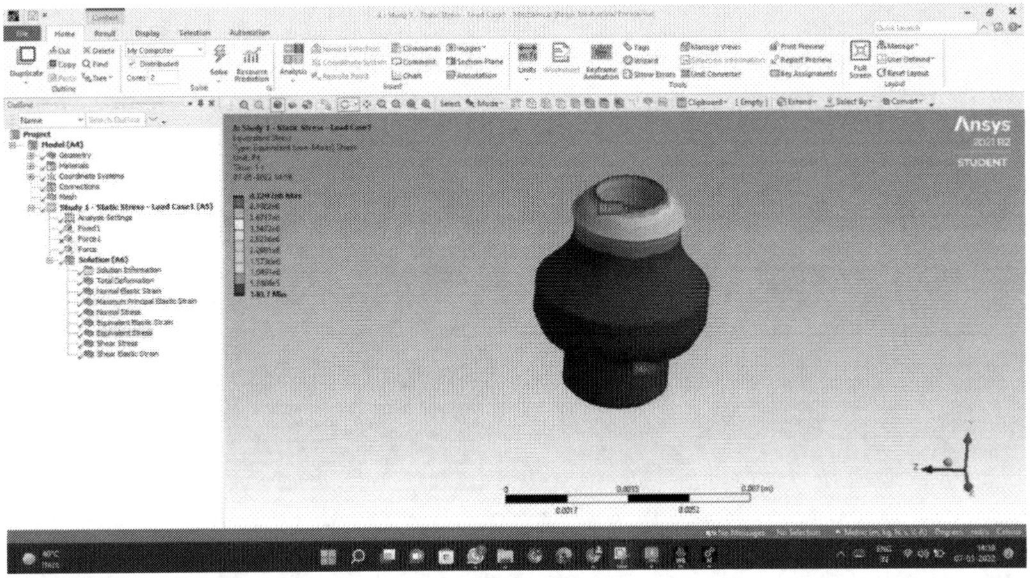

Fig.6 Equivalent Stress

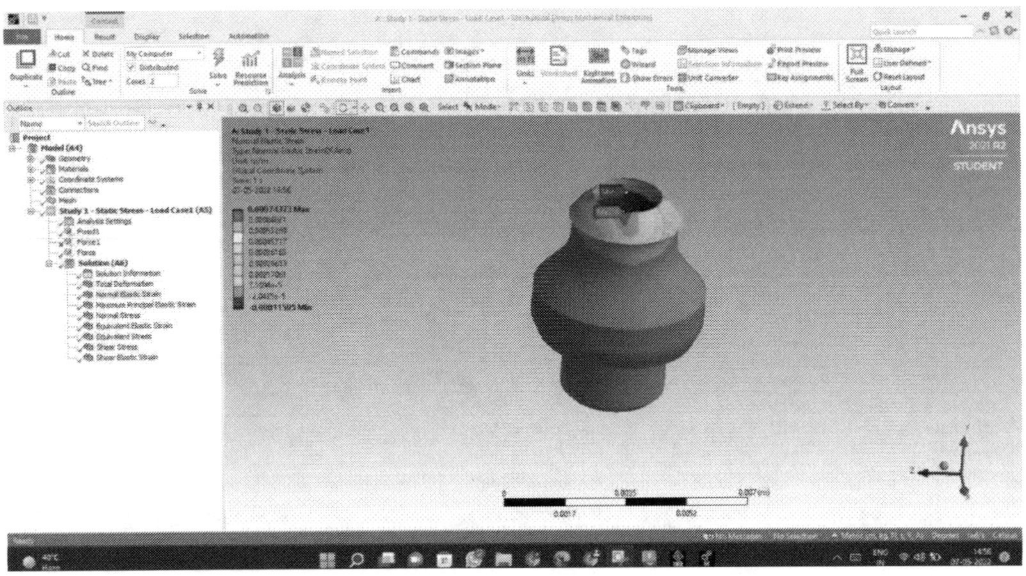

Fig.7 Normal Strain

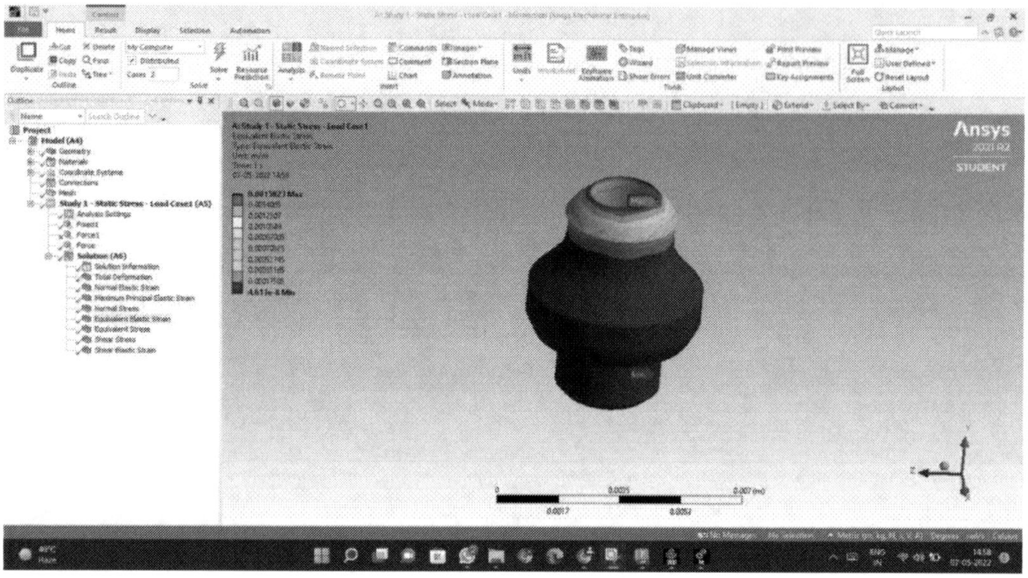

Fig.8 Equivalent Strain

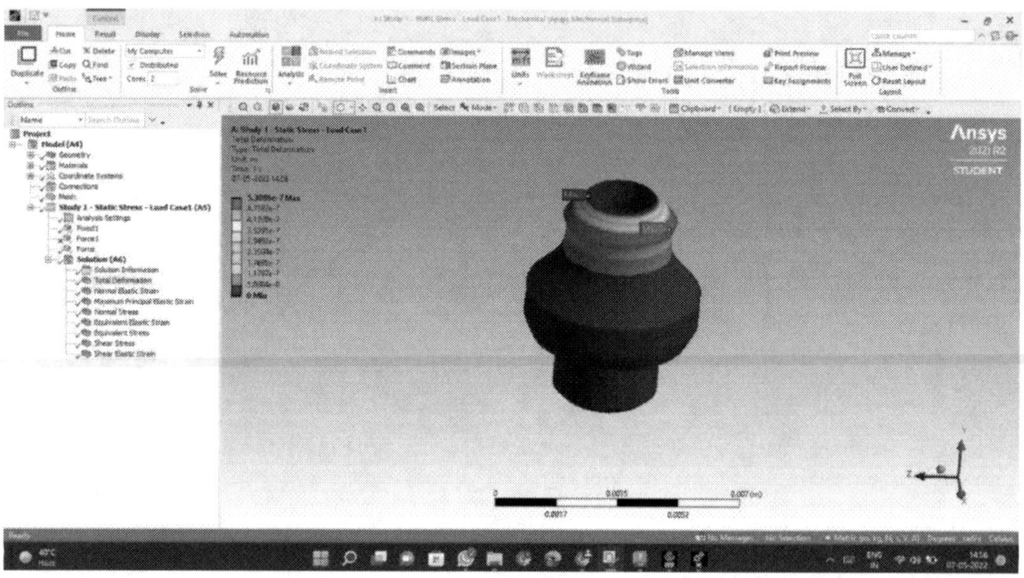

Fig.9 Total Deformation

Test Type	*Total Deformation*	*Normal Elastic Strain*	*Principal ElasticStrain*	*Normal Stress*	*Equivalent ElasticStrain*	*Equivalent Stress*	*Shear Stress*	*Elastic Strain*
Minimum	0 m	-1.1595×10^{-4} m/m	4.3677×10^{-8} m/m	-7.7681×10^{5}Pa	4.617×10^{-8}m/m	140.7 Pa	-2.5374×10^{6}Pa	-2.1852×10^{-3} m/m
Maximum	5.3086×10^{-7} m	7.4373×10^{-4} m/m	1.4033×10^{-3} m/m	3.2703×10^{6} Pa	1.5823×10^{-3}m/m	4.7207×10^{6}Pa	2.2614×10^{6}Pa	1.9475×10^{-3} m/m
Average	9.7136×10^{-8} m	4.2949×10^{-5} m/m	1.1243×10^{-4} m/m	-12864 Pa	1.8382×10^{-4}m/m	5.4803×10^{5}Pa	3616.8 Pa	3.1148×10^{-6} m/m

Table: 1 Ansys Results

9. Comparison of Mechanical properties of different Dental implant materials

Material Name	**Modulus of elasticity (GPa)**	**Poisson's ratio**	**Tensile Strength(MPa)**
PEEK	3.7-4.0 GPa	0.3779	99 MPa
Titanium	116 GPa	0.34	220 MPa
Zirconia	210 GPa	0.34	330 MPa

Table: 2 The Elastic modulus, Poisson's ratios and Tensile Strength of the materials [18]–[20]

10. Conclusion

The stresses on the abutment part were more affected by the change of the abutment material than the change of the crown material. PEEK abutments can help to reduce stress on the abutment compared to other materials. When the crown stresses are studied through various papers, the results stated that by using a PEEK crown, the stress is decreased and the overall lifespan increases. The PEEK material has also encouraged the remodeling procedure of the bone and it has been noted that PEEK material is a suitable choice over titanium in abutment structure. Since titanium has a high elastic modulus, it is the reason chewing loads cannot be absorbed efficiently. PEEK has shown great mechanical properties making it an ideal material to be used in dentistry.

A rigid structure can easily transfer the loads revealed by the implant to the bone structure, it can also cause bone resorption. As the elastic modulus and Tensile strength of the PEEK material and cortical bone is very close, it undervalues the stresses on the bone by absorbing the incoming forces. It has been found that PEEK has the edge of shielding the bone structure by absorbing many of the stresses. By coating PEEK with veneer materials as well as it can also be produced with monolithic materials and aesthetics can be increased by this process. Because PEEK is lighter, it may be a practical choice to metal supported ceramics.

11. References

[1] R. N. Ionescu *et al.*, "Prosthetic Materials Used for Implant-Supported Restorations and Their Biochemical Oral Interactions: A Narrative Review," *Materials (Basel).*, vol. 15, no. 3, 2022, doi:10.3390/ma15031016.

[2] R. Gupta, N. Gupta, and K. K. Weber, "Dental Implants," *Bioceram. Calcium Phosphate*, pp.115–130, Aug. 2021, doi: 10.1201/9781351070133.

[3] L. Gaviria, J. P. Salcido, T. Guda, and J. L. Ong, "Current trends in dental implants," *J. KoreanAssoc. Oral Maxillofac. Surg.*, vol. 40, no. 2, p. 50, 2014, doi: 10.5125/JKAOMS.2014.40.2.50.

[4] S. Hulbert, J. B.-J. of dental research, and undefined 1975, "State of the art in dental implants,"*journals.sagepub.com*, Accessed: Apr. 10, 2022. [Online]. Available: https://journals.sagepub.com/doi/abs/10.1177/00220345750540021001

[5] J. Steigenga, K. Al-Shammari, F. N.-... dentistry, and undefined 2003, "Dental implant design and its relationship to long-term implant success," *journals.lww.com*, Accessed: Apr. 10, 2022. [Online]. Available: https://journals.lww.com/implantdent/fulltext/2003/12040/dental_implant_design_and_its_relati o nship_to.11.aspx

[6] R. Yanase, R. Bodine, J. Tom, S. W.-J. of P. Dentistry, and undefined 1994, "The mandibular subperiosteal implant denture: A prospective survival study," *Elsevier*, Accessed: Apr. 10, 2022.[Online]. Available: https://www.sciencedirect.com/science/article/pii/0022391394900965

[7] L. Y. Jr, J. Michel, D. M.-J. of P. Dentistry, and undefined 1983, "A twenty-year evaluation ofsubperiosteal implants.," *europepmc.org*, Accessed: Apr. 10, 2022. [Online]. Available: https://europepmc.org/article/med/6574238

[8] M. McCracken, "Dental implant materials: Commercially pure titanium and titanium alloys," *J.Prosthodont.*, vol. 8, no. 1, pp. 40–43, 1999, doi: 10.1111/J.1532-849X.1999.TB00006.X.

[9] Z. Özkurt, E. K.-J. of oral implantology, and undefined 2011, "Zirconia dental implants: aliterature review," *meridian.allenpress.com*, doi: 10.1563/AAID-JOI-D-09-00079.

[10] K. Kim, M. Eo, ... T. N.-... of implant, and undefined 2019, "General review of titanium toxicity," *journalimplantdent.springeropen ...*, Accessed: Apr. 10, 2022. [Online]. Available:https://journalimplantdent.springeropen.com/articles/10.1186/s40729-019-0162-x?_ga=2.226012953.2029240304.1572958783-776814952.1572958783

[11] "ZIRCONIA DENTAL IMPLANTS - Pros and Cons." https://blog.ddslab.com/zirconia-dental-implants-pros-and-cons (accessed Apr. 10, 2022).

[12] A. Al-Radha, D. Dymock, C. Younes, D. O.-J. of dentistry, and undefined 2012, "Surfaceproperties of titanium and zirconia dental implant materials and their effect on bacterial adhesion," *Elsevier*, Accessed: Apr. 10, 2022. [Online]. Available: https://www.sciencedirect.com/science/article/pii/S0300571211003125

[13] F. Rahmitasari, Y. Ishida, K. Kurahashi, T. Matsuda, M. Watanabe, and T. Ichikawa, "PEEK withReinforced Materials and Modifications for Dental Implant Applications," *Dent. J.*, vol. 5, no. 4, Dec. 2017, doi: 10.3390/DJ5040035.

[14] G. Skirbutis, A. Dzingutė, V. Masiliūnaitė, G. Šulcaitė, and J. Žilinskas, "A review of PEEKpolymer's properties and its use in prosthodontics," *Stomatologija*, vol. 19, no. 1, pp. 19–23,2017.

[15] "What Is A PEEK Implant?" https://monib-health.com/en/post/15-what-is-a-peek-implant(accessed Apr. 11, 2022).

[16] T. P. Verma, K. I. Kumathalli, V. Jain, and R. Kumar, "Bite Force Recording Devices - AReview," *J. Clin. Diagn. Res.*, vol. 11, no. 9, p. ZE01, Sep. 2017, doi: 10.7860/JCDR/2017/27379.10450.

[17] "[The load carrying capacity of the teeth of the human jaw] - PubMed." https://pubmed.ncbi.nlm.nih.gov/150164/ (accessed Apr. 11, 2022).
[18] "Zirconium - Mechanical Properties And Material Applications." https://www.azom.com/article.aspx?ArticleID=7645 (accessed Apr. 11, 2022).
[19] "Applications that will Benefit from PEEK." https://www.azom.com/article.aspx?ArticleID=17671 (accessed Apr. 11, 2022).
[20] "Titanium (Ti) - The Different Properties and Applications." https://www.azom.com/article.aspx?ArticleID=9118 (accessed Apr. 11, 2022).
[21] P.M. Solanki, D.S. Deshmukh, V.R. Diware, M.S. Deshmukh, Experimental investigation of thermoelectric generator system, Materials Today: Proceedings, Volume 47, Part 11, 2021, Pages 3012-3017, ISSN 2214-7853, https://doi.org/10.1016/j.matpr.2021.05.478.
[22] Sheetal D Deshmukh, Shubham Gabhane and Dheeraj S Deshmukh, published a research paper titled "Osmotic dehydration of carrot strips and modelling" , Published under licence by IOP Publishing Ltd, Journal of Physics: Conference Series, Volume 1913, International Conference on Research Frontiers in Sciences (ICRFS 2021) 5th-6th February 2021, Nagpur, India, doi:10.1088/1742-6596/1913/1/012093, Journal of Physics: Conference Series, 1913 (2021) 012093, Page No. 1-10, June 2021
[23] Mahesh.V. Kulkarni, D.S. Deshmukh, S.P Shekhawat, An innovative design approach of hot water storage tank for solar water heating system using artificial neural network, Materials Today: Proceedings, Volume 46, Part 11, 2021, Pages 5400-5405, ISSN 2214-7853, https://doi.org/10.1016/j.matpr.2020.09.058.
[24] Patil, P.P. and .Deshmukh, Dr.D.S and .Vaidya,, Dr.A.M and Paul, I.D., Experimental Investigation of New Innovative Flat Plate Collector for Solar Water Heater (May 17, 2019). Proceedings of the Second International Conference on Emerging Trends in Science & Technologies For Engineering Systems (ICETSE-2019), SSRN: https://ssrn.com/abstract=3509071 or http://dx.doi.org/10.2139/ssrn.3509071
[25] Pundlik N. Patil, Dheeraj S. Dehmukh and Vilas S. Patil, "Analysis of the effect of LPG on the Performance and Frictional Power Loss for SI Engine", International Journal of Innovative Technology and Exploring Engineering (IJITEE), ISSN: 2278-3075, Volume-8, Issue-9S3, July 2019, Page No. 1406 -1409, Retrieval Number: I33000789S319/2019©BEIESP DOI: 10.35940/ijitee.I3300.0789S319, Published By: Blue Eyes Intelligence Engineering & Sciences Publication.
[26] Shailesh Pimpale, Manish Deshmukh, Rajesh Shelke and Dheeraj Deshmukh, published a research paper titled " Experimental Investigation on Hip Implant Materials Development through Analytical and Finite Element Analysis: 3D Modelled Computed Tomography", Biointerface Research in Applied Chemistry, Platinum Open Access Journal (ISSN: 2069-5837, Volume 12, Issue 3, 2022, 4103 – 4125, online available 15 August. 2021, https://biointerfaceresearch.com/, https://doi.org/10.33263/BRIAC123.41034125
[27] Shailesh S. Pimpale, Manish S. Deshmukh, Rajesh T. Shelke and Dheeraj S. Deshmukh, published a research paper titled "Biomaterial Properties of Femur Implant on Acetabulum Erosion: A Review", Journal of Biomimetics, Biomaterials and Biomedical Engineering, ISSN: 2296- 9845, Volume 51, Page No. 39-62, June 2021, Online available 14 June 2021, @ Trans Tech Publications Ltd., Switzerland, ESCI Indexed, https://doi.org/10.4028/www.scientific.net/jbbbe.51.39
[28] Dr. B. S. Bhaskar, Dr. M. S. Deshmukh, Dr. S.K. Choudhary and Dr. D. S. Deshmukh, published a research paper titled "Heat Transfer Through Porous Materials (Aluminum Foam) Empirical Optimization of a Heat Exchanger", International Journal of Mechanical Engineering, ISSN: 0974-5823, Vol. 7 No. 1, Page No. 47 – 55, January, 2022, Copyrights @Kalahari.

Detailed study on basic methodology of dental implant and surface modification techniques

**Aditya Alagatu [1], Darshan Dhapade[1], Mayuresh Gajbhiye[1]
Ritik Panjrekar [1], Ashish Raut [1], Nilesh Awate[1], Praneeta Priya [2]**

[1] Mechanical Engineering Department, G H Raisoni College of Engineering Nagpur, India-440016.
[2] Advanced Dental Hospital Nagpur, India-440025.

alagatu_aditya.me@ghrce.raisoni.net , ashish.raut@raisoni.net .

Abstract: Dental implant can be defined as the process of inserting an artificial tooth in the place of natural tooth. This artificial tooth consists of: a) abutment b) Custom made crown and c) an implant. Since it is replacing the natural tooth, it should resemble and work in the same manner. It should not break under the chewing load. Dental implant shows great success rate but over the years. They are regarded as a significant contribution to dentistry since they have changed the process of replacing lost teeth. The capacity of the implant material to integrate with the surrounding tissue is critical to its success. Dental implants are artificial dental roots that are used to replace missing teeth. Dental implant is slightly expensive and patient has to go through plenty of processes like surgery. Moreover, even after taking all the preventive measures there are still chances of failing the implant and hence material selection and surface modification techniques used in Dental implant plays a crucial role. In this review paper we will see the different material used in dental implant and different surface modification techniques.

Keywords: Osseointegration, Biocompatibility, fibronectin.

1. Introduction

It has been observed that everyone is having different tooth orientation and some of us faced issue with our internal tooth structure and tooths are very essential from the starting of our life till the end. People who lost their tooth due to any incident for them dental implant is proving to be lifesaving method. Though one can opt for dental bridges, full or partial dentures but dental implant is the most effective way to get artificial tooth. Inserting the implant properly is also one of the most important parameter and study is still going on. Drilling techniques and equipment used in this process plays a crucial role for success rate of dental implant [1–5].

2. Osseointegration

Osseointegration is the direct connection between living bone and surface of implant. Few studies shows that there is an amorphous layer of proteoglycans and collagen between bone and implant surface. While placing an implant, cell adhesion is most important factor. Some adhesive proteins such as fibronectin, vitronectin, fibrinogen, etc. supports the cell adhesion process. it is observed that many cells show different behaviour to different surfaces. Development of the interlocking between bone and implant is complex and depends on several factors. These includes not only implant affiliated factors similar as geomorphology of implant, material of implant, shape of implant and surface chemistry but also surgical styles, patients' health, mechanical loading [6–13].

3. Factors for Successful Osseointegration

1. Biocompatibility of the implant material which we are using.
2. Microscopic and macroscopic nature of implant surface.
3. Implant surface (implant material should be rough to achieve the better bone to implant contact).
4. Surgical techniques.
5. Loading conditions.
6. Implant design characteristics.

4. Implant Material

4.1 Metal

Due to good machinability and showing good biocompatibility titanium is most widely and effective dental implant material. Day by day it is over shadowing other materials in every aspect. It is having the success rate of more than 98%. Titanium and its alloys have an elasticity modulus that is comparable to that of bone. Titanium is considered to be the most widely used implant material owing to its superior corrosion resistance property. Comparing to other implant material like cobalt-chromium alloy, stainless steel, etc titanium is having the excellent biocompatibility[14–20]. For various biomedical purposes 316SS and 316 VM, the alloys of this materials are used. 316L SS consist of Iron, chromium and nickel. Chromium is best for protecting the material from the corrosion and hence it is used to protect the implant from corrosion by the process of passivation. An oxide coating get develops on the metal surface by the chromium in order to prevent corrosion. The property of 316L of having low carbon content do not corrode in the human body. However, when stainless steel is exposed to chlorine ions it can corrode. Stainless-steel implants are used for temporary purpose but it can be used in various biomedical applications[21-22].

4.2 Ceramics

Ceramic is the material used for dental implants as an alternative or substitute material. As titanium is the most common implant material used in the market from past few years. Ceramic implants are designed for the patients who are facing the allergy due to metal implants, having autoimmune disease, have galvanic reaction to metal. The most common ceramic implant used in the dental surgery is "Zirconia implants". Zirconia implants are made from the tooth-coloured ceramic material known as "Zirconium oxide". Zirconia is biocompatible material and supports the osseointegration process by integrating with the jawbone to provide strong base of your replacement tooth. The most common ceramic material used for the coating is plasm- sprayed hydroxyapatite (HA) ceramic. Ceramic implants are being developed to address the aforementioned disadvantages. As a result, zirconia is now being employed as a dental insert material. When compared to metallic components, zirconia emits the fewest particles and is thought to be inactive in the body. This material provides high quality, crack resistance, and biocompatibility[23–25].

4.3 Polymers

The early experiments with methyl methacrylate resin implants were mainly unsuccessful. Hodosh, on the other hand, claimed in 1969 that polymers were biologically acceptable chemicals. Milton Hodosh developed the polymer dental implant concept after conducting research on polymethacrylate tooth-replica implants. The polymer copy proved to be suitable for restoring function and look when replacing a natural tooth [26].

Table 1 Material used for fabrication of dental implants [27]

Implant Material	Common Name
1. Metals: Titanium	C_pTi
Titanium Alloys	Ti-6Al-4 V extra low interstitial (ELI) Ti-6Al-4 V Ti-6Al-7Nb Ti-5Al-2.5Fe Ti-15 Zr-4Nb-2Ta 0.2Pd Ti-29Nb-13Ta-4.6Zr Roxolid (83% -87% Ti- 13% - 17%Zr)
Stainless Steel	SS, 316 LSS
Cobalt Chromium Alloy Vitallium	Co-Cr-Mo
2. Ceramics	
Alumina	Al_2O_3, polycrystalline alumina
Hydroxyapatite	HA, $Ca_{10}(PO_4)_{10,}(OH)_2$
Carbon-Silicon	C- Si
Biogas	$SiO_2/CaO/Na_2O/P_2O_5$

5. Surface Modification Techniques

5.1 Sand Blasting

The main objective of this method is the improve the surface roughness of the dental implants. This roughness increases the surface area of implants resulting in effective osseointegration. Sand blasting is done with the help of alumina Al_2O_3 and titanium dioxide (TiO_2) at the high pressure. The effectiveness of sand blasting depends on following factors: - Number and size of particles and Pressure and speed of rotation to which the implant is fixed. Blasting process leaves some residual particles on the surface of implant that could lead to modification of bone healing process. Some resorbable blast media can also be used like calcium phosphate in the form of hydroxyapatite (HA), beta tricalcium phosphate[28].

5.2 Acid Etching

Strong acids are used in acid etching methods for dental implant surface modifications like Nitric acid (HNO_3), Sulphuric acid (H_2SO_4), Hydrofluoric acid (HF), and their combinations. This method can be used to increase the bonding between the cells, osseointegration and they can also enhance the bone formation. This process treats the surface of the implant with strong acids. After applying this procedure as a result, it produces uniform surface roughness, it also activates more amount of surface area [29–32].

5.3 Anodization

Anodization can be done by keeping the dental implant in the one side which is the called as the anode side of the galvanic cell which consists of phosphoric acid electrolyte. This process treats the surface implants with strong acids like phosphoric acid (H_3PO_4) Sulphuric acid (H_2SO_4), Hydrofluoric acid (HF), Nitric acid (HNO_3) at high current density around (200 A/m^2) or either voltage potential of 100V.

5.4 Grit Blasting
In this method, the machine turned implant are grit-blasted with a titanium dioxide particle with the size of 0.25 to 0.5 μm. This method is used to give the rough surface to the implant material as compared to turned implant surface. Several studies show that in a human study using micro implants, the osseointegration will increase in mandible but not in maxilla. By studying TiO_2 from last 10 years it is observed that the success rate of grit blasted material is 97% which is the higher success rate as compared to machine turned implants and un-blasted implants. There will be more predictable clinical results with the grit blasting technique as compared to other techniques.

6. Conclusion
No doubt there are plenty of advancement has been made in the field of medical science but plenty of parameters still needs to be considered. When it comes to dental implant it is very necessary to take preventive measure so that during the entire process patient will not cause any damage to health. Material selection plays a crucial role. "Osseointegration" is one the key factor for long term stability of dental implant. Titanium and its alloy are considered to be best implant material owing to the properties it possesses. But it shows allergic reaction in few cases. Zirconia is also best option but it is slightly expensive as compared to titanium. Material selection will get guaranteed the success of implant. Surface modification techniques also played a crucial role for long term stability and success of dental implant.

7. References
[1] G. Giro *et al.*, "The effect of simplifying dental implant drilling sequence on osseointegration: An experimental study in dogs," *International Journal of Biomaterials*, vol. 2013, 2013, doi: 10.1155/2013/230310.

[2] R. v. Dahibhate, S. B. Jaju, and R. I. Sarode, "Design Improvements in Conventional Drilling Machine to Control Thermal Necrosis during Orthopaedic Surgeries," in *IOP Conference Series: Materials Science and Engineering*, Dec. 2020, vol. 1004, no. 1. doi: 10.1088/1757-899X/1004/1/012015.

[3] R. v Dahibhate, S. B. Jaju, and R. I. Sarode, "Inventive Methods Used to Study and Control Thermal Necrosis: A Review," *JOURNAL OF CLINICAL AND DIAGNOSTIC RESEARCH*, 2020, doi: 10.7860/jcdr/2020/45110.14052.

[4] R. v. Dahibhate, S. B. Jaju, and R. I. Sarode, "Development of mathematical model for prediction of bone drilling temperature," in *Materials Today: Proceedings*, 2020, vol. 38, pp. 2732–2736. doi: 10.1016/j.matpr.2020.08.537.

[5] R. v. Dahibhate and S. B. Jaju, "Study of influence of tool geometry and temperature on bone substructure to reduce bone drilling injury," in *Journal of Physics: Conference Series*, Jun. 2021, vol. 1913, no. 1. doi: 10.1088/1742-6596/1913/1/012114.

[6] Mihalchenko DV, Sirak SV, Yarigina EN, Khvostov SN, and Zhidovinov AV, "THE ISSUE OF A METHOD OF STIMULATING OSTEOITEGRATSII DENTAL IMPLANTS."

[7] Y. Liu, B. Rath, M. Tingart, and J. Eschweiler, "Role of implants surface modification in osseointegration: A systematic review," *Journal of Biomedical Materials Research - Part A*, vol. 108, no. 3. John Wiley and Sons Inc., pp. 470–484, Mar. 01, 2020. doi: 10.1002/jbm.a.36829.

[8] J. Li, J. A. Jansen, X. F. Walboomers, and J. J. van den Beucken, "Mechanical aspects of dental implants and osseointegration: A narrative review," *Journal of the Mechanical Behavior of Biomedical Materials*, vol. 103. Elsevier Ltd, Mar. 01, 2020. doi: 10.1016/j.jmbbm.2019.103574.

[9] R. S. Jayesh and V. Dhinakarsamy, "Osseointegration," *Journal of Pharmacy and Bioallied Sciences*, vol. 7. Wolters Kluwer Medknow Publications, pp. S226–S229, Apr. 01, 2015. doi: 10.4103/0975-7406.155917.

[10] G. R. M. Matos, "Surface Roughness of Dental Implant and Osseointegration," *Journal of Maxillofacial and Oral Surgery*, vol. 20, no. 1. Springer, Mar. 01, 2021. doi: 10.1007/s12663-020-01437-5.

[11] A. Barfeie, J. Wilson, and J. Rees, "Implant surface characteristics and their effect on osseointegration," *British Dental Journal*, vol. 218, no. 5. Nature Publishing Group, Mar. 13, 2015. doi: 10.1038/sj.bdj.2015.171.

[12] R. Smeets *et al.*, "Impact of Dental Implant Surface Modifications on Osseointegration," *BioMed Research International*, vol. 2016. Hindawi Limited, 2016. doi: 10.1155/2016/6285620.

[13] Z. G. M. Azzawi, T. I. Hamad, S. A. Kadhim, and G. A. H. Naji, "Osseointegration evaluation of laser-deposited titanium dioxide nanoparticles on commercially pure titanium dental implants," *Journal of Materials Science: Materials in Medicine*, vol. 29, no. 7, Jul. 2018, doi: 10.1007/s10856-018-6097-6.

[14] M. T. Mohammed, Z. A. Khan, and A. N. Siddiquee, "Surface Modifications of Titanium Materials for developing Corrosion Behavior in Human Body Environment: A Review," *Procedia Materials Science*, vol. 6, pp. 1610–1618, 2014, doi: 10.1016/j.mspro.2014.07.144.

[15] K. Subramani, R. T. Mathew, and P. Pachauri, "Titanium surface modification techniques for dental implants-From microscale to nanoscale," in *Emerging Nanotechnologies in Dentistry: Second Edition*, Elsevier, 2018, pp. 99–124. doi: 10.1016/B978-0-12-812291-4.00006-6.

[16] M. Herrero-Climent *et al.*, "Relevant design aspects to improve the stability of titanium dental implants," *Materials*, vol. 13, no. 8, Apr. 2020, doi: 10.3390/MA13081910.

[17] L. Damiati *et al.*, "Impact of surface topography and coating on osteogenesis and bacterial attachment on titanium implants," *Journal of Tissue Engineering*, vol. 9, Jul. 2018, doi: 10.1177/2041731418790694.

[18] Y. Doe *et al.*, "Titanium surface treatment by calcium modification with acid-etching promotes osteogenic activity and stability of dental implants," *Materialia (Oxf)*, vol. 12, Aug. 2020, doi: 10.1016/j.mtla.2020.100801.

[19] Y. Shibata and Y. Tanimoto, "A review of improved fixation methods for dental implants. Part I: Surface optimization for rapid osseointegration," *Journal of Prosthodontic Research*, vol. 59, no. 1. Elsevier Ltd, pp. 20–33, Jan. 01, 2015. doi: 10.1016/j.jpor.2014.11.007.

[20] D. Duraccio, F. Mussano, and M. G. Faga, "Biomaterials for dental implants: current and future trends," *Journal of Materials Science*, vol. 50, no. 14. Kluwer Academic Publishers, pp. 4779–4812, Jul. 01, 2015. doi: 10.1007/s10853-015-9056-3.

[21] B. Priyadarshini, M. Rama, Chetan, and U. Vijayalakshmi, "Bioactive coating as a surface modification technique for biocompatible metallic implants: a review," *Journal of Asian Ceramic Societies*, vol. 7, no. 4. Taylor and Francis Ltd., pp. 397–406, Oct. 02, 2019. doi: 10.1080/21870764.2019.1669861.

[22] D. Anna Abraham *et al.*, "Different Surface Modifications of Titanium Implant: A Review," *IOSR Journal of Dental and Medical Sciences (IOSR-JDMS) e-ISSN*, vol. 20, pp. 49–53, 2021, doi: 10.9790/0853-2006014953.

[23] D. Velmurugan, A. Masilamani Santha, and S. Gaurishankar Sarate, "DENTAL IMPLANT MATERIALS, IMPLANT DESIGN, AND ROLE OF FEA- A BRIEF REVIEW," *Journal of Evolution of Medical and Dental Sciences*, vol. 6, no. 44, pp. 3487–3492, Jun. 2017, doi: 10.14260/jemds/2017/753.

[24] A. Siddiqi, A. S. Khan, and S. Zafar, "Thirty years of translational research in zirconia dental implants: A systematic review of the literature," *Journal of Oral Implantology*, vol. 43, no. 4. Allen Press Inc., pp. 314–326, Aug. 01, 2017. doi: 10.1563/aaid-joi-D-17-00016.

[25] S. Gupta, "A Recent Updates on Zirconia Implants: A Literature Review," 2016. [Online]. Available: https://www.researchgate.net/publication/334376146

[26] M. Saini, "Implant biomaterials: A comprehensive review," *World Journal of Clinical Cases*, vol. 3, no. 1, p. 52, 2015, doi: 10.12998/wjcc.v3.i1.52.

[27] S. Nimbalkar, P. Dhatrak, C. Gherde, and S. Joshi, "A review article on factors affecting bone loss in dental implants," in *Materials Today: Proceedings*, 2020, vol. 43, pp. 970–976. doi: 10.1016/j.matpr.2020.07.428.

[28] A. El-Banna, M. W. Bissa, Z. Khurshid, S. Zohaib, F. Y. I. Asiri, and M. S. Zafar, "Surface modification techniques of dental implants," in *Dental Implants*, Elsevier, 2020, pp. 49–68. doi: 10.1016/b978-0-12-819586-4.00004-4.

[29] Y. Sasikumar, K. Indira, and N. Rajendran, "Surface Modification Methods for Titanium and Its Alloys and Their Corrosion Behavior in Biological Environment: A Review," *Journal of Bio- and Tribo-Corrosion*, vol. 5, no. 2. Springer International Publishing, Jun. 01, 2019. doi: 10.1007/s40735-019-0229-5.

[30] J. Li, P. Zhou, S. Attarilar, and H. Shi, "Innovative surface modification procedures to achieve micro/nano-graded ti-based biomedical alloys and implants," *Coatings*, vol. 11, no. 6. MDPI AG, Jun. 01, 2021. doi: 10.3390/coatings11060647.

[31] G. Asensio, B. Vázquez-Lasa, and L. Rojo, "Achievements in the topographic design of commercial titanium dental implants: Towards anti-peri-implantitis surfaces," *Journal of Clinical Medicine*, vol. 8, no. 11. MDPI, Nov. 01, 2019. doi: 10.3390/jcm8111982.

[32] R. v. Dahibhate and S. B. Jaju, "Effect of irrigation mode, volume, medium and drill on heat generation during implant surgery: A review," *Materials Today: Proceedings*, vol. 50, pp. 1799–1803, Jan. 2022, doi: 10.1016/J.MATPR.2021.09.206.

IOP Conf. Series: Materials Science and Engineering 1259 (2022) 012047 doi:10.1088/1757-899X/1259/1/012047

UTILIZATION OF NANOMATERIALS IN BIOIMAGING AND CANCER THERANOSTICS

Prachi Dabhade[1,5], Dr. Ashutosh Bagde[2], Dr. Zahiruddin Quazi Syed[3], Dr. Punit Fulzele[4]

[1,5] Research Associate, Bio-Innovation Centre, Jawaharlal Nehru Medical College, Datta Meghe Institute of Medical Sciences, Sawangi, Wardha (MS)- 442004
[2] Research Scientist, Bio-Innovation Centre, Jawaharlal Nehru Medical College, Datta Meghe Institute of Medical Sciences, Sawangi, Wardha (MS)- 442004
[3] Head of Department, Bio-Innovation Centre, Jawaharlal Nehru Medical College, Datta Meghe Institute of Medical Sciences, Sawangi, Wardha (MS)- 442004
[4] Associate Professor, Sharad Pawar Dental College, Datta Meghe Institute of Medical Sciences Sawangi, Wardha (MS)- 442004

Abstract. Cancer therapy has evolved in recent times on diagnostic and therapeutic grounds. Nanotechnology is the latest technique used to design nanoparticles having therapeutic applications in diseases like cancer. There are various such nanoparticles types made up of gold, silver, iron, and other compounds. Much recent are the bioinspired and theragnostic nanoparticles which exhibit good biocompatibility and much more effective drug delivery compared to the traditional drug delivery mechanisms. Bioimaging is a visualization process that aids to locate the target cancer cell or tumor-targeted delivered nanoparticles to understand and study the actual status of the disease conditions. In this review, we have discussed these various nanomaterials and its counterparts having applications in cancer.

1. Introduction

Material science in the medical field has an important application in the form of Biomaterials. These biomaterials are involved in the treatment strategies like the tissues and organ replacement and regeneration. They are micro-environmentally sensitive and interacts with damaged or injured tissue regeneration. Nanotechnology has ensured the expansion of tissue engineering by creating new assemblies similar to the bone structure size and shape. Organic and inorganic Nano biomaterials are becoming more popular, particularly in bone tissue engineering [1]. Nanomedicine involves the non-invasive tracking and transplantation of the delivered drug. The different nanoprobes developed are used for bioimaging the tissues and predictive treating ailments ranging from cardiovascular and inflammatory diseases to cancer and further to facilitate multimodal bioimaging and targeted therapy. The significant advantage of developing this technique is the better understanding diverse etiologies enabling diagnosis, finding illness and evaluating therapeutic effects and drug discovery [2]. Nanomaterials in the form of nanoparticles ranges of 1- 1000 nm, small in size, flexible with high surface area-to-volume ratio resulting in an ideal drug delivery system. Nanoparticles could be synthesized from the metals, polysaccharides, proteins. Other biological materials like silk, collagen, soy protein, corn zein, keratin, elastin. Many protein nanoparticles are easy to fabricate and thus reshaped into desirable

IOP Conf. Series: Materials Science and Engineering 1259 (2022) 012047 doi:10.1088/1757-899X/1259/1/012047

size, weight and morphology. Protein nanoparticles are a good substitute to the non- biocompatible and hence has a good applications in the targeted drug delivery and biomedical field [3].

Bioimaging is the visualization process of the biological structures. Many techniques are found to fulfil the clinical and laboratory needs. The imaging probes are designed using metal or organic compounds having applications in labelling target molecules or organs to provide enhanced visibility and with structural and functional details [4]. Fluorescent dyes as imaging probes are which are employed for optical imaging, undergoes photobleaching and the contrast agents used in MRI, made up of Gd3+ chelates exhibits weak contrast effect on account of low magnetic moment. In the above examples, they are found to have short circulation time in vivo thus showing weak target efficiency and lower imaging enhancement. In this context, Nanotechnology has played a vital role in developing imaging probes like inorganic nanoparticles, having unique chemical and physical properties originating from nanoscale dimensions [5]. Examples like magnetic nanoparticles, superparamagnetic iron oxide nanoparticles are found to be strong T2 MRI contrast agents, with high sensitivity which is not that evident in conventional MRI contrast agents. Despite the advantages, inorganic nanoparticles have certain drawbacks in the clinical use like magnetic susceptibility artifacts of the contrast agents, photoinduced tissue damage, toxicity of heavy metal in the nanoparticles. Multimodal and simultaneous imaging and therapy can be enabled by targeting ligands along with certain functional molecules viz. radioisotopes, fluorescent dyes, nanoparticles attached drugs [6]. In comparison to traditional imaging modes, recently developed techniques such as Magnetic Resonance Imaging, Photoacoustic Imaging, Computed Tomography, and a variety of other imaging modes that play an important role in disease early detection and treatment are much more effective in terms of imaging time and noise reduction, as well as disease diagnosis efficacy. The inorganic nanomaterial is used in the designing the novel imaging systems. The nanomaterials used in targeting the diseases can be improved via surface modification or fusing with functional materials for betterment of the targeting property [7]. Fluorescent probes are utilized in fluorescence imaging (FL) by labeling specific cells or molecules to generate fluorescence signal via excitation of external light. It is involved in varied applications owing to less expenses, convenient operation, intrinsic biological safety of the imaging and quantitative sensitivity on the biological imaging grounds [7]. The existing contrast agents used in the imaging technique have the certain drawbacks like weak and short life, difficulty in maintaining high MRI resolution and fast weakening of MRI enhancement signal in the body, side effects due to improper dose. Hence vast research is going on combat this problem and find new ways and preparation technique for developing new effective contrast agents with good relaxation rate and to meet its basic requirement [7]. Recently, nanomaterials can be surface functionalized which improves its biocompatibility and also changes the charge distribution on various nanomaterials surface and also achieves the targeting of the other functional effects. The inorganic nanomaterial is used as the material to depict the imaging application of metal and non-metal based nanomaterials in various diseases[7]. Computed Tomography (CT) considers or scans the specific required part of the human body using X-ray beams, ultrasound, and gamma rays, among other things, with the assistance of a scanning detector. It facilitates fast scanning and clear images to inspect the diseases. The contrasting agents used for CT imaging, ensures clear and accurate images. For ensuring the proper functioning the high density and atomic number is considered. Small molecule CT contrasts agents face the problem of weak targeting and rapid renal excretion in the course of clinical use. As a result, CT imaging contrast agents that are both safe and effective are being developed. [8].

2. Types of nanomaterials

Silicon-Based Nanomaterials: Silicon is a trace element that is necessary for the formation of connective tissue and cartilage in the human body. It has a stable structure and good biocompatibility, so it is widely used in medicine and diagnostics [1]. Nanomaterials like gold nanoclusters are the promising emerging agent used in diagnosis and cancer treatment. As far as its small size is concerned, significant quantization is required to the conduction band to lead to the significant photonic property and fading away the plasmonic response which is seen in the large gold nanoparticle.

Gold based nanoparticles: AuNCs exhibits luminescent properties, peptides, proteins, and other biomolecules used as ligands or structural scaffold improves biocompatibility and stability of AuNCs with retained distinct optical properties. The utilization of AuNCs is thus ensured in fluorescent and multimodal imaging combined with CT and MRI. The active passive tumor targeting strategies develops AuNCs based cancer diagnostics application [9]. In most of the research conducted it is preferred to obtain the stabilized fluorescent AuNCs with enhanced yield of luminescent emission (QY) or pharmacokinetic factors by using capping ligands like thiols, bovine serum albumin, glutathione [9]. The luminescent emission of the AuNCs is an important characteristic of the light emission properties. Few surface modification is done for the enhancing the fluorescence from AuNCs by employing electron rich atoms like Nitrogen, Phosphorous, Oxygen, Sulphur and groups like amines and carboxylic acid [10].

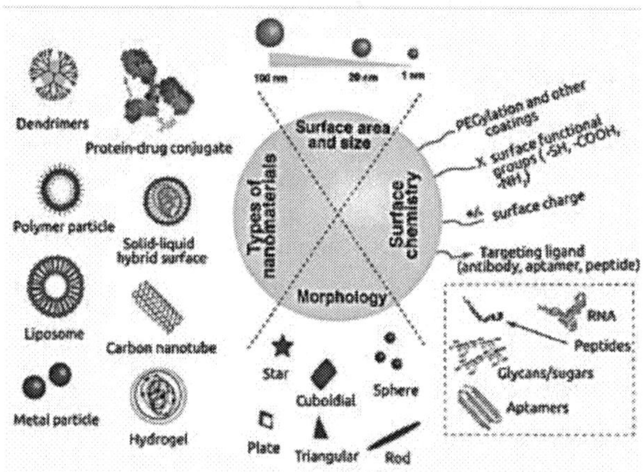

Figure 1. Types of nanomaterials used in cancer therapy with crucial surface chemistry and physical properties for drug delivery [5].

AuNCs are designed using polymers like capping ligands, peptides, proteins, DNA to increase the biocompatibility [8]. Gold nanomaterials are inert in nature compared to other metals and is easily modifiable and turned into nanorods, nanoclusters, nano stars. Also, optical property of Gold nanomaterials possess surface plasmon resonance, characteristics enables it use in cell imaging, drug delivery, detection, and photothermal therapy [11]. Multimodal imaging is a combined multiple imaging technique and thus having more than one application over using the single technique. For the cancer cells imaging, the imaging agent quantity for cancer cells should be high than the other organs. Owing to this, the imaging platform should facilitate the imaging agent to the tumor cells via either active or passive targeting [12].

2.1. Active targeting

Combination of the targeting ligands and nanomaterials (nanoclusters) and providing cellular uptake through receptor mediated endocytosis.

2.2 Passive targeting

Involves enhanced permeability and retention i.e., EPR effect at the body region exhibiting high inflammation and hypoxia or a tumor microenvironment. The extravasation of the AuNCs in the tumor tissue is enabled due to the gaps in endothelium on account of abnormal growth of the tumor vasculature [9].

IOP Conf. Series: Materials Science and Engineering 1259 (2022) 012047 doi:10.1088/1757-899X/1259/1/012047

3. Bioinspired nanomaterials

These are the type of nanomaterials which are easy to use, versatile tool in bioimaging and biosensing, bio catalysis, and biotherapy thus exhibiting utility in biomedical challenges. Chemical modification and biological interaction on the surface can be utilized to construct functional nanomaterials to recognize specific site at the in vitro and in vivo level. This could be highly useful than other biological systems who depends on lock and key approach for molecular interaction [13]. Hence it is evident that surface engineering is an alternative technique for developing specificity of bifunctional nanomaterials. The advantages of the advanced nanomaterials are that it is highly compatible, provides low toxicity, shows specific stimulus responses like light, electricity, magnetism which adds on to its features. Additionally, biological molecules like antibodies, proteins, nucleic acids attached onto the nanomaterials surface is a vital process extending the biosafety of the nanomaterials. Target triggered nanomaterials are also the potential drug delivery system and has achieved the antitumor drugs delivery, ensured decrease in toxicity, and enhanced therapeutic efficacy [14].

Table 1. Examples of Bioinspired nanomaterials

Nanomaterial used	Mechanism used	Targeted type of cancer
Polypyrrole (PPy)-modified Fe3O4 nanoparticles (PPy@ Fe3O4 NPs)	Inhibiting growth and metastasis in lung cancer using combination of photothermal therapy	Non-small cell lung cancer
Glucose oxidase (GOx)-loaded hydrogel	Hydrogel mediated starvation therapy using combination of alginate solution with pH sensitive charge transfer nanoparticles as Near Infrared (NIR) photothermal agents and GOx	Lung cancer, Breast cancer

4. Designing of the nanomaterials for the cancer therapeutics

The drug delivery system at the nanoscale based on liposomes, polymeric micelles, and inorganic nanoparticles has distinct advantages and has been a fascinating topic of study in cancer research. [15]. At the cancer proliferation and survival, this nanomaterial ranging in nanometres sizes involves intracellular interactions with the biomolecules, thus manipulating the target signals of the signal pathway. Nanomaterials with a high surface area to volume ratio transport imaging or therapeutic agents. Polymeric nanoparticles have an average diameter of 70 nm and carry approximately 2000 drug molecules, whereas polymer drug conjugates only carry nine drug molecules. This property of the nanomaterial of high drug loading capacity is much advantageous in the field of cancer medicine [16]. Along with this they exhibit the surface modification with multiple target moieties like peptides, antibodies, and other small molecules. Researchers have proven that the nanocarrier system having number of targeting ligands (also called as Ligand-functionalized nano vehicles) overcomes the limitations of conventional chemotherapeutic agents with the property of enhanced intracellular uptake by target cancer cell through multivalent binding on the cell surface [17]. For Example, Paclitaxel, promoting strong apoptotic effects, a water insoluble anti-cancer drug, is largely used in ovarian, breast and other cancer. But due to its insoluble nature it is difficult to prepare its formulations for anticancer therapy. Encapsulation using nano vehicles like liposomes and polymeric micelles are found to improve the water-soluble properties of these hydrophobic anticancer drugs by adjusting them to the hydrophobic

IOP Conf. Series: Materials Science and Engineering 1259 (2022) 012047 doi:10.1088/1757-899X/1259/1/012047

environment. This procedure results in increased bioavailability and the resulting therapeutic efficacy in the bloodstream [18]. Certain nanomaterials are recently studied which responds to the external signals like light, temperature, pH, tumor environment and thus these nano systems is expected to enhance therapeutic efficacy of anticancer agents providing effective drug delivery in tumor tissues [19]. In another example, spherical structured polymeric micelles are nanoscale colloidal carriers formed by amphipathic copolymer in the aqueous solution. These drug carriers embody hydrophobic anti-cancer drugs like camptothecin and paclitaxel within their hydrophobic core. The cancerous tissues can be targeted by the micelles via EPR effect. The hydrophilic cell layer on the micelles prevents the non-specific uptake mediated by macrophages in the entire body and also prevents enzyme degradation [20].

Apart from the useful desirable properties, it is required to enhance the stability of the polymeric micelles in vivo. The polymeric micelles structure is maintained by noncovalent interactions, thus it tends to dissociate in bloodstream and further excreting out through kidney. The inner core domain or outer shell possess chemical cross linking via various technical approaches which ensures the stability of polymeric micelles. Polyelectrolyte complex micelles is another type of polymeric micelles for carrying out the effective drug delivery of charged therapeutic agents. These are created by facilitating electrostatic interaction in between the anionic macromolecules and di-block copolymers made up of cationic and hydrophilic segments [21]. The Polyelectrolyte complex micelles have spherical core shell architecture where a charge neutralized inner core is generated by oppositely charged molecules and the outer shell is composed of hydrophilic segments facing to the aqueous solution. The intravenous and intracellular bioactive macromolecules delivery like peptides, proteins and nucleic acids is achieved by polymeric micelles and polyelectrolyte complex micelles [22]. Micelles couples with ligands like transferrin, folic acid, antibodies or peptides enforces effective tumor targeting and penetration. Liposomes are another drug delivery vehicle consisting of concentric lipid bilayers with inner aqueous compartment. The hydrophilic therapeutic agents are fetched in the aqueous interior and hydrophobic domain with lipid bilayers are filled with anti-cancer drugs which are water insoluble [23]. Liposomes has relatively high drug loading capacity and are biocompatible and in involved in minimum antigenic, toxic and allergic reactions, since they are usually composed of naturally derived phospholipids. Liposomes are excellent therapeutic and diagnostic agent which carry out intracellular drug delivery to the tumor tissues. However, liposome has few limitations in the in vivo stability since is recognized and cleared off by macrophages in the reticuloendothelial systems owing to it large size (50-400 nm). Compared to the unmodified liposome, the inculsion of polyethylene glycol (PEG) onto the liposome surface which forms a protective layer around the liposome thus promoting stability and prolonged plasma half-life [24].

5. Development of theragnostic nanomaterials

Multifunctional nanomaterials enable to visualize and take therapeutic steps in disease diagnosis and its therapy and is studied extensively in the recent years. The contrast agents like siRNAs, iron oxide nanoparticles, small molecules, etc constitutes the theragnostic nanoparticles. The early detection of the tumors can be achieved by designing the system to localize at the target disease site. Magnetic nanoparticles made of iron oxide, manganese oxide and doped composite nanomaterials are widely studied for cancer imaging on account of their impressive contrast effects by Magnetic resonance Imaging (MRI) and are widely studied as carriers of anticancer drugs drug delivery[25] . However, there are few limitations in case of iron oxide nanoparticles and it can be corrected by altering the surface properties by hydrophilic coating or by designing the shell layer. The chemical modification of iron oxide nanoparticles has extended the scope of employing it for drug delivery purpose. Recent studies have reported designing the multifunctional polymeric iron oxide embedded nanoparticles along with anti-cancer drug e.g., doxorubicin. The polymeric matrix exhibits a reservoir for delivering hydrophobic drug molecules into the cancer cells with high MRI contrast.

IOP Conf. Series: Materials Science and Engineering 1259 (2022) 012047 doi:10.1088/1757-899X/1259/1/012047

5.1. Nano system

siRNA can also be facilitated inside the target cancer cells by adsorbing or conjugating it onto the iron oxide nanoparticles. The MRI of a nano system for siRNA delivery was created by coating Mg-doped iron oxide nanoparticles with Bovine serum albumin and chemically modifying them with siRNA and RGD peptide. This Nano system also ensures highly effective delivery into the cancer cells and shows excellent MR imaging and optical imaging techniques contrast and specific gene silencing efficiency. The trafficking of the particles are studied by the concept of Quantum dots both at in vitro and in animal model and facilitates sharp fluorescence and high photostability [26].

6. Nanoparticles (NPS) role in cancer immunotherapy

Immunotherapy has widened the area of cancer treatment. Nanoparticles play important role in drug delivery in chemotherapy. Cancer Immunotherapy comes into the picture by triggering anti-tumor immune response. Nanoparticle associated immunotherapy involves nano vaccines, artificial antigen presenting cells (aAPCs), and functions in targeting immunosuppressed tumor microenvironment [27]. Nano vaccines deliver tumor-associated antigen and adjuvants to APCs like dendritic cells. The activation of anti-tumor function T cells can be resulted due to enhanced APC antigen presentation and dendritic cells maturation utilizing nanoparticles. NPs like liposomes, PLGA and Gold NPs, dendrimers, micelles have the capacity to enhance immune response against tumor cells. Polymers like acetylated dextran (AcDEX), inorganic NPs like mesoporous silica, have involved in the role as an adjuvant in immunotherapy, thus stimulating the immune response. Artificial APC and MHC-antigen complexes and costimulatory molecules binding to T cell receptors, results into T cell activation. Immunosuppressive Tumor microenvironment is targeted by macrophages, T-cells and Myeloid derived suppressor cells (MDSCs). The concept of chemoimmunotherapy involving co-delivery monoclonal antibodies and chemotherapeutics into porous silicon nanoparticles are prone to trigger immune response against cancer cells by competent activation and antibody dependent cell cytotoxicity (ADCC) including other signalling cascades [28].

7. Recent findings

7.1. Quantum Dots

The quantum dots are the recently used nanomaterials in cancer therapeutics. They are the type of small particles or nano crystals made up of semiconductor materials in the range of 2 and 10 nm. Ratio of height of the surface to volume of the particles ensures QD, an intermediate electron property which lies in between mass semiconductor and discrete atom. QD immunostaining works at the low protein expression level and comparatively accurate than traditional immunochemical techniques. Quantum dots have applications in cancer diagnosis by detecting various tumor biomarkers like cell protein or other components in tumor samples. It aids in drug delivery by gathering at the internal organ and facilitating drugs at the target site. The surface modification of QDs is achieved by combining with the biomolecules, peptides, antibodies, in vivo and is used for targeting tumors and their subsequent utilization in cancer diagnostics. For example, in other cases QD with prostate specific antigen to label cancer and biomarkers are used to speed up the process which has stable intensity of light compared to traditional fluorescent biomarkers. The quantum dots has more stable, concreate and brighter light stability [29].

7.2. Fenton/Fenton like chemistry

Tumor microenvironment is been looked up in the cancer diagnostics and chemo dynamics therapy (CDT). Fenton like chemistry reactions encounters the mild and overexpressed H_2O_2 tumor microenvironment by generating more cytotoxic hydroxyl radicals. Reactive oxygen species (ROS) exhibits targeted and desirable anti-cancer effects without providing external stimulus and drug

IOP Conf. Series: Materials Science and Engineering 1259 (2022) 012047 doi:10.1088/1757-899X/1259/1/012047

resistance and is thus found to be promising strategy in cancer therapy. However, the efforts are being undertaken to fulfil the treatment efficacy of chemo dynamic therapy in TME on account of the limited efficacy of CDT agents. More CDT agents are need to be designed and fabricated by the researchers enriched with excellent catalytic activity and remodeling tumor microenvironment to ensure optimal CDT assisted by Fenton chemistry [30].

7.3. mAb nanoparticles
mAbs having vast immunological applications are used to design novel anti-tumor nanoplatforms, conjugated with cytotoxic drugs i.e., antibody-drug conjugates (ADCs) and it is ensured that specificity and less toxicity is achieved. For example: Trastuzumab (Herceptin) (Tmab) is a mAb having application in treating breast cancer with positive expression of HER2- Human epithelial growth factor receptor. The comparative study of the Tmab alone with Tmab in ADC shows that the later one is more therapeutically effective [13].

7.4. Aptamer functionalized non biological materials
On the advent of biotechnology, physical and chemical nanosized materials have arisen to showcase varied biomedical applications. Researchers using the recent sophisticated synthetic methodologies and techniques has designed and manufactured non-biological materials having customized shape, size and dispersion. Inherent features like large surface area, high loading capacity, easy modification, enhanced permeability, retention effects attract researchers to explore drug delivery and bioimaging. Few materials possess unique physiological properties like magnetism or surface plasma resonance (SPR). The targeting moieties are incorporated onto non biological materials like hydrogels, micelles, liposomes, gold and silica nanoparticles and exhibit the enhanced therapeutic efficiency [31].

7.5. CRISPR/Cas9 system
CRISPR/Cas9 system is a highly specific laboratory genome editing tool with other applications requiring delivery to specific cells. Research have developed aptamer functionalized liposome - CRISPR/Cas9 in counter to Polo like kinase 1 in prostrate tumor cells using post insertion approach. Osteosarcoma cells specific aptamer conjugated with PEG-PEI Cholesterol (PPC) lipopolymer containing plasmid coding for CRISPR/Cas9 and VEGFA gRNA sequences. These showcased gene editing and tumor suppression in cell specific manner. The efficacy of aptamers functionalized CRISPR/Cas9 has been revealed in animal models. Apart from this, many other synthetic delivery platforms are found to be promising and function driven, still, possess potential toxicity and triggers immune response and has got low specificity limiting its use in wide range [31].

8. Overview on nanomaterial approach in cancer treatment
In the cancer treatment various methods are being undertaken moreover, in spite of different working platforms, functional and mechanism diversity between the nanomaterials, most researchers target tumor cells and Tumor microenvironment including immune system related to tumor. Enhanced permeability and retention (EPR) effect and targeting, dendrimers, modified nanocarriers such as NPs, carbon nanomaterials, or drug delivery system (DDS) approach cancer cells and release drugs or biomaterials. The overexpressed antigens on the cancer cells are widely targeted by the antibodies [13].

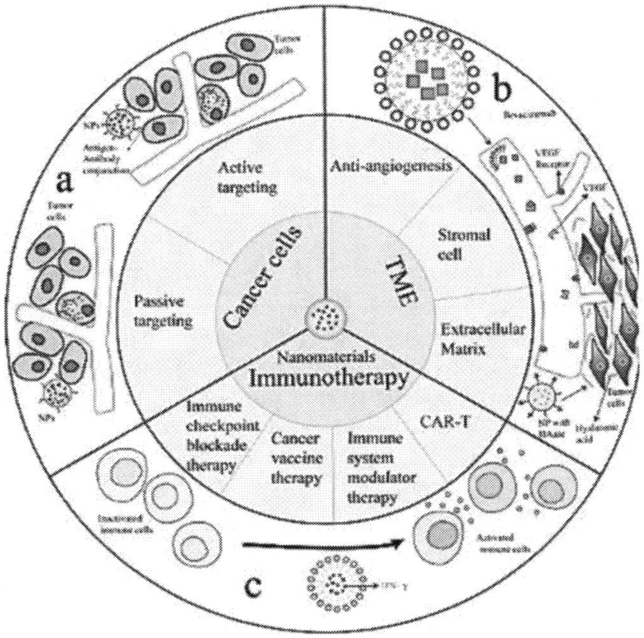

Figure 2. Nanomaterial approach in Cancer Treatment (a) Active or Passive targeting of cancer cells (b) Targeting Tumour microenvironment (c) liposome mediated immune modulation via activated immune cells in cancer immunotherapy [13].

Post endocytosis by cancer cells, encapsulated chemical biomolecules or drugs forces cytotoxicity or cell apoptosis is promoted by the nucleic acid. It is crucial to target the tumour micro environment and the involved uncontrolled angiogenesis by overactive cell proliferation. Extra cellular matrix functions as a scaffold in cancer proliferation, metastasis, invasion, abnormal vasculature and angiogenesis. Collagen is the primary structural protein of ECM. Collagen along with HA and other enzymes facilitates the migration for the tumour cells. It is evident that it contributes to the high interstitial fluid pressure to ensure no drug delivery and penetration. Also, enzymes like Matrix metalloprotein (MMPs) regulates the TME by controlling and manipulating non-ECM molecules, cytokines and other growth factors [13].

9. Conclusion

In the recent advances in drug delivery and therapeutics, nanotechnology has offered an innovative approach to deal with the existing chemotherapeutic agents and has thus gained impetus in the area of cancer diagnostics and therapeutics. For example, Nonvehicle platforms with size around 100-200 nm can efficiently promote endocytic intracellular uptake, specific targeting and high drug delivery. The effective designing of these artificially engineered nanomaterials ensures the possibility of safe and effective cancer treatment. On the same lines, recently nano systems for cancer diagnostics and therapy are developed and is envisioned that the tumors can be provided with the intervention using targeted drug delivery and monitor their response via bioimaging tumor lesions in the body. It is envisioned that the nanotechnology and biological sciences together can carry out significant purpose of cancer theragnostics. The varied nanomaterials used in the cancer cell imaging and therapy has unique characteristics exhibiting high cellular uptake, lengthen circulation after surface modification with efficient approaching towards the targeted tumor sites. Nano particulate formulations depending on the polyelectrolyte complex micelles and inorganic nanoparticles are highly explored for targeted siRNA delivery to execute desired functions. It is further envisioned to develop nano systems that can visualize and simultaneously treat diseases. The developments in nanotechnology and biomedical science would widen the designing of nanomaterials for cancer theragnostic and imaging applications. Nanoparticles used as an alternative to the traditional drugs, it's delivery system is associated with enhanced biocompatibility, pharmacokinetics, tumor targeting, stability, reduction in the systemic toxicity and overcoming drug resistance. The nanoparticles-based drugs have wide application in chemotherapy, radiotherapy, targeted therapy, hyperthermia, gene therapy and combination therapy which aids in overcoming mechanism of drug resistance including overexpression of efflux transporter, hypoxia tumor microenvironment and defective apoptotic pathway. Hybrid nanoparticles are also widely used and has effective drug delivery mechanism in cancer cells. It is quite noted that the interaction between the NPs and immune system are complex and various factors like NP shape, size, composition, surface characteristics are the crucial to consider for the interaction of Nanoparticles with the immune system. It is envisioned that studying the Tumor Microenvironment and exploring the crosstalk between drug delivery based on nanoparticle and tumor immunity are crucial for drug designing and its utilization.

10. Challenges in nano drug delivery

On account of the growing advances in the nanomaterials field with specific properties and latest advances in the drug delivery system it has been anticipated as an effective cancer theragnostic tool. It is envisioned that nanomaterials would tremendously upgrade the health care system depending on the progression of the targeted drug delivery phenomenon in the recent times. The physiochemical properties like biocompatibility, toxicity in the biological system is crucial to consider for efficiency of nanoparticles and its drug delivery mechanism should be away from any toxicity to the healthy cells. The pharmacological performance is influenced by nanomaterials storage and its stability. Nanotoxicology is a branch of medicine and is an essential research field, paving the way to assess the toxicity of the nanoparticles. Due to lack of detailed information and understanding of nano bio interfacial interaction, it can face a setback in commercialization of nanomedicine. A large-scale production of the nanomedicine products for releasing it on the commercial scale is a quite challenging task and hence amongst the effective cancer therapeutics only few nano formulations has made it to the clinical trials and its approval. Regulatory approval of nanomedicines is another key issue since due to lack of FDA specific guidelines for products with the nanomaterials. Individual assessment of nano formulated drugs and its regulatory permission and verdicts are quite time-consuming procedures. In the overall process deciphering the complications entailed in cancer cell and its microenvironment, pharmacokinetics, regulatory guidelines and risks involved in administrating to the patient case study and their prior investigations are all mandatory to overlook.

11. References

[1] Kumar P, Saini M, Dehiya B S, Sindhu A, Kumar V, Kumar R, Lamberti L, Pruncu C I and Thakur R 2020 Comprehensive Survey on Nanobiomaterials for Bone Tissue Engineering Applications *Nanomater. (Basel, Switzerland)* **10** 1–60

[2] Yang Y, Wang L, Wan B, Gu Y and Li X 2019 Optically Active Nanomaterials for Bioimaging and Targeted Therapy *Front. Bioeng. Biotechnol.* **7** 320

[3] De Frates K, Markiewicz T, Gallo P, Rack A, Weyhmiller A, Jarmusik B and Hu X 2018 Protein Polymer-Based Nanoparticles: Fabrication and Medical Applications *Int. J. Mol. Sci.* **19**

[4] Rong P, Huang P, Liu Z, Lin J, Jin A, Ma Y, Niu G, Yu L, Zeng W, Wang W and Chen X 2015 Protein-based photothermal theranostics for imaging-guided cancer therapy *Nanoscale* **7** 16330–6

[5] Navya P N, Kaphle A, Srinivas S P, Bhargava S K, Rotello V M and Daima H K 2019 Current trends and challenges in cancer management and therapy using designer nanomaterials *Nano Converg. 2019 61* **6** 1–30

[6] Kim D, Kim J, Park Y Il, Lee N and Hyeon T 2018 Recent Development of Inorganic Nanoparticles for Biomedical Imaging *ACS Cent. Sci.* **4** 324–36

[7] Du W, Zhou L, Zhang Q, Liu X, Wei X and Li Y 2021 Inorganic Nanomaterial for Biomedical Imaging of Brain Diseases *Molecules* **26**

[8] Harish V, Tewari D, Gaur M, Yadav A B, Swaroop S, Bechelany M and Barhoum A 2022 Review on Nanoparticles and Nanostructured Materials: Bioimaging, Biosensing, Drug Delivery, Tissue Engineering, Antimicrobial, and Agro-Food Applications *Nanomater. (Basel, Switzerland)* **12**

[9] Van de Looij S M, Hebels E R, Viola M, Hembury M, Oliveira S and Vermonden T 2022 Gold Nanoclusters: Imaging, Therapy, and Theranostic Roles in Biomedical Applications *Bioconjug. Chem.* **33** 4–23

[10] Krug P, Kwiatkowska M, Mojzych I, Głowala P, Dorant S, Kępińska D, Chotkowski M, Janiszewska K, Stolarski J, Wiktorska K, Kaczyńska K and Mazur M 2019 Polypyrrole microcapsules loaded with gold nanoparticles: Perspectives for biomedical imaging *Synth. Met.* **248** 27–34

[11] Ricciardi L, Martini M, Tillement O, Sancey L, Perriat P, Ghedini M, Szerb E I, Yadav Y J and La Deda M 2014 Multifunctional material based on ionic transition metal complexes and gold–silica nanoparticles: Synthesis and photophysical characterization for application in imaging and therapy *J. Photochem. Photobiol. B Biol.* **140** 396–404

[12] Wang Y, Li P, Tran T T D, Zhang J and Kong L 2016 Manufacturing Techniques and Surface Engineering of Polymer Based Nanoparticles for Targeted Drug Delivery to Cancer *Nanomater. (Basel, Switzerland)* **6**

[13] Cheng Z, Li M, Dey R and Chen Y 2021 Nanomaterials for cancer therapy: current progress and perspectives *J. Hematol. Oncol. 2021 141* **14** 1–27

[14] Yang Y J and Gao Z F 2022 Editorial: Bio-Inspired Nanomaterials in Surface Engineering and Bioapplications *Front. Chem.* **10**

[15] Kesse S, Boakye-Yiadom K O, Ochete B O, Opoku-Damoah Y, Akhtar F, Filli M S, Farooq M A, Aquib M, Mily B J M, Murtaza G and Wang B 2019 Mesoporous Silica Nanomaterials: Versatile Nanocarriers for Cancer Theranostics and Drug and Gene Delivery *Pharmaceutics* **11**

[16] Xie J, Liu G, Eden H S, Ai H and Chen X 2011 Surface-engineered magnetic nanoparticle platforms for cancer imaging and therapy *Acc. Chem. Res.* **44** 883–92

[17] Lin G, Chen S and Mi P 2018 Nanoparticles Targeting and Remodeling Tumor Microenvironment for Cancer Theranostics *J. Biomed. Nanotechnol.* **14** 1189–207

[18] McCarthy J R and Weissleder R 2008 Multifunctional magnetic nanoparticles for targeted imaging and therapy *Adv. Drug Deliv. Rev.* **60** 1241–51

[19] Duncan R 2003 The dawning era of polymer therapeutics *Nat. Rev. Drug Discov. 2003 25* **2** 347–60

[20] Kataoka K, Harada A and Nagasaki Y 2001 Block copolymer micelles for drug delivery: design, characterization and biological significance *Adv. Drug Deliv. Rev.* **47** 113–31

[21] Lee Y, Lee H, Kim Y B, Kim J, Hyeon T, Park H, Messersmith P B and Park T G 2008 Bioinspired Surface Immobilization of Hyaluronic Acid on Monodisperse Magnetite Nanocrystals for Targeted Cancer Imaging *Adv. Mater.* **20** 4154–7

[22] Park T G, Jeong J H and Kim S W 2006 Current status of polymeric gene delivery systems *Adv. Drug Deliv. Rev.* **58** 467–86

[23] Liu C G, Han Y H, Kankala R K, Wang S Bin and Chen A Z 2020 Subcellular Performance of Nanoparticles in Cancer Therapy *Int. J. Nanomedicine* **15** 675

[24] Klibanov A L, Maruyama K, Beckerleg A M, Torchilin V P and Huang L 1991 Activity of amphipathic poly(ethylene glycol) 5000 to prolong the circulation time of liposomes depends on the liposome size and is unfavorable for immunoliposome binding to target *Biochim. Biophys. Acta* **1062** 142–8

[25] Liang G, Wang H, Shi H, Wang H, Zhu M, Jing A, Li J and Li G 2020 Recent progress in the development of upconversion nanomaterials in bioimaging and disease treatment *J. Nanobiotechnology 2020 181* **18** 1–22

[26] Bae K H, Chung H J and Park T G 2011 Nanomaterials for Cancer Therapy and Imaging *Mol. Cells* **31** 295

[27] Zang X, Zhao X, Hu H, Qiao M, Deng Y and Chen D 2017 Nanoparticles for tumor immunotherapy *Eur. J. Pharm. Biopharm.* **115** 243–56

[28] Yao Y, Zhou Y, Liu L, Xu Y, Chen Q, Wang Y, Wu S, Deng Y, Zhang J and Shao A 2020 Nanoparticle-Based Drug Delivery in Cancer Therapy and Its Role in Overcoming Drug Resistance *Front. Mol. Biosci.* **7** 193

[29] Jin C, Wang K, Oppong-Gyebi A and Hu J 2020 Application of Nanotechnology in Cancer Diagnosis and Therapy - A Mini-Review *Int. J. Med. Sci.* **17** 2964–73

[30] Cao C, Wang X, Yang N, Song X and Dong X 2022 Recent advances of cancer chemodynamic therapy based on Fenton/Fenton-like chemistry *Chem. Sci.* **13** 863–89

[31] Liu M, Wang L, Lo Y, Shiu S C C, Kinghorn A B and Tanner J A 2022 Aptamer-Enabled Nanomaterials for Therapeutics, Drug Targeting and Imaging *Cells* **11**

IOP Conf. Series: Materials Science and Engineering 1259 (2022) 012048 doi:10.1088/1757-899X/1259/1/012048

Experimental Studies on TIG Welding of Advanced Material under various parameters

Safraj Ansari[1], Amrit Anand Dosar[2] and Sunil kumar[2]

[1]Research Scholar, Department of Mechanical Engineering, Babu Banarsi Das University, Lucknow, U.P., India.
[2]Assistant Professor, Department of Mechanical Engineering, Babu Banarsi Das University, Lucknow, U.P., India.

Abstract: The material stainless steel (AISI 309) is extensively used in almost all types of thermal power plant, petroleum industries, waste treatment industries etc. The material AISI 309 stainless steel is used as sample which is prepared by wire cut EDM process. The sample dimension is tested according to the ASTM standard. Tungsten inert gas (TIG) welding is used to improve mechanical properties and to reduce the manufacturing costs, by the taguchi`s experimental design method which is used for optimization. The three different welding process parameters such as GFR, WS and WC have opted to perform the tests using L9 orthogonal array. The tensile strength and bending strength test are conducted to check the final weld quality. The maximization of ultimate tensile strength, impact strength and bending strength of weld joints results has obtained by input parameters such as welding current 170amp, welding speed 2 mm/sec and gas flow rate 15 lit/min. It is observed that current and welding speed influence to maximize mechanical properties of weld joints.

1. Introduction

1.1 Material AISI 309 Stainless Steel

Stainless steel (AISI 309) is extensively used in almost all types of thermal power plant, petroleum industries, waste treatment industries etc. The material is AISI 309 has high strength, weldability, toughness, ductility, excellent oxidization resistance and formability. AISI 309 stainless steel applications includes jet engine parts, evaporators, heat exchangers, furnace parts, automotive exhaust part, chemical processing equipment, tanks, fire box sheet and other high temperature containers. The chemical composition of AISI309 is show in table 1.

Table 1. The chemical constituent of AISI309

Grade	%C	%Si	%Mn	%P	%S	%N	%Cr	%Ni
309	0.067	0.26	1.73	0.036	0.003	-	22.15	12.19

1.2 TIG welding of AISI 309 stainless steel

Tungsten inert gas (TIG) welding method is applied to increase mechanical properties in stainless steel. TIG welding provides better quality, accurate weld and absence of slag, with less heat effect other than welding processes so it can be used widely. TIG welding is abbreviated GTAW (gas tungsten arc welding) which is non consumable tungsten electrode [2-5].
The preparation of a good quality weld is considered to be a complex task. So welding quality can be controlled by input parameters like WC (welding current), WS (welding speed), GFR (gas flow rate), gas pressure, etc. TIG welding process generates an electric arc between a tungsten electrode and base metal. In TIG welding process, inert gas (argon-helium) shield is used to protect the weldpool [6-7]. TIG welding setup show in Figure.3.

2. Literature Review

Shanmugasundar et al., (2019) reviewed that TIG welding processes improve the quality of a welded material of austenitic stainless steel (AISI 304l). Tungsten inert gas (TIG) has been used in this welding process and optimized the outcome on the basis of welding current, gas flow rate and nozzle to work piece distance.

Natrayan et al., (2020) The taguchi approach was used to maximise the quality of the wielded joint using control parameters such as welding current, weld speed, and filler diameter for tensile strength. The experiments were carried out with the help of L9 orthogonal array. A correlation between welding input parameters and tungsten inert gas penetration was discovered using a regression model. This study found that ANOVA determined the exceptional process parameters for tensile strength at level 1 (a1), weld speed at level 3 (b3), filler diameter at level 3 (c3), and the outcomes, revealing that wielding speed is the most important component in improving tensile strength.

Vinoth et al., (2021) focused on Taguchi's method was utilised to optimise input parameters such as welding speed, current, and gas flow rate for TIG welding on 316 stainless steels. Tensile strength, impact strength, and corrosion rate are the response parameters. Using the taguchi based grey relational analysis approach, the values of input parameters were optimised and assessed to obtain the target mechanical properties of 190 mm/min welding speed, 150-amp welding current, and 15 ltr/min gas flow rate.

Kumar et al., (2019) described to lower manufacturing costs, the mechanical qualities of the weld structure and the welding parameters of the TIG welding process must be optimised. The current study focuses on optimising the TIG joining technique's procedure parameters. As a base metal, AISI 304 materials were chosen. The sample's dimensions are 200 mm x 50 mm x 3 mm. TIG welding was used to create the sample. The input parameters were chosen based on their impact on the mechanical properties of the weld sample. Voltage, current, gas flow, and root gap are the input parameters chosen. Minitab-17 software was used to conduct 27 tests, which were developed using Taguchi's L27 orthogonal array with four components and three levels. The tensile strength test verifies the integrity of the material.

Avinash et al., (2019) worked on Using ernicrmo-3 as a filler, a pcgta welding method is used to combine two dissimilar plates made of steel grade 304 and monel 400. Taguchi approach was used to develop and analyse experimental results in order to improve welding quality attributes. An l-9 orthogonal array is used to determine process parameters for maximising the mechanical characteristics and heat affected region of base metals. The analysis of variance (ANOVA) method was also utilised to compile and describe the results in mathematical terms.

Ahmad et al., (2019) focused on Optimization of tungsten inert gas process parameters (TIG). Taguchi method is very effective and efficient to solve the process parameters optimization problem to find optimum processing parameters, but it has been discovered that the traditional taguchi method is insufficient to solve the problem with multi response problem to get optimum outcomes and to meet the global challenges of the producers because it is impractical to obtain more than one exceptional processing parameter using taguchi. It is vital to apply updated and customised optimization approaches in order to thrive in the global market.

Jayashree et al., (2019) worked on Because of its strength, 6061al alloy is employed in harsh environments such as aerospace applications. During the welding of aluminium alloy, numerous problems such as porosity, hot cracking, and stress corrosion cracking must be avoided. Tungsten inert gas (TIG) welding of 6061al alloy has been completed using Taguchi's l9 orthogonal array design of tests.

Mishra et al., (2019) TIG welded using single and double 'v' butt joints two distinct sheets of metal of different grades of stainless steel 310 and 304 of 3 mm thickness. With a current of 110 amps and a voltage of 19 volts, two gas tungsten arc weld joints were created. When the current was kept at 80 amps and the voltage at 19 volts, two shielded metal arc weld junctions were created. To identify weld faults, dye penetrant inspection was used on the welded samples. Weld samples were determined to be defect-free, with no fractures, porosity, or lack of penetration, among other things. The gas tungsten arc welding procedure produces 556 MPa in the case of a double 'v' butt joint. The findings will aid in

IOP Conf. Series: Materials Science and Engineering 1259 (2022) 012048 doi:10.1088/1757-899X/1259/1/012048

the selection of an appropriate welding procedure for different alloys of stainless-steel grades 310 and 304 with single or double 'v' edges in 3 mm thickness for practical usage.

Ramana et al., (2019) focussed on Robotic tungsten inert gas welding of work components using high frequency technology. The process parameters play a crucial role in successful dissimilar material welding. Changes in weld control settings were employed to optimise the response surface methodology. The impact strength of each weld was tested. The control settings must be tuned to maximise the impact strength of weld joints. The study concluded that a current of 180 amps, a moving speed of 0.08 mm/min, and a wire feed rate of 0.82 mm/min are the best conditions for maximising impact strength.

Chandrasekaran et al., (2017) investigated Inconel 825 weld strength utilising the gas tungsten arc welding (gtaw) technique. The development of high-strength, corrosion-resistant materials has always been a difficulty for researchers. There were 27 trial runs in a box-behnken configuration. To investigate different welding responses, comprehensive quadratic models were created. The model's predictive performance indicates a 6.72 percent average percentage error for uts, 2.79 percent for ys, and 17.75 percent for percent e. For each answer, welding current I was determined to be the most important process parameter, followed by welding speed (v). The process parameters were optimised using desirability graphical analysis, yielding a maximum desirability value of 0.7089.

Modenesi et al., (2000) In applications where the weld bead shape and metallurgical parameters must be controlled, gas tungsten arc welding is essential. ATIG welding uses a small coating of an active ux to boost weld penetration dramatically. This effect is caused by high electronegativity elements capturing electrons in the arc's outer sections, causing the arc to constrict. The electrical signals from the arc and the arc shape were compared to the changes in weld geometry. The effect of the ux on the microstructure of the weld was also investigated. The results show that even the most basic ux can significantly increase the penetration of the weld bead.

Manurunget al., (2012), reviewed an alternative strategy for optimising resistance spot welding (RSW) process parameters for weld zone development Using the multi-objective Taguchi technique, the optimization method was attempted to account several quality criteria, namely heat affected zone (HAZ) and weld nugget. The test was carried out on 1.5 mm thick plates with various weld currents, hold times, and weld times. MTM was used to determine the ideal value, which included calculating total normalised quality loss and multi signal to noise ratio. The analysis of variance was used to produce a significant level of welding parameters (ANOVA). Using response surface methods, a first-order model for predicting weld zone development was also developed (RSM). The proposed method can be used to predict the size of the weld zone based on the experimental results.

Chakraborty al., One of the oldest non-traditional machining technologies is electrical discharge machining (EDM). The thermoelectric energy between the work piece and an electrode is used in the EDM process. The working fluid affects the material removal rate and the qualities of the machined surface in electrical dis-charge machining (EDM), a procedure that uses the removal phenomenon of electrical discharge in dielectric. For successful operations, selecting the correct dielectric fluid is crucial. This paper includes a review of the literature on the usage of dielectric fluids and their implications on the features of electrical discharge machining.

3. Experimental Set-up

The material is AISI309 stainless steel use as sample prepared by wire cut EDM process. EDM cutting process show in figure 1(a) and figure 1(b). Testing sample dimension standardised by ASTM standard as show in figure 2 (a) and figure 2 (b). Then implementation of taguchi optimisation, L9 taguchi's orthogonal array in design of experimental method utilise 3 levels of WC, 3 level of WS and 3 level of GFR as show in table no.2. The experimental procedure of L9 orthogonal array have been developed by Minitab L9 software.

Figure.1(a) EDM Cutting

Figure.1(b) EDM Cutting

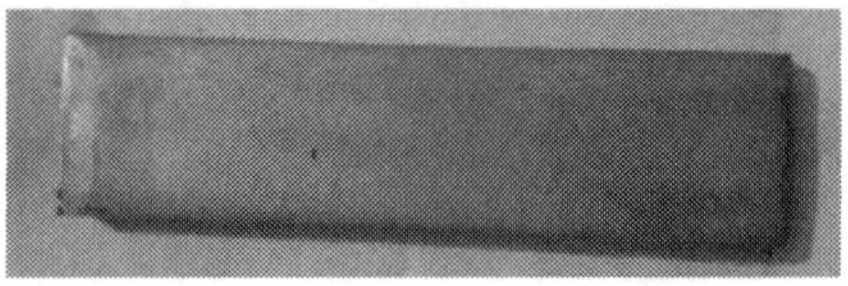

Figure.2 (a) Before EDM Cut

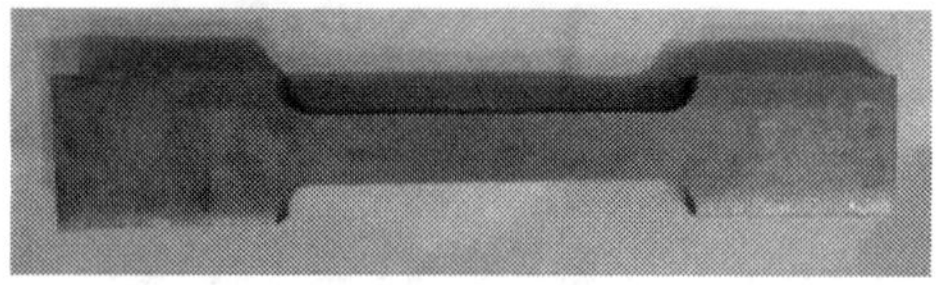

Figure.2 (b) After EDM Cut

Table 2. Process parameters

Factor	Levels 1	Levels 2	Levels 3
WC (amp)	150	160	170
WS (mm/sec)	2	3.5	5
GFR (ltr/min)	10	12	15

After preparing the sample, TIG welding were applied on the following selected parameters such as: (1) WC, (2) WS, (3) GFR show in figure.3 and using boistig200 machine for welding as show in Figure.4(a), figure.4(b) and figure.4(c).

Figure.3 TIG Welding setup

Figure.4 (a) Welded Sample for Tensile Test

Figure.4 (b) Welded Sample for impact Test

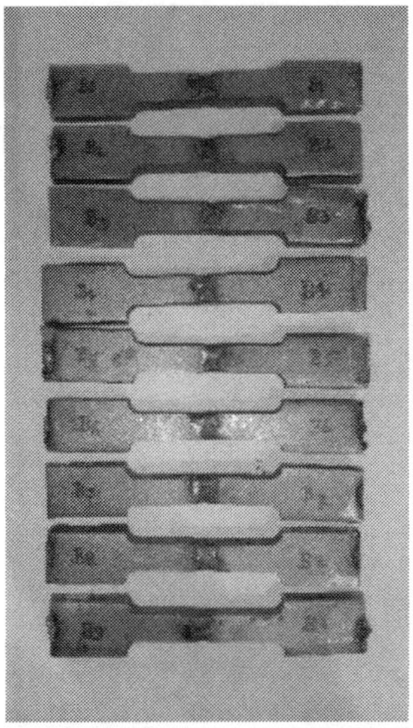

Figure.4 (c) Welded Sample for bending Test

After that performing the test, tensile test on utm and impact test on izod impact and bending test on utm show in figure.5(a), figure.5(b) and figure.5(c). The outcomes achieved from all the three tests performed are mentioned below in table no. 3.

Figure.5 (a) Sample tensile Test

Figure.5 (b) Sample impact Test

Figure.5 (c) Sample bending Test

IOP Conf. Series: Materials Science and Engineering 1259 (2022) 012048 doi:10.1088/1757-899X/1259/1/012048

Table 3. Experimental results using L9 (OA)

Exp. No.	Input parameters WC (amp)	WS (mm/sec)	GFR (lit/min)	Output parameters (UT) Ultimate tensile strength (MPa)	(IT) Impact strength (joule)	(BT) Bending strength (MPa)
1	150	2	10	311.85	90	436.52
2	150	3.5	12	339.43	95	439.22
3	150	5	15	385.22	99	512.33
4	160	2	12	362.06	112	447.23
5	160	3.5	15	394.34	139	126.32
6	160	5	10	343.97	103	441.95
7	170	2	15	416.42	154	578.22
8	170	3.5	10	361.62	143	470.51
9	170	5	12	405.66	140	525.34

4. Result

According to ASTM standards, L9 taguchi orthogonal array test were applied on the samples. Hence, the tensile test, impact test and bending test were carried out and results occur in table 3. In the table three the best outcome is achieved from test no. 7 (ultimate tensile strength = 416.42 mpa, impact strength = 154 j, bending strength = 578.22 mpa.) At input parameters (WC = 170 amp, WS = 2 mm/sec and GFR = 15 ltr/min) and the smallest outcome is achieved from test no. 1 (ultimate tensile strength = 311.85 mpa, impact strength = 90 j, bending strength = 436.52 mpa.) At input parameters (WC = 150 amp, WS = 2 mm/sec and GFR = 10 ltr/min).

4.1 taguchi optimization

The optimum TIG welding parameters values have been achieved by the taguchi method. In this analysis the TIG welding characteristics parameters like WC, WS and GFR were set as objective function. The ratio of mean and standard deviation in described as (signal-to-noise) S/N ratio and these are of three types. Maximization of the three tests (UT, impact strength, and bending strength) are the essential. So larger the better has been use. The equation shown in below

$$S/N = -10 \times \log_{10}\left(\frac{1}{y_i^2}\right) \tag{1}$$

The multiple S/N ratios (msrn) have been calculated. These results are listed in table 4.
Multiple S/N ratios (msrn) are listed in table 5. The parameters with msrn values for responses (WC, WS and GFR) have been plotted.

Table 4. Multiple S/N ratios (msrn)

Snra1	Snra2	Snra3	Msnr
49.8789	39.0849	52.8001	0.879831
50.6150	39.5545	52.8536	0.913743
51.7142	39.9127	54.1910	0.908343
51.1756	40.9844	53.0106	0.831797
51.9174	42.8603	42.0294	0.891516
50.7304	40.2567	52.9075	0.990000
52.3906	43.7504	55.2419	0.944057
51.1650	43.1067	53.4514	0.964893
52.1632	42.9226	54.4088	0.879831

Table 5 and Figure. 6. Show that for WC maximum mean s/n ratio value is achieved at level 3. And WS maximum mean S/N ratio value is achieved at level 3. Also, GFR maximum mean S/N ratio value

IOP Conf. Series: Materials Science and Engineering 1259 (2022) 012048 doi:10.1088/1757-899X/1259/1/012048

is achieved at level 3. Hence the optimum value is WC 170 amp (level 3), WS 2mm/sec (level 1) and GFR 15 lit/min (level 3).

Table 5. Response for S/N ratios

Symbol	Factors	Mean of multiple S/N ratio		
		Level 1	Level 2	Level 3
A	WC (amp)	43.80	44.50	47.18*
B	WS (mm/sec)	45.39*	44.89	45.21
C	GFR (lit/min)	44.90	45.29	45.30*

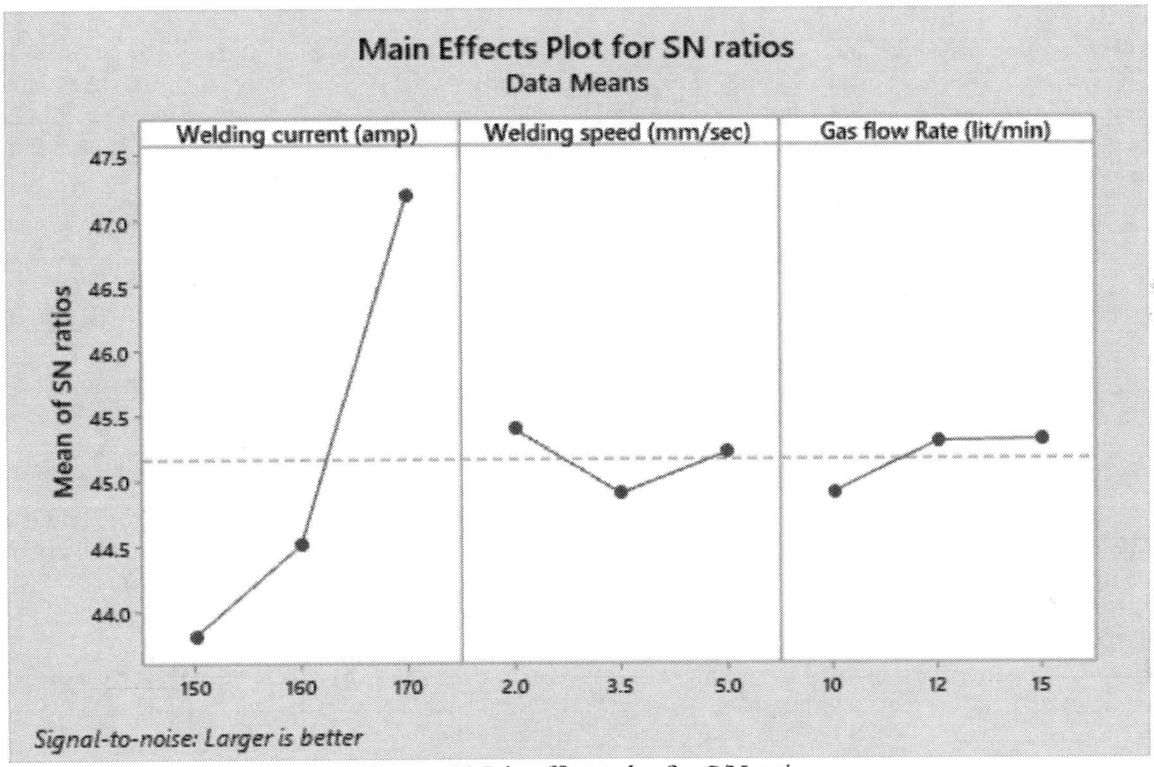

Figure.6 Main effects plot for S/N ratio

4.2 Analysis if variance

The analysis of variance (ANOVA) has been applied.

Correlative factor (c.f.) = $\frac{(GT)^2}{n}$ (2)

Where,

GT = grand total of response

n = no. Of observations

Sum of square = variance of response

Total sum of square = $\sum$sum of square – cf (3)

Mean square = $\frac{ss^2}{DOF}$ (4)

Percentage contribution (pc) = $\frac{ss^2}{\text{total sum of square}}$ x 100 (5)

F-value = $\frac{MS}{MS_{error}}$ (6)

The Analysis of variance analysis and their results are listed in table 6.

IOP Conf. Series: Materials Science and Engineering 1259 (2022) 012048 doi:10.1088/1757-899X/1259/1/012048

Table 6. Analysis of variance for S/N ratios

Source	Df	Ss	Pc (%)	Ms	F
WC (amp)	2	19.1054	92.68	9.5527	23.41
WS (mm/sec)	2	0.3784	1.84	0.1892	0.46
GFR (lit/min)	2	0.3146	1.53	0.1573	0.39
Residual error	2	0.8162	3.96	0.4081	
Total	8	20.6145			

The percentage contribution (pc) of each parameter on responses are as WC=92.68%, WS=1.84% and GFR=1.53% positively.

5. Experimental validation

The higher value of mean msnr (table 5) represents the exceptional value. The confirmation tests have been executed by supervise a test at central value of the factor and the levels that were calculated previously. The motive is to prove the optimum level suggested by taguchi technique. The experimental results and the predicted values at initial setting of parameters are listed in table 7.

Table 7. Confirmation test using tm technique

Response	Initial setting	Optimum values Test	% improvement
Level	A_1-b_1-c_1	A_3-b_3-c_3	
(UT) Ultimate tensile strength	311.85	415.90	25.02%
(IT) Impact strength	90.00	153.00	41.18%
(BT) Bending strength	436.52	579.01	24.61%

Table 7 shows that with the experimental values of WC (170 amp), WS (2 mm/sec) and GFR (15 ltr/min), the improvement of 25.02%, 41.18%, 24.61% in UT, IT and BT is found respectively on comparing it with the initial values of WC (150 amp), WS (2 mm/sec) and GFR (10 ltr/min).

6. Conclusion

The best outcome has been achieved for the test no.7. as shown in Table 8. The lowest output parameters from the tensile testing have been achieved for the test no. 1. as shown in Table 9.

The optimization by taguchi method provides best conditions which get applicable through obtained exceptional responses in confirmatory tests.

The best parametric combination was discovered using mean main effect plots and S/N ratio plots. Parameter (welding current=170-amp, gas flow rate=15l/min, and speed=2mm/sec) is the best factor setting.

According to the ANOVA, all of the Process factors, including WC, WS, and GFR, contributed significantly with 92.68 percent, 1.84 percent, and 1.53 percent, respectively, and Error Contribution was 3.96 percent.

Table 8. Best Outcome

TEST	Result	WC	WS	GFR
Ultimate tensile strength	416.42 MPa	170Amp	2mm/sec	15lit/min
Impact strength	154 Joule	170Amp	2mm/sec	15lit/min
Bending strength	578.22 MPa	170Amp	2mm/sec	15lit/min

Table 9. Lowest Outcome

TEST	Result	WC	WS	GFR
Ultimate tensile strength	311.85 MPa	150Amp	2mm/sec	10lit/min
Impact strength	90 Joule	150Amp	2mm/sec	10lit/min
Bending strength	436.52 MPa	150Amp	2mm/sec	10lit/min

References

[1] Shanmugasundar, g., karthikeyan, b., ponvell, p.s. and vignesh, v., 2019. Optimization of process parameters in TIG welded joints of AISI 304l-austenitic stainless steel using taguchi's experimental design method. Materials today: proceedings, 16, pp.1188-1195.

[2] Natrayan, l., anand, r. And kumar, s.s., 2021. Optimization of process parameters in TIG welding of AISI 4140 stainless steel using taguchi technique. Materials today: proceedings, 37, pp.1550-1553.

[3] Vinoth, v., sudalaimani, r., ajay, c.v., kumar, c.s. and prakash, k.s., 2021. Optimization of mechanical behaviour of TIG welded 316 stainless steels using taguchi based grey relational analysis method. Materials today: proceedings, 45, pp.7986-7993.

[4] Kumar, s., jena, s., lahoty, v., paswan, m.k., sharma, b., patel, d., prasad, s.b. and sharma, v.k., 2020. Experimental investigation on the effect of welding parameters of tig welded joints using ANOVA. Materials today: proceedings, 22, pp.3181-3189.

[5] Avinash, s., balram, y., babu, b. S., & venkatramana, g. (2019). Multi-response optimization of pulse tig welding process parameters of welds AISI 304 and monel 400 using grey relational analysis. Materials today: proceedings, 19, 296-301.

[6] Ahmad, a. And alam, s., 2019. Integration of rsm with grey based taguchi method for optimization of pulsed tig welding process parameters. Materials today: proceedings, 18, pp.5114-5127.

[7] Jayashree, p.k., sharma, s.s., shetty, r., mahato, a. And gowrishankar, m.c., 2018. Optimization of tig welding parameters for 6061al alloy using taguchi's design of experiments. Materials today: proceedings, 5(11), pp.23648-23655.

[8] Mishra, d. And dakkili, m., 2020. Gas tungsten and shielded metal arc welding of stainless steel 310 and 304 grades over single and double 'v'butt joints. Materials today: proceedings, 27, pp.772-776.

[9] Ramana, m.v., kumar, b.r., krishna, m., rao, m.v. and kumar, v.s., 2020. Optimization and influence of process parameters of dissimilar ss304l–ss430 joints produced by robotic tig welding. Materials today: proceedings, 23, pp.479-482.

[10] Modenesi, P.J., Apolinario, E.R. and Pereira, I.M., 2000. TIG welding with single component fluxes. Journal of materials processing technology, 99(1-3), pp.260-265.

[11] Ghosh, n., pal, p. K., & nandi, g. (2017). Parametric optimization of gas metal arc welding process by pca-based taguchi method on ferritic stainless steel AISI409. Materials today: proceedings, 4(9), 9961-9966.

[12] Muhammad, n., manurung, y.h., hafidzi, m., abas, s.k., tham, g. And haruman, e., 2012. Optimization and modeling of spot-welding parameters with simultaneous multiple response consideration using multi-objective taguchi method and rsm. Journal of mechanical science and technology, 26(8), pp.2365-2370.

[13] Chakraborty, S., Dey, V. and Ghosh, S.K., 2015. A review on the use of dielectric fluids and their effects in electrical discharge machining characteristics. Precision Engineering, 40, pp.1-6.

Surface Functionalization of Jute Fibers In The Presence Of Sodium Hydroxide and Graphene Oxide in the Preparation of Bio Composite

Yogendra Kumar Verma[1], **Amit Kumar Singh**[1] **and M. K. Paswan**[2], **Puneet Kumar Sonker**[3]

[1'3] National Institute of Technology Nagaland, Dimapur, Nagaland, India

[2] National Institute of Technology Jamshedpur, Jharkhand, India
amit.kumar965@yahoo.com

Abstract. The natural fibers (jute, bamboo, banana etc.) has high strength to weight ratios, inexpensiveness, low densities, good mechanical properties, environment friendly, ease of manufacturing and availability due to this properties natural fibers are now days vitally popular. Jute fibers have good mechanical properties. The interfacial strength (23.6%) and tensile strength (96%) of the coated jute fibers increases more than untreated jute fibers. The morphologies of treated and untreated jute fibers surface are display very different in the analysis of SEM. Which jute fibers are untreated that smooth and it contains hemicellulose, lignin and pectin and they are covered with lots of lump materials. Which are removed after functionalization of the jute fibers in the presence of NaOH. This work presented the improvement of the epoxy based composite with the functionalized jute fibers as filler used for increased the mechanical (tensile, flexural strength) properties of the composites. The flexural strength of the prepared composites increases from 259.9MPa to 325.5MPa and the flexural modulus increases from 16.7GPa to 21.5GPa in the prepared composites by using functionalized jute fibers in the comparison of untreated jute fibers composites. These jute fibers based composites very useful for the environment so it can reduce the synthetic fibers based composites.

Keyword: Jute fibers, functionalization, SEM, bio composites, NaOH, GO.

1. Introduction

The natural fibers (jute, bamboo, banana etc.) has high strength to weight ratios, inexpensiveness, low densities, good mechanical properties, environment friendly, ease of manufacturing and availability due to this properties natural fibers are now days vitally popular[1]. Not only during growth stage of natural fibers but also after life time natural fibers are environment friendly because they are the biodegradable[2]. Fabrication and development of polymer composites natural redeemable materials are using for the better environment in place of the soaring price of raw material for the engineering and standard plastics[3]. The cell walls of the plants are typically described as macromolecular composites and multi layered wall structure. The plants have three major parts which are respectively as middle lamella, primary cell walls and secondary cell walls[4]. Two types of parenchyma cells in the ground tissues of the bamboo, woody monocot and in contracts to rice has single type of parenchyma cell is present[5]. For the better environment now days biodegradable polymers are vitally used in place of plastics which are not good for the better environment[6]. Jute fibers brought alternative into the spotlight and as a good replacement of the synthetic fibers in the textile field of innovation[7]. For the preparation of natural fibers composites, natural fibers such as jute, banana, bamboo, pine apple are mainly used as filler for the good strength of the composites. The mechanical properties such as tensile and flexural strength of the composites increases as filler contents increases in the composites preparation[8]. The subsequent swelling and high moisture absorption and degradation, poor chemical and fire resistance, high dispersion of mechanical properties type limitation are present in the natural fibers. So for the overcome this limitation some modification are done with some chemical[9]. For the replacing of the synthetic composites at lower cost with better environment friendly natural fibers are now days very attractive for better environment. The interfacial properties and mechanical properties (strength, toughness) of natural fiber based composites suffer. For the improvement of the properties (mechanical strength, toughness) of the natural fibers composites, fibers are coated with the graphene oxide and by graphene flakes. The interfacial strength (23.6%) and tensile strength (96%) of the coated jute fibers increases more than untreated jute fibers[10]. For the high performance and sustainability of the composites jute fibers are used in the composites and by this dimensional a stability compromised, and dimensional stability directly related with the shrinkage[11]. The attentions of the researchers are attracted towards the bio polymers composites due to the environment friendly, environment sustainability and the various sustainable applications[12]. Natural fibers has high mechanical properties and ease production so they can used as filler in the preparation of bio degradable based composites and the efficiency of the reinforcements of the plants fibers crystallinity and on type of cellulose are defined[13]. The density of the cellulose fibers (1.5g/cm^3) are lower than the glass fibers (2.5 g/cm^3) and talc fillers (2.8 g/cm^3). The stiffness of the axial tissue and density of the tissue fibers effected by the differential thickness cell wall[14]. That's why this can used in the automotive industries for the meeting the light weight and the standard of the fuel economy. The fuel economy of the vehicle can improve by 3% to 7%, by the reducing the vehicle weight by 10%[15]. Parts which are used in the field of the aerospace and automotive identified the desirable properties such as lightweight, high strength and high stiffness[16]. The mechanical properties such as heat resistance and impact strength of jute fibers is improve when the jute fibers mix with the polylactic acid[17].

2. Materials and Methods

2.1 Materials

The fibers of jute collected from the Gloster jute mill, India in the grade of TD4. The jute fibers which are untreated obtained in golden color and its diameter (o.062mm) and average length (2.5mm). NaOH, Graphene oxide, epoxies resin.

2.2 Methodology

2.2.1 Functionalization of jute fibers with NaOH and with GO

Jute fibers socked in the solution of 5 wt. % NaOH for 5hrs at room temperature. The mass ratio (30:1) of water and jute fibers was taken. For the maintaining the fibers pH value in nutral form washed the treated fibers with distilled water in several times. Then functionalized jute fibers put in the vacuum oven for the curing till 20h at 60°C.

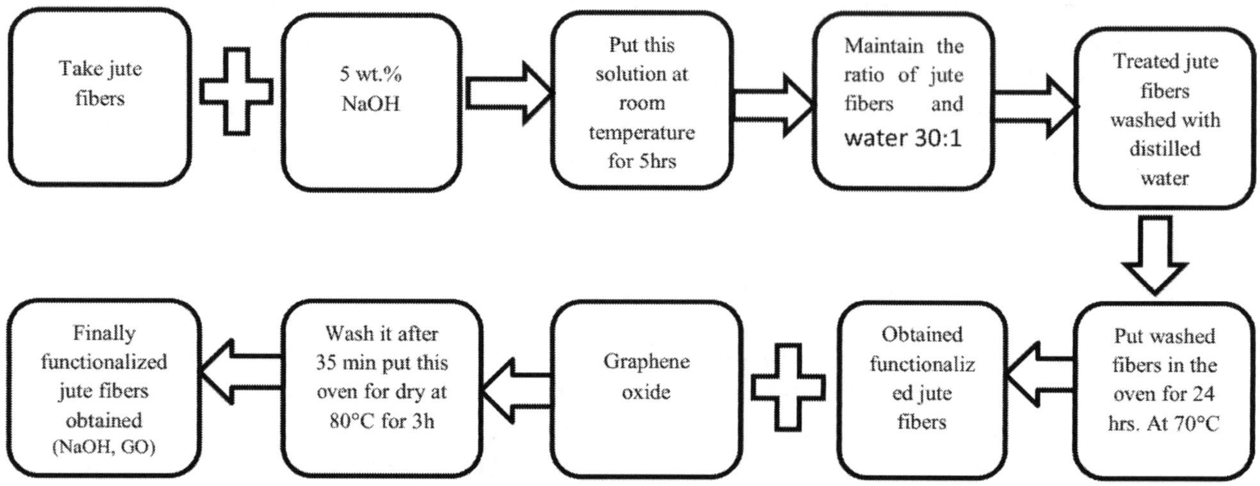

Fig. 1. Functionalization of jute fibers with NaOH and with GO

2.2.2 Preparation of bio composites

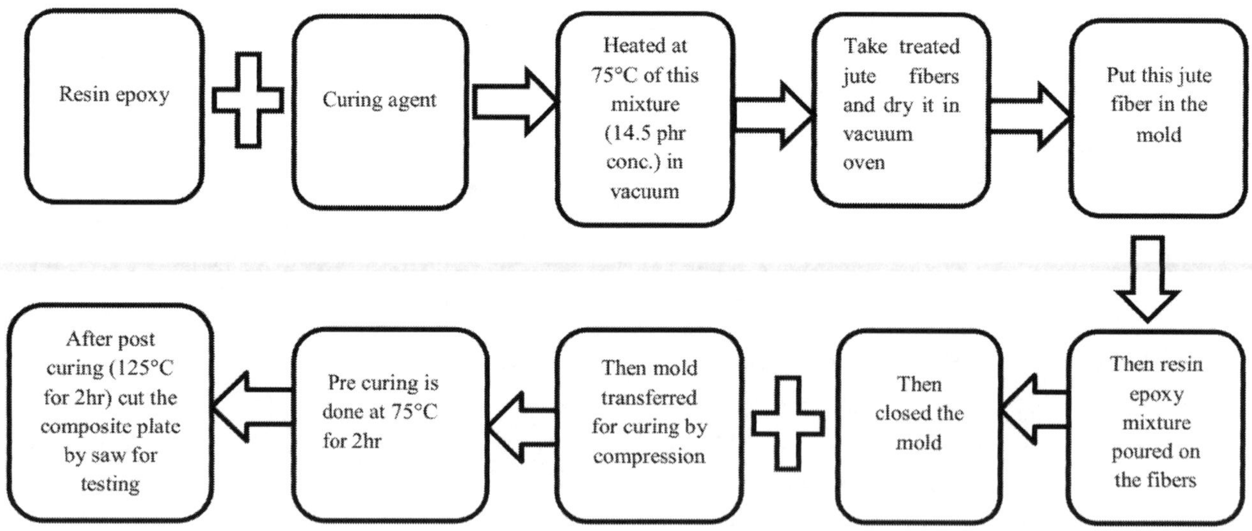

Fig. 2. Preparation of bio composites

IOP Conf. Series: Materials Science and Engineering 1259 (2022) 012049 doi:10.1088/1757-899X/1259/1/012049

3. Results and discussion

3.1 Characterizations of functionalized jute fibers

The morphologies of treated and untreated jute fibers surface are display very different in the SEM analysis. The untreated jute fibers are smooth and it contains hemicellulose, lignin and pectin and they are covered with lots of lump materials. Which are removed after functionalization of the jute fibers in the presence of NaOH [18]. At the temperatures range from 170°C to 210°C thermal degradation of jute fibers done. The fibers properties affected below the 170°C temperature. The degree of polymerization and tenacity of jute fibers decreases when the temperature above the 170°C[19]. Due to the removal of surface materials by the treatment of the jute fibers jute fibers surface become rougher which are very important for mechanical properties of the composite interlocking between jute fibers and epoxy resin. And also improved jute fibers adhesive properties[20]. Excessive concentration of NaOH treatment of fibers can decrease the fiber strength[21].

Table 1. TGA summary for jute fibers, neat epoxy and jute/epoxy composites [18][22].

Samples	$T_{5\%}$ (°C)[1]	$T_{10\%}$ (°C)[1]	T_{max} (°C)[2]	Y_c (°C)[3]
UJF	236	289	368	16.7
TJF	264	299	374	13.1
P(epoxy)	363	372	392	6.0
P(epoxy)/UJF	330	356	386	9.7
P(epoxy)/TJF	334	358	387	8.9

3.2 Improvement in the Mechanical properties of jute fiber based composites

The length of fiber and the content of the jute fibers affect the (tensile, flexural) mechanical properties of the prepared bio based composites. At the 15mm (Fig.3) fibers length strength of the composites increases[23]. The flexural properties of the treated jute based bio composites increases when the the percentage of jute fibers increase in comparison with untreated jute fibers composites. Due to void formation in the prepared composites if fibers volume increased more than 50% prepared composites are not more than 50% of the fiber volume[24]. The crystallinity of prepared composite in the presence of jute fibers increases in the addition of jute fibers. The bending properties of prepared composites increases in the steam exploded filament because it has highest interfacial strength between this[25].

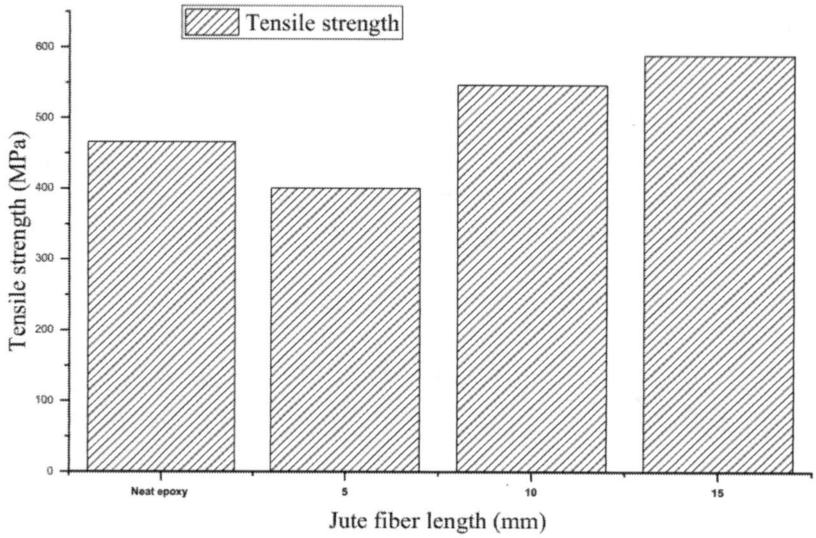

Fig.3 Tensile strength of jute fiber

3.3 Variations in the Mechanical properties of functionalized (with NaOH and GO) based jute fibers composites

The flexural properties of the prepared composites increases by using functionalized jute fibers in the presence of NaOH and GO compared than which jute fibers are untreated used for preparation of neat composites. This will happened due to the treatment of the jute fibers surface remove the impurities, and due to the free hydroxyl group from the treated fibers so they become more chemical reactive. And they also increases the interfacial bonding and mechanical interlocking in the epoxy matrix composites because it has the greater surface roughness of the treated jute fibers[26].

4. Conclusions

This work presented the improvement of the epoxy based composite with the functionalized jute fibers as filler used for increased the mechanical properties (tensile, flexural) of the bio based composites. The flexural strength of the prepared composites increases from 259.9MPa to 325.5MPa and the flexural modulus increases from 16.7GPa to 21.5GPa in the prepared composites by using functionalized jute fibers in the comparison of untreated jute fibers composites. These jute fibers based composites very useful for the environment so it can reduce the synthetic fibers based composites. And this natural fibers (jute) based composites used in the sector of household and also used in the automotive industry.

5. References

[1] T. K. Mulenga, A. U. Ude, and C. Vivekanandhan, "Techniques for modelling and optimizing the mechanical properties of natural fiber composites: A review," *Fibers*, vol. 9, no. 1, pp. 1–17, 2021, doi: 10.3390/fib9010006.

[2] L. Osorio, E. Trujillo, A. W. Van Vuure, and I. Verpoest, "Morphological aspects and mechanical properties of single bamboo fibers and flexural characterization of bamboo/ epoxy composites," *J. Reinf. Plast. Compos.*, vol. 30, no. 5, pp. 396–408, 2011, doi: 10.1177/0731684410397683.

[3] H. P. S. Abdul Khalil, I. U. H. Bhat, M. Jawaid, A. Zaidon, D. Hermawan, and Y. S. Hadi, "Bamboo fibre reinforced biocomposites: A review," *Mater. Des.*, vol. 42, no. June 2018, pp. 353–368, 2012, doi: 10.1016/j.matdes.2012.06.015.

[4] C. Lian *et al.*, "Ultrastructure of parenchyma cell wall in bamboo (Phyllostachys edulis) culms," *Cellulose*, vol. 27, no. 13, pp. 7321–7329, 2020, doi: 10.1007/s10570-020-03265-9.

[5] X. Q. He, K. Suzuki, S. Kitamura, J. X. Lin, K. M. Cui, and T. Itoh, "Toward understanding the different function of two types of parenchyma cells in bamboo culms," *Plant Cell Physiol.*, vol. 43, no. 2, pp. 186–195, 2002, doi: 10.1093/pcp/pcf027.

[6] H. Zaki, S. M. Sapuan, and R. A. Ilyas, "processing technique of natural fiber-reinforced polylactic acid (PLA)," no. January, 2021.

[7] L. Nayak and S. P. Mishra, "Prospect of bamboo as a renewable textile fiber , historical overview , labeling , controversies and regulation," *Fash. Text.*, 2016, doi: 10.1186/s40691-015-0054-5.

[8] H. Banga, V. K. Singh, and S. K. Choudhary, "Fabrication and Study of Mechanical Properties of Bamboo Fibre Reinforced Bio-Composites," *Innov. Syst. Des. Eng.*, vol. 6, no. 1, pp. 84–99, 2015.

[9] J. Cruz and R. Fangueiro, "Surface Modification of Natural Fibers: A Review," *Procedia Eng.*, vol. 155, pp. 285–288, 2016, doi: 10.1016/j.proeng.2016.08.030.

[10] F. Sarker, N. Karim, S. Afroj, V. Koncherry, K. S. Novoselov, and P. Potluri, "High-Performance Graphene-Based Natural Fiber Composites," *ACS Appl. Mater. Interfaces*, vol. 10, no. 40, pp. 34502–34512, 2018, doi: 10.1021/acsami.8b13018.

[11] A. M. Davis, L. E. Hanzly, B. L. DeButts, and J. R. Barone, "Characterization of dimensional stability in flax fiber reinforced polypropylene composites," *Polym. Compos.*, vol. 40, no. 1, pp. 132–140, 2019, doi: 10.1002/pc.24614.

[12] R. Siakeng, M. Jawaid, H. Ariffin, S. M. Sapuan, M. Asim, and N. Saba, "Natural fiber reinforced polylactic acid composites: A review," *Polym. Compos.*, vol. 40, no. 2, pp. 446–463, 2019, doi: 10.1002/pc.24747.

[13] P. Zakikhani, R. Zahari, M. T. H. Sultan, and D. L. Majid, "Extraction and preparation of bamboo fibre-reinforced composites," *Mater. Des.*, vol. 63, pp. 820–828, 2014, doi: 10.1016/j.matdes.2014.06.058.

[14] "Cell wall structure and formation of maturing fibres of moso bamboo (Phyllostachys pubescens) increase buckling resistance Cell wall structure and formation of maturing fibres of moso bamboo (Phyllostachys pubescens) increase buckling resistance Refer," no. May, 2012, doi: 10.1098/rsif.2011.0462.

[15] M. D. Reale Batista, L. T. Drzal, A. Kiziltas, and D. Mielewski, "Hybrid cellulose-inorganic reinforcement polypropylene composites: Lightweight materials for automotive applications," *Polym. Compos.*, vol. 41, no. 3, pp. 1074–1089, 2020, doi: 10.1002/pc.25439.

[16] M. D. Reale Batista and L. T. Drzal, "Carbon fiber/epoxy matrix composite interphases modified with cellulose nanocrystals," *Compos. Sci. Technol.*, vol. 164, pp. 274–281, 2018, doi: 10.1016/j.compscitech.2018.05.010.

[17] R. Tokoro, D. M. Vu, K. Okubo, T. Tanaka, T. Fujii, and T. Fujiura, "How to improve mechanical properties of polylactic acid with bamboo fibers," *J. Mater. Sci.*, vol. 43, no. 2, pp. 775–787, 2008, doi: 10.1007/s10853-007-1994-y.

[18] R. E. Composites, "Thermal and Mechanical Properties of Bamboo Fiber Reinforced Epoxy Composites," 2018, doi: 10.3390/polym10060608.
[19] J. Gassan and A. K. Bledzki, "Thermal degradation of flax and jute fibers," *J. Appl. Polym. Sci.*, vol. 82, no. 6, pp. 1417–1422, 2001, doi: 10.1002/app.1979.
[20] K. V, C. DP, K. S, and K. K, "Study on the Performance of Bamboo Fibre Modified with Different Concentrations of Sodium Hydroxide and Chlorine Containing Agents," *J. Text. Sci. Eng.*, vol. 08, no. 03, 2018, doi: 10.4172/2165-8064.1000362.
[21] M. I. Ibrahim, M. Z. Hassan, R. Dolah, M. Z. M. Yusoff, and M. S. Salit, "Tensile behaviour for mercerization of single kenaf fiber," *Malaysian J. Fundam. Appl. Sci.*, vol. 14, no. 4, pp. 437–439, 2018, doi: 10.11113/mjfas.v14n4.1099.
[22] D. Ray, B. K. Sarkar, R. K. Basak, and A. K. Rana, "Study of the thermal behavior of alkali-treated jute fibers," *J. Appl. Polym. Sci.*, vol. 85, no. 12, pp. 2594–2599, 2002, doi: 10.1002/app.10934.
[23] B. Biswas, N. R. Bandyopadhyay, G. Mandal, and A. Sinha, "Effect of alkaline treatment on mechanical properties of composites: Unsaturated polyester reinforced ZrO2/jute and sisal," *Polym. Polym. Compos.*, vol. 29, no. 9_suppl, pp. S1000–S1008, 2021, doi: 10.1177/09673911211033949.
[24] M. M. Thwe and K. Liao, "Effects of environmental aging on the mechanical properties of bamboo-glass fiber reinforced polymer matrix hybrid composites," *Compos. - Part A Appl. Sci. Manuf.*, vol. 33, no. 1, pp. 43–52, 2002, doi: 10.1016/S1359-835X(01)00071-9.
[25] H. P. S. Abdul Khalil, I. U. H. Bhat, M. Jawaid, A. Zaidon, D. Hermawan, and Y. S. Hadi, "Bamboo fibre reinforced biocomposites: A review," *Mater. Des.*, vol. 42, pp. 353–368, 2012, doi: 10.1016/j.matdes.2012.06.015.
[26] A. Roy, S. Chakraborty, S. P. Kundu, R. K. Basak, S. Basu Majumder, and B. Adhikari, "Improvement in mechanical properties of jute fibres through mild alkali treatment as demonstrated by utilisation of the Weibull distribution model," *Bioresour. Technol.*, vol. 107, pp. 222–228, 2012, doi: 10.1016/j.biortech.2011.11.073.

A review on wear characterization of electrodeposited nickel matrix composites

Atul Dhakne [1], Santosh Jaju[1], Sourabh Shukla[1*]

[1]Department of Mechanical Engineering, G H Raisoni College of Engineering Nagpur, India

*Corresponding Author Email: sourabh.shukla@raisoni.net

Abstract. Corrosion resistance across a wide temperature range, increased physical, mechanical, and tribological capabilities, variable surface hydrophobicity, and an attractive appearance are all advantages of electrodeposited metal matrix composite coatings (MMCs). As a result, several studies have been conducted to evaluate the various physical, mechanical, biochemical, electrochemistry, and thermodynamics characteristics of nanocrystalline composite material, generated on a variety of substrates with numerous electrodeposition settings. In this review paper, the current literatures on wear rate of electrode-posited Ni-based coatings made of composite materials are being reviewed. More specifically, the research papers on weight management, wear resistance, coefficient of friction, surface roughness, hardness and associated deterioration methods has been identified and reviewed. The observations showed that nano-composite films offer substantially superior wear resistance than magnesium alloys and pure nickel coatings due to the diffraction stiffening and grain refining effects. Also, abrasion is the main wear process for nano-composite coatings, whereas adhesion wear is the primary wear arrangement for magnesium alloys and major wear mechanism for pure nickel coatings is exfoliation wear.

Key words: Wear; Composite coating; Electrodeposition; Magnesium alloy

1. Introduction

Metal matrix composite coatings (MMCs) can be made using an electrodeposition process. Initially, electrodeposited coats were designed for decorative and protection purposes. These coatings were later discovered to be suitable for wear-resistant applications. In a metal/alloy matrix system, inert ceramic particles of micron to nanoscale are inserted, and electrodeposited composite metal matrix coatings were created [1]. Copper (Cu), aluminum (Al), cobalt (Co), nickel (Ni), chromium (Cr), and other graphene-enhanced metal matrix composite coatings are currently being electrodeposited. In terms of solid lubricating capabilities, the Ni based graphene composite coatings stand out. Ni as well as its compounds can be used in a variety of nanocrystalline coverings, with strong strength, organic stability, good electrical and thermal conductance, corrosion resistance, outstanding reflectance of light, and ability to layer finish in a variety of colours and patterns [2]. Ni coatings have a number of major advantages such as their ability to create hybrid versions which results in substantial improvements in resistance to corrosion in a number of media including watery, salinity, chemical, alkali, and acidic minerals. To further improve the qualities of manufactured

IOP Conf. Series: Materials Science and Engineering 1259 (2022) 012050 doi:10.1088/1757-899X/1259/1/012050

coatings, abrasive particles such as silicon carbide are added to electrodeposited Ni-based coatings. Because of their toughness and elastic modulus, as well as outstanding corrosion and wear resistance, Ni-based ceramic materials are commonly used in construction, military, aerospace, automobile, marine, instrumentation, healthcare, oilfield, biochemical, and petroleum industries [3]. Ni matrix coatings can be manufactured in a variety of ways, with good reproducibility, low price, and reduced energy consumption. Hard micron-sized ceramic carbides, oxides, oleaginous polymeric compounds, and graphene are all popular reinforcing phases. Second phases can now be used in electrodeposits to a greater extent to incorporate nanotechnology reinforcements, according to recent research. Compared to standard micron-sized reinforcement, nanoparticles render a number of important benefits, for example nanoparticles such as carbon nanotubes and tungsten disulphide offer exceptional chemical characteristics due to their unique properties more uniformly scattered nanoparticles can enhance a metal matrix's reinforcing; and a metal matrix containing hard nanoparticles has a stronger boosting impact [4].So far, several approaches have been used to create composite coatings with reinforced particles.

Dynamic chemical plating, sol-gel procedures, these methods include electroless and electroforming processes, as well as electrodeposition. Numerous studies and investigations have been conducted on the impact of manufacturing criteria with respect to the morphological features, morphological characteristics, physical quality, toughness, corrosion resistance, and wear behaviour of Ni-based coatings in various deposition techniques, as well as their impacts on morphological characteristics, texture of the structure, toughness, wear corrosion resistant behaviour of Ni-based coatings [5]. It is generally known that composite coatings via electrodeposition are a cost-effective way to provide increased qualities of self-lubrication, resistance to abrasion, resistance to corrosion, and chemical resistance. Over the last few years, there has been a constant pursuit of lowering the friction coefficient in Ni based coatings. Pilot tubes, automotive parts, hydraulic systems, gearboxes, actuation, clutches, and other aircraft components typically use pure Ni coatings, as well as Ni-based alloy or composite coatings [6]. Commercialized Ni/Si-C composites have been used to secure frictional parts, internal combustion, and cast moulds. In the late 1990s, the accessibility of nano-size particles, as well as the improved coating properties they provided, that sparked an interest in nickel-based nano-composite coatings [7]. Composite electroplating is the preferred process for deposition and co-depositing microscopic metal and non-metal particles, and also polymeric compounds, in protective coatings. Durability, high hardness, lubrication resistance, corrosion and oxidation resistance are all improved by including such nanoparticles into the coating layer [8]. Electrodepositions are a viable option for commercial processes because they can be performed at a minimal price, at a low temperature, at air pressure, and with simple, low-maintenance equipment. In recent years, electrodeposition has been used to create massive super-hydrophobic surfaces with exceptional hardness, erosion and wear resistance, and conscience characteristics [9]. Irresolvable particles dispersed in a plating solution get captured in a developing metallic surface during electrodeposition of composite coatings, often electrostatic repulsion. These particles are widely used in composite coatings to greatly improve the coating's resistance to abrasion. The pin-on-disc method is a popular and versatile method for evaluating composite coating wear behaviour as shown in Fig. 1 [10].

Electrodeposition is a film building method that involves the electrochemical reduction of metal ions from an electrolyte to generate a metallic coating on a base material. Electroplating is the common name for the equivalent technique. Electrochemical metal reduction is also utilised for the extraction of metals from their ores (electrometallurgy) and the reproduction of moulds to create items directly in their final shape, in addition to the creation of metallic coatings (electroforming) [11]. Because the metallic deposit formed in most cases is crystalline, this process is also known as electrocrystallization, a name used by Russian scientist V. Kistiakovski in the early twentieth century.

Chemical species containing the metal of interest are dissolved in a suitable solvent or brought to a liquid state to produce a molten salt, which is the electrolyte. The most common solvent is water, however for some electroplating techniques, chemical molecules and other ionic liquids have lately been utilised. Chemical species containing the metal of interest are dissolved in a suitable solvent or brought to a liquid state to produce a molten salt, which is the electrolyte. The most common solvent is water, however for some electroplating techniques, chemical molecules and other ionic liquids have lately been utilised [12].

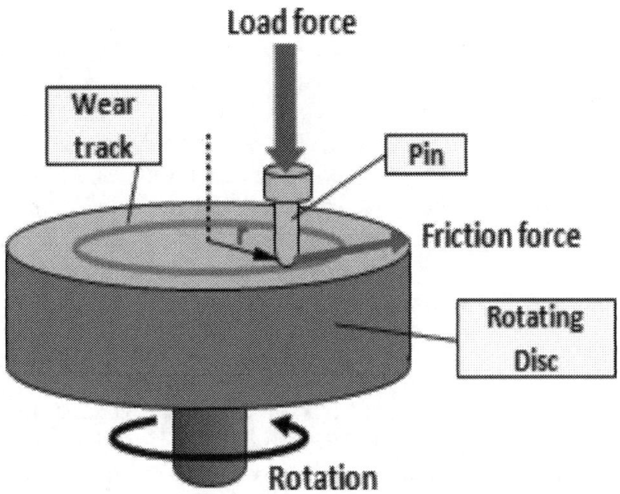

Figure 1. Pin on disc method

2. Recently update in wear and tribology on electrodeposition

As the graph shows that study of wear and tribology on electrodeposition of Ni composite coating gradually increased.

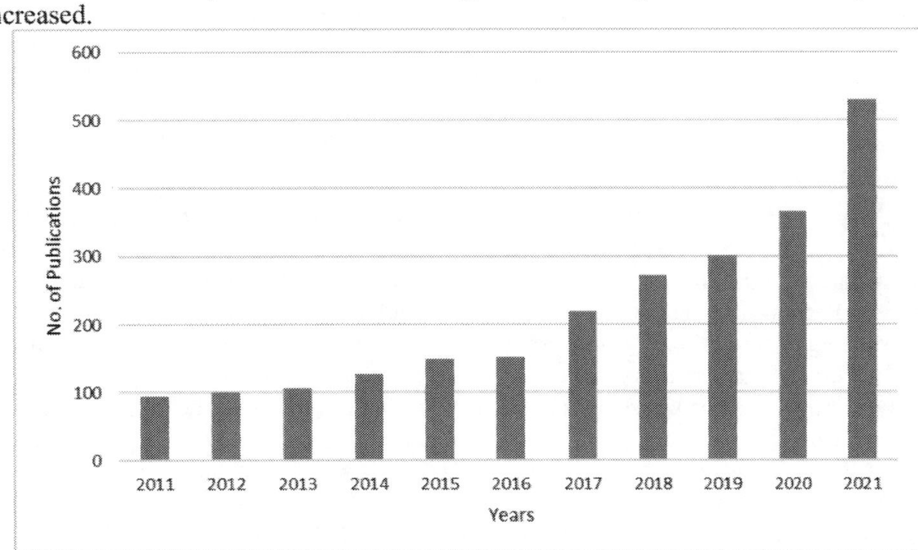

Figure 2. Graph between no. of publication and year

In a very recent work, Bajwa et al. [11] studies on the tribological performance of electrodeposited coatings with nano-Al_2O_3 and graphene, as well as Ni/nano-Al_2O_3 composite coatings subjected to nano-lubricants. A pulse electrodeposition approach was used to create electrodeposited coatings. Tribological experiments were performed with the assistance of on a flat sliding tribometer, a linear reciprocating ball is used. Nano effects loading in lubricating oils and composite coatings considerably improved the wear and friction resistance qualities, according to experimental data. The greatest tribological qualities were identified when Al_2O_3 nanomaterials were introduced to the lubricating oil, followed by Ni–Al_2O_3 composite material and graphite nano-oil. Li et al. [12] works on the use of nano-crystalline zinc on steel electrodes to improve corrosion rate. Corrosive wear tests in a modelled seawater environment were carried out to compare the durability of nano-crystalline zinc coatings with grain sizes of 40 nm to poorly graded zinc coats (grain size ~ 5 μm). The coarse-grained zinc coating showed material loss to be 39 times more than that of the nanocrystalline one. Liang et al. [13] discussed the work on Ni–$SrSO_4$ nanocomposites which were successfully deposited on the superalloy Inconel 718 and was fabricated using a watt's electrolyte containing a $SrSO_4$ solvent. A scanning electron microscope, a microstructural sample, and a friction coefficient sample were used to investigate Ni–$SrSO_4$ coatings and sliding against with a bearing specimen in an unlubricated scenario. When contrasted with unadulterated Ni coatings and the substrate, the Ni-$SrSO_4$ composite covering has a low contact, low grating and wear rate.

Xia et al. [14] work on this paper for the construction of Ni/Ti-N nanocomposites (NCs), in order to form a co-deposit with Ni ions, as a reinforcement phase, and therefore TiN nanoparticles were utilised. The coatings were made with a pulsation current density near about 60 mA/cm^2 that exhibited a low deposition amount and the smallest Ni embryos, measuring 141.6 nm on averaged. The thickness of Ni/Ti-N NCs inserted at 100, 60, and 20 mA/cm^2 was 78.3, 62.8, and 50.7 m, accordingly. With a Ra value of 29.57 nm, the Ni/Ti-N NC produced on 60 mA/cm2 had finest then compressed surface properties of entirely Ni/Ti-N coatings. Through a microhardness of 693.2 HV, the Ni/Ti-N NC deposited on 20 mA/cm2 has lowest microhardness among three Ni/Ti-N NCs. Furthermore, among these NCs, the coating generated at 60 mA/cm^2 had the maximum micro-hardness rate of 922.1 HV. Wasekar et al. [15] studied electrodeposited Ni-W/SiC nanocomposite coatings dry sliding wear behaviour after heat treatment. Pulsed current electrodeposition was used to deposit Ni (14 at percent)-W/SiC composite coatings which were heat-treated in vacuum at 500°C with different amounts of applied SiC particles. A pin-on-disc configuration was used to conduct sliding wear tests against WC-Co discs in dry situations (at room temperature). The mixing rule relates to nanocomposite coatings mechanical characteristics (ROM). The wear rate of the nanocomposite system is determined by the inter-particle spacing of SiC particles implanted in the Ni-W matrix.

Xiang et al. [16] the nickel graphene composite covering was investigated with the help of XRD, SEM, and EDS. The graphene disperses uniformly throughout the composite covering. Although the Ni granules are polished, the alignment of the crystallographic planes stays unchanged. The Ni-graphene coating has a substantially higher hardness and wear resistance than a pure Ni coating. Liu et al. [17] studies on pulsed electrodeposition which was used to development of Cobalt (Co) and graphene oxide/cobalt (GO/Co) composite coatings and have two metals with similar properties made from an aqueous solution containing cobalt sulphate and graphene oxide, among many other components. The GO nanosheets scatter equally in the composite coating, as seen by SEM and energy dispersed X-ray imaging, and the extension of GO alters the deposit's shape from circular to trying to project. The co-distribution of GO with Co ions in the composite coating favours the creation of hcp (1 0 0), (0 0 2), and (1 0 1) textures, as well as grain refinement and hardness enhancement functions. Wear studies demonstrate that including GO into the coating improves the deposit's wear resistance and reduces friction. The GO/Co composite coating has superior corrosion resistance than a pure Co coating, according to electrochemical corrosion experiments employing potentiodynamic polarisation and electrochemical impedance spectroscopy.

IOP Conf. Series: Materials Science and Engineering 1259 (2022) 012050 doi:10.1088/1757-899X/1259/1/012050

Xue et al. [18] works on electrodeposition that was used to improve tribological properties of Ni–CeO_2 composite coatings produced from CeO_2 nanoparticles in nickel sulfamate solvent. Changing the number of CeO_2 particles in a solution could regulate the level of Co-deposited CeO_2 in composite material. The Co-deposited CeO_2 nanoparticles reduce the friction coefficient and wear weight, according to the current study. Furthermore, in terms of friction and wear, the concentration of CeO2 has an impact on the properties of Ni–CeO_2 composite coatings. Despite the fact that the pure Ni coating has somewhat better wear resistance than the composite coating with a lesser CeO_2 concentration of 2.3 wt% but has substantially superior wear resistance than the composite coating with a higher CeO_2 level. Luo et al. [19] works on a unique functional organic–inorganic composite covering with good adhesion and barrier performance was devised in this work to ensure concrete serving in severe and corrosive environments. The prepared composite coating was constructed using ultrasonic mixing procedures to include kaolinite nanosheets into permeable epoxy resin. The distributed nanosheets provided stronger cross-linking, high temperature stability, and improved barrier properties to the epoxy. The composite coatings had good corrosion resistance and a stable bond with the concrete matrix. The composite coating with 1 wt% nanosheets had the best overall performance. It could penetrate concrete to a depth of 2376 metres, and the coating's dehydration reaction temperature was 168.1 °C. The maximum oxygen transmission and total charge passed were lowered to 63°C and 4.438 cc/m^2 respectively. The composite covering with concrete had a pull-off strength of 5.68 MPa, which was 32% greater than the permeable epoxy. The composite coatings produced strong adhesion and a high barrier to the concrete structure, indicating that they might be used in anti-corrosion engineering and construction. Cheng et al. [20] studied on metal surface quality and performance must be upgraded to keep up with the rapid expansion of modern industries. Composite electrodeposition (co-deposition) has become a popular method for increasing metal material surface performance. Co-deposition was used to successfully prepare a new type of graphene oxide (GO)-reinforced nickel–boron (Ni-B) composite coating on a 7075 aluminium (Al) alloy. The addition of GOs improved the composite's mechanical and anti-corrosion properties significantly, according to the results. When compared to the Ni-B alloy deposit (878 HV, 9.64 x 10^{-5} $mm^3.N^{-1}.m^{-1}$, and 116.64 105 $mm.y^{-1}$, the composite had a rougher, compact, cauliflower-like morphology with finer grains, a greater hardness (1532 HV), a lower rate of wear (5.20 x 10^{-5} $mm^3.N^{-1}.m^{-1}$), and a lower corrosion rate (33.66 x 10^{-5} $mm.y^{-1}$).

3. Discussion

3.1 Effect of wear in SiO2 particles are reinforced in a Ni matrix

On carbon steel, stainless steel, and magnesium alloys, SiO2-based synthetic amazingly areas with superior mechanical, corrosive and nanocomposite coatings with remarkable form, tenacity, and permeability have recently been produced to improve corrosion protection and wear properties of the material [21].

Ni–SiO2 successfully deposited composite coating on such an aluminium alloy surface with 3 preparations to improve adherence between the coatings and substrate. According to their findings, the Ni–SiO2 composite coating was tougher (65 HV) than the coatings generated could withstand fast changes in temperature (from 100 °C to -35 °C) and were made of pure nickel (50 HV). Coating's remarkable performance was attributed to low surface energy and micro/nanostructures [22]. To optimise the even SiO2 nanocrystals distributed throughout the Ni matrix, researchers introduced a new in-situ production of Nanocomposites (s-SiO2) in ChCl: 2EG ionic liquid solvent (DES). As a result, friction factor and wear rate were greatly enhanced by Nanostructures of SiO2 in a Ni matrix [23]. Carbon-based components were introduced along with composite coatings metallic matrix organizational structure using DES and with a nano-SiO2 particle content of 10 g/l. Using AZ91HP magnesium alloy, the maximum wear durability of Ni-SiO2 composite material may be attained, and the strengthening action in the Ni matrix nanomaterials was attributed with enhanced wear rate of Ni-nano-SiO2 composite coatings put over whitish pure Ni coated substrates [24].

Table 1: SiO2 particles in nickel matrix coatings. N.A. denotes instances where specific data is not available [23].

Coating composition	Size of the particles (nm)	Rate of wear (mm^3 /N.mm)	Coefficient of friction	Volume loss due to wear	Roughness of the surface (µm)	Microhardness (HV)
Ni-SiO_2	25	N.A.	N.A.	N.A.	3-4.5	Pure Ni: 30-50 / Ni-SiO_2: 50-65
Ni-SiO_2	30	Pure Ni: 2.4×10^{-2}, , Ni/s-SiO_2: 8×10^{-3}, Ni/c-SiO_2: 1.0-1.4×10^{-2}	Pure Ni: ~ 0.75, , Ni/s-SiO_2: ~ 0.35, Ni/c-SiO_2: 0.4-0.5	N.A.	N.A.	Pure Ni: 280-290 / Ni/s-SiO_2: 830-850, / Ni/c-SiO_2: 630- 830
Ni-SiO_2	20+5/ 20-5	N.A.	0.3 - 0.47 (F:30N, SiO_2 content: 5, 10, 15, 20 g/l)	At SiO_2 content of 5, 10, 15, and 20 g/l, wear track width at F=10N is 1680, 1500, 1540, and 1570 m.	N.A.	80 / Ni/ nano-SiO2: 270 to 350 (0 to 10 g/l) / 350 to 320 (10 to 20 g/l) Substrate (AZ91 Mg alloy):

3.2 Effect of wear on Ni-SiO_2 composite coating on Mg alloys

At low loads, AZ91HP mg alloy wear morphology, pure nickel coatings, and composites coating are shown in Fig. 3 whereas at high load, AZ91HP mg alloy wear morphologies, pure nickel covering, & nanocomposite coating are shown in Fig. 4. Adhesive wear is major wear mechanism of the AZ91HP magnesium alloy surface [26]. Friction pairs furrow the substrate surface during low-load (10 N) wear as shown in Fig. 3(a). Ploughing on the surface of AZ91HP mg alloy deepens and widens as the load (30 N) increases [27]. Elasto-plastic deformation of the surface on the material, followed by recurrent plastic deformation, causes surface fractures and increase wear rate. The temperature rises as the friction persists, and the wear on the surface adhesion becomes visible. The frictional adhesion sites in the friction course have been cut, resulting in adhesive wear as shown in Fig. 4 (a) [28].

At low loads (10 N), the pure nickel coating's wear surface is flat, despite the fact that certain shallow grooves run parallel to the direction of frictional couple signal as shown in Fig. 3(b). As during slide wear test, the contact parts would be the first to experience fatigue harm as a result of repetitive shear stress. Microcracks arise when fatigue effect accumulates to some point [29]. Microcracks expand and link under cyclic load, resulting in macrocrack formation. The macrocracks spread towards the misstatement at about the same instant. The macrocracks continue to expand as the load (30 N) rises. Once the crack tip stress intensity factor (Kmax) reaches material fracture toughness, the cracks develop an abrupt local buckling that behaves like being sliced. As a result, the fractures in the pure nickel coating show considerable exfoliation, which is a normal exfoliation wear pattern [30].

Figure 3. Low-load wear surface morphologies (10 N) (a) Magnesium alloy; (b) Pure nickel coating; (c) Nanocomposite coating [31]

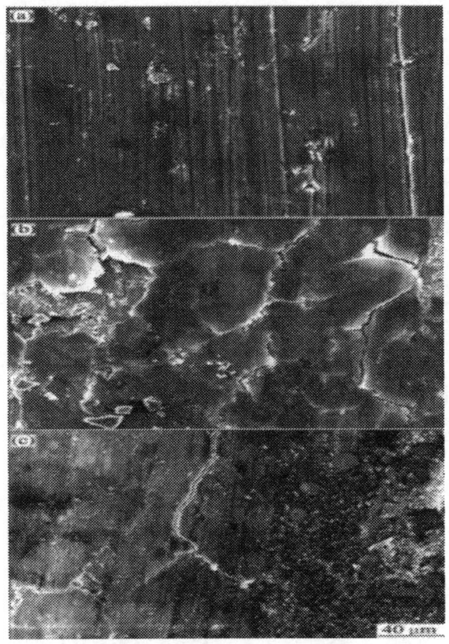

Figure 4. High-load wear surface morphologies (30 N) (a) Magnesium alloy; (b) Pure nickel coating; (c) Nanocomposite coating [31]

Nano-composite coatings wear surfaces are not only flat under low loads, grooves are also parallel to the direction of frictional pair motion, with shallow grooves, identical to pure nickel coating materials. However, compared to pure nickel plating, number of cracks is clearly lower [32]. As the load increases, grooves on the composite coating's wear surface deepen and wear particles emerged. However, with a conventional abrasive wear mechanism, exfoliation will not occur in vast areas, and wear debris will be incorporated into the wear test [33]. As a result, the composite coating's wear resistance has greatly increased. Nanoparticulate grain refinement and dispersive strengthening effects are the key explanations. The nanoparticles are evenly distributed throughout the composite covering, and there is significant distortion around the nanoparticles [34]. The movement and distortion of the nickel matrix dislocations are hampered as imperfections accumulate. As a result, the coatings hardness and strength dramatically improve which provides better wear resistance. The contact area eventually exposes the composite coating's nanoparticles throughout the friction process and serve as load-bearing elements, preventing adhesion areas from expanding and flaking and increasing adhesive wear resistance [35].

4. Conclusion

Depending on the coating's quantitative qualities, a correlation between electrodeposition parameters and the produced coatings tribological qualities are detailed. Ni-SiO2 nanocomposites with a homogenous, thick texture were produced on the AZ91HP mg alloy. The effectiveness of nanocomposite coatings & mg alloy together is revealed by form of the contact. Nanoparticulates grain refinement and dispersive strengthening effects improves wear resistance.

References

[1] Raghavendra, C.R., S. Basavarajappa, and I. Sogalad, Electrodeposition of Ni-nano composite coatings: a review. *Inorganic and Nano-Metal Chemistry*, 2018. **48**(1-2): p. 583-598.

[2] Akbulut H, Hatipoglu G, Algul H, Tokur M, Kartal M, Uysal M, Cetinkaya T. Co-deposition of Cu/WC/graphene hybrid nanocomposites produced by electrophoretic deposition. *Surface and Coatings Technology*. 2015 Dec 25;**284**:344-52.

[3] Cui W, Wang K, Wang K, Wang P. Effects of jet rate on microstructure, microhardness, and wear behavior of jet electrodeposited Ni–SiC composites. *Ceramics International*. 2018 Apr 15;**44**(6):7214-20.

[4] Carpenter CR, Shipway PH, Zhu Y. Electrodeposition of nickel-carbon nanotube nanocomposite coatings for enhanced wear resistance. *Wear*. 2011 Jul 29;**271**(9-10):2100-5.

[5] Omidvar H, Sajjadnejad M, Stremsdoerfer G, Meas Y, Mozafari A. Characterization of NiBP-graphite composite coatings deposited by dynamic chemical plating. *Anti-Corrosion Methods and Materials*. 2015 Mar 2.

[6] Maharana HS, Basu A, Mondal K. Structural and tribological correlation of electrodeposited solid lubricating Ni-WSe_2 composite coating. *Surface and Coatings Technology*. 2018 Sep 15;**349**:328-39.

[7] Vaezi MR, Sadrnezhaad SK, Nikzad L. Electrodeposition of Ni–SiC nano-composite coatings and evaluation of wear and corrosion resistance and electroplating characteristics. *Colloids and Surfaces A: Physicochemical and Engineering Aspects*. 2008 Feb 15;**315**(1-3):176-82.

[8] Sajjadnejad M, Mozafari A, Omidvar H, Javanbakht M. Preparation and corrosion resistance of pulse electrodeposited Zn and Zn–SiC nanocomposite coatings. *Applied Surface Science*. 2014 May 1;**300**:1-7.

[9] Wasekar NP, Haridoss P, Seshadri SK, Sundararajan G. Sliding wear behavior of nanocrystalline nickel coatings: Influence of grain size. *Wear*. 2012 Aug 30;**296**(1-2):536-46.

[10] Shafiei M, Alpas AT. Friction and wear mechanisms of nanocrystalline nickel in ambient and inert atmospheres. *Metallurgical and materials transactions A*. 2007 Jul;**38**(7):1621-31.

[11] Gamburg YD, Zangari G. Introduction to electrodeposition: Basic terms and fundamental concepts. In *Theory and Practice of Metal Electrodeposition* 2011 (pp. 1-25). Springer, New York, NY.

[12] Walsh FC, Wang S, Zhou N. The electrodeposition of composite coatings: Diversity, applications and challenges. *Current Opinion in Electrochemistry*. 2020 Apr 1;**20**:8-19.

[13] Bajwa R, Khan Z, Nazir H, Chacko V, Saeed A. Wear and friction properties of electrodeposited Ni-based coatings subject to nano-enhanced lubricant and composite coating. *Acta Metallurgica Sinica*. 2016 Oct;**29**(10):902-10.

[14] Dhakne A, Jaju S, Shukla S. Review on analysis of enhancing wear properties through thermo-mechanical treatment and grain size. *Materials Today: Proceedings*. 2022 Jan 1;**60**:2270-2.

[15] Liang XS, Ouyang JH, Li YF, Wang YM. Electrodeposition and tribological properties of Ni–SrSO4 composite coatings. *Applied Surface Science*. 2009 Jan 15;**255**(7):4316-21.

[16] Xia F, Li C, Ma C, Li Q, Xing H. Effect of pulse current density on microstructure and wear property of Ni-TiN nanocoatings deposited via pulse electrodeposition. *Applied Surface Science*. 2021 Feb 1;**538**:148139.

[17] Wasekar NP, Bathini L, Ramakrishna L, Rao DS, Padmanabham G. Pulsed electrodeposition, mechanical properties and wear mechanism in Ni-W/SiC nanocomposite coatings used for automotive applications. *Applied Surface Science*. 2020 Oct 15;**527**:146896.

[18] Xiang L, Shen Q, Zhang Y, Bai W, Nie C. One-step electrodeposited Ni-graphene composite coating with excellent tribological properties. *Surface and Coatings Technology*. 2019 Sep 15;**373**:38-46.

[19] Liu C, Su F, Liang J. Producing cobalt–graphene composite coating by pulse electrodeposition with excellent wear and corrosion resistance. *Applied Surface Science*. 2015 Oct 1;**351**:889-96.

[20] Xue YJ, Jia XZ, Zhou YW, Ma W, Li JS. Tribological performance of Ni–CeO2 composite coatings by electrodeposition. *Surface and Coatings Technology*. 2006 May 22;**200**(20-21):5677-81.

[21] Luo S, Wei J, Xu W, Chen Y, Huang H, Hu J, Yu Q. Design, preparation, and performance of a novel organic–inorganic composite coating with high adhesion and protection for concrete. *Composites Part B: Engineering*. 2022 Apr 1;**234**:109695.

[22] Cheng D, Zhang L, Zhu Y, Xia H, Li N, Song W, Bai H, Ma H. Preparation and Properties of Electrodeposited Ni-B-Graphene Oxide Composite Coatings. *Materials*. 2022 Mar 20;**15**(6):2287.

[23] She Z, Li Q, Wang Z, Li L, Chen F, Zhou J. Researching the fabrication of anticorrosion superhydrophobic surface on magnesium alloy and its mechanical stability and durability. *Chemical Engineering Journal*. 2013 Jul 15;**228**:415-24.

[24] Li X, Yin S, Luo H. Fabrication of robust superhydrophobic Ni–SiO2 composite coatings on aluminum alloy surfaces. *Vacuum*. 2020 Nov 1;**181**:109674.

[25] Sajjadnejad M, Haghshenas SM, Badr P, Setoudeh N, Hosseinpour S. Wear and tribological characterization of nickel matrix electrodeposited composites: A review. *Wear*. 2021 Dec 15;**486**:204098.

[26] Du L, Xu B, Dong S, Yang H, Tu W. Study of tribological characteristics and wear mechanism of nano-particle strengthened nickel-based composite coatings under abrasive contaminant lubrication. *Wear*. 2004 Nov 1;**257**(9-10):1058-63.

[27] Shukla, Sourabh, et al. Effect of grain refinement on sensitization of high manganese austenitic stainless steel. *Materials Today: Proceedings*, 2021, **44**: 2802-2807.

[28] Wang L, Tu JP, Chen WX, Wang YC, Liu XK, Olk C, Cheng DH, Zhang XB. Friction and wear behavior of electroless Ni-based CNT composite coatings. *Wear*. 2003 Nov 1;**254**(12):1289-93.

[29] Shi L, Sun CF, Zhou F, Liu WM. Electrodeposited nickel–cobalt composite coating containing nano-sized Si3N4. *Materials Science and Engineering*: A. 2005 Apr 25;**397**(1-2):190-4.

[30] Borkar, S., Dabhekar, K., Khedikar, I., Jaju, S., Analysis of Flat Slab Structures in Comparison with Conventional Slab Structures, *IOP Conference Series: Earth and Environmental Science*, 2021, **822**(1), 012049.

[31] Yan L, Yu SR, Liu JD, Han ZW, Yuan DS. Microstructure and wear resistance of electrodeposited Ni-SiO_2 nano-composite coatings on AZ91HP magnesium alloy substrate. *Transactions of Nonferrous Metals Society of China*. 2011 Aug 1;**21**: s483-8.

[32] Jaju, S.B., Charkha, P.G., Kale, M., Gas metal arc welding process parameter optimization for AA7075 T6, *Journal of Physics: Conference Series*, 2021, **1913**(1), 012122.

[33] Ravekar K, Jaju S. Design and analysis of material discharge plate of microsurfacing machine. *Materials Today: Proceedings*. 2021 Jan 1;**38**:2385-91.

[34] Ravekar K, Jaju SB, Narsingpurwala AZ. Comparative analysis of material for material discharge plate of microsurfacing machine. *Materials Today: Proceedings*. 2021 Jan 1;**46**:249-55.

[35] Ravekar K, Jaju S. Performance improvement of microsurfacing machine: a case study. In *Advances in Mechanical and Materials Technology* 2022 (pp. 93-101). Springer, Singapore.

IOP Conf. Series: Materials Science and Engineering 1259 (2022) 012051 doi:10.1088/1757-899X/1259/1/012051

Flexural strength of Epoxy resin and Bagasse fiber reinforced composite materials

Anurag Saroj[1], Ravi Vishwakarma[2], Chandra Bhushan[2]

[1]Research Scholar, Department of Mechanical Engineering, Babu Banarsi Das University, Lucknow, UP, India
[2]Assistant Professor, Department of Mechanical Engineering, Babu Banarsi Das University, Lucknow, UP, India

Abstract. In this research, we are going to take out to investigate flexural strength and tensile strength properties of composite material with the using of epoxy resin and bagasse fibre in the powder form. These composites are bonded using epoxy resin and hardener mixed in appropriate volume. This composite was prepared with using bagasse fibre in powder form and the bagasse fibre powder used with different ratio in each material which are prepared and we found the high strength at the using 5gm bagasse fibre powder in both of Tensile and Flexural strength. Here for preparing specimen using Hand layup method. It can be use in more way because rare future of the environmentally friendly, waste utilization, good strength, bio-degradability and alternative for plastic.

1. Introduction

The sugarcane bagasse fiber is highly generated in the agriculture area and it goes waste after extracting the juice of sugarcane. Nowadays use of agriculture crop is located as substances improvement of fiber composite reinforcement and it is a simple and secure recyclable product at the end of that provider and its maximum appropriate aspect to use of this agricultural crop like as cotton, wheat, banana, bagasse, etc. it's far low price and mild in weight and it additionally has true properties. In this research we study on the mechanical behavior of bagasse fiber reinforced epoxy composite, different type of force such as tensile and compressive force test perform on this product, which is prepared by us, in which apply different amount of force on those prepared materials also we see what is the maximum wearing capacity of force. These studies, look at approximately the mechanical conduct of bagasse fiber reinforced epoxy composite material, an exceptional form of energy together with Flexural strength take a look at that reinforced composite. The traits in composite fabric after assembly the demanding situations of aerospace zone have deluge down for serve to home and business applications. Composites, the surprise fabric with light-weight; excessive strength-to-weight ratio and stiffness homes have come a protracted manner in changing the traditional substances like metals, wooden etc. The fabric scientists everywhere in the global centred their interest on herbal composites bolstered with Jute, Sisal, Coir, Pineapple etc. mainly to put down the rate of raw materials. and it is Dry pulpy stalky substances that remain after crushing sugarcane to bring out their juice. It is used as biofuel for the manufacturing of heat, energy, etc and It has good mechanical properties and it's also easily recyclable, and eco-friendly it is low cost, light in weight with good strength And manufacture of building materials. Its residences consisting of Cellulose: 45-55%, Hemicellulose: 20-25%, Lignin: 18-24%, Ash: 1-4%. Sandesh S Nayak et al, Issue 3 March 2020 Investigated on the use of bagasse fiber with polymer (epoxy resin) to make matrix composite and study its mechanical properties and performance for automobile (structural) application. Good bonding with epoxy resin. flexural strength. [1] V. Vidyashri et al. December 2019, waste of bagasse fibre using epoxy chemically ($KMnO_4$ + NaOH) treated to improve the capability and adhesion Scanning Electron Microscopy (SEM), X-Ray Diffraction (XRD) and Thermo Gravimetric Analysis (TGA), also find out mechanical properties by using tensile test. roughness of the surface improved. mechanical properties increase after using chemically treated with the help of reinforcement. [2]

1.1 Bagasse Fibre

It is dry mushy stalky substances that remains after crushing sugarcane to bring out their juice. It is used as clean-energy for the manufacturing of heat, energy etc. and manufacture of constructing materials. Its residences which include Cellulose: 45-55%, Hemicellulose: 2025%, Lignin: 18-24%, Ash: 1-4%. It additionally satisfies the greening necessities with the aid of using being biodegradable, recyclable reusable and additionally eco-friendly. [3]

Figure 1. Bagasse fibre

1.2 Epoxy Resin

Epoxy is the own circle of relatives of simple additives or cured cease merchandise of epoxy resin. Epoxy resin is likewise called poly-epoxides. It is very viscous liquid.[4]

Epoxy resin properties:

High strength.
Low shrinkage.
Excellent adhesion to various substrates.
Low cost.
Low toxicity.

Figure 2. Epoxy resin and Hardener

2. Literature Review

Sandesh S Nayak et al. investigated that uses of bagasse fibre with polymer to make matrix composite (epoxy resin) and study its mechanical properties and performance for automobile (structural) application. The fabricated composite sample were cut according to the ASTM standards for different experiment like as: density, hardness, tensile strength and impact strength. The work is successful and the result which obtained are suitable and comparable for the application of few automobile sectors. It has good bonding with epoxy resin. Maximum tensile strength and flexural strength observed for the prepared composite with the weight percentage of the bagasse fibre being 3.75%.[1]

V. Vidyashri et al. In this work use of waste of bagasse fibre to make a useful thing, in which the using epoxy with sugarcane bagasse fibre chemically treated to improve the capability and adhesion with the epoxy polymer. Treated and untreated fibres analysis such as Scanning Electron Microscopy (SEM), X-Ray Diffraction (XRD) and Thermo Gravimetric Analysis (TGA) and also find out mechanical properties by using tensile test. While chemically treatment performed under goes on using alkaline treatment, KMnO4 for improve mechanical and thermal properties. The result shows that using chemical these fibres to surface analysis using SEM which revealed the roughness of the surface improved. After all test mechanical properties showed improved.[2]

Sebastin joyal J et al. research on the analysis of composite material using sugarcane bagasse with epoxy resin HY951 resin and LY551 hardener suitably mixed in appropriate volume. The result found from the sugarcane bagasse reinforced epoxy-based hybrid composite is usefulness. These tests performed on the UTM machine. The composite become less ductile as sugarcane content is increased. Further work done on the fibre surface treatment and residue elimination in recycled HDPE to improve mechanical properties.[3]

Lalta Prasad et al. this research about the physical and mechanical behaviour of sugarcane bagasse fibre-reinforced epoxy composite. In which the test will performed on the various type of length for the test like as: Density test, wear test, tensile test, impact strength test, hardness test. The result of this test show on the base of study that minimum void obtained 5mm length of all varieties of fibre reinforced is better from other composite formed. Maximum ultimate tensile strength obtained in composite variety A and minimum tensile strength obtained C variety the optimum length was 10mm. All test is good to improve its strength.[4]

3. Method

In this method, first of all make the pattern by using of the wooden ply with the dimension of 100mm*30mm*10mm and made 6 pieces of wooden pattern for flexural test. Each of the workpiece have different ratio of the epoxy resin, hardener and sugarcane bagasse fiber. The measurement of each parameter with the measuring instrument, after that mixed every one with measuring weight in glass till 3 to 5 min then filled in the pattern with prepared raw material, and kept them for 1 to 2 days in the pattern for dry.

3.1 Flexural strength

The three-point bending flexural take a look at offers values for the modulus of elasticity in bending Ef, flexural stress σf, flexural strain εf and the flexural strain–stress reaction of the material. This take a look at is accomplished on a general checking out system (tensile checking out system or tensile tester) with a three-point or four-point bend fixture. The most important gain of a three-factor flexural take a look at is the convenience of the specimen preparation and checking out.

Figure 3. Testing

Table 1. Ratio of raw materials

S.N	Specimen designation	bagasse fiber (grams)	epoxy resin (grams)	Hardener (grams)
1	0	1	18	10
2	1	2	19	10
3	2	3	20	10
4	3	3	21	10
5	4	4	22	10
6	5	5	23	10

Figure 4. workpiece of flexural strength

3.1.1 *Calculations For flexural strength*

- For F = 0.285kN (1)
- Flexural stress = $\frac{3bd}{2bd^2}$ (2)

Put the eq. no. (1) in eq. no (2), then we get;

$$= \frac{3*285*100}{2*30*10^2}$$

$$= 14.25 \text{ MPa}$$

- For F = 0.425kN (3)

Put the eq. no. (3) in eq. no. (2), then we get;

Flexural stress = $\frac{3*425*100}{2*30*10^2}$

$$= 21.25\text{MPa}$$

- For F = 0.328kN (4)

Put the eq. no. (4) in eq. no. (2), then we get;

Flexural stress = $\frac{3*328*100}{2*30*10^2}$

$$= 16.4 \text{ MPa}$$

- For F = 0.320kN (5)

Put the eq. no. (5) in eq. no.(2), then we get;

Flexural stress = $\frac{3*320*100}{2*30*10^2}$

$$= 16\text{MPa}$$

- For F = 0.455kN (6)

Put the eq. no. (6) in eq. no. (2), then we get;

Flexural stress = $\frac{3*455*100}{2*30*10^2}$

$$= 22.75\text{MPa}$$

- For F = 0.565kN (7)

Put the eq. no (7) in eq. no (2), then we get;

$$\text{Flexural stress} = \frac{3*565*100}{2*30*10^2}$$

$$= 28.25\text{MPa}$$

Table 2. Result of flexural strength

S.N	Specimen designation	Flexural Strength (MPa)
1	0	14.25
2	1	21.25
3	2	16.4
4	3	16
5	4	22.75
6	5	28.25

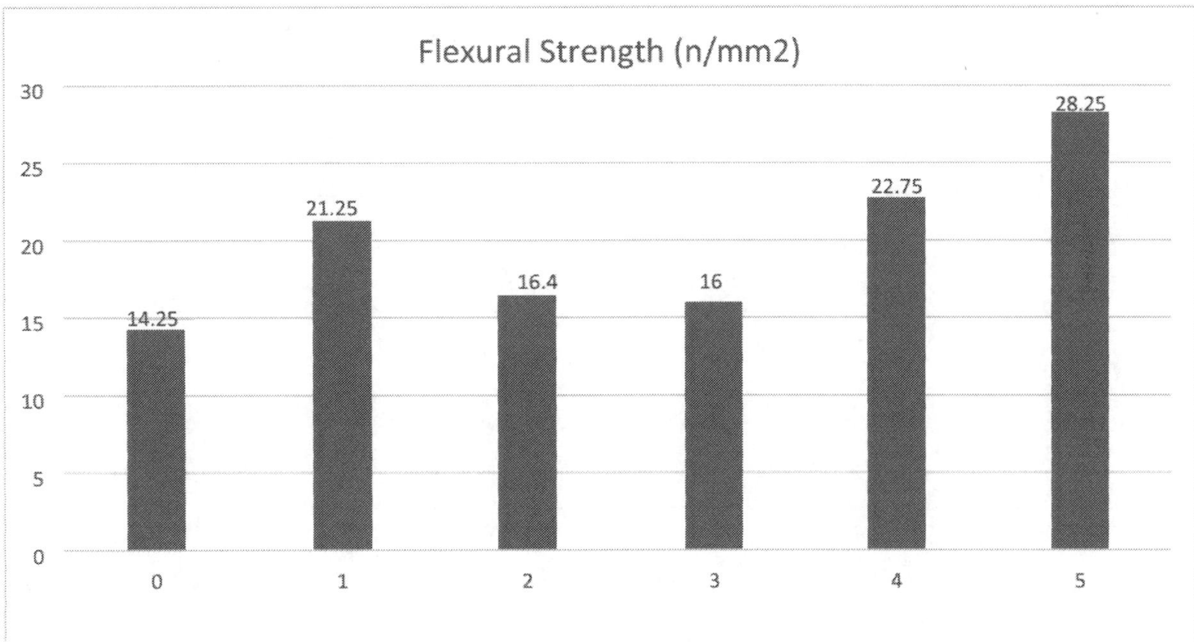

Figure 5. Chart for Flexural strength

3.2 Tensile Strength

Tensile testing, additionally referred to as tension testing, is an essential substances technology and engineering take a look at wherein a pattern is subjected to a managed anxiety till failure. Properties which might be without delay measured through a tensile take a look at are breaking strength, ultimate tensile strength, most elongation and discount in area. From those measurements the subsequent properties also can be determined; load at peak and yield strength, traits. The dimension of workpiece is 165mm*15mm*10mm. These workpieces have different ratio of raw materials.

Figure 6. Specimen "B"

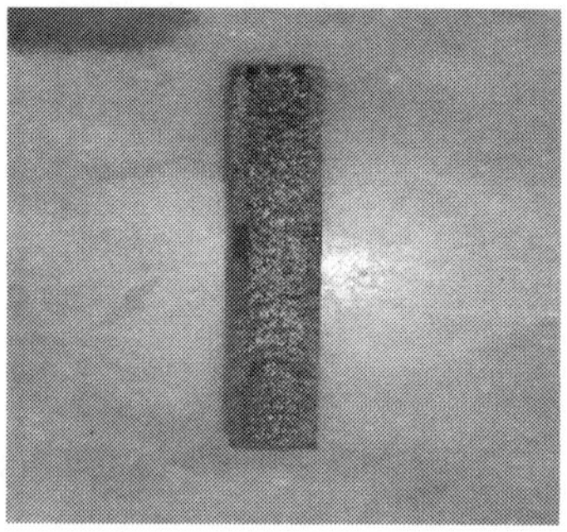

Figure 7. Specimen "C"

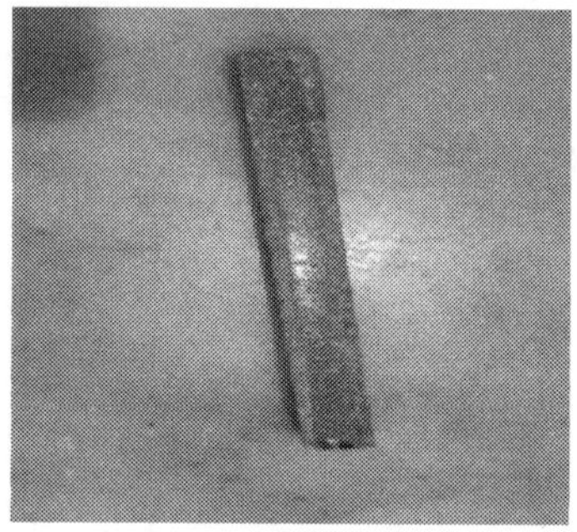

Figure 8. Specimen "A"

Table 3. Raw materials

S. N	Specimen designation	bagasse fibre (grams)	epoxy resin (grams)	Hardener (grams)
1	A	4	32	4
2	B	5	30	5
3	C	3	34	3

Table 4. Tensile strength

S. N	Specimen designation	Yield stress (MPa)	Tensile strength (MPa)	Load at peak (kN)
1	A	8.213	11.115	1.79
2	B	18.293	19.453	1.93
3	C	6.523	8.314	1.3

3.2.1 Graph for Tensile strength

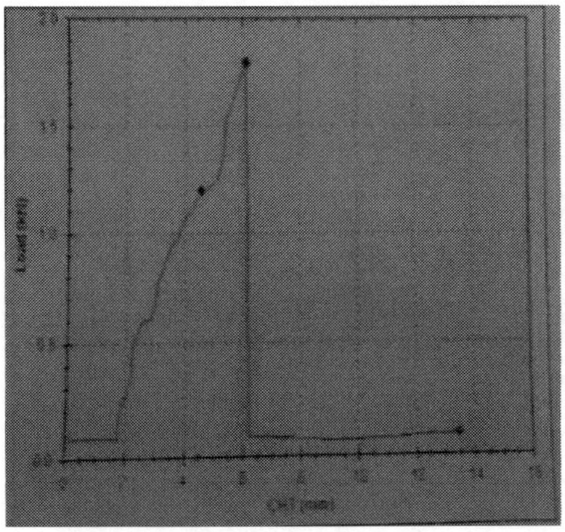

Figure 9. Graph of w/p "A"

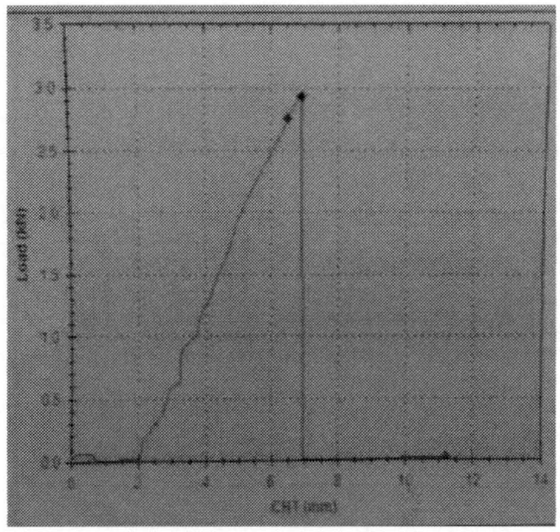

Figure 10. Graph of w/p "B"

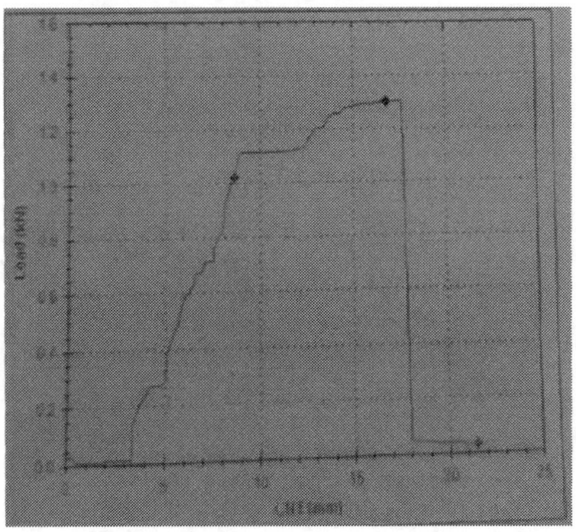

Figure 11. Graph of w/p "C"

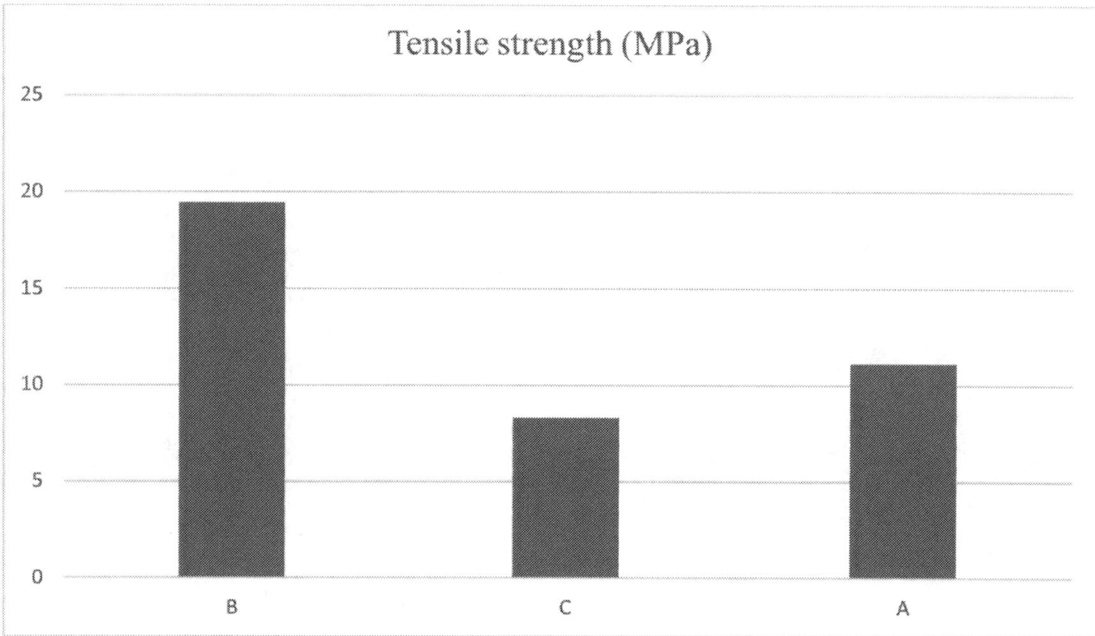

Figure 12. Chart for tensile strength

4. Results and discussion

Hence, the value change with their variation in mixture of epoxy resin and bagasse fiber and result shown following:

- The maximum value of flexural strength is found on workpiece "6" which is 28.25 MPa
- The minimum value of flexural strength is found on workpiece "0" which is 14.25 MPa
- The maximum value of tensile strength is found on workpiece "B" which is 19.453 MPa
- The minimum value of tensile strength is found on workpiece "A" which is 8.314 MPa
- The percentage of elongation on each w/p has same 5%
- The flexural strength of this reinforced composite materials is got by 3-point bend test using UTM machine. The flexural strength is varying with using the different type ratio of each other (epoxy resin, bagasse fiber) and hardener put same in each mixture. The flexural strength varies from 14.25 MPa to 28.25 MPa and maximum result found in 5gm of bagasse fibre.
- In this study, observed the result of fibre reinforced composite materials.it is observed that the tensile strength varies from 8.314 MPa to 19.453 MPa. The result varies with variation in the raw materials, the tensile strength increases with the increase in bagasse fibre up to 5grams. It has good bonding between epoxy and bagasse fibre.

5. Conclusion

The investigation on flexural and tensile behaviour of bagasse fibre with epoxy resin reinforced composite materials with different ratio of each other. The conclusion drawn from this work are here; The flexural strength of the composite varies from 14.25 MPa to 28.25 MPa. the maximum flexural strength is obtained for the composite with 5grams. This is the water resistance materials. In this study, observed the result of fibre reinforced composite materials.it is observed that the tensile strength varies from 8.314 MPa to 19.453 MPa. Here we found the maximum result of both (flexural and tensile) test on 5gm bagasse fibre with different ratio of epoxy resin, but when increase the bagasse fibre then the result will increase and epoxy resin also varies. It can be use in more way because unique future of the eco-friendly, waste utilization, good strength, bio-degradability, alternative for plastic and also recyclable. It can be used in the place of plastic, wood. In the future, work to do some more research on it to make it fireproof. When it will be used in the place of wood, then deforesting would be reduced and use of the waste of bagasse fiber to make some wooden thing.

Save The Tree

6. References

[1] Sandesh S Nayak, K Reuben Joseph2020. Evaluation of Sugarcane Bagasse Polymer Composite for Structural Applications Volume: 07 Issue: 01 | Jan 2020, e-ISSN: 2395-0056 p-ISSN: 2395-0072

[2] V. Vidyashri, Henrita Lewis, P. Narayanaswamy, K.Subrahmanya Bhat 2017. Preparation of chemically treated sugarcane bagasse fibre reinforced epoxy composites and their characterization.

[3] 0072Sebastin joyal J, Karthick A, Mahendran K, Palanisamy K, Ramkumar R 2020. analysis of composite material formed by sugarcane bagasse and epoxy resin Volume 8, Issue 3 March 2020 | ISSN: 2320-2882.

[4] Lalta Prasad , Shiv Kumar , Raj Vardhan Patel , Anshul Yadav , Virendra Kumar 2020.Jerzy Winczek Physical and Mechanical Behaviour of Sugarcane Bagasse Fibre-Reinforced Epoxy Bio-Composites Received: 14 October 2020; Accepted: 26 November 2020; Published: 27 November 2020 Materials 2020, 13, 5387.

[5] Vinay K Singh 2012. Evaluation of mechanical properties of bagasse-glass fiber reinforced composite September 2012

[6] Deepa G. Devadiga1, K. Subrahmanya Bhat GT Mahesha 2020. Sugarcane bagasse fiber reinforced composites: Recent advances and application, Article: 1823159 | Received 24 Jun 2020, Accepted 03 Sep 2020, Published online: 21 Sep 2020.

[7] Y.R.LohaD.SujanaM.E.RahmanaC.A.Dasb 2013. Sugarcane bagasse—The future composite material, Received 9 July 2012, Revised 14 February 2013, Accepted 6 March 2013, Available online 24 April 2013.

[8] Sajjad Ali Mangi, N Jamaluddin1, M H Wan Ibrahim, Abd Halid Abdullah, A S M Abdul Awal1, Samiullah Sohu and Nizakat Ali ,2017. Utilization of sugarcane bagasse ash in concrete as partial replacement of cement, 2017 IOP Conf. Ser.: Mater. Sci. Eng. 271 012001.

[9] Marwa A. Abd El-baky, Mona Megahed, Hend H. El-Saqqa, A. E. Alshorbagy, 2021. Mechanical properties of natural-synthetic hybrid composites, April 2021 DOI:10.21608/eijest.2021.58084.103.

A review of effect on metallurgical properties of friction stir welding using high entropy alloy

M D Shende[1], A S Bombate[1], M N Kumbhare[1], C R Dhabale[1], and S S Shukla[1]

[1]Department of Mechanical Engineering, G.H. Raisoni College of Engineering, Nagpur, India

Email: sourabh.shukla@raisoni.net

Abstract. The work summarizes the review of use of High Entropy Alloy in the Friction Stir Welding. When High Entropy Alloy is utilized in the Friction Stir Welding the resultant mechanical, metallurgical as well as corrosive properties show the influence of HEA base material properties. The Process used to obtain HEA provides base properties affecting the microstructural impact due to heat generation in FSW. Taking into account HEA, FSW too improve the welding properties of the material. The grain size was affected by heat generation and pressure of rotational as well as transverse speed, and could be differentiated in different zones (SZ, HAZ, and TMAZ) and sides (advancing, retreating, top, and bottom). Results of misorentation, HABs, and LABs were influences by the type of recrystallization. GGAE are associated to HABs and SFE to heating phenomenon, so overall it is HEA reaction to the heat generation in different areas and its effects seen in recrystallization of grains.

Keywords: Friction Stir Welding, High Entropy Alloy, Stacking Fault Energy, Grain Growth Activation Energy

1. Introduction

In 1991 Friction Stir welding (FSW) was introduced at The Welding Institute (TWI), UK, and patented in the USA, Europe, Japan, and Australia by TWI in 1992[1-2]. Initially, the welding method used lightweight metal. Aluminum and its alloys were the first metal used in Friction Stir Welding [4]. Aluminum and most of its alloys carrying 1000x-7000x series are widely researched and studied in FSW [4-10]. Friction Stir Welding is solid-phase welding using a rotational tool having a cylindrical shoulder and pin within it. Tool's friction produces heat, deforming the contact area and then with the transverse moment creating mechanical pressure and pin moment mixing the two-metal result in joining of the two work-pieces [3]. The working temperature of FSW is lower than the melting point of the material. Result of which is no solidification defect [3]. Till now, Friction Stir Welding has evolved a lot, from machine design to the work piece's joint strength, hardness, corrosion resistance, microstructure, and FSW effect on different alloys and metals affecting grain properties [9-10]. FSW has undertaken various similar or dissimilar alloys welding providing efficient joints. [11-16]

Researchers have found Friction Stir Welding with High Entropy Alloys as the welding material strengthens the overall result of the weld seam. Research on multicomponent alloys started in 1981, but in 1995 concept of HEA came into the limelight. Later in 2004, two simultaneous pieces of research by Jien-Weh Yeh [17] and Brian Corton [18] created the base for High Entropy Alloy (HEA). High Entropy Alloy is the mixture of five or more principal elements in an equiatomic proportion. The concentration of these elements should be between 5 to 35% and can contain some minor elements below 5% concentration. The name High Entropy alloy deduces that the liquid or random solid Solution has significantly higher mixing entropies. Studies suggest High Entropy Alloy can develop many phases and intermetallic compounds [19]. HEA shows high strength, wear resistance, temperature strength, good structure stability, corrosion resistance, and more [19-21].

Welding acts as a necessary parameter for any newly introduced structured material. Several materials, when introduced, have been experimented with fusion welding. The HEA, when subjected to fusion welding [21], shows significant results. FSW joint shows improved properties, and it can be achieved by HEA as welding material. FSW's when joined with the help of FSW, provides weld with significant mechanical and microstructural properties [30-37]. First, Zhu et al. [38-41] used FSW in welding dissimilar HEA at different welding speeds, resulting in the study of microstructure and recrystallization of joints.

In this paper, we review how High Entropy Alloy, when used in Friction Stir Welding reacts to process and provides output properties. This paper discusses the various aspect of welding joint of HEA material when used in the FSW process. This paper will provide various metallurgical aspects of FS-Welded HEA covering its grain recrystallization, refinement, formation of different zone and bands to parameters of FSW causing changes in its microstructure.

2. Current Scenario

Currently, Friction Stir Welding of High Entropy Alloy is done for the evolution of the metallurgical and mechanical properties of HEA. The changes in microstructure, grain structure, strength, fatigue, and various other aspects of the mechanical and metallurgical properties are studied.

Some of the recent studies on the Friction Stir Welding of HEA are discussed. Friction Stir Welding of metastable Cu HEA [33] was tested under different scanning calorimeter, mechanical and microstructural characteristics. The result obtained showed a high tensile strength of weld seam and also heat-affected zone (HAZ) provided higher ductility and strength comparable to Base Metal (BM). Al0.3CoCrCu0.3FeNi HEA was Friction Stir Welded and the observation was focused on the Stir Zone (SZ). The SZ saw increased tensile strength of 920 MPa affecting 37% elongation. Microstructurally, grains were partially recrystallized during FSW which resulted in the synergetic strengthening of SZ of the weld seam. FeCoCrNiMn HEA was welded with 1060Al alloy to study intermetallic compounds (IMCs) as seen in many dissimilar metals welding. Microstructural observation showed the interfacial layer was continuous, but the thickness of the layer varied from 1.3 to 1.7μm. The interfacial layer was partitioned into three out of which two were close to Al showing the same monoclinic Al13Fe4 type IMC but differences in composition and depletion of the element were seen. And the third had parallelly columnar grain of orthorhombic Al5Fe2 type IMC. CoCrFeNi HEA was Friction stir welded [22] and under different processing parameters differently affected zone was observed. Thermo-mechanically affected zone (TMAZ) showed mixed microstructure with elongation in grain, whereas Stir Zone (SZ) was equiaxed until the change in rotation speed caused an increase in grain size and change in the shape of grain with

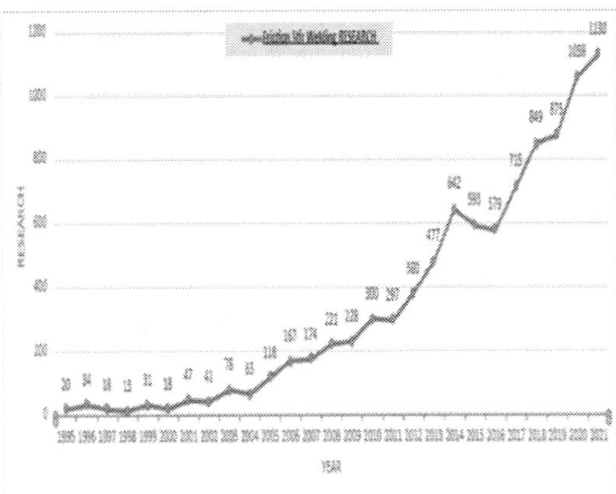

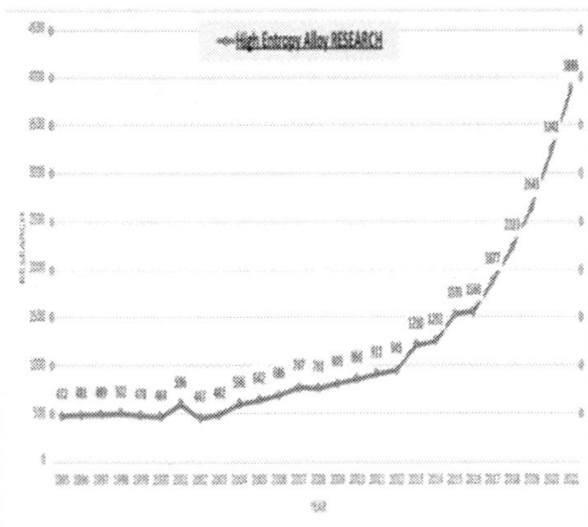

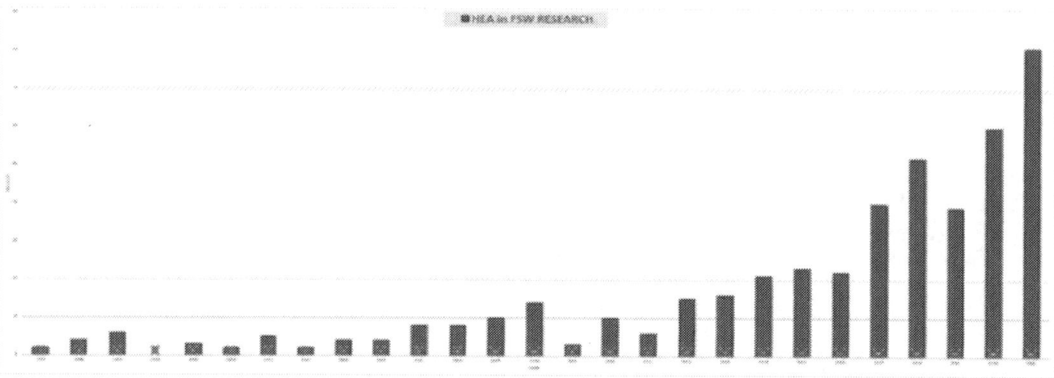

Figure 1. (a) graph represent development in friction Stir Welding, (b) graph represents development

change in various parameters. SZ hardness was higher than base metal (BM) observed due to dynamic recrystallization. The ductility of the joint seemed to be close to the base metal.

3. Metallurgical Properties

There is a lot to understand when it comes to metallurgical properties, we haven't had a clear idea of some of the phenomenon which occurs exceptionally. But there are lots of studies that clear mostly all aspects of metallurgical necessity during welding and study of FS-Weld High Entropy Alloy. In this section, we tried to explore the various methodologies used to study the metallurgical aspect of weld and also the result that was found because of the process parameter. We will explore various findings in Electron Backscattered Diffraction (EBSD), Scanning Electron Microscopy (SEM), Transmission Electron Microscopy (TEM), X-ray Crystallography (XRD), Energy Dispersive X-ray Spectroscopy (EDS/EDX). It will be carrying texture formation, Banded structure details, affected grain size and boundary, Stacking Fault Energy (SFE), Grain Growth Activation Energy (GGAE), Geometrically Necessary Dislocations (GND), and misorientation.

3.1 Methodologies Used

Let's get through the technologies that were used to observe the result. It's not the base we want to explore but let us get an understanding of some of the widely used methodologies in the research field to observe metallurgical specifically microstructural properties of the weld during Friction Stir Welding. EDX/EDS, SEM, EBSD, and TEM contain the technologies that were used with the proper method during the process to get the resulting observations. The observation can be said to be observed for the research paper so there might be a slight difference concerning the same observation for the practical purpose. Let's start with the EDX or EDS which stands for Energy Dispersive X-ray Spectroscopy [23-3 2] is mostly used alone to get the chemical composition properties of the weld. It is an analysis technique to produce an observational graph on the fundamental basis that each element has a unique atomic structure that produced relative peaks when observed. There are four major components used for the technique that is such as x-ray source, the x-ray detector, the pulse processor, and the analyzer. Apart from chemical composition, multilayer thickness of various alloys. This might give a double thickness peak also the accuracy of EDS [25, 28, 29-33] depends on the nature of x-ray deflection by atom. It is often used with Scanning Electron Microscopy (i.e., SEM) to analyze considered surface area samples. SEM on its individual is practiced to get the high-resolution imaging of the sample which could act as high informative details of the various surface properties like flaws, containment, fracture, or corrosion. It uses a focused beam of electrons to produce a highly magnified image of a complex surface topography structure. The use of SEM is especially for getting visual aspects of various small details that could be seen and could also be observed when

SEM is used with EDX analysis. When both are used together add a proper understanding of each other details which can be said to be complete chemical and elemental characteristics. To be more specific most of the metallurgical observation is found through their combined test. Some used EBSD [23-33] or Electron Backscattered Diffraction as we know it, used to measure grain size and boundaries, yet it is a type of SEM that provides data in the form of crystallography representing different crystal orientations to define different metals. In this technique, stationary electrons are exposed to the crystalline sample tilted at an angle. The electron scattering throughout the same produced a different pattern that is detected in the fluorescent screen. It helps in getting various details with the study of texture and orientation analysis, yet it has a drawback that it produces only geometry of the same yet ignoring various physical parameters that could be of importance. Kernel average misorientation (KAM) under the influence of EBSD provides a measure of local grain misorientation. XRD stands for X-ray Crystallography [23, 25-33] used to get the pattern of crystal also the phase and orientation. These are the various processing unit that compiles data of the weld before and after the Friction Stir Welding of HEA, be more broadly these techniques are used with various others to get more detailed data. In this, mainly a beam of x-ray is passed through the specimen and the diffraction by the crystalline structure is used to gather the density of electrons present in the structure. In short, the atom's position, atomic bond, and disorder details could be obtained. These were the basic techniques that were used mainly for the metallurgical aspect of the FS-welded HEA. There might be other processes you might be aware of yet these were the some mostly mentioned in the study of this topic.

3.2 Formation Of Different Regions

Base Metal (BM) after friction stir welding goes through various processes that when noticed in the metallurgy can cause the formation of various zones. This zone may be the work of welding working parameter and the heat transfer that may lead to various other parameters differentiation. These differentiating parameters can be studied in the group by separating BM into various zones. These zones as we may know are Stir Zone (SZ), Heat Affected Zone (HAZ), [22-24], and Thermo-mechanically Affected Zone (TMAZ) [25-28, 30-32]. This region was registered on the area starting from the center of the weld and then moving away from it. The area like SZ is the mid-section where the main welding tool acts, later on, is the HAZ which from its name suggests the area after the SZ which is mostly affected by the heat transfer that takes place while welding. This heat register differentiation in different microstructural parameters. Moving forward is the TMAZ which is a Thermo-mechanically Affected Zone that is the work of mechanical movement with the heat transfer that takes place through the same zone resulting in changing parameters that can be noticed differentiating itself into a different zone [25-30]. Potentially heat affected zone and Thermomechanically affected zone are based on the welding process and how the metal undergoes the process affecting its properties. The three zones were differentiated in the microstructural properties with slightly change or having a high difference in some of the grain structure, grain size, grain boundaries or the surface structure, etc. In some of the researchers both HAZ and TMAZ were noticed. The heat-affected zone (HAZ) is the non-melted area of the metal that changed the material properties when exposed to the friction stir welding rotational high temperature or we can say that the heat produced by the tool's rotation in the area after the Stir Zone. Whereas TMAZ [23-28] is the zone formed in the welding especially frictions stir welding that acts as a transition zone between SZ and heat-affected zone. Stir zone as from the name we can conclude it is the zone which was under the friction stir welding tool rotation the simply the affected zone that underwent the rotation of the tool and where we can see the onion rings.

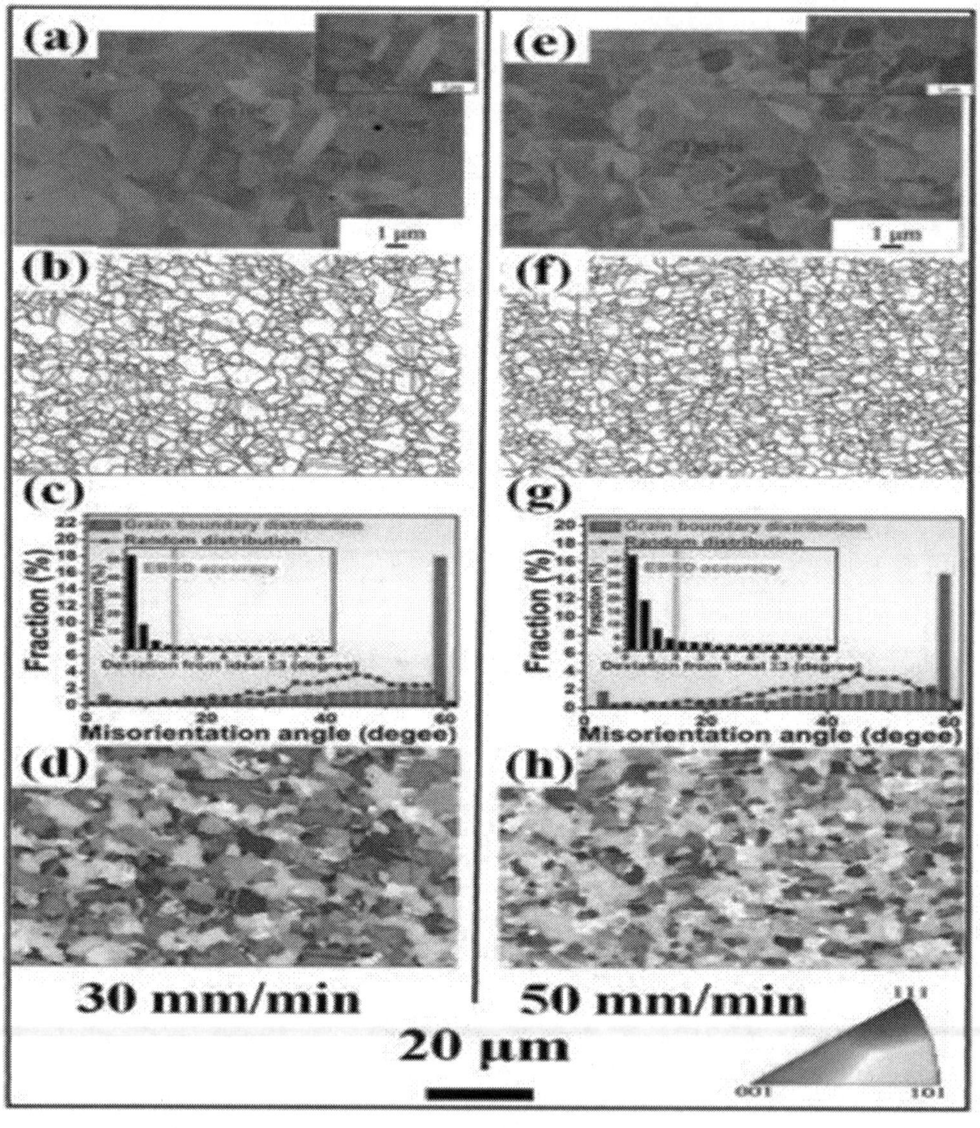

Figure 2. (a-d) BSE images, EBSD grain boundary map, misorientation angle distribution, IPF map ofthe center of the SZ in the retreating side with welding speed of 30 mm/min. (e-h) Corresponding images of the center of the SZ in the retreating side with a welding speed of 50 mm/min. LAB, HAB, and the Σ3 boundaries are shown as green, black, and red lines. The deviations of the real twin misorientation from the ideal Σ3 relationship were shown in the insets of (c) and (g) used from the [29]

3.3 Affected Grain Parameter

Grain is the fundamental unit of the microstructure that varies both the microstructural as well as mechanical properties of the material. It is the most important foundation for a material's properties. Every microstructural research starts with the grain parameters including grain size, grain boundary, and various others. The research of FSW in CoCrFeNiMn [23] resulted in the formation of different regions that can be divided based on grain size varying from smallest (2.0+-0.9) to (4.6+-2.4). It was the work of [25] to provide small grain surrounded with LABs that were not fully recrystallized with higher residual strain. It was also observed that the grain in banded structures had less deformed grain when compared to SZ. In welding especially in FSW, the grain is affected due to the penetrating tool, we need to understand that we can improve the weld properties of the HEA higher than the original plates by improving the conditions of the welding and its parameters. In FSW, the speed both transverse and rotational is the working parameter that is put into consideration. It is with the conclusion of various research that the rotational and transverse speed should be calculated with keeping the actual working cycle in mind. As it has been observed that increased rotational as well as transverse speed results in increased grain size. In one of the research, it was understood that grain size varied on the retreating and advancing side. To explain this advancing side had a slightly higher grain size than the retreating side. It's because the flow to the material the causing advancing side to have a higher temperature than the retreating. Also, when considering the top and bottom, the top appears to have a higher temperature resulting in higher grain than the bottom part. This is also the result of having a backing plate at the bottom reducing its temperature. Also, this was one [26] that showed microstructure with a fine equiaxed recrystallized grain of FCC phase. Whereas increased rotational speed caused coarsening of grain on the other side higher transverse speed caused smaller grain. It was also noted that the temperature of the surface also affected grain size. It was also observed that e-martensitic during deformation in the longitudinal specimen. The research by Sangwan park et al. [27] was about finding differences in rolled and cast material weld that had SZ with refined grain and TMAZ with severely deformed grain. Rolled SZ had the smallest grain size compared to cast SZ, Discontinuous recrystallization ratio of rolled SZ and cast SZ to BM increased from 60-70% to 90%.In [26] SZ was composed of severe deformation, it reformed grain from 124 μm to 100 μm.

3.4 Banded Structure

Banded structure act as the result of the wear of the rotating tool. The wear may not be considered physical wear yet it needs to be understood that the wear that created the banded structure is caused due to the chemical wear that can be understood due to high temperature and high penetrating rotation of the tool. [26] Found a narrow structure of 30 μm. Whereas a researcher in [27] suggests three-banded structures were distinguished as Bright Band (BB), Dark Band (DB), and White Band (WB). He observed that the rotational speed of the tool was the cause of the banded structure because increasing the speed reduced the fraction of the banded structure. Yet the white band was mostly related to the tool wear. A similar view was pointed out by [28] while testing defect pores uneven distribution that can be connected to the temperature rise during welding. Oxides and Nitrides can be observed in the banded structure [29] mostly on the advancing side rather than the retreating side. The white band is the result of tool wear that can be observed through it being carbon-rich which is associated with the tool and to add on SZ with WB has fined refined grain when compared to normal SZ.

Table 1. Observation for grain recrystallization

Authors	Observation
[20]	Welded region grain size decreased compared to BM, finest grain was found in White0 band whereas opposite to most result stir zone had coarse grain comparably. LABs in weld seam (11-14%) were 1% less than BM. High percentage of HABs gave rise to misorentation of 37.2°–41.8 in welded region.
[21]	SZ have the smallest fine grains comparably to another region. BM have largest grains. Increased transverse sped decrease grain size due to the heat production on the surface.
[22]	The refined grain sizes in the SZ at different rotational speed are: 200rpm – 1.7, 300rpm – 4.1, 400rpm – 4.7, 500rpm – 6.1. Increased rotational speed increases grain size
[23]	Grain size increases with increasing rotational speed and decreasing transverse speed.

3.5 Analysis Of Misorientation And Dislocation

Misorientation is the change in orientation of the two respective grains concerning each other. Majorly twinning and Coincidence Site Lattice (CSL) are the two aspects of misorientation. It is the smallest angle between any pair of grains within all the possible limits. While this is the base we observe misorientation, in the case of major lattice, plane coincides it can be said to be twinning misorientation. Also if the misorientation is found between two-phase it is said to be CSL. In this study, twinning was found in some cases and when it comes to CSL no such misorientation could be observed as the HEA exhibited only a single phase before and after FSW. In the work by Po Ting et al. on the effect of rotational speed in FSW observation in [25], it was noted that Low Angle Boundaries (LABs) of Base Metal had slightly greater misorientation than Weld Zone by the average of 1°-4°. The High Angle Boundaries misorientation result observed base metal with the highest misorientation of 43.2% compared to 37.2° - 41.8° in the weld zone. It can be seen from this observation that misorientation decreased after the weld was performed. In addition to this, no misorientation in SZ with deformed and substructure grain orientation uniformity was observed in [28]. The misorientation can be the result of the HEA's original composition and structure. The dislocation density of recrystallized layer was low compared to the deformed layer with high dislocation density, it was the work of [30] that also observed recrystallization sandwiched between two dislocation layers. But to the misorientation in the weld zone [31] notice LABs fractions were greater at the weld center and decreased while moving away from the center in the SZ. While dislocation density may be lower in severely distorted lattice it made it difficult to eliminate complete dislocation of grain while performing FSW. Most all the research indicates decreased misorientation and dislocation density after the use of FSW for HEA. This ideal work can be investigated through the work [32] that tells us that it was the strain effect at the extremity at the SZ that resulted in grain reorientation, which in turn led to the formation of HABs. LABs to HABs arose due to the continuous recrystallization. SZ contained a high fraction of HABs in the sense of 90% which was said to be the work of discontinuous recrystallization that acted as the main dominant in the structure reconstruction. They also observed HABs and twins increased with the increase in speed.

3.6 Basic result of structure Texture

Texture act as the reformation after the weld was performed. The initial texture depends on the material properties as well as the process it has been through for its formation. In this context of FSW of HEA, The texture is the important aspect which is the distribution of crystallographic orientation of the sample. the texture is observed in most of the welds to see if the output of the weld is good enough for their utilization. It is also one of the influencing factors in the microstructural property. The texture is said to be used for the observational study in the sense of quantity or quality. On the above basis, the texture is observed by various appropriate techniques according to the requirement, XRD, EBSD, and SEM are mostly used in such observation which was found in the studies. Some of the case texture was represented in the pole figure which is the crystallographic representation. The result of texture help in getting the result of Orientation Distribution Function (ODF). Properties whether mechanical or microstructural depend on the texture of the weld. We will be studying the texture of the weld formed during or after FSW. In Po Ting et al. in the effect of rotational speed in FSW observation in [25], FSW resulted from irregular in the base metal to low Stacking Fault Energy which was thought of as the work of material flow and texture formation due to tool rotation and geometry. The basic differentiation of the texture varying with different zones was observed in [26-28], [26] One had a columnar structure with the TMAZ having deformed grain concerning the SZ that consisted of refined grain due to the dynamic recrystallization. [27] Observation noted a special boundary between SZ and TMAZ on the retreating side, with grain bulging observed at this particular side which can be called the result of discontinuous recrystallization. Although the overall structure was thought to be a shear structure. whereas [28] microstructural was found to be bimodal due to the small grain in the SZ and coarser grain found in the TMAZ, the coarser grain can be said to be the result of insufficient temperature and incomplete recrystallization. There was no change in phase although 0.2 to 5% of hcp was identified there was no connection found between the FSW and hcp formation. In [29] it deals with the simple shear deformation, where the shear plane and shear direction do not match the sample geometry. This can be seen as they were parallel to the pin shoulder and pin respectively. Also varying the speed showed similar shear orientation FCC metal. The texture depends on the various factor from initial texture to the working parameter, yet speed change didn't affect much in some, but it can be said that texture can be correlated to the stress-strain, heating temperature and not to forget it can be also due to Stacking Fault Energy (SFE).

3.7 Energy Aspect (SFE, GGAE)

As we know HEA mostly are FCC structured and it is noticeable that the SFE of FCC structure metal was too low. [30] Explained this behavior was due to a multicomponent mixture. It influences the recrystallization mechanism as it can be caused due to discontinuous recrystallization. SFE can be related to the heating phenomenon of the dynamical recrystallization of the thermomechanical processing of low Stacking fault energy metal that can increase the working temperature up to 80-90% of the melting temperature of the weld. GGAE can be seen as the energy that acts as force energy that takes part in the growth of the grain. This energy can be associated with the HABs which are mostly stored energy while recrystallization occurs during strain formation which can be suggested as crystalline defects. GGAE affects mostly all the mechanical properties as grain size acts as the medium for the same. Hardness, yield strength, tensile strength, and fatigue increase with decreasing temperature. While [30] investing FS-weld found fine grain with partial recrystallization in SZ which was the work of the low heat index and GGAE. As we know GGAE is dependent on atomic size and in [30] it was 3.65%. High atomic size difference can lead to a high degree of lattice distortion which increases strain energy and reduces energy between normal lattice and defect area.

4. Discussion

The whole finding of friction stir welding utilization for welding high entropy alloy is the positive evolvement that needs to be enriched within the coming year. We have been exploring different parameters of the metallurgical as well as the mechanical aspect of the weld seam. The findings by the different authors provide evidence of enhancement of the produced weld life and its mechanical and metallurgical properties. As discussed above metallurgical properties when grouped seems to prove that the properties of the weld are more close to the actual base metal. Not to forget some welds in fact in some cases can be showing higher results in the respective properties than the base metal. Let's start with the different zones mainly the stir zone, heat-affected zone and thermo-mechanically affected zone were the zones formed concerning base metal and this formation was said to be the result of parameters such as transverse speed, rotational speed, heat transfer, tool dimensions, tool shoulder, and tool pin design, material flow and others. So, we can say the overall result of the different zone can be due to the working parameters and the effects of the same on the base metal. The differentiation of these zones was because of different metallurgical properties that could be observed in the microstructure. The main difference that everyone observed was the grain size and other grain-related parameters. The base metal was seen to be the original size of the grain as the application of FSW resulted in the formation of zones in which all the authors noted SZ as having the least size range of the grain. Not to be surprised but the HAZ and TMAZ were also among the zones that exhibited lower grain size compared to the base metal. Yet to be clear about a point mainly researchers differentiated the group of zones into BM, SZ, and TMAZ or SM, SZ, and HAZ. This drop-in size was said to be the result of the heat produced by the rotation of the tool with the effect of transverse speed, this produced heat was transferred throughout the base metal forming zone with increasing grain size from midsection to the end part. The relation of grain size with the transverse speed is such that increasing speed increases grain size and vice versa. A similar result could be obtained in the case of rotational speed. Grain size increased from the advancing side to the retreating side and top to bottom this was because of material and heat flow during welding. A single author found three-banded structures in the FSW of HEA was Bright band, the white band, and the Dark band. Yet mostly all the authors came with the white band and it is the result of chemical wear of the tool because tungsten was found to be a part of the band, whereas the other band was the result of rotational speed changes. One of the interesting facts about the FSW of HEA was the formation of a single phase. The texture of the weld was quite impressive and finding through the texture misorientation of grain was observed LAB's misorientation is slightly lesser in weld than the base metal, one such observation found no change in orientation in SZ. LAB's fraction increased from center to moving away from it in the case of SZ. The overall result of the observation can be concluded to FSW decreased misorientation and dislocation density. Some cases of LABs To HABs arises due to the continuous recrystallization, whereas SZ exhibits higher HAB which can be said to be the result of discontinuous recrystallization. The energy aspect of the weld in this case was stacking fault energy and grain growth energy. SFE was the result of the heating phenomenon of dynamic recrystallization whereas GGAE is said to be the energy that increased grain size and the reason for such was the HABs that had stored energy extracted and the process of recrystallization.

5. Conclusion

Friction Stir Welding had a noticeable effect on various lightweight alloys .Here HEA as the base metal have influences result at somewhat extent. Grain parameters are affected due to the effect of rotational speed on the heat generation affecting different areas distinctively. Transverse speed affected the pressure as well as the time to be influenced by rotational movement. Increased grain size was the result of increased rotational speed or decreased transverse speed. The overall heat produced at particular point and the flow of material resulted the type of recrystallization that a particular area went into. The dynamic recrystallization whether continuous , discontinuous or partial were observe and each affected a particular zone differently and hence distinct grain size , misorentation , dislocation, texture are observed. We can say that at particular rotational and transverse speed a particular HEA can show higher metallurgical properties in weld region compared to base metal.

References:

[1] W. M. Thomas, E. D. Nicholas, J. C. Needham, M.G. Murch, P. Temple-Smith And C. J. Dawes- *'Friction Stir Butt Welding'*, 1993.

[2] W. M. Thomas, E. D. Nicholas, J. C. Needham, M.G. Murch, P. Temple-Smith And C. J. Dawes- *'Improvements Relating To Friction Welding'*, 1993

[3] prof. Bharart Raj Singh- *'A handbook on Friction Stir Welding'*, 2012.

[4] O.V. Flores, C. Kennedy, L.E. Murr, D. Brown, S. Pappu, B.M. Nowak, And J.C. McClure- *'Microstructural Issues In A Friction-Stir-Welded Aluminum Alloy'*, 1998.

[5] S Benavides, Y. Li, L.E. Murr, D. Brown And J.C. McClure- *'Low-Temperature Friction-Stir Welding Of 2024 Aluminum'*, 1999.

[6] K.V. Jata, S.L AND Semiatin- *'Continuous Dynamic Recrystallization During Friction Stir Welding Of High Strength Aluminum Alloys'*, 2000.

[7] M.J. Jones, P. Heurtier, C. Desrayaud, F. Montheillet, D. Allehaux and J.H. Driver- *'Correlation Between Microstructure And Microhardness In A Friction Stir Welded 2024 Aluminum Alloy'*, 2005.

[8] C Genevois, A Deschamps, A Denquin And B. Doisneau-Cottignies- *'Quantitative Investigation Of Precipitation And Mechanical Behavior For Aa2024 Friction Stir Welds'*, 2005.

[9] Kalinenkoigor, I. Vysotskiy S. Malopheyev S. Mironov And R. Kaibyshev- *'New Insight Into The Phenomenon Of The Abnormal Grain Growth In Friction-Stir Welded Aluminum'*, 2021.

[10] R.P. Mahto, M. Rout and S.K. Pal- *'Mechanism Of Microstructure Evolution And Grain Growth In Friction Stir Welding Of Aa6061-T6 And Aisi304 In Air And Water Media'*, 2021.

[11] M. Verma, S. Ahmed and P. Saha- *'Challenges, Process Requisites/Inputs, Mechanics And Weld Performance Of Dissimilar Micro-Friction Stir Welding (Dissimilar μFSW)'*, 2021.

[12] Yibo Sun, Ping Li, yuan Zhang, Li Zou AND Xinhua Yang- *'Study On The Thermal-Assisted Friction Stir Lap Welding For Dissimilar Materials Of 6061-T6 Al Alloy And Thermoplastic Pc'*, 2021.

[13] Omid Karimi-Dermani, Alireza Abbasi, Ghasem Azimi Roeen AND Mohammad Javad Nayyeri- *'Dissimilar Friction Stir Lap Welding of Aa7075 To Az31b In The Presence Of Sn Interlayer'*, 2021.

[14] S.P. Shrivas, G.K. Agrawal, S. Nagpal, A.K. Vishvakarma And A.K. Khandelwal- *'Dissimilar Aluminum Alloy Joint Strength Is Affected By Heat Addition In Friction Stir Welding (FSW)'*, 2020.

[15] M. Makeshkumar, S.R. Surender, S. Arunprakash, R. MADESH, M.S. Kumar, K. Sudharsan- *'Microstructural and Mechanical Properties Evaluation Of Dissimilar Aluminum Alloy And Bronze Joints Using Friction Stir Welding'*, 2021.

[16] R. Sokkalingam, B. Pravallika, K. Sivaprasad, V. Muthupandi, K.G. Prashanth- *'Dissimilar Welding Of High-Entropy Alloy To Inconel 718 Superalloy For Structural Applications'*, 2022.

[17] J.W. Yeh, S.K. Chen, S.J. Lin , J.Y. Gan, T.S. Chin, T.T. Shuh, C.H. Tsa, And, S.Y. Chang - *'Nanostructured High-Entropy Alloys With Multiple Principal Elements: Novel Alloy Design Concepts And Outcomes'*, 2004.

[18] B. Cantor, I.H.T. Chang, P. Knight, And A.J.B. Vincent- *'Microstructural Development In Equiatomic Multicomponent Alloys'*, 2003.

[19] R.K. Mishra And Rohit R. Shahi- *'Effect Of Annealing On Phase Formation And Their Correlation With Magnetic Characteristics OF Tifenicrco Hea'*, 2019.

[20] Xu Chen, Di Gao, Jia Xuan Hu, YE LIU, AND Chang Ping Tang- *'Evolution Of Microstructures And Properties In Alxcrfemn0.8ni2.1 HEAs'*, 2019.

[21] W. Muftah, John Allport, And Vladimir Vishnyakov.- *'Corrosion Performance And Mechanical Properties Of Fecrsinb Amorphous Equiatomic Hea Thin Film'*, 2021.

[22] D. Shaysultanov, N. Stepanov, S. Malopheyev, I. Vysotskiy, V. Sanin, S. Mironov, R. Kaibyshev, G. Salishchev, S. Zherebtsov- *'Friction Stirs Welding Of A Carbon-Doped CoCrFeNiMn High-Entropy Alloy'*, 2018.

[23] Z.G. Zhu, Y.F. Sun, M.H. Goh, F.L. Ng, Q.B. Nguyen, H. Fujii, S.M.L. Nai, J. Wei, And C.H. Shek- *'Friction Stir Welding Of A Cocrfenial0.3 High Entropy Alloy'*,2017.

[24] Po-Ting Lin, Chan-Sheng Wu, Chun-Hao Peng, Che-Wei Tsai, And Yutaka S. Sato- *'Effects Of Rotational Speed On The Al0.3cocrcu0.3feni High-Entropy Alloy By Friction Stir Welding'*, 2020.

[25] Dmitry Shaysultanov, Kazimzhon Raimov, Nikita Stepanov And Sergey Zherebtsov- *'Friction Stir Welding Of A Trip Fe49mn30cr10co10c1 High Entropy Alloy'*, 2021.

[26] Sangwon Park, Hyunbin Nam, YOUNGSANG Na, Hyoungseop Kim, Younghoon Moon, AND Namhyun Kang- *'Effect Of Initial Grain Size On Friction Stir Weldability For Rolled And Cast CoCrFeNiMn High-Entropy Alloys'*, 2020.

[27] Po-Ting Lin, Hung-Chi Liu, Po-Ying Hsieh, Cheng-Yu Wei, Che-Wei Tsai, Yutaka S. Sato, Shih-Che Chen, Hung-Wei Yen, Nian-Hu Lu, Chih-Hsuan Chen- *'Heterogeneous Structure-Induced Strength- Ductility Synergy By Partial Recrystallization During Friction Stir Welding Of A High-Entropy Alloy'*, 2020.

[28] Min-Gu Jo, Han-Jin Kim, Minjung Kang, Phaniraj P. Madakashira, Eun Soo Park, Jin-Yoo Suh, Dong-Ik Kim, Sung-Tae Hong, And Heung Nam Han- *'Microstructure And Mechanical Properties Of Friction Stir Welded And Laser Welded High Entropy Alloy CrMnFeCoNi'*, 2018.

[29] Z.G. Zhu, Y.F. Sun, M.H. Goh, F.L. Ng, Q.B. Nguyen, H. Fujii, S.M.L. Nai, J. Wei, Y. XU, P.K. LIAW And C.H. Shek - *'Friction-Stir Welding Of A Ductile High Entropy Alloy: Microstructural Evolution And Weld Strength'*, 2017.

[30] Sergey Zherebtsov, Nikita Stepanov, Dmitry Shaysultanov, Sergey Malopheyev, Igor Vysotskiy, Vladimir Sanin, Nikolay Kashaev And Rustam Kaibyshev- *'Use Of Novel Welding Technologies For High-Entropy Alloys Joining'*, 2018.

[31] Xiaomeng Qin, Yi Xu, Yufeng Sun, Hidetoshi Fujii, Zhiguang Zhu, Chan Hung SheK- *'Effect Of Process Parameters On Microstructure And Mechanical Properties Of Friction Stir Welded CoCrFeNi High Entropy Alloy'*, 2020.

[32] Peng Li, Haotian Sun, Shuai Wang, Xiaohu Hao, Honggang Dong- *'Rotary Friction Welding Of Alcocrfeni2.1 Eutectic High Entropy Alloy'*,2019.

[33] Jing Guo, Cong Tang, Glynn Rothwell, Lisa Li, Yun-Che Wang, Qingxiang Yang And Xuejun Ren- *'Welding Of High Entropy Alloys—A Review'*, 2019.

[34] Sanya Gupta, Priyanka Agrawal, Saurabh S. Nene, Rajiv S. Mishra- *'Friction Stirs Welding Of γ-Fcc Dominated Metastable High Entropy Alloy: Microstructural Evolution And Strength'*, 2021.

IOP Conf. Series: Materials Science and Engineering 1259 (2022) 012052 doi:10.1088/1757-899X/1259/1/012052

[35] Haining Yao, Hongyuan Wen, Ke Chen, Muyang Jiang, Kolan Madhav Reddy, Katsuyoshi Kondoh, Min Wang, Xueming Hua, Aidang Shan- *'Interfacial Phases Formed In Friction Stir Lap Welding High Entropy Alloy To Al Alloy'*,2021.

[36] Ankur V Bansod, Awanikumar Patil, Abhijeet Moon, Sourabh Shukla- *'Microstructural And Electrochemical Evaluation Of Fusion Welded Low Nickel And 304 Ass At Different Heat Input'*, 2017.

[37] Ankur V Bansod, Awanikumar P. Patil, Sourabh Shukla- *'Effect Of Heat On Microstructural, Mechanical, And Electrochemical Evaluation Of Tungsten Inert Gas Welding Of Low-Nickel Ass'*, 2018.

[38] Sourabh Shukla, Awanikumar Patil, Ashlesha P. Kawale, Rakesh Haldkar- *'Effect Of Grain Refinement On Sensitization Of High Manganese Austenitic Stainless Steel'* 2021.

[39] Sourabh Shukla, Awanikumar Patil, Ashlesha Kawale, Inayat Ullah, *'Effect Of Recrystallization On The Degree Of Sensitization In Nickel-Free Austenitic Stainless Steel'*, 2021.

[40] Sourabh Shukla, Awanikumar Patil, Ankur V Bansod- *'The Effect Of Cold Work And Thermal Aging On Intergranular And Transgranular Corrosion Failure Of High Manganese Austenitic Stainless Steel'*, 2022

AUTHOR INDEX

Institute of Physics
Dirac House, Temple Back
Bristol BS1 6BE UK

ISSN: 1757-8981
ISBN 978-1-7138-6422-6

6th International Forum on Industrial Design (IFID 2018)

MATEC Web of Conferences Volume 176 (2018)

Luoyang, China
18 - 20 May 2018

edp sciences